T0181852

Lecture Notes in Computer Science　13981

The series Lecture Notes in Computer Science (LNCS), including its subseries Lecture Notes in Artificial Intelligence (LNAI) and Lecture Notes in Bioinformatics (LNBI), has established itself as a medium for the publication of new developments in computer science and information technology research, teaching, and education.

LNCS enjoys close cooperation with the computer science R & D community, the series counts many renowned academics among its volume editors and paper authors, and collaborates with prestigious societies. Its mission is to serve this international community by providing an invaluable service, mainly focused on the publication of conference and workshop proceedings and postproceedings. LNCS commenced publication in 1973.

Jaap Kamps · Lorraine Goeuriot · Fabio Crestani ·
Maria Maistro · Hideo Joho · Brian Davis ·
Cathal Gurrin · Udo Kruschwitz ·
Annalina Caputo
Editors

Advances in Information Retrieval

45th European Conference on Information Retrieval, ECIR 2023
Dublin, Ireland, April 2–6, 2023
Proceedings, Part II

Springer

Editors
Jaap Kamps ⓘ
University of Amsterdam
Amsterdam, Netherlands

Fabio Crestani ⓘ
Università della Svizzera Italiana
Lugano, Switzerland

Hideo Joho ⓘ
University of Tsukuba
Ibaraki, Japan

Cathal Gurrin ⓘ
Dublin City University
Dublin, Ireland

Annalina Caputo ⓘ
Dublin City University
Dublin, Ireland

Lorraine Goeuriot ⓘ
Université Grenoble-Alpes
Saint-Martin-d'Hères, France

Maria Maistro ⓘ
University of Copenhagen
Copenhagen, Denmark

Brian Davis ⓘ
Dublin City University
Dublin, Ireland

Udo Kruschwitz ⓘ
Universität Regensburg
Regensburg, Germany

ISSN 0302-9743   ISSN 1611-3349 (electronic)
Lecture Notes in Computer Science
ISBN 978-3-031-28237-9   ISBN 978-3-031-28238-6 (eBook)
https://doi.org/10.1007/978-3-031-28238-6

Preface

The 45th European Conference on Information Retrieval (ECIR 2023) was held in Dublin, Ireland, during April 2–6, 2023, and brought together hundreds of researchers from Europe and abroad. The conference was organized by Dublin City University, in cooperation with the British Computer Society's Information Retrieval Specialist Group (BCS IRSG).

These proceedings contain the papers related to the presentations, workshops, and tutorials given during the conference. This year's ECIR program boasted a variety of novel work from contributors from all around the world. In total, 489 papers from authors in 52 countries were submitted to the different tracks. The final program included 65 full papers (29% acceptance rate), 41 short papers (27% acceptance rate), 19 demonstration papers (66% acceptance rate), 12 reproducibility papers (63% acceptance rate), 10 doctoral consortium papers (56% acceptance rate), and 13 invited CLEF papers. All submissions were peer-reviewed by at least three international Program Committee members to ensure that only submissions of the highest relevance and quality were included in the final program. The acceptance decisions were further informed by discussions among the reviewers for each submitted paper, led by a senior Program Committee member. In a final PC meeting all the final recommendations were discussed, trying to reach a fair and equal outcome for all submissions.

The accepted papers cover the state of the art in information retrieval: user aspects, system and foundational aspects, machine learning, applications, evaluation, new social and technical challenges, and other topics of direct or indirect relevance to search. As in previous years, the ECIR 2023 program contained a high proportion of papers with students as first authors, as well as papers from a variety of universities, research institutes, and commercial organizations.

In addition to the papers, the program also included 3 keynotes, 7 tutorials, 8 workshops, a doctoral consortium, the presentation of selected papers from the 2022 issues of the Information Retrieval Journal, and an industry day. Keynote talks were given by Mounia Lalmas (Spotify), Tetsuya Sakai (Waseda University), and this year's BCS IRSG Karen Spärck Jones Award winner, Yang Wang (UC Santa Barbara). The tutorials covered a range of topics including conversational agents in health; crowdsourcing; gender bias; legal IR and NLP; neuro-symbolic representations; query auto completion; and text classification. The workshops brought together participants to discuss algorithmic bias (BIAS); bibliometrics (BIR); e-discovery (ALTARS); geographic information extraction (GeoExT); legal IR (Legal IR); narrative extraction (Text2story); online misinformation (ROMCIR); and query performance prediction (QPP).

The success of ECIR 2023 would not have been possible without all the help from the team of volunteers and reviewers. We wish to thank all the reviewers and meta-reviewers who helped to ensure the high quality of the program. We also wish to thank: the short paper track chairs: Maria Maistro and Hideo Joho; the demo track chairs: Liting Zhou and Frank Hopfgartner; the reproducibility track chair: Leif Azzopardi; the workshop track chairs: Ricardo Campos and Gianmaria Silvello; the tutorial track chairs: Bhaskar Mitra and Debasis Ganguly; the industry track chairs: Nicolas Fiorini and Isabelle Moulinier; the doctoral consortium chair: Gareth Jones; and the awards chair: Suzan Verberne. We thank the students Praveen Acharya, Chinonso Osuji and Kanishk Verma for help with preparing the proceedings. We would like to thank all the student volunteers who helped to create an excellent experience for participants and attendees. ECIR 2023 was sponsored by a range of research institutes and companies. We thank them all for their support.

Finally, we wish to thank all the authors and contributors to the conference.

April 2023

Lorraine Goeuriot
Fabio Crestani
Jaap Kamps
Maria Maistro
Hideo Joho
Annalina Caputo
Udo Kruschwitz
Cathal Gurrin

Organization

General Chairs

Annalina Caputo   Dublin City University, Ireland
Udo Kruschwitz   Universität Regensburg, Germany
Cathal Gurrin   Dublin City University, Ireland

Program Committee Chairs

Jaap Kamps   University of Amsterdam, Netherlands
Lorraine Goeuriot   Université Grenoble Alpes, France
Fabio Crestani   Università della Svizzera Italiana, Switzerland

Short Papers Chairs

Maria Maistro   University of Copenhagen, Denmark
Hideo Joho   University of Tsukuba, Japan

Demo Chairs

Liting Zhou   Dublin City University, Ireland
Frank Hopfgartner   University of Koblenz-Landau, Germany

Reproducibility Track Chair

Leif Azzopardi   University of Strathclyde, UK

Workshop Chairs

Ricardo Campos   Instituto Politécnico de Tomar/INESC TEC, Portugal
Gianmaria Silvello   University of Padua, Italy

Tutorial Chairs

Bhaskar Mitra   Microsoft, Canada
Debasis Ganguly   University of Glasgow, UK

Industry Day Chairs

Nicolas Fiorini   Algolia, France
Isabelle Moulinier   Thomson Reuters, USA

Doctoral Consortium Chair

Gareth Jones   Dublin City University, Ireland

Awards Chair

Suzan Verberne   Leiden University, Netherlands

Publication Chairs

Brian Davis   Dublin City University, Ireland
Joachim Wagner   Dublin City University, Ireland

Local Chairs

Brian Davis   Dublin City University, Ireland
Ly Duyen Tran   Dublin City University, Ireland

Senior Program Committee

Omar Alonso   Amazon, USA
Giambattista Amati   Fondazione Ugo Bordoni, Italy
Ioannis Arapakis   Telefonica Research, Spain
Jaime Arguello   University of North Carolina at Chapel Hill, USA
Javed Aslam   Northeastern University, USA

Krisztian Balog	University of Stavanger & Google Research, Norway
Patrice Bellot	Aix-Marseille Université - CNRS (LSIS), France
Michael Bendersky	Google, USA
Mohand Boughanem	IRIT University Paul Sabatier Toulouse, France
Jamie Callan	Carnegie Mellon University, USA
Ben Carterette	Spotify, USA
Charles Clarke	University of Waterloo, Canada
Bruce Croft	University of Massachusetts Amherst, USA
Maarten de Rijke	University of Amsterdam, Netherlands
Arjen de Vries	Radboud University, Netherlands
Giorgio Maria Di Nunzio	University of Padua, Italy
Laura Dietz	University of New Hampshire, USA
Shiri Dori-Hacohen	University of Connecticut, USA
Carsten Eickhoff	Brown University, USA
Tamer Elsayed	Qatar University, Qatar
Liana Ermakova	HCTI, Université de Bretagne Occidentale, France
Hui Fang	University of Delaware, USA
Nicola Ferro	University of Padova, Italy
Ingo Frommholz	University of Wolverhampton, UK
Norbert Fuhr	University of Duisburg-Essen, Germany
Debasis Ganguly	University of Glasgow, UK
Nazli Goharian	Georgetown University, USA
Marcos Goncalves	Federal University of Minas Gerais, Brazil
Julio Gonzalo	UNED, Spain
Jiafeng Guo	Institute of Computing Technology, China
Matthias Hagen	Friedrich-Schiller-Universität Jena, Germany
Martin Halvey	University of Strathclyde, UK
Allan Hanbury	TU Wien, Austria
Donna Harman	NIST, USA
Faegheh Hasibi	Radboud University, Netherlands
Claudia Hauff	Spotify, Netherlands
Ben He	University of Chinese Academy of Sciences, China
Jiyin He	Signal AI, UK
Dietmar Jannach	University of Klagenfurt, Austria
Adam Jatowt	University of Innsbruck, Austria
Hideo Joho	University of Tsukuba, Japan
Gareth Jones	Dublin City University, Ireland
Joemon Jose	University of Glasgow, UK
Jaap Kamps	University of Amsterdam, Netherlands

Paul Thomas	Microsoft, Australia
Nicola Tonellotto	University of Pisa, Italy
Theodora Tsikrika	Information Technologies Institute, CERTH, Greece
Julián Urbano	Delft University of Technology, Netherlands
Suzan Verberne	LIACS, Leiden University, Netherlands
Gerhard Weikum	Max Planck Institute for Informatics, Germany
Marcel Worring	University of Amsterdam, Netherlands
Andrew Yates	University of Amsterdam, Netherlands
Jakub Zavrel	Zeta Alpha, Netherlands
Min Zhang	Tsinghua University, China
Shuo Zhang	Bloomberg, Norway
Justin Zobel	University of Melbourne, Australia
Guido Zuccon	University of Queensland, Australia

Program Committee

Shilpi Agrawal	Linkedin, India
Qingyao Ai	Tsinghua University, China
Dyaa Albakour	Signal AI, UK
Mohammad Aliannejadi	University of Amsterdam, Netherlands
Satya Almasian	Heidelberg University, Germany
Omar Alonso	Amazon, USA
Ismail Sengor Altingovde	Middle East Technical University, Turkey
Giuseppe Amato	ISTI-CNR, Italy
Enrique Amigó	UNED, Spain
Sophia Ananiadou	University of Manchester, UK
Linda Andersson	Artificial Researcher IT GmbH, TU Wien, Austria
Vito Walter Anelli	Politecnico di Bari, Italy
Negar Arabzadeh	University of Waterloo, Canada
Arian Askari	Leiden Institute of Advanced Computer Science, Leiden University, Netherlands
Giuseppe Attardi	Università di Pisa, Italy
Maurizio Atzori	University of Cagliari, Italy
Sandeep Avula	Amazon, USA
Mossaab Bagdouri	Walmart Global Tech, USA
Ebrahim Bagheri	Ryerson University, Canada
Georgios Balikas	Salesforce Inc, France
Krisztian Balog	University of Stavanger & Google Research, Norway
Alvaro Barreiro	University of A Coruña, Spain

Célia da Costa Pereira	Université Côte d'Azur, France
Duc Tien Dang Nguyen	University of Bergen, Norway
Maarten de Rijke	University of Amsterdam, Netherlands
Arjen de Vries	Radboud University, Netherlands
Yashar Deldjoo	Polytechnic University of Bari, Italy
Gianluca Demartini	University of Queensland, Australia
Amey Dharwadker	Meta, USA
Emanuele Di Buccio	University of Padua, Italy
Giorgio Maria Di Nunzio	University of Padua, Italy
Gaël Dias	Normandie University, France
Laura Dietz	University of New Hampshire, USA
Vlastislav Dohnal	Faculty of Informatics, Masaryk University, Czechia
Zhicheng Dou	Renmin University of China, China
Antoine Doucet	University of La Rochelle, France
Pan Du	Thomson Reuters Labs, Canada
Tomislav Duricic	Graz University of Technology, Austria
Liana Ermakova	HCTI, Université de Bretagne Occidentale, France
Ralph Ewerth	L3S Research Center, Leibniz Universität Hannover, Germany
Guglielmo Faggioli	University of Padova, Italy
Anjie Fang	Amazon.com, USA
Hossein Fani	University of Windsor, Canada
Yue Feng	UCL, UK
Marcos Fernández Pichel	Universidade de Santiago de Compostela, Spain
Juan M. Fernández-Luna	University of Granada, Spain
Nicola Ferro	University of Padova, Italy
Komal Florio	Università di Torino, Italy
Thibault Formal	Naver Labs Europe, France
Ophir Frieder	Georgetown University, USA
Ingo Frommholz	University of Wolverhampton, UK
Maik Fröbe	Friedrich-Schiller-Universität Jena, Germany
Norbert Fuhr	University of Duisburg-Essen, Germany
Michael Färber	Karlsruhe Institute of Technology, Germany
Petra Galuščáková	Université Grenoble Alpes, France
Debasis Ganguly	University of Glasgow, UK
Dario Garigliotti	No affiliation, Norway
Eric Gaussier	LIG-UJF, France
Kripabandhu Ghosh	Indian Institute of Science Education and Research (IISER) Kolkata, India
Anastasia Giachanou	Utrecht University, Netherlands

Lorraine Goeuriot	Univ. Grenoble Alpes, CNRS, Grenoble INP, LIG, France
Nazli Goharian	Georgetown University, USA
Marcos Goncalves	Federal University of Minas Gerais, Brazil
Julio Gonzalo	UNED, Spain
Michael Granitzer	University of Passau, Germany
Adrien Guille	ERIC Lyon 2, EA 3083, Université de Lyon, France
Nuno Guimaraes	CRACS - INESC TEC, Portugal
Chun Guo	Pandora Media LLC., USA
Dhruv Gupta	Norwegian University of Science and Technology, Norway
Christian Gütl	Graz University of Technology, Austria
Matthias Hagen	Friedrich-Schiller-Universität Jena, Germany
Lei Han	University of Queensland, Australia
Preben Hansen	Stockholm University, Sweden
Donna Harman	NIST, USA
Morgan Harvey	University of Sheffield, UK
Maram Hasanain	Qatar University, Qatar
Claudia Hauff	Spotify, Netherlands
Mariya Hendriksen	University of Amsterdam, Netherlands
Daniel Hienert	GESIS - Leibniz Institute for the Social Sciences, Germany
Orland Hoeber	University of Regina, Canada
Frank Hopfgartner	Universität Koblenz, Germany
Gilles Hubert	IRIT, France
Juan F. Huete	University of Granada, Spain
Bogdan Ionescu	University Politehnica of Bucharest, Romania
Radu Tudor Ionescu	University of Bucharest, Faculty of Mathematics and Computer Science, Romania
Adam Jatowt	University of Innsbruck, Austria
Faizan Javed	Kaiser Permanente, USA
Renders Jean-Michel	Naver Labs Europe, France
Tianbo Ji	Dublin City University, Ireland
Noriko Kando	National Institute of Informatics, Japan
Nattiya Kanhabua	SCG CBM, Thailand
Sarvnaz Karimi	CSIRO, Australia
Sumanta Kashyapi	NIT Hamirpur, USA
Makoto P. Kato	University of Tsukuba, Japan
Abhishek Kaushik	Dundalk Institute of Technology, Ireland
Mesut Kaya	Aalborg University Copenhagen, Denmark
Roman Kern	Graz University of Technology, Austria

Salvatore Orlando	Università Ca' Foscari Venezia, Italy
Iadh Ounis	University of Glasgow, UK
Pooja Oza	University of New Hampshire, USA
Özlem Özgöbek	Norwegian University of Science and Technology, Norway
Deepak P.	Queen's University Belfast, UK
Panagiotis Papadakos	Information Systems Laboratory - FORTH-ICS, Greece
Javier Parapar	IRLab, University of A Coruña, Spain
Pavel Pecina	Charles University, Czechia
Gustavo Penha	Delft University of Technology, Brazil
Maria Soledad Pera	TU Delft, Netherlands
Vivien Petras	Humboldt-Universität zu Berlin, Germany
Giulio Ermanno Pibiri	Ca' Foscari University of Venice, Italy
Francesco Piccialli	University of Naples Federico II, Italy
Karen Pinel-Sauvagnat	IRIT, France
Florina Piroi	TU Wien, Institue of Information Systems Engineering, Austria
Marco Polignano	Università degli Studi di Bari Aldo Moro, Italy
Martin Potthast	Leipzig University, Germany
Ronak Pradeep	University of Waterloo, Canada
Xin Qian	University of Maryland, USA
Fiana Raiber	Yahoo Research, Israel
David Rau	University of Amsterdam, Netherlands
Andreas Rauber	Vienna University of Technology, Austria
Gábor Recski	TU Wien, Austria
Weilong Ren	Shenzhen Institute of Computiing Sciences, China
Zhaochun Ren	Shandong University, China
Chiara Renso	ISTI-CNR, Pisa, Italy, Italy
Thomas Roelleke	Queen Mary University of London, UK
Kevin Roitero	University of Udine, Italy
Haggai Roitman	eBay Research, Israel
Paolo Rosso	Universitat Politècnica de València, Spain
Stevan Rudinac	University of Amsterdam, Netherlands
Anna Ruggero	Sease Ltd., Italy
Tony Russell-Rose	Goldsmiths, University of London, UK
Ian Ruthven	University of Strathclyde, UK
Sriparna Saha	IIT Patna, India
Tetsuya Sakai	Waseda University, Japan
Eric Sanjuan	Laboratoire Informatique d'Avignon—Université d'Avignon, France
Maya Sappelli	HAN University of Applied Sciences, Netherlands

Jacques Savoy   University of Neuchatel, Switzerland
Harrisen Scells   Leipzig University, Germany
Philipp Schaer   TH Köln (University of Applied Sciences),
                                     Germany
Ferdinand Schlatt   Martin-Luther Universität Halle-Wittenberg,
                                     Germany
Jörg Schlötterer   University of Duisburg-Essen, Germany
Falk Scholer   RMIT University, Australia
Fabrizio Sebastiani   Italian National Council of Research, Italy
Christin Seifert   University of Duisburg-Essen, Germany
Ivan Sekulic   Università della Svizzera italiana, Switzerland
Giovanni Semeraro   University of Bari, Italy
Procheta Sen   University of Liverpool, UK
Mahsa S. Shahshahani   Accenture, Netherlands
Eilon Sheetrit   Technion - Israel Institute of Technology, Israel
Fabrizio Silvestri   University of Rome, Italy
Jaspreet Singh   Amazon, Germany
Manel Slokom   Delft University of Technology, Netherlands
Mark Smucker   University of Waterloo, Canada
Michael Soprano   University of Udine, Italy
Laure Soulier   Sorbonne Université-ISIR, France
Marc Spaniol   Université de Caen Normandie, France
Damiano Spina   RMIT University, Australia
Andreas Spitz   University of Konstanz, Germany
Torsten Suel   New York University, USA
Kazunari Sugiyama   Kyoto University, Japan
Dhanasekar Sundararaman   Duke University, USA
Irina Tal   Dublin City University, Ireland
Lynda Tamine   IRIT, France
Carla Teixeira Lopes   University of Porto, Portugal
Joseph Telemala   University of Cape Town, South Africa
Paul Thomas   Microsoft, Australia
Thibaut Thonet   Naver Labs Europe, France
Nicola Tonellotto   University of Pisa, Italy
Salvatore Trani   ISTI-CNR, Italy
Jan Trienes   University of Duisburg-Essen, Germany
Johanne R. Trippas   RMIT University, Australia
Andrew Trotman   University of Otago, New Zealand
Theodora Tsikrika   Information Technologies Institute, CERTH,
                                     Greece
Kosetsu Tsukuda   National Institute of Advanced Industrial Science
                                     and Technology (AIST), Japan

Yannis Tzitzikas	University of Crete and FORTH-ICS, Greece
Md Zia Ullah	Edinburgh Napier University, UK
Kazutoshi Umemoto	University of Tokyo, Japan
Julián Urbano	Delft University of Technology, Netherlands
Ruben van Heusden	University of Amsterdam, Netherlands
Aparna Varde	Montclair State University, USA
Suzan Verberne	LIACS, Leiden University, Netherlands
Manisha Verma	Amazon, USA
Vishwa Vinay	Adobe Research, India
Marco Viviani	Università degli Studi di Milano-Bicocca - DISCo, Italy
Ellen Voorhees	NIST, USA
Xi Wang	University College London, UK
Zhihong Wang	Tsinghua University, China
Wouter Weerkamp	TomTom, Netherlands
Gerhard Weikum	Max Planck Institute for Informatics, Germany
Xiaohui Xie	Tsinghua University, China
Takehiro Yamamoto	University of Hyogo, Japan
Eugene Yang	Human Language Technology Center of Excellence, Johns Hopkins University, USA
Andrew Yates	University of Amsterdam, Netherlands
Elad Yom-Tov	Microsoft, Israel
Ran Yu	University of Bonn, Germany
Hamed Zamani	University of Massachusetts Amherst, USA
Eva Zangerle	University of Innsbruck, Austria
Richard Zanibbi	Rochester Institute of Technology, USA
Fattane Zarrinkalam	University of Guelph, Canada
Sergej Zerr	Rhenish Friedrich Wilhelm University of Bonn, Germany
Fan Zhang	Wuhan University, China
Haixian Zhang	Sichuan University, China
Min Zhang	Tsinghua University, China
Rongting Zhang	Amazon, USA
Ruqing Zhang	Institute of Computing Technology, Chinese Academy of Sciences, China
Mengyisong Zhao	University of Sheffield, UK
Wayne Xin Zhao	Renmin University of China, China
Jiang Zhou	Dublin City University, Ireland
Liting Zhou	Dublin City University, Ireland
Steven Zimmerman	University of Essex, UK
Justin Zobel	University of Melbourne, Australia
Lixin Zou	Tsinghua University, China

Guido Zuccon   University of Queensland, Australia

Additional Reviewers

Ashkan Alinejad
Evelin Amorim
Negar Arabzadeh
Dennis Aumiller
Mohammad Bahrani
Mehdi Ben Amor
Giovanni Maria Biancofiore
Ramraj Chandradevan
Qianli Chen
Dhivya Chinnappa
Isabel Coutinho
Washington Cunha
Xiang Dai
Marco de Gemmis
Alaa El-Ebshihy
Gloria Feher
Yasin Ghafourian
Wolfgang Gritz
Abul Hasan
Phuong Hoang
Eszter Iklodi
Andrea Iovine
Tania Jimenez
Pierre Jourlin

Anoop K.
Tuomas Ketola
Adam Kovacs
Zhao Liu
Daniele Malitesta
Cataldo Musto
Evelyn Navarrete
Zhan Qu
Saed Rezayi
Ratan Sebastian
Dawn Sepehr
Simra Shahid
Chen Shao
Mohammad Sharif
Stanley Simoes
Matthias Springstein
Ting Su
Wenyi Tay
Alberto Veneri
Chenyang Wang
Lorenz Wendlinger
Mengnong Xu
Shuzhou Yuan

Contents – Part II

Short Papers

Full Papers

Extractive Summarization of Financial Earnings Call Transcripts
Or: When GREP Beat BERT

Tim Nugent[1], George Gkotsis[2], and Jochen L. Leidner[3,4(✉)] (iD)

[1] GSR Markets, London, UK
[2] Kailua Labs, Patras, Greece
[3] Coburg University of Applied Sciences, Friedrich-Streib-Straße 2,
96450 Coburg, Germany
[4] University of Sheffield, Regents Court, 211 Portobello,
Sheffield S1 4DP, UK
leidner@acm.org

Abstract. To date, automatic summarization methods have been mostly developed for (and applied to) general news articles, whereas other document types have been neglected. In this paper, we introduce the task of summarizing financial earnings call transcripts, and we present a method for summarizing this text type essential for the financial industry. Earnings calls are briefing events common for public companies in many countries, typically in the form of conference calls held between company executives and analysts that consist of a spoken monologue part followed by moderated questions and answers.

We show that traditional methods work less well in this domain, we present a method suitable for summarizing earnings calls. Our large-scale evaluation on a new human-annotated corpus of summary-worthy sentences shows that this method outperforms a set of strong baselines, including a new one that we propose specifically for earnings calls. To the best of our knowledge, this is the first application of summarization to financial earnings calls transcripts, a primary source of information for financial professionals.

Keywords: Automatic document summarization · Finance applications · Natural language processing (NLP) · Applied Machine Learning (ML) · Information Retrieval (IR)

1 Introduction

Automatic document summarization has long been part of information retrieval as well as natural language processing. Text summarization or abstracting has a long history [1,9,25,30], going back to Luhn's heuristic sentence scoring [20]. However, most recent research has been conducted on *agency news*, a text type

---

Most of this research was conducted while all three authors were at Refinitiv Ltd., 5 Canada Square, London E14 5AQ, United Kingdom.

© The Author(s), under exclusive license to Springer Nature Switzerland AG 2023
J. Kamps et al. (Eds.): ECIR 2023, LNCS 13981, pp. 3–15, 2023.
https://doi.org/10.1007/978-3-031-28238-6_1

that by its very design mostly contains very short documents, where the first sentence often summarizes the core message, which is then elaborated further.[1] By contrast, financial earnings call transcripts are very important and *long* documents: they are one of three document types, together with SEC filings and news, that analysts rely on regularly to assess the investment-worthiness of public companies on a regular basis. Earnings calls are regular, quarterly or annual, pre-scheduled conference calls held between company executives and financial analysts and investors. They consist of a (transcribed) spoken presentation part followed by moderated questions to company executives by analysts and their answers [4,5]. Earnings call transcripts may amount to 30-60 pages in print, so the case for summarization research can arguably be more easily made than for news summarization, especially where single-document summarization is concerned. Financial earnings call transcripts also pose an interesting target for automatic summarization research because (i) they are transcripts produced from originally spoken language, (ii) they contain redundant parts and (iii) because of their two-part nature comprising a CEO-CFO "duolog", i.e. a presentation conducted by two people, followed by an interactive Q&A part, in which analysts probe the contents of the presentation or ask for omitted information. Earnings calls are mostly held in English, on which we therefore focus. Because our system was developed for industry deployment where questions of misleading investors through wrongly-generated abstractive text are prohibitive, we subscribe to a sentence-level extractive paradigm.

In this paper, we explore three research hypotheses:

H1 Typical summarization methods developed for news will not perform well on financial earnings call transcripts.
H2 A small set of simple features can capture well what is essential information from financial earnings call transcripts.
H3 A large-scale, general-purpose pre-trained neural language model outperforms a set of human-devised features.

We present ECSumm, our implementation that forms the basis for our studies of these three hypotheses. Our contributions include:

– a **new baseline**, devised for financial earnings calls or other financial report, which is simple to implement and replicate (code in Listing 1.1);
– **several novel methods** for the automatic summarization of financial earnings calls transcript documents, including heuristic/unsupervised, traditional supervised machine-learning based and deep-learning based methods;
– the description of our **system ECSumm**, which implements all of them;
– an **experimental evaluation** on a new, 5-way annotated gold standard corpus, which includes a detailed comparison of several baselines (old and new) with our new methods.

Our unsupervised and unsupervised methods for the task, which is framed as a single-document extractive (sentence selection) summarization task, are

---

[1] The so-called *inverted pyramid structure*, a property that has been exploited in supervised learning for summarization [28].

Good day, ladies and gentlemen, and welcome to the SAP Quarter Three 2019 Earnings Conference Call. (Operator Instructions) At this time, I would like to turn the conference over to Mr. Stefan Gruber, Head of Investor Relations. Please go ahead, sir.
Stefan Gruber – Head of Investor Relations
Thank you. Good morning or good afternoon. This is Stefan Gruber, Head of Investor Relations. Thank you for joining us to discuss our results for the third quarter 2019.
...

Look in April we promised a stronger focus on profits and we are clearly delivering. I'm very pleased to say that our operational excellence measures allowed us to achieve double digit operating profit growth and a substantial operating margin expansion. And just as important we achieved this result with continued strong top line momentum.
Let me now provide you with some background on the key drivers of the third quarter. Both cloud and software revenue as well as total revenue grew 13% this quarter, cloud revenue was again a big driver of this growing 37%. New cloud bookings were up 39% and up 51% excluding our infrastructure-as-a-service business.

Mohammed Moawalla – Goldman Sachs – Analyst Great, thank you very much. And Jennifer and Christian, my congratulations as well on your new roles. I had a couple. Firstly, Luka, you obviously have some pretty strong gross margin tailwinds continuing into next year, but also some of the big opex benefits kicking in. You've also talked about sort of reinvestments back in the business, can you talk about sort of the flexibility you have around delivering the margin and perhaps the shape of the margin expansion over the next couple of years, in the event that the top line potentially faces risks with a macro or anything else.
Luka Mucic – Chief Financial Officer Yeah, sure. Thank you. So, first of all, obviously the benefits from the restructuring program this year has not been significant, because it takes a while for the program to take effect, so the bigger impact will come
next year. What you see this year in terms of progress on the gross margin side is really around the replatforming and increased operational efficiency through consolidation of data center operations and infrastructure operations. And therefore, I'm very, very confident that we can continue to scale this business with increasing gross margin contributions also next year and even beyond next year.

Fig. 1. Excerpt from an earnings call transcript for SAP SE (triple dots and horizontal lines indicate editorially cut material for reasons of space)

evaluated by comparing their performance against each other and against several well-known baselines.

There are not many summarization approaches that have been directed towards finance [9,17], and we are not aware of any previous work on summarization applied to text type of financial earnings calls transcripts, one of the primary sources of evidence for investors dealing with public companies.

2   Related Work

General Summarization. The earliest summarizer by Luhn at IBM was built for business communications [20], a heuristic sentence selection method. [30] and [25] are monograph-length general and comprehensive surveys that cover the history of summarization and seminal methods until just before the arrival of deep learning methods.

News Summarization. See [11] for a discussion of typical news summarization baselines. Recent approaches like [23] or [29] are representative examples of state of the art neural models for news summarization.

Financial Summarization. [10] propose a method for financial summarization that uses a variant of TF IDF weighting in the relevance weighting for sentences where the inverse document frequency is conditioned specifically on company-relevant documents, which penalizes words common to company information (such as words like "company", "CEO" but also more specifically "Apple", "computer", "iPad" for Apple Inc., for instance). They assume the (professional) user is already familiar with the company (e.g., a financial investment analyst whose job it is to study one particular oil company on a daily does not need to be told the name of its CEO). Their objective is to provide actionable information for near-term trading of the company (inference from news to stock price movement within a day). They also introduce a novel query expansion method based on the company's name. [13] present a neural model for Japanese financial reports (already summarized by humans) and news. [17] provides a survey of financial summarization work. More recently, [14] use a Longformer-Encoder-Decoder (LED) model that they fine-tuned in two rounds, first on scientific summaries from ArXiv and then on British financial annual reports, to summarize financial reports as part of the Financial Narrative Summarization (FNS) shared task [7,8]. Unlike our work, they did not do dedicated pre-training on financial language, and they reported that their fine-tuning did not improve over a zero-shot approach (i.e., just running the model pre-trained on news without further fine-tuning). Their work is complementary to ours as financial reports typically get published around the same time earnings calls are held, i.e. quarterly and/or annually.[2]

3   Data Set and Annotation

As our gold data, we randomly sampled $N = 50$ English-language documents from the two-year period 2017-2018 of a commercially available multi-year dataset of financial earnings call transcripts (Anonymized) and stripped off leading boilerplate text like the cast of characters on the call (Table 1); $k = 5$ human annotators, all financial information professionals that work with transcripts on a daily basis, and each of whom have several years of financial data experience were tasked to classify each sentence for binary relevance using a Web interface. To reduce arbitrariness, they were instructed to aim for a soft target summary size of 20% of sentences compared to the original document length: annotators judged all 20,463 sentences of our sample (102,315 total judgments). Annotators processed all sentences of all sample documents (complete overlap), leading to 5 binary judgments for each sentence (i.e., whether a sentence is essential or not). We chose many judgments per data point because of the known difficulty of the task and because even partial agreement can be integrated during training and evaluation (e.g. [26]). Our data set contains two labels. The first label is a binary label that is true if $k \geq 3$ annotators have identified a sentence belonging to the summary. We chose 3, since this lets us do absolute majority voting, and sentences picked up by at least 3 annotators also correspond to 25% of the overall corpus (deviation of the length of the gold summary is 6%), which also agrees with coder guidance. The second label contains the number of times a sentence has been identified as part of the summary. The latter is used

---

[2] We are grateful to one anonymous reviewer that pointed out to us a recent pre-print on ArXiv at https://arxiv.org/pdf/2210.12467v1.pdf (uploaded on October 22, 2022 – after the ECIR submission deadline), which is about the release of a freely available dataset, also in the financial earnings call space.

Table 1. Earnings call transcript corpus: summary statistics.

Corpus text size (in word units)	312 k	Number of unique sentences (annot.)	20,463
Corpus text size (in MB)	1.5	Annotations per sentence	5
Number of documents	50	Number of companies covered	50

Listing 1.1. 2GREP: A one-line UNIX baseline summarizer (assumes one sentence per line input format; split into two physical lines here for formatting reasons only).

```
grep -E '[0-9]+' | \
   grep -i -E '(profit|loss|revenue|EBITDA|margin|EPS|dividends)'
```

for the Pyramid score (cf. below). We measured the inter-annotator agreement using Krippendorff's α ([15]), and found it to be low (0.36), consistent with past observations that humans find it hard to agree in sentence selection tasks ([21]). Unfortunately, due to the commercial nature of our project, the dataset cannot be released; however, our approach towards creating the gold data as described here can be replicated in principle.

4 Methods

4.1 Baselines Used for Comparison

We will first lay out a set of baseline methods for comparison.

Random. We evaluate a random baseline, drawing from a uniform distribution until the number of sentences equals 20% of the expected summary length.

2GREP. A good question to pose is *what is the simplest conceivable baseline that is actually useful?* In the context of news summarization, it was found that simply taking the first three sentences of a story is a rather good summary (so-called "LEAD3-baseline" [23,24]). It can be implemented in UNIX by the simple command `head -n 3`.[3] Inspired by the quest for simplicity in times dominated by more and more complex models [6], we define the "2GREP-baseline" for summarizing financial earnings calls in Listing 1.1. 2GREP's summaries are very short, so there was no need to control size by imposing a 20% cut-off compared to other methods. The first part retains sentences that contain numbers, since monetary amounts are important, and the second part fishes for key company performance indicators in the hope that the numeric information pertains to these (note the cascaded filter implements an implied Boolean "AND" operator). Any more sophisticated method should at least be able to outperform 2GREP to command our attention.

---

[3] This and the next command assume a one sentence per line format.

Luhn. We also compare our method to [20]'s due to its simplicity, familiarity, as it has long served as a reference. The Luhn algorithm is a simplified version of a TF-IDF calculation of words within sentence-units and takes into account density. Note the fundamental difference between the Luhn and 2GREP baselines, despite their simplicity: whereas Luhn's method is based on information retrieval metrics like term frequency applied to a token window of fixed size, 2GREP uses finite-state pattern matching techniques, closer to those applied in information extraction, to find sentences with number-dimension pairs, such as "profit increased by 5%".

LexRank. LexRank is an unsupervised, graph-based approach inspired by the random walker model that is also behind PageRank [3]; it uses centrality scoring of sentences. We used the Sumy library for our experiments.[4]

BertSum. BertSum [18] is simple variant of BERT [6] with inter-sentence Transformer layers. In BertSum, as well as BERT, position embeddings have a maximum length of 512 tokens. In the majority of summarization data sets, this length is sufficient and longer documents can be truncated without loss of information, at least with regards to the gold-standard summary. A follow-up paper by the same authors [19] allows extending the length of the document further to 800 tokens, in order to accommodate the NYT dataset. In our case, we faced 2 limitations: (a) the documents in our labelled data are not enough for re-training BertSum on our corpus, and (b) our documents are significantly larger (avg. number of tokens: 7,339). In fact, even the mean length of the expected summaries is longer than 512 tokens. In order to overcome the above limitations, we break the documents into chunks of 512 tokens, allowing for one sentence overlap (stride of size 1). For each chunk, we collect the score of the sentences. We aggregate all scores for the whole document, and keep the highest scored sentences to generate a summary of size equal to gold summary length.

4.2   Our Novel Methods

We now describe our novel methods, which treat each document as a sequence of sentences (including the question & answer section) for which a decision (i.e., whether to include a sentence in the summary or not) has to be made. **Heuristic Approach.** Our first extractive summarizer, called *ECSumm/Rul*, uses features as heuristics to estimate the salience of any sentence. The feature vector $f = (f_1; \ldots ; f_9)$ has the elements or feature functions shown in Table 2.

A simple scoring mechanism then adds one bonus point if digit sequences and currency designators are seen together. To avoid favoring extremely short sentences, a bonus is awarded for sentences exceeding 50 characters in length. If $|sentiment\_polarity| > 0.5$ (based on the Vader sentiment lexicon described in [12]), another point is awarded. Yet another bonus point is given for each financial lexicon match and each capitalized word (except at the beginning of a

---

[4] https://pypi.org/project/sumy/ (cited 2020-01-10).

Table 2. Feature functions used by ECSumm/Rul, ECSumm/Bin and ECSumm/Reg.

No	Feature name	Type	Description
f_1	DIGSEQS	Int	Number of disconnected digit sequences in a sentence
f_2	LENGTH	Int	The sentence length in number of characters
f_3	HASPERC	Bool	Whether this sentence contains at least one "%"
f_4	HASCURR	Bool	Whether a currency word or symbol is present
f_5	SENTIM	Real	Absolute value of the sentiment polarity [12]
f_6	FINLEX	Int	Number of matches from a tiny financial lexicon
f_7	CAPTOK	Int	Number of capitalized tokens
f_8	FLUFF	Int	Number of fluff phrases that match in the sentence
f_9	DISTFSENT	Real	Distance from the beginning of the text (in percent)

sentence). A "fluff" lexicon is a short list of phrases to identify boilerplate language or politeness protocol utterances (greetings, thanks etc.) used to exclude such sentences. Our definition of a "fluff phrase" is "a sentence containing them may be removable from a summary without substantial loss". ECSumm/Rul penalizes matches from it with 3 penalty points subtracted from a sentence's score. The sentences are then re-ranked based on their heuristic score values, after which the top-k are selected that constitute the summary. The financial and fluff lexicons include the 2GREP keywords but extend them with additional as well as adversarial signals in the hope to beat the baseline; this was done by inspecting a few transcripts outside our corpus and based on human intuition.

Learning Approaches. Throughout this section, and for all methods that include supervised learning, we perform the training and evaluation using a uniform, document-level, 3-fold cross-validation. *Decision Tree Regressor and Binary Classifier.* We induced our second summarizer, a decision tree regressor pruned to 8 levels max. to avoid overfitting, and call it ECSumm/Reg. This model is trained from the regression data wherein label values are derived from the number of votes received. A third, binary tree classifier called ECSumm/Bin is also implemented from our data with the binary labels. During training, standard deviation of the accuracy was less than 0.002 (binary classifier) whereas for Mean Square Error of the regressor was less than 0.028. Finally, our summarizers produce an output that has 20% of the number of input sentences, in line with the guidance for the annotation task.

Two Neural Language Model-Based Approaches. Our next extractive summarizer is called ECSumm-Bert-Base, and is based on BERT [6], which leverages large-scale pre-training and a multi-headed attention architecture to learn complex features from text. We modify the standard BERT-based architecture, consisting of 12 transformer layers each with 12 attention heads, 768 hidden units, and a total of 110 million parameters, by adding a single linear layer. Inputs are single sentences from the transcript, and a predicted score is generated for each by passing the classification token (CLS) representation through the linear layer and a sigmoid function. We fine-tune the model using our gold data corpus, starting with the cased version of the BERT-base checkpoint files, for 40 epochs using a learning rate of 1e-5, a dropout rate of 0.1, a learning rate warm-up proportion of 0.1, a batch size of 64 and a mean squared error (MSE) loss function. We also developed a variant of this model by applying additional pre-training on top of the BERT-base checkpoint before commencing fine-tuning, which we call ECSumm/Bert-Tran. For the pre-training protocol, we used a large corpus of earnings calls, consisting of 390,000 transcripts totaling 2.9 billion tokens. We converted this corpus into TensorFlow record format[5] for BERT's masked language model (LM) and next sentence prediction loss function tasks, at sequence lengths of 128 and 512 tokens with a duplication factor of 10 and masked LM probability of 0.15. We performed sentence boundary disambiguation using spaCy[6]. Pre-training was run using a batch size of 512 for 2.5 million steps at a maximum sequence length of 128 tokens, and then 500,000 steps at a maximum sequence length of 512 tokens, since the long sequences are mostly needed to learn positional embeddings which can be learned fairly quickly. This additional pre-training results in a domain-specific version of BERT, which has been demonstrated to yield significant performance gain when fine-tuned on in-domain downstream tasks [2]. After pre-training, we fine-tuned the model in exactly the same way as ECSumm/Bert-Base. After establishing that our human-crafted features f1-f9 did not lead to an improvement for our two BERT-based models, we removed them from the ECSumm/Bert-Base and ECSumm-Bert-Tran pipelines.

5   Evaluation and Discussion

Methodology: Metrics and Protocol. We will evaluate whether all assessed summarization methods can retrieve the sentences marked as "relevant" in the gold data reference corpus, and report Precision, Recall and F-score. We also report a variant of the Pyramid score [26], a consensus-based metric that scores a proposed summary against a pool of human-generated summaries which has text spans marked as *Summary Content Units (SCUs)*. The Pyramid score is usually used to evaluate abstractive summarization tasks; we borrow it to evaluate extractive summarization based on sentence level units by making our SCUs full sentences. Furthermore, in our implementation, SCUs are sentences which are assigned a rank based on the number of appearances in human summaries. A summary is scored based on the sum of its ranked units, normalized by the optimal score it would have achieved for its length. This ensures that the 5-fold redundant annotation of each of our gold data sentences gets put to the best use. The exact formula we use is computed as follows: let v_i be the number of votes a sentence with index i has received. For a given summary S of length n, we define its weight as follows: $D = \sum_{i \in S} v_i$. We assign the optimal weight for this summary as the weight you could have acquired for its given length: $w_{Max} = \sum_{i \in S_{MAX}} v_i$, where $|S_{Max}| = |S|$ and

---

[5] https://www.tensorflow.org/tutorials/loaddata/tfrecord (accessed 2020-05-28).
[6] https://spacy.io (accessed 2020-05-28).

$S_{Max} = \arg\max_{x}\{v_x\} = \{x | \forall y : v_y \leq v_x \land y \notin w_{Max}\}$. Hence, the pyramid score is defined as $P = \frac{D}{w_{Max}}$.

Quantitative Evaluation. We conducted a component-based evaluation using our gold standard corpus for assessing the quality of a range of summarizers and measured precision (P), recall (R), and F1-score (cf. Table 3). We report both macro- and micro-average results because the task is document-centric (macro) but the unit of decision is a sentence (micro). A gold-standard is derived automatically using a voting regime that required the agreement of at least 3 annotators in order to assign a "relevant sentence" gold label to a sentence based on three human annotators' judgements (Fig. 2).

Table 3. Evaluation results (best scores in bold; note that comparison methods (rows 1-4) are arranged roughly in order of increasing complexity, and that the Pyramid score broadly decreases (rows 2-5) rather than increases with method complexity).

Method	Pyram.	Macro (documents)			Micro (sentences)		
		Precision	Recall	F1	Precision	Recall	F1
Random baseline	0.36	0.25	0.20	0.22	0.24	0.20	0.22
2GREP baseline	0.74	**0.79**	0.12	0.19	**0.78**	0.11	0.19
Luhn	0.49	0.31	0.35	0.32	0.28	0.35	0.31
LexRank	0.42	0.25	0.31	0.27	0.23	0.31	0.27
BertSum	0.38	0.25	0.25	0.24	0.23	0.25	0.24
ECSumm/Rul	0.49	0.36	0.36	0.35	0.34	0.36	0.35
ECSumm/Bin	0.61	0.52	0.42	0.46	0.49	0.41	0.45
ECSumm/Reg	0.78	0.65	0.53	0.57	0.62	0.52	0.57
ECSumm/Bert-Base	**0.86**	0.74	**0.61**	**0.66**	0.71	**0.60**	0.65
ECSumm/Bert-Tran	**0.86**	0.74	**0.61**	**0.66**	0.72	**0.60**	**0.66**

Qualitative Evaluation. Speaking qualitatively, we have observed the following types of errors made by *all* models:

– the exclusion of relevant sentences;
– the inclusion of non-relevant or redundant sentences or fragments;
– suboptimal ordering of the sentences selected to make up a summary;
– the lack of cohesion of a sequence of sentences part of a summary.

Most of these phenomena may merit specific mechanisms to be developed in order to improve the resulting summaries.

6 Summary, Conclusions and Future Work

Here, we introduced the task of automatic financial earnings call transcript summarization for the English language and from a single-document, extractive perspective. We presented one new, unsupervised (heuristic) method and several

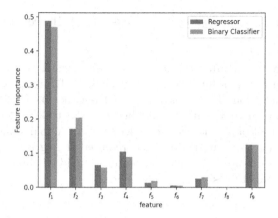

Fig. 2. Individual feature contribution of the 9 features in ECSumm/Reg (left) and ECSumm/Bin (right)

supervised learning models for the task, which were evaluated on a new gold data corpus in an automatic component evaluation based on sentence-level gold data. We described our industry-strength ECSumm system implementation, which supports these methods. We also described a variant of the Pyramid metric borrowed from abstractive summarization that makes use of this overlap to provide a robust automatic evaluation, and used it on a corpus with 5 judgments per sentence.

How do our findings support or refute our three hypotheses?

/H1/. Our experiments show: traditional news summarization methods are indeed insufficient for this document type, as a range of news summarizers fail when tested on earnings transcripts.

/H2/. We devised a set of simple features for the financial earnings call domain, and found them to work reasonably well, somewhat supporting our second hypothesis. As we expected, supervised learning mostly outperforms a heuristic approach. Compared to our best model overall, and in line with our expectations informed by the success stories in other application areas ([27]), our deep learning based neural models especially pre-trained from scratch for this task on financial English outperform approaches based on human-engineered features.

/H3/. Indeed, our two best models, ECSumm-Bert-Base and ECSumm-Bert-Tran, outperform all other methods. In evaluation, 2GREP performs best in terms of precision; however, its recall is extremely low as it is unable to make use of the expected summary length according to our evaluation protocol that sought a 20% summary length. **Still, it is perhaps our most remarkable finding that a single-line UNIX shell script can outperform BERT on any measure.** ECSumm/Bert-Tran and ECSumm/Bert-Base are the overall top approaches, with Bert-Tran being marginally better. Future work could explore modeling the discourse structure explicitly. in ways that inform existing summarization models with discourse knowledge, e.g. via a Markov chain integrated as graph embeddings [31]. The notion of an *update summary* [22], i.e. what changed since the last earnings call, should also be investigated in the context of earnings calls.

7   Limitations

Our experiments were carried out based on a large quantity of sentences but still a small number of 50 documents (due to the great length of each of them). Our findings should be re-affirmed by additional experiments on more earnings calls. Our experiments were all in English, which is the main language for earnings calls worldwide, but there are a smaller number of companies that report e.g. in Japanese, for which our methods are not suitable without customization. One limitation of the non-transformer methods presented is the inability to find a globally good solution due to sentence selection based on local evidence only; the limitations of BERT regarding length and remedies were already discussed above. Finally, earnings reports call recordings and their transcripts are subject to copyright and database rights. Our research was done under commercial license agreements, and Fig. 1 constitutes Fair Use for education and research. They also contain personal information, but the individuals mentioned are in official functions of publicly traded companies. Bert-based models overfit to news articles: they have a tendency to favour sentences in the beginning of the text, are short and very similar to each other. In our error analysis, they are also weakly correlated with human judgement, consistent with past findings [16]. An overall finding of our study is a limitation of many summarization methods to date, namely they do not generalize across domains. Special adaptation to each domain (like financial earnings calls in this case) is very important to achieve good quality outcomes.

Acknowledgements. We would like to thank Geoffrey Horrell and Diana Serbu for supporting this project. The authors would like to particularly acknowledge the contributions of Michelle Scott, in person as well as a leader of her team, for numerous discussions, for contributing invaluable domain expertise and for managing and contributing to the annotation of our gold data corpus. We also thank three anonymous reviewers for helpful comments that improved the presentation of this paper.

References

1. Alterman, R.: Text summarization. In: Shapiro, S.C. (ed.) Encyclopedia of Artificial Intelligence, vol. 2, pp. 1579–1587. Wiley (1992)
2. Beltagy, I., Lo, K., Cohan, A.: SciBERT: A pretrained language model for scientific text. Tech. rep., https://arxiv.org/abs/1903.10676, arXiv pre-print
3. Brin, S., Page, L.: The anatomy of a large-scale hypertextual Web search engine. Comput. Netw. ISDN Syst. **30**(1–7), 107–117 (1998). https://doi.org/10.1016/S0169-7552(98)00110-X

4. Crawford Camiciottoli, B.: Earnings calls: Exploring an emerging financial reporting genre. Dis. Commun. **4**(4), 343–359 (2010). https://doi.org/10.1177/1750481310381681
5. Crawford Camiciottoli, B.: Ethics and ethos in financial reporting: Analyzing persuasive language in earnings calls. Bus. Commun. Q. **74**(3), 298–312 (2011). https://doi.org/10.1177/1080569911413810
6. Devlin, J., Chang, M.W., Lee, K., Toutanova, K.: BERT: Pre-training of deep bidirectional transformers for language understanding. In: Proceedings of the 2019 Conference of the North American Chapter of the Association for Computational Linguistics: Human Language Technologies, vol. 1, pp. 4171–4186. Association for Computational Linguistics, Minneapolis, MN, USA (2019). https://doi.org/10.18653/v1/N19-1423
7. El-Haj, M., AbuRa'ed, A., Litvak, M., Pittaras, N., Giannakopoulos, G.: The financial narrative summarisation shared task (FNS 2020). In: Proceedings of the 1st Joint Workshop on Financial Narrative Processing and MultiLing Financial Summarisation, pp. 1–12. COLING, Barcelona, Spain (Online) (2020), https://aclanthology.org/2020.fnp-1.1
8. El-Haj, M., et al.: The financial narrative summarisation shared task (FNS 2022). In: Proceedings of the 4th Financial Narrative Processing Workshop held at LREC 2022, pp. 43–52. European Language Resources Association, Marseille, France (2022)
9. El-Kassas, W.S., Salama, C.R., Rafea, A.A., Mohamed, H.K.: Automatic text summarization: A comprehensive survey. Expert Syst. Appl. **165**, 113679 (2021). https://doi.org/10.1016/j.eswa.2020.113679
10. Filippova, K., Surdeanu, M., Ciaramita, M., Zaragoza, H.: Company-oriented extractive summarization of financial news. In: Proceedings of the 12th Conference of the European Chapter of the ACL, EACL 2009, ACL, pp. 246–254. Association for Computational Linguistics, Athens, Greece (2009)
11. Hong, K., Conroy, J.M., Favre, B., Kulesza, A., Lin, H., Nenkova, A.: A repository of state of the art and competitive baseline summaries for generic news summarization. In: Proceedings of the 9th International Conference on Language Resources and Evaluation, LREC 2014, Reykjavik, Iceland, 26–31 May 2014, pp. 1608–1616 (2014)
12. Hutto, C.J., Gilbert, E.: Vader: A parsimonious rule-based model for sentiment analysis of social media text. In: Proceedings of the International AAAI Conference on Web and Social Media (2014)
13. Isonuma, M., Fujino, T., Mori, J., Matsuo, Y., Sakata, I.: Extractive summarization using multi-task learning with document classification. In: Proceedings of the 2017 Conference on Empirical Methods in Natural Language Processing, pp. 2101–2110. Association for Computational Linguistics, Copenhagen, Denmark (2017). https://doi.org/10.18653/v1/D17-1223
14. Khanna, U., Ghodratnama, S., Molla, D., Beheshti, A.: Transformer-based models for long document summarisation in financial domain. In: Proceedings of the The 4th Financial Narrative Processing Workshop held at LREC 2022, pp. 73–78. European Language Resources Association, Marseille, France (2022)
15. Krippendorff, K.K.: Bivariate agreement coefficients for reliability data. In: Sociological Methodology, pp. 139–150. Jossey, San Francisco, CA, USA (1970)
16. Kryscinski, W., Keskar, N.S., McCann, B., Xiong, C., Socher, R.: Neural text summarization: A critical evaluation. CoRR abs/ arXiv: 1908.08960 (2019)

17. Leidner, J.L.: Summarization in the financial and regulatory domain. In: Fiori, A. (ed.) Trends and Applications of Text Summarization Techniques, chap. 7. IGI Global, Hershey, PA, USA (2019)
18. Liu, Y., Lapata, M.: Fine-tune BERT for extractive summarization (2019). https://arxiv.org/abs/1903.10318
19. Liu, Y., Lapata, M.: Text summarization with pretrained encoders. In: Proceedings of the 2019 Conference on Empirical Methods in Natural Language Processing and the 9th International Joint Conference on Natural Language Processing (EMNLP-IJCNLP), pp. 3730–3740. Association for Computational Linguistics, Hong Kong, China (2019). https://doi.org/10.18653/v1/D19-1387
20. Luhn, H.P.: The automatic creation of literature abstracts. IBM J. Res. Developm. **2**(2), 159–165 (1958)
21. Mani, I.: Automatic Summarization. John Benjamins Publishing Company, Amsterdam/Philadephia (2001)
22. Mani, I., Bloedorn, E.: Multi-document summarization by graph search and matching. In: Proceedings of the 14th National Conference on Artificial Intelligence, pp. 622–628. Providence, Rhode Island (1997)
23. Nallapati, R., Zhai, F., Zhou, B.: SummaRuNNer: A recurrent neural network based sequence model for extractive summarization of documents. In: AAAI Conference on Artificial Intelligence (2017)
24. Nallapati, R., Zhou, B., Ma, M.: Classify or select: Neural architectures for extractive document summarization. ArXiv preprint - CoRR abs/ arXiv: 1611.04244 (2016)
25. Nenkova, A., McKeown, K.: Automatic Summarization, Foundations and Trends in Information Retrieval, vol. 5. Now, Delft, The Netherlands (2011)
26. Nenkova, A., Passonneau, R.: Evaluating content selection in summarization: The pyramid method. In: Proceedings of the Human Language Technology Conference of the North American Chapter of the Association for Computational Linguistics: HLT-NAACL 2004, 2 May–7 May, pp. 145–152. Association for Computational Linguistics, Boston, Massachusetts, USA (2004). https://www.aclweb.org/anthology/N04-1019
27. Qiu, X., Sun, T., Xu, Y., Shao, Y., Dai, N., Huang, X.: Pre-trained Models for Natural Language Processing: A Survey. arXiv e-prints arXiv:2003.08271 (Mar 2020)
28. Schilder, F., Kondadadi, R.: FastSum: Fast and accurate query-based multi-document summarization. In: Proceedings of ACL-08: HLT, Short Papers, pp. 205–208. Association for Computational Linguistics, Columbus, OH, USA (2008)
29. See, A., Liu, P.J., Manning, C.D.: Get to the point: Summarization with pointer-generator networks. In: Proceedings of the 55th Annual Meeting of the Association for Computational Linguistics, ACL 2017, vol. 1, pp. 1073–1083. Association for Computational Linguistics (2017). https://doi.org/10.18653/v1/P17-1099
30. Torres-Moreno, J.M.: Automatic Text Summarization. Cognitive Science and Knowledge Management Series, Wiley, Hoboken, NJ, USA (2014)
31. Xinyi, Z., Chen, L.: Capsule graph neural network. In: International Conference on Learning Representations, ICLR 2019, New Orleans, LA, USA, 6–9 May (2019)

Parameter-Efficient Sparse Retrievers and Rerankers Using Adapters

Vaishali Pal[1,2]([✉]) [ID], Carlos Lassance[2] [ID], Hervé Déjean[2] [ID],
and Stéphane Clinchant[2] [ID]

[1] IRLab, University of Amsterdam, Amsterdam, Netherlands
v.pal@uva.nl
[2] Naver Labs Europe, Meylan, France

Abstract. Parameter-Efficient transfer learning with Adapters have been studied in Natural Language Processing (NLP) as an alternative to full fine-tuning. Adapters are memory-efficient and scale well with downstream tasks by training small bottle-neck layers added between transformer layers while keeping the large pretrained language model (PLMs) frozen. In spite of showing promising results in NLP, these methods are under-explored in Information Retrieval. While previous studies have only experimented with dense retriever or in a cross lingual retrieval scenario, in this paper we aim to complete the picture on the use of adapters in IR. First, we study adapters for SPLADE, a sparse retriever, for which adapters not only retain the efficiency and effectiveness otherwise achieved by finetuning, but are memory-efficient and orders of magnitude lighter to train. We observe that Adapters-SPLADE not only optimizes just 2% of training parameters, but outperforms fully fine-tuned counterpart and existing parameter-efficient dense IR models on IR benchmark datasets. Secondly, we address domain adaptation of neural retrieval thanks to adapters on cross-domain BEIR datasets and TripClick. Finally, we also consider knowledge sharing between rerankers and first stage rankers. Overall, our study complete the examination of adapters for neural IR. (The code can be found at: https://github.com/naver/splade/tree/adapter-splade.)

Keywords: Adapters · Information Retrieval · Sparse neural retriever

1 Introduction

Information Retrieval (IR) systems often aim to return a ranked list of documents ordered with respect to their relevance to a user query. In modern web search engines, there is, in fact, not a single retrieval model but several ones specialized in diverse information needs such as different search verticals. To add to this complexity, multi-stage retrieval considers effectiveness-efficiency trade-off where first stage retrievers are essential for fast retrieval of potentially relevant candidate documents from a large corpus. Further down the pipeline, rerankers are added focusing on effectiveness.

J. Kamps et al. (Eds.): ECIR 2023, LNCS 13981, pp. 16–31, 2023.
https://doi.org/10.1007/978-3-031-28238-6_2

With the advent of large Pretrained Language Models (PLM), recent neural retrieval models have millions of parameters. Training, updating and adapting such models implies significant computing and storage cost calling for efficient methods. Moreover, generalizability across out-of-domain datasets is critical and even when effectively adapted to new domains, full finetuning often comes at the expense of large storage and catastrophic forgetting. Fortunately, such research questions have already been studied in the NLP literature [1,2,9,10] with parameter-efficient tuning. In spite of very recent work exploring parameter-efficient techniques for neural retrieval, the use of adapters in IR has been overlooked. Previous work on dense retriever had mixed results [11] and successful adaptation was achieved for cross lingual retrieval [17]. Our study aims to complete the examination of adapters for neural IR and investigates it with neural sparse retrievers. We study ablation of adapter layers to analyze whether all layers contribute equally. We examine how adapter-tuned neural sparse retriever SPLADE [5] fares on benchmark IR datasets MS MARCO [21], TREC DL 2019 and 2020 [3] and out-of-domain BEIR datasets [30]. We explore whether generalizability of SPLADE can be further improved with adapter-tuning on BEIR and out-of-domain dataset such as TripClick [26]. In addition, we examine knowledge transfer between first stage retrievers and rerankers with full fine-tuning and adapter-tuning. To the best of our knowledge, this is the first work which studies adapters on sparse retrievers, focuses on sparse models' generalizability and explores knowledge transfer between retrievers in different stages of the retrieval pipeline. In summary, we address the following research questions:

1. RQ1: What is the efficiency-accuracy trade-off of parameter-efficient finetuning with adapters on the sparse retriever model SPLADE?
2. RQ2: How does each adapter layer ablation affect retrieval effectiveness?
3. RQ3: Are adapters effective for adapting neural sparse neural retrieval in a new domain?
4. RQ4: Could adapters be used to share knowledge between rerankers and first stage rankers?

2   Background and Related Work

Parameter efficient transfer learning techniques aim to adapt large pretrained models to downstream tasks using a fraction of training parameters, achieving comparable effectiveness to full fine-tuning. Such methods [9,10,15,25,28] are memory efficient and scale well to numerous downstream tasks due to the massive reduction in task specific trainable parameters. This makes them an attractive solution for efficient storage and deployment compared to fully fine-tuned instances. Such methods have been successfully applied to language translation [25], natural language generation [16], Tabular Question Answering [22], and on the GLUE benchmark [7,28], In spite of all its advantages and a large research footprint in NLP, parameter-efficient methods remain under-explored in IR.

A recent comprehensive study [4] categorises parameter efficient transfer learning into 3 categories: 1) Addition based 2) Specification based 3) Reparameterization based. Addition based methods insert intermediate modules into the pretrained model. The newly added modules are adapted to the downstream task while keeping the rest of the pretrained model frozen. The modules can be added vertically by increasing the model depth as observed in Houlsby Adapters [9] and Pfeiffer Adapters [25]. Houlsby Adapters insert small bottle-neck layers after both the multi-head attention and feed-forward layer of the each transformer layer which are optimized for NLP tasks on GLUE benchmark. Pfeiffer Adapter inserts the bottle-neck layer after only the feed-forward layer and has shown comparable effectiveness to fine-tuning on various NLP tasks. Prompt-based adapter methods such as Prefix-tuning [15] prepend continuous task-specific vectors to the input sequence which are optimized as free-parameters. Compacter [20] hypothesizes that the model can be optimized by learning transformations of the bottle-neck layer in a low-rank subspace leading to less parameters.

Specification based methods fine-tune only a subset of pretrained model parameters to the task-at-hand while keeping the rest of the model frozen. The fine-tuned model parameters can be only the bias terms as observed in BitFit [2], or only cross-attention weights as in the case of Seq2Seq models with X-Attention [6]. Re-parameterization methods transform the pretrained weights into parameter efficient form during training. This is observed in LoRA [10] which optimises rank decomposition matrices of pretrained layer while keeping the original layer frozen.

Recent studies exploring parameter efficient transfer learning for Information Retrieval show promising results of such techniques for dense retrieval models [11,17,19,29]. [11] studies parameter efficient prefix-tuning [15], and LoRA [10] on bi-encoder and cross-encoder dense models. Additionally, they combine the two methods by sequentially optimizing one method for m epochs, freezing it and optimizing the other for n epochs. Their studies show that while cross-encoders with LoRA and LoRA+(50% more parameters compared to LoRA) outperform fine-tuning with TwinBERT [18] and ColBERT [13], parameter-efficient methods *do not outperform fine-tuning* for bi-encoders across all datasets. [17] uses parameter-efficient techniques such as Sparse Fine-Tuning Masks and Adapters for multilingual and cross-lingual retrieval tasks with rerankers. They train language adapters with Masked Language Modeling (MLM hereafter) task and then task-specific retrieval adapters. This enables the fusion of reranking adapter trained with source language data together with the language adapter of the target language. Concurrent to our work, [29] studies parameter-efficient prompt tuning techniques such as Prefix tuning and P-tuning v2, specification based methods such as BitFit and adapter-tuning with Pfeiffer Adapters on late interaction bi-encoder models such as Dense Passage Retrieval [12] and ColBERT. They are motivated by cross-domain generalization of dense retrievals and achieve better results with P-tuning compared to fine-tuning on the BEIR benchmark. [19] studies various parameter-efficient tuning procedures at both retrieval and reranking stages. They conduct a comprehensive study of parameter-efficient techniques such as BitFit, Prefix-tuning, Adapters, LoRA, MAM adapters with dense bi-encoders and cross-encoders with BERT-base as the backbone model. Their parameter-efficient techniques achieve comparable effectiveness to fine-tuning on top-20 retrieval accuracy and marginal gains on top-100 retrieval accuracy.

Compared to prior works, our experiments first study the use of adapters for state of the art sparse models such as SPLADE, contrary to previous work that studied dense bi-encoder models[1]. Furthermore, our results show improvements compared to the previous studies. We also studied the case of using distinct adapters for query and document encoders in a "bi-adapter" setting where the same pretrained backbone model is used by both the query and the document encoder but different adapters are trained for the queries and documents. Secondly, we address another research questions ignored by previous work, which is efficient domain adaptation[2] for neural first stage rankers. We start from a trained neural ranker and study adaptation with adapters on a different domain, such as the ones present in the BEIR benchmark. Finally, we also study parameters sharing between rerankers and first stage rankers using adapters, which to our knowledge has not been studied yet.

3   Parameter-Efficient Retrieval with Adapters

In this section, we first present the self-attention used in transformers and how the adapters we use for our experiments interact with them. We then introduce the models used for first stage ranking and reranking.

3.1   Self-attention Transformer Layers

Large pretrained language models are based on the transformer architecture composed of N stacked transformer layers . Each transformer layer comprises of a fully connected feed-forward module and a multi-headed self attention module. Each attention layer has a function of query matrix ($Q \in R^{nXd_k}$), a key matrix and a value matrix. The attention can be formally written as:

$$A(Q, K, V) = softmax(\frac{QK^T}{\sqrt{d_k}})V \tag{1}$$

where the query Q, key K and value V are parameterized by weight matrices $W_q \in R^{nXd_k}$, $W_k \in R^{nXd_k}$, and $W_v \in R^{nXd_v}$, as $Q = XW_q$, $K = XW_k$ and $V = XW_v$. Each of the N heads has its respective Q_i, V_i and K_i weights and its corresponding attention A_i. The feed-forward layer takes as input a transformation of the concatenation of the N attentions as:

$$FFN(x) = \sigma(XW_1 + b_1)W_2 + b_2 \tag{2}$$

where $\sigma(.)$ is the activation function. A residual connection is further added after each attention layer and feed-forward layer.

---

[1] To the best of our knowledge the only work involving SPLADE and adapters/freezing layers is [32], which found that freezing the embeddings improves effectiveness.

[2] Here we use adaptation as further finetuning on the target domain.

3.2 Adapters

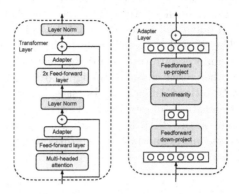

Fig. 1. Houlsby Adapter, image from the original paper [9]

In this paper, we focus on the Houlsby adapter [9], which as described in Sect. 3 can be considered an additive adapter and is depicted in Fig. 1. An additive adapter inserts trainable parameters in addition to the aforementioned transformer layers. The added modules form a bottle-neck architecture with a down-projection, an up-projection and a non-linear transformation. The size of the bottle-neck controls the number of training parameters in an adapter layer. Additionally, a residual connection is applied across each adapter layers. Finally, a layer normalization is added after each transformer sublayer. Formally, this is defined as:

$$x = f(hW_{down})W_{up} + x \qquad (3)$$

where $x \in R^d$ is the input to the adapter layer, $W_{down} \in R^{dXr}$ is the down projection matrix transforming input x into bottle-neck dimension d, $W_{up} \in R^{rXd}$ is the up projection matrix transforming the bottle-neck representation back to the d-dimensional space. Each adapter layer is initialized with a near-identity weights to enable stable training.

3.3 Neural Sparse First Stage Retrievers

Neural sparse first stage retrievers learn contextualized representations of documents and queries in a sparse high-dimensional latent space. In this work, we focus on SPLADE sparse retriever [5,14], which uses both L_1 and *FLOPS* regularizations to force sparsity. We freeze the pretrained language model while training the adapter layers. SPLADE predicts term weights of each vocabulary token j with respect to an input token i as:

$$w_{ij} = transform(h_i)^T E_j + b_j \qquad j \in 1, ..., |V| \qquad (4)$$

where E_j is the jth vocabulary token embedding, b_j is it's bias, h_i is i^{th} input token embedding, $transform(.)$ is a linear transformation followed by GeLU activation and LayerNorm. The final term importance for each vocabulary term j is obtained by taking the maximum predicted weights over the entire input sequence of length n, after applying a log-saturation effect:

$$w_j = \max_n \ \log(1 + ReLU(w_{ij})) \tag{5}$$

Given a query q_i, the ranking score s of a document d is defined by the degree to which it is relevant to q obtained as a dot product $s(q,d) = w(q).w(d)$. The learning objective is to discriminate representations obtained from Eq. 5 of a relevant document d^+ and non-relevant hard-negatives d^- obtained from BM25 and in-batch negatives $d^-_{i,j}$ by minimizing the contrastive loss:

$$L = -log \frac{e^{s(q_i,d^+)}}{e^{s(q_i,d^+_i)} + e^{s(q_i,d^-_i)} + \sum_j e^{s(q_i,d^-_{i,j})}} \tag{6}$$

SPLADE can be further improved with distillation. The learning objective here is to minimize the MarginMSE [5] loss: mean-squared-error between the positive negative margins of a cross-encoder teacher and the student:

$$L = MSE(M_s(q_i,d^+) - M_s(q_i,d^-), M_t(q_i,d^+) - M_t(q_i,d^-)) \tag{7}$$

where MSE is mean-squared error, M_t is the teacher's margin and M_s is the student's margin. The final objective optimizes either of the objective in Eq. 6 or 7 with regularization losses:

$$\mathcal{L}_{SPLADE} = L + \lambda_q \, \mathcal{L}_1 + \lambda_d \, \mathcal{L}_{FLOPS} \tag{8}$$

$$where \ \mathcal{L}_{FLOPS} = \sum_{j \in V} \widehat{a}^2_j = \sum_{j \in V} (\frac{1}{N} \sum_{i=1}^{N} w_j^{d_i}) \tag{9}$$

The Flops regularizer is a smooth relaxation of the average number of floating-point operations necessary to compute the score of a document, and hence directly related to the retrieval time. It is defined using as a continuous relaxation of the activation (i.e. the term has a non zero weight) probability a_j for token j, and estimated for documents d in a batch of size N by $\widehat{a}^2_j$.

Retrieval Flops: SPLADE also reports the retrieval flops (noted R-FLOPS), i.e., the number of floating point operations on the inverted index to return the list of documents for a given query. The R-FLOPS metric is defined by an estimation of the average number of floating-point operations between a query and a document which is defined as the expectation $\mathbb{E}_{q,d} \left[\sum_{j \in V} p_j^{(q)} p_j^{(d)} \right]$ where p_j is the activation probability for token j in a document d or a query q. It is empirically estimated from a set of approximately 100k development queries, on the MS MARCO collection. It is thus an indication of the inverted index sparsity and of the computational cost for a sparse model (which is different from the inference i.e. forward cost of the model)

3.4   Cross-Encoding Rerankers

Another way to use PLMs for neural retrieval is to use what is called "cross-encoding" [33]. In this case, both query and document are concatenated before being provided to the network and the score is directly computed by the network. The cross-encoding procedure allows for networks that are much more effective, but this effectiveness comes with a cost on efficiency as the retrieval procedure now has to go through the entire network for each query document pair, instead of being able to precompute document representations and only go through the network for the query representation. The models are trained with a contrastive loss as seen in Eq. (6) that aims to maximize the score of the true query/document pair compared to a BM25 negative query/document pair, without using in-batch negatives.

4   Experimental Setting and Results

We use the SPLADE github repository[3] to implement our modifications and followed the standard procedure to train SPLADE models. We implement our SPLADE models using an L_1 regularization for the query, and $FLOPS$ regularization for the document following [14]. Unless otherwise stated, the document regularization weight λ_d is set to 9e−5 and the query regularization weight λ_q to 5e−4 to train all variants of Adapters-SPLADE. In order to mitigate the contribution of the regularizer at the early stages of training, we follow [23] and use a scheduler for λ, quadratically increasing λ at each training iteration, until the 50k step. We use a learning rate of 8e−5, a batch size of 128, a linear scheduler and warmup step of 6000. We set the maximum sequence length to 256. We train for 300k iterations and keep the best checkpoint using MRR@10 on the validation set. We use a bottle-neck reduction factor of 16 (i.e. 16 times smaller) for all adapter layers. We use PyTorch [24], Huggingface Transformers [31] and AdapterHub [1] to train all models on 4 T V100 GPUs with 32GB memory. We compute statistical significance with $p \leq 0.05$ using the Student's t-test and use superscripts to identify statistical significance for almost all measures safe for metrics related to BEIR[4].

4.1   RQ1: Adapters-SPLADE

We study 2 different settings of encoding with adapters. The first called `adapter`, is a mono-encoder setup where the query and document shares a single encoder. The adapter layers are optimized with both the input sequences keeping the PLM frozen. The second setting inspired by the work on [14], is a bi-encoder setup which separates query and document encoders by training distinct query and document adapters on a shared frozen PLM. We call this setting `bi-adapter`. This setting not only benefits from optimizing exclusive adapters for input

---

[3] https://github.com/naver/splade.
[4] Due to lack of standard procedure.

sequence type (different lengths of query/document, etc.), it is also possible to use smaller PLMs for the queries instead of sharing PLM weights. We explore different backbone PLMs: `DistilBERT` and `CC+MLM Flops`, a pretrained PLM of cocondenser trained on the masked language model (MLM) task using the FLOPS regularization in order to make it easier to work with SPLADE, introduced in [14]. We trained and evaluated Adapter-SPLADE models on the MS MARCO passage ranking dataset [21] in full ranking setting. The results for finetuning with BM25 triplets are available in Table 1, whereas in Table 2 we make available the results of training models with distillation. For distillation, we use hard-negatives and scores generated by a cross-encoder reranker[5] and the MarginMSE loss as described in [5] and set λ_d to 1e−2 and λ_q to 9e−2.

Table 1. Finetuning and adapter-tuning comparison using BM25 triplets for training.

Model	#	Method	MS MARCO dev		TREC DL 2019	TREC DL 2020	R-Flops	Training params
			MRR@10	R@1000	NDCG@10	NDCG@10		
DistilBERT	a	Finetuning	0.346	0.963	0.692	0.677	1.43	100%
	b	Adapter	0.351	0.968^a	0.711	0.676	1.44	2.23%
	c	Bi-adapter	0.352	0.967^a	0.690	0.666	0.74	2.23%
CC + MLM FLOPS	d	Finetuning	0.366^{abc}	0.977^{abc}	**0.712**	0.684	1.09	100%
	e	Adapter	$\mathbf{0.376}^{abcd}$	$\mathbf{0.980}^{abcdf}$	0.712	0.688	0.8	2.23%
	f	Bi-adapter	0.372^{abc}	0.976^{abc}	0.701	**0.700**	0.37	2.23%

Table 2. Finetuning and adapter-tuning comparison using distillation training.

Model	#	Method	MS MARCO dev		TREC DL 2019	TREC DL 2020	R-Flops	Training params
			MRR@10	R@1000	NDCG@10	NDCG@10		
DistilBERT	a	Finetuning	0.371	0.979^b	0.727	0.711	3.93	100%
	b	Adapter	0.373	0.975	0.728	0.716	1.86	2.16%
CC + MLM FLOPS	c	Finetuning	0.388^{ab}	0.982^{ab}	0.734	**0.732**	4.38	100%
	d	Adapter	$\mathbf{0.390}^{ab}$	$\mathbf{0.983}^{ab}$	**0.740**	0.729	2.34	2.16%

To study efficiency-effectiveness trade-off of Adapters-SPLADE, we compare effectiveness, R-FLOPS size and number of training parameters of adapter-tuned models with their baseline finetuned counterparts having the same backbone PLM. [23] first showed that R-FLOPs reduction is a reasonable measure of retrieval speed. R-FLOPS measure the average number of floating-point operations needed to compute a document score during retrieval. A sparse embedding and subsequently lower FLOP achieves a retrieval speedup of the order of $1/p^2$ over an inverted index where p is the probability of each document embedding dimension being non-zero.

---

[5] https://huggingface.co/cross-encoder/ms-marco-MiniLM-L-6-v2.

Overall, we observe, from Table 1 and 2, all variants of adapter-tuned SPLADE outperform all baseline fine-tuned counterparts on MS MARCO and TREC DL 2019. The distilled cocondenser with MLM mono-encoder model is the highest performing with an MRR@10 score of 0.390 and R@100 of 0.983. The difference in effectiveness between the mono-encoder and bi-encoder adapter-tuning is marginal and depends on the PLM. Most noteworthy, we also observe that the R-FLOPS are lower for adapter-tuned models indicating sparser representation than the fine-tuned counterparts. This is more pronounced in the adapter-tuned models with distillation. Finally, the **bi-adapter** models have even lower R-FLOPS than the mono-encoder settings, which shows that for the same effectiveness the bi-adapters models are more efficient and sparse. We also observe that the number of training parameters is only 2.23% of the total model parameters for *triplets* training (1.5M/67M for mono-adapter DistilBERT, 3M/135M for bi-adapter DistilBERT, 2M/111M for CC + MLM FLOPS) and 2.16% for the distillation process (1.5M/67M for mono-adapter DistilBERT, 2M/111M for CC + MLM FLOPS). This has direct consequence in low-hardware setting where adpaters with lower number of number of training parameters and gradients can be trained on a smaller GPU(such as 24GB P40) but full finetuning is infeasible. Overall, there is a clear advantage in using Adapter-SPLADE over finetuning, which differs from the previous results on dense adapters [11].

We also evaluate with the full BEIR benchmark [41] comprising of 18 different datasets to measure generalizability of IR models with zero-shot effectiveness on out-of-domain data. The results are listed in Table 3. We observe from that in the mono-adapter Triplets training, adapter outperforms finetuning on mean nDCG@10 with the highest gap in arguana. With CC+MLM Flops as the backbone model, finetuning and adapter-tuning performs similarly. However, adapter scores drop on models trained with distillation. This can be attributed to the adapter representations being sparser compared to the finetuned models. As depicted by the R-FLOPS in Table 1, adapter-tuned DistilBERT has less than half the number of R-FLOPS than its finetuned counterpart whereas CC+MLM Flops finetuned model has approximately 1.87 times the number of R-FLOPS of the adapter-tuned model. This reflects in model representation capacity in 0-shot setting in Table 3. However, as discussed in Sect. 4.3, adapters are well suited for domain adaptation when trained on out-of-domain datasets keeping the backbone retriever intact and free from catastrophic forgetting.

4.2   RQ2: Adapter Layer Ablation

Furthermore, we perform extensive adapter layer ablation by progressively removing adapter layers from the early layers of the encoder. Doing so results in n separate models for each layer ablation setting. The frozen pretrained model for our ablation studies is DistilBERT in a mono-encoder setting where the same instance of the encoder is used to encode both the document and the query, which is the same configuration as the adapter method in Table 1. This results in a total of 6 configurations for the ablation study corresponding to the 6 adapter layers after each pretrained transformer layer. The final experimental setting removes all 6 adapter layers $(0 - 5)$ and fine-tunes only the language model head.

Table 3. nDCG@10 score comparison on the BEIR zero-shot evaluation

Datasets	Triplets training				Distillation training			
	DistilBERT		CC + MLM FLOPS		DistilBERT		CC + MLM FLOPS	
	Finetuning	Adapter	Finetuning	Adapter	Finetuning	Adapter	Finetuning	Adapter
arguana	0.298	0.364	0.427	0.388	0.513	0.443	0.463	0.433
climate-fever	0.167	0.172	0.180	0.187	0.202	0.197	0.229	0.202
dbpedia-entity	0.379	0.392	0.388	0.401	0.419	0.417	0.438	0.432
Fever	0.730	0.734	0.724	0.722	0.773	0.757	0.792	0.773
fiqa	0.295	0.289	0.317	0.320	0.332	0.314	0.342	0.337
hotpotqa	0.626	0.647	0.650	0.603	0.687	0.670	0.687	0.629
nfcorpus	0.318	0.321	0.331	0.333	0.335	0.335	0.340	0.344
nq	0.481	0.482	0.506	0.523	0.522	0.508	0.539	0.544
quora	0.819	0.810	0.821	0.806	0.825	0.722	0.841	0.552
scidocs	0.143	0.150	0.151	0.153	0.154	0.147	0.152	0.153
scifact	0.614	0.611	0.658	0.669	0.687	0.658	0.690	0.673
trec-covid	0.694	0.684	0.668	0.689	0.703	0.728	0.700	0.713
webis-touche2020	0.270	0.255	0.277	0.274	0.260	0.258	0.294	0.290
mean	0.449	0.455	0.469	0.467	0.493	0.473	0.500	0.467

Table 4. Adapter layer Ablation with `adapters` on `DistilBERT` PLM.

#	Adapters Removed	MS MARCO dev		TREC DL 2019	TREC DL 2020	BEIR	R-Flops	Training params	Training Time (Hrs)
		MRR@10	R@1000	NDCG@10	NDCG@10	NDCG@10			
a	None	**0.351**^{cdefg}	**0.968**^{defg}	**0.711**^{fg}	0.676^g	0.455	1.44	2.23%	34.42
b	0	0.348^{defg}	0.967^{efg}	0.708^{fg}	0.674^g	0.458	1.27	2.01%	32.23
c	0–1	0.344^{fg}	0.968^{fg}	0.709^{fg}	**0.699**^{abdefg}	0.459	1.34	1.80%	28.55
d	0–2	0.341^{efg}	0.966^{fg}	0.703^{fg}	0.665^g	0.459	1.36	1.59%	26.70
e	0–3	0.325^{fg}	0.962^{fg}	0.689	0.660^g	0.455	1.50	1.37%	24.18
f	0–4	0.318^g	0.956	0.659	0.663^g	0.455	1.27	1.15%	22.51
g	0–5	0.312	0.955	0.660	0.617	0.449	2.78	0.90%	21.35

We note that such an experiment (dropping adapter layers from transformer models) has been studied in NLP [28] and was shown to improve both training and inference time while retaining comparable effectiveness. We report the effectiveness of each adapter ablation setting on MS MARCO, TREC DL 2019 and TREC DL 2020 in Table 4. We actually observe gradual performance drop for MS MARCO and TREC DL datasets as the training parameters decrease with the progressive removal of adapter layers as shown in Table 4. The drop is significantly higher (a drop of 0.25 MRR score) when layers are removed from the second half of the model ($\geq 0 - 3$). This phenomenon is consistent with studies in NLP [22,28] that task-specific information is stored in the later layers of the adapters. For the BEIR datasets, this effectiveness drop is not as evident until all adapters but the language model head is removed (configuration $0 - 5$). The last configuration also has less sparsity as observed from the R-FLOPS size of 2.78 compared to the other configurations. We also observe that the training time drops proportional to the drop in adapter layers. The training time for adapter-tune without any drop in adapter layers is 34.42 h on 4 T V100 GPUS for 150,000 iterations, and it drops to 26.70 h with only 1% drop in MRR with the first 0 − 2 adapter layers dropped. The lowest training time is 21.35 h with a drop of 3.2% in MRR for the configuration with all adapters dropped but the language model head.

4.3   RQ3: Out-of-Domain Dataset Adaptation

For the next research question, we want to check how adapters compare to full finetuning when adapting a model trained on MSMARCO on a smaller out-of-domain dataset. We evaluate this question under two scenarios: i) BEIR and ii) TripClick.

BEIR: On the beir benchmark we use 3 datasets (FEVER, FiQA and NFCorpus) that have training, development and test sets and aim for very different domains and tasks (fact checking , financial QA and bio-medical IR). We start from a pre-finetuned SPLADE model called "splade-cocondenser-ensembledistil" made available in [5]. We verify the effectiveness of the models in zero shot and get a first set of hard negatives. These hard negatives are then used to train either via finetuning of all parameters or via the introduction of adapters. The networks are trained for either 10 (FEVER) or 100 epochs (FiQA and NFCorpus), and at the end of each epoch we compute the development set effectiveness. We use the models with the best development set to compute the 1st round test set effectiveness and generate hard negatives that are used for another round of training that we call 2nd round (which repeats the 1st round, starting from the best network of the 1st round and using negatives from the 1st round).

Results are available in Table 5. While finetuning is not always able to improve the results over the zero-shot, mostly due to overfitting on the training/dev sets. For example, on fever fine-tuning first makes all representations as it can easily overfit to the training even without using many words and only on the second round of training started using more dimensions. On the other hand, adapter tuning is able to consistently improve the effectiveness over the zero shot and first rounds (even if it does not always perform the best, as is the case on NFCorpus). Overall, we conclude that adapters are more stable than finetuning when finetuning on these specific domains.

Table 5. Domain adaptation comparison on BEIR Datasets

Dataset	Training	Zero Shot		1st round		2nd round	
		NDCG@10	Recall@100	NDCG@10	Recall@100	NDCG@10	Recall@100
Fever	Finetuning	0.793	0.954	0.692	0.866	0.851	0.959
	Adapter			0.841	0.960	**0.881**	**0.964**
FiQA	Finetuning	0.348	0.632	0.371	0.678	0.356	0.694
	Adapter			0.373	0.675	**0.393**	**0.711**
NFCorpus	Finetuning	0.348	0.285	0.384	0.466	**0.403**	**0.484**
	Adapter			0.362	0.435	0.371	0.428

TripClick: Given that in the BEIR benchmark the adapters underperformed finetuning on bio-medical data, we decided to further experiment on a larger bio-medical dataset called TripClick. The TripClick collection [27] contains approximately 1.5 millions MEDLINE documents (title and abstract), and 692,000 queries. The test set is divided into three categories of queries: Head, Torso and Tail (according to their decreasing frequency), which contain 1,175 queries each. For the Head queries, a DCTR click model was employed to created relevance signals, otherwise raw clicks were used. We use the triplets released by [8]. Similarly to the BEIR experiments, we start from the "splade-cocondenser-ensembledistil" SPLADE model and fine-tune or adapt-tune it over 100,000 iterations (batch size equal to 100). As shown in Table 6, adapter-tuning shows very competitive results, on par with finetuning for head categories (frequent queries), and achieving even better results for the less frequent queries (torso and tail).

Table 6. Performance of mono-encoder on out-of-domain Tripclick Dataset

#	Training	HEAD (dctr)		HEAD		Torso		Tail	
		NDCG@10	Recall@100	NDCG@10	Recall@100	NDCG@10	Recall@100	NDCG@10	Recall@100
a	Finetuning	0.218	**0.579**	**0.302**	0.523	0.219	**0.679**	0.238	**0.722**
b	Adapter	**0.219**	0.578	0.299	**0.526**	**0.229**[a]	0.679	**0.253**[a]	0.720

4.4 RQ4: Knowledge Sharing Between Rerankers and First Stage Rankers

The final research question explores sharing knowledge between rerankers and first-stage rankers. We explore this with transforming first stage rankers into rerankers. First, we tune the pretrained `DistilBERT` for reranking task as a baseline for both finetuning and adapter-tuning. We then test transforming both sparse (`splade-cocondenser`) and dense (`tct_colbert-v2-msmarco`) first stage rankers into rerankers, using either fine-tuning or adapter-tuning. To be clear, the cross-encoder is initialized with the weights of the aforementioned first stage models, but the reranker classification head on the CLS token is randomly initialized. Also note that we rerank the top-1k returned from "splade-cocondenser-ensembledistil" (represented by "first stage" on table).

We compare adapter-tuning with finetuning and display the results in Table 7. We observe that finetuning the baseline model (`DistilBERT`) is better than adapter-tuning. When using first stage rankers, results are varied. Dense first stage rerankers were able to learn similarly with both adapter and fine-tuning. However, this was not the case for sparse first stage rankers (`splade-cocondenser-ensembledistil`). We posit that this may come from two different reasons: i) The SPLADE model does not focus on the CLS representations, but on the MLM head representations of all tokens, thus needing more flexibility; ii) The model has been trained multiple times (initial BERT training, then condenser, then cocondenser and finally SPLADE), while not always using

Table 7. Knowledge Sharing between first stage rankers and rerankers comparison between finetuning and adapter-tuning.

Base model	#	Training	MS MARCO dev	TREC DL 2019	TREC DL 2020
			MRR@10	NDCG@10	NDCG@10
First stage	a	None	0.383^e	0.732	0.721
DistilBERT	b	Finetune	0.396^{ace}	$\mathbf{0.764}^e$	0.736
	c	Adapter	0.388^e	0.737	0.727
SPLADE++	d	Finetune	$\mathbf{0.408}^{abceg}$	0.753	$\mathbf{0.743}$
	e	Adapter	0.358	0.723	0.707
TCT Colbert v2	f	Finetune	0.404^{abce}	0.749	0.731
	g	Adapter	0.400^{ace}	0.740	0.739

the same precision (fp16 or fp32), which under preliminary analysis seems to have made some parts of the model unusable for cross-encoding without full finetuning. Overall, there is slight gain in using the first stage model for the reranker. However, there's no increase in effectiveness of using adapters, we actually see worse effectiveness on all settings.

5   Conclusion

Retrieval models, based on PLM, require finetuning millions of parameters which makes them memory inefficient and non-scalable for out-of-domain adaptation. This motivates the need for efficient methods to adapt them to information retrieval tasks. In this paper, we examine adapters for sparse retrieval models. We show that with approximately 2% of training parameters, adapters can be successfully employed for SPLADE models with comparable or even better effectiveness on benchmark IR datasets such as MS MARCO and TREC. We further analyze adapter layer ablation and see a further reduction in training parameters to 1.8% retains effectiveness of full finetuning. For domain adaptation, adapters are more stable and outperform finetuning, which is prone to overfitting, On Tripclick dataset, adapters outperform on precision metrics Torso and Tail queries and performs comparably on Head queries. We explore knowledge transfer between first stage rankers and rerankers as a final study. Adapters underperform full finetuning when trying to reuse sparse model to rerankers. Dense first stage rankers perform similarly for adapters and finetuning while sparse first stage rankers is less effective compared to finetuning. We leave this as future work. As memory-efficient adapters are effective for Splade, we leave for future studying larger sparse models and their generalizability. Finally, an interesting scenario could also be to tackle unsupervised domain adaptation with adapters.

References

1. Beck, T., et al.: Adapterhub playground: Simple and flexible few-shot learning with adapters. In: ACL (2022)
2. Ben Zaken, E., Goldberg, Y., Ravfogel, S.: BitFit: Simple parameter-efficient fine-tuning for transformer-based masked language-models. In: Proceedings of the 60th Annual Meeting of the Association for Computational Linguistics, vol. 2: Short Papers, pp. 1–9. Association for Computational Linguistics, Dublin, Ireland (May 2022). https://doi.org/10.18653/v1/2022.acl-short.1, https://aclanthology.org/2022.acl-short.1
3. Craswell, N., Mitra, B., Yilmaz, E., Campos, D., Voorhees, E.M., Soboroff, I.: Trec deep learning track: reusable test collections in the large data regime. In: Proceedings of the 44th International ACM SIGIR Conference on Research and Development in Information Retrieval, pp. 2369–2375 (2021)
4. Ding, N., et al.: Delta tuning: A comprehensive study of parameter efficient methods for pre-trained language models. ArXiv abs/ arXiv: 2203.06904 (2022)
5. Formal, T., Lassance, C., Piwowarski, B., Clinchant, S.: From distillation to hard negative sampling: Making sparse neural ir models more effective. In: Proceedings of the 45th International ACM SIGIR Conference on Research and Development in Information Retrieval, SIGIR 2022, pp. 2353–2359. Association for Computing Machinery, New York, NY, USA (2022). https://doi.org/10.1145/3477495.3531857
6. Gheini, M., Ren, X., May, J.: Cross-attention is all you need: Adapting pretrained Transformers for machine translation. In: Proceedings of the 2021 Conference on Empirical Methods in Natural Language Processing, pp. 1754–1765. Association for Computational Linguistics, Online and Punta Cana, Dominican Republic (Nov 2021). https://doi.org/10.18653/v1/2021.emnlp-main.132
7. Han, W., Pang, B., Wu, Y.: Robust transfer learning with pretrained language models through adapters (2021). https://doi.org/10.48550/ARXIV.2108.02340, https://arxiv.org/abs/2108.02340
8. Hofstätter, S., Althammer, S., Sertkan, M., Hanbury, A.: Establishing strong baselines for tripclick health retrieval (2022)
9. Houlsby, N., et al.: Parameter-efficient transfer learning for nlp. In: Chaudhuri, K., Salakhutdinov, R. (eds.) ICML. Proceedings of Machine Learning Research, vol. 97, pp. 2790–2799. PMLR (2019)
10. Hu, E., et al.: Lora: Low-rank adaptation of large language models (2021)
11. Jung, E., Choi, J., Rhee, W.: Semi-siamese bi-encoder neural ranking model using lightweight fine-tuning. In: Proceedings of the ACM Web Conference 2022, WWW 2022, pp. 502–511. Association for Computing Machinery, New York (2022). https://doi.org/10.1145/3485447.3511978
12. Karpukhin, V., et al.: Dense passage retrieval for open-domain question answering. In: Proceedings of the 2020 Conference on Empirical Methods in Natural Language Processing (EMNLP), pp. 6769–6781. Association for Computational Linguistics, Online (Nov 2020). https://doi.org/10.18653/v1/2020.emnlp-main.550, https://aclanthology.org/2020.emnlp-main.550
13. Khattab, O., Zaharia, M.: Colbert: Efficient and effective passage search via contextualized late interaction over bert. In: Proceedings of the 43rd International ACM SIGIR Conference on Research and Development in Information Retrieval, SIGIR 2020, pp. 39–48. Association for Computing Machinery, New York (2020). https://doi.org/10.1145/3397271.3401075

14. Lassance, C., Clinchant, S.: An efficiency study for splade models. In: Proceedings of the 45th International ACM SIGIR Conference on Research and Development in Information Retrieval, pp. 2220–2226 (2022)

15. Li, X.L., Liang, P.: Prefix-tuning: Optimizing continuous prompts for generation. In: Proceedings of the 59th Annual Meeting of the Association for Computational Linguistics and the 11th International Joint Conference on Natural Language Processing, vol. 1: Long Papers, pp. 4582–4597. Association for Computational Linguistics, Online (Aug 2021). https://doi.org/10.18653/v1/2021.acl-long. 353, https://aclanthology.org/2021.acl-long.353

16. Lin, Z., Madotto, A., Fung, P.: Exploring versatile generative language model via parameter-efficient transfer learning. In: Findings of the Association for Computational Linguistics: EMNLP 2020, pp. 441–459. Association for Computational Linguistics, Online (Nov 2020). https://doi.org/10.18653/v1/2020.findings-emnlp. 41, https://aclanthology.org/2020.findings-emnlp.41

17. Litschko, R., Vulic, I., Glavas, G.: Parameter-efficient neural reranking for cross-lingual and multilingual retrieval (2022). https://doi.org/10.48550/arXiv.2204. 02292. CoRR abs/ arXiv: 2204.02292

18. Lu, W., Jiao, J., Zhang, R.: Twinbert: Distilling knowledge to twin-structured compressed bert models for large-scale retrieval. In: Proceedings of the 29th ACM International Conference on Information & Knowledge Management, CIKM 2020, pp. 2645–2652. Association for Computing Machinery, New York (2020). https:// doi.org/10.1145/3340531.3412747

19. Ma, X., Guo, J., Zhang, R., Fan, Y., Cheng, X.: Scattered or connected? an optimized parameter-efficient tuning approach for information retrieval (2022). https://doi.org/10.48550/arXiv.2208.09847, arXiv: 2208.09847

20. Mahabadi, R.K., Ruder, S., Dehghani, M., Henderson, J.: Parameter-efficient multi-task fine-tuning for transformers via shared hypernetworks. In: ACL (2021)

21. Nguyen, T., et al.: Ms marco: A human generated machine reading comprehension dataset. In: CoCo@ NIPs (2016)

22. Pal, V., Kanoulas, E., Rijke, M.: Parameter-efficient abstractive question answering over tables or text. In: Proceedings of the Second DialDoc Workshop on Document-grounded Dialogue and Conversational Question Answering, pp. 41–53. Association for Computational Linguistics, Dublin, Ireland (May 2022). https://doi.org/10. 18653/v1/2022.dialdoc-1.5, https://aclanthology.org/2022.dialdoc-1.5

23. Paria, B., Yeh, C.K., Yen, I.E., Xu, N., Ravikumar, P., Póczos, B.: Minimizing flops to learn efficient sparse representations. In: International Conference on Learning Representations (2019)

24. Paszke, A., et al.: Pytorch: An imperative style, high-performance deep learning library. In: Advances in Neural Information Processing Systems 32 (2019)

25. Pfeiffer, J., Vulic, I., Gurevych, I., Ruder, S.: Mad-x: An adapter-based framework for multi-task cross-lingual transfer. In: EMNLP (2020)

26. Rekabsaz, N., Lesota, O., Schedl, M., Brassey, J., Eickhoff, C.: Tripclick: the log files of a large health web search engine. In: Proceedings of the 44th International ACM SIGIR Conference on Research and Development in Information Retrieval, pp. 2507–2513 (2021)

27. Rekabsaz, N., Lesota, O., Schedl, M., Brassey, J., Eickhoff, C.: Tripclick: The log files of a large health web search engine. In: Proceedings of the 44th International ACM SIGIR Conference on Research and Development in Information Retrieval, pp. 2507–2513 (2021). https://doi.org/10.1145/3404835.3463242

28. Rücklé, A., et al.: AdapterDrop: On the efficiency of adapters in transformers. In: Proceedings of the 2021 Conference on Empirical Methods in Natural Language Processing, pp. 7930–7946. Association for Computational Linguistics, Online and Punta Cana, Dominican Republic (Nov 2021). https://doi.org/10.18653/v1/2021.emnlp-main.626

29. Tam, W.L., et al.: Parameter-efficient prompt tuning makes generalized and calibrated neural text retrievers (2022). https://doi.org/10.48550/ARXIV.2207.07087, https://arxiv.org/abs/2207.07087

30. Thakur, N., Reimers, N., Rücklé, A., Srivastava, A., Gurevych, I.: Beir: A heterogeneous benchmark for zero-shot evaluation of information retrieval models. In: Thirty-fifth Conference on Neural Information Processing Systems Datasets and Benchmarks Track (Round 2) (2021)

31. Wolf, T., et al.: Transformers: State-of-the-art natural language processing. In: Proceedings of the 2020 Conference On Empirical Methods in Natural Language Processing: System Demonstrations, pp. 38–45 (2020)

32. Yang, J.H., Ma, X., Lin, J.: Sparsifying sparse representations for passage retrieval by top-k masking. arXiv preprint arXiv:2112.09628 (2021)

33. Yates, A., Nogueira, R., Lin, J.: Pretrained transformers for text ranking: Bert and beyond. In: Proceedings of the 14th ACM International Conference on Web Search and Data Mining, pp. 1154–1156 (2021)

Feature Differentiation and Fusion for Semantic Text Matching

Rui Peng, Yu Hong$^{(\boxtimes)}$, Zhiling Jin, Jianmin Yao, and Guodong Zhou

School of Computer Science and Technology, Soochow University, Suzhou, China
rpeng124@gmail.com, tianxianer@gmail.com, zhljinjackson@gmail.com,
{jyao,guodongzhou}@suda.edu.cn

Abstract. Semantic Text Matching (STM for short) stands for the task of automatically determining the semantic similarity for a pair of texts. It has been widely applied in a variety of downstream tasks, e.g., information retrieval and question answering. The most recent works of STM leverage Pre-trained Language Models (abbr., PLMs) due to their remarkable capacity for representation learning. Accordingly, significant improvements have been achieved. However, our findings show that PLMs fail to capture task-specific features that signal hardly-perceptible changes in semantics. To overcome the issue, we propose a two-channel Feature Differentiation and Fusion network (FDF). It utilizes a PLM-based encoder to extract features separately from the unabridged texts and those abridged by deduplication. On this basis, gated feature fusion and interaction are conducted across the channels to expand text representations with attentive and distinguishable features. Experiments on the benchmarks QQP, MRPC and BQ show that FDF obtains substantial improvements compared to the baselines and outperforms the state-of-the-art STM models.

Keywords: Semantic Text Matching · Deep neural networks · Natural Language Processing

1 Introduction

STM is a fundamental and well-studied task of Natural Language Processing (NLP). It is defined as the task of determining the semantic consistency between texts. It has been applied for a wide range of downstream tasks. For example, in Community Question Answering (CQA), the STM model can be employed to retrieve the historical questions that are semantically equivalent to the queries.

Recently, the transformer-based PLMs have been leveraged to STM [24], playing the role of encoding sentences with attention mechanism (e.g., BERT [4], RoBERTa [14]). In general, they possess multi-layer transformers and learn to perceive and represent semantics from large corpora via well-designed self-supervised tasks. Transforming and fine-tuning PLMs have been proven effective in enhancing the current neural STM models.

© The Author(s), under exclusive license to Springer Nature Switzerland AG 2023
J. Kamps et al. (Eds.): ECIR 2023, LNCS 13981, pp. 32–46, 2023.
https://doi.org/10.1007/978-3-031-28238-6_3

Table 1. Examples of sentence pairs from the QQP corpus. The shared words of two sentences are marked in bold. A and B are the original sentences, while A' and B' denote the masked ones by replacing the shared words with a special token [MASK]. "Label" denotes the ground-truth label for matching, including *Paraphrase* (i.e., *matched*) and *Non − paraphrase*. "*Predict*" indicates the prediction of a PLM-based STM model (BERT is used here), where the percentage numbers represent the prediction confidence levels.

A	What is the best course for learning **data structures**?
B	How can I learn **data structures** effectively?
A'	What is the best course for learning [MASK][MASK]?
B'	How can I learn [MASK][MASK] effectively?
Label	Non-paraphrases
Predict	Paraphrases (79.5%) → Non-paraphrases (100%)
A	**What are the requirements to** get into a **German** university?
B	**What are the requirements to** apply for **German** universities?
A'	[MASK][MASK][MASK][MASK][MASK] get into a [MASK] university?
B'	[MASK][MASK][MASK][MASK][MASK] apply for [MASK] universities?
Label	Paraphrases
Predict	Non-paraphrases (79.1%) → Non-paraphrases (54.6%)

The apparent contributions of PLMs for STM can be attributed to their profound perception of linguistic phenomena, as well as their awareness of a broader range of commonsense knowledge. However, our findings show that PLMs fail to address the most challenging issue—anti-distraction. Distraction is caused by the repetitive contents occurring in a pair of sentences, which may easily distract an STM model. For example, the two instances in Table 1 separately provide a pair of sentences A and B, and each pair contains a large block of duplicated contents, such as "*learn data structures*" and "*what are the requirements to*". Such contents cause a close similarity between sentences, and therefore they are extra-distracting for determining semantic consistency.

In order to alleviate the distraction problem, we propose a two-channel Feature Differentiation and Fusion network (FDF). It is designed to highlight the features of non-repetitive contents in a sentence pair, with less information loss of attentive features in the repetitive contents. Specifically, we conduct deduplication for the sentence pair by masking the shared tokens in them (see A' and B' in Table 1). This produces a pair of seemingly abridged sentences that merely possess non-repetitive contents. We utilize PLMs to extract token-level context-aware features for the original (i.e., unabridged) and abridged sentence pairs, through two separate channels. The resultant features are referred to as "shareable features" and "exclusive features" respectively. On this basis, Graph Convolutional Network (GCN for short) [11] is used to model interactions between shareable and exclusive features. Conditioned on the interaction strength, we additionally apply a gated layer to capture the important information of exclusive features, fusing that into shareable features. The goal is to produce distinguishable features with less information loss of attentive shareable features.

In our experiments, we combine the aforementioned shareable and exclusive features, and follow the conventional approaches to perform self-attention computation over them as well as pooling. Using the encoded features as reliance, we conduct binary classification (paraphrase or non-paraphrase) by a fully-connected linear layer with Softmax. Experiments are carried out over different benchmark corpora, including English MRPC [5] and QQP [9], as well as Chinese BQ [1]. Experimental results show that our method (FDF) yields substantial improvements compared to the PLM baselines BERT [4] and RoBERTa [14], where the most significant improvement is up to 2.1% accuracy rate (*Acc.*). Besides, FDF outperforms the state-of-the-art STM models.

The main contributions of this paper are concluded as follows:

- We propose to highlight the exclusive features under the condition that attentive information of shareable features is perceived and preserved.
- We construct a new PLM-based STM model (FDF) whose distinct components lay in the part of feature differentiation and fusion. Experiments show that FDF outperforms the existing STM models, over English and Chinese benchmark corpora.

2 Related Work

The previous work can mainly be divided into two categories: representation-based [10] and interaction-based approaches [4]. Representation-based models are generally constructed with the Siamese architecture, which encodes the considered sentences into embeddings separately, and decodes their relationship like semantic consistency by similarity estimation over embeddings. By contrast, interaction-based models straightforwardly involve interaction characteristics of sentences into the feature representation during encoding, instead of in the decoding phase. Our FDF can be sorted into the family of interaction-based models.

In order to perceive and represent deep features of texts for semantic matching, a variety of neural networks have been utilized at the earlier time, including CNN [8,12], RNN [2,17] and attention mechanism [10,22]. Recently, PLMs have been further leveraged for SMT due to their remarkable success in boosting performance and versatility.

Specifically, Zhang et al. [23] incorporate a relation of relation classification task into their method to fully exploit the pairwise relation information. Their proposed method obtains the performance of 84.3% *Acc* on the MRPC corpus. Zou et al. [24] construct STM models merely using PLMs. Though, they develop a sophisticated and effective training strategy (namely divide-and-conquer training), where different losses are considered for optimizing the STM models, including KL-divergence loss, binary classification loss and distant supervision loss. Such training strategy contributes to feature differentiation and well-directed matching of concrete and abstract contents.

In addition, external knowledge has been used to enhance the PLM-based STM models. Liu et al. [13] train a semantic labeler over the external dataset

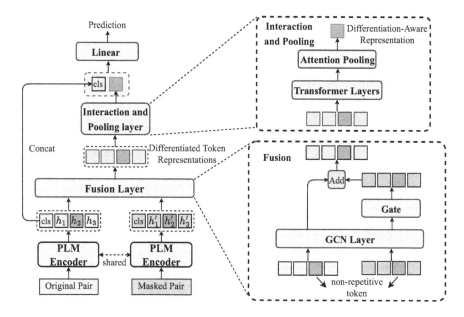

Fig. 1. Overview of the FDF network.

CONLL-2005, which provides semantic role embeddings for tokens. On this basis, they integrate semantic role embeddings with context-aware embeddings output by BERT. Xia et al. [21] incorporate synonym knowledge into BERT, enhancing word similarity perception at the self-attention computation stage.

3 Approach

First, let us give a formal definition of the STM task. Given two sentences $S_a = \{t_{a_1}, t_{a_2}, \cdots, t_{a_m}\}$ and $S_b = \{t_{b_1}, t_{b_2}, \cdots, t_{b_n}\}$ as the input, an STM model is required to output the binary decision about whether S_a and S_b are semantically consistent. Thus, STM can be boiled down to a binary classification task, grounded on the understanding of sentence semantics.

Our work concentrates on the encoding of sentences, providing reliable semantic features and representations for linear classification. The architecture of our model (FDF) is shown in Fig. 1.

3.1 Input Layer

We concatenate the sentences S_a and S_b to form the original input sequence S_{ori}, i.e., $S_{ori} = \{[CLS], t_{a_1}, \cdots, t_{a_m}, [SEP], t_{b_1}, \cdots, t_{b_n}, [SEP]\}$, where $[CLS]$ and $[SEP]$ are specified as the special tokens. We regard S_{ori} as the unabridged sentence pair. Duplication detection is conducted to recognize the shared tokens in S_a and S_b (i.e., mutually-repetitive contents). Further, we uniformly replace

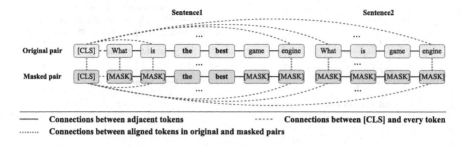

Fig. 2. An example of our proposed graph.

the shared tokens with the [MASK] tokens. This results in the production of the masked abridged sentence pair, which is denoted with S_{mask}:

$$S_{mask} = \{[CLS], t'_{a_1}, \cdots, t'_{a_m}, [SEP], t'_{b_1}, \cdots, t'_{b_n}, [SEP]\}$$

$$t'_i = \begin{cases} [MASK], & if \ t_i \in S_a \wedge t_i \in S_b \\ t_i, & Otherwise \end{cases} \tag{1}$$

Subsequently, we employ a PLM-based encoder to separately extract token-level features from the unabridged S_{ori} and abridged S_{mask}: $H = \text{PLM}(S)$. In this way, we obtain the shareable features H_{ori} and exclusive features H_{mask}.

3.2 Fusion Layer

We suggest that some exclusive features may signal the most distinguishable difference between the sentences S_a and S_b. Though, they cannot be used solely but cooperatively with the shareable features. Therefore, we fuse the two kinds of features (i.e., H_{ori} and H_{mask}), so as to avoid information loss throughout the semantics representation process, meanwhile preserving the positive effects of exclusive features.

We employ GCN [11] to fuse the features, in terms of the local interactive relationships of tokens in a predefined graph. Specifically, we construct a graph $G = (\mathcal{V}, \mathcal{E})$ using each token t_i ($t_i \in S_a \cup S_b$) as a node v_i ($v_i \in \mathcal{V}$). The edges $\mathcal{E}$ connecting the nodes are defined as follows: (1) Every node is connected to itself; (2) If two tokens in the unabridged sentence pair S_{ori} or the abridged sentence pair S_{mask} are adjacent, we connect them with an edge; (3) The special token $[CLS]$ of S_{ori} is connected with all the tokens in S_{ori} itself, while the special token $[CLS]$ of S_{mask} is connected with all the tokens (including [MASK]s) in S_{mask} itself; (4) Each token in S_{ori} is connected with the corresponding token (including [MASK]) in S_{mask}. We provide an example of $G = (\mathcal{V}, \mathcal{E})$ in Fig. 2, where the lines indicate the connections between nodes.

We build the graph in this manner primarily for the following reasons. Employing multi-layer transformer architecture, PLMs are able to capture the critical words in sentences. However, they are deficient in perceiving local information [23], which can also be instrumental to the matching process. To better model the local information, we adopt the second condition to gather the adjacent representations of each token. Besides, we design the third condition to enhance each token representation by global representation $[CLS]$ and vice versa.

The last condition enables the GCN module to model interactions between H_{ori} and H_{mask}. On the one hand, without the impact of shared tokens, the representations of non-repetitive tokens in H_{mask} only carry their semantic information. As a result, the message propagation between the nodes of non-repetitive tokens in H_{ori} and H_{mask} can enhance the representations of non-repetitive contents. On the other hand, during the encoding phase of PLM, the [MASK] tokens can only gather the contextual information from the tokens which are not masked (i.e., non-repetitive tokens). Therefore, their representations can weaken the corresponding representations of the repetitive tokens in H_{ori}, which can also be seen as an enhancement of non-repetitive token representations.

After constructing Graph G, we introduce its adjacency matrix $A \in \mathbb{R}^{2N \times 2N}$, where N is the sequence length of both S_{ori} and S_{mask}. Then we apply GCN to get the updated node features $\tilde{H}_{ori}$ and $\tilde{H}_{mask}$ as follows:

$$\tilde{H}_{ori}, \tilde{H}_{mask} = ReLU(\tilde{A}[H_{ori}; H_{mask}]W) \qquad (2)$$

Here, $[\cdot; \cdot]$ denotes the concatenation operation, $\tilde{A} = D^{-\frac{1}{2}}AD^{-\frac{1}{2}}$ is the normalized symmetric matrix. D is the degree matrix of A, $D_{ii} = \sum_j A_{ij}$. $ReLU(\cdot)$ is the activation function, and $W \in \mathbb{R}^{d \times d}$ is a trainable weight matrix.

After the GCN module, each node is fully updated by aggregating the representations of its neighbors. Then we design a gated module to dynamically integrate the token representations $\tilde{h}_i^{ori}$ and their counterparts $\tilde{h}_i^{mask}$. Specifically, we first compare the two representations and calculate the score g to decide how to combine them, which can typically be conducted as follows:

$$\tilde{h}_i = [\tilde{h}_i^{ori}; \tilde{h}_i^{mask}] \qquad (3)$$

$$g_i = tanh(w_g \tilde{h}_i + b_g) \qquad (4)$$

where $tanh(\cdot)$ is the tanh activation function, $w_g \in \mathbb{R}^{d \times 1}$ and b_g are trainable parameters. Note that the g in Eq. (4) is a scalar score. Thus if we use it to integrate the two representations, all the dimensions are treated equally. Since the representation space is anisotropic [6] and each dimension represents different information, it is more reasonable to assign different scores to each dimension to achieve better fusion. Inspired by Shen et al. [18], we propose a multi-dimensional gated module. Instead of calculating a single scalar score, we calculate a feature-wise score matrix G_i, which has the same length as h_i. Accordingly, the Eq. (4) can be revised as follows:

$$G_i = tanh(W_g \tilde{h}_i + b_g) \qquad (5)$$

where $W_g \in \mathbb{R}^{d \times d}$ is the weight matrix, b_g is the bias term. Then we apply the score vector G_i to integrate $\tilde{H}_{ori}$ and $\tilde{H}_{mask}$ to get the distinguishable features C, which is calculated as follows:

$$C = \tilde{H}_{ori} + G \odot \tilde{H}_{mask} \tag{6}$$

Here, $\odot$ denotes the element-wise multiplication.

3.3   Interaction and Pooling Layer

To fully exploit the distinguishable features C, we devise an interaction layer to further compare each token in two sentences:

$$\hat{C} = TransformerBlock(C) \tag{7}$$

By performing interaction on the enhanced features C, the model can perceive the differentiated information of the sentence pair and enable better refinement of the features from both sequences. To obtain the high-level differentiation-aware representation for the sentence pair, we then employ an additional attention layer to aggregate the output $\hat{C} = \{\hat{c}_1, ..., \hat{c}_N\}$ of interaction layer:

$$\alpha_i = \frac{exp(w_a^T \hat{c}_i / \sqrt{d})}{\sum_{j=1}^{N} exp(w_a^T \hat{c}_j / \sqrt{d})} \tag{8}$$

$$h_{dif} = \sum_{i=1}^{N} \alpha_i \hat{c}_i \tag{9}$$

where $w_a \in \mathbb{R}^{d \times 1}$ is the trainable parameter. Then we feed the final output h_{dif} of this layer to the relation classifier module for the final prediction.

3.4   Relation Classifier

The semantic information of the original sequence S_{ori} is completely preserved in the global representation h_{cls}^{ori}, while h_{dif} is a differentiation-aware representation with attentive and distinguishable features. Combining them enables the model better to determine the semantic relation between the two sentences. Therefore, we concatenate them to make the final classification:

$$p(y|S_a, S_b) = FFN([h_{cls}^{ori}; h_{dif}]) \tag{10}$$

where $FFN(\cdot)$ is a feed forward network with one layer. During the training stage, the training object is to minimize the binary cross-entropy loss.

4   Experimentation

4.1   Corpora and Hyperparameter Settings

Corpora. We conduct experiments on three STM benchmarks: two English corpora QQP [9] and MRPC [5], and one Chinese corpus BQ [1]. Both QQP

Table 2. Statistics of three corpora QQP, MRPC and BQ. "Avg. words" denotes the average number of words of all sentences, and "Avg. shared words" is the average number of shared words of all sentence pairs.

Corpora	Size	Avg. words	Avg. shared words	Domain
QQP	404,276	11.06	10.08	open-domain
MRPC	5,801	21.89	31.39	open-domain
BQ	120,000	11.64	7.82	bank

and MRPC are open-domain corpora collected from online websites, while BQ is a domain-specific corpus derived from bank service logs. Each sentence pair in these corpora is associated with a binary label indicating whether they are the same in semantics. Data statistics are shown in Table 2, where we report the details of instances that contain repetitive content.

Hyperparameters. We use the pre-trained BERT and RoBERTa released by the huggingface community[1], and fine-tune them on each corpus. The hyperparameters are set as follows. We use AdamW [15] ($\beta_1 = 0.9$, $\beta_2 = 0.999$, $\epsilon = $ 1e-8) with weight decay of 0.01 to fine-tune the parameters. We set the initial learning rate to 2e-5 and decrease its value with linear scheduling as the model training. As for batch size, we use 64 for QQP and BQ, and 16 for MRPC. We fine-tune the model for five epochs and evaluate the model after every 200 steps. Checkpoints with the best performance on the development set are evaluated on the test set to report the performance. All of our experiments are conducted on a single Nvidia Tesla V100-16GB GPU.

4.2 Main Results

The main results are shown in Table 3 and Table 4. The reported results are average scores using five different seeds, and all the improvements over baselines are statistically significant ($p - value < 0.05$ in the statistical significance test).

English STM. We compare our FDF to the previous STM models on QQP and MRPC. The results are shown in Table 3. All the models in Table 3 can be divided into three groups. The first group contains traditional neural matching models without pre-training, while the second and third groups comprise PLMs and the ones which employ PLMs as their backbones.

Table 3 shows that PLMs have salient performance gains over the traditional method due to pre-training on large-scale corpora. From the results of the MRPC corpus, we can observe that PLMs show their superiority, especially when performed on the small-scale corpus. It is worth mentioning that DRCN is capable of preserving both the original and the co-attentive feature information, while

---

[1] https://huggingface.co/.

Table 3. Results (*Acc.*) on QQP and MRPC.   **Table 4.** Results on BQ.

Models	QQP	MRPC
BiMPM [20]	88.2	–
DIIN [7]	89.1	–
DRCN [10]	90.2	82.5
DRr-Net [22]	89.8	82.9
BERT [4]	90.9	82.7
-*large version*	91.0	85.9
SS-BERT [13]	91.4	–
R^2-Net [23]	91.6	84.3
DC-Match [24]	91.2	83.8
FDF (BERT-base)	**91.6**	**84.8**
RoBERTa [14]	91.4	87.2
-*large version*	92.0	88.3
DC-Match (RoBERTa-large)	92.2	88.9
FDF (RoBERTa-large)	**92.4**	**89.3**

Models	BQ	
	Acc	F1
Text-CNN [8]	68.5	69.2
BiLSTM [17]	73.5	72.7
Lattice-CNN [12]	78.2	78.3
BiMPM [20]	81.9	81.7
ESIM [2]	81.9	81.9
LET [16]	83.2	83.0
BERT-wwm [3]	84.9	84.3
BERT-wwm-ext [3]	84.7	83.9
ERNIE [19]	84.7	84.2
BERT [4]	84.5	84.0
LET-BERT [16]	85.3	85.0
FDF (BERT)	**85.4**	**85.4**

DRr-Net repeatedly reads the important words to understand the sentences better. The performance of both DRCN and DRr-Net is close to that of BERT-base, which is impressive for models without pre-training.

The second group shows the performance of solely utilizing BERT [4], and the ones expanding BERT in different ways, such as the most representative R^2-Net. Benefiting from the pairwise relation learning processing, R^2-Net is able to make full use of the relation information. It achieves the best performance among the BERT-based models at the earlier time. By contrast, FDF outperforms R^2-Net on MRPC and obtains comparable performance on QQP.

For a better comparison, we also conduct experiments using RoBERTa-large [14] as baseline, and the results are shown in the third group in Table 3. Within the ones using RoBERTa-large as the backbone, DC-Match achieves the best performance on both MRPC and QQP corpora. Instead of performing text comparison by processing each word uniformly, DC-Match matches the intents and keywords under different levels of granularity. However, this approach could lead to incomplete semantic information, and thus may affect the model performance. In contrast, FDF is able to better preserve the semantics through the fusion and interaction layers. It can be observed that FDF outperforms DC-Match on the two corpora. All the results from Table 3 prove the necessity of reducing distractions by highlighting distinguishable features.

Chinese STM. Furthermore, we evaluate FDF on the Chinese benchmark BQ, which is a domain-specific corpus for bank question matching. Following Lyu et al. [16], we also report the F1-score besides the accuracy rate (*Acc.*). Table 4 shows the comparison results of different models. Note that the models in Table 4 can be divided into two categories: BERT-free models and BERT-based models.

Within the models, LET-BERT uses word lattices and introduces HowNet's knowledge to solve word sense disambiguation. Therefore, it achieves impressive

Table 5. Ablation performance (*Acc.*) of FDF network.

Model	QQP	MRPC	BQ
FDF	**91.55**	**84.83**	**85.37**
w/o masking	91.35	84.16	84.80
simple gate	91.43	84.19	84.77
w/o fusion	91.43	83.13	84.70
w/o interaction	91.40	83.97	84.73
BERT-base	90.91	82.70	84.50

performance among all the models, including both BERT-free and BERT-based ones. Although both LET-BERT and FDF utilize graph neural networks to extract and represent features of local structures. Though, FDF does not utilize external information. Briefly, FDF is more straightforward but outperforms LET-BERT. The experimental results of FDF on BQ indicate that it performs effectively in different languages and domain-specific scenarios.

4.3   Ablation Study

We verify the possible contributions of different components in FDF by ablation study. The results are shown in Table 5. To validate the effectiveness of the masking strategy, we replace the masked sequence with the original sequence in Fig. 1 and remain the rest of the FDF network unchanged to eliminate the effect of the number of parameters. It can be observed that the accuracy decreased by 0.2, 0.67 and 0.57 on QQP, MRPC and BQ respectively, when the shared-token masking is disabled. This demonstrates the effectiveness of exclusive feature extraction by deduplication. Further, we replace the multi-dimensional gate module with a simplified gate module. Specifically, we merely calculate a single scalar score to integrate two representations. It can be found that accuracy decreased to 91.43, 84.19 and 84.77 on the three datasets. This proves that treating each feature differently can obtain a better integration of representations.

Besides, when we remove the fusion layer and integrate the representations by directly adding them together, the performance also decreases. This implies that the fusion layer is helpful for producing distinguishable features. Moreover, the MRPC dataset possesses a longer text length (which can be observed from Table 2), thus making it more challenging for the STM model to determine whether the given pairs are semantically equivalent. In this case, the direct summation of the results from two channels cannot yield useful token representations for supporting the subsequent modules to predict the semantic relations. Therefore, the performance of FDF on the MRPC benchmark drops dramatically by 1.7 when the fusion layer is removed. In the last ablation, we remove the interaction layer and directly aggregate the outputs of the fusion layer. It can be observed that performance drops on all corpora. This demonstrates that the enhanced representations obtained by the fusion layer are not fully exploited, and the interaction layer enables further comparison between the sentence pair.

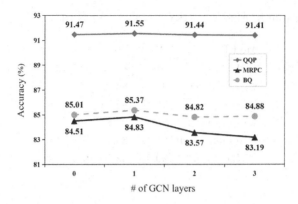

Fig. 3. Performance (*Acc.*) of FDF with different GCN layers.

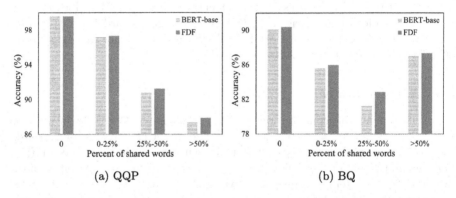

(a) QQP   (b) BQ

Fig. 4. Comparison between Baselines and FDFs on QQP and BQ with different proportions of shared words. The accuracy metric (*Acc.*) is considered.

4.4 Effect of GCN Layers

We validate the performance of FDF with different numbers of GCN layers. Figure 3 shows the accuracy curves over three corpora. It can be found that FDF approaches the best performance on all corpora when we set the number to 1.

By contrast, both disabling GCN and utilizing more GCN layers lead to performance degradation. Frankly, when GCN is disabled, FDF cannot effectively model interactions between shareable and exclusive features. This most probably causes performance reduction. When we use a larger number of GCN layers, the model suffers from less relevant information or even the noises introduced by GCN, which can hamper the model from making final predictions.

Table 6. Examples of sentence pairs from the QQP and BQ corpora. Words in **bold** are the distinctions between the two sentences. **PLM** and **FDF** represent the prediction of the BERT-base and FDF, respectively.

ID	Sentence pair	Label	PLM	FDF
1	Who is the best singer **now**? Who is the best singer **of all time**?	0	1	0
2	What **should I do** to sleep better? What **is the best way** to sleep better?	1	0	1
3	我什么时候可以使用微利贷 (When can I use the loan app) 什么时候可以**再次**使用微粒贷 (When can I use the loan app **again**)	0	1	0

4.5 Effectiveness Analysis

To verify the effectiveness of FDF when it deals with the sentence pairs possessing different proportions of shared words, we split the validation sets into quarters and validate the performance of models trained on the original training set. Figure 4 shows the comparison of FDF and baseline on QQP and BQ.

It can be observed that when the proportion increases, the performance of the baseline usually decreases. As shown in Fig. 4(a), BERT achieves nearly 100% accuracy when no shared word occurs in the sentence pair. Though, when the proportion is more than 50%, the accuracy drops dramatically. By contrast, FDF consistently yields improvements over the baseline when different proportions are considered. Specifically, when the proportion is higher than 25%, FDF achieves a substantial improvement compared to the baseline. When the proportion lies between 25% and 50%, FDF gains the improvements of 1.57% and 0.47% *Acc* on BQ and QQP, respectively. When the proportion increases to above 50%, improvements slightly reduce. The results imply that FDF can alleviate the distraction of repetitive contents effectively though incompletely, when the duplication is overly severe.

4.6 Case Study

This section presents several sample cases with predicted labels of FDF and the fine-tuned BERT in Table 6. These cases show that the original PLMs are confused by the words shared by both sentences and therefore fail to identify their semantic relations. For example, in the NO.1 and NO.3 cases, two sentences differ only in a few words (e.g., *"now"* and *"of all time"* in the No.1 case). The PLMs fail to highlight the non-repetitive contents of the pair and thus make the wrong prediction. By contrast, FDF is capable of producing distinguishable features to support the STM model to determine semantic consistency.

In addition, the baseline model is prone to classify sentence pairs that are partially different but contain the same meaning as negative cases. As shown in the NO.2 case, two sentences differ in *"should I do"* and *"is the best way"*, the baseline fails to identify their relation. During the training stage, FDF is able to highlight the representations of these phrases and employ the ground-truth labels to guide the model to learn that these phrases actually express the same meaning in this scenario.

5   Conclusion

We propose a Feature Differentiation and Fusion network (FDF) to enhance the current PLMs-based STM models. It is designed to alleviate the distraction of repetitive contents between sentences, conducting separate extraction of shareable and exclusive features and gated information fusion by GCN. Experiments on the benchmark corpora demonstrate the effectiveness of FDF.

In the future, we will carry out the study of the general multilingual FDF networks for STM. It is motivated by the findings in this paper that some linguistic features and commonsense knowledge are shareable among different languages for signaling semantic consistency. This implies the possibility of utilizing the Parent-Child learning model to transform experiences (parameters of neurons) across different languages. Progressive and contrastive learning will be used.

Acknowledgements. We thank all reviewers for their insightful comments, as well as the great efforts our colleagues have made so far. This work is supported by National Key R&D Program of China (2020YFB1313601) and National Science Foundation of China (62076174, 62076175).

References

1. Chen, J., Chen, Q., Liu, X., Yang, H., Lu, D., Tang, B.: The BQ corpus: A large-scale domain-specific Chinese corpus for sentence semantic equivalence identification. In: Proceedings of the 2018 Conference on Empirical Methods in Natural Language Processing, pp. 4946–4951
2. Chen, Q., Zhu, X., Ling, Z.H., Wei, S., Jiang, H., Inkpen, D.: Enhanced LSTM for natural language inference. In: Proceedings of the 55th Annual Meeting of the Association for Computational Linguistics (Volume 1: Long Papers). pp. 1657–1668
3. Cui, Y., Che, W., Liu, T., Qin, B., Yang, Z.: Pre-training with whole word masking for chinese BERT. IEEE ACM Trans. Audio Speech Lang. Process. **29**, 3504–3514 (2019)
4. Devlin, J., Chang, M.W., Lee, K., Toutanova, K.: BERT: Pre-training of deep bidirectional transformers for language understanding. In: Proceedings of the 2019 Conference of the North American Chapter of the Association for Computational Linguistics: Human Language Technologies, Volume 1 (Long and Short Papers), pp. 4171–4186 (2019)
5. Dolan, W.B., Brockett, C.: Automatically constructing a corpus of sentential paraphrases. In: Proceedings of the Third International Workshop on Paraphrasing (IWP2005)

6. Gao, J., He, D., Tan, X., Qin, T., Wang, L., Liu, T.: Representation degeneration problem in training natural language generation models. In: 7th International Conference on Learning Representations, ICLR 2019, New Orleans, LA, USA, May 6–9 (2019)

7. Gong, Y., Luo, H., Zhang, J.: Natural language inference over interaction space. In: 6th International Conference on Learning Representations, ICLR 2018, Vancouver, BC, Canada, April 30 - May 3, 2018, Conference Track Proceedings

8. He, T., Huang, W., Qiao, Y., Yao, J.: Text-attentional convolutional neural network for scene text detection. IEEE Trans. Image Process. **25**(6), 2529–2541 (2016)

9. Iyer, S., Dandekar, N., Csernai, K.: First quora dataset release: Question pairs (2017), https://quoradata.quora.com/First-Quora-Dataset-Release-Question-Pairs

10. Kim, S., Kang, I., Kwak, N.: Semantic sentence matching with densely-connected recurrent and co-attentive information. In: The Thirty-Third AAAI Conference on Artificial Intelligence, AAAI 2019, Honolulu, Hawaii, USA, January 27 - February 1, pp. 6586–6593 (2019)

11. Kipf, T.N., Welling, M.: Semi-supervised classification with graph convolutional networks. In: 5th International Conference on Learning Representations, ICLR 2017, Toulon, France, April 24–26, 2017, Conference Track Proceedings

12. Lai, Y., Feng, Y., Yu, X., Wang, Z., Xu, K., Zhao, D.: Lattice cnns for matching based chinese question answering. In: The Thirty-Third AAAI Conference on Artificial Intelligence, AAAI 2019, Honolulu, Hawaii, USA, January 27 - February 1, pp. 6634–6641 (2019)

13. Liu, T., Wang, X., Lv, C., Zhen, R., Fu, G.: Sentence matching with syntax- and semantics-aware BERT. In: Proceedings of the 28th International Conference on Computational Linguistics, pp. 3302–3312

14. Liu, Y., et al.: Roberta: A robustly optimized bert pretraining approach. arXiv preprint arXiv:1907.11692

15. Loshchilov, I., Hutter, F.: Decoupled weight decay regularization. In: 7th International Conference on Learning Representations, ICLR 2019, New Orleans, LA, USA, May 6–9 (2019)

16. Lyu, B., Chen, L., Zhu, S., Yu, K.: LET: linguistic knowledge enhanced graph transformer for chinese short text matching. In: Thirty-Fifth AAAI Conference on Artificial Intelligence, AAAI 2021, Virtual Event, February 2–9, pp. 13498–13506 (2021)

17. Mueller, J., Thyagarajan, A.: Siamese recurrent architectures for learning sentence similarity. In: Proceedings of the Thirtieth AAAI Conference on Artificial Intelligence, February 12–17, Phoenix, Arizona, USA. pp. 2786–2792 (2016)

18. Shen, T., Zhou, T., Long, G., Jiang, J., Pan, S., Zhang, C.: Disan: Directional self-attention network for rnn/cnn-free language understanding. In: Proceedings of the Thirty-Second AAAI Conference on Artificial Intelligence, AAAI-18, New Orleans, Louisiana, USA, February 2–7, pp. 5446–5455 (2018)

19. Sun, Y., et al.: Ernie: Enhanced representation through knowledge integration. arXiv preprint arXiv:1904.09223

20. Wang, Z., Hamza, W., Florian, R.: Bilateral multi-perspective matching for natural language sentences. In: Proceedings of the Twenty-Sixth International Joint Conference on Artificial Intelligence, IJCAI 2017, Melbourne, Australia, August 19–25, pp. 4144–4150 (2017)

21. Xia, T., Wang, Y., Tian, Y., Chang, Y.: Using prior knowledge to guide bert's attention in semantic textual matching tasks. In: WWW '21: The Web Conference 2021, Virtual Event / Ljubljana, Slovenia, April 19–23, pp. 2466–2475 (2021)

22. Zhang, K., et al.: Drr-net: Dynamic re-read network for sentence semantic matching. In: The Thirty-Third AAAI Conference on Artificial Intelligence, AAAI 2019, Honolulu, Hawaii, USA, January 27 - February 1, pp. 7442–7449 (2019)
23. Zhang, K., Wu, L., Lv, G., Wang, M., Chen, E., Ruan, S.: Making the relation matters: Relation of relation learning network for sentence semantic matching. In: Thirty-Fifth AAAI Conference on Artificial Intelligence, AAAI 2021, Virtual Event, February 2–9, pp. 14411–14419 (2021)
24. Zou, Y., et al.: Divide and conquer: Text semantic matching with disentangled keywords and intents. In: Findings of the Association for Computational Linguistics: ACL 2022,D pp. 3622–3632 (2022)

Multivariate Powered Dirichlet-Hawkes Process

Gaël Poux-Médard$^{(\boxtimes)}$ (ID), Julien Velcin (ID), and Sabine Loudcher (ID)

Université de Lyon, Lyon 2, ERIC UR 3083, 5 Avenue Pierre Mendès France,
69676 Bron Cedex, France
{gael.poux-medard,julien.velcin,sabine.loudcher}@univ-lyon2.fr

Abstract. The publication time of a document carries a relevant information about its semantic content. The Dirichlet-Hawkes process has been proposed to jointly model textual information and publication dynamics. This approach has been used with success in several recent works, and extended to tackle specific challenging problems –typically for short texts or entangled publication dynamics. However, the prior in its current form does not allow for complex publication dynamics. In particular, inferred topics are independent from each other –a publication about finance is assumed to have no influence on publications about politics, for instance.

In this work, we develop the Multivariate Powered Dirichlet-Hawkes Process (MPDHP), that alleviates this assumption. Publications about various topics can now influence each other. We detail and overcome the technical challenges that arise from considering interacting topics. We conduct a systematic evaluation of MPDHP on a range of synthetic datasets to define its application domain and limitations. Finally, we develop a use case of the MPDHP on Reddit data. At the end of this article, the interested reader will know how and when to use MPDHP, and when not to.

Keywords: Dirichlet process · Multivariate Hawkes process · Clustering · Information spread · Sequential data

1 Introduction

Understanding the data publication mechanisms on online platforms is of utmost importance in computer science. The amount of user-generated content that flows on social networks (e.g. Reddit) daily appeals for efficient and scalable approaches; they should provide us detailed insights within these mechanisms. A favoured approach to this problem is to cluster published documents according to their semantic content [2,3,18].

In the specific case of data flowing on social networks, time also carries a valuable information about the underlying data flow generation process [4,8,16]. Typically, we expect given publications to trigger ulterior publications within a

© The Author(s), under exclusive license to Springer Nature Switzerland AG 2023
J. Kamps et al. (Eds.): ECIR 2023, LNCS 13981, pp. 47–61, 2023.
https://doi.org/10.1007/978-3-031-28238-6_4

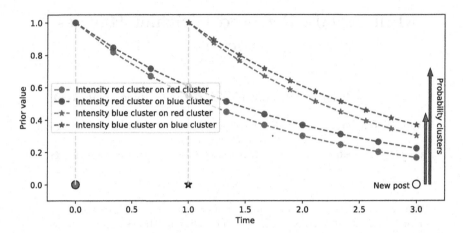

Fig. 1. Illustration of the multivariate powered Dirichlet-Hawkes process prior — A new event appears at time $t = 3$ from a cluster which is yet to be determined. The *a priori* probability that this event belongs to a given cluster c_{red} depends on the sum of the red intensity functions at time $t = 3$. Similarly, the *a priori* probability that this event belongs to a cluster c_{blue} depends on the sum of the blue lines at time $t = 3$. In previous models, this prior probability depends on each cluster self-stimulation only (Color figure online).

short time range. This effect has been studied by considering spreading agents, who are individually influenced by contacts' publications [5,6,15].

In previous works, the understanding of large data flows boils down to sorting data pieces (documents) into independent topics (clusters). However, it has been underlined on several occasions that online publication mechanisms are more complex than that. Typically, a correct description should involve clusters that interact with each other [9,11,19]. We illustrate the implications of this claim in Fig. 1. In most existing works that explicitly model both text and time, a given topic is assumed to only trigger observations from the same topic [4,13] –the red cluster can only trigger observations from the red cluster. Instead, we must allow clusters to trigger publications in any other cluster.

Therefore, we extend a previous class of models (DHP [4] and PDHP [13]) to account for cluster interaction mechanisms. We show that technical challenges arise when considering topical interaction, and solve them. This results in the Multivariate Powered Dirichlet-Hawkes Process (MPDHP). We conduct systematic experiments to test the limits of MPDHP and define its application domain. In particular, we show that it performs well in cases when textual data is scarce and when the number of coexisting clusters is large. Finally, we investigate a real-world use case on a Reddit dataset.

2   Background

The original Dirichlet-Hawkes process (DHP) [4] merges Dirichlet processes and Hawkes processes. It is used as a prior in Bayesian clustering along with a main model –typically a language model. The prior expresses the assumption that a new event from a given topic appears conditionally on the presence of older events from this same topic. The conditional probability is encoded into the intensity function of a Hawkes process. One such Hawkes process is associated to each topic. The temporal (Hawkes) intensity of a topic c is written $\lambda_c(t|\mathcal{H}_c)$; it depends on the history of all previous events associated to topic c, written $\mathcal{H}_c$. If no Hawkes intensity manages to explain well enough the presence of a new observation happening further in time, the DHP *a priori* guess is that the new observation belongs to a new cluster. DHP have first been used for automated summary generation [4]. A list of textual documents appear in chronological order and are treated as such; the DHP infers clusters of documents that are based both on their textual content and their publication date, and studies their auto-stimulated publication dynamics. This process knew several developments, that essentially consider alternative Dirichlet-based priors combined with Hawkes processes –Hierarchical DHP [8], Indian Buffet Hawkes process [16] and powered DHP [13].

However, in [13,18], the authors underline several limits of the standard Dirichlet-Hawkes processes and of the extensions mentioned earlier. For instance, DHP fails in cases where publications content carry few information: when textual content is short (e.g. tweets [18]) or when vocabularies overlap significantly (e.g., topic-specific datasets). Similarly, when each topic's temporal dynamics are hard to distinguish from each other, relying too much on the temporal information in the prior leads the model on the wrong track. In [13], the problem is alleviated by considering a Powered Dirichlet process [12] instead of a standard Dirichlet process. This process is merged with a univariate Hawkes process to make the Powered Dirichlet-Hawkes process. The authors retrieve better results in challenging clustering situations (large temporal and textual overlaps).

However, none of these works allow clusters to interact with each other, despite clues pointing in that direction [9,11,19]. Indeed in [4,8,13,16], the considered Hawkes processes are univariate: a cluster can only be used to trigger events within itself. Exploring how clusters interact with each other would significantly extend our comprehension of the publication mechanisms at stake in various datasets –such as social media or scientific articles. Identifying which topics trigger the publication of other seemingly unrelated topics might be interesting in the study of fake news spreading. Understanding the dynamics at stake may help to surgically inhibit the spread of such topics using the right refutation. Another possible use case would be nudging users towards responsible behaviours regarding environment, health, tobacco, etc.

In this paper, we extend the (univariate) Powered Dirichlet-Hawkes Process to its multivariate version. There are several reasons why it has not been done in prior works: first of all, the adaptation to the multivariate case is not trivial and poses several technical challenges. As a **first contribution**, we detail the challenges that arise when developing the Multivariate Powered Dirichlet-Hawkes Process (MPDHP). We propose methods to overcome them while retaining a linear time complexity $\mathcal{O}(N)$. Doing so, we also relax the near-critical Hawkes process hypothesis made in [4,8,13]. The second reason why the multivariate extension has not been developed in prior works, is that it greatly raises the number of parameters to estimate. The inference task might become harder, and the results irrelevant. Our **second contribution** consists in a systematic evaluation of the MPDHP on a variety of synthetic situations. Our goal is to identify the limits of MPDHP regarding textual overlap, computation time, the amount of available data, the number of co-existing clusters, etc. We show that MPDHP is perfectly fit for solving a variety of challenging situations. Finally, we illustrate the new insights on topical interaction obtained by running MPDHP on a real-world Reddit dataset.

3   The Multivariate Powered Dirichlet-Hawkes Process

3.1   Powered Dirichlet Process

The Dirichlet process can be expressed using the Chinese Restaurant Process metaphor. Consider a restaurant with an infinity of empty tables. A first client enters the restaurant and sits to any of the empty tables with a probability proportional to α –the *concentration parameter*. Another client then enters the restaurant, and sits either at one of the occupied tables with a probability linearly proportional to the number of clients already sat at the table, or to any of the empty tables with a probability proportional to α. The process is then iterated for an infinite number of clients. The resulting clients distribution over the tables is equivalent to a draw from a Dirichlet distribution. The Powered Dirichlet process is intended as a generalisation of the Dirichlet process [12].

The probability for a new client to sit at one of the occupied tables is now proportional to the number of clients at the power r. Let C_i be the table chosen by the ith client and $\mathcal{H}$ the history of table allocation. Formally, the probability for the ith to choose a table reads:

$$PDP(C_i = c|\alpha, r, \mathcal{H}) = \begin{cases} \frac{N_c^r}{\alpha + \sum_c N_c^r} & \text{if } c = 1, ..., K \\ \frac{\alpha}{\alpha + \sum_c N_c^r} & \text{if } c = K+1 \end{cases} \quad (1)$$

where N_c is the number of people that already sat at table c, K is the total number of tables, and r a hyper-parameter. Note that when $r = 1$ we recover the regular Dirichlet process, and when $r = 0$ we recover the Uniform Process [17].

3.2   Multivariate Hawkes Process

A Hawkes process is a temporal point process where the appearance of new events is conditional on the realisation of previous events. It is fully characterised by an intensity function, noted $\lambda(t|\mathcal{H})$ that depends on the history of previous events. It is interpreted as the instantaneous probability of a new observation appearing: $\lambda(t)dt = P(e \in [t, t + dt])$ with e an event and dt an infinitesimal time interval. For simplicity of notation, we omit the term $\mathcal{H}$ which is implicit anytime the intensity function λ is mentioned.

As in the DHP [4], we define one Hawkes process for each cluster. However in DHP each of them is associated to a **univariate** Hawkes process, that depends only on the history of events comprised in this cluster. In our case, instead, we associate each cluster to a **multivariate** Hawkes process that depends on all the observations previous to the time being. Let t_i^c be the time of realisation of the ith event belonging to cluster c. We write the intensity function for cluster c at all times as:

$$\lambda_c(t) = \sum_{t_i^{c'} < t} \alpha_{c,c'} \cdot \kappa(t - t_i^{c'}) \quad ; \quad \kappa_l(\Delta t) = \frac{e^{-\frac{(\Delta t - \mu_l)^2}{2\sigma_l^2}}}{\sqrt{2\pi\sigma_l^2}} \tag{2}$$

In Eq. 2, $\alpha_{c,c'}$ is a vector of L parameters to infer, and $\kappa(t - t_i^{c'})$ is a vector of L temporal kernel functions depending only on the time difference between two events. In our case, we consider a Gaussian RBF kernel, that allows to model a range of different intensity functions.

The log-likelihood of a multivariate Hawkes process for all observations up to a time T reads:

$$\log \mathcal{L}(\alpha|\mathcal{H}) = \sum_c \int_0^T \lambda_c(t)dt + \sum_{t_i^c} \lambda_c(t_i^c) \tag{3}$$

3.3   Multivariate Powered Dirichlet-Hawkes Process

The Multivariate Powered Dirichlet-Hawkes Process (MPDHP) arises from the merging of the Powered Dirichlet Process (PDP) and the Multivariate Hawkes Process (MHP), described in the previous sections. As in [4,8,13], the counts in PDP are substituted with the intensity functions of a temporal point-process, here MHP. The *a priori* probability that a new event is associated to a given cluster no longer depends on the population of this cluster, but on its temporal intensity at the time the new observation appears. This is illustrated in Fig. 1, where two events from two different clusters c_{red} and c_{blue} have already happened at times $t_0 = 0$ and $t_1 = 1$. A new event appears at time $t = 3$. The *a priori* probability that this event belongs to the cluster c_{red} depends on the sum of the intensity functions of observations at t_0 and t_1 on cluster c_{red} at time $t = 3$ –sum of the red dotted lines. Similarly, the *a priori* probability that this event belongs to the cluster c_{blue} depends on the sum of the blue dotted lines at time $t = 3$.

Let t_i be the time at which the ith event appears. The resulting expression reads:

$$P(C_i = c | t_i, r, \lambda_0, \mathcal{H}) = \begin{cases} \frac{\lambda_c^r(t_i)}{\lambda_0 + \sum_{c'} \lambda_{c'}^r(t_i)} & \text{if } c \leq K \\ \frac{\lambda_0}{\lambda_0 + \sum_{c'} \lambda_{c'}^r(t_i)} & \text{if } c = K+1 \end{cases} \quad (4)$$

In Eq. 4, λ_c in defined as in Eq. 2, and the parameter λ_0 is the equivalent of the concentration parameter described in Eq. 1. Taking back the illustration in Fig. 1, this parameter corresponds to a time-independent intensity function. It has a chance to get chosen typically when the other intensity functions are below it (meaning they do not manage to explain the dynamic aspect of a new event). In this case, a new topic is opened, and gets associated to its own intensity function.

3.4 Language Model

Similarly to what has been done in [4,13], the MPDHP must be associated to a Bayesian model given it is a prior on sequential data. Since we study applications on textual data, we choose to side the MPDHP prior with the same Dirichlet-Multinomial language model as in previous publications [4,13]. According to this model, the likelihood of the ith document belonging to cluster c reads:

$$\mathcal{L}(C_i = c | N_{<i,c}, n_i, \theta_0) = \frac{\Gamma(N_c + \theta_0)}{\Gamma(N_c + n_i + \theta_0)} \prod_v \frac{\Gamma(N_{c,v} + n_{i,v} + \theta_{0,v})}{\Gamma(N_{c,v} + \theta_0)} \quad (5)$$

where N_c is the total number of words in cluster c from observations previous to i, n_i is the total number of words in document i, $N_{c,v}$ the count of word v in cluster c, $n_{i,v}$ the count of word v in document i and $\theta_0 = \sum_v \theta_{0,v}$. Note that for any empty cluster, the likelihood is computed using $N_{c_{empty}} = 0$ for every empty cluster c_{empty}.

4 Implementation

4.1 Base Algorithm

SMC Algorithm. We use a Sequential Monte Carlo (SMC) algorithm for the optimisation. The base algorithm is the same as in [4,8,13] – a graphical representation of the SMC algorithm is provided in [13]-Fig. 1 and as Supplementary Material. The goal of the SMC algorithm is to jointly infer textual documents' clusters and the dynamics associated with them. It runs as follows. First, the algorithm computes each cluster's posterior probability for a new observation by multiplying the temporal prior on cluster allocation (see Eq. 4, illustrated Fig. 1) with the textual likelihood (see Eq. 5). It results in an array of $K+1$ probabilities, where K is the number of non-empty clusters. A cluster label is then sampled from this probability vector. If the empty $(K + 1)^{th}$ cluster is chosen, the new observation is added to this cluster, and its dynamics are randomly initialised (i.e. a $(K+1)^{th}$ row and a $(K+1)^{th}$ column are added to the parameters matrix α). If a non-empty cluster is chosen, its dynamics are updated by maximising the new likelihood Eq. 3. The process then goes on to the next observation.

This routine is repeated N_{part} times in parallel. Each parallel run is referred to as a *particle*. Each particle keeps track of a series of cluster allocation hypotheses. After an observation has been treated, we compute the particles likelihood given their respective cluster allocations hypotheses. Particles that have a likelihood relative to the other particles' one below a given threshold ω_{thres} are discarded and replaced by a more plausible existing particle.

Sampling the Temporal Parameters. The parameters α are inferred using a sampling procedure. A number N_{sample} of precomputed vectors is drawn from a Dirichlet distribution with probability $P(\alpha|\alpha_0)$, with α_0 a concentration parameter. As the SMC algorithm runs, within each existing cluster, each of these candidate vectors is associated to a likelihood computed from Eq. 3, noted $P(\mathcal{H}|\alpha)$, where $\mathcal{H}$ represents the data. The sampling procedure returns the average of each of the N_{sample} precomputed α, weighted by the posterior distribution associated to them $P(\alpha|\mathcal{H}) \propto P(\mathcal{H}|\alpha)P(\alpha|\alpha_0)$. The so-returned matrix is guaranteed to be a good statistical approximation of the optimal matrix, provided the number of sample matrices N_{sample} is large enough.

Limits. This algorithm described here works well for the univariate case, but fails for the multivariate case. In particular, updating the multivariate intensity function of each cluster requires knowing the number of already existing clusters, which vary over time. Therefore, we cannot precompute the sample matrices in advance –they must be updated as the algorithm runs to account for the right number of non-empty clusters. Moreover, the number of parameters to estimate evolves linearly with the number of active clusters K, instead of remaining constant as in DHP and variants [4,13]. Because the number of parameters is not constant anymore, their candidate values cannot be sampled from a Dirichlet distribution anymore. In the following, we review these challenges and present our solutions to overcome them. We manage to preserve a constant time complexity for each observation.

4.2   Optimisation Challenges

Temporal Horizon. A first problem that has been answered in [4] is that, for each new observation, the algorithm has to run through the whole history of events to compute the DHP prior. However, carefully choosing the kernel vector $\kappa(\cdot)$ described in Eq. 2 allows to perform this step in constant time. If the chosen kernel vanishes as time goes, it happens a point where old events have a near-zero chance to have any influence on new observations, according to our model. These events can be discarded from the computation for new events.

Updating the Triggering Kernels. In the univariate case [4,8,13], the coefficients $\alpha_c \in \mathbb{R}^L$ are sampled from a collection of existing sample vectors computed at the beginning of the algorithm (where L is the size of the kernel vector). However, we must now infer a matrix instead. We recall that matrix α_c represents the weights given to the temporal kernel vector of every cluster influence on c –see Eq. 2. The likelihood Eq. 3 can be updated incrementally for each sample matrix. A given cluster c has a likelihood value associated to each of those N_{sample} sample matrices, which represents how fit one sample matrix is to explain one cluster's dynamics. The final value of the parameters matrix is sampled simply by averaging the samples matrices weighted by their likelihood for a given cluster times the prior probability of these vectors being drawn in the first place.

Such sampling was possible in the univariate case, where each sample matrix was in fact a vector of fixed length. In our case, because Hawkes processes are multivariate, each entry $\alpha_c \in \mathbb{R}^{K \times L}$ is now a matrix. Moreover, the number of existing clusters K increases over with time, and can grow very large. Each time a new cluster is added to the computation, a row is appended to the α_c matrix –it accounts for the influence of this new cluster regarding c.

However, some older events can be discarded from the computation. When an event is older than 3σ with respect to the longest range entry of the RBF kernel, it can be safely discarded. Clusters whose last observation has been discarded thus have a near-zero chance to get sampled once again. These clusters' contribution to the likelihood Eq. 3 will not change anymore. Therefore, they do not have a role in the computation of the parameters matrix α_c. The row corresponding to each of these clusters in the parameters matrix can then be discarded in every sample matrix. Put differently, the last sampled value for their influence on c will remain unchanged for the remaining of the algorithm. The dimension of α_c only depends on the number of *active* clusters, whose intensity function has not faded to zero. For a given dataset, the number of active clusters typically fluctuates around a constant value, making one iteration running in constant time $\mathcal{O}(1)$.

A Beta Prior on Parameters. Another problem inherent to the multivariate modelling is the prior assumption on sample vectors. In [4,13], each sample vector is sampled from a Dirichlet distribution. This choice is to infer Hawkes processes that are nearly-unstable: the spectral radius of the temporal kernel function $\lambda_c(t)$ is close to 1. However in our case, such assumption is not possible because the size of each sample matrix can vary as the number of active clusters evolve. Drawing one Dirichlet vector for each entry $\alpha_{c,c'}$ would force the spectral radius of α_c to equal $K = |\alpha_c|$, which transcribes a highly-unstable Hawkes process. Our solution is to consider the parameters as completely independent from each other. Each entry of the matrix α is drawn from an independent β distribution of parameter β_0. In this way, we make no assumption on the spectral radius of the Hawkes process, and samples rows/columns corresponding to new clusters can be generated one after the other.

5   Experiments

5.1   Setup

We design a series of experiments to determine the use cases of the Multivariate Powered Dirichlet-Hawkes Process[1]. We list the parameters we consider in our experiments. When a parameter does not explicitly vary, it takes a default value given between parenthesis. These parameters are: the textual overlap (0) and the temporal overlap (0) discussed further in the text, the temporal concentration parameter λ_0 (0.01), the strength of temporal dependence r (1), the number of synthetically generated clusters K (2), the number of words associated to each document n_{words} (20), the number of particles N_{part} (10) and the number of sample matrices used for sampling α, noted N_{sample} (2 000). For the detail of these parameters, please refer to Eq. 4. The interplay between the parameters that are not part of MPDHP (N_{sample} and N_{part}) is studied in Supplementary Material.

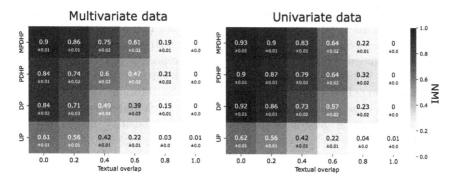

Fig. 2. Numerical results on synthetic data — MPDHP consistently outperforms other baselines designed for the univariate case on both univariate and multivariate data. The standard error has been computed using 100 independent runs.

Note that the overlap $o(f_1, ..., f_N)$ between N functions is defined as the sum over each function f_i of its intersecting area with the largest of the $N-1$ other functions, divided by the sum of each function's total area [13]. This value is bounded between 0 (perfectly separated functions) and 1 (identical functions):

$$o(f_1, ..., f_N) = \sum_i \int_{\mathbb{R}} min(f_i(x), max(\{f_j(x)\}_{j \neq i})) dx$$

For each combination of parameters considered, we generate 10 different datasets. In all datasets, we consider a fixed size vocabulary $V = 1000$ for each cluster.

---

[1] Data and implementations are available at https://github.com/GaelPouxMedard/ MPDHP.

All datasets are made of 5 000 observations. Observations for each cluster c are generated using a RBF temporal kernel $\kappa(t)$ weighted by a parameter matrix α_c. We set $\kappa(t) = [\mathcal{G}(3; 0.5); \mathcal{G}(7; 0.5); \mathcal{G}(11; 0.5)]$ where $\mathcal{G}(\mu; \sigma)$ is a Gaussian of mean μ and standard deviation σ. We note $L = 3$ the number of entries of κ. The inferred entries of α determine the amplitude (weight) of each kernel entry.

The generation process is as follows. First, we draw a random matrix $\alpha \in \mathbb{R}^{K \times L}$ and normalise it so that its spectral radius equals 1 –near unstable Hawkes process. We repeat this process until we obtain the wanted temporal overlap. Then, we simulate the multivariate Hawkes process using the triggering kernels $\alpha \cdot \kappa(t)$, where $\kappa(t)$ is the RBF kernel as defined earlier. Given the Hawkes process is multivariate, each event is associated to its class it has been generated from among K possible classes. For each event, we draw n_{words} words from a vocabulary of size V. The vocabularies are drawn from a multinomial distribution and shifted over this distribution so that they overlap to a given extent.

5.2   Baselines

We evaluate our clustering results in terms of Normalised Mutual Information score (NMI). This metric is standard when evaluating non-parametric clustering models. It compares two cluster partitions (i.e. the inferred and the ground truth ones); it is bounded between 0 (each true cluster is represented to the same extent in each of the inferred ones) and 1 (each inferred partition comprises 100% of one true cluster). The standard error is computed on 100 runs. We compare our approach to 3 closely related baselines. **Powered Dirichlet-Hawkes process (PDHP)** [13]: in this model, clusters can only self replicate. It means that the intensity function of a cluster c Eq. 2 only considers past events that happened in the same cluster c. r is set to 1. **Dirichlet process (DP)**: this prior is standard in clustering problems. It corresponds to a special case of Eq. 1 where $r = 1$. The prior probability for an observation to belong to a cluster depends on its population. **Uniform process (UP)** [17]: this prior corresponds to a special case of Eq. 1 where $r = 0$. It assumes that the prior probability for an observation to belong to a cluster does not depend on any information about this cluster (neither population nor dynamics).

5.3   Results on Synthetic Data

MPDHP Outperforms State-of-the-Art. In Fig. 2, we plot our results for datasets that have been generated using a Multivariate Hawkes process (clusters have an influence on each other) and a Univariate Hawkes process (clusters can only influence themselves). We compare MPDHP to our baselines for various values of textual overlap –we provide a similar study that considers various temporal overlaps as Supplementary Material. **MPDHP** systematically outperforms the baselines on multivariate data –when clusters interact with each other. Considering that clusters interacts with each other improves our description of the datasets. **MPDHP** performs at least as good as PDHP on univariate data – when clusters can only self-stimulate. The complexity of MPDHP does not make it unfit to simpler tasks. **PDHP** performs better than MPDHP when the textual overlap is large (textual overlap of 0.8) due to its reduced complexity. Increasing the number of observations would fix this.

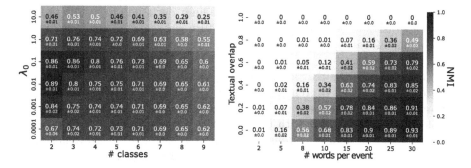

Fig. 3. MPDHP can handle a large number of coexisting clusters and scarce textual information — MPDHP yields good results when a large number of clusters coexist simultaneously (left) and when texts are short or little informative (right). It is also robust against variations of λ_0 over 5 order of magnitude.

Highly Interacting Processes. We test when a large number of clusters coexist simultaneously. The rate at which new clusters get opened is mainly controlled by the λ_0 hyperparameter (see Eq. 4), which we vary to see whether MPDHP is robust against it. In Fig. 3 (left), we plot the performances of MPDHP according to these two parameters. We draw two conclusions: MPDHP can handle a large number of coexisting clusters and still correctly identify to which one each document belongs, and MPDHP is robust against large variations of λ_0. In this case, results are similar for λ_0 varying over 5 orders of magnitude. It means MPDHP does not have to be fine-tuned according to the number of expected clusters in cases where this number is not known in advance. An extended discussion on the choice of the parameter λ_0 is provided as Supplementary Material.

Handling Scarce Textual Information. In this paragraph, we determine how much data should be provided to MPDHP to get satisfying results. In Fig. 3 (right), we plot the performances of MPDHP with respect to the number of words generated by each observation and to the clusters' vocabulary overlap. MPDHP needs a fairly low number of words to yield good results over 5 000 observations. For reference, the overlap between topics can be estimated around 0.25 ([10], in Spanish). Similarly, we can estimate an average of ~10-20 named entities per Twitter post (240 characters). These results support the application of MPDHP to model real-world situations.

58   G. Poux-Médard et al.

5.4 Real-world Application on Reddit

Data. We conclude this work with an illustration of MPDHP in a real-world situation. We investigate the interplay between topics on news subreddits, that is how much influence a topic can exert on other ones. The dataset is collected from the Pushshift Reddit repository [1]. We limit our study to headlines from popular English news subreddits in January 2019: inthenews, neutralnews, news, nottheonion, offbeat, open news, qualitynews, truenews, worldnews. From these, we remove posts that have a popularity (difference between upvotes and down-votes) lesser than 20, as they are of lesser influence in the dataset and only add noise to the modeling. We remove headlines that contain less than 3 words as they only add noise to the modelling. After curating the dataset in the way described above, we are left with roughly 8,000 news headlines, which makes a total of 65,743 tokens drawn from a vocabulary of size 7,672. Additional characteristics of the dataset are provided as Supplementary Material.

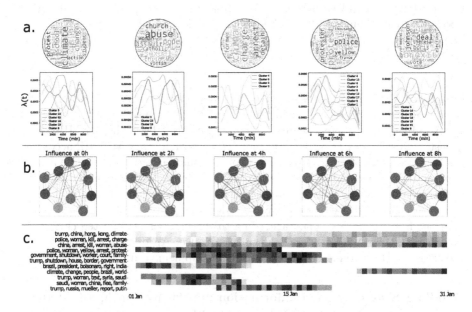

Fig. 4. Real-world application on Reddit – **a.** Examples of clusters along with their inferred reproduction dynamics. **b.** Visualisation of interaction patterns at different times as a network; each dot is a cluster, each edge accounts for the value of $\lambda(t)$ at a given time $t \in [0; 2; 4; 6; 8]$ hours. **c.** Most used clusters represented over real time.

Parameters. We run our experiments using a RBF kernel made of Gaussians centred around $[0, 2, 4, 6, 8]$ hours, with a standard deviation σ of 1 h, $\lambda_0 = 0.001$, and $r = 1$. We use a Dirichlet-Multinomial language model as in the synthetic experiments with $\theta_0 = 0.01$. As for the SMC algorithm, we set $N_{samples} = 100000$.

From our observations, the number of coexisting clusters remains around 10 coexisting clusters (roughly 1,000 parameters), allowing sampling each parameter from approximately 100 candidate values. Each sample parameter is drawn from an identical Beta distribution of concentration parameter $\alpha_0 = 2$. We consider 8 particles for the SMC algorithm, similarly to [4].

Results. We present the results of MPDHP on real-world data in Fig. 4. Figure 4a. illustrates a typical output from the model. The transparency in the representation of $\lambda(t)$ accounts for the number of times such interaction has effectively been observed; transparency of the intensity function $\lambda_{c,c'}(t)$ of c' on c is proportional to $\sum_{t^c} \sum_{t^{c'} < t^c} \lambda_{c,c'}(t)$. We can make two interesting observations from this figure. Firstly, the interaction strength between clusters seems to fade as time passes (Fig. 4b.). Cluster interactions are more likely to happens within short time ranges. Secondly, the first two clusters seem to be consistently used across the whole month (Fig. 4c.). When we look at their composition, we notice that the first cluster is made of 75% of articles from the subreddit r/worldnews, which is +20% from what one would expect from chance (55% of the corpus is from r/worldnews, see Supplementary Material). Similarly, the second cluster comprises 46% of r/news articles, which is also roughly +20% from expected at random. These two clusters therefore significantly account for publications from either of these subreddits, independently from the textual content. Both are general news forums with a large audience; an article that gets posted on other subreddits is likely to also appear on these. Other clusters follow a bursty dynamic, which concurs with [7]. More details on this experiment can be found in [14].

6   Conclusion

In this paper, we extended the Powered Dirichlet-Hawkes process so that it can consider multivariate processes, resulting in the Multivariate Powered Dirichlet-Hawkes process (MPDHP). This new process can infer temporal clusters interaction networks from textual data flow. We overcome several optimisation challenges to preserve a time complexity that scales linearly with the dataset.

We showed that MPDHP outperforms existing baselines when clusters interact with each other, and performs at least as well as the PDHP baseline when clusters do not (which PDHP is designed to model). MPDHP can handle cases where textual content is lesser informative better than other baselines. It is robust against tuning of the temporal concentration parameter λ_0, which allows to handle highly intricate processes. We finally showed that MPDHP performs well with scarce textual data. Our results suggest that MPDHP can be applied in a robust way to a broad range of problems, which we illustrate on a real-world application, that provides insights in topical interactions mechanisms between news published on Reddit.

References

1. Baumgartner, J., Zannettou, S., Keegan, B., Squire, M., Blackburn, J.: The pushshift reddit dataset. Proc. Int. AAAI Conf. Web Social Media **14**(1), 830–839 (2020)
2. Blei, D.M., Lafferty, J.D.: Dynamic topic models. In: Proceedings of the 23rd International Conference on Machine Learning. pp. 113–120. ICML '06, Association for Computing Machinery, New York, NY, USA (2006). https://doi.org/10.1145/1143844.1143859
3. Blei, D.M., Ng, A.Y., Jordan, M.I.: Latent dirichlet allocation. J. Mach. Learn. Res. **3**, 993–1022 (2003)
4. Du, N., Farajtabar, M., Ahmed, A., Smola, A., Song, L.: Dirichlet-hawkes processes with applications to clustering continuous-time document streams. In: 21th ACM SIGKDD International Conference on Knowledge Discovery and Data Mining (2015). https://doi.org/10.1145/2783258.2783411
5. Gomez-Rodriguez, M., Balduzzi, D., Schölkopf, B.: Uncovering the temporal dynamics of diffusion networks. In: ICML. pp. 561–568 (2011)
6. Gomez-Rodriguez, M., Leskovec, J., Schölkopf, B.: Modeling information propagation with survival theory. In: ICML. vol. 28, pp. III-666–III-674 (2013)
7. Haralabopoulos, G., Anagnostopoulos, I.: Lifespan and propagation of information in on-line social networks: A case study based on reddit. JNCA 56 (03 2014). https://doi.org/10.1016/j.jnca.2015.06.006
8. Mavroforakis, C., Valera, I., Gomez-Rodriguez, M.: Modeling the dynamics of learning activity on the web. In: Proceedings of the 26th International Conference on World Wide Web. pp. 1421–1430. WWW '17 (2017)
9. Myers, S.A., Leskovec, J.: Clash of the contagions: Cooperation and competition in information diffusion. In: 2012 IEEE 12th International Conference on Data Mining, pp. 539–548 (2012)
10. Posadas Duran, J., Gomez Adorno, H., Sidorov, G., Moreno, J.: Detection of fake news in a new corpus for the spanish language. Journal of Intelligent and Fuzzy Systems vol. 36, pp. 4869–4876 (05 2019). https://doi.org/10.3233/JIFS-179034
11. Poux-Médard, G., Velcin, J., Loudcher, S.: Interactions in information spread: quantification and interpretation using stochastic block models. RecSys'21 (2021). https://doi.org/10.1145/3460231.3474254
12. Poux-Médard, G., Velcin, J., Loudcher, S.: Powered dirichlet process for controlling the importance of "rich-get-richer" prior assumptions in bayesian clustering. ArXiv (2021)
13. Poux-Médard, G., Velcin, J., Loudcher, S.: Powered hawkes-dirichlet process: challenging textual clustering using a flexible temporal prior. ICDM (2021). https://doi.org/10.1109/ICDM51629.2021.00062
14. Poux-Médard, G., Velcin, J., Loudcher, S.: Properties of reddit news topical interactions. In: Complex Networks & Their Applications XI (under press). Springer International Publishing (2022)
15. Poux-Médard, G., Velcin, J., Loudcher, S.: Dirichlet-survival process: Scalable inference of topic-dependent diffusion networks. ECIR (2023)
16. Tan, X., Rao, V.A., Neville, J.: The indian buffet hawkes process to model evolving latent influences. In: UAI (2018)
17. Wallach, H., Jensen, S., Dicker, L., Heller, K.: An alternative prior process for nonparametric bayesian clustering. In: Proceedings of the Thirteenth International Conference on Artificial Intelligence and Statistics, pp. 892–899. JMLR (2010)

18. Yin, J., Chao, D., Liu, Z., Zhang, W., Yu, X., Wang, J.: Model-based clustering of short text streams. In: Proceedings of the 24th ACM SIGKDD International Conference on Knowledge Discovery and Data Mining. pp. 2634–2642. KDD '18, Association for Computing Machinery, New York, NY, USA (2018). https://doi. org/10.1145/3219819.3220094
19. Zarezade, A., Khodadadi, A., Farajtabar, M., Rabiee, H.R., Zha, H.: Correlated cascades: Compete or cooperate. In: Proceedings of the Thirty-First AAAI Conference on Artificial Intelligence, pp. 238–244 (2017)

Fragmented Visual Attention in Web Browsing: Weibull Analysis of Item Visit Times

Aini Putkonen[1]($\boxtimes$)(iD), Aurélien Nioche[2](iD), Markku Laine[1](iD),
Crista Kuuramo[3](iD), and Antti Oulasvirta[1](iD)

[1] Department of Information and Communications Engineering, Aalto University,
Espoo, Finland
{aini.putkonen,markku.laine,antti.oulasvirta}@aalto.fi
[2] School of Computing Science, University of Glasgow, Glasgow, UK
aurelien.nioche@glasgow.ac.uk
[3] Department of Psychology and Logopedics, University of Helsinki, Helsinki, Finland
crista.kuuramo@helsinki.fi

Abstract. Users often browse the web in an exploratory way, inspecting what they find interesting without a specific goal. However, the temporal dynamics of visual attention during such sessions, emerging when users gaze from one item to another, are not well understood. In this paper, we examine how people distribute visual attention among content items when browsing news. Distribution of visual attention is studied in a controlled experiment, wherein eye-tracking data and web logs are collected for 18 participants exploring newsfeeds in a single- and multi-column layout. Behavior is modeled using Weibull analysis of item (article) visit times, which describes these visits via quantities like durations and frequencies of switching focused item. Bayesian inference is used to quantify uncertainty. The results suggest that visual attention in browsing is fragmented, and affected by the number, properties and composition of the items visible on the viewport. We connect these findings to previous work explaining information-seeking behavior through cost-benefit judgments.

Keywords: Eye-tracking · Web browsing · Visual attention · Newsfeeds

1 Introduction

A large proportion of people's time engaged with computers gets devoted to web browsing [14]. In considerable proportions, this activity includes exploration [4,46,47], characterized by lack of an explicit informational goal. Unlike focused search tasks, exploration permits users to review the available content freely and engage with anything they find interesting. Given the information-rich nature of many modern browsing environments (news feeds, social media, catalogues, etc.), understanding how users choose what to attend to and for how long during exploration is a key challenge for behavioral and psychological research.

J. Kamps et al. (Eds.): ECIR 2023, LNCS 13981, pp. 62–78, 2023.
https://doi.org/10.1007/978-3-031-28238-6_5

This paper presents new empirical data and modeling results on visual attention when browsing feed-format news. We focus on "open-ended" browsing tasks, wherein information is gathered without a specific goal. The literature has described this type of browsing with several terms [6], such as "undirected" [13,19], "unstructured" [13], "casual" [28], "serendipitous" [9,13], "capricious" [43], and "hedonic" [27], denoting a contrast against directed or semi-directed browsing, which assumes a specified or somewhat specified goal.

At present, it is not well known *how* users spread their attention across content items when browsing. Previous work has used static spatial representations such as heatmaps for eye movement patterns during browsing. The "F-pattern" [30,33] and the Golden Triangle [21,22] are well-known examples. In addition, commercial tools have been developed to predict visual attention to visual stimuli (e.g., Attention Insight[1] and 3M Visual Attention Software[2]). Some research has examined how users distribute attention to a page's various HTML elements [25], while other studies focused on how people look at a single element type, such as specific features of images [12,20,23]. Still, the temporal dynamics of this behavior remain relatively unknown. Exceptions to this are work by Liu et al. [26] and Luo et al. [27]. Modeling page-level dwell times via Weibull analysis, Liu et al. concluded that general browsing exhibits a *screening pattern*: only some web pages pass an initial screening. They used data on page-visit times without considering visual attention. Similarly, Luo et al. modeled page-level dwell times, using an inverse Gaussian distribution.

This paper presents new findings on the temporal dynamics of visual attention when browsing news. We model gaze behavior in a task where users were given a long newsfeed to explore, and were asked to read what they find interesting. As depicted in Fig. 1, we examined the temporal dynamics of visual attention over a page in terms of two concepts: *an item* offers a clickable preview of content, here consisting of textual (title and description) and visual (picture) elements, and *a visit* consists of a continuous sequence of fixations on an item.

We report on the distributions of visit times and examine them with Weibull analysis, a technique employed for analyzing time-to-event data [39], with special attention to the parameters of the Weibull distribution. We use Bayesian inference to quantify the uncertainty in the parameter estimates. Presenting how visit durations are affected by two independent variables – layout type (single- vs. multi-column) and item content (a picture and/or a description with a title), we look in particular at *survival* (how long the visit lasts, or "survives") and *hazard* (the rate at which visits end). Our main finding is that visual attention in news browsing is fragmented. That is, visits are very brief on average, and items may be visited more than once.

In summary, this paper offers two contributions:

- We show that the distribution of visual attention in news browsing is fragmented and depends on the properties of the items and the environment.
- We quantify this fragmentation by means of Weibull analysis, extending the model presented by Liu et al. [26] to item-level dynamics.

---

[1] https://attentioninsight.com/.

[2] https://vas.3m.com/.

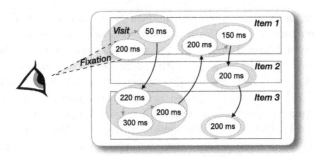

Fig. 1. We model the temporal dynamics of visual attention in browsing of feed-format news. The paper examines "visits" to content "items" (see text for definitions).

We conclude with synthesis, discussing these results and future work. In particular, we analyze the findings in relation to existing theories of information-seeking that take cost–benefit judgments as a basis for behavior.

2   Related Work

Below, we provide a brief overview of literature on visual attention and of empirical results related to browsing. We then discuss theories specifically addressing how people seek information on the web.

2.1   Visual Attention and Web Browsing

Eye-tracking work in information-search studies commonly associates eye movements with attention [7,15,25]. Several eye-tracking studies have analyzed the distribution of visual attention over a website during a browsing session, many of these focusing on search-engine result pages (SERPs) [15,21,22,25]. Among the well-known outputs pertaining to web browsing are the depiction of how people "scan" some web pages in a pattern resembling the letter F [30] and Google's Golden Triangle, wherein the upper-left corner of a SERP attracts the majority of eye-tracking activity [21,22] (similar patterns have been found for other types of pages too [7]). Most often, static representations of eye movements describe the findings, typically represented via heatmaps. In addition, previous research has examined visual attention in relation to specific content displayed on a website – for instance, tendencies in ignoring advertisement banners [36,37].

2.2   Empirical Understanding of Browsing Without a Specific Goal

In comparison, browsing without a specific goal has gained little empirical attention. Work using similar methodology to ours consists primarily of the aforementioned studies of general and hedonic browsing, by Liu et al. [26] and Luo et al.

[27], both using statistical models. While Liu et al.'s work identified a screening pattern in web browsing, the inverse Gaussian dwell-time distribution found by Luo and colleagues in hedonic content systems reinforces the law-of-surfing results proposed by Huberman et al. [24] in more structured tasks. The "law" suggests that the number of pages visited on a given website follows strong regularities and this can be captured with distinct probability distributions.

2.3   Theories of Browsing Behavior

While there are multiple theoretical models of browsing, views on its fundamental mechanisms differ. Cove and Walsh describe browsing as "an art" wherein individuals know what they want only as they come across it [13]. In fact, many theoretical accounts address this observation, whether focusing on directed, semi-directed, or undirected browsing. White and Roth [47] distinguish *exploratory browsing* as a part of exploratory search (as opposed to focused search) undertaken to 1) specify information needs and 2) encourage information discovery [40, 47]. Using the "berrypicking" model, Bates [3, 4] draws on the analogy of picking berries in a forest: browsing is an activity comprising a series of glimpses that may lead to closer inspection and acquisition of an object, connecting it to curiosity and exploratory behavior in humans. In this analogy, browsing is undirected behavior [40]. In contrast, information-foraging theory (IFT) [34] is based on biology and anthropology's theory of optimal foraging; this posits that individuals weigh the costs of performing an action against the potential information gain. The IFT notion of browsing is of a dynamic activity wherein the individual is guided between information items by "information scent," a subjective measure of item value [10]. Work in SERP and other settings has suggested that cost–benefit judgments may drive users' browsing behavior [2]. Under the models developed for these settings, the effort and time required to complete the task are the costs while the relevance of the information discovered constitutes the benefits [2]. An alternative is to interpret the benefits as utility, though such notions are seldom used in IR research, due to difficulties in measuring it [2, 11, 41].

3   Method

We report on a controlled experiment where participants were given a newsfeed to explore in two different layouts. They were allowed to inspect, click, and scroll as they found natural, without being asked to perform particular actions during the browsing session. We present aggregated results obtained from 18 participants, with multiple data sources: eye tracking, web logging, and a questionnaire.

3.1   Experimental Design

Participants were advised to read news items they found interesting for an unlimited amount of time. They were told that the experiment is not designed to assess them. All participants were shown a single-column (mobile-like) and multi-column (desktop-like) layout, where news items in various categories were presented with different levels of detail visible (see Fig. 2). The conditions' presentation order was counterbalanced across the participants: every other person saw the single-column condition first followed by the multi-column condition, and the tasks had the opposite order for the other half of the participants. The logs captured interaction with an item if at least 60% of it was visible on the screen and it remained visible for at least 300 ms. In addition to eye-tracking and log data for each participant, we recorded participants' interests, via the questionnaire form. Our analysis examined the effects of two independent variables on item (article) visit times: layout (single- or multi-column) and level of detail (accompanying the title, always presented, with an image and/or description).

3.2   Participants

In total, 24 participants (9 male and 15 female university students, mean age 26.13 years with $SD=4.27$) were enrolled in the study, between December 12 and 20, 2018, from a student mailing list at Aalto University. Five of them wore glasses, and one used contact lenses. On a five-point Likert scale, from participants' self-reporting, the mean level of their knowledge of the English language was 3.71 ($SD=0.62$) and of their interest in news related to North America was 3.67 ($SD=0.64$). All participants received a movie ticket as compensation. The study was conducted in accordance with the principles stated in the Declaration of Helsinki and a local procedure for ethics approval. Each participant signed an informed-consent form before taking part.

3.3   Apparatus and Setup

A custom news-aggregator web application called *WebNews* was created for the experiment. This application's purpose was twofold: 1) to present stored news

Fig. 2. Participants browsed news in a single (left) and a multi-column (right) layout.

items to participants in a single-column and multi-column layout (see Fig. 2) and 2) to log the participants' browsing-behavior data. On average, the single-column condition presented three items per viewport, and the multi-column condition had eight. The application was implemented with standard web technologies (front end) and with Python and MongoDB (back end). The experiment was carried out via the Chrome 71 web browser, running on Windows 8.1, with an Intel Core i7-5930K CPU @3.50 GHz and 64 GB of RAM. Other hardware used in the experiment included a 24-inch LCD monitor and a Logitech M100 optical mouse with scroll wheel. The participants' eye movements were tracked by means of a Tobii EyeX eye-tracker attached to the bottom edge of the monitor. The tracker was calibrated with Tobii Eye Tracking Core Software v2.13.4 for each participant individually, once at the beginning of the experiment and then a second time, between conditions. We collected fixation data by using a custom C# program with Tobii Interaction Library SDK 0.7.3. A video of the participants' browsing behavior with overlaid eye positions was recorded via Tobii Ghost v1.4 and OBS Studio v22.0.2, while eye positions and fixation data was captured through a custom C# program.

3.4   Materials

Headlines of the top live news articles from the US were obtained each morning of the experiment via News API[3] and stored locally in an empty database. Each news item belonged to a single, specified topic category ("business," "entertainment," "health," "science," "sports," or "technology") and contained data such as title, teaser, publication date, and URL to both an actual news article and a related image. Four categories were presented in the single-column layout ("business," "entertainment," "health," and "technology"), while the multi-column layout covered all six. Out of the stored news articles (approximately 400 pieces), 64 + 64 (no duplicates) news items were randomly sampled for each participant to be shown in the experiment. The previews of the articles varied in their level of detail: all items were presented with a title, but some featured a picture and/or description in addition. In the single-column condition, levels of detail within the given layout were determined randomly. The template applied for the multi-column layout displayed the levels of detail in the same order for all participants, but different articles were assigned to each item position. This design choice was intended to generate realistic-looking layouts in the multi-column condition: had the items' detail level been allowed to vary randomly, the page may have looked unrealistic, with items of differing size shown side by side. Participants could freely choose to browse the WebNews app or visit the external sites where the articles were hosted. Upon visiting an external site, participants were instructed to return to the WebNews app and not follow any further links. Additionally, they rated how interesting they found the news in each category generally.

---

[3] https://newsapi.org/.

3.5 Data Pre-processing

We considered a visit to consist of a continuous dwell on an item. When a participant fixated on an area outside the item and then returned to it, we deemed the subsequent dwelling as a *revisit* and regarded the item as having been visited twice so far. Visits were calculated from fixation data, obtained using the Tobii software's fixation filter. For six users, the beginnings and endings of fixations that it calculated were ambiguous, likely on account of a logging error (that is, either a "Begin" or an "End" tag being missing for the gaze points' associated event type). For consistency in fixation calculations (i.e., comparability across all fixations included in the modeling), we omitted these users' data from consideration. Roughly following earlier work's approach [18], we filtered out fixation outliers, which we defined as fixations of below 50 ms or longer than 1500 ms, or 22% of all fixations, across conditions. Fixation duration may depend on the type of activity (e.g., reading [45] or visual search, with varying difficulty [35]), so we used sensitivity analysis with outliers included, to be sure the definition chosen for outliers did not affect the qualitative modeling results.

We considered only those fixations taking place within the WebNews app, to focus on browsing internal to the newsfeed (rather than on external sites). Likewise, we excluded fixations on areas in the margins or on items that were not included in the logs (i.e., those visible for below 300 ms or with less than 60% of their area visible). Our final dataset consisted of 18 participants' data, for 7,446 fixations (with a mean of 0.33 s, $SD=0.29$) in the single-column condition and 7,122 (mean: 0.33 s, $SD=0.28$) in the multi-column layout. Since the participants were allowed to sit 45–100 cm from the monitor and move their head back and forth (our calculations used a mean of 72.5 cm), we assumed the foveal area to correspond to a diameter of 2.53 cm, or 96 px. If any part of an item fell within the foveal area, the calculation of visits took it into account. We carried out sensitivity analysis to test the effect of different foveal areas (diameters of 56 px, 96 px, and 132 px) and concluded that the qualitative modeling results are not affected by these choices. Our dataset covers 2,200 visits, to 794 articles, in the single-column condition and 3,178 visits, to 898, in the multi-column one.

4 Modeling Browsing Behavior

We use Weibull analysis to examine browsing behavior. The Weibull distribution has been used in different contexts as it can fit data from a number of different applications (e.g., biology, engineering and economics) [39]. We draw an analogy between system failure and web browsing, in a manner similar to Liu et al.'s [26] but with item-visit times rather than page-dwell times as the time-to-event-data. Inherent to this approach is that visiting is considered a random process.

4.1 Visiting as a Random Process

In web browsing, a user can visit an item for one or more fixations, then shift the focus of attention somewhere else. Consider a user who is examining a screen displaying three items as in Fig. 1. In this example, there are five visits (labeled from top to bottom in the figure: item 1 → item 3 → item 1 → item 2 → item 3). We model visiting as a random process. Formalizing phenomena as a random process has proven suitable for application in such fields as general browsing [26], gaming [5,44], and medical research [8], with survival analysis of time-to-event data. It is plausible that item visits in browsing are affected by multiple latent variables, introducing randomness. For instance, a door suddenly closing during browsing may draw the user's attention away from the screen, interrupting a visit. We assume that, alongside the random component, browsing behavior is affected by the properties of the items and the browsing environment.

Similarly to Liu et al., we assume that visit durations follow a Weibull distribution, as its parameters can be interpreted with respect to user behavior. We consider a two-parameter Weibull distribution with a shape k and a scale λ.

- The distribution's shape parameter (k) determines whether a process follows *negative aging* (i.e., the immediate probability of the process ending decreases over time) or *positive aging* (i.e., the immediate probability of it ending rises over time). Positive aging is associated with $k > 1$, no aging with $k = 1$, and negative aging with $k < 1$.
- The scale parameter denotes where 63.2% of the processes have ended [31].

One can analyze these parameters by applying two concepts from Weibull analysis: the *survival function* and the *hazard function*, for which we use the following formulations. The former, $S(t)$, describes the proportion of processes (visits) that exceed a given duration t. It is the inverse of the cumulative distribution function, $F(t)$, which can be written as follows for the Weibull distribution:

$$S(t) = 1 - F(t) = e^{-(t/\lambda)^k} \tag{1}$$

The hazard function $h(t)$ at time t of a process (instantaneous failure probability) is calculated thus [26]:

$$h(t) = \frac{k}{\lambda^k} t^{k-1} \tag{2}$$

Here, $h(t)$ gives the instantaneous rate at which a visit to an item ends.

4.2   Weibull Model Specification

To obtain more robust estimates of the Weibull model parameters (shape and scale), we use Bayesian inference to obtain their posterior distributions. This allows quantifying the uncertainty in these estimates. Two models are considered: 1) a separate model for single- and multi-column environments and 2) an extension of this that takes into account properties of an item as covariates.

Separate Model. The separate model accounts for single- and multi-column environments having distinct, or "separate," shape and scale parameters. We assume that k and λ both have a weakly informative prior distribution in the positive domain. Hence, the data y can be modeled via the Weibull distribution:

$$y \sim \text{Weibull}(k, \lambda) \tag{3}$$

Separate Model with Covariates. We also can consider adding item properties to the separate model as covariates. The properties we examine are whether the item contains a picture (p), a description (d), or both. All items contain a title in our setting. We assume that all parameters have weakly informative prior distributions. Again, dataset y is assumed to follow a Weibull distribution:

$$y \sim \text{Weibull}\left(k, \exp\left(\frac{-(\beta_0 + \beta_p x_p + \beta_d x_d)}{k}\right)\right) \tag{4}$$

where y denotes the data, β_0 is an intercept, β_p and β_d are coefficients, and x is a Boolean indicating whether an item preview included a picture (p) or a description (d). Hence, the covariates x are included as a linear combination for the scale λ parameter. This way of adding covariates to the Weibull distribution is referred to as the accelerated life model or the proportional hazard model [39], and the implementation chosen is based on one from prior work [29, 32].

4.3   Model Fitting

To obtain posterior samples for the parameters, a sampler implemented in PyStan3 (Hamiltonian Monte Carlo with No-U-Turn) [38] was run with four chains for 1,000 iterations, with 500 iterations being discarded as warm-up in line with recommendations [16, p. 282]. We achieved good convergence (measured as rank-normalized $\hat{R} < 1.01$ [42]). Model fit was evaluated via comparison of the observations to data produced under the posterior. We used a posterior predictive p-value, which is the probability of data drawn from the posterior being more extreme than the observations, as measured by a test quantity [16] (note that this metric is not the commonly used frequentist p-value). We performed a prior sensitivity analysis too, concluding that the qualitative results hold also for both an uninformative and an informative prior.

5   Results

We report both statistics describing the visit durations and the results of model fitting. Since uncertainty is quantified in the Bayesian model (see Subsect. 5.2), we do not provide related measurements for the descriptive results.

5.1   Descriptive Results

The participants were allowed to browse for an unlimited amount of time. On average, the participants spent more time browsing in the single-column than the multi-column condition: 18:01.80 $(SD{=}599\,\text{s})$ vs. 13:15.94 $(SD{=}257\,\text{s})$. Visits to items were, in general, short, and the distribution of visit durations was right-skewed: most visits were very brief, with some extended visits creating a long right tail for the distribution. Mean visit durations were longer in the single-column condition (1.54 s, with $SD{=}1.90$, vs. 0.84 s, with $SD{=}1.04$). Total dwell times on items (the sums of all visit durations) were higher in the single-column condition, with means of 3.38 s (SD=3.09) and 2.50 s (SD=2.17), respectively, and exceed the mean visit durations, thus reflecting that items frequently received several visits. That is, users seemed to engage in the following pattern: observe an item, look elsewhere on the screen, then return to an item they had already examined. In the single-column condition, approximately 58% of the items were visited more than once, while 76% of the items in the multi-column condition received several visits. The mean number of visits was higher in the multi-column condition, at 1.25 (SD=1.69) as opposed to 2.01 (SD=2.02).

5.2   Modeling Results

To analyze the strategies users may adopt during web browsing, we look at survival and hazard functions for the two fitted models.

Model Fitting Results. We begin by describing the fitting results for the Weibull models' shape and scale parameters. In the single-column condition, the k parameter suggests that visits follow negative aging, since k is below 1 for both models (90% credibility intervals of $k \in [0.89, 0.93]$ and $k \in [0.90, 0.95]$ for, respectively, the separate model and the separate model with covariates). That is, the immediate probability of a user glancing away from an item decreases over time. For the multi-column condition, k is higher (with corresponding 90% credibility intervals of $k \in [0.96, 1.00]$ and $k \in [0.97, 1.01]$). This translates to behavior wherein the immediate probability that a user switches between items is more stable in the multi-column condition. The 90% credibility intervals for the two conditions do not overlap for the k parameter. The scale parameter's value is lower in the multi-column condition for both models (means: $\lambda \approx 0.8$ vs. $\lambda \approx 1.5$) and for items with less information. This result can be interpreted as follows: 63.2% of visits end before reaching a duration of approximately 1.5 s (single-column) or 0.8 s (multi-column). Posterior predictive p-values calculated with the mean as the test statistic indicate a good model fit [16,17] ($p \approx 0.43$ for the separate model and $p \approx 0.47$ for the separate model with covariates), though the fitted model underestimates standard deviation ($SD \approx 1.3$ vs. $SD \approx 1.5$).

Survival. Next, we turn to the survival functions evaluated for the fitted shape and scale parameters. Both models estimate that visits frequently last less than a second, suggesting that visits to items are brief. The models also estimate that visits are longer in the single-column condition. For instance, the separate model estimates that 47–53% of them last over a second in the single-column condition while the equivalent figure for the other condition is only 28–33% (see Fig. 3, pane A). These estimates seem consistent with the empirical observations of the proportions of visits above a given duration (see the dotted lines in Fig. 3, A). The fitted model also shows that visits to items that have less information (e.g., only a title) tend to be shorter (see Fig. 3, B and C).

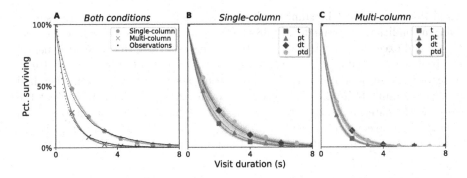

Fig. 3. The percentage of visits surviving to a given duration in both conditions (A) and for different levels of detail (B and C, where t=title, p=picture, and d=description). Visit durations are shorter in the multi-column condition and when the items have less detail. The marked line is drawn from the posterior means of the models' parameters. Uncertainty is indicated with 100 functions for parameters sampled from the posterior.

Hazard. The model fitting's results suggest that the hazard functions for the two conditions differ in shape. Users seem to move between items frequently, with the switching rate being higher in the multi-column condition. The instantaneous rate of switching one's focus of attention per second (the hazard rate) decreases over time in the single-column condition (the slope of the hazard function is steeper for that condition in Fig. 4's pane A). On the other hand, in the multi-column condition, this probability stays more stable. Users in the multi-column setting were approximately 50% more likely to switch their focus of attention upon landing on an item than users in the single-column condition (with roughly 0.8 vs. 1.2 switches per second in the first fixation). Users move their attention away quicker from items with fewer details (as Fig. 4's panes B and C attest). This pattern is more distinct for the single- than the multi-column layout, where the average hazard rate decreases as the amount of detail increases (e.g., compare the hazard functions for items with title only vs. with image, title, and description in Fig. 4, B). However, for the multi-column condition, the hazard rate is similar between items with a title only and ones with an additional image (t and pt in Fig. 4, C). In addition, items with descriptions in the multi-column condition show similar hazard rates (dt and ptd in Fig. 4, C). These observations arise from the different estimates for the shape k and the scale λ parameters for the different models. The hazard rate in the single-column condition exhibits negative aging (a hazard rate that falls over time), which roughly corresponds to a screening pattern wherein most visits are brief, with some items passing this initial test [26]. A less prominent effect is visible in the multi-column condition.

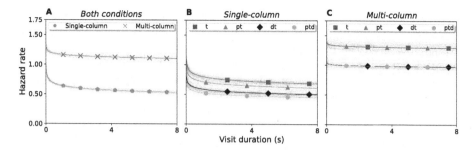

Fig. 4. Hazard functions for the fitted models in both conditions (A) and for different levels of detail (B and C, where t=title, p=picture, and d=description). Users switch focus of attention more often in the multi-column condition and from items with fewer details. The marked line is drawn from the posterior means of the models' parameters. Uncertainty is indicated with 100 functions for parameters sampled from the posterior.

6 Discussion and Conclusions

The main findings of this paper are the following:

- People distribute their attention to items on a screen in a fragmented manner. Instead of making a single, focused visit to an item, users gather information in a sequence of visits.
- We found the "fragmentation" to be more prominent in desktop (multi-column) than mobile-like (single-column) environments in our setting.

We measured fragmentation of attention as the frequency of gaze shifts between items and formalized it by modeling visit durations via Weibull analysis in line with prior work [26]. These results could inform design of content feeds, commonly used in social media and news applications.

Weibull analysis presents the advantage of having parameters (scale and shape) that can be interpreted with respect to user behavior. Our results suggest that mobile-like environments with a single-column layout are more effective at maximizing the attention a user directs toward any single item. If the goal is instead a maximal number of items attended to, desktop-like environments with multi-column layouts are better. For example, the fitted model suggests that when user gaze shifts to a target item, the rate of switching one's focus of attention (the hazard rate) is higher in a desktop-like environment. The properties of the item matter also: items that contain a title, a description, and an image are given attention longer than those with just a title. This observation is sensible, since items with only a title offer less information – hence its processing is quicker. In addition, we found that a screening pattern wherein items are quickly scanned is more prevalent in the single-column condition, suggested by the lower value of the Weibull distribution's shape parameter. This observation parallels that of Liu et al., who find a similar pattern when analyzing page-level data.

One way to interpret the model proposed here is that a user samples item-visit times from Weibull distributions. Our results point to these distributions diverging between the two conditions (mobile- and desktop-like) and with the level of detail visible in an item (picture, title, and/or description). Longer visits to items that are richer in detail may be a natural consequence of there being more information to explore. Similarly, the shorter visits in the multi-column condition may stem from the a more complex layout and the larger number of items presented. Additionally, with fewer items being visible on the viewport in the single-column layout than the multi-column one at any given time, longer visits may be explained by the effort it would take to switch viewport.

We hypothesize that a cost–benefit (or utility) lens [2,11,41] may aid in interpreting these results. In previous work, the notion of costs and benefits has been used in reference to browsing behavior in more structured search tasks (e.g., with SERPs [1,2]). Some of this work, building on IFT, suggests that search behavior is determined by a judgment of whether the information sources are relevant for the information diet. We suggest that our results can be viewed through this lens (i.e., in relation to cost–benefit analysis) even though we concern ourselves with an unstructured task. Browsing the newsfeed brings a cost to the user in the form of time and effort. In addition, users may choose items to attend to by gauging some utility to be gained from the activity, even when a specific information need is not specified. This approach ties in with our finding that visit times were lower in the desktop- than the mobile-like environment and with items showing a title only. The switching cost of glancing at another item may be lower when the target displays only text and in multi-column layouts that position items near one another and make more items visible without a need for scrolling. When switching costs are lower, moving between items more frequently may offer strategic benefits. Related work sometimes characterizes undirected browsing tasks as oriented toward randomness [4,9,13]. Were browsing purely random selection and sampling, however, we would not expect the inter-condition differences observed in our study to emerge.

Future work could address certain limitations of the study. Running a similar experiment in a different geographical region, with participants who are not primarily students and using another commercial eye-tracker, could aid in assessing whether the results generalize to the population at large. In addition, future efforts should aim to explore visit durations in more complex conditions, such as the richer interaction scenarios emerging when users follow links.

Acknowledgements. This work was supported by the Finnish Center for Artificial Intelligence (FCAI), Business Finland (MINERAL project), the Academy of Finland (projects Human Automata – ID: 328813, and BAD – ID: 318559), as well as the Technology Industries of Finland (project SOWP). We would also like to thank the reviewers for their feedback.

Data Availability Statement. The data and code for the Weibull analysis is available through the project page: https://userinterfaces.aalto.fi/browsing/.

References

1. Azzopardi, L., Thomas, P., Craswell, N.: Measuring the utility of search engine result pages: An information foraging based measure. In: The 41st International ACM SIGIR Conference on Research & ; Development in Information Retrieval, pp. 605–614. SIGIR '18, Association for Computing Machinery, New York, NY, USA (2018). https://doi.org/10.1145/3209978.3210027

2. Azzopardi, L., Zuccon, G.: An analysis of the cost and benefit of search interactions. In: Proceedings of the 2016 ACM International Conference on the Theory of Information Retrieval, pp. 59–68. ICTIR '16, Association for Computing Machinery, New York, NY, USA (2016). https://doi.org/10.1145/2970398.2970412

3. Bates, M.J.: The design of browsing and berrypicking techniques for the online search interface. Online Rev. **13**(5), 407–424 (1989). https://doi.org/10.1108/eb024320

4. Bates, M.J.: What is browsing - really? a model drawing from behavioural science research. Inform. Res. **12**(4), (2007)

5. Bauckhage, C., Kersting, K., Sifa, R., Thurau, C., Drachen, A., Canossa, A.: How players lose interest in playing a game: an empirical study based on distributions of total playing times. In: 2012 IEEE Conference on Computational Intelligence and Games (CIG), pp. 139–146 (2012). https://doi.org/10.1109/CIG.2012.6374148

6. Bawden, D.: Encountering on the road to Serendip? Browsing in new information environments, pp. 1–22. Facet (2011). https://doi.org/10.29085/9781856049733.003

7. Buscher, G., Cutrell, E., Morris, M.R.: What do you see when you're surfing? Using eye tracking to predict salient regions of Web pages, pp. 21–30. Association for Computing Machinery, New York, NY, USA (2009). https://doi.org/10.1145/1518701.1518705

8. Carroll, K.J.: On the use and utility of the Weibull model in the analysis of survival data. Control. Clin. Trials **24**(6), 682–701 (2003). https://doi.org/10.1016/S0197-2456(03)00072-2

9. Catledge, L.D., Pitkow, J.E.: Characterizing browsing strategies in the world-wide web. Computer Networks and ISDN Systems **27**(6), 1065–1073 (1995). https://doi.org/10.1016/0169-7552(95)00043-7, proceedings of the Third International World-Wide Web Conference

10. Chi, E.H., Pirolli, P., Chen, K., Pitkow, J.: Using information scent to model user information needs and actions and the Web. In: Proceedings of the SIGCHI Conference on Human Factors in Computing Systems, pp. 490–497. CHI '01, Association for Computing Machinery, New York, NY, USA (2001). https://doi.org/10.1145/365024.365325

11. Cooper, W.S.: On selecting a measure of retrieval effectiveness. J. Am. Soc. Inform. Sci. **24**(2), 87–100 (1973). https://doi.org/10.1002/asi.4630240204

12. Couture Bue, A.C.: The looking glass selfie: instagram use frequency predicts visual attention to high-anxiety body regions in young women. Comput. Hum. Behav. **108**, 106329 (2020). https://doi.org/10.1016/j.chb.2020.106329

13. Cove, J., Walsh, B.: Online text retrieval via browsing. Inform. Process. Manage. **24**(1), 31–37 (1988). https://doi.org/10.1016/0306-4573(88)90075-1

14. Crichton, K., Christin, N., Cranor, L.F.: How do home computer users browse the web? ACM Trans. Web **16**(1) (2021). https://doi.org/10.1145/3473343

15. Dumais, S.T., Buscher, G., Cutrell, E.: Individual differences in gaze patterns for web search. In: Proceedings of the Third Symposium on Information Interaction in Context, pp. 185–194. IIiX '10, Association for Computing Machinery, New York, NY, USA (2010). https://doi.org/10.1145/1840784.1840812
16. Gelman, A., Carlin, J., Stern, H., Dunson, D., Vehtari, A., Rubin, D.: Model checking. Chapman Hall/CRC (2013). https://doi.org/10.1201/b16018
17. Gelman, A.: Two simple examples for understanding posterior p-values whose distributions are far from uniform. Electroni. J. Stat. **7**, 2595–2602 (2013). https://doi.org/10.1214/13-EJS854
18. Henderson, J.M., Choi, W., Luke, S.G., Schmidt, J.: Neural correlates of individual differences in fixation duration during natural reading. Quart. J. Exp. Psychol. **71**(1), 314–323 (2018). https://doi.org/10.1080/17470218.2017.1329322
19. Herner, S.: Browsing. In: Kent, A., Lancour, H., Nasri, W. (eds.) Encyclopedia of Library and Information Science, vol. 3, pp. 408–415 (1970)
20. Ho, H.F.: The effects of controlling visual attention to handbags for women in online shops: Evidence from eye movements. Comput. Hum. Behav. **30**, 146–152 (2014). https://doi.org/10.1016/j.chb.2013.08.006
21. Hotchkiss, G.: Google's Golden Triangle - Nine Years Later (2014). https://outofmygord.com/2014/10/09/googles-golden-triangle-nine-years-later/
22. Hotchkiss, G., Alston, S., Edwards, G.: Eye Tracking Study: An In Depth Look at Interactions with Google Using Eye Tracking Methodology. Enquiro Search Solutions Incorporated (Jun 2005)
23. Huang, Y.T.: The female gaze: content composition and slot position in personalized banner ads, and how they influence visual attention in online shoppers. Comput. Hum. Behav. **82**, 1–15 (2018). https://doi.org/10.1016/j.chb.2017.12.038
24. Huberman, B.A., Pirolli, P.L.T., Pitkow, J.E., Lukose, R.M.: Strong regularities in world wide web surfing. Science **280**(5360), 95–97 (1998). https://doi.org/10.1126/science.280.5360.95
25. Lagun, D., Agichtein, E.: Inferring searcher attention by jointly modeling user interactions and content salience. In: Proceedings of the 38th International ACM SIGIR Conference on Research and Development in Information Retrieval, pp. 483–492. SIGIR '15, Association for Computing Machinery, New York, NY, USA (2015). https://doi.org/10.1145/2766462.2767745
26. Liu, C., White, R.W., Dumais, S.: Understanding web browsing behaviors through weibull analysis of dwell time. In: Proceedings of the 33rd International ACM SIGIR Conference on Research and Development in Information Retrieval. pp. 379–386. SIGIR '10, Association for Computing Machinery, New York, NY, USA (2010). https://doi.org/10.1145/1835449.1835513
27. Luo, P., Zhou, G., Tang, J., Chen, R., Yu, Z., He, Q.: Browsing regularities in hedonic content systems. In: Proceedings of the Twenty-Fifth International Joint Conference on Artificial Intelligence, pp. 3811–3817. IJCAI'16, AAAI Press (2016)
28. Marchionini, G.: Information Seeking in Electronic Environments. Cambridge Series on Human-Computer Interaction, Cambridge University Press (1995). https://doi.org/10.1017/CBO9780511626388
29. (to mi), T.P.: stan-survival-shrinkage (2015). https://github.com/to-mi/stan-survival-shrinkage
30. Nielsen, J.: F-Shaped Pattern for Reading Web Content (original study) (Apr 2006). https://www.nngroup.com/articles/f-shaped-pattern-reading-web-content-discovered/
31. Pasha, G., Khan, M.S., Pasha, A.: Empirical analysis of the weibull distribution for failure data. J. Stat. **13**, 33–45 (2006)

32. Peltola, T., Havulinna, A.S., Salomaa, V., Vehtari, A.: Hierarchical Bayesian survival analysis and projective covariate selection in cardiovascular event risk prediction. In: Proceedings of the Eleventh UAI Conference on Bayesian Modeling Applications Workshop - Volume 1218, pp. 79–88. BMAW'14, CEUR-WS.org, Aachen, DEU (2014)
33. Pernice, K.: Text Scanning Patterns: Eyetracking Evidence (Aug 2019). https://www.nngroup.com/articles/text-scanning-patterns-eyetracking/
34. Pirolli, P., Card, S.: Information foraging in information access environments. In: Proceedings of the SIGCHI Conference on Human Factors in Computing Systems, pp. 51–58. CHI '95, ACM Press / Addison-Wesley, USA (1995). https://doi.org/10.1145/223904.223911
35. Reingold, E.M., Glaholt, M.G.: Cognitive control of fixation duration in visual search: the role of extrafoveal processing. Vis. Cogn. **22**(3–4), 610–634 (2014). https://doi.org/10.1080/13506285.2014.881443
36. Resnick, M., Albert, W.: The impact of advertising location and user task on the emergence of banner ad blindness: An eye-tracking study. Int. J. Human-Comput. Interact. **30**(3), 206–219 (2014). https://doi.org/10.1080/10447318.2013.847762
37. Resnick, M.L., Albert, W.: The influences of design esthetic, site relevancy and task relevancy on attention to banner advertising. Interact. Comput. **28**(5), 680–694 (2016). https://doi.org/10.1093/iwc/iwv042
38. Riddell, A., Hartikainen, A., Carter, M.: pystan (3.0.0). PyPI (Mar 2021)
39. Rinne, H.: Related distributions. In: The Weibull Distribution. CRC Press (2008). https://doi.org/10.1201/9781420087444.ch3
40. Savolainen, R.: Berrypicking and information foraging: comparison of two theoretical frameworks for studying exploratory search. J. Inf. Sci. **44**(5), 580–593 (2018). https://doi.org/10.1177/0165551517713168
41. Varian, H.R.: Economics and search. ACM. SIGIR Forum **33**(1), 1–5 (1999)
42. Vehtari, A., Gelman, A., Simpson, D., Carpenter, B., Bürkner, P.C.: Rank-normalization, folding, and localization: an improved $\hat{R}$ for assessing convergence of mcmc (with discussion). Bayesian Analysis 16(2) (Jun 2021). https://doi.org/10.1214/20-ba1221
43. Vickery, J.: A note in defense of browsing. BLL Rev. **5**(3), 110 (1977)
44. Viljanen, M., Airola, A., Heikkonen, J., Pahikkala, T.: Playtime measurement with survival analysis. IEEE Trans. Games **10**(2), 128–138 (2018). https://doi.org/10.1109/TCIAIG.2017.2727642
45. Vitu, F., McConkie, G.W., Kerr, P., O'Regan, J.: Fixation location effects on fixation durations during reading: An inverted optimal viewing position effect. Vision. Res. **41**(25), 3513–3533 (2001). https://doi.org/10.1016/S0042-6989(01)00166-3
46. White, R.W., Drucker, S.M.: Investigating behavioral variability in web search. In: Proceedings of the 16th International Conference on World Wide Web. pp. 21–30. WWW '07, Association for Computing Machinery, New York, NY, USA (2007). https://doi.org/10.1145/1242572.1242576
47. White, R.W., Roth, R.A.: Exploratory search: beyond the query-response paradigm. Synthesis Lect. Inform. Concepts, Retrieval Serv. **1**(1), 1–98 (2009). https://doi.org/10.2200/S00174ED1V01Y200901ICR003

Topic-Enhanced Personalized Retrieval-Based Chatbot

Hongjin Qian and Zhicheng Dou[✉]

Gaoling School of Artificial Intelligence, Renmin University of China, Beijing, China
{ian,dou}@ruc.edu.cn

Abstract. Building a personalized chatbot has drawn much attention recently. A personalized chatbot is considered to have a consistent personality. There are two types of methods to learn the personality. The first mainly model the personality from explicit user profiles (*e.g.*, manually created persona descriptions). The second learn implicit user profiles from the user's dialogue history, which contains rich, personalized information. However, a user's dialogue history can be long and noisy as it contains long-time, multi-topic historical dialogue records. Such data noise and redundancy impede the model's ability to thoroughly and faithfully learn a consistent personality, especially when applied with models that have an input length limit (*e.g.*, BERT). In this paper, we propose deconstructing the long and noisy dialogue history into topic-dependent segments. We only use the topically related dialogue segment as context to learn the topic-aware user personality. Specifically, we design a **Top**ic-enhanced personalized **Re**trieval-based **C**hatbot, TopReC. It first deconstructs the dialogue history into topic-dependent dialogue segments and filters out irrelevant segments to the current query via a Heter-Merge-Reduce framework. It then measures the matching degree between the response candidates and the current query conditioned on each topic-dependent segment. We consider the matching degree between the response candidate and the cross-topic user personality. The final matching score is obtained by combining the topic-dependent and cross-topic matching scores. Experimental results on two large dataset show that TopReC outperforms all previous state-of-the-art methods.

Keywords: Personalization · Dialogue systems

1 Introduction

Developing an open-domain chatbot is a long-lasting task in the AI domain. The main reason is that an open-domain chatbot enables human-machine interactions via text from any domain irrespective of any constraints, which is considered as an ultimate goal of AI [3]. Methods for building an open-domain chatbot can be divided into two categories: generation-based and retrieval-based. The former leverages models (*e.g.*, encoder-decoder) to generate a new response [9,16,24].

© The Author(s), under exclusive license to Springer Nature Switzerland AG 2023
J. Kamps et al. (Eds.): ECIR 2023, LNCS 13981, pp. 79–93, 2023.
https://doi.org/10.1007/978-3-031-28238-6_6

The latter retrieves a set of response candidates and chooses the most matched one as the output [10,12]. In this paper, we focus on the retrieval-based chatbot.

For an open-domain chatbot, a consistent personality is crucial as personality inconsistency might bring a sense of unpredictability and untrustworthiness [3]. To this end, many works seek to develop personalized chatbots that have consistent personalities. Previous works about personalized chatbots can be divided into three groups: (1) Early works assign a trainable user embedding to each user, which is updated during training and can be used to guide response retrieval or generation [8]; (2) some works model the user personality from explicit user profiles which are usually persona descriptions or attributes [15,19]; (3) recent works propose learning implicit user profiles from the user's dialogue history [11,13,23]. As discussed in [13], learning implicit user profiles from the dialogue history is advantageous regarding flexibility and effectiveness. First, a user's dialogue history is easy to obtain and update. Second, a user's dialogue history contains rich personalized information, such as the user's preferences and preferred expressions, which are essential for personality modeling.

However, a user's dialogue history contains long-time and multi-topic dialogue records, which might be redundant and noisy. Directly modeling the raw dialogue history has two challenges: (1) the redundant dialogue history contains a large number of historical dialogues. Feeding the whole dialogue history into a neural model might lead to model capacity overflow, especially when applying pre-trained language models with token length limits (e.g., BERT has 512 length limits); (2) a user might have dynamic preferences over different topics. Modeling such topical preference dynamism is challenging to maintain the consistent personality of a personalized chatbot.

Most previous methods that learn implicit user profiles fail to overcome the two challenges. They usually learn several user representations directly from the whole dialogue history to guide response selection or generation. In this paper, we instead propose **deconstructing the long and noisy dialogue history into topic-dependent dialogue segments from which we learn the topic-aware implicit user profiles for personalized chatbot**. Modeling the implicit user profiles from the topic-dependent dialogue segments has three advantages: (1) as the long dialogue history is split into short dialogue segments, we can model the implicit user profile from each segment separately, which greatly reduces the required model capacity; (2) the topic-dependent dialogue segments are less noisy than the whole dialogue history. The reason is that the data noise in the dialogue history is primarily caused by its varied topics. And a user might have different personal preferences over various topics. For example, regarding the topic of organic vegetables, a vegetarian is likely to show great interest while a meatatarian would not; (3) given an input query, we can further measure the topical relevance between the topic-dependent dialogue segments and the query and filter out the irrelevant dialogue segments.

We design TopReC, which learns topic-aware user personality from topic-dependent dialogue segments. TopReC comprises two modules, the Topic-dependent Context Deconstruction module, and the Personalized Topic Matching module.

In the Topic-dependent Context Deconstruction module, the dialogue history is first reassembled into topic-dependent segments concerning the topical inter-relations among the historical dialogues. We then filter out the topic segments according to their relevance to the current query and only keep the topically related dialogue segments to model personality. When deconstructing the dialogue history into topic-dependent dialogue segments, one challenge is that the number of topics in each user's dialogue history is dynamic. To tackle such dynamism, we propose a Heter-Merge-Reduce method that can flexibly deconstruct the dialogue history into topic-dependent segments without deciding the topic number in advance. **In the Personalized Topic Matching module**, we measure the matching degree between the response candidate and each topic-dependent topic-dependent dialogue segment to obtain topic-dependent matching scores. Besides, we also measure the relevance between the response candidate and the cross-topic user profile to get the cross-topic matching score. The final matching scores are obtained by fusing the topic-dependent and cross-topic matching scores.

To verify the effectiveness of the proposed model TopReC, we conduct extensive experiments on two publicly available datasets for personalized response selection. The empirical results show that our model achieves the best performance overall baseline models. Our contributions are three-fold: (1) We point out that a user's dialogue history might reflect multi-faceted user interests, which indicates that a user's personalized preferences can be dynamic in the dialogue history; (2) We propose TopReC that deconstructs the user's dialogue history into topic-dependent segments via the Herter-Merge-Reduce method and performs personalized response selection by learning topic-aware implicit user profile from the topic-dependent dialogue segments; (3) Comprehensive experiments show that our model outperforms the state-of-the-art models.

2   Related Work

2.1   Retrieval-Based Chatbot

A retrieval-based chatbot aims to select a proper response from the response candidates given the current query. Early works mainly focus on single-turn dialogue, which takes the current query as the dialogue context. Afterwards, many works turn attention to the multi-turn dialogue, which takes a series of follow-up dialogues as the context [10]. To model the multi-turn dialogues, early works directly encode the multi-turn dialogues into hidden states via RNN and use the last hidden states to perform matching [10]. Later works mainly improve the multi-turn dialogue task by either obtaining deep context representation (*e.g.*, DAM [18]) or selecting useful dialogue context (*e.g.*, MSN [24]). With the huge success of the pre-trained language model (*e.g.*, BERT), recent works further improve the effectiveness of the multi-turn dialogue model. For example, Han et al. design self-supervised tasks to continue training BERT [6] and Xu et al. split the dialogue context into segments and feed them into BERT to compute relevance scores [17].

2.2   Personalized Chatbot

For open-domain chatbots, inconsistent personalities bring unpredictability and untrustworthiness to the end-user. Maintaining a consistent personality is the ultimate goal for the domain. To endow consistent personality to the open-domain chatbots, early works assign a user embedding to each user, which can be updated during training [8]. Inspired by the PERSONA-CHAT dataset [20], which contains user descriptions for each user, many works explore directly modelling the explicit user profile (e.g., personality descriptions or user attributes). For example, DGMN [22] lets the dialogue context and the user profile interact with each other to learn a user representation. Some works also claim that the explicit user profile contains noise which might undermine the user modeling. Hence, models like CSN [25] and RSM-DCK [7] propose context selection to denoise the explicit user profile. Besides, Gu et al. concatenate the dialogue context, user profile, and the current query into a long sequence to feed into BERT to obtain the matching representation [5]. Though the explicit user profile can partly reflect the user's personality, it suffers from inflexibility and limited personalized information. Therefore, recent works propose learning implicit user profiles from the user's dialogue history [11,13]. In this paper, we argue that the user's dialogue history might be long and noisy. Directly learning the implicit user profile from the whole dialogue history might limit the model's performance. Therefore, we propose deconstructing the dialogue history into topic-dependent segments and learn topic-dependent user representations from the topic-dependent segment.

3   Methodology

3.1   Preliminary

For a retrieval-based chatbot, the major goal is to return the best response from a response repository given an input query. Formally, let $g(\cdot, \cdot)$ be a scoring model evaluating the matching degree of a candidate response r for an input query q under the context C. The chatbot will choose the response r^* with the highest scores of g from a repository of responses $\mathcal{R}$ as the output. Hence, following [13], we have:

$$r^* = \arg\max_{r \in \mathcal{R}} g(q, r, C),$$

where the context C can be versatile. Taking the personalized chatbot as an example, C is the user profile that portrays the personality of the user. As mentioned in Sect. 1, the personalized chatbot can either learn the user personality from explicit user descriptions [7,19,25] or implicitly learn the personality from the dialogue history [11,13]. Inspired by recent works that highlight the effectiveness and availability of learning implicit user profiles from the user's dialogue history, we can define a mapping function $\mathcal{F}(\cdot)$ that learns the implicit user profile from the dialogue history. Formally, we have $C = \mathcal{F}(H)$, where $H = \{(p_j, r_j)\}, j \in [1, t]$ represents the dialogue history of the user and (p_j, r_j) refers to the j-th historical post-response pair.

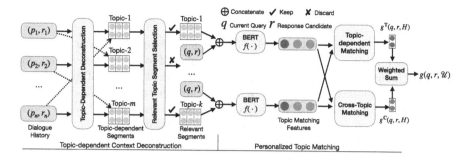

Fig. 1. The overview of TopReC.

3.2 The Proposed Model: TopReC

When learning the implicit user profile from the user's dialogue history, data noise and redundancy of dialogue history are two major issues we need to address. The data noise undermines the faithfulness of the learned user personality. And the data redundancy might lead to model capacity overflow (*e.g.*, BERT has 512 length limits). Our TopReC proposes deconstructing the long dialogue history into topic-dependent dialogue segments and filtering out dialogue segments that are irrelevant to the current query. Afterward, TopReC performs relevance matching between the response candidates and the topically-related dialogue segments to obtain the relevance scores.

Figure 1 shows the overview of TopReC. Specifically, TopReC comprises two modules: the Topic-dependent Context Deconstruction module and the Personalized Topic Matching module. The former deconstruct the dialogue history $H = \{(p_j, r_j)\}, j \in [1, t]$ into topic-dependent segments $\{H_1, \cdots, H_m\}, m \leq t$ and filter out irrelevant ones to get $\{\tilde{H}_1, \cdots, \tilde{H}_k\}, k \leq m$ that are topically-related to the current query. The latter applies a pre-trained encoder (*e.g.*, BERT) to perform topic-dependent matching and cross-topic matching to obtain the topic-dependent feature $g^{\mathbf{T}}(q, r, H)$ and the cross-topic matching feature $g^{\mathbf{C}}(q, r, H)$, respectively. The final matching score is computed by fusing the topic-dependent and cross-topic matching features.

3.3 Topic-Dependent Context Deconstruction

The Topic-dependent Context Deconstruction module obtains the topic-dependent dialogue segments via three steps: Heter-Merge-Reduce. We illustrate the procedures in Fig. 2. For a user's dialogue history $H = \{(p_j, r_j)\}, j \in [1, t]$ and the current query q, the goal of the module is to first deconstruct the dialogue history H into m topic-dependent segments $\{H_1, \cdots, H_m\}, m \leq t$ and then select k topic segments $\{\tilde{H}_1, \cdots, \tilde{H}_k\}, k \leq m$ that are topically-related to the current query q. We then will explain the details of each step of Heter-Merge-Reduce.

In the Heter step, we seek to decide the number of topics of a user's dialogue history. The difficulty of this step is that the number of topics in a user's dialogue history is changeable. Therefore, we cannot preset a fixed number of topics for all users. TopReC applies a soft margin to dynamically control the number of topics of each user's dialogue history. We achieve the goal by choosing m historical posts $P^{\text{topic}} = \{\hat{p}_1, \cdots, \hat{p}_m\}$ as the topic centers in which the mutual similarities of any two posts are smaller than a threshold γ. Specifically, we feed all the historical posts $\{p_1, \cdots, p_t\}$ into a pretrained encoder (e.g., BERT). And we use the [CLS] token's hidden states of the i-th post p_i as its sentence representation $\mathbf{p}_i$. We then compute the point-wise similarities M of the historical posts:

$$M = \{\theta(p_i, p_j)\}, i, j \in [1, t], i \neq j, \tag{1}$$

$$\theta(p_i, p_j) = \frac{\mathbf{p}_i \cdot \mathbf{p}_j}{\|\mathbf{p}_i\|_2 \cdot \|\mathbf{p}_j\|_2}, \tag{2}$$

$$\mathbf{p} = \text{Pool}_{cls}(\text{BERT}(p)), \tag{3}$$

where $\theta(\cdot, \cdot)$ is the cosine similarity function.

After obtaining the point-wise similarities of all historical posts, we can choose the topic center posts $P^{\text{topic}} = \{\hat{p}_1, \cdots, \hat{p}_m\}$ by:

$$P^{\text{topic}} = \{\hat{p}_m\}, \theta(\hat{p}_m, \hat{p}_j) < \gamma, \hat{p}_m \neq \hat{p}_j. \tag{4}$$

In the merge step, we assign each historical dialogue (p_j, r_j) to a topic segment $H_n, n \in [1, m]$ of which the topic center $\hat{p}_n$ is the most similar to p_j:

$$(p_j, r_j) \rightarrow H_n; n = \text{argmax}_{n \in [1, m]} \theta(p_j, \hat{p}_n). \tag{5}$$

In the reduce step, we remove the negative impact of the irrelevant topic segments. Thus, we prune the topic segments $\{H_1, \cdots, H_m\}$ to $\{\tilde{H}_1, \cdots, \tilde{H}_k\}, k \leq m$ by measuring the similarity $\theta(q, \hat{p}_n), n \in [1, m]$ between the topic center $\hat{p}_n, n \in [1, m]$ and the current query q. We keep the k topic segments with the highest similarity score $\theta(q, \hat{p}_n)$. We will discuss the impact of the choice of k in Sect. 4.5.

Taking Fig. 2 as an example, we explain the Heter-Merge-Reduce method. In the Heter step, we choose the historical posts $\{p_1, p_5, p_6\}$ as the topic centers. We assign historical post-response pairs to the most similar topic segment in the Merge step. In the Reduce step, we keep $k = 2$ topic segments and filter out the segments centered by p_3.

3.4   Personalized Topic Matching

As mentioned in Sect. 1, a user's personal preferences can be dynamic over topics. Therefore, instead of modelling the user personality from the whole dialogue history, we propose modelling the topic-aware personality from topic-dependent

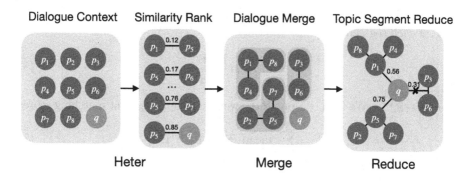

Fig. 2. The Heter-Merge-reduce method

dialogue segments, which can greatly avoid the personality bias brought by such preference discrepancy. In the Topic-dependent Context Deconstruction module, we obtain k topic-dependent segments $\{\tilde{H}_1, \cdots, \tilde{H}_k\}$ that are relevant to the current query q. We then perform matching between the current query q and the response candidate r under the context of each topic-dependent segment $\tilde{H}$ via the Personalized Topic Matching module. Formally, given the k topic-aware segments $\{\tilde{H}_1, \cdots, \tilde{H}_k\}$, the current query q and response candidate r, we seek to compute k topic-dependent matching features $\{\mathbf{e}_1, \cdots, \mathbf{e}_k\}$ which represent the topic-dependent relevance between the topic segments and the response candidate given the current query q. Taking the k-th topic segments $\tilde{H}_k = \{(p_{k,1}, r_{k,1}), \cdots, (p_{k,n_k}, r_{k,n_k})\}$ as an example, we first concatenate the topic-dependent segment with the current query q and the response candidate r into a token sequence S_k:

$$S_k = [\text{CLS}], p_{k,1}, r_{k,1}[\text{SEP}], \cdots, [\text{SEP}], q, [\text{SEP}], r \qquad (6)$$

And we then feed the token sequence S_k into a pretrained encoder $\phi(\cdot)$ (*e.g.*, BERT) to get the token representations. We use the representation of the first token ([CLS]) as the sequence representation which is fed into a Multi-layer Perceptron (MLP) to obtain the k-th matching representations $\mathbf{e}_k$:

$$\mathbf{e}_k = \text{MLP}_1(\text{Pooling}_{\text{CLS}}(\phi(S_k))), \qquad (7)$$

where $\text{MLP}_1 \in \mathbb{R}^{d \times d}$ and d is the hidden size.

Likewise, we perform the topic-dependent matching over each topic segments respectively and obtain k topic-dependent matching representations $\mathbf{E} = \{\mathbf{e}_1, \cdots, \mathbf{e}_k\}$, $\mathbf{E} \in \mathbb{R}^{k \times d}$. The k matching representations measure the matching degree between the response candidate r and the k topic-dependent segments given the current query q.

Furthermore, we think that the impact of the topic-aware segments is different as their topical relatedness to the current query is not the same. The topic-dependent segments with larger relevance scores to the current query should be

more important. Thus, we perform self-attention to compute the relative importance of each segment by:

$$\mathbf{S} = \text{softmax}(\frac{\mathbf{E} \cdot \mathbf{E}^\top}{\sqrt{d}}), \tag{8}$$

where $\mathbf{S} \in \mathbb{R}^{k \times k}$. We then computed the weighted topic segment matching scores
by:

$$g^\mathbf{T} = \sum \underset{dim=-1}{\text{mean}} (\mathbf{S}) \cdot \text{MLP}_2(\mathbf{E}), \tag{9}$$

where $\text{MLP}_2 \in \mathbb{R}^{d \times 1}$.

Besides the matching signal among topic-dependent segments, we also want
to model the matching signal from the cross-topic user profile. Therefore, we
compute the cross-topic matching representation by using a residual connection
with an MLP to get a fused representation $\tilde{\mathbf{E}}$:

$$\tilde{\mathbf{E}} = \text{MLP}_3(\hat{\mathbf{E}}) + \hat{\mathbf{E}}, \quad \hat{\mathbf{E}} = \mathbf{E} \cdot \mathbf{S} + \mathbf{E}, \tag{10}$$

where $\text{MLP}_3 \in \mathbb{R}^{d \times d}$. We then pool the weighted matching representation $\tilde{\mathbf{E}}$ to
obtain the cross-topic matching feature and feed it into a MLP to obtain the
matching score of the cross-topic user profile.

$$g^\mathbf{C} = \text{MLP}_2(\underset{dim=-1}{\text{mean}} (\tilde{\mathbf{E}})). \tag{11}$$

We combine the two scores by:

$$g = \alpha \cdot g^\mathbf{C} + (1 - \alpha) \cdot g^\mathbf{T}, \tag{12}$$

where α is a trainable parameter and is initialized by 0.5.

We use cross-entropy loss to train the model:

$$\mathcal{L}(\theta) = -\frac{1}{|D|} \sum_D [y \log(g) + (1 - y) \log(1 - g)]. \tag{13}$$

4   Experiments

4.1   Dataset and Evaluation

We explore learning implicit user profiles from the user's dialogue history. Therefore, we require datasets with user identifications to construct users' dialogue
history. We use two public datasets: Weibo and Reddit. Specifically, the Weibo
dataset is derived from the PChatbotW dataset, in which all posts and responses
have timestamps and user IDs [14]. The Reddit dataset is released by [21],
which is crawled from the Reddit forum from Dec. 1, 2015, to Oct. 30, 2018.
By traversing the chain-like responses, we can obtain post-response pairs with
timestamps and user IDs. We first aggregate the user's dialogue history for the

two datasets and then filter out users who have less than fifteen historical dialogues. Besides, we limit the length of all utterances by 50 tokens. Following previous works [10,13], we use the latest post as the current query and create a list of ten response candidates in which the negative samples are mined via a BM25 engine. The candidate list contains: (1) the ground-truth response made by the user; (2) other user's responses under the same post (non-personalized response); (3) retrieved response candidates via a retrieval engine (hard negative samples). The statistic information of the two datasets is shown in Table 1.

Table 1. The statistics of the two datasets.

	Weibo	Reddit
Number of users	420,000	280,642
Average history length	32.3	85.4
Average length of post	24.9	10.5
Average length of response	10.1	12.4
Number of response candidates	10	10
Number of training samples	3,000,000	2,000,000
Number of validation samples	600,000	403,210
Number of testing samples	600,000	403,210

To evaluate our proposed TopReC and all baseline models, we use $\mathbf{R}_n@\mathbf{k}$ (recall at position k in n candidates) and **MRR** (Mean Reciprocal Rank) as evaluation metrics. As the ground-truth response is the personalized response, the two metrics can directly evaluate the model's ability to output a response that is consistent with the user's personality.

4.2   Baseline Models

In the task, the user's dialogue history comprises many single-turn dialogues. Besides, the dialogue history can be considered as the multi-turn context. Hence, except for the two types of personalized baseline models, we also consider the single-turn and multi-turn models as the baseline: (1) Single-turn models: **Conv-KNRM** [1]: The model utilizes a kernel-based ranking method with CNN to learn soft n-gram matches for ad-hoc matching; **BERT-adhoc** [2]: We fine-tune the BERT model with single-turn dialogue data. (2) Multi-turn models: **DAM** [24]: The model stacks multiple attentive modules to extract deep semantic interactive semantics. **IOI** [16]: The model designs a chain of deep interactive blocks to perform semantic interactions. **MSN** [18]: The model filters irrelevant dialogue context and performs matching at multi-grained. (3) Explicit user profile-based models: **DIM** [4]: The model separately encodes the context, user profile, and response candidates and then performs interactions; **RSM-DCK** [7]: The model performs context selection over dialogue context and then perform response selection; **CSN** [25]: The model uses a content selection network to select relevant dialogue context and then perform matching; (4) Implicit user profile-based models: **IMPChat** [13]: The model proposes learning implicit user profile from the user's dialogue history. **BERT** [2]: We fine-tune the BERT model with all users' dialogue history.

Table 2. Evaluation results of all models on both Weibo and Reddit corpus. "†" denote the TopReC is significantly better than all baselines in t-test with $p < 0.05$ level. The best results are in bold.

	Weibo Corpus				Reddit Corpus			
	$R_{10}@1$	$R_{10}@2$	$R_{10}@5$	MRR	$R_{10}@1$	$R_{10}@2$	$R_{10}@5$	MRR
(1) Conv-KNRM	0.323	0.520	0.893	0.538	0.576	0.711	0.917	0.712
(1) BERT (adhoc)	0.342	0.545	0.966	0.561	0.668	0.797	0.991	0.787
(2) DAM	0.438	0.644	0.966	0.635	0.605	0.748	0.965	0.741
(2) IOI	0.442	0.651	0.969	0.639	0.620	0.764	0.974	0.753
(2) MSN	0.355	0.554	0.931	0.567	0.555	0.733	0.977	0.715
(3) DIM	0.388	0.557	0.835	0.571	0.678	0.813	0.979	0.794
(3) RSM-DCK	0.428	0.627	0.947	0.623	0.615	0.753	0.972	0.748
(3) CSN	0.387	0.560	0.842	0.572	0.681	0.807	0.976	0.794
(4) IMPChat	0.460	0.665	0.963	0.651	0.691	0.820	0.982	0.804
(4) BERT	0.445	0.653	0.967	0.641	0.727	0.849	0.991	0.830
(4) TopReC	**0.486**†	**0.695**†	**0.972**†	**0.677**†	**0.750**†	**0.868**†	**0.992**	**0.852**†

4.3   Implementation Details

We employ the *bert-base-uncased* and *chinese-bert-wwm-ext* as the backbone of TopReC for the Reddit and Weibo datasets, respectively. The codes are implemented based on the PyTorch-Lightning[1] and Transformers[2] libraries. We train the TopReC on 4 T V100 16GB GPUs for 3 epochs. We set the batch size as 128, and the learning rate as 1e-5. For the number of topic segments k and the sequence length l of each topic segment, we set $k = 3, l = 256$, and $k = 4, l = 128$ for the Weibo and Reddit dataset, respectively. The reason that we keep a longer sequence length for the Weibo dataset is that dialogues on Weibo are usually longer than on Reddit (see Table 1). The further analysis of the choice of k and l can refer to Sect. 4.5. We use the history length of 15 for all baseline models and TopReC. The detailed analysis of the choice of history length can refer to Sect. 4.5. We tune TopReC and all baseline models on the dev set and evaluate the models on the test set. The codes will be released at https://github.com/qhjqhj00/ECIR23-TopReC.

---

[1] https://github.com/PyTorchLightning/pytorch-lightning.
[2] https://github.com/huggingface/transformers.

4.4   Experimental Results

Table 2 shows the experiment results from which we have the following findings: **First, the proposed TopReC outperforms all baseline models regarding all evaluation metrics.** And TopReC lead statistically significant improvement regarding all metrics on the Weibo dataset and most metrics on the Reddit dataset (t-test with $p < 0.05$). It proves the effectiveness of TopReC's ability to find the most proper response that is consistent with the user's personality. **Second**, all models perform worse in the Weibo dataset than the Reddit dataset, which implies that the dialogue history in the Weibo dataset might contain more noise than the Reddit dataset. Impacted by such noise, in the Weibo dataset, the pre-trained model BERT performs worse than the IMPChat, which does not benefit from the pre-trained language model. The reason might be that IMPChat conduct reweighs the importance of the historical dialogues, which alleviate the impact of data noise. Compared to BERT, TopReC models the user's personality concerning the topical inter-relations inside the dialogue history and prunes the topically irrelevant dialogue history. As a result, TopReC can be partially immune to the negative effect of the data noise and therefore booster the performances; **Third**, regarding the model types, we find that the models learning implicit user profile from the dialogue history perform better than the rest types of models. It demonstrates the superior effectiveness of learning implicit user profiles from the dialogue history. A fundamental problem of learning implicit user profiles is how to use the dialogue history properly. In this paper, TopReC uses the dialogue history from a topic-aware perspective which is empirically effective. Future works might explore more promising perspectives to use the dialogue history and provide better performances and explainability.

4.5   Discussion

Ablation Study. To verify the effect of the Topic-dependent Context Deconstruction module, we randomly deconstruct the dialogue history into the same number of dialogue segments for comparison. To study the effect of the Personalized Topic Matching module, we respectively remove the topic-dependent matching scores and the cross-topic matching score. Table 3 shows the results. We find: (1) removing any module of TopReC would bring performance decline, implying that any module of TopReC captures orthogonal information that is indispensable to the overall model; (2) randomly deconstructing the dialogue history lead to big performance decline, verifying the validity of our idea that models the user personality from topic-dependent dialogue segments; (3) removing any of the topic-dependent matching scores and the cross-topic matching score would lead to performance decline, which implies that the two matching scores capture the personalized information from different perspectives (*e.g.*, local and global).

Impact of History Length. We conduct experiments with our TopReC and previous SOTA model IMPChat to study the impact of history length. Figure 3 shows the results: (1)the model performances show an increasing tendency with

Table 3. Ablation results on the Reddit dataset.

	$R_{10}@1$	$R_{10}@2$	$R_{10}@5$	MRR
TopReC	**0.750**	**0.868**	**0.992**	**0.852**
w/o cross	0.739	0.858	0.991	0.838
w/o topic	0.736	0.857	0.990	0.836
Random segment	0.732	0.852	0.986	0.833
BERT	0.727	0.849	0.991	0.830

longer dialogue history, which indicates that longer dialogue history can provide more personalized information; (2) our TopReC outperforms IMPChat after the history length of 15, which proves that TopReC is more effective when modeling user personality from long dialogue history. Before the history length of 15, TopReC is more sensitive to data insufficiency than IMPChat, as the latter is designed to learn multi-grained user representations from the whole dialogue history. Such saturated fitting is effective for short dialogue history but also limits the model capacity for longer dialogue history; (3) the increasing tendency slow down after the history length of 15, the reason might be that the experiments setting[3] limits TopReC's capacity for longer dialogue history, which indicates that longer dialogue history contains more dialogue topics, and correspondingly, we should increase the choice of the max number of topics. Figure 3 middle shows the impact of the max number of topics (the k value in Sect. 3.3). The model performance peaks at $k = 4$ and then decreases, verifying the effectiveness of using topically-related dialogue segments as context. And it also proves that less relevant topic segments might undermine the model performance.

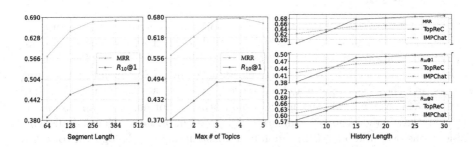

Fig. 3. Left shows the impact of segment length, middle shows the impact of max number of topics to keep, and right shows the impact of history length.

---

[3] For TopReC, we set the max segment length as 256 and the max number of topics as 4. For IMPChat, we feed all dialogue history into the model without truncation.

Impact of Sequence Length. Figure 3 left shows the impact of the length of topic segment (the l value in Sect. 4.3). We find that the model performance steadily increase with longer segment length. The reason is that less context would be truncated when using longer segment length. But in the meantime, the required computing resources greatly increase with longer segment length (*e.g.*, BERT's complexity exponentially increases with the sequence length), for which we choose to use relatively small l value[4] in this paper.

5   Conclusion

This paper explores learning implicit user profiles from dialogue history for a personalized chatbot. We observe that a user's dialogue history might be long and noisy as the dialogue history contains the user's long-term, multi-topic dialogue records. To reduce the data noise and increase the model's capacity to adapt long dialogue history, we propose deconstructing the user's dialogue history into topic-dependent segments and filtering out irrelevant dialogue segments. We design a model TopReC, which first performs dialogue history deconstructions via a Heter-Merge-Reduce method and learns the topic-aware personality from each topic-dependent segment. Besides, TopReC also explores a cross-topic personalized matching feature that measures the matching degree of the response candidate from a general perspective. The final response is selected by fusing the topic-dependent and cross-topic matching scores. Experimental results verify the effectiveness of the proposed TopReC. The limitations of this work are: (1) we conduct experiments on datasets that come from social media, which might not reflect how people usually talk; (2) we prune noisy topical segments by measuring the similarities, which might be biased by data noise and therefore lack interpretability. In the future, we will further explore how TopReC performs in more dialogue datasets and how to better learn an implicit user profile from the dialogue history to enhance personalized response selection regarding both effectiveness and interpretability.

Acknowledgments. Zhicheng Dou is the corresponding author. This work was supported by the National Natural Science Foundation of China No. 62272467 and No. 61872370, Beijing Outstanding Young Scientist Program NO. BJJWZYJH012019100020098, the Fundamental Research Funds for the Central Universities, the Research Funds of Renmin University of China NO. 22XNKJ34, Public Computing Cloud, Renmin University of China, Intelligent Social Governance Platform, Major Innovation & Planning Interdisciplinary Platform for the "Double-First Class" Initiative, Renmin University of China, and the Outstanding Innovative Talents Cultivation Funded Programs 2023 of Renmin Univertity of China. The work was partially done at Engineering Research Center of Next-Generation Intelligent Search and Recommendation, Ministry of Education, and Beijing Key Laboratory of Big Data Management and Analysis Methods.

---

[4] $l = 128$ and $l = 256$ for Reddit and Weibo.

References

1. Dai, Z., Xiong, C., Callan, J., Liu, Z.: Convolutional neural networks for soft-matching n-grams in ad-hoc search. In: Proceedings of the 11th WSDM. ACM (2018). https://doi.org/10.1145/3159652.3159659
2. Devlin, J., Chang, M.W., Lee, K., Toutanova, K.: Bert: pre-training of deep bidirectional transformers for language understanding. arXiv preprint arXiv:1810.04805 (2018)
3. Gao, J., Galley, M., Li, L.: Neural Approaches to Conversational AI: Question Answering. Task-oriented Dialogues and Social Chatbots. Now Foundations and Trends (2019)
4. Gu, J.C., Ling, Z.H., Zhu, X., Liu, Q.: Dually interactive matching network for personalized response selection in retrieval-based chatbots. In: Proceedings of the EMNLP-IJCNLP 2019 (2019)
5. Gu, J.C., Liu, H., Ling, Z.H., Liu, Q., Chen, Z., Zhu, X.: Partner matters! an empirical study on fusing personas for personalized response selection in retrieval-based chatbots. In: Proceedings of the 44th International ACM SIGIR Conference on Research and Development in Information Retrieval, pp. 565–574 (2021)
6. Han, J., Hong, T., Kim, B., Ko, Y., Seo, J.: Fine-grained post-training for improving retrieval-based dialogue systems. In: Proceedings of NAACL 2021. ACM (2021). https://www.aclweb.org/anthology/2021.naacl-main.122
7. Hua, K., Feng, Z., Tao, C., Yan, R., Zhang, L.: Learning to detect relevant contexts and knowledge for response selection in retrieval-based dialogue systems. In: Proceedings of the 29th ACM International Conference on Information & Knowledge Management, pp. 525–534 (2020)
8. Li, J., Galley, M., Brockett, C., Spithourakis, G.P., Gao, J., Dolan, W.B.: A persona-based neural conversation model. In: Proceedings of the ACL 2016. ACL (2016)
9. Liu, Y., Qian, H., Xu, H., Wei, J.: Speaker or listener? the role of a dialog agent. In: Findings of the Association for Computational Linguistics: EMNLP 2020. pp. 4861–4869. Association for Computational Linguistics (2020). https://doi.org/10.18653/v1/2020.findings-emnlp.437, https://aclanthology.org/2020.findings-emnlp.437
10. Lowe, R., Pow, N., Serban, I.V., Pineau, J.: The ubuntu dialogue corpus: a large dataset for research in unstructured multi-turn dialogue systems. In: SIGDIAL 2015, pp. 285–294 (2015). https://doi.org/10.18653/v1/w15-4640
11. Ma, Z., Dou, Z., Zhu, Y., Zhong, H., Wen, J.R.: One chatbot per person: creating personalized chatbots based on implicit user profiles. In: SIGIR 2021 (2021)
12. Mao, K., Dou, Z., Qian, H.: Curriculum contrastive context denoising for few-shot conversational dense retrieval. In: Proceedings of the 45th International ACM SIGIR Conference on Research and Development in Information Retrieval, SIGIR 2022, , pp. 176–186. Association for Computing Machinery, New York (2022). https://doi.org/10.1145/3477495.3531961
13. Qian, H., Dou, Z., Zhu, Y., Ma, Y., Wen, J.R.: Learning implicit user profile for personalized retrieval-based chatbot. In: Proceedings of the 30th CIKM, pp. 1467–1477 (2021)
14. Qian, H., et al.: Pchatbot: a large-scale dataset for personalized chatbot. In: Proceedings of the 44th SIGIR. ACM, Virtual Event (2021). https://doi.org/10.1145/3404835.3463239, https://doi.org/10.1145/3404835.3463239

15. Qian, Q., Huang, M., Zhao, H., Xu, J., Zhu, X.: Assigning personality/profile to a chatting machine for coherent conversation generation. In: IJCAI, pp. 4279–4285 (2018)
16. Tao, C., et al.: One time of interaction may not be enough: Go deep with an interaction-over-interaction network for response selection in dialogues. In: Proceedings of the 57th ACL, pp. 1–11 (2019). https://doi.org/10.18653/v1/P19-1001, https://www.aclweb.org/anthology/P19-1001
17. Xu, Y., Zhao, H., Zhang, Z.: Topicaware multi-turn dialogue modeling. In: The Thirty-Fifth AAAI Conference on Artificial Intelligence (AAAI-2021) (2021)
18. Yuan, C., et al.: Multi-hop selector network for multi-turn response selection in retrieval-based chatbots. In: Proceedings of EMNLP-IJCNLP 19. ACL (2019). https://doi.org/10.18653/v1/D19-1011, https://www.aclweb.org/anthology/D19-1011
19. Zhang, S., Dinan, E., Urbanek, J., Szlam, A., Kiela, D., Weston, J.: Personalizing dialogue agents: i have a dog, do you have pets too? In: Proceedings of the ACL 2018, pp. 2204–2213. ACL (2018)
20. Zhang, S., Dinan, E., Urbanek, J., Szlam, A., Kiela, D., Weston, J.: Personalizing dialogue agents: i have a dog, do you have pets too? In: Proceedings of the 56th Annual Meeting of the Association for Computational Linguistics, vol. 1: Long Papers, pp. 2204–2213. Association for Computational Linguistics, Melbourne (2018). https://doi.org/10.18653/v1/P18-1205, https://aclanthology.org/P18-1205
21. Zhang, Y., Sun, S., Galley, M., Chen, Y., Brockett, C., et al.: DIALOGPT: large-scale generative pre-training for conversational response generation. In: Proceedings of the ACL 2020 (2020)
22. Zhao, X., Tao, C., Wu, W., Xu, C., Zhao, D., Yan, R.: A document-grounded matching network for response selection in retrieval-based chatbots (2019). https://doi.org/10.48550/ARXIV.1906.04362, https://arxiv.org/abs/1906.04362
23. Zhong, H., Dou, Z., Zhu, Y., Qian, H., Wen, J.R.: Less is more: learning to refine dialogue history for personalized dialogue generation. In: Proceedings of the 2022 Conference of the North American Chapter of the Association for Computational Linguistics: Human Language Technologies, pp. 5808–5820. Association for Computational Linguistics, Seattle (2022). https://doi.org/10.18653/v1/2022.naacl-main.426, https://aclanthology.org/2022.naacl-main.426
24. Zhou, X., et al.: Multi-turn response selection for chatbots with deep attention matching network. In: Proceedings of the 56th ACL, pp. 1118–1127 (2018). https://doi.org/10.18653/v1/P18-1103, https://www.aclweb.org/anthology/P18-1103
25. Zhu, Y., Nie, J.-Y., Zhou, K., Du, P., Dou, Z.: Content selection network for document-grounded retrieval-based chatbots. In: Hiemstra, D., Moens, M.-F., Mothe, J., Perego, R., Potthast, M., Sebastiani, F. (eds.) ECIR 2021. LNCS, vol. 12656, pp. 755–769. Springer, Cham (2021). https://doi.org/10.1007/978-3-030-72113-8_50

Improving the Generalizability of the Dense Passage Retriever Using Generated Datasets

Thilina C. Rajapakse$^{(\boxtimes)}$ and Maarten de Rijke

University of Amsterdam, Amsterdam, The Netherlands
{t.c.r.rajapakse,m.derijke}@uva.nl

Abstract. Dense retrieval methods have surpassed traditional sparse retrieval methods for open-domain retrieval. While these methods, such as the Dense Passage Retriever (DPR), work well on datasets or domains they have been trained on, there is a noticeable loss in accuracy when tested on out-of-distribution and out-of-domain datasets. We hypothesize that this may be, in large part, due to the mismatch in the information available to the context encoder and the query encoder during training. Most training datasets commonly used for training dense retrieval models contain an overwhelming majority of passages where there is only one query from a passage. We hypothesize that this imbalance encourages dense retrieval models to *overfit* to a single potential query from a given passage leading to worse performance on out-of-distribution and out-of-domain queries. To test this hypothesis, we focus on a prominent dense retrieval method, the dense passage retriever, build generated datasets that have multiple queries for most passages, and compare dense passage retriever models trained on these datasets against models trained on single query per passage datasets. Using the generated datasets, we show that training on passages with multiple queries leads to models that generalize better to out-of-distribution and out-of-domain test datasets.

1 Introduction

Recently, a number of transformer-based dense retrieval models have achieved state-of-the-art results on various benchmark datasets [13,14,28]. The Dense Passage Retriever (DPR) architecture consists of two encoder models, typically BERT models [8], which encode the query and the passages separately. A simple similarity metric, such as the inner product or cosine distance, is then used to compute the relevance of a passage for a query.

An advantage of the DPR architecture is that passage representations can be pre-computed offline and built into an index with relatively small computational cost, making it a preferred model over recent proposals such as, e.g., ColBERT [14] and ANCE [28] with higher computational cost for training and/or retrieval. At runtime, the query encoder is used to compute a dense representation for the query and approximate nearest neighbor methods are used to find the most relevant passage.

© The Author(s), under exclusive license to Springer Nature Switzerland AG 2023
J. Kamps et al. (Eds.): ECIR 2023, LNCS 13981, pp. 94–109, 2023.
https://doi.org/10.1007/978-3-031-28238-6_7

A disadvantage of this approach is that a mismatch may exist between the information available to the passage encoder and the information available to the query encoder. As the training objective forces the passage and query encoders to generate representations that are similar, we hypothesize that the passage encoder (which has access to more information) learns to discard information that is not relevant to the query in a given training query-passage pair. The issue is exacerbated by the fact that most retrieval datasets and benchmarks contain far more passages with only one query from a given passage than passages with multiple queries per passage (see Table 1). In such situations, the model is not sufficiently penalized against learning to discard information that is not relevant to the (single) query that is asked from a given passage.

We hypothesize that a DPR model trained on datasets where a given passage typically has one associated query generalizes poorly to other datasets, new types of queries or topics, or both. We investigate this hypothesis by testing the zero-shot performance of the pretrained DPR model (from [13], which is trained on NQ [16]) in both out-of-distribution and out-of-domain settings. Here, we define *out-of-distribution* to be datasets that share the same passage corpus but with queries collected at different times and/or using different methods, and *out-of-domain* to be datasets with their own unique passage collection typically focused on a particular domain (see Sect. 4.1).

Having established that a DPR model trained on datasets where a given passage typically has one associated query, generalizes poorly, we propose a treatment to help improve out-of-distribution and out-of-domain performance. We synthetically generate training datasets where the passages typically have multiple queries from any given passage. The generation pipeline consists of a NER model to tag entities, a sequence-to-sequence model to generate queries, and a question answering model to filter out bad queries (see Sect. 3.1).

Our results show that training on data with multiple queries per passage leads to a DPR model with better generalizability to both out-of-distribution and out-of-domain data. In both settings, our DPR model trained on multiple queries per passage data easily outperforms the baseline DPR model trained on mostly single query per passage data (NQ).

In summary, then, we answer the following research question:

RQ Does training a DPR model on data containing multiple queries per passage improve the generalizability of the model?

In the out-of-distribution setting, the pre-trained DPR model [13], serving as the baseline, and our DPR model trained on generated queries with multiple queries per passage are tested, zero-shot, on six datasets. Our model achieves higher retrieval accuracy on five out of the six datasets demonstrating that training data containing multiple queries per passage does improve the generalizability of dense retrievers to out-of-distribution queries.

The picture becomes even clearer in the out-of-domain setting where our model outperforms the pretrained DPR model on 12 out of 13 datasets. Training DPR models on passages with multiple associated queries prevents the context encoder from (exclusively) focusing on a specific detail or piece of information in the passage, leading to a better generalized retrieval model.

Our analysis of increasing the size of the set of generated queries with multiple queries per passage as a way to improve the generalizability of dense retrievers indicates a subtle balance. While the model trained on the largest training dataset does achieve higher scores compared to the others, the improvements are relatively minor. But, these relatively minor improvements come at a significantly higher costs in terms of compute and training time. Even the smallest generated dataset with multiple queries per passage performs competitively with larger generated datasets and handily outperforms the pre-trained model trained on mostly single query per passage data.

2   Related Work

Passage Retrieval. Passage retrieval has classically been performed using sparse retrieval methods such as BM25 [25]. Recently, transformer-based dense retrieval methods have garnered interest as the performance of dense retrieval methods surpasses that of traditional sparse methods [13,14,28]. A dense passage retriever indexes a collection of passages in a low-dimensional and continuous space, such that the top-k passages are relevant to a given query [13]. Here, the size of the passage collection is typically very large (21M passages in this work and in [13]) and k is very small (e.g., 20–100). Going beyond *in-distribution* and *in-domain* testing, we focus on generalizability to new data which can be *out-of-distribution* and *out-of-domain*.

Test Collections. The Benchmarking-IR (BEIR) [22] test collection was introduced to facilitate the effectiveness of retrieval models in out-of-domain settings. It provides a collection of 18 datasets (13 of which are readily available) from diverse retrieval tasks and domains. Thakur et al. [22] also highlight considerable room for improvement in the generalization capabilities of dense retrieval models. Our work aims to improve the generalizability of dense retrievers by using synthetic datasets with specially chosen composition of data (multiple queries per passage).

Automatically Generated Collections. Automatically generating training, development and test collections for retrieval has a long history in information retrieval. Examples include test collections for bibliographic systems [21], known-item test collections [2], desktop search [15], web search [1], test collections for academic search [3]. Berendsen et al. [4] focus on test collection generation to improve robustness for tuning and learning. A comprehensive approach to simulated test collection building with considerable attention to privacy preservation is offered in [11]. What we add on top of this is test collection building with a specific focus on generalizability by preventing overfitting.

3   Methodology

We train DPR models on generated query datasets and compare their retrieval performance against the pre-trained model on the test datasets.

3.1   Dataset Generation Process

For our dataset generation process, we follow the steps below:

(1) Identify potential answers to questions to be generated;
(2) Generate queries that are answered by one of the potential answers; and
(3) Filter out bad queries, that is, queries that are unanswerable or do not end with a question mark.

Identifying Potential Answers. We train a token classification model to identify words or phrases from a passage that could serve as potential answers to queries. The trained model is then used to tag potential answers for each passage in a dataset. This process enables us to find all potential answers in a passage, which is critical to ensure that there are sufficient queries from any given passage.

Generating Queries. The passages, along with the tagged answers, are fed to a sequence-to-sequence model that generates a query for each passage-answer pair. Each passage can have multiple associated answers, resulting in multiple queries from the same passage. This ensures that there are queries related to most, if not all, entities found in a given passage.

Filtering Queries. The generated queries are filtered to remove potentially unanswerable queries (from the originating passage). To find such queries, we feed the passages and queries to a question answering (QA) model and discard queries where the QA model answer does not match the original tagged answer. We also discard queries that contain more than one sentence or do not end with a question mark (?). This is to ensure that all the generated queries used for training are reasonable queries (see Sect. 4.2) and provide a good training signal for the model being trained on them.

3.2   Training the Retriever

We build training datasets by generating queries following the procedure given in Sect. 3.1. The generation process ensures that most passages in the training datasets have multiple queries associated with them. We train bi-encoder retrieval models on these training datasets.

4 Experimental Setup

4.1 Datasets

Most popular open-domain retrieval datasets contain a much larger number of passages with only a single query originating from it than passages with multiple queries. Table 1 shows the frequency of passages with a given number of queries originating from the passage for the five datasets used in [13] as well as the five datasets that were generated. The Wikipedia collection and five of the datasets used (NQ, Trivia QA, Curated TREC, Web Questions, and SQuAD) are the same versions provided by [13] available on GitHub.[1]

Table 1. Frequency of passages with a given number of queries originating from the passage.

Dataset	Number of queries/passage		
	1	2	≥2
Natural questions	32,155	4,973	3,542
Trivia QA	43,401	5,308	1,793
Curated TREC	990	41	16
Web Questions	2,019	148	46
SQuAD	8,468	6,056	11,790
Generated from NQ train	2,784	3,418	30,120
Wikipedia passages (˜58k) single	58,880	0	0
Wikipedia passages (˜58k) multi	16,634	19,641	985
Wikipedia passages (˜236k)	19,487	18,061	41,308
Wikipedia passages (˜786k)	62,264	60,472	137,266

Out-of-Distribution Test Datasets. To test the models on out-of-distribution data, we use the four datasets available from [13] that were not used in training the baseline model, namely Trivia QA, Curated TREC, Web Questions, and SQuAD. In addition to these four, we include two generated test datasets. The first of these is generated from the NQ *dev* passages and the second is generated from randomly selected Wikipedia passages. This results in a total of six out-of-distribution test datasets. As these datasets use the same passage collection but contain queries collected or generated using different approaches, we consider the datasets to be *out-of-distribution* but *in-domain*.

---

[1] https://github.com/facebookresearch/DPR.

Out-of-Domain Test Datasets. We use the 13 readily available datasets from [22], each with their own distinct passage collection, to test the models on out-of-domain data. The datasets are as follows: TREC-COVID [24], NFCorpus [6], HotpotQA [29], FiQA-2018 [18], ArguAna [26], Touché-2020 [5], CQADupStack [12], Quora, DBPedia [10], SCIDOCS [7], FEVER [23], Climate-FEVER [9], and SciFact [27]. These datasets cover multiple domains, including bio-medical, Wikipedia/general, finance, news, and scientific domains.

4.2 Generation Pipeline

Named Entity Recognition Model for Tagging Answers. The named entity recognition model is a RoBERTa [17] model trained on the large NER dataset (1 million sentences) from Naman Jaswani on Kaggle,[2] with the tags: *Organization, Person, Location, Date, Time, Money, Percent, Facility,* and Geo-Political Entity (GPE). The RoBERTa model, trained on a large NER dataset, ensures that we find all the entities in a passage.

MACAW Model for Query Generation. The pretrained *MACAW* [20] model (3 billion parameters) is used to generate the queries. It is a strong sequence-to-sequence question generation model (among other tasks) based on the T5 model [19]. This model is capable of generating queries for each entity found in the passage such that they are relevant to the context of the passage.

Table 2. Examples of generated queries and answers for a randomly sampled passage.

Passage	Generated query	Generated answer	Related	Answerable
Sirocco (play) Sirocco is a play, in four acts, by Noël Coward. It originally opened at Daly's Theatre, on November 24, 1927. The production was directed by Basil Dean. Ivor Novello was part of the original cast. The plot told a tale of free love among the wealthy. The London opening of "Sirocco" met with violently unfavorable audience reaction and a very harsh critical reception. Coward was later asked whether he had ever despaired when faced with a failure like "Sirocco". He replied, "Well, if I'm going to have a flop, I like it to be a rouser. I didn't	Sirocco was first performed at which theater in London?	Dalys Theatre	Yes	Yes
	When did the first performance of Sirocco take place?	November 24 1927	Yes	Yes
	Which actor played the role of Sirocco in the original production?	Ivor Novello	Yes	No
	Who wrote the play Sirocco?	Noël Coward	Yes	Yes
	Who directed the first production of Sirocco?	Basil Dean	Yes	Yes

---

[2] https://www.kaggle.com/namanj27/ner-dataset.

Question Answering Model for Query Filtering. A RoBERTa [17] model trained on the SQuAD dataset is used to filter out potential bad queries in the generated datasets. The RoBERTa model is a question answering model that is good at extractive question answering. We can reasonably assume that the questions the model is incapable of answering are most likely flawed.

This generation pipeline results in queries that are typically relevant and answerable from their passages of origin. We found 92% of queries to be relevant, and 86% to be answerable from their passages of origin, based on a randomly sampled set of 50 queries (example shown in Table 2).

4.3 Retrieval Pipeline

The architecture of the retrieval model is identical to [13], i.e., a bi-encoder architecture consisting of two BERT [8] encoders, one for encoding the passages/contexts and the other for encoding the queries. We also use the same hyperparameters as [13] except for the batch size, where we use a batch size of 80 vs. a batch size of 120 due to resource limitations.

We choose the DPR [13] model as our architecture of choice to avoid introducing any confounding factors in our analysis. Other architectures, notably the late interaction based ColBERT [14] architecture, has demonstrated superior retrieval accuracy over the original DPR [13] architecture. However, ColBERT has higher latency and much larger space footprints for indices. As our work is focused on the composition of data, the simpler and more straightforward architecture of DPR is better suited to our analysis. Furthermore, the higher resource demands and complexity of ColBERT makes it a less viable option compared to DPR in any setting with even moderate computational resource constraints.

We build five training datasets by generating queries following the procedure given in Sect. 3.1. One dataset is built by generating queries from the same passages used in the NQ train set, while the other four are from randomly selected Wikipedia passages. A bi-encoder DPR model, starting from the pretrained BERT [8] weights, is trained on each of these five datasets.

While positive training examples (matching query and document pairs) are available directly in retrieval datasets, negative training examples must be selected from the set of all documents. The original DPR model is trained using a combination of in-batch negatives (the positive documents of all other queries in the batch used as negatives for a given query) and BM25 selected negatives (highest ranked document retrieved by BM25, which does not contain the answer to the query). In our work, we simply use the in-batch negatives as the negative examples leaving improvements from more complex negative selection strategies for future work as our results demonstrate improved generalizability even without using hard negatives.

4.4 Experiment

We use two models trained on two different datasets to compare the generalizability of DPR models trained on data with multiple queries per passage versus DPR models trained on data with mostly a single query per passage. The pre-trained DPR model from [13], trained on NQ with mostly single query per passage data, is used as the baseline model to be compared against our model trained 58,880 generated queries containing mostly multiple queries per passage data (*58k generated*).

The two models are tested in both the out-of-distribution (6 datasets) and the out-of-domain settings (13 datasets). *Top-100 accuracy* is used as the evaluation metric for the out-of-distribution setting while *recall@100* is used as the evaluation metric for the out-of-domain setting. The decision to use two different metrics is motivated by the fact that the set of all relevant passages is only available for the out-of-domain datasets, which is necessary to calculate recall. Only the true answers are available for the out-of-distribution datasets, so we calculate top-100 accuracy by checking whether the true answer is present in any of the top-100 retrieved documents. In addition to this, we also report MRR@100 (Mean Reciprocal Rank) for all experiments.

5   Results

We report results from the baseline pretrained model trained on NQ (58,880 queries) against our model trained on 58,880 generated queries for the two generalizability settings; out-of-distribution and out-of-domain. Here, the generated query dataset contains mostly passages with multiple queries per passage.

5.1   Out-of-Distribution Generalizability

Table 3 shows the top 100 accuracy scores obtained by the baseline DPR model (trained on NQ) and our DPR model, trained on the 58k generated query dataset with multiple queries per passage (*58k generated*), on the out-of-distribution datasets. We also include the scores on the NQ dataset itself for completeness, but it should be noted that this dataset is an in-distribution dataset for the baseline model.

Table 3. Top 100 accuracy scores for the model trained on *58k generated* and the baseline DPR model trained on NQ for out-of-distribution datasets. The highest score is in **bold** and ‡ indicates in-domain performance. Statistical significance with paired t-test: * indicates $p < 0.05$ and ** indicates $p < 0.01$.

Model	Standard datasets					Generated datasets	
	NQ	TriviaQA	TREC	WebQ	SQuAD	NQ dev.	Wikipedia
Baseline DPR	**84.9**‡**	78.7	**90.7**	77.6	63.5	81.5	56.7
58k generated (ours)	75.0	**80.0****	89.6	**78.3**	**69.4****	**85.3****	**79.2****

The model trained on *58k generated* (our model) outperforms the baseline DPR model on 5 out of 6 out-of-distribution datasets, with the Curated TREC

Table 4. MRR@100 scores for the model trained on *58k generated* and the baseline DPR model trained on NQ for out-of-distribution datasets. Same notational conventions as in Table 3.

Model	Standard datasets					Generated datasets	
	NQ	TriviaQA	TREC	WebQ	SQuAD	NQ dev.	Wikipedia
Baseline DPR	$0.512^{\ddagger**}$	0.437^{**}	0.583^{**}	0.389^{**}	0.234	0.449^{**}	0.240
58k generated (ours)	0.313	0.426	0.507	0.358	0.258^{**}	0.426	0.415^{**}

dataset being the sole exception. However, the difference in accuracy between the two models on Curated TREC and WebQ are not statistically significant. Our model generalizes better in all four datasets (out of six) where the difference is statistically significant. The baseline DPR model does better on the NQ test dataset (in-distribution) compared to the our model trained on generated queries (out-of-distribution).

Interestingly, the baseline DPR model trails our model trained on *58k generated* even on the queries generated from the NQ passages despite being trained on fairly similar data. This indicates that the performance of DPR models trained on data with mostly a single query from each passage deteriorates rapidly when tested on new queries. This observation may be explained by our initial hypothesis. If a model trained on data with a single query per passage learns to discard information, it is logical that the model would struggle when dealing with multiple queries from a passage as this requires the context encoder to encode all information available in the passage in order to correctly match all the queries from that passage. These results indicate that training a model on data with multiple queries per passage results in improved generalizability in the out-of-distribution setting.

The baseline model outperforms the model trained on *58k generated* on 4 out of 6 out-of-distribution datasets when considering MRR@100 scores (Table 4). However, the *58k generated* model performs slightly better on average.

5.2   Out-of-Domain Generalizability

Table 5 shows the recall@100 scores obtained by the baseline DPR model (trained on NQ) and our DPR model trained on *58k generated*. The model trained on *58k generated* outperforms the baseline DPR model achieving higher recall@100 scores in 12 out of 13 out-of-domain datasets. Considering only the statistically significant results ($p < 0.05$), our model trained on multiple query per passage data outperforms the baseline DPR model on all 10 out of 10 datasets.

The MRR@100 scores (Table 5) follow a similar pattern, with the model trained on *58k generated* outperforming the baseline in 9 out of 10 out-of-domain datasets where the results are statistically significant.

The model trained with data containing multiple queries per passage (our model trained on *58k generated*) dominates the baseline DPR model, trained on mostly single query per passage data, in both the out-of-distribution and out-of-domain setting. This clearly superior zero-shot generalization performance when

Table 5. Recall@100 and MRR@100 scores for the baseline DPR model trained on NQ and the model trained on 58k generated queries for out-of-domain datasets. Same notational conventions as in Table 3.

Dataset	Recall@100		MRR@100	
	Baseline DPR	58k generated	Baseline DPR	58k generated
ArguAna	0.480	**0.919****	0.051	**0.213****
Climate FEVER	**0.410**	0.405	**0.258****	0.220
CQA dup stack	0.109	**0.139****	0.041	**0.068****
DBPedia	0.310	**0.335***	0.559	**0.564**
FEVER	0.748	**0.805****	**0.497**	0.492
FiQa	0.313	**0.369****	0.131	**0.195****
HotpotQA	0.493	**0.502**	0.419	**0.559****
NFCorpus	0.170	**0.238**	0.306	**0.377****
Quora	0.566	**0.880****	0.279	**0.590****
SciDocs	0.196	**0.253****	0.136	**0.207****
SciFact	0.581	**0.704****	0.247	**0.372****
Touche	0.276	**0.344****	0.234	**0.386****
TREC-COVID	0.096	**0.177****	0.287	**0.354**

a DPR model is trained on data with multiple queries per passage answers our research question (RQ) demonstrating that training a DPR model on data with multiple queries per passage does result in a better generalized model.

6   Analysis

6.1   Generation Versus Data Composition

We conduct a further analysis to confirm that the improvements in generalizability shown in Sect. 5 is due to the composition of the dataset, specifically the number of queries per passage, rather than any artifact of the query generation process. Here, we compare the generalizability to out-of-distribution and out-of-domain data of two models trained on generated queries. The first model is trained on generated queries with multiple queries per passage (same as in Sect. 5) and the second model is trained on generated queries with only a single query from each passage.

Table 6 shows the top-100 accuracy scores obtained by the two models on the out-of-distribution datasets. The model trained on *58k generated (multi)* outperforms the model trained *58k generated (single)* on 5 out of 7 datasets (one loss and one tie). Four of these results are statistically significant with the model trained on *58k generated (multi)* generalizing better in all four cases. Similarly, the model trained on *58k generated (multi)* outperforms the model trained on *58k generated (single)*, in terms of MRR@100 scores (Table 7), on all six out-of-distribution datasets with four of the results being statistically

Table 6. Top 100 accuracy scores for the models trained on *58k generated (single)* and *58k generated (multi)* for out-of-distribution datasets. Same notational conventions as in Table 3.

Model	Standard datasets					Generated datasets	
	NQ	TriviaQA	TREC	WebQ	SQuAD	NQ dev	Wikipedia
58k generated (single)	**75.0**	78.4	**90.2**	77.5	67.9	81.9	74.5
58k generated (multi)	**75.0**	**80.0**[**]	89.6	**78.3**	**69.4**[**]	**85.3**[**]	**79.2**[**]

Table 7. MRR@100 scores for the models trained on *58k generated (single)* and *58k generated (multi)* for out-of-distribution datasets. Same notational conventions as in Table 3.

Model	Standard datasets					Generated datasets	
	NQ	TriviaQA	TREC	WebQ	SQuAD	NQ dev	Wikipedia
58k generated (single)	0.309	0.397	0.489	0.350	0.247	0.394	0.366
58k generated (multi)	**0.313**	**0.426**[**]	**0.507**	**0.358**	**0.258**[**]	**0.426**[**]	**0.415**[**]

significant. These results clearly show that having multiple queries per passage in the training data helps the model generalize better to out-of-distribution queries, as the only difference between the two models is the composition of the training data.

Table 8 shows the recall@100 scores obtained by the two models on the out-of-domain datasets. Again, the model trained with multiple queries per passage outperforms the model trained on single query per passage data and generalizes

Table 8. Recall@100 and MRR@100 scores for the models trained on *58k generated (single)* and *58k generated (multi)* for out-of-domain datasets. Same notational conventions as in Table 3.

Dataset	Recall@100		MRR@100	
	58k generated (single)	58k generated (multi)	58k generated (single)	58k generated (multi)
ArguAna	0.885	**0.919**[**]	0.208	**0.213**
Climate FEVER	0.378	**0.405**[**]	0.188	**0.220**[**]
CQA Dup Stack	0.134	**0.139**[**]	**0.068**	**0.068**
DBPedia	0.312	**0.335**[**]	0.545	**0.564**
FEVER	0.722	**0.805**[**]	0.415	**0.492**[**]
FiQa	0.358	**0.369**	0.189	**0.195**
HotpotQA	0.430	**0.502**[**]	0.460	**0.559**[**]
NFCorpus	0.185	**0.238**	0.376	**0.377**
Quora	**0.909**[**]	0.880	**0.658**[**]	0.590
SciDocs	0.246	**0.253**	0.202	**0.207**
SciFact	0.685	**0.704**	0.346	**0.372**
Touche	**0.371**	0.344	0.343	**0.386**
TREC-COVID	**0.181**	0.177	0.300	**0.354**

better to 10 out of 13 out-of-domain datasets. Looking at the statistically significant results, the model trained on *58k generated (multi)* does better on 6 out of 7 datasets. The results on the remaining six datasets are likely not statistically significant as they contain a very small number of queries.

Overall, the model trained on *58k generated (multi)* generalizes better, in both out-of-distribution and out-of-domain settings, compared to the model trained on *58k generated (single)* when all other factors are kept constant. This confirms that the composition of training data, specifically the number of queries per passage, is an important factor to consider when training dense retrieval models and that training on data with multiple queries per passage leads to a model that is capable of generalizing better to out-of-distribution and out-of-domain queries.

6.2   Effect of Dataset Size

We also investigate the effect of the total number of generated queries in a training dataset on the generalizability of DPR models. For this analysis we compare three DPR models trained on three generated query datasets, where each dataset contains 58,880 (*58k generated*), 236,444 (*236k generated*), and 786,312 (*786k generated*) queries respectively. Note that all three of these datasets contain data with multiple queries per passage. Again, we report zero-shot scores in both the out-of-distribution and out-of-domain settings.

Table 9 shows the top-100 accuracy scores obtained by each model on the out-of-distribution datasets. The model trained on *786k generated* generalizes better to all seven datasets, with five of the results being statistically significant. In terms of MRR@100 (Table 10), the model trained on *786k generated* obtains higher scores on 5 out of 6 datasets, with four being statistically significant. These results indicate that training on larger datasets, containing data with multiple queries per passage, does yield better results on out-of-distribution datasets in a zero-shot setting.

Table 9. Top 100 accuracy scores for the models trained on the three generated query datasets *58k*, *236k*, and *786k* for out-of-distribution datasets. Same notational conventions as in Table 3.

Model	Standard datasets					Generated datasets	
	NQ	TriviaQA	TREC	WQ	SQuAD	NQ dev	Wikipedia
58k Generated	75.0	80.0	89.6	78.3	69.4	85.3	79.2
236k Generated	79.5	82.5	91.7	80.6	71.6	90.1	85.4
786k Generated	**80.5***	**83.2****	**92.2**	**80.7**	**72.9****	**92.4****	**89.4****

Table 11 shows the recall@100 scores obtained by each model on the out-of-domain datasets. Overall, the model trained on the largest dataset, *786k generated*, does marginally better than the other two models, obtaining the highest recall@100 score for seven out of thirteen out-of-domain datasets. The other

Table 10. MRR@100 scores for the models trained on the three generated query datasets *58k*, *236k*, and *786k* for out-of-distribution datasets. Same notational conventions as in Table 3.

Model	Standard Datasets					Generated Datasets	
	NQ	TriviaQA	TREC	WQ	SQuAD	NQ dev	Wikipedia
58k Generated	0.313	0.426	0.507	0.358	0.258	0.426	0.415
236k Generated	0.339	0.467	0.515	**0.381**	0.274	0.493	0.488
786k Generated	**0.360****	**0.492****	**0.526**	0.379	**0.283****	**0.522****	**0.542****

two models, trained on *236k generated* and *58k generated*, achieve the highest scores in four out of thirteen and two out of thirteen, respectively. Only three of these results are statistically significant with the model trained on *786k generated* doing better on two and the model trained on *58k generated* performing better on the other. The MRR@100 scores (Table 11) are even more mixed, with the model trained on *236k genrated* performing better in 2 out of 4 statistically significant results while the other two models perform better on one each.

Table 11. Recall@100 and MRR@100 scores for the model trained on the three generated query datasets *58k generated*, *236k generated*, and *786k generated* for the out-of-domain datasets. Same notational conventions as in Table 3.

Dataset	Recall@100			MRR@100		
	58k generated	236k generated	786k generated	58k generated	236k generated	786k generated
ArguAna	0.919	0.939	**0.940**	**0.213**	0.209	0.202
Climate FEVER	0.405	**0.406**	0.371	0.220	**0.224**	0.198
CQA Dup Stack	0.139	**0.154**	0.153	0.068	**0.072****	0.069
DBPedia	0.335	0.362	**0.364**	**0.564**	**0.564**	**0.564**
FEVER	0.805	0.853	**0.856**	0.492	**0.508****	0.476
FiQa	0.369	**0.385**	0.377	**0.195**	0.190	0.171
HotpotQA	0.502	0.557	**0.572****	0.559	0.598	**0.603**
NFCorpus	**0.238**	0.216	0.216	0.377	**0.387**	0.382
Quora	0.880	0.897	**0.929****	0.590	0.613	**0.636****
SciDocs	0.253	0.253	**0.261**	0.207	**0.212**	0.198
SciFact	0.704	0.737	**0.790**	0.372	0.373	**0.374**
Touche	0.344	**0.366**	0.325	**0.386**	0.325	0.314
TREC-COVID	**0.177****	0.124	0.119	**0.354***	0.219	0.166

While larger training datasets help with zero-shot performance on out-of-distribution datasets, the benefit of more generated data is less clear with regard to zero-shot performance on out-of-domain datasets. Although the model trained on *786k generated* generalizes better than the other two models, the increase in recall scores are marginal, especially compared to the increased cost of training which increases linearly with dataset size. Overall, training DPR models on more generated queries with multiple queries per passage can improve the generalizability of the model, but with sharply diminishing gains. This is likely due to the fact that increasing the size of the training dataset does not necessarily increase the diversity of the training data.

7   Conclusion and Future Work

We have shown that the generalizability of dense passage retrievers may suffer from learning to discard information from passages during training. This problem can be mitigated by using training data containing a sufficient number of passages with multiple associated queries. By exposing the dense retriever to multiple facets of information contained in the same passage, we ensure that the model does not learn to discard potentially useful information, leading to improved retrieval accuracy for out-of-domain topics and queries and a better-generalized model overall.

As a general lesson, when training a dense retrieval model, it is important to consider the number of queries per passage, or more generally, how much of the information contained in a given passage is covered by the queries. Training datasets with a large number of queries per passage can be automatically generated for training dense retrievers resulting in a better generalized model.

As to limitations, we did not use hard negative mining [28] or late interaction [14], which are known to improve the generalizability of dense retrievers. We leave their integration to future work but note that our method is trivially compatible with such techniques and is also independent of the actual dense retriever architecture that is used.

Finally, it would be interesting to use our proposed dataset generation method on a full collection of Wikipedia passages to train a DPR model. While our analysis of the effect of dataset size (Sect. 6.2) did not demonstrate meaningful gains in generalizability, a sufficiently large query collection (a generated query dataset of the full Wikipedia collection would be several orders of magnitude larger) containing diverse topics may generalize very well to most domains.

Acknowledgements. This research was supported by the Dreams Lab, a collaboration between Huawei, the University of Amsterdam, and the Vrije Universiteit Amsterdam, and by the Hybrid Intelligence Center, a 10-year program funded by the Dutch Ministry of Education, Culture and Science through the Netherlands Organisation for Scientific Research, https://hybrid-intelligence-centre.nl. All content represents the opinion of the authors, which is not necessarily shared or endorsed by their respective employers and/or sponsors.

References

1. Asadi, N., Metzler, D., Elsayed, T., Lin, J.: Pseudo test collections for learning web search ranking functions. In: Proceedings of the 34th International ACM SIGIR Conference on Research and Development in Information Retrieval, pp. 1073–1082. ACM (2011)

2. Azzopardi, L., de Rijke, M.: Automatic construction of known-item finding test beds. In: Proceedings of the 29th Annual International ACM SIGIR Conference on Research & Development on Information Retrieval, pp. 603–604. ACM (2006)

3. Berendsen, R., Tsagkias, M., de Rijke, M., Meij, E.: Generating pseudo test collections for learning to rank scientific articles. In: Catarci, T., Forner, P., Hiemstra, D., Peñas, A., Santucci, G. (eds.) CLEF 2012. LNCS, vol. 7488, pp. 42–53. Springer, Heidelberg (2012). https://doi.org/10.1007/978-3-642-33247-0_6

4. Berendsen, R., Tsagkias, M., Weerkamp, W., de Rijke, M.: Pseudo test collections for training and tuning microblog rankers. In: Proceedings of the 36th International ACM SIGIR Conference on Research and Development in Information Retrieval, pp. 53–62. ACM (2013)

5. Bondarenko, A., et al.: Overview of touché 2022: argument retrieval. In: International Conference of the Cross-Language Evaluation Forum for European Languages, pp. 311–336. Springer, Heidelberg (2022). https://doi.org/10.1007/978-3-031-13643-6_21

6. Boteva, V., Gholipour, D., Sokolov, A., Riezler, S.: A full-text learning to rank dataset for medical information retrieval. In: Ferro, N., et al. (eds.) ECIR 2016. LNCS, vol. 9626, pp. 716–722. Springer, Cham (2016). https://doi.org/10.1007/978-3-319-30671-1_58

7. Cohan, A., Feldman, S., Beltagy, I., Downey, D., Weld, D.: SPECTER: Document-level representation learning using citation-informed transformers. In: Proceedings of the 58th Annual Meeting of the Association for Computational Linguistics, pp 2270–2282. Association for Computational Linguistics (2020). https://doi.org/10.18653/v1/2020.acl-main.207, https://aclanthology.org/2020.acl-main.207

8. Devlin, J., Chang, M.W., Lee, K., Toutanova, K.: Bert: pre-training of deep bidirectional transformers for language understanding (2018). arXiv preprint arXiv:1810.04805

9. Diggelmann, T., Boyd-Graber, J., Bulian, J., Ciaramita, M., Leippold, M.: Climate-fever: a dataset for verification of real-world climate claims (2020). https://doi.org/10.48550/ARXIV.2012.00614, https://arxiv.org/abs/2012.00614

10. Hasibi, F., et al.:Dbpedia-entity v2: a test collection for entity search. In: Proceedings of the 40th International ACM SIGIR Conference on Research and Development in Information Retrieval, SIGIR 2017, pp. 1265–1268. Association for Computing Machinery, New York (2017) . https://doi.org/10.1145/3077136.3080751

11. Hawking, D., Billerbeck, B., Thomas, P., Craswell, N.: Simulating Information Retrieval Test Collections. Morgan & Claypool (2020)

12. Hoogeveen, D., Verspoor, K.M., Baldwin, T.: Cqadupstack: a benchmark data set for community question-answering research. In: Proceedings of the 20th Australasian Document Computing Symposium, pp. 1–8 (2015)

13. Karpukhin, V., et al.: Dense passage retrieval for open-domain question answering. In: Proceedings of the 2020 Conference on Empirical Methods in Natural Language Processing (EMNLP), pp. 6769–6781. Association for Computational Linguistics (2020)

14. Khattab, O., Zaharia, M.: Colbert: efficient and effective passage search via contextualized late interaction over bert. In: Proceedings of the 43rd International ACM SIGIR Conference on Research and Development in Information Retrieval, pp. 39–48 (2020)

15. Kim, J., Croft, W.B.: Retrieval experiments using pseudo-desktop collections. In: Proceedings of the 18th ACM Conference on Information and Knowledge Management, pp. 1297–1306. ACM (2009)

16. Kwiatkowski, T., et al.: Natural questions: a benchmark for question answering research. Trans. Assoc. Comput. Linguist. **7**, 453–466 (2019)
17. Liu, Y., et al.: Roberta: a robustly optimized bert pretraining approach. arXiv preprint arXiv:1907.11692 (2019)
18. Maia, M., et al.: Www'18 open challenge: financial opinion mining and question answering. In: Companion Proceedings of the The Web Conference 2018, International World Wide Web Conferences Steering Committee, Republic and Canton of Geneva, CHE, WWW 2018, pp. 1941–1942 (2018). https://doi.org/10.1145/3184558.3192301
19. Raffel, C., et al.: Exploring the limits of transfer learning with a unified text-to-text transformer. arXiv preprint arXiv:1910.10683 (2019)
20. Tafjord, O., Clark, P.: General-purpose question-answering with macaw. arXiv preprint arXiv:2109.02593 (2021)
21. Tague, J., Nelson, M., Wu, H.: Problems in the simulation of bibliographic retrieval systems. In: Proceedings of the 3rd Annual ACM Conference on Research and Development in Information Retrieval, Butterworth & Co., pp. 236–255 (1980)
22. Thakur, N., Reimers, N., Rücklé, A., Srivastava, A., Gurevych, I.: BEIR: a heterogeneous benchmark for zero-shot evaluation of information retrieval models. In: Thirty-fifth Conference on Neural Information Processing Systems Datasets and Benchmarks Track (Round 2) (2021). https://openreview.net/forum?id=wCu6T5xFjeJ
23. Thorne, J., Vlachos, A., Christodoulopoulos, C., Mittal, A.: FEVER: a large-scale dataset for fact extraction and VERification. In: Proceedings of the 2018 Conference of the North American Chapter of the Association for Computational Linguistics: Human Language Technologies, vol. 1 (Long Papers), pp. 809—819. Association for Computational Linguistics, New Orleans (2018). https://doi.org/10.18653/v1/N18-1074, https://aclanthology.org/N18-1074
24. Voorhees, E., et al.: Trec-covid: constructing a pandemic information retrieval test collection. SIGIR Forum **54**(1) (2021). https://doi.org/10.1145/3451964.3451965
25. Voorhees, E.M.: The trec-8 question answering track report. In: Proceedings of TREC-8, pp. 77–82 (1999)
26. Wachsmuth, H., Syed, S., Stein, B.: Retrieval of the best counterargument without prior topic knowledge. In: Proceedings of the 56th Annual Meeting of the Association for Computational Linguistics, vol. 1: Long Papers, pp. 241–251. Association for Computational Linguistics, Melbourne (2018). https://doi.org/10.18653/v1/P18-1023, https://aclanthology.org/P18-1023
27. Wadden, D., et al.: Fact or fiction: verifying scientific claims. In: Proceedings of the 2020 Conference on Empirical Methods in Natural Language Processing (EMNLP), pp. 7534–7550. Association for Computational Linguistics (2020). https://doi.org/10.18653/v1/2020.emnlp-main.609, https://aclanthology.org/2020.emnlp-main.609
28. Xiong, L., et al.: Approximate nearest neighbor negative contrastive learning for dense text retrieval (2020). arXiv preprint arXiv:2007.00808
29. Yang, Z., et al.: HotpotQA: a dataset for diverse, explainable multi-hop question answering. In: Proceedings of the 2018 Conference on Empirical Methods in Natural Language Processing, pp. 2369–2380. Association for Computational Linguistics, Brussels (2018). https://doi.org/10.18653/v1/D18-1259, https://aclanthology.org/D18-1259

SegmentCodeList: Unsupervised Representation Learning for Human Skeleton Data Retrieval

Jan Sedmidubsky[1]([✉])[iD], Fabio Carrara[2][iD], and Giuseppe Amato[2][iD]

[1] Masaryk University, Brno, Czech Republic
xsedmid@fi.muni.cz
[2] ISTI-CNR, Pisa, Italy

Abstract. Recent progress in pose-estimation methods enables the extraction of sufficiently-precise 3D human skeleton data from ordinary videos, which offers great opportunities for a wide range of applications. However, such spatio-temporal data are typically extracted in the form of a continuous skeleton sequence without any information about semantic segmentation or annotation. To make the extracted data reusable for further processing, there is a need to access them based on their content. In this paper, we introduce a universal retrieval approach that compares any two skeleton sequences based on temporal order and similarities of their underlying segments. The similarity of segments is determined by their content-preserving low-dimensional code representation that is learned using the Variational AutoEncoder principle in an unsupervised way. The quality of the proposed representation is validated in retrieval and classification scenarios; our proposal outperforms the state-of-the-art approaches in effectiveness and reaches speed-ups up to 64x on common skeleton sequence datasets.

Keywords: 3D skeleton sequence · Segment similarity · Unsupervised feature learning · Variational AutoEncoder · Segment code list · Action retrieval

1 Introduction

The rapid development of pose-estimation methods [6] enables more and more precise detection of human body keypoints (2D or even 3D) in individual frames of a standard video-camera recording. The detected keypoints are then used to simplify human motion using the spatio-temporal *skeleton sequence* representation. Since such skeleton sequences can nowadays be extracted from a common video, the analysis of human motion is becoming very popular in a broad spectrum of application domains, ranging from computer animation through robotics, security, autonomous driving, to healthcare and sports [25].

J. Sedmidubsky and F. Carrara—These authors contributed equally.

J. Kamps et al. (Eds.): ECIR 2023, LNCS 13981, pp. 110–124, 2023.
https://doi.org/10.1007/978-3-031-28238-6_8

The skeleton sequences may generally appear in different forms – short or long, segmented or continuous, labeled or unlabeled. Current research in skeleton-data processing mainly focuses on recognizing classes of short and labeled *actions* [5,19,22] or detecting such actions [21,27] or anomalies [1] in *continuous sequences*. These tasks require examples of actions to be defined in advance, so that action classifiers or detectors can be trained in a supervised way. However, supervised training is not applicable to environments where examples of actions are not known in advance. In environments where skeleton data are extracted as long continuous sequences without any information about their annotation or semantic partitioning, unsupervised content-based processing methods are the only possibility to make the recorded data searchable and thus reusable. One of the most general principles is to partition continuous sequences into short *segments* and access the data based on similarities of the underlying segments.

In this paper, we adopt such general segment-based processing principle by partitioning a continuous sequence into fixed-size segments and extracting the content-preserving segment representation – in the form of a low-dimensional *code* – in an unsupervised way. The most desirable property is that two codes are similar in terms of the cosine distance if their corresponding segments exhibit similar movement characteristics and vice-versa. In this way, we can represent a skeleton sequence of any length by the list of codes and determine the similarity of any two code lists based on the time-warping principle. This allows the proposed approach to be integrated within any retrieval-based operation.

2   Related Work and Our Contributions

Related works almost exclusively learn a skeleton-data representation on the level of *pre-segmented* actions, that are commonly provided by benchmark datasets [16,17,20]. Since the individual actions constitute standalone semantic units, the action representation can be straightforwardly learned in a *supervised* or *self-supervised* way. However, representation learning is not an easy task for the continuous (unsegmented) sequences whose content is generally unpredictable.

Approaches for Pre-segmented Actions. Plenty of papers propose various architectures of *supervised* neural-network classifiers that trade between classification accuracy and the number of network parameters, pretty much influencing the training time. Such approaches are usually based on transformers [5], convolutional [19], recurrent [27], graph-convolutional [22] networks, or their combinations including attention-based mechanisms. However, they are limited to scenarios where both segmented actions and their labeling are provided in advance. Recently, *self-supervised* learning, where action labeling is not known, has become increasingly popular. In such cases, the action representation can be learned using reconstruction-based or contrastive-learning-based methods. The former group applies the encoder-decoder principle to reconstruct the original skeleton data of an action and uses the learned intermediate representation as the action feature. The latter group [12,14,29] aims at learning a meaningful metric that sufficiently reflects semantic similarity to discriminate actions belonging to different classes in the validation step. Still, all these methods are applied to scenarios where the actions are pre-segmented (known) in advance.

Approaches for Unsegmented Sequences. Compared to the previous research, a limited number of approaches provide content-based access to unsegmented skeleton data. In [26], the continuous sequences are synthetically partitioned into many overlapping and variable-size segments that are represented by 4,096D deep features. However, such features are learned in a supervised way by exploiting supplementary knowledge about labeled actions, and indexing is very difficult due to both high feature dimensionality and a large number of segments. To move towards more efficient processing, the approaches in [2,18] quantize high-dimensional segment features into low-dimensional codes using k-means clustering. However, it is not generally possible to partition a given segment-feature space in such way that all pairs of similar segments are in the same partition. Some pairs of similar segments thus get separated by partition borders and become non-matching, which decreases the effectiveness of applications with an increasing number of clusters (i.e., the vocabulary size). This problem is partly solved in [4,24] by applying soft quantization; nevertheless, limited effectiveness is still achieved as the quantization process employs a numerical distance function for comparison of segments. Such function can not principally partition segment data based on their semantics.

Contributions of This Paper

We propose an effective representation of unsegmented and unlabeled skeleton sequences using a list of compact codes learned in an unsupervised way. Compared to existing methods, the proposed codes are very compact (in contrast to high-dimensional features in [2]) and preserve motion semantics (in contrast to hand-crafted segment features in [18,24]). Specifically,

- we propose a lightweight residual neural-network architecture to effectively process short segments of spatio-temporal skeleton data;
- we apply the reconstruction-based Variational AutoEncoder approach in combination with the proposed architecture to learn semantic information of segment data in the form of a compact code;
- we propose to adopt the time-warping principle to determine a similarity between two code lists representing pairs of motions of any lengths;
- we verify the effectiveness and efficiency of the proposed code lists in the context of two retrieval-based application scenarios.

3   Code List Representation

In this section, we describe the whole process of transformation of a continuous skeleton sequence into a list of codes. In particular, we formally define the skeleton data domain together with the retrieval-based principle using k-nearest neighbor queries. Then, we present how continuous sequences are partitioned and how semantic codes are learned from such unlabeled segment data. Finally, we propose how to compare any two sequences represented by the lists of codes.

3.1 Problem Definition

We represent skeleton data as a continuous sequence $(P_1, \ldots, P_n)$ of n consecutive 3D *poses* P_i, where the i-th pose $P_i \in \mathbb{R}^{j \cdot 3}$ is captured at time moment i ($1 \leq i \leq n$) and consists of xyz-coordinates of j tracked *joints*. In this paper, we use two different body models with $j = 31$ joints for the HDM05 dataset [20] and $j = 25$ joints for the PKU-MMD dataset [15]. The sequence of n poses is then partitioned into a list of segments $(S_1, \ldots, S_m)$, where $m \ll n$. Each segment $S_i = (P_1, \ldots, P_f)$ consists of a fixed number of f poses and is further transformed using the $codeTrans(S_i)$ function into a low-dimensional *code* representation $C_i \in \mathbb{R}^d$. The appropriate code dimensionality d typically ranges between $d \in 2^{[3..6]}$. Thus, the original high-dimensional skeleton-data sequence $(P_1, \ldots, P_n)$ is transformed into a short *code list* $(C_1, \ldots, C_m)$, so-called *SegmentCodeList* (SCL), consisting of m low-dimensional codes (e.g., on the HDM05 dataset, a 744-dimensional segment – consisting of eight 93D poses – is transformed into a 32D code). The whole transformation process is schematically illustrated in Fig. 1.

We evaluate the SCL representation on classification and retrieval scenarios using the k-nearest neighbor approach. Having a set $\{D_1, D_2, \ldots\}$ of *database sequences* $D_i = (P_1, \ldots, P_{n_i})$ and a *query sequence* $Q = (P_1, \ldots, P_n)$, the objective is to find such k database sequences that are the most similar to the query sequence Q. The similarity between the query Q and any database sequence D_i is quantified using a distance function $dist(Q, D_i)$ that operates over their corresponding SCLs. Our approach does not anyhow limit the length of query or database sequences, so they can correspond to long skeleton recordings, short pre-segmented actions, or their combinations.

3.2 Partitioning Skeleton Sequences

For efficient content-based management of especially longer skeleton sequences, it is necessary to partition them into meaningfully-sized segments. The segment-level representation constitutes the smallest processing unit that preserves a reasonable volume of spatio-temporal information and is much better manageable in comparison with either many only-spatial poses, or hardly-processable continuous sequences. In addition, processing on the level of segments can be utilized in a broad variety of tasks.

The most straightforward way is to apply a mechanical slicing of a skeleton sequence into *fixed-size* segments. There is no optimal length of segments, but the rule of thumb suggests that such length should be upper-bounded by the length of the shortest retrievable query. Besides the segment length, the problem is that the segments originating from the query need not be perfectly aligned

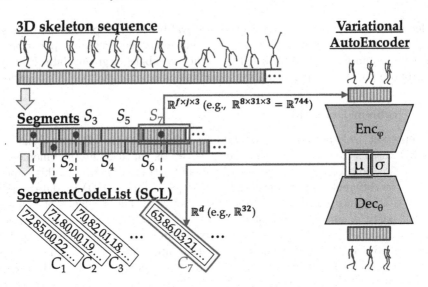

Fig. 1. A schematic illustration of the proposed transformation process of a continuous 3D skeleton sequence into the SCL representation.

with the segments covering query-relevant parts within database sequences – thus, such relevant database sub-sequences can become unfindable. To overcome this problem, we apply the *overlapping* segment principle, which balances the trade-off between data findability at the price of increased data redundancy. The appropriate overlap between two consecutive segments is often set to between 50–80 %. For example, the 50 % segment overlap is illustrated in Fig. 1.

Formally, we partition a continuous skeleton sequence $(P_1, \ldots, P_n)$ into a list of segments $(S_1, \ldots, S_m)$. The i-th segment S_i $(i \in [1, m])$ is represented by the following sub-sequence of f poses:

$$S_i = (P_{(i-1)\cdot ss+1}, \ldots, P_{(i-1)\cdot ss+f}),$$

where $f \in \mathbb{N}$ is the fixed *segment length* and $ss \in \mathbb{N} \, (1 \leq ss \leq f)$ is the fixed *segment shift* (in number of poses), determining that two consecutive segments overlap in $(1 - \frac{ss}{f}) \cdot 100\%$ frames. For the skeleton sequence of n poses, this segmentation policy generates $\lfloor \frac{n-f}{ss} \rfloor = m$ segments in total.

3.3   Learning Codes for Segment Data

Given a set $\{S_i\}$ of unlabelled segments $S_i \in \mathbb{R}^{f \times j \cdot 3}$ that are extracted from training sequences, we want to learn an encoding function $codeTrans : \mathbb{R}^{f \times j \cdot 3} \to \mathbb{R}^d$ that maps segments $\{S_i\}$ to small semantic codes $\{C_i\}, C_i \in \mathbb{R}^d$. The similarity between codes (e.g., their cosine similarity) should reflect the semantic similarity between the original segment data. To learn $codeTrans(\cdot)$ in an unsupervised way, we apply the reconstruction-based principle by adopting a deep generative model; specifically a Variational AutoEncoder (VAE) [13].

VAE Formulation. We assume that the segment space $\mathbf{S} \subset \mathbb{R}^{f \times j \cdot 3}$ is induced by a latent code space $\mathbf{C} \subset \mathbb{R}^d$. Following the commonly used VAE terminology, in our formulation, the encoder network Enc_ϕ takes as input a segment $S \in \mathbf{S}$ and produces a distribution $q_\phi(C|S)$ over the latent code space $\mathbf{C}$ describing the codes that could have generated S. Specifically,

$$\left(\mu_C, \sigma_C^2\right) = \text{Enc}_\phi(S) \tag{1}$$
$$q_\phi(C|S) \sim \mathcal{N}(\mu_C, \sigma_C^2 I), \tag{2}$$

where $q_\phi(C|S)$ is defined as a Gaussian distribution whose parameters (the mean $\mu_C \in \mathbb{R}^d$ and diagonal covariance matrix whose diagonal values are $\sigma_C^2 \in \mathbb{R}^d$) are produced by $\text{Enc}_\phi(S)$. Similarly, the decoder network Dec_θ takes a code C and defines $p_\theta(S|C)$ (the distribution of sequences in $\mathbf{S}$ corresponding to the latent code C) by providing $\mu_S, \sigma_S^2 \in \mathbb{R}^{f \times j \cdot 3}$:

$$\left(\mu_S, \sigma_S^2\right) = \text{Dec}_\phi(C) \tag{3}$$
$$p_\theta(S|C) \sim \mathcal{N}(\mu_S, \sigma_S^2 I). \tag{4}$$

The parameters of the encoder and decoder networks ϕ, θ are jointly optimized via mini-batch gradient descent by maximizing the evidence lower bound (ELBO):

$$\text{ELBO}(\theta, \phi, S_i) = \mathbb{E}_{C_i \sim q_\phi(C|S_i)} \left[\log p_\theta(S_i|C_i)\right] - \beta \cdot D_{KL}\left(q_\phi(C|S_i) \,\|\, p(C)\right), \tag{5}$$

where β is a hyperparameter that controls the trade-off between reconstruction accuracy and latent code disentanglement [9]. When assuming a normal prior for $p(C)$ (i.e., $C \sim \mathcal{N}(0, I)$), maximizing Eq. 5 reduces to minimizing the following loss function for a sample sequence S:

$$\mathcal{L}(\theta, \phi, S) = \sum_{k=1}^{f \times j \cdot 3} \left(\frac{\left(S^{(k)} - \mu_S^{(k)}\right)^2}{2\sigma_S^{2(k)}} + \log \sigma_S^{(k)} \right)$$
$$+ \frac{\beta}{2} \sum_{k=1}^{d} \left(\mu_C^{(k)} + \sigma_C^{2(k)} - 1 - \log \sigma_C^{(k)} \right), \tag{6}$$

where the notation $^{(k)}$ indicates the k-th component of a vector. The first term in Eq. 6 represents the negative log-likelihood of the sample S given the reconstruction mean μ_S and variance σ_S^2 produced by the decoder. The variance σ_S^2 can be interpreted as the uncertainty of the reconstruction [11]; the decoder is pushed to minimize uncertainty (via the $\log \sigma_S^{2(j)}$ term) but is discouraged to output a low uncertainty when the predicted mean μ_S deviates too much from the original sample S (via the $(S^{(j)} - \mu_S^{(j)})^2/2\sigma_S^{2(j)}$ term). The second term of Eq. 6 is a regularization term that pushes codes produced by the encoder to be normally distributed, thus reducing code overfitting.

Encoder and Decoder Architectures. The encoder and decoder networks are implemented as residual 1D convolutional networks [8]. The encoder is comprised of a single-layer 64-channels convolutional stem followed by three residual blocks and a last fully connected layer. In the second and third residual blocks, the time dimension is halved by 2-stride convolutions, while channels are doubled. The decoder network is comprised of four residual blocks and a final convolutional layer; before each residual block, the input is upsampled by a factor 2 in the time dimension until it matches the correct segment size, while the channel dimension is halved by each block. A residual block is implemented as BN-ReLU-Conv-BN-ReLU-Conv, where BN and Conv are 1-dimensional batch normalization and convolutional layers, respectively. A convolutional layer is added in the shortcut path of the residual block when output and input dimensionalities do not match.

Segment Encoding. Once trained, we adopt the encoder network to transform the skeleton data of each segment into a code. Specifically, for each segment S_i, we take the mean parameter μ_C produced by $\text{Enc}_\phi(S_i)$ as code C_i:

$$C_i = codeTrans(S_i) = \text{Enc}_\phi(S_i)[0]\,, \tag{7}$$

where [0], with abuse of notation, indicates the selection of only the first output of the encoder. The similarity between codes is quantified using the cosine distance. The code for the extraction of segment features is available at: https://github.com/fabiocarrara/mocap-vae-features.

3.4   Determining Similarity of SCL Representations

For the purpose of k-nearest neighbor retrieval, there is a need to determine a similarity between the query Q and any database sequence D_i. Let us recall that the query and database sequences have to be first transformed into the SCL representation by partitioning a given sequence into segments and transforming each segment into the code using the $codeTrans(\cdot)$ function. Thus, the query Q is then represented by its SCL as $(C_1, \ldots, C_m)$ and the database sequence D_i as $(C'_1, \ldots, C'_{m'})$. Since the lengths of SCLs can be generally different, i.e., $m \neq m'$, the time-warping or bag-of-words principle constitutes possible candidates for similarity-based comparison. To respect the temporal order of codes, we have decided to apply the Dynamic Time Warping (DTW) distance function to compare two SCLs, where the similarity of two particular codes C_i and C'_j inside DTW is quantified using the cosine distance (as stated in Sect. 3.3):

$$codeDist(Q, D_i) = \frac{1}{\overline{m}} \cdot DTW\left((C_1, \ldots, C_m), (C'_1, \ldots, C'_{m'})\right). \tag{8}$$

We further normalize the DTW distance by the length of the warping path $\overline{m}$ ($\max\{m, m'\} \leq \overline{m} \leq m + m'$) inside the DTW matrix (i.e., by the number of identified mappings between pairs of codes) so that shorter sequences are not favored at the expense of longer ones with respect to the same query.

The disadvantage of DTW is its quadratic time complexity. On the other hand, the SCLs are typically quite short. In case long SCLs are needed to be compared, several DTW enhancements can possibly be applied to decrease processing time up to linear complexity [23].

4 SCL in Retrieval Applications

We experimentally verify that the SCL representation preserves important characteristics of 3D skeleton segment data in the context of two popular applications: *action retrieval* and *action classification*. The evaluation on the level of pre-segmented actions allows us to demonstrate that the SCL approach trained without the information about pre-segmented actions or their labels can achieve high effectiveness even when compared to the purposely trained classifiers.

4.1 Datasets

Even though there is a variety of 3D skeleton datasets, they usually provide only pre-segmented actions used for supervised or self-supervised learning tasks (such as NTU RGB+D 60/120 [17] or Kinetics 400 [10]). To properly evaluate our approach, which does not assume anything about pre-segmentation, we need datasets that provide continuous (unsegmented) skeleton sequences. The suitable possibilities are the following two datasets.

- **HDM05** dataset [20] is captured by a marker-based motion capture technology with a 31-joint body model. The dataset contains up to 324 continuous skeleton sequences with the total length of about 3.5 h, which corresponds to 1.5 M frames with the frame-per-second rate (FPS) of 120 Hz. The dataset also provides a fine-grained annotation of 241 (out of 324) continuous sequences in which 2,345 actions belonging to 130 classes are labeled.
- **PKU-MMD** dataset [15] is captured by Kinect with a 25-joint body model. The dataset provides 860 single-subject continuous sequences with the total length of 20 h captured with the FPS rate of 30 Hz. Such sequences contain almost 20 K labeled single-subject actions that are categorized in 43 classes. The dataset defines the cross-view (**CV**) and cross-subject (**CS**) evaluation scenarios that specifically divide the actions as well as sequences into training and test batches.

As recommended in most of the papers, we also pre-process the datasets by downsampling the skeleton data (downsampling HDM05 10 times from 120 to 12 and PKU-MMD 3 times from 30 to 10) and applying the position, orientation, and skeleton-size normalization.

4.2 Evaluation Methodology of Retrieval Applications

To support unsupervised segment-code learning, we use only continuous (unsegmented) skeleton sequences without any information about pre-segmented

actions or their labels. We train one model for the whole HDM05 dataset and
two models for the PKU-MMD dataset corresponding to the CV and CS scenarios. As discussed in Sect. 3.2, we synthetically partition each skeleton sequence
into a series of fixed-size overlapping segments. As recommended in [2,24], we
fix the segment length to 0.666 s with the segment shift of 0.133 s, which corresponds before downsampling to 80 and 24 frames with the segment shift of
16 and 5 frames for the HDM05 and PKU-MMD dataset, respectively. After
downsampling, this results in 8-frame segments for both datasets. In total, 70 K
segments were generated from the 241 HDM05 sequences and 1.2 M from the 860
PKU-MMD sequences. All the HDM05 segments were used to train the HDM05
model, while the subsets of PKU-MMD segments originating from the training
sequences specified for the CV/CS evaluation scenarios were only used to train
the models for the CV and CS scenarios. For training purposes, such segments
were randomly split in the 80:20 manner to define the sets for the training and
validation phases of each model.

To study the effectiveness of the proposed approach, we need a ground truth
that is, however, defined only on the level of actions. Therefore, we focus on
traditional action-retrieval and action-classification applications which we evaluate on the skeleton-based modality using k-nearest neighbor (kNN) *queries*.
In both applications, the model trained on unsegmented skeleton sequences is
used to extract the SCL representation for each dataset action. Then, each kNN
query is evaluated in a sequential way by computing the normalized DTW distance between the specific *query-action* SCL and each *database-action* SCL (see
Eq. 8). In particular, on the PKU-MMD dataset, the queries correspond to the
test actions and the database actions to the *training actions* defined on the
CS/CV scenarios. On the HDM05 dataset, there is no standard evaluation protocol, so the leave-one-out approach is applied over all the 2,345 actions.

For the *action-retrieval* application, we quantify *effectiveness* as the average
precision (*Precision@k*) of all the kNN queries, where the query precision is
computed as a ratio of correctly retrieved actions. An action is considered as
correctly identified if it belongs to the same class as the query action. For the
action-classification application, we evaluate different values of k and apply the
kNN classifier as adopted in [24]. We measure the application effectiveness as
the average *classification accuracy* (*Accuracy@k*) over all the queries. For both
scenarios, we measure *efficiency* as the average time (in milliseconds) needed to
evaluate a single query on the collection of database actions.

4.3 Effectiveness and Efficiency Results

We evaluate effectiveness and efficiency results for varying the SCL dimensionality d, which is experimentally set to $d \in \{8, 16, 32, 64, 128, 256\}$. Thus, the
original segment dimensionality – 744/600-dimensional segment data consisting of eight 93D/75D poses for the HDM05/PKU-MMD datasets – is decreased
roughly from 2 times (for $d = 256$) to 75 times (for $d = 8$). Figure 2 reports
the effectiveness-efficiency trade-off when varying such SCL dimensionality d

(blue lines) in both action-retrieval and action-classification scenarios. As a reference, we also report the results of the *baseline* configuration (green cross) that uses DTW to determine the similarity of actions on the level of individual poses; the distance between poses inside DTW is implemented as the sum of the Euclidean distances between corresponding raw joint coordinates. SCL representations deliver a much improved effectiveness in both scenarios (with $d = 256$, a Precision@1 of 87.42% vs 75.22% on HDM05, 90.03% vs 83.58% on PKU-MMD (CV), 74.41% vs 62.34% on PKU-MMD (CS), and an Accuracy@5 of 87.59% vs 77.31% on HDM05, 90.44% vs 81.19% on PKU-MMD (CV), 77.80% vs 59.43% on PKU-MMD (CS)). These results indicate that the learned SCL representations preserve semantic information. From the efficiency point of view, the comparison is quite fast as the DTW function is applied to relatively short SCLs – a single action contains 12 and 24 codes on average for the HDM05 and PKU-MMD datasets, respectively. As a result, we reduce the average query processing time by more than an order of magnitude with respect to operating on raw joint coordinates.

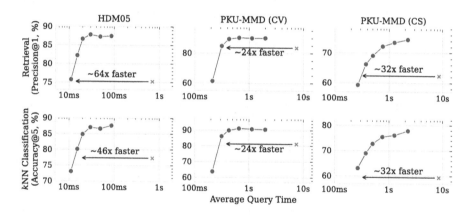

Fig. 2. Effectiveness-Efficiency trade-off of SCL. Effectiveness is measured in the kNN-based retrieval and classification scenarios as Precision@1 and Accuracy@5, respectively. Efficiency is measured as the average query-processing time needed to evaluate a single kNN query, and plotted on a logarithmic scale. (Color figure online)

Figure 3 shows the effectiveness in both scenarios when considering different numbers k of nearest neighbors during query processing. The best effectiveness is often reached when $k \in [3, 10]$, with SCL representations being more robust to the choice of k in the classification scenario with respect to the baseline. A very high effectiveness is already achieved when the code dimensionality d equals to 64 (red line), which means that the dimensionality of original segment data (i.e., 744/600 dimensions for the HDM05/PKU-MMD) is reduced by about ten times.

Regarding the training phase of SCL representations, a coarse grid search over the parameter β (see Table 1) showed that $\beta = 1$ delivers an optimal or

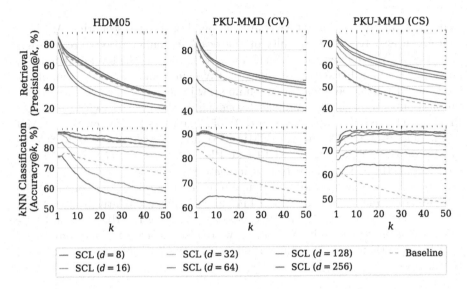

Fig. 3. kNN effectiveness of SCL for both retrieval and classification scenarios, when varying the number k of nearest neighbors during query processing. Dashed lines represent the baseline effectiveness. (Color figure online)

comparable effectiveness most of the time. There is an exception for small SCL dimensionalities ($d = 8$) where tuning β led to slight improvements. Besides evaluation of effectiveness and efficiency, we also employed the trained VAE decoder to reconstruct original skeleton data of a segment purely from its latent code representation. In Fig. 4, we illustrate examples of reconstructed segment data for two specific HDM05 segments.

Table 1. Precision@1 (%) of SCL representations for different settings of latent code dimensionality d and hyperparameter β controlling the trade-off between segment-reconstruction accuracy and latent-code disentanglement.

	(a) HDM05					(b) PKU-MMD (CV)					(c) PKU-MMD (CS)				
	β					β					β				
d	0	.01	.1	1	10	0	.01	.1	1	10	0	.01	.1	1	10
8	75.1	77.0	74.4	75.8	**77.4**	61.6	62.5	63.2	61.3	**63.7**	55.0	58.6	**60.3**	59.3	55.4
16	80.6	81.2	81.1	**82.2**	82.2	74.1	75.6	78.4	**84.8**	74.5	63.1	62.8	65.1	**66.2**	56.0
32	85.1	84.4	85.5	**86.7**	82.1	78.7	78.5	80.8	**89.4**	74.7	65.4	66.6	68.3	**69.0**	53.9
64	87.2	87.5	87.6	**87.8**	80.9	82.7	81.3	83.9	**90.3**	73.0	67.9	68.1	70.0	**72.1**	51.5
128	86.8	87.6	**87.9**	87.3	83.1	85.5	85.2	88.0	**90.0**	69.7	71.7	70.9	72.3	**73.5**	56.6
256	87.5	87.2	**87.9**	87.4	82.8	84.5	84.2	88.6	**90.0**	79.0	70.4	69.4	72.4	**74.4**	58.5

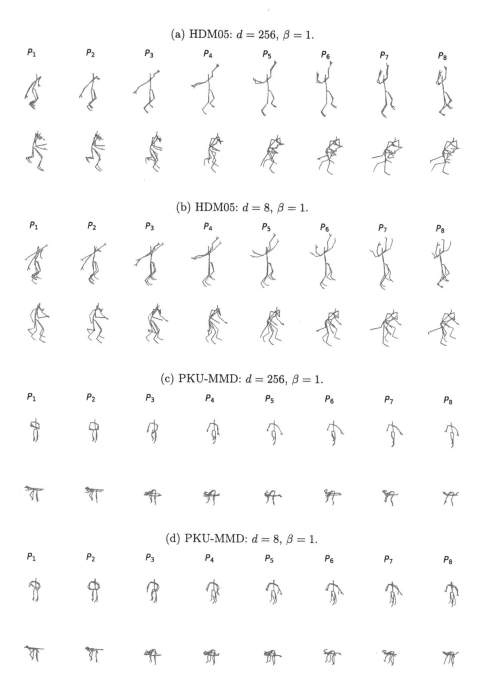

Fig. 4. Examples of original (blue) and reconstructed (red) segment poses from the HDM05 and PKU-MMD datasets. In each panel, the top (bottom) row depicts a success (failure) example of reconstruction. (Color figure online)

4.4 State-of-the-Art Comparison

The state-of-the-art skeleton-data processing primarily focuses on classifying pre-segmented and labeled actions. For this reason, we compare the results of our classification approach to the results of existing classifiers evaluated on the HDM05 and PKU-MMD datasets. Specifically, we adopt our kNN retrieval approach by fixing k to 4 as this setting reaches high classification accuracy. In Table 2, we demonstrate that our approach achieves superior accuracy on both datasets compared to existing unsupervised classifiers. Let us also emphasize that our solution is approaching the accuracy of supervised classifiers, even if no information about the pre-segmentation nor labels of actions was available in the training phase.

Table 2. Comparison of our approach with the existing supervised/unsupervised classifiers trained on the pre-segmented actions. The values of classification accuracy are taken from the referenced papers.

	HDM05	PKU (CV)	PKU (CS)
Supervised approaches			
Activity images + CNN [28]	–	92.00	85.00
DSwarm-Net [3]	**90.67**	–	–
BiLSTM [7]	89.26	**92.11**	**84.73**
Unsupervised approaches			
Baseline: raw skeleton data + DTW	75.22	83.58	62.34
Motion words + DTW [24]	80.30	–	–
LSTM + triplet-loss [12]	83.76	–	–
MS^{2}L [14]	–	–	64.86
Our approach ($\beta = 1, d = 256, k = 4$)	**87.80**	**90.53**	**77.20**

5 Conclusions

We have proposed a new skeleton-data representation that is learned using a unique combination of the β-VAE approach and a lightweight convolutional neural network. Such representation has several advantages in contrast to related approaches. First, the representation can be extracted for skeleton sequences of any length on the level of short segments. Second, the segment feature is learned from continuous skeleton sequences completely in an unsupervised way, without the requirement of knowledge of pre-segmented actions or their labels. Third, the learned segment feature preserves semantic information of the underlying skeleton data, which is confirmed by reaching much higher effectiveness in retrieval-based scenarios compared to the baseline approach. Fourth, the segment feature is very compact and efficiently comparable with the cosine distance, which supports indexing possibilities for the future. In addition, the universality of the proposed approach enables its high applicability in many tasks, e.g., not only for action recognition or detection but also for sub-sequence search.

Acknowledgements. This research was supported by ERDF "CyberSecurity, CyberCrime and Critical Information Infrastructures Center of Excellence" (No. CZ.02.1.01/0.0/0.0/16_019/0000822), by AI4Media - A European Excellence Centre for Media, Society, and Democracy (EC, H2020 n. 951911), and by SUN - Social and hUman ceNtered XR (EC, Horizon Europe n. 101092612).

References

1. Acsintoae, A., et al.: Ubnormal: new benchmark for supervised open-set video anomaly detection. In: IEEE/CVF Conference on Computer Vision and Pattern Recognition (CVPR), pp. 20143–20153 (2022)
2. Aristidou, A., Cohen-Or, D., Hodgins, J.K., Chrysanthou, Y., Shamir, A.: Deep motifs and motion signatures. ACM Trans. Graph. **37**(6), 187:1–187:13 (2018)
3. Basak, H., Kundu, R., Singh, P.K., Ijaz, M.F., Wozniak, M., Sarkar, R.: A union of deep learning and swarm-based optimization for 3D human action recognition. Sci. Rep. **12**(5494), 1–17 (2022)
4. Budikova, P., Sedmidubsky, J., Zezula, P.: Efficient indexing of 3D human motions. In: International Conference on Multimedia Retrieval (ICMR), pp. 10–18. ACM (2021)
5. Cheng, Y.B., Chen, X., Chen, J., Wei, P., Zhang, D., Lin, L.: Hierarchical transformer: Unsupervised representation learning for skeleton-based human action recognition. In: IEEE International Conference on Multimedia and Expo (ICME), pp. 1–6 (2021)
6. Dubey, S., Dixit, M.: A comprehensive survey on human pose estimation approaches. Multimedia Syst. **29**, 1–29 (2022)
7. Elias, P., Sedmidubsky, J., Zezula, P.: Understanding the limits of 2D skeletons for action recognition. Multimedia Syst. **27**(3), 547–561 (2021)
8. He, K., Zhang, X., Ren, S., Sun, J.: Identity mappings in deep residual networks. In: Leibe, B., Matas, J., Sebe, N., Welling, M. (eds.) ECCV 2016. LNCS, vol. 9908, pp. 630–645. Springer, Cham (2016). https://doi.org/10.1007/978-3-319-46493-0_38
9. Higgins, I., et al.: BETA-VAE: learning basic visual concepts with a constrained variational framework. In: 5th International Conference on Learning Representations (ICLR), pp. 1–22. OpenReview.net (2017)
10. Kay, W., et al.: The kinetics human action video dataset. arXiv (2017)
11. Kendall, A., Gal, Y., Cipolla, R.: Multi-task learning using uncertainty to weigh losses for scene geometry and semantics. In: IEEE Conference on Computer Vision and Pattern Recognition (CVPR), pp. 7482–7491 (2018)
12. Kico, I., Sedmidubsky, J., Zezula, P.: Towards efficient human action retrieval based on triplet-loss metric learning. In: 33rd International Conference on Database and Expert Systems Applications (DEXA), pp. 234–247. Springer, Cham (2022). https://doi.org/10.1007/978-3-031-12423-5_18
13. Kingma, D.P., Welling, M.: Auto-encoding variational bayes. arXiv preprint arXiv:1312.6114 (2013)
14. Lin, L., Song, S., Yang, W., Liu, J.: MS2L: multi-task self-supervised learning for skeleton based action recognition. In: 28th ACM International Conference on Multimedia (MM), pp. 2490–2498. ACM, New York (2020)

15. Liu, C., Hu, Y., Li, Y., Song, S., Liu, J.: PKU-MMD: a large scale benchmark for skeleton-based human action understanding. In: Workshop on Visual Analysis in Smart and Connected Communities (VSCC@MM), pp. 1–8. ACM (2017)
16. Liu, J., Song, S., Liu, C., Li, Y., Hu, Y.: A benchmark dataset and comparison study for multi-modal human action analytics. ACM Trans. Multimedia Comput. Commun. Appl. **16**(2), 1–24 (2020)
17. Liu, J., Shahroudy, A., Perez, M., Wang, G., Duan, L., Kot, A.C.: NTU RGB+D 120: a large-scale benchmark for 3D human activity understanding. IEEE Trans. Pattern Anal. Mach. Intell. **42**, 2684–2701 (2019)
18. Liu, X., He, G., Peng, S., Cheung, Y., Tang, Y.Y.: Efficient human motion retrieval via temporal adjacent bag of words and discriminative neighborhood preserving dictionary learning. IEEE Trans. Human-Mach. Syst. **47**(6), 763–776 (2017)
19. Lv, N., Wang, Y., Feng, Z., Peng, J.: Deep hashing for motion capture data retrieval. In: IEEE International Conference on Acoustics, Speech and Signal Processing (ICASSP), pp. 2215–2219. IEEE (2021)
20. Müller, M., Röder, T., Clausen, M., Eberhardt, B., Krüger, B., Weber, A.: Documentation Mocap Database HDM05. Technical RepORT CG-2007-2, Universität Bonn (2007)
21. Papadopoulos, K., Ghorbel, E., Baptista, R., Aouada, D., Ottersten, B.: Two-stage RGB-based action detection using augmented 3D poses. In: Vento, M., Percannella, G. (eds.) CAIP 2019. LNCS, vol. 11678, pp. 26–35. Springer, Cham (2019). https://doi.org/10.1007/978-3-030-29888-3_3
22. Peng, W., Hong, X., Zhao, G.: Tripool: graph triplet pooling for 3d skeleton-based action recognition. Pattern Recogn. **115**, 107921 (2021)
23. Rakthanmanon, T., et al.: Searching and mining trillions of time series subsequences under dynamic time warping. In: 18th ACM SIGKDD International Conference on Knowledge Discovery and Data Mining (KDD), pp. 262–270. ACM (2012)
24. Sedmidubsky, J., Budikova, P., Dohnal, V., Zezula, P.: Motion words: a text-like representation of 3D skeleton sequences. In: Jose, J.M., et al. (eds.) ECIR 2020. LNCS, vol. 12035, pp. 527–541. Springer, Cham (2020). https://doi.org/10.1007/978-3-030-45439-5_35
25. Sedmidubsky, J., Elias, P., Budikova, P., Zezula, P.: Content-based management of human motion data: Survey and challenges. IEEE Access **9**, 64241–64255 (2021). https://doi.org/10.1109/ACCESS.2021.3075766
26. Sedmidubsky, J., Elias, P., Zezula, P.: Searching for variable-speed motions in long sequences of motion capture data. Inf. Syst. **80**, 148–158 (2019)
27. Song, S., Lan, C., Xing, J., Zeng, W., Liu, J.: Spatio-temporal attention-based LSTM networks for 3D action recognition and detection. IEEE Trans. Image Process. **27**(7), 3459–3471 (2018)
28. Vernikos, I., Koutrintzes, D., Mathe, E., Spyrou, E., Mylonas, P.: Early fusion of visual representations of skeletal data for human activity recognition. In: 12th Hellenic Conference on Artificial Intelligence (SETN). ACM (2022)
29. Yang, Y., Liu, G., Gao, X.: Motion guided attention learning for self-supervised 3D human action recognition. IEEE Trans. Circ. Syst. Video Technol. **32**, 1–13 (2022)

Knowing What and How: A Multi-modal Aspect-Based Framework for Complaint Detection

Apoorva Singh[1] , Vivek Gangwar[1], Shubham Sharma[2],
and Sriparna Saha[1(✉)]

[1] Indian Institute of Technology Patna, Patna, India
{apoorva_1921cs19,vivek_2111mc14,sriparna}@iitp.ac.in
[2] Panjab University, Chandigarh, India

Abstract. With technological advancements, the proliferation of e-commerce websites and social media platforms has created an avenue for customers to provide feedback to enterprises based on their overall experience. Customer feedback serves as an independent validation tool that could boost consumer trust in the brand. Whether it is a recommendation or review of a product, it provides insight allowing businesses to understand what they are doing right or wrong. By automatically analyzing customer complaints at the aspect-level enterprises can connect to their customers by customizing products and services according to their needs quickly and deftly. In this paper, we introduce the task of Aspect-Based Complaint Detection (ABCD). ABCD identifies the aspects in the given review about a product and also finds if the aspect mentioned in the review signifies a complaint or non-complaint. Specifically, a task solver must detect duplets (What, How) from the inputs that show WHAT the targeted features are and HOW they are complaints. To address this challenge, we propose a deep-learning-based multi-modal framework, where the first stage predicts what the targeted aspects are, and the second stage categorizes whether the targeted aspect is associated with a complaint or not. We annotate the aspect categories and associated complaint/non-complaint labels in the recently released multi-modal complaint dataset (CESAMARD), which spans five domains (books, electronics, edibles, fashion, and miscellaneous). Based on extensive evaluation our methodology established a benchmark performance in this novel aspect-based complaint detection task and also surpasses a few strong baselines developed from state-of-the-art related methods (Resources available at: https://github.com/appy1608/ECIR2023_Complaint-Detection).

Keywords: Complaint detection · Aspect-based complaint detection · Aspect category detection · Multi-modal learning · Deep learning

1 Introduction

Multimodality has managed to bridge the divide between the various branches of artificial intelligence, such as natural language processing and computer vision.

J. Kamps et al. (Eds.): ECIR 2023, LNCS 13981, pp. 125–140, 2023.
https://doi.org/10.1007/978-3-031-28238-6_9

The supplementary sources of images, video, and audio help develop robust end-to-end frameworks that offer the user a comprehensive understanding. Multimodal inputs, i.e., a combination of text and other nonverbal cues (images), offer additional information that can assist in establishing efficient downstream application modules such as chatbot systems, highlighting the need to add multi-modal inputs into the process. With the help of multi-modal systems, customers can examine products and make educated choices about the items to be purchased. Customer reviews also directly affect the company image (for better or worse), increase or decrease sales, and could be the ultimate cue that either gets converted or persuades a customer never to consider the company again.

However, the volume of information generated daily presents a variety of obstacles to effective maintenance and analysis. The unpredictable nature of user-generated texts and the lack of many essential resources and techniques for the handling of such kinds of multi-modal reviews are some of the obstacles. Thus, it has been a matter of significance for researchers to develop suitable tools and methodologies for accurately and effectively analyzing consumer content. One such task is complaint detection [15,23,32], which entails identifying breach of expectation (complaint) from customer reviews posted on social media platforms.

Prior studies on complaint detection [12,26,27] identify the overall document or sentence-level complaints. Although such research findings provide vital information, they can sometimes be insufficient. For example, in complaint detection at the sentence level, the overall complaint does not reflect the specific attribute or feature that the user disapproves of. For an instance, when a customer expresses a complaint regarding an electronic product purchased online, it is unclear which feature or attribute the customer dislikes (e.g., whether it is the design, price, software, etc.). Such instances can be misleading for the complaint detection system and may not be of much assistance to the companies in making an informed complaint-related decision. In such a scenario, aspect-based complaint detection (ABCD) provides a solution. However, to the best of our knowledge, neither multi-modal nor unimodal complaint detection frameworks have been developed that incorporate aspect information. ABCD identifies complaints at a fine-grained level by associating a complaint/non-complaint label for each attribute present in the sentence. It consists of two sub-tasks:

(a) Aspect Category Detection (ACD), and
(b) Aspect Category Complaint detection (ACC)

The first sub-task is concerned with categorizing an attribute/aspect term into one of the predefined categories. The second sub-task involves classifying reviews with respect to the aspect categories as complaint or non-complaint.

Table 1 shows a few example review texts and images with the aspect categories and the associated labels. It is observable that the review instance relies on both text and images to provide a complete understanding of the product being discussed and to aid a complaint detection system in a better manner. The first review instance contains two aspect categories 'Taste' and 'Packaging'. The

Table 1. Examples scenarios for the aspect based complaint detection.

Review	Image	Aspect Terms	Labels
Taste of lentil was very good		Taste	Non-Complaint
but received torn package.		Packaging	Complaint
Ordered navy blue colour		Colour	Complaint
but got red colour tshirt.			

labels for the aspect categories are 'Non-Complaint' and 'Complaint', respectively. However, the second review instance contains only one aspect category 'Colour' and the associated label is 'Complaint'. In this paper, we identify not only the various aspect categories in a supervised end-to-end multi-modal framework but also the complaint/non-complaint classes associated with each aspect. We believe that identifying complaints at the fine-grained aspect level could provide organizations with a deeper understanding of the complaints expressed in online product reviews, thereby enabling new research avenues in the complaint detection domain.

The major attributes of our current work are as follows:

- *We propose the task of aspect-guided complaint classification in a multi-modal setup.*
- *We extend the recently released multi-modal complaint dataset (CESAMARD) [26] by annotating the aspect categories and associated complaint/non-complaint labels.*
- *We propose a deep learning-based multi-modal bitransformer framework that uses local and global attributes and relates them to the textual context for aspect-guided complaint classification.*
- *The proposed model is a benchmark setup for aspect-based complaint detection (ABCD), which also surpasses a few strong baselines developed from state-of-the-art related methods.*

2   Related Work

In computational linguistics, previous works on complaint detection only pivoted on identifying complaints using feature-based machine learning models [6,23], transformer network-based deep learning models [11,12]. Recently multitask complaint analysis models have been developed that leveraged polarity and affect information for enhancing the complaint mining task [27–29]. Besides complaint classification, studies have focused on product hazards and risks [3], and the propensity of escalation [36]. In pragmatics, [20] classified complaints into the following five categories based on their straightforwardness and intensity: (a) below reproach; (b) statement of disapproval; (c) explicit complaint; (d) allegation; and (e) warning. Quite recently, the research reported in [31], grouped complaints into four granular severity levels: (a) no specific reproach; (b) disapproval; (c) accusation; and (d) blame. In the work [12], authors have evaluated the severity level of complaints by training several transformer-based networks paired with linguistic information to predict severity levels in complaints.

In the related field of emotion and sentiment analysis, the study presented in [22,24,25] has helped bridge the divide between vision and language. The research in [26] proposed a binary complaint classification model based on multi-modal information without taking into consideration the specific features or aspects regarding which the user is complaining. With the release of the multi-modal complaint dataset (CESAMARD) [26], a collection of customer-posted reviews and images of the purchases made from the e-commerce website Amazon[1], has facilitated further research in complaint detection in multi-modal setup.

Our current study differs from the previous work on multi-modal complaint detection in that we concentrate on the task of identifying complaints based on different aspects of product reviews in accordance with the text and image associated with the review instance. We extend the CESAMARD dataset, which contains both textual and visual information, by annotating the aspect categories and associated complaint/non-complaint labels. To demonstrate how the extended dataset can be used effectively, we propose a deep-learning-based multi-modal framework to address two main problems: aspect category detection and complaint classification.

Table 2. Aspect categories and the total number of instances corresponding to different domains present in the *CESAMARD-Aspect* dataset.

Domains	Instances	Aspect categories
Books	690	Content, Packaging, Price, Quality
Edibles	450	Taste, Smell, Packaging, Price, Quality
Electronics	1507	Design, Software, Hardware, Packaging, Price, Quality
Fashion	1275	Colour, Style, Fit, Packaging, Price, Quality
Miscellaneous	40	Miscellaneous, Packaging, Price, Quality

---

[1] https://www.amazon.in.

3   Corpus Extension

The existing complaint datasets [23,27] deal with text-based complaints only. For this work, we utilize the *CESAMARD* dataset[2] published in [26]. We selected this dataset because it is the only publicly available multi-modal complaint dataset. The CESAMARD dataset comprises 3962 reviews, with 2641 reviews in the non-complaint category (66.66%) and 1321 reviews in the complaint category (33.34%). Each record in the dataset consists of the image URL, review title, review text, and corresponding complaint, polarity, and emotion labels. The instances in the CESAMARD dataset have also been grouped according to various domains, such as electronics, edibles, fashion, books, and miscellaneous. This categorization of instances domain-wise motivates us to utilize this dataset for fine-grained aspect-level complaint detection.

We take it a step forward by including the pre-defined set of aspect categories for each of the 5 domains with the associated complaint/non-complaint labels based on the text and image information available. In order to do so, we re-annotate the CESAMARD dataset. These steps are detailed in the following subsection.

3.1   Annotation Specifications

We define and compile a list of aspect categories for the various domains (books, edibles, electronics, fashion, and miscellaneous) present in the CESAMARD dataset. All domains share three common aspect categories, namely packaging, price, and quality because these aspects are relevant for products purchased online. We follow a similar scheme in line with works in the related area of aspect-based sentiment analysis [1] and SemEval shared tasks for selecting and annotating the aspect categories. Table 2 shows the different aspect categories and the number of instances across the 5 domains in the extended CESAMARD dataset (CESAMARD-Aspect).

We annotate every instance in the dataset with relevant aspect categories. Secondly, a complaint/non-complaint label is chosen based on whether the user is expressing a complaint or non-complaint with respect to the aspect category. Three annotators (one doctoral and two undergraduate students in the computer science discipline) with adequate domain knowledge and expertise in developing supervised corpora were entrusted with annotating each review instance in the dataset. We observe a Fleiss-Kappa [8] agreement ratio of approximately 69% which can be considered reliable [2]. Note that if the review in its entirety has been labelled as non-complaint, the annotations at the aspect level will also be non-complaint. But in the case of a few review-level complaints, there are some aspects that were annotated as non-complaints. Table 1 shows some examples of aspect terms and label annotations.

---

[2] https://www.iitp.ac.in/~ai-nlp-ml/resources.html#CESAMARD.

4   Proposed Methodology

In this section, we describe the problem at hand and discuss the details of the proposed framework. In the following subsections, we discuss the major components of the architecture.

4.1   Problem Definition

We address the aspect-based complaint detection model as a task solver that detects duplets (What, How) from the inputs, consisting of the aspect categories (WHAT) and the complaint/non-complaint label concerning the respective aspects (HOW). For a given review instance R, is represented as $\{[T, I, W, A_k, c_k]_i\}_{i=1}^{N}$, where T denotes the review text, I is the review image, and W is the web entity tag, A_k denotes the aspect categories, c_k is the associated complaint/non-complaint labels for every aspect category present in the review instance. The first task is to identify the aspect categories, A_k and the second task involves detecting the complaint/non-complaint labels, c_k, for each of the identified aspect categories present in A_k. The Transformer network [33] acts as the foundation of our proposed architecture, as can be seen in Fig. 1. We propose a multi-modal bi-transformer-based architecture that combines text-only self-supervised representations with the strength of state-of-the-art convolutional neural network architecture from computer vision, using web entities as a higher-level image concept.

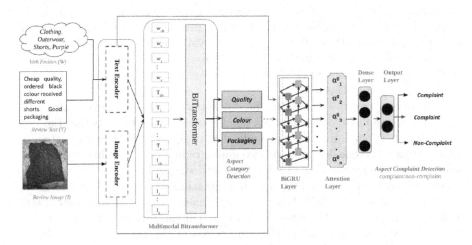

Fig. 1. Architectural diagram of the proposed ABCD framework.

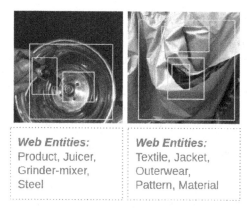

Web Entities:
Product, Juicer,
Grinder-mixer,
Steel

Web Entities:
Textile, Jacket,
Outerwear,
Pattern, Material

Fig. 2. Detected web entities for a few sample images from the *CESAMARD-Aspect* dataset.

4.2 Text Encoder

RoBERTa: Bidirectional Encoder Representations from Transformers (BERT) is a multi-layer bidirectional transformer encoder [7] based on the primary work described in [34]. RoBERTa [18] is a modification of BERT that has demonstrated improved performance on social media analytics [16,17] via training on a larger number of datasets with changing hyperparameters. We use the *RoBERTa-base* model[3] (12-layer, 768-hidden, 12-heads, 125M parameters) to embed the words of each review instance.

Web Entities: Motivated by [4], we propose web entities[4] as a higher-level image concept for our model. It can be observed from the review images shown in Fig. 2 that entities such as textile, juicer, grinder-mixer, outerwear, etc., provide background context corresponding to every review image, which could help in identifying complaints. We use Google Cloud Vision API to foreground various web entities[5]. Assuming for a given input review image, the entities are $\{P_1, P_2, ..., P_k\}$. We encode each web entity with a *RoBERTa* encoder to generate 512-dimensional features. We represent these features as $O_f = \{f_1, f_2, ..., f_k\}$, where $O_f \in R^{k \times 512}$.

Image Encoder: We employ a ResNet-152 [10] with average pooling over X×K grids in the image, generating $V = X \times K$ output vectors with 2048 dimensions per image, where V is the number of image embeddings. The images are resized, normalized, and cropped to the center. Please note that pooling operation over feature maps in the final fully-connected layer is not required for multimodal bitransformers, as they can handle an arbitrary number of dense inputs.

---

[3] https://huggingface.co/roberta-base.
[4] The foreground items are enclosed in rectangular bounding boxes.
[5] https://cloud.google.com/vision/docs/detecting-web.

4.3 Aspect Category Detection (ACD)

Motivated by the work [13], we utilize the multi-modal bitransformer for the current study as it has outperformed a number of other competitive fusion techniques. One of the reasons behind this could be the capability of the multi-modal bitransformer to concurrently apply self-attention to both modalities, resulting in a more fine-granular level multi-modal fusion. Additionally, the multi-modal bitransformer architecture is easily generalizable to an arbitrary number of modalities.

Text and image networks are merged to create a shared representation before the classification layer in the multimodal network. The text network is BERT [34], whereas the image network is ResNet152 [10]. In our study, feature representation of the web entities obtained from RoBERTa model is concatenated with the review text features and passed through the text network. During training, the model simultaneously learns the token embedding spaces of BERT and the image embeddings. For training the model, we use the Adam optimizer with a minibatch size of 32.

4.4 Aspect Category Complaint Detection (ACC)

BiGRU Layer: Subsequently, for each of the instances, representations of the encoded textual, visual, and web-entities, combined with the aspect category outputs obtained from the multimodal bitransformer module are passed through a Bi-directional Gated Recurrent Units (BiGRUs) [5] layers (256 neurons) to obtain hidden states with complementary semantic relations between the modalities. The BiGRU layer retains contextual information from both forward ($\overrightarrow{GRU}$) and backward ($\overleftarrow{GRU}$) time steps and produces a hidden representation (h_i) of each word in the sentence. The shared BiGRU layer's final hidden state matrix is $H \in \mathbb{R}^{n \times 2d_l}$, where $n \in \mathbb{R}^{(n_T + n_I)}$.

Self Attention: We take a similar approach to that outlined in [34], in which the authors proposed that attention should be calculated by mapping a query and a set of key-value pairs to an output. We estimate the output by the weighted sum of values, where weights are assigned by calculating the compatibility function of a query with its corresponding key. We use self-attention over l review texts and k web entities to assign weights to the most relevant ones. The resulting self-attended representations can be written as: $SA_i = softmax(Q_i K_i^T)V_i$, where $i \in \{WebEntities, ReviewText\}$. The attended representation is then passed to a fully-connected layer (100 units).

4.5 Loss Function

For Aspect Category Detection, a multilabel task with multiple correct answers, we use a sigmoid on the logits and train each output class with a binary cross-entropy loss. For the Aspect Category Complaint detection task, the softmax function and categorical-cross entropy loss are utilized.

5 Experiments, Results, and Analysis

In this section, we discuss the experiments, results, and analysis.

5.1 Baselines

We compare the proposed ABCD model against both unimodal baseline and more advanced state-of-the-art multimodal fusion techniques. We discuss each baseline individually:

- **Unimodal baselines:** For the text-only (Text) model, we feed the classifier the initial output of the final layer of a base-uncased RoBERTa model that has been pre-trained. For the image-only model (Image), we classify the output of a standard, pretrained ResNet-152 with average pooling, which yields a 2048-dimensional vector for each image.
- **Text&Image:** For this baseline, the outputs of the Text and Image baselines are concatenated, with the input to the classifier being of size 2048 + 768 dimensions. This baseline is significant as it incorporates the encoder for both modalities, with the classifier having full access to the encoder outputs.
- **ViLBERT:** We employ Vision and Language BERT (ViLBERT) [19], a collaborative methodology for acquiring task-agnostic visual foundation for paired visual-linguistic data. It has distinct visual and linguistic processing streams that interact with one another via co-attentional transformer layers. Co-attentional layers allow cooperation between modalities at different representational levels. It has surpassed the single-stream unified model across various tasks.

5.2 Experimental Setup

We utilized Python-based libraries Pytorch[6] and Tensorflow[7] and Scikit-learn[8] [21]. All our experiments were conducted on a hybrid cluster of multiple GPUs comprised of GTX 1080Ti. We report the macro-F1 and micro-F1 scores for aspect category detection results. For the aspect category complaint detection task, we report the macro-F1 score and accuracy results of all the models. We divided the *CESAMARD-Aspect* dataset into 70% training data, 10% for validation, and the rest 20% was used as testing data on all the experimental models. A seed value of 32 was chosen for a fair comparison. We use *ReLU* activation [9] for the dense layer (100 neurons each) to avoid overfitting and apply *dropout* [30] of 0.5, *dropout* ϵ {0.2, 0.3, 0.5} following the dense layer output. We utilize early stopping on micro-F1 for the multilabel aspect category detection task and on validation loss for the binary complaint classification. To fully utilize the GPU, we kept the batch size at 8. The models are optimised using Adam [14], with learning rate, $lr = 3e\text{-}5$ lr ϵ {1e-4, 3e-5, 5e-5}, $\beta_1 = 0.9$ and $\beta_2 = 0.999$.

---

[6] https://pytorch.org.
[7] https://www.tensorflow.org/.
[8] https://scikit-learn.org/stable/.

Table 3. Experimental results in terms of Micro-F1 score and Macro-F1 score for aspect category detection task and for aspect category complaint detection task in terms of Accuracy and Macro-F1 score metrics. M-F1, F1, and A metrics are given in %. Bold-faced values represent the maximum scores achieved. ACD: Aspect Category Detection, ACC: Aspect Category Complaint detection, †: Signifies statistically significant findings.

Domain	Model	ACD		ACC	
		Micro-F1	Macro-F1	Accuracy	Macro-F1
Books	Text	60.45	52.89	73.61	72.19
	Image	31.25	29.97	47.77	45.57
	Text&Image	66.04	60.31	74.78	73.05
	SOTA [26]	62.09	57.88	77.42	76.28
	ViLBERT	71.34	**68.41**	77.84	76.78
	ABCD	**71.54†**	68.18†	**78.96†**	**78.03†**
Edibles	Text	59.08	55.87	74.38	72.02
	Image	33.47	29.87	48.52	47.22
	Text&Image	61.03	57.89	78.95	77.67
	SOTA [26]	63.78	59.98	78.73	78.41
	ViLBERT	65.79	61.05	81.28	79.18
	ABCD	**65.98†**	**62.09†**	**81.94†**	**80.03†**
Electronics	Text	67.45	59.87	77.51	76.76
	Image	35.55	31.89	50.78	49.17
	Text&Image	68.88	65.49	79.48	78.12
	SOTA [26]	69.88	63.56	81.34	78.27
	ViLBERT	71.89	65.87	82.46	80.28
	ABCD	**72.56†**	**68.25†**	**84.57†**	**84.08†**
Fashion	Text	65.56	59.14	76.59	74.77
	Image	32.43	30.12	46.56	44.06
	Text&Image	66.45	61.51	78.08	77.62
	SOTA [26]	65.78	59.08	81.23	80.04
	ViLBERT	70.48	65.67	83.37	82.07
	ABCD	**70.84†**	**69.32†**	**84.27†**	**83.25†**

5.3 Results and Discussion

Table 3 shows the outcomes of the proposed framework and all the baselines discussed in the previous section. The table clearly indicates that the proposed framework surpasses all existing baseline models (in unimodal as well as multimodal setup). In comparison to all unimodal and multimodal baselines, the proposed network vastly enhances performance for both ACD and ACC tasks. The efficiency of unimodal baselines does not match that of the proposed framework. This could be credited to the supplemental information provided by the images. Additionally, the knowledge of web entities aids the performance of the proposed ABCD model. We also observe that for the text-only model, the highest attained results for ACD task are 67.45% and 59.87% on micro-F1 and macro-F1 metrics, respectively; this implies that textual information plays a crucial role in correctly identifying the fine-grained aspect categories[9]. It is also worth noting the domain-wise best performance attained by the proposed network is for

---

[9] Kindly note we do not report the results for miscellaneous domain as it consists of 40 instances, which is insufficient for training a deep learning model.

the Electronics domain, which has the highest number of instances (38%) in the CESAMARD-Aspect dataset. All of the results presented here are statistically significant[10] [35].

Comparison with State-of-the-art Technique (SOTA): We also compare our proposed approach with the existing state-of-the-art technique for complaint detection task, as we are unaware of any existing multi-modal aspect-based complaint detection model. The SOTA technique [26] (SOTA), uses an attention-based adversarial multi-task deep neural network framework for complaint detection in multimodal scenario. We re-implement the SOTA model for the aspect-based complaint detection task, keeping the experimental setup the same as our current work. It is evident from Table 3 that the proposed model, ABCD is able to outperform the scores of SOTA model for both the sub-tasks (ACD, ACC).

5.4   Qualitative Analysis

We discovered that correct classifications are biased towards the *Non-Complaint* class since it has a considerably higher number of instances in the dataset.

Table 4 shows a qualitative analysis of the predictions made by the proposed model in comparison to other baselines. The table also illustrates that integrating text and images with web entity information enhanced the predictions of the ABCD model over the next best-performing baseline, ViLBERT, which did not include all of these features.

From the predictions for sentence 1, the user expresses no explicit declaration of disapproval. Still, the visual modality and the web entity information in combination help the proposed multimodal model in correctly predicting the aspect categories and the associated labels. Whereas in sentence 2, even though the ViLBERT model correctly identifies one of the aspect categories and label pair, only the proposed model is able to strongly correlate the visual modality and web entity information with the assertion of wrongdoing by the seller.

5.5   Error Analysis

Here we discuss some of the reasons the proposed model falters while classifying the aspect categories and the associated labels:

Incomprehensible Images: In a few of the reviews, the user-uploaded images are obscure and arbitrary, based on which the extracted web entities are generic,

---

[10] The results are found to be statistically significant when testing the null hypothesis (p-value < 0.05).

Table 4. Qualitative study of the predictions from the proposed model and few other baselines. Bold-faced labels indicate the true labels of the task. W.E: Web Entities, Com: Complaint; Non Com: Non-Complaint

Review	Image	ABCD	ViLBERT
Dull Color, Average Product. Quality is fine. W.E: Jeans, Textile, Denim		**Colour-Com** **Quality-Non Com**	Style-Com
Pathetic product quality, rotating plastic blade broken. It's overpriced item. W.E: Cookware, Mixer, Plastic		**Quality-Com** **Price-Com**	**Quality-Com** Design-Non Com

thereby providing little to no assistance to the model. As a result, some instances rely solely on the text portion of the review, which may not always contain sufficient information for accurate classification.

Misclassified Aspect Category: The model misclassifies the aspect categories and, as a result, the associated complaint/non-complaint labels in the electronics and fashion domains, which have a higher number of aspect categories and fewer training samples per aspect class. For example, *'Its good but not very good product. Its working well for any pc, but it's wire is not durable its very thin.'*. The ABCD model predicts 'hardware-complaint', but the actual aspect category and label pair is 'quality-complaint'. One of the possible reasons for misclassification could be the less number of training samples for the 'quality' aspect in the electronics domain.

Superficial Feedback: Such instances where the overall tone is neutral, but the instance is a complaint, the proposed model wrongly classifies such instances as non-complaint. For example, *'Received in this condition. All the red ants were enjoying their feast.'*. For the given sentence, the actual class is 'complaint' but the models predict it as 'non-complaint', which could be due to the absence of explicit terms to express dissatisfaction.

6   Conclusion and Future Work

In this paper, we have proposed a benchmark setup for aspect-based complaint detection. ABCD identifies the aspects in the given review about a product and also finds if the aspect mentioned in the review signifies a complaint or non-complaint. We extended the recently released multi-modal complaint dataset by annotating 3962 review instances across 5 domains. Based on this extended dataset we proposed a deep-learning-based multi-modal framework, where the first stage predicts what the targeted aspects are, and the second stage categorizes whether the targeted aspect is associated with a complaint or not. Based on our present study, we believe that automatically detecting complaints cannot be viewed solely as a coarse-grained classification task rather it should be analyzed at the fine-grained aspect-level. To the best of our knowledge, this is the very first attempt at solving these two specific problems (ACD, ACC) together, in a multi-modal setup.

In the future, we would like to work on finding rationales for complaints at the aspect-level, in addition to working on aspect-based complaint detection in code-mixed scenarios.

Acknowledgement. This publication is an outcome of the R&D work undertaken in the project under the Visvesvaraya Ph.D. Scheme of Ministry of Electronics & Information Technology, Government of India, being implemented by Digital India Corporation (Formerly Media Lab Asia).

References

1. Akhtar, M.S., Ekbal, A., Bhattacharyya, P.: Aspect based sentiment analysis: category detection and sentiment classification for hindi. In: Gelbukh, A. (ed.) CICLing 2016. LNCS, vol. 9624, pp. 246–257. Springer, Cham (2018). https://doi.org/10.1007/978-3-319-75487-1_19
2. Artstein, R., Poesio, M.: Inter-coder agreement for computational linguistics. Comput. Linguist. **34**(4), 555–596 (2008)
3. Bhat, S., Culotta, A.: Identifying leading indicators of product recalls from online reviews using positive unlabeled learning and domain adaptation. In: Proceedings of the International AAAI Conference on Web and Social Media, vol. 11 (2017)
4. Cai, Y., Cai, H., Wan, X.: Multi-modal sarcasm detection in twitter with hierarchical fusion model. In: Proceedings of the 57th Annual Meeting of the Association for Computational Linguistics, pp. 2506–2515 (2019)
5. Cho, K., van Merrienboer, B., Bahdanau, D., Bengio, Y.: On the properties of neural machine translation: encoder-decoder approaches. In: Wu, D., Carpuat, M., Carreras, X., Vecchi, E.M. (eds.) Proceedings of SSST@EMNLP 2014, Eighth Workshop on Syntax, Semantics and Structure in Statistical Translation, Doha, Qatar, 25 October 2014, pp. 103–111. Association for Computational Linguistics (2014). https://doi.org/10.3115/v1/W14-4012, https://www.aclweb.org/anthology/W14-4012/
6. Coussement, K., Van den Poel, D.: Improving customer complaint management by automatic email classification using linguistic style features as predictors. Decis. Supp. Syst. **44**(4), 870–882 (2008)
7. Devlin, J., Chang, M., Lee, K., Toutanova, K.: BERT: pre-training of deep bidirectional transformers for language understanding. In: Burstein, J., Doran, C., Solorio, T. (eds.) Proceedings of the 2019 Conference of the North American Chapter of the Association for Computational Linguistics: Human Language Technologies, NAACL-HLT 2019, Minneapolis, MN, USA, 2–7 June 2019, vol. 1 (Long and Short

Papers), pp. 4171–4186. Association for Computational Linguistics (2019). https://doi.org/10.18653/v1/n19-1423

8. Fleiss, J.L.: Measuring nominal scale agreement among many raters. Psychol. Bull. **76**(5), 378 (1971)

9. Glorot, X., Bordes, A., Bengio, Y.: Deep sparse rectifier neural networks. In: Proceedings of the Fourteenth International Conference on Artificial Intelligence and Statistics, pp. 315–323 (2011)

10. He, K., Zhang, X., Ren, S., Sun, J.: Deep residual learning for image recognition. In: Proceedings of the IEEE Conference on Computer Vision and Pattern Recognition, pp. 770–778 (2016)

11. Jin, M., Aletras, N.: Complaint identification in social media with transformer networks. In: Scott, D., Bel, N., Zong, C. (eds.) Proceedings of the 28th International Conference on Computational Linguistics, COLING 2020, Barcelona, Spain (Online), 8–13 December 2020, pp. 1765–1771. International Committee on Computational Linguistics (2020). https://doi.org/10.18653/v1/2020.coling-main.157, https://doi.org/10.18653/v1/2020.coling-main.157

12. Jin, M., Aletras, N.: Modeling the severity of complaints in social media. In: Toutanova, K., et al. (eds.) Proceedings of the 2021 Conference of the North American Chapter of the Association for Computational Linguistics: Human Language Technologies, NAACL-HLT 2021, 6–11 June 2021, pp. 2264–2274. Association for Computational Linguistics (2021). https://doi.org/10.18653/v1/2021.naacl-main.180

13. Kiela, D., Bhooshan, S., Firooz, H., Perez, E., Testuggine, D.: Supervised multimodal bitransformers for classifying images and text. arXiv preprint arXiv:1909.02950 (2019)

14. Kingma, D.P., Ba, J.: Adam: a method for stochastic optimization. arXiv preprint arXiv:1412.6980 (2014)

15. Lailiyah, M., Sumpeno, S., Purnama, I.E.: Sentiment analysis of public complaints using lexical resources between Indonesian sentiment lexicon and sentiwordnet. In: 2017 International Seminar on Intelligent Technology and Its Applications (ISITIA), pp. 307–312. IEEE (2017)

16. Liao, W., Zeng, B., Yin, X., Wei, P.: An improved aspect-category sentiment analysis model for text sentiment analysis based on roberta. Appl. Intell. **51**(6), 3522–3533 (2021)

17. Liu, Y., Liu, H., Wong, L.-P., Lee, L.-K., Zhang, H., Hao, T.: A hybrid neural network RBERT-C based on pre-trained RoBERTa and CNN for user intent classification. In: Zhang, H., Zhang, Z., Wu, Z., Hao, T. (eds.) NCAA 2020. CCIS, vol. 1265, pp. 306–319. Springer, Singapore (2020). https://doi.org/10.1007/978-981-15-7670-6_26

18. Liu, Z., Lin, W., Shi, Y., Zhao, J.: A robustly optimized BERT pre-training approach with post-training. In: Li, S., et al. (eds.) CCL 2021. LNCS (LNAI), vol. 12869, pp. 471–484. Springer, Cham (2021). https://doi.org/10.1007/978-3-030-84186-7_31

19. Lu, J., Batra, D., Parikh, D., Lee, S.: Vilbert: pretraining task-agnostic visiolinguistic representations for vision-and-language tasks. Adv. Neural Inf. Process. Syst. **32**, 1–11 (2019)

20. Olshtain, E., Weinbach, L.: Complaints: a study of speech act behavior among native and nonnative speakers of hebrew. the prag-matic perspective (1985)

21. Pedregosa, F., et al.: Scikit-learn: machine learning in python. J. Mach. Learn. Res. **12**(Oct), 2825–2830 (2011)

22. Poria, S., Hazarika, D., Majumder, N., Naik, G., Cambria, E., Mihalcea, R.: Meld: a multimodal multi-party dataset for emotion recognition in conversations. arXiv preprint arXiv:1810.02508 (2018)
23. Preotiuc-Pietro, D., Gaman, M., Aletras, N.: Automatically identifying complaints in social media. In: Korhonen, A., Traum, D.R., Màrquez, L. (eds.) Proceedings of the 57th Conference of the Association for Computational Linguistics, ACL 2019, Florence, Italy, 28 July–2 August 2019, vol. 1: Long Papers, pp. 5008–5019. Association for Computational Linguistics (2019). https://doi.org/10.18653/v1/p19-1495
24. Saha, T., Patra, A.P., Saha, S., Bhattacharyya, P.: Towards emotion-aided multimodal dialogue act classification. In: Jurafsky, D., Chai, J., Schluter, N., Tetreault, J.R. (eds.) Proceedings of the 58th Annual Meeting of the Association for Computational Linguistics, ACL 2020, 5–10 July 2020, pp. 4361–4372. Association for Computational Linguistics (2020). https://doi.org/10.18653/v1/2020.acl-main.402
25. Saha, T., Upadhyaya, A., Saha, S., Bhattacharyya, P.: Towards sentiment and emotion aided multi-modal speech act classification in twitter. In: Proceedings of the 2021 Conference of the North American Chapter of the Association for Computational Linguistics: Human Language Technologies, pp. 5727–5737 (2021)
26. Singh, A., Dey, S., Singha, A., Saha, S.: Sentiment and emotion-aware multi-modal complaint identification. In: Thirty-Sixth AAAI Conference on Artificial Intelligence, AAAI 2022, Thirty-Fourth Conference on Innovative Applications of Artificial Intelligence, IAAI 2022, The Twelveth Symposium on Educational Advances in Artificial Intelligence, EAAI 2022 Virtual Event, 22 February–1 March 2022, pp. 12163–12171. AAAI Press (2022). https://ojs.aaai.org/index.php/AAAI/article/view/21476
27. Singh, A., Nazir, A., Saha, S.: Adversarial multi-task model for emotion, sentiment, and sarcasm aided complaint detection. In: Hagen, M., Verberne, S., Macdonald, C., Seifert, C., Balog, K., Nørvåg, K., Setty, V. (eds.) ECIR 2022. LNCS, vol. 13185, pp. 428–442. Springer, Cham (2022). https://doi.org/10.1007/978-3-030-99736-6_29
28. Singh, A., Saha, S.: Are you really complaining? a multi-task framework for complaint identification, emotion, and sentiment classification. In: Lladós, J., Lopresti, D., Uchida, S. (eds.) ICDAR 2021. LNCS, vol. 12822, pp. 715–731. Springer, Cham (2021). https://doi.org/10.1007/978-3-030-86331-9_46
29. Singh, A., Saha, S., Hasanuzzaman, M., Dey, K.: Multitask learning for complaint identification and sentiment analysis. Cogn. Comput., 1–16 (2021)
30. Srivastava, N., Hinton, G., Krizhevsky, A., Sutskever, I., Salakhutdinov, R.: Dropout: a simple way to prevent neural networks from overfitting. J. Mach. Learn. Res. **15**(1), 1929–1958 (2014)
31. Trosborg, A.: Interlanguage Pragmatics: Requests, Complaints, and Apologies, vol. 7. Walter de Gruyter (2011)
32. Vásquez, C.: Complaints online: the case of tripadvisor. J. Pragmat. **43**(6), 1707–1717 (2011)
33. Vaswani, A., et al.: Attention is all you need. CoRR abs/1706.03762 (2017). http://arxiv.org/abs/1706.03762
34. Vaswani, A., et al.: Attention is all you need. arXiv preprint arXiv:1706.03762 (2017)

35. Welch, B.L.: The generalization of 'student's'problem when several different population varlances are involved. Biometrika **34**(1–2), 28–35 (1947)
36. Yang, W., et al.: Detecting customer complaint escalation with recurrent neural networks and manually-engineered features. In: Loukina, A., Morales, M., Kumar, R. (eds.) Proceedings of the 2019 Conference of the North American Chapter of the Association for Computational Linguistics: Human Language Technologies, NAACL-HLT 2019, Minneapolis, MN, USA, 2–7 June 2019, vol. 2 (Industry Papers), pp. 56–63. Association for Computational Linguistics (2019). https://doi. org/10.18653/v1/n19-2008

What Is Your Cause for Concern? Towards Interpretable Complaint Cause Analysis

Apoorva Singh[1]([✉])[iD], Prince Jha[1][iD], Rohan Bhatia[2], and Sriparna Saha[1][iD]

[1] Indian Institute of Technology Patna, Bihta, India
{apoorva_1921cs19,princekumar_1901cs42,sriparna}@iitp.ac.in
[2] Delhi Technological University, Delhi, India
rohanbhatia_2k19ep079@dtu.ac.in

Abstract. The abundance of information available on social media and the regularity with which complaints are posted online emphasizes the need for automated complaint analysis tools. Prior study has focused chiefly on complaint identification and complaint severity prediction: the former attempts to classify a piece of content as either complaint or non-complaint. The latter seeks to group complaints into various severity classes depending on the threat level that the complainant is prepared to accept. The complainant's goal could be to express disapproval, seek compensation, or both. As a result, the complaint detection model should be interpretable or explainable. Recognizing the cause of a complaint in the text is a crucial yet untapped area of natural language processing research. We propose an interpretable complaint cause analysis model that is grounded on a dyadic attention mechanism. The model jointly learns complaint classification, emotion recognition, and polarity classification as the first sub-problem. Subsequently, the complaint cause extraction and the associated severity level prediction as the second sub-problem. We add the causal span annotation for the existing complaint classes in a publicly available complaint dataset to accomplish this. The results indicate that existing computational tools can be repurposed to tackle highly relevant novel tasks, thereby finding new research opportunities (Resources available at: https://bit.ly/Complaintcauseanalysis).

Keywords: Complaint detection · Cause analysis · Severity level classification · Sentiment analysis · Emotion recognition · Explainable AI · Multi-task learning · Deep learning

1 Introduction

Complaining is a speech act wherein negative emotions are expressed in consequence of a contradiction between actuality and expectations [17]. In linguistics, complaints have been classified into fine-grained severity levels, the extent of forewarning that the complainant is willing to accept, as well as the complainant's motivation [10, 26]. Identifying complaints and corresponding severity levels in everyday language is critical for downstream software developers, such as customer relations virtual assistants [31] and corporate entities, who seek to improve their customer support skills by recognizing and resolving complaints [27].

A. Singh and P. Jha—Denotes equal contribution.

© The Author(s), under exclusive license to Springer Nature Switzerland AG 2023
J. Kamps et al. (Eds.): ECIR 2023, LNCS 13981, pp. 141–155, 2023.
https://doi.org/10.1007/978-3-031-28238-6_10

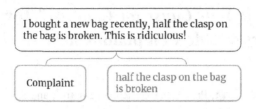

Fig. 1. Example of complain cause analysis

The emotional state of an entity has a substantial influence on the intended content in any form of communication [13]. When coupled with polarity knowledge, emotion offers a more in-depth insight of the customer's cognition. A statement, could entail only negative sentiment, whereas emotion could be sadness or even disgust. Earlier studies on multitask complaint analysis [23], encourage us to integrate polarity and emotion information when studying complaints.

Table 1. Comparative study on the extension of *Complaints* dataset.

Dataset	Labels				
	Complaint	Severity	Sentiment	Emotion	**Cause**
Complaints [21]	✓	✗	✗	✗	✗
Complaints [8]	✓	✓	✗	✗	✗
Complaints-ESS [23]	✓	✓	✓	✓	✗
Com_Cause (current work)	✓	✓	✓	✓	✓

Recognizing the causal span of the expressed emotions is a fundamental research development in automatic reasoning about human emotions [20], [4]. Drawing cues from these related studies, we aim to determine the cause (reason or stimuli) of the opinion (complaints/non-complaints) expressed on the social web. Specifically, we look for the rationale or experience for the evoked complaint in the discourse act. Figure 1 provides an example of a complaint instance about a handbag; the complaint detection model requires not only recognizing the voiced concern but also identifying the cause-in this case, "*half the clasp on the bag is broken*".

Previous works [8] categorized complaints into different severity classes which merely recognize the level of dissatisfaction of the complainee, whereas identifying the cause of the complaint gives insight into the source or reason for an event or action that produced the complaint as a result. Furthermore, determining the cause of the complaint presents additional obstacles. For example, (i) data from media platforms are fragmented text excerpts with a word limit and random acronyms, and (ii) a sentence indicative of a complaint may not always contain the associated cause in it and it could be present somewhere else in the review text. Our study tries to make headway in this direction by identifying the source of the complaint using social media data. The existing complaint corpora are invaluable for training complaint detection models, yet they consist of only binary complaint labels and severity levels. Such frameworks must be interpretable and investigated further to demonstrate their trustworthiness and applicability. Our work's major contributions are as follows:

- This is the first study on explainable complaint identification where the focus is to identify the rationale/causes which are responsible for categorizing social media data having complaints.
- We extend a publicly available dataset with manual annotation of causal spans for complaint/non-complaint labeled sentences.
- We establish a benchmark framework for complaint cause recognition, emphasizing cause detection and extraction.
- We propose a hierarchical attention-based framework that jointly learns (a) binary complaint classification, (b) emotion recognition, and (c) polarity classification as the first sub-problem, and the second sub-problem involves (d) cause extraction and (e) severity level classification. The main tasks in the joint-learning frameworks are Complaint Identification (CI), Severity Classification (SC), and Cause Extraction (CE), whereas, Polarity Classification (PC) and Emotion Recognition (ER) are the supporting tasks.

2   Related Studies

Earlier works on complaint detection majorly focused on distinguishing complaints from non-complaints using hand-crafted feature-based machine learning models [21], [1], transformer network-based deep learning models [7], and a few multitask models that leveraged polarity and affect information for the complaint mining task [23,24]. The authors in [21] employed logistic regression with a diverse set of features to identify complaints on Twitter. Quite recently, [7] investigated an array of transformer-based models for complaint detection from social media. In pragmatics theory, [26] introduced four basic complaint severity levels: (a) no explicit reproach, (b) disapproval, (c) accusation, and (d) blame. In computational linguistics, [8] evaluated the severity level of complaints by training several transformer-based networks paired with linguistic information to predict severity levels in complaints.

The extraction of certain stimuli behind an emotion in an utterance is known as emotion cause extraction (ECE). The authors of the work [12] was the first to investigate the task of emotion cause extraction using rule-based approaches. The emotion cause pair extraction (ECPE) model as a variation to the ECE task proposed in [30], recognizes both emotions and their accompanying causes. Furthermore, [20] recently extended ECE work in conversations to improve the explainability of emotion-aware models.

Despite numerous studies on cause detection and extraction in the related domain, no such study on complaint cause detection and extraction utilizing computational techniques exist. We introduce an interpretable complaint cause analysis model grounded on a dyadic attention mechanism at the word and sentence levels, allowing it to pay attention to more and less important information in differing respects while exploiting the multi-task model to tackle this challenging problem. We consider a Twitter-based complaint corpus for this work and enrich it with the causal span of complaint/non-complaint.

Table 2. Example instances of causal span annotation. Label: Class labels for complaint, sentiment and emotion tasks

Tweet text	Cause
$$ incredibly beautiful day, No wind at all and the power is out. Fix it fast **Label:** *complaint, sadness, negative*	The power is out. Fix it fast
Just had a fantastic customer service from $<USER>$ - Cameron in claims was so helpful and understanding **Label:** *non-complaint, happiness, positive*	had a fantastic customer service
$<USER>$ Any explanation on the dozens of other purple base cards yet? **Label:** *complaint, other, neutral*	no cause

3   Corpus Extension

For this work, we utilize the *Complaints* dataset[1] published in [21], which includes 2,214 non-complaints and 1,235 complaints in English. We selected this dataset because it is openly available and comprises annotated complaints from the social-networking site Twitter, a popular choice for data analysis. In the work [8], the *Complaints* dataset has been augmented with five severity levels (*no explicit reproach, disapproval, accusation, blame, and non-complaints*). Recently, [23] enriched the *Complaints* dataset with the sentiment (*negative, neutral, positive*) and emotion (*anger, disgust, fear, happiness, sadness, surprise and other*) classes, the 'other' emotion class depicts tweets that do not fall under the scope of Ekman's six basic emotions [3]. Our work focuses on strengthening the available benchmark setup by extending the Complaints dataset with the manual annotation of causal spans for complaint/non-complaint labeled sentences to provide scope for multi-faceted research. Hence, we utilize this extended dataset annotated with severity levels, sentiment, and emotion classes for our current work. Table 1 provides a comparative study on the diversification of *Complaints* dataset in different works.

3.1   Cause Extraction Method

Task Definition. The study's primary objective is to define the concept of causal span detection and extraction, specifically for complaint causes.

Complaint Cause is a portion of the text that expresses why the user feels compelled to file a complaint. It is the speech act used by the individual to describe the circumstances in which their expectations have been violated.

---

[1] https://github.com/danielpreotiuc/complaints-social-media.

Annotations. Three annotators (one doctoral and two undergraduate students of computer science discipline) with adequate domain knowledge and expertise in developing supervised corpora were entrusted with annotating the causal span identification task for each sample in the dataset.

Annotators were directed to identify the causal span, X(I), that appropriately represented the basis of the complaint (C) or non-complaint (NC) for each instance (I) in the *Complaints* dataset. If there was no explicit X(I) for C/NC in I, the annotators marked the sentence as 'no cause'. Note that 94% of the cases in the dataset have only one complaint/non-complaint cause, while instances with no cause are 5%. We employ the macro-F1 [22] metric to assess inter-annotator agreement based on earlier studies on span extraction and obtain a 0.77 F1 score, indicating that the annotations are of decent quality. The extended dataset (*Com_Cause*) used in this work consists of the tweet text, domain, complaint label, severity level, sentiment and emotion label, and the corresponding annotated causal span for each record. Table 2 shows few example instances of causal span annotation.

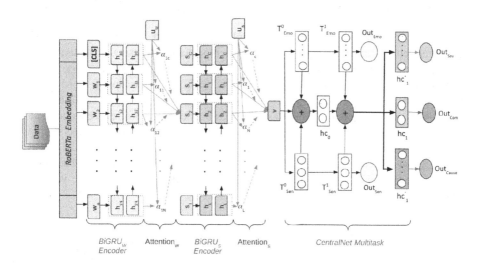

Fig. 2. The architectural diagram of proposed hierarchical attention multitask framework. W: word-level, S: sentence-level

4 Methodology

This section outlines the tasks at hand before delving into the details of the proposed architecture. The overall framework is shown in Fig. 2.

4.1 Problem Definition

We approach the task of extracting the cause and the associated severity classification as two sub-problems: (a) complaint identification using a multitask architecture where polarity and emotion classification are supplementary tasks; (b) cause extraction and severity prediction.

We define the cause extraction task for any instance as follows: Given a text t with a complaint label x, sentiment label s, emotion label e, and severity level m, determine the causal span $c(t)$ in t that is relevant to the label x.

4.2   Text Features

Bidirectional Encoder Representations from Transformers (BERT) is a multi-layer bidirectional transformer encoder [2] based on the primary work described in [28]. RoBERTa [16] is a modification of BERT that has demonstrated improved performance on social media analytics [14,15] via training on a larger number of datasets with changing hyperparameters. To embed the words of each tweet instance, we employ the *RoBERTa-base-uncased* model.

4.3   Hierarchical Attention Network (HAN)

To make the representation of data samples context-rich, we expand our model to a hierarchical one; we compute sentence representation using a sentence-level encoder. The input is a collection of text instances, $T = [S_1, S_2, S_3,....., S_n]$ from the source data, where a given sentence, S_i, is made up of word tokens $[W_{l-k+1}, W_{l-k+2}, W_{l-k+3},...,W_l]$. These word tokens from the entire tweet are considered as a single sequential input to a Bi-GRU encoder as a word-level encoder. Let's say for each time step, output of the input word token w_t is h_t^w (we use superscript w to indicate word level and s for sentence level). Since not all words contribute equally to the sentence meaning, word-level attention (WLA) has been introduced. The aggregate representation of those words with attention is used to form a sentence vector as follows:

$$h_{s_i}^w = \sum_{t=1}^{k} a_{l-k+t}^w h_{l-k+t}^w \tag{1}$$

Thus, we obtain word level sentence embedding of the instance, $h_T^w = [h_{s_1}^w, h_{s_2}^w, h_{s_3}^w,...,h_{s_n}^w]$. It is then fed into Bi-GRU as a sentence-level encoder. Subsequently, we obtain final sentence level embedding $h_T^s = [h_{s_1}^s, h_{s_2}^s, h_{s_3}^s,...,h_{s_n}^s]$ which is aggregated with attention to obtain a tweet representation as follows:

$$h_s^T = \sum_{t=1}^{n} a_{s_t}^s h_{s_t}^s \tag{2}$$

We feed these representations of tweets to the CentralNet Multitask module in order to investigate the performance of complaint, severity, and cause extraction with the help of sentiment and emotion information in multitask settings.

4.4 Multitask Framework

CentralNet [6] is a multimodal data fusion framework. We modify the CentralNet framework into a multitask architecture as part of our work. CentralNet multitask is a neural network paradigm in which task-specific networks are divided into n separate networks with one central network. Emotion and sentiment are the task-specific layers in this case. The central network is a standard layer for identifying complaints, predicting severity levels, and extracting causes. The central network considers the weighted summation of task-specific networks and their preceding layers to aggregate the features acquired from various single tasks (sentiment, emotion). The multitask layers can be defined as follows:

$$MTL_{i+1} = \alpha n MTL_i + \sum_{k=1}^{n} \alpha s_i^k ST_i^k \tag{3}$$

where n is the number of task-specific networks, αs are scalar trainable weights, ST_i^k is the hidden representation of kth task-specific network at i^{th} layer, and MTL_i is the central hidden representation of the main task. MT_{i+1} is the final layer, which is fed to an operating layer (dense layer accompanied by an activation function).

4.5 Post-processing for CE Task

The post-processing of CE task is done based on the 0/1 label predicted by our model corresponding to each word token where $0 \in non - complaint$ and $1 \in complaint$. After that $label \in \{0, 1\}$, will be mapped corresponding to each word token in a given tweet text. This way the beginning and end of the casual span can be directly decoded from the token length in a sentence. For example:
The/[0] $< USER >$/[0] this/[1] stinks/[1], 10mins/[1] to/[1] take/[1] my/[1] order/[1] and/[1] another/[1] 15/[1] to/[1] get/[1] it/[1]. And/[0] stop/[0] asking/[0] my/[0] name/[0] like/[0] we're/[0] friends/[0].

4.6 Loss Function

- The categorical-cross entropy (J_{CE}) losses are calculated for the complaint (CI), emotion (ER), polarity (PC), and severity (SC) tasks. The integrated loss function ($J(\theta)$) is realized as follows:

$$J(\theta) = J_{CE}^{CI}(\theta) + J_{CE}^{ER}(\theta) + J_{CE}^{PC}(\theta) + J_{CE}^{SC}(\theta) \tag{4}$$

All the model parameters to be optimized are denoted by θ.
- We use Binary Cross-entropy to evaluate the loss in the cause extraction task, where each instance may belong to more than one class (J_{BE}). It enables the model to determine whether or not a given instance belongs to a specific class.
- The combined loss function (L) of our proposed framework is:

$$L = J(\theta) + J_{BE} \tag{5}$$

4.7 Model Details

This section provides the details of the sentence-level and word-level attention sub-modules of the proposed model we used to evaluate this dataset.

HAN + CentralNet. We pass the features obtained from RoBERTa to the HAN module, which generates attention-rich word level and sentence level encodings of the tweet instances. The context-rich instances are passed to the emotion and sentiment-aided CentralNet multitask module for complaint detection, severity level prediction, and cause extraction.

WLA + CentralNet. We also develop another related framework where we pass the RoBERTa features only through WLA as described in Sect. 4.3. The obtained hidden representation is further fed to emotion and sentiment-aided CentralNet multitask for prediction of complaint and severity level and cause extraction.

5 Experiments, Results, and Analysis

The findings of several variants of our proposed model tested on the *Complaints* corpus are shown in this section.

5.1 Baselines

- **Multitask systems:** Based on recent work in CI in multitasking framework, we develop Baseline$_1$ [23] model as a multitask baseline. Baseline$_1$ is a BERT-based adversarial multitask model for complaint, emotion, sentiment, and sarcasm classification. We re-implement the Baseline$_1$ model for the joint learning of CI, SC, and CE with PC and ER as additional tasks, keeping the experimental setup the same as our current work. At the final stage for the binary and multi-class prediction, we add four separate output layers with softmax activation function for each CI, SC, ER, and PC task and a sigmoid function for the CE task.
- **Baselines for Cause Extraction Task:** As the cause extraction is a novel task in the domain of complaint analysis, we took motivation from the works of [20] and [4] in the emotion recognition domain and utilized pre-trained SpanBERT base model that is fine-tuned on SQuAD 2.0 [22] dataset. The SpanBERT model is specifically used for the CE task.
- **Ablation models:** We implement the proposed model as both HAN + CentralNet model and WLA + CentralNet model. Additionally, to understand the impact of emotion and sentiment classification individually on the complaint identification (CI), severity classification (SC), and cause extraction (CE) tasks, we develop multiple ablation models keeping CI, SC, and CE tasks fixed. The architecture is similar to the proposed system in other aspects.

Table 3. Results of all the ablation studies performed on the proposed models for CI and SC tasks in terms of macro-F1 score (F1) and Accuracy (A) values. JS: Jaccard Similarity, HD: Hamming distance, and ROS: Ratcliff-Obershelp Similarity. All the metrics are given in %. Bold-faced values represent the maximum scores achieved. The † denotes statistically significant findings.

Model	Complaint (CI)		Severity (SC)		Cause (CE)		
	F1	A	F1	A	HD	JS	ROS
SOTA [8]	86.6 ± .03	87.6 ± .03	59.4 ± .03	55.5 ± .02	–	–	–
HAN+CentralNet$_{CI+SC+CE}$	84.08 ± 0.04	84.35 ± 0.03	63.23 ± 0.02	67.23 ± 0.02	71.87	86.24	92.64
HAN+CentralNet$_{CI+SC+CE+PC}$	84.87 ± 0.02	84.93 ± 0.02	65.25 ± 0.02	68.99 ± 0.02	72.23	86.5	93.14
HAN+CentralNet$_{CI+SC+CE+ER}$	84.52 ± 0.04	84.80 ± 0.03	63.94 ± 0.02	68.85 ± 0.02	72.4	86.62	92.88
HAN+CentralNet$_{All}$	85.7 ± 0.02	85.8 ± 0.02	65.02 ± 0.01	70.12 ± 0.02	**72.87**	**86.75**	**93.29**
WLA+CentralNet$_{CI+SC+CE}$	86.81 ± 0.03	86.96 ± 0.03	64.63 ± 0.02	70.14 ± 0.02	61.96	72.94	83.12
WLA+CentralNet$_{CI+SC+CE+PC}$	88.05 ± 0.02	88.08 ± 0.02	66.66 ± 0.01	70.93 ± 0.01	62.36	73.21	83.54
WLA+CentralNet$_{CI+SC+CE+ER}$	87.09 ± 0.02	87.25 ± 0.01	65.18 ± 0.02	70.14 ± 0.02	62.31	73.26	83.62
WLA+CentralNet$_{All}$	**88.11**† ± 0.02	**88.12**† ± 0.02	**66.93**† ± 0.03	**71.22**† ± 0.02	62.76	73.54	83.73
Baseline$_1$	81.41 ± 0.14	82.84 ± 0.03	61.38 ± 0.05	62.87 ± 0.13	61.53	72.28	81.67
SpanBERT [9]	–	–	–	–	62.18	72.51	82.66

5.2 Experimental Setup

We implemented our proposed framework and all the baselines on Python-based libraries, PyTorch[2] [18], Tensorflow[3] and Scikit-learn[4] [19]. All our experiments were conducted on a hybrid cluster of multiple GPUs comprised of RTX 2080Ti. We keep our training setup similar to that of [8]. All the models are executed using a nested 10-fold cross-validation approach, which comprises two nested loops. In the outer loop, 9-folds are used for training and one for testing, whereas in the inner loop, data from the nine folds (from the outer loop) is utilized for 3-fold cross-validation, with 2-folds used for training and one fold for validation. We select the model with the smallest validation loss over 30 epochs during training. We set the maximum length of the input sequence as 20. We train our model with batch size of 32 using Adam optimizer [11] with learning rate of $lr = 1e-4$, $lr \in \{1e-1, 1e-2, 1e-3, 1e-4\}$. We utilize $ReLU$ [5] activation to prevent overfitting and a $dropout$ [25] of 30%, $dropout \in \{20\%, 30\%, 50\%\}$ following the dense layers. $Softmax$ activation with 2, 5, 7, and 3 neurons are used for the output layers for complaint, severity, emotion, and sentiment classification tasks, respectively. For the cause extraction task, we use the $Sigmoid$ activation function with 20 neurons in the output layer. The pre-trained baseline models (SpanBERT, RoBERTa) used for the cause extraction task are from the open-source repositories[5] huggingface transformer. The mean accuracy and macro-F1 across 10-folds are used to evaluate predictive performance (we also report the standard deviations).

---

[2] https://pytorch.org/.

[3] https://www.tensorflow.org/.

[4] https://scikit-learn.org/stable/.

[5] https://huggingface.co/model and https://tfhub.dev/google/collections/bert/1.

Table 4. Qualitative study of the CI and SC predictions by the SOTA [8] and the proposed (WLA+CentralNet) model. 'Actual Label': true labels for CI and SC tasks, the bold text indicates the causal span annotation of the sentence.

Tweet text	SOTA	Proposed	Actual label
The < *USER* > **this stinks, 10mins to take my order and another 15 to get it**. And stop asking my name like we're friends	complaint blame	complaint disapproval	complaint disapproval
< *USER* > I love this product featured on < *USER* > today **but I cannot find the price?** Help a girl out?	complaint disapproval	complaint no explicit reproach	complaint no explicit reproach

5.3   Results and Discussion

In this work, we define a new task named cause extraction (CE) in interpretable complaint analysis. Furthermore, we intend to improve the performance of SC when jointly learned with CI and CE tasks. Therefore, we state the results and analysis with CI, SC, and CE as key tasks in all the task combinations.

This work introduces two different versions of the proposed architecture, HAN+CentralNet and WLA+CentralNet multitask models. The primary motivation behind this is that not all portions of a text are equally significant for addressing a problem. Identifying which portions are relevant requires modeling the associations of the words instead of their existence in isolation. Table 3 depicts the classification results from the various experiments for the CI, SC, and CE tasks.

As can be observed, the proposed model WLA+CentralNet$_{All}$ outperforms all the other baselines for CI and SC tasks. *Complaints* dataset is a Twitter-based dataset having fixed character constraints, due to which the word-level attention model is able to capture more information in comparison to the sentence-level attention model, which requires more contextual information to perform better. Kindly note the average sentence length of the dataset used is 15. Sample sentences from the dataset, such as *'Thank you', 'I need help'* depict a lack of contextual information. Moreover, WLA+CentralNet$_{CI+SC+CE+PC}$ outperforms WLA+CentralNet$_{CI+SC+CE+ER}$ model. It can be driven by the fact that emotion itself, a fine-grained task, is highly context-dependent and unable to contribute much to the multitask model alone.

For the quantitative assessment of the CE task, we used the Jaccard Similarity (JS), Hamming Distance (HD), and Ratcliff-Obershelp Similarity (ROS) metrics which are based on token distance, edit distance, and sequence distance metrics, respectively. The ROS metric which relies on the longest sequence matching approach is appropriate for the training objective of the CE task and provides a more accurate indication of the framework's effectiveness. It can be observed from Table 3, with a ROS score of 93.29, the HAN+CentralNet$_{All}$ model exceeds all the other models for the cause extraction task. *All of the results are statistically significant*[6] [29].

---

[6] For the significance test, we used the Student's t-test (p-value < 0.04).

Comparison with State-of-the-Art Technique (SOTA): We also compare our proposed approach with the existing state-of-the-art technique [8] for CI and CS tasks. SOTA utilizes an array of neural language models boosted by a pre-trained transformer network with additional linguistic information for the CI task. The best performing SOTA model initially uses a combination of textual and multimodal information and is then fine-tuned for the current dataset, which gives an advantage to the model, unlike in our case where the RoBERTa model is not pre-trained with multimodal information. For a fair comparison, we compare our model with the next best performing model in their work, i.e., the RoBERTa base model. For the SC task, both the proposed multitask models outperform the SOTA technique. This validates the proposed architecture's efficient usage of interactions amongst associated tasks, mainly CI and CE, for severity classification.

Kindly note we have not made any assumption on the length of the span in the proposed approach. Thus the proposed approach can handle spans of any length.

Table 5. Example instances comparing the cause predicted by human annotators and the HAN+CentralNET and WLA+CentralNet Multitask models. The span with bold face was selected by the human annotator and the models to be essential for the prediction. The text in italics indicates tokens relevant to the model but not to the human annotators.

Model	Text	Severity
Human Annotator	$< USER >$ again with their **crappy customer service! Make u wait on hold for hours then ask for u to call** back later. WHAT?!	Accusation
$HAN + CentralNET_{All}$	$< USER >$ again with *their* **crappy customer service! Make u wait on hold for hours then ask for u to call** back later. WHAT?!	Accusation
$WLA + CentralNet_{All}$	$< USER >$ *again with their* **crappy customer service! Make u wait on hold for hours then ask for u to call** back later. WHAT?!	Accusation

5.4 Analysis

Qualitative Analysis: Since the *Non-Complaint* class has a significantly higher number of instances in the dataset, we noticed correct classifications are relatively biased towards it. Tweets including strong complaint indications, such as expression of accusation, or blame-related words, are less misclassified.

The qualitative study of the complaint severity predictions obtained by the SOTA [8] and the best performing WLA + CentralNet multitask system on a few sample test instances are shown in Table 4. The table shows that the CE task combined with CI led to improved predictions than the SOTA system that lacks this element. One of the possible reasons could be the removal of unnecessary information in the cause extraction task, which leads to better prediction of fine-grained severity levels. It can be observed from Table 4 that for both the example instances, both the models correctly predict them as a complaint, but the severity level is correctly predicted only by the proposed model.

We also perform a qualitative analysis for the CE task as shown in Table 5. The first row represents the causal span that human annotators picked as relevant for classification. The spans obtained by the proposed models are shown in the following two rows. The rationale for this varies between models, even though the model makes an accurate prediction (complaint/non-complaint cause).

Error Analysis: In addition, we investigate the errors that the proposed model faces.

- Disproportionate dataset: The *Complaints* dataset's skewed class distribution impacts the proposed model's predictions. The complaint class makes up only 35.8% of the whole dataset, due to which the proposed model is biased towards non-complaint instances.
- Scattered Causes: The proposed model is not able to identify multiple causes spread across a tweet instance. For example, *The display is scratched on the front and left side. Product delivery was quite fast. Packaging was okay. Not able to contact support.* The causal span predicted: *The display is scratched on the front and left side.* In the current work, the causes are annotated based on the first encounter with a strong expression of complaint reason in the tweet. Causes scattered across the complete tweet cannot be identified by the proposed model.
- Concealed Intentions: The model predicts concealed intent sentences inaccurately. When a user voices a complaint without conveying the actual reason, the model classifies it as non-complaint based on the text's literal meaning. For example, $< USER >$ *congratulations. You have reached popular status and the spamming has begun.* The model predicts the example as non-complaint. The correct class is complaint. The complainant's absence of straightforward disapproval or accusation could be one of the reasons behind this.
- Neighbouring Severity Level Errors: Instances that belong to neighboring severity levels tend to be semantically similar, which leads to misclassification. For example, *Real disappointed in $< USER >$ leaving me high and dry. Ordered some new Iowa gear Tues with 1 day shipping and it has not even shipped yet.* The predicted severity level is disapproval. The actual severity level for the given instance is accusation. However, the proposed model misclassifies it owing to the inclusion of negative terms like *disappointed*, which generally appear in the disapproval severity class.

6   Conclusion and Future Work

Current work discusses the task of complaint cause detection from social media data by jointly analyzing complaint cause detection, severity level categorization, polarity, and emotion recognition in a hierarchical attention-based multitask setting. As explainable AI systems help improve trustworthiness and confidence while deploying models in real-time, companies can improve the quality of different products/services by generating rationales behind decisions taken by CI models. To facilitate the research on explainable complaint identification, we enrich the existing open-source complaint dataset (*Com_Cause*) by providing a gold standard cause annotation for each instance. Based on our present study, we believe that automatically detecting complaints cannot be viewed solely as a classification task. The algorithm's role and the significance to be assigned to its prediction must be carefully considered, as the goal is to mitigate the breach of expectation scenario and not only to detect one. On this point, our proposed approach has a considerable advantage due to its interpretability.

Future research will focus on recognizing multiple causes and using multimodal cues in causal span detection and extraction to make the proposed cause analysis approach more explanatory.

Acknowledgement. This publication is an outcome of the R&D work undertaken in the project under the Visvesvaraya Ph.D. Scheme of Ministry of Electronics & Information Technology, Government of India, being implemented by Digital India Corporation (Formerly Media Lab Asia).

References

1. Coussement, K., Van den Poel, D.: Improving customer complaint management by automatic email classification using linguistic style features as predictors. Dec. Support Syst. **44**(4), 870–882 (2008)
2. Devlin, J., Chang, M., Lee, K., Toutanova, K.: BERT: pre-training of deep bidirectional transformers for language understanding. In: Burstein, J., Doran, C., Solorio, T. (eds.) Proceedings of the 2019 Conference of the North American Chapter of the Association for Computational Linguistics: Human Language Technologies, NAACL-HLT 2019, Minneapolis, MN, USA, June 2–7, 2019, Volume 1 (Long and Short Papers). pp. 4171–4186. Association for Computational Linguistics (2019). https://doi.org/10.18653/v1/n19-1423, https://doi.org/10.18653/v1/n19-1423
3. Ekman, P., et al.: Universals and cultural differences in the judgments of facial expressions of emotion. J. Pers. Soc. Psychol. **53**(4), 712 (1987)
4. Ghosh, S., Roy, S., Ekbal, A., Bhattacharyya, P.: CARES: CAuse recognition for emotion in suicide notes. In: Hagen, M., et al. (eds.) ECIR 2022. LNCS, vol. 13186, pp. 128–136. Springer, Cham (2022). https://doi.org/10.1007/978-3-030-99739-7_15
5. Glorot, X., Bordes, A., Bengio, Y.: Deep sparse rectifier neural networks. In: Proceedings of the Fourteenth International Conference on Artificial Intelligence and Statistics, pp. 315–323 (2011)
6. Jha, P., et al.: Combining vision and language representations for patch-based identification of Lexico-semantic relations. In: Proceedings of the 30th ACM International Conference on Multimedia, pp. 4406–4415 (2022)
7. Jin, M., Aletras, N.: Complaint identification in social media with transformer networks. In: Scott, D., Bel, N., Zong, C. (eds.) Proceedings of the 28th International Conference on Computational Linguistics, COLING 2020, Barcelona, Spain (Online), December 8–13, 2020, pp. 1765–1771. International Committee on Computational Linguistics (2020). https://doi.org/10.18653/v1/2020.coling-main.157, https://doi.org/10.18653/v1/2020.coling-main.157
8. Jin, M., Aletras, N.: Modeling the severity of complaints in social media. In: Proceedings of the 2021 Conference of the North American Chapter of the Association for Computational Linguistics: Human Language Technologies, pp. 2264–2274 (2021)

9. Joshi, M., Chen, D., Liu, Y., Weld, D.S., Zettlemoyer, L., Levy, O.: Spanbert: improving pre-training by representing and predicting spans. Trans. Assoc. Comput. Linguist. **8**, 64–77 (2020)
10. Kakolaki, L.N., Shahrokhi, M.: Gender differences in complaint strategies among Iranian upper intermediate EFL students. Stud. English Language Teach. **4**(1), 1–15 (2016)
11. Kingma, D.P., Ba, J.: Adam: a method for stochastic optimization. arXiv preprint arXiv:1412.6980 (2014)
12. Lee, S.Y.M., Chen, Y., Huang, C.R.: A text-driven rule-based system for emotion cause detection. In: Proceedings of the NAACL HLT 2010 Workshop on Computational Approaches to Analysis and Generation of Emotion in Text, pp. 45–53 (2010)
13. Lewis, M., Haviland-Jones, J.M., Barrett, L.F.: Handbook of Emotions. Guilford Press, New York (2010)
14. Liao, W., Zeng, B., Yin, X., Wei, P.: An improved aspect-category sentiment analysis model for text sentiment analysis based on Roberta. Appl. Intell. **51**(6), 3522–3533 (2021)
15. Liu, Y., Liu, H., Wong, L.-P., Lee, L.-K., Zhang, H., Hao, T.: A hybrid neural network RBERT-C based on pre-trained RoBERTa and CNN for user intent classification. In: Zhang, H., Zhang, Z., Wu, Z., Hao, T. (eds.) NCAA 2020. CCIS, vol. 1265, pp. 306–319. Springer, Singapore (2020). https://doi.org/10.1007/978-981-15-7670-6_26
16. Liu, Z., Lin, W., Shi, Y., Zhao, J.: A robustly optimized BERT pre-training approach with post-training. In: Li, S., et al. (eds.) CCL 2021. LNCS (LNAI), vol. 12869, pp. 471–484. Springer, Cham (2021). https://doi.org/10.1007/978-3-030-84186-7_31
17. Olshtain, E., Weinbach, L.: Complaints: a study of speech act behavior among native and nonnative speakers of hebrew. the prag-matic perspective (1985)
18. Paszke, A., et al.: Automatic differentiation in pytorch. In: 31st Conference on Neural Information Processing Systems (2017)
19. Pedregosa, F., et al.: Scikit-learn: machine learning in python. J. Mach. Learn. Res. **12**, 2825–2830 (2011)
20. Poria, S., et al.: Recognizing emotion cause in conversations. Cogn. Comput. **13**(5), 1317–1332 (2021)
21. Preotiuc-Pietro, D., Gaman, M., Aletras, N.: Automatically identifying complaints in social media. arXiv preprint arXiv:1906.03890 (2019)
22. Rajpurkar, P., Zhang, J., Lopyrev, K., Liang, P.: Squad: 100,000+ questions for machine comprehension of text. arXiv preprint arXiv:1606.05250 (2016)
23. Singh, A., Nazir, A., Saha, S.: Adversarial multi-task model for emotion, sentiment, and sarcasm aided complaint detection. In: Hagen, M., et al. (eds.) ECIR 2022. LNCS, vol. 13185, pp. 428–442. Springer, Cham (2022). https://doi.org/10.1007/978-3-030-99736-6_29
24. Singh, A., Saha, S.: Are you really complaining? A multi-task framework for complaint identification, emotion, and sentiment classification. In: Lladós, J., Lopresti, D., Uchida, S. (eds.) ICDAR 2021. LNCS, vol. 12822, pp. 715–731. Springer, Cham (2021). https://doi.org/10.1007/978-3-030-86331-9_46
25. Srivastava, N., Hinton, G., Krizhevsky, A., Sutskever, I., Salakhutdinov, R.: Dropout: a simple way to prevent neural networks from overfitting. J. Mach. Learn. Res. **15**(1), 1929–1958 (2014)
26. Trosborg, A.: Interlanguage Pragmatics: Requests, Complaints, and Apologies, vol. 7. Walter de Gruyter (2011)
27. Van Noort, G., Willemsen, L.M.: Online damage control: The effects of proactive versus reactive webcare interventions in consumer-generated and brand-generated platforms. J. Interact. Mark. **26**(3), 131–140 (2012)
28. Vaswani, A., et al.: Attention is all you need. arXiv preprint arXiv:1706.03762 (2017)
29. Welch, B.L.: The generalization of 'student's' problem when several different population variances are involved. Biometrika **34**(1–2), 28–35 (1947)

30. Xia, R., Ding, Z.: Emotion-cause pair extraction: a new task to emotion analysis in texts. arXiv preprint arXiv:1906.01267 (2019)
31. Xu, A., Liu, Z., Guo, Y., Sinha, V., Akkiraju, R.: a new chatbot for customer service on social media. In: Proceedings of the 2017 CHI Conference on Human Factors in Computing Systems, pp. 3506–3510 (2017)

DeCoDE: Detection of Cognitive Distortion and Emotion Cause Extraction in Clinical Conversations

Gopendra Vikram Singh[1], Soumitra Ghosh[1], Asif Ekbal[1(✉)],
and Pushpak Bhattacharyya[2]

[1] Indian Institute of Technology Patna, Bihta, India
{gopendra_1921cs15,asif}@iitp.ac.in
[2] Indian Institute of Technology Bombay, Mumbai, India
pb@cse.iitb.ac.in

Abstract. Despite significant evidence linking mental health to almost every major development issue, individuals with mental disorders are among those most at risk of being excluded from development programs. We outline a novel task of detection of *Cognitive Distortion* and *Emotion Cause* extraction of associated *emotions* in conversations. Cognitive distortions are inaccurate thought patterns, beliefs, or perceptions that contribute to negative thinking, which subsequently elevates the chances of several mental illnesses. This work introduces a novel multi-modal mental health conversational corpus manually annotated with *emotion*, *emotion causes*, and the presence of *cognitive distortion* at the utterance level. We propose a multitasking framework that uses multi-modal information as inputs and uses both external commonsense knowledge and factual knowledge from the dataset to learn both tasks at the same time. This is because commonsense knowledge is a key part of understanding how and why emotions are implied. We achieve commendable performance gains on the *cognitive distortion* detection task (+3.91 F1%) and the *emotion cause* extraction task (+3 ROS points) when compared to the existing state-of-the-art model.

Keywords: Cognitive distortion · Emotion cause · Mental health · Angular momentum · Multi-modal · Multi-task · Attention · Conversations

1 Introduction

The World Health Organization (WHO) estimated a cost of \$1 trillion per year in lost productivity due to depression and anxiety disorders[1]. The COVID-19

---

[1] https://www.who.int/teams/mental-health-and-substance-use/promotion-prevention/mental-health-in-the-workplace.

G. V. Singh and S. Ghosh—These authors contributed equally to this work and are joint first authors.

© The Author(s), under exclusive license to Springer Nature Switzerland AG 2023
J. Kamps et al. (Eds.): ECIR 2023, LNCS 13981, pp. 156–171, 2023.
https://doi.org/10.1007/978-3-031-28238-6_11

pandemic's trauma has also exacerbated the world's mental health crises. Negatively biased errors in thinking, also known as *Cognitive Distortion* [8] is a major contributor to the development of many different mental illnesses. Short-term use may help with stress and boost confidence, but chronic use can lead to mental decline and the onset of feelings of depression and anxiety [33]. Cognitive distortion manifests itself in a variety of ways, and the study in [4] found ten main manifestations of the same. Mindreading, catastrophizing, all-or-nothing thinking, emotive reasoning, labeling, mental filtering, overgeneralization, personalization, should statements, and diminishing or rejecting the positive are examples of these. Given the relevance of the interpersonal context in the start and progression of several mental health conditions [8,16], early detection of cognitive distortions among individuals may play an important role in the symptomatology of such illnesses.

Humans often exercise some restraint while interacting with one another. But people were more likely to talk about their thoughts and feelings with a virtual therapist than with a real one. In order to create efficient and low-cost interactive systems (like chatbots), which are often quick to install and may be utilized in combination with a human therapist, understanding human emotional states is vital. In order to identify how to best avoid acts of self-harm (such as suicide), it is important to recognize not just the emotional states but also the cause(s) of those feelings. This will allow for a deeper knowledge of the mental health of those involved. In such a situation, the Emotion cause extraction (ECE) task, which seeks to identify the possible causes behind a certain emotion expression in the text, might be useful.

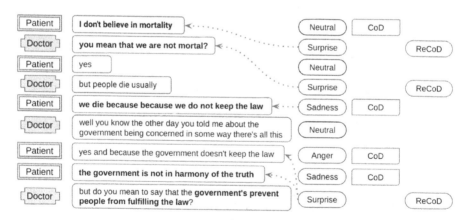

Fig. 1. Sample snapshot of our CoDeC Dataset. CoD: Cognitive Distortion; ReCoD: Response to CoD. The font highlighted in bold is the causal span.

Over the years, several conversational datasets have been introduced on various domains such as TV shows, social media, news, etc., but no such dataset exists, to our knowledge, related to mental health. Also, there is a big shortage of multi-modal datasets that can be used in clinical conversation settings.

The task of emotion-cause extraction in conversations is very nascent, and the existing studies [12,28] on this topic provide baseline systems that are majorly fine-tuned language models on emotion-cause annotated datasets. Certain features distinguish mental health discussions from other conversational datasets. We see a general trend in the flow of a typical doctor-patient interview. The doctor inquires about numerous elements and situations concerning the patient's well-being, and the patient relates the scenario they have been experiencing. At the end of the session, the doctor provides his diagnosis or remedy for the patient's ailment. In order to identify cognitively warped phenomena in any patient statement, information from future time steps may be required in addition to the context history. Figure 1 illustrates a conversation snippet describing the phenomenon of cognitive distortion, its response to it, and the various types of association of causes for emotions.

We take this opportunity to introduce a high-quality multi-modal clinical conversation dataset and a task-specific framework, especially for the task of emotion cause extraction. The current study analyzes emotion, emotion causes, and cognitive distortion in videos of dyadic conversations between psychiatrists and mental illness patients. The dataset and code is open-sourced to aid research[2].

The main contributions of this work are summarized below:

1. We propose the novel task of Detection of Cognitive Distortion and Emotion cause extraction in clinical conversations.
2. We introduce the first Cognitive Distortion and Emotion Cause (*CoDEC*) annotated multi-modal clinical conversation dataset comprising doctor-patient interactions on the premise of mental health interviews. We also provide manual annotations for Cognitive Distortion and Emotion at the utterance level.
3. We develop an emotion-aware multi-modal multi-task framework for the Detection of Cognitive Distortion and Emotion cause extraction (*DeCoDE*) in Clinical Conversations.
4. We also hypothesize that the performance of the above two tasks can be enhanced by the incorporation of information from future time steps.

The remainder of the paper is structured as follows. Section 2 summarises some previous works in this area. Following that, we go into the dataset preparation in depth in Sect. 3. We address our suggested methodology for multimodal multitask experiments in Sect. 4. In Sect. 5, we discuss the experiments, their results, and their outcomes. Finally, in Sect. 6, we bring our effort to a close and define the scope of future work.

2 Related Work

A few studies using computational methods have focused on the detection of various mental health issues, but none of them have concentrated on identifying

---

[2] https://www.iitp.ac.in/~ai-nlp-ml/resources.html#DeCoDE-CoDEC.

the core cause of such difficulties, which is a cognitive distortion in general. The emotion-cause extraction (ECE) problem has also received a lot of attention in several studies, but none of them have focused on the domain of clinical conversations or mental health, leaving a major gap that calls for more research in this area. In this section, we go through some of the earlier studies on mental health as well as emotion-cause extraction techniques.

2.1   Mental Health Studies

Mental health illnesses in general, being a major public health issue [25], have gained attention in past research, including computational studies. Although depression has received the most attention, other mental illnesses such as anxiety disorder, schizophrenia, post-traumatic stress disorder, suicide risk, and self-harm have also been studied [31]. Furthermore, psychological research supports the use of multimodal data for developing automated systems to recognize human emotion [1,26]. The impact of online content for doctor-patient interactions on patient satisfaction was investigated in [2]. The authors in [7] developed a deep learning method to categorize a variety of detrimental mental-health emotions, including addiction, anxiety, despair, stress, etc. While there is growing interest in the subject of the explainability of machine learning models in NLP [13], there is less such research for mental health condition identification. ECE tasks may be the initial step in making any automated system that can aid with mental health condition identification intelligible.

2.2   Emotion Cause Extraction

Due to the inherent long-term dependencies present in the utterances, determining the causes of emotions in a conversational situation is a challenging task. First proposed by Lee et al. [20] as a word-level sequence labeling problem, the ECE task was re-formalized in [14] as a clause-level extraction problem. End-to-end networks, such as the one shown in [29], have been proven to provide additional advantages over multi-stage approaches by leveraging the interdependence between the extracted emotion words and cause clauses. Li et al. [22] developed a context-aware co-attention model for the extraction of emotion cause pair. The authors in [3] suggested a strategy that narrows the search field and enhances productivity by matching emotions and causes concurrently utilizing the local search. Emotion-cause recognition in a conversation scenario was initially introduced in [28], which provided an emotion-cause annotated conversation corpus and evaluated it using a pair of deep learning-based systems. In a similar way, Ghosh et al. [12] introduced an emotion-cause annotated suicide note corpus and solved the emotion-cause identification and extraction independently. In this study, we consider the works in [12,28] as baselines to evaluate our suggested approach and mental health conversation dataset.

3   Dataset

In the following subsections, we discuss the various aspects of the developed *CoDEC* dataset.

3.1   Data Collection

Mental health-related content is scarcely available in the public domain, mainly due to its sensitive nature of it and also the associated stigma in sharing such content. YouTube is one of the most popular social media sites for sharing videos. It has a wide range of content about mental health, most of which is meant to promote and support educational needs. Applying a certain combination of keywords and phrases[3], we collected 30 doctor-patient conversation sessions[4] where the patients suffer from some form of cognitive distortions (such as polarized thinking, catastrophizing, over-generalization, etc.). Among the collected videos, 13 are from female patients and 17 are from male patients. Twenty of them are genuine interviews with psychiatrists and patients. The remaining 10 interviews are case studies/tutorial films with actual psychiatrists and actors conversing (posing as mental illness patients of various types of mental illness). Because mental health is sensitive and stigmatized, easily available relevant data in the public domain is limited. As a result, we chose to generate the dataset by considering both actual and enacted doctor-patient exchanges. The average number of utterances in the conversations is 125.1 and the average sentence length is 11.41 words.

3.2   Data Annotation

Each utterance of the conversations is marked with a start and end time stamp, which is essential to extract video extracts per utterance during multi-modal training. Also, each utterance is marked with speaker information (doctor or patient), the presence of factual information (fact), and a response to cognitive distortion (ReCoD). The annotations for emotion, emotion causes, and cognitive distortion are performed by three annotators[5], and final labels are obtained by performing a majority vote among the individual annotations. Text transcripts of some videos were already available from the uploading source. For the rest, we first collected the auto-generated transcripts in English and manually validated them to correct any inherent errors and produce good-quality transcripts for each utterance of the conversations.

---

[3] mental health, psychiatric interview, psychotic, paranoia, hallucination, etc.

[4] Links to some sample videos: https://www.youtube.com/watch?v=P7qMfG-yNfA
https://www.youtube.com/watch?v=Ii2FHbtVJzc.

[5] 2 Ph.D. linguistics degree holders and 1 undergraduate student from the computer science discipline.

Annotating Cognitive Distortion. Identifying the utterances with cognitive distortion is a challenging task. With a sound understanding of the phenomenon of cognitive distortion and its various forms, the annotators identified utterances as cognitively distorted if they presented biasedness and/or depicted irrational ways of perceiving real-world situations. Doctor responses to patient utterances at various junctures of the conversations presented vital clues to anticipate CoD utterances. We compute the Fleiss-Kappa (κ) score for the overall inter-rater agreement [30], as it is a popular choice when more than two raters are involved. The cognitive distortion task yielded a score of 0.83 which is considered to be 'almost perfect agreement'.

Annotating Emotion. Each utterance is marked with one of Ekman's [9] six basic emotions: anger, disgust, fear, joy, sadness, and surprise. We add a *neutral* class to accommodate all other utterances that are out of scope of the Ekman's emotions. Table 1b shows the distribution of utterances over the various emotion classes. We also observe that the dataset has an over-representation of the *others* class. The average Fleiss-Kappa [30] score obtained for the Emotion task is 0.77 which signifies 'substantial agreement' among the annotators.

Table 1. Dataset details

(a) Frequency of utterances over various attributes. CoD: Cognitive Distortion; ReCoD: Response to CoD

Attribute	Count
CoD	743
ReCoD	410
One Cause	410
Two Causes	179
Three Causes	36

(b) Emotion and Cause distribution.

Class	Count	# Causes
Anger	184	One: 101; Two: 42; Three: 10
Disgust	77	One: 49; Two: 22; Three: 2
Fear	169	One: 96; Two: 32; Three: 6
Joy	128	One: 28; Two: 7; Three: 2
Sadness	503	One: 198; Two: 80; Three: 10
Surprise	176	One: 78; Two: 24; Three: 2
Neutral	2516	No causal spans exists

Annotating Emotion Cause. Following the work in [12,28], we marked the causal spans (cs) for an emotion of each utterance in the dataset. We mark at most 3 causal spans for each utterance as we observed most utterances have single causes and few of them have two or more causes. The final causal span for an utterance U_t is marked using the span-level aggregation approach detailed in [15]. For a target utterance U_t, C_t denotes the set of causal spans ($C_t = \{cs_1, cs_2, cs_3\}$) for U_t. The causal spans are marked from $v+1$ utterances, where v denotes the number of context utterances of U_t and $v+1^{th}$ utterance is the target utterance U_t itself. We quantify the inter-rater agreement using the macro-F1 metric based on earlier work on span extraction [12,28], and we get an F1-score of 0.81, indicating that the annotations are of very high quality.

Table 1 shows the various details of the *CoDEC* dataset. The distribution of utterances over the various emotion classes and the number of causal spans per emotion class is shown in Table 1b.

4 Methodology

In this section, the *CoDEC* framework is illustrated for Cognitive Distortion detection and Emotion cause extraction from conversations. The system leverages the utterance-level emotion information for which the causes are to be extracted. The overall architecture of the proposed method is shown in Fig. 2.

4.1 Problem Definition

Given a document $D = [u_1, \cdots u_i \cdots, u_p]$ composed of a sequence of utterances (u), and each utterance can be further decomposed into a sequence of words, represented as $u_i = [word_{i,1}, \cdots term_{i,j} \cdots, term_{i,q}]$, where p indicates the number of utterances in the document, and q denotes the length of the word sequence contained in the utterance. Let $E = [e_1, \cdots e_i \cdots, e_p]$, denotes the utterance-level emotions in the document D. For a target utterance u_t, the task objective is to detect whether the utterance is cogntive distortion or not (0 or 1) and extract all possible causal spans for the given emotion e_t.

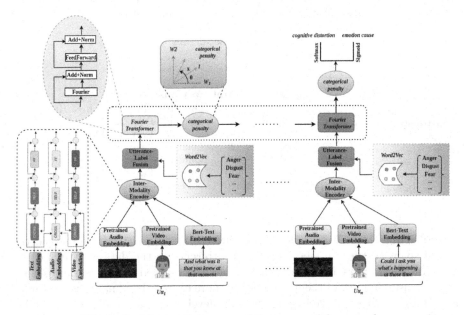

Fig. 2. Architectural diagram of the proposed framework

4.2 Detection of Cognitive Distortion and Emotion Cause Extraction (*DeCoDE*)

We illustrate the various components of the *DeCoDE* method below.

Input Feature Representation. We generate textual features for the utterances in the conversations using a pre-trained Bidirectional Encoder Representations from Transformers (BERT) [6] due to its strong ability to learn context-sensitive information and its ability to generalize on various downstream tasks. We utilize openSMILE[6] [10] tonal low-level features group to extract the acoustic features. We employ 3D-ResNeXt-101[4][7] [17] to extract visual features from the video snippets at the utterance level. We fetch the word vectors for the input emotion labels from the Word2Vec [24] pre-trained word embeddings.

Inter-Modality Encoder (IME). The inter-modality encoder comprises primarily three self-attention sub-layers, two bi-directional cross-attention sub-layers, and three feed-forward sub-layers. The output of the rth layer is used as the input to the $(r-1)^{th}$ layer, therefore $N *$ inter-modality layers are stacked together in the encoder. Here, we capture the inter-relatedness of the audio and visual counterparts with respect to the textual representation as it is obvious that textual utterance is more contributing than the other modalities. Subsequently, in the rth layer, the bi-directional cross-attention sub-layer is designed to interchange the knowledge and align the features between the said modalities in order to learn the inter-modality representations. In addition, to build internal connections the self-attention sub-layers are then applied to the output of the cross-attention sub-layer. Finally, the rth layer output $\hat{h}_i^k$ and $\hat{a}_i^k$ are produced by feed-forward layers. Residual connections and layer normalization are added after each sub-layer, in a similar manner as the single-modality encoders.

Utterance-Label Fusion (ULF). We combine the utterance level semantic features from BERT and the emotion features from Word2Vec through self-attention [32]. The separate feature vectors are passed through independent dense layers to make them the same length. Unlike linear concatenation of emotion labels with corresponding utterances [28], our strategy enables to learn the importance of the emotion for a particular utterance and dynamically update its initial weights during training.

Contextual Fourier Transformer (CFT). To model the contextual information among the utterances, we develop Contextual Fourier Transformer. The Fourier Transformer encoder [21] is an efficient method to the regular transformer encoder [32] where the self-attention sublayers in the transformer encoder are replaced with simple linear transformations that 'mix' the input tokens to create the FNet. Each utterance from the ULF module in the input sequence is passed through the CFT module. Each passing utterance acts as a context for the next utterance up to the target utterance.

---

[6] https://github.com/audeering/opensmile.

[7] https://github.com/kaiqiangh/extracting-video-features-ResNeXt.

Categorical Penalty. To aid the model in understanding how an emotion label and its related utterance are linked, we add a Categorical Penalty word to the intermediate CFT units. This will enable better prediction ability of the start and end tokens. For this purpose, we first represent softmax and sigmoid by the equations below.

$$\mathcal{L} = -\frac{1}{b_s} \sum_{i=1}^{b_s} \log \frac{\exp^{\mathcal{W}l_i + b_i}}{\sum_{j=1}^{N} \exp^{\mathcal{W}l_j + b_j}} \qquad \mathcal{L} = -\frac{1}{b_s} \sum_{i=1}^{b_s} \frac{1}{\exp^{\mathcal{W}l_i + b_i}} \qquad (1)$$

where $l_i \in \mathbb{R}^d$ is the feature of ith sample. b_s is batch size. b_i and b_j denote the bias. $\mathcal{W} \in \mathbb{R}^{d*n}$ denotes the weight matrix. It is known for information extraction tasks, that finding the decision boundary for the start and end markers of a span is challenging, and a simple softmax/sigmoid classifier will not be able to handle this distinction effectively. Because of this, some samples can fall into the wrong region due to the ambiguity of the classification boundary. This can call for a higher convergence rate. To handle this we use the strategy used in Insightface loss [5] which normalizes the feature l_i and the weight matrics $\mathcal{W}$ to measure the similarity of feature by the difference of angle by which it maps the vector more closely. It adds a penalty value x into the angle to force the feature to converge.

$$\mathcal{L}_{u1} = -\frac{1}{b_s} \sum_{i=1}^{b_s} \log \frac{\exp^{a(cos(\theta + x))}}{\exp^{a(cos(\theta + x))} + \sum_{j=1}^{N} \exp^{a(cos(\theta))}} \qquad (2)$$

$$\mathcal{L}_{u2} = -\frac{1}{b_s} \sum_{i=1}^{b_s} \frac{1}{\exp^{a(cos(\theta + x))} + \exp^{a(cos(\theta))}} \qquad (3)$$

where $\mathcal{L}_{u1}$ and $\mathcal{L}_{u2}$ is updated loss function for softmax and sigmoid respectively, θ denotes the angle between weight $\mathcal{W}$ and feature l and a denotes the amplifier function.

Task-Specific Layers: The output from the last CFT unit which corresponds to the target utterance U_t is passed to two task-specific dense layers and the following output layers for the CoD and ECE tasks. The output layer for the ECE task is a linear layer to calculate span start and end logits which employ sigmoid activation in which the threshold value is set at 0.4. This results in the output of the probability of three first tokens and three last tokens, which signifies the capability to output three causal spans at most.

Calculation of Loss. The model is trained using a unified loss function as shown in Eq. 4. We employ categorical cross-entropy loss and binary cross-entropy loss for the CoD and ECE tasks, respectively.

$$L = \sum_{\omega} W_\omega L_\omega \qquad (4)$$

Here, ω denotes the two tasks, CoD and ECE. The weights (W_ω) are updated using back-propagation for specific losses for each task.

5 Experiments and Results

This section discusses the experiments performed, the results, and the analysis.

5.1 Experimental Setup

Since the *CoDEC dataset* is having skewed class proportion, we report both the accuracy and macro-F1 scores for the CoD task. Following the work in [12], we report the full match (FM), partial match (PM), Hamming Distance (HD), Jaccard Similarity (JS) and Ratcliffe-Obershelp Similarity (ROS) measures to evaluate the ECE task. We use PyTorch[8], a Python-based deep learning package, to develop our proposed model. We experiment with the base version of BERT imported from the huggingface transformers[9] package. To determine the optimal value of the additive angle x, which influences performance, we tested five values ranging from 0.1 to 0.5. The default value of x is set at 0.3. We set amplification value a as 64. For openSMILE, voice normalization and voice intensity threshold are used to discriminate between samples with and without speech. Z-standardization is used for voice normalizing. ResNext has been pre-trained on Kinetics at 1.5 features per second and a resolution of 112. All experiments are carried out on an NVIDIA GeForce RTX 2080 Ti GPU. We perform 80–20 split of the *DeCoDE* dataset for training and testing purposes. The best model is saved on the performance of the validation set. We run our experiments for 200 epochs and report the averaged scores after 5 runs of the experiments to account for the non-determinism of Tensorflow GPU operations.

Baselines: For the comprehensive evaluation of our proposed *DeCoDE* method and the introduced *CoDEC* dataset, we consider the following systems as baselines in this study: RoBERTa [23], SpanBERT [19], MT-BERT [27] and Cascaded Multitask System with External Knowledge Infusion (CMSEKI) [11]. Similar to the *DeCoDE* method, to adapt the baselines to our multi-task scenario, we add a linear layer on top of the hidden-states output in the output layer of the ECE task to calculate span start and end logits. The output layer for the ECE task employs sigmoid activation in which the threshold value is set as 0.4.

5.2 Results and Discussion

We investigate the contribution of multi-modal aspects to the tasks at hand. The results of our *DeCoDE* method on the *CoDEC* dataset are shown in Table 2. The trimodal configuration yields the best results, followed by the bimodal and the unimodal networks. This may be due to the fact that texts have less background noise than audio-visual sources, yet when the three are compared, textual modality outperforms the others. For similar jobs, our results are consistent with prior research [18]. We performed experiments using the *DeCoDE* method varying the context length on the *CoDEC* dataset and observe that the best results

---

[8] https://pytorch.org/.
[9] https://huggingface.co/docs/transformers/index.

are obtained when the context length is set as 5. Figure 3 illustrates the detailed results of the *DeCoDE* method on various context length sizes.

Table 2. Experimental results of *DeCoDE* on various modalities

Modality	Cognitive distortion		Emotion cause				
	F1%	Acc. %	FM	PM	HD	JF	ROS
T	66.68	68.71	21.98	28.31	0.45	0.58	0.69
A	62.69	64.11	20.74	24.46	0.41	0.53	0.68
VZ	55.96	52.13	18.29	19.31	0.37	0.48	0.61
T+V	68.31	69.59	25.19	29.78	0.49	0.61	0.71
T+A	69.74	71.11	27.31	31.91	0.51	0.63	0.72
A+V	66.22	67.63	24.33	27.58	0.47	0.59	0.70
T+V+A	**73.48**	**75.91**	**29.43**	**33.24**	**0.53**	**0.65**	**0.74**

Table 3. Results from our proposed model and the various baselines. Values in bold are the maximum scores attained.

Models	Cognitive distortion		Emotion cause				
	F1 (%)	Acc. (%)	FM	PM	HD	JF	ROS
Baselines							
RoBERTa [19]	67.16	69.24	25.73	25.51	0.46	0.59	0.69
SpanBERTa [23]	65.79	66.83	23.58	21.12	0.44	0.57	0.67
MTL-BERT [27]	66.93	69.79	25.11	23.67	0.47	0.58	0.69
CMSEKI [11]	70.31	71.47	27.11	28.59	0.50	0.62	0.71
Proposed							
DeCoDE	73.48	75.91	29.43	33.24	0.53	0.65	0.74
DeCoDE$_{-[CP]}$	71.25	72.35	27.22	30.89	0.50	0.62	0.72
DeCoDE$_{-[EMO]}$	71.76	73.17	28.18	31.17	0.51	0.63	0.71
DeCoDE$_{-[EMO+CP]}$	69.47	70.85	25.91	29.33	0.49	0.60	0.69
DeCoDE$_{+[ReCoD]}$	**74.21**	**76.31**	**30.15**	**34.31**	**0.54**	**0.66**	**0.74**

Comparison with Prior Works: Table 3 shows that CMSEKI is the best-performing baseline, which is not surprising given that it uses common-sense knowledge from external knowledge sources to grasp the input information. However, the proposed *DeCoDE* method outperforms the performances of CMSEKI for all metrics, specifically by 3.17% F1 for CoD task and 3 ROS points. Low performances by RoBERTa [23] and SpanBERT [19] shows the difficulty of powerful language models in perceiving critical tasks as emotion cause extraction, more so, especially in clinical situations where training data may be insufficient. We also observe that harnessing the information from future utterances (doctor's responses) enhances the performances of the *DeCoDE* method for both the tasks

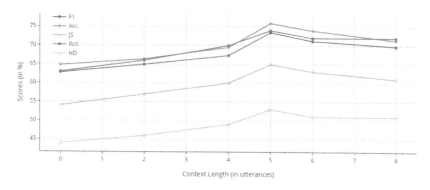

Fig. 3. Results on varying context length

(as shown by $DeCoDE_{+[ReCoD]}$). This shows the relevance of the information available at future time steps, particularly in the case of clinical conversations, in comprehending mental health-related discussion.

Ablation Experiments: As shown in Table 3, we performed an ablation study on the $DeCoDE$ dataset to analyze the performance of the different modules in our proposed strategy. The values of all the metrics over both the CoD and ECE tasks are shown to decrease when either the categorical penalty factor ($DeCoDE_{-[CP]}$) or the emotion task ($DeCoDE_{-[EMO]}$) is removed. The decrease is more profound when we remove both the penalty factor and the emotion task from the $DeCoDE$ method ($DeCoDE_{-[EMO+CP]}$). This confirms that the inclusion of the categorical penalty factor and the emotion information of the utterances is an integral contributor to the performances of the cognitive distortion and emotion-cause extraction tasks.

Qualitative Analysis: We performed an extensive analysis of the predictions from the various systems. It is observed that the proposed DeCoDE performs comparatively better than MTL-BERT and CMSEKI systems in generating correct predictions for both the CoD and ECE tasks. Some sample instances are shown in Table 4. In the first two instances, we can see that the DeCoDE method correctly predicts both the causal spans and the CoD label. The MTL-BERT system extracts an incomplete span for the first example whereas it is unable to extract any part of the cause in the second case. In the third example, we see that both the baselines incorrectly classified the utterance as CoD, however, the DeCoDE method correctly categorized it as non-CoD. Lengthier utterances seem to cause difficulty for all systems, as can be seen from the last example. Although CMSEKI and the proposed DeCoDE are able to predict the CoD label correctly, it manages to extract a part of the causal span fully.

Table 4. Sample predictions from the various systems. Color Coding: Blue- Correct, Red: Incorrect; Teal: Incomplete. [Y] and [N] indicate Yes and No predictions for the CoE task, respectively.

DeCoDE	CMSEKI	MTL-BERT
Actual: *she might be reporting back to them [Y]*		
she might be reporting back to them [Y]	might be reporting back to [Y]	she might be reporting back to them [Y]
Actual: *i am a lord god jehovah [Y]*		
i am a lord god jehovah [Y]	No Cause [N]	a lord god jehovah [Y]
Actual: *you started carrying guns [N]*		
you started carrying guns [N]	you started carrying guns [Y]	you started carrying guns [Y]
Actual: *try to ensure somehow that they are being raised properly, from a distance [N]*		
to ensure somehow that they are being raised [N] properly	try to ensure somehow that they are being raised properly [Y]	that they are being raised properly, from a distance [N]

6 Conclusion

In this work, we present the first multi-modal, emotion-cause annotated clinical conversation dataset, consisting of conversations between doctors and patients in the context of mental health interviews. Additionally, we present sentence-level manual annotations for cognitive distortion and emotion. In order to extract the emotional causes of cognitive distortions in clinical conversations, we develop, *DeCoDE*, a multi-modal, multi-task framework that takes into account the inherent speaker's emotions present in utterances of any conversations. To the best of our knowledge, the *DeCoDE* framework is the first task-specific system to address the emotion-cause extraction task in conversations. We demonstrate the efficacy of our technique by comparing it to different state-of-the-art baselines.

Even if a negative emotional context has little to do with how the patient feels about other people or things, the patient's behaviours and judgments may be negatively impacted. Future research would concentrate on creating techniques to educate people about the cognitive biases brought on by cognitive distortions in order to provide fresh treatment approaches. It is also important to pay attention to how to properly capture the implicit aspects of complex causation.

Ethical Consideration. This study has been evaluated and approved by our Institutional Review Board (IRB). The videos used to create the dataset for this study do not have any copyright clauses attached to them. Furthermore, the videos are shared via various channels for the main purpose of facilitating research and educational purposes.

References

1. Aviezer, H., Trope, Y., Todorov, A.: Body cues, not facial expressions, discriminate between intense positive and negative emotions. Science **338**(6111), 1225–1229 (2012)

2. Chen, S., Guo, X., Wu, T., Ju, X.: Exploring the online doctor-patient interaction on patient satisfaction based on text mining and empirical analysis. Inf. Process. Manage. **57**(5), 102253 (2020)

3. Cheng, Z., Jiang, Z., Yin, Y., Yu, H., Gu, Q.: A symmetric local search network for emotion-cause pair extraction. In: Proceedings of the 28th International Conference on Computational Linguistics, pp. 139–149. International Committee on Computational Linguistics, Barcelona, Spain (Online), December 2020. https://doi.org/ 10.18653/v1/2020.coling-main.12, https://aclanthology.org/2020.coling-main.12

4. David, B., Burns, M.: Feeling good-the new mood therapy. NY: Signet Books. Chin, Richard. (1995) p. 3 (1980)

5. Deng, J., Guo, J., Xue, N., Zafeiriou, S.: Arcface: additive angular margin loss for deep face recognition. In: Proceedings of the IEEE/CVF Conference on Computer Vision and Pattern Recognition, pp. 4690–4699 (2019)

6. Devlin, J., Chang, M., Lee, K., Toutanova, K.: BERT: pre-training of deep bidirectional transformers for language understanding. In: Proceedings of the 2019 Conference of the North American Chapter of the Association for Computational Linguistics: Human Language Technologies, NAACL-HLT 2019, Minneapolis, MN, USA, 2–7 June 2019, Volume 1 (Long and Short Papers), pp. 4171–4186. Association for Computational Linguistics (2019). https://doi.org/10.18653/v1/n19-1423

7. Dheeraj, K., Ramakrishnudu, T.: Negative emotions detection on online mental-health related patients texts using the deep learning with MHA-BCNN model. Expert Syst. Appl. **182**, 115265 (2021)

8. Dozois, D.J., Beck, A.T.: Cognitive schemas, beliefs and assumptions. Risk factors in Depression, pp. 119–143 (2008)

9. Ekman, P.: An argument for basic emotions. Cogn. Emot. **6**(3-4), 169–200 (1992)

10. Eyben, F., Wöllmer, M., Schuller, B.W.: Opensmile: the munich versatile and fast open-source audio feature extractor. In: Proceedings of the 18th International Conference on Multimedia 2010, Firenze, Italy, October 25–29, 2010, pp. 1459–1462. ACM (2010). https://doi.org/10.1145/1873951.1874246

11. Ghosh, S., Ekbal, A., Bhattacharyya, P.: A multitask framework to detect depression, sentiment and multi-label emotion from suicide notes. Cogn. Comput. **14**(1), 110–129 (2021). https://doi.org/10.1007/s12559-021-09828-7

12. Ghosh, S., Roy, S., Ekbal, A., Bhattacharyya, P.: CARES: CAuse recognition for emotion in suicide notes. In: Hagen, M., et al. (eds.) ECIR 2022. LNCS, vol. 13186, pp. 128–136. Springer, Cham (2022). https://doi.org/10.1007/978-3-030-99739-7_15

13. Gilpin, L.H., Bau, D., Yuan, B.Z., Bajwa, A., Specter, M., Kagal, L.: Explaining explanations: An overview of interpretability of machine learning. In: 2018 IEEE 5th International Conference on data science and advanced analytics (DSAA), pp. 80–89. IEEE (2018)

14. Gui, L., Wu, D., Xu, R., Lu, Q., Zhou, Y.: Event-driven emotion cause extraction with corpus construction. In: Proceedings of the 2016 Conference on Empirical Methods in Natural Language Processing, pp. 1639–1649 (2016)

15. Gui, L., Wu, D., Xu, R., Lu, Q., Zhou, Y.: Event-driven emotion cause extraction with corpus construction. In: Proceedings of the 2016 Conference on Empirical Methods in Natural Language Processing, Austin, Texas, pp. 1639–1649. Association for Computational Linguistics, November 2016. https://doi.org/10.18653/v1/ D16-1170, https://aclanthology.org/D16-1170

16. Hammen, C.L., Shih, J.: Depression and interpersonal processes (2014)
17. Hara, K., Kataoka, H., Satoh, Y.: Can spatiotemporal 3D CNNs retrace the history of 2d CNNs and imagenet? In: 2018 IEEE Conference on Computer Vision and Pattern Recognition, CVPR 2018, Salt Lake City, UT, USA, 18–22 June 2018, pp. 6546–6555. Computer Vision Foundation/IEEE Computer Society (2018). https://doi.org/10.1109/CVPR.2018.00685, http://openaccess.thecvf.com/content_cvpr_2018/html/Hara_Can_Spatiotemporal_3D_CVPR_2018_paper.html
18. Hazarika, D., Poria, S., Zadeh, A., Cambria, E., Morency, L., Zimmermann, R.: Conversational memory network for emotion recognition in dyadic dialogue videos. In: Proceedings of the 2018 Conference of the North American Chapter of the Association for Computational Linguistics: Human Language Technologies, NAACL-HLT 2018, New Orleans, Louisiana, USA, 1–6 June 2018, Volume 1 (Long Papers), pp. 2122–2132. Association for Computational Linguistics (2018). https://doi.org/10.18653/v1/n18-1193
19. Joshi, M., Chen, D., Liu, Y., Weld, D.S., Zettlemoyer, L., Levy, O.: Span-BERT: improving pre-training by representing and predicting spans. Trans. Assoc. Comput. Linguist. **8**, 64–77 (2020). https://doi.org/10.1162/tacl_a_00300, https://aclanthology.org/2020.tacl-1.5
20. Lee, S.Y.M., Chen, Y., Huang, C.R.: A text-driven rule-based system for emotion cause detection. In: Proceedings of the NAACL HLT 2010 Workshop on Computational Approaches to Analysis and Generation of Emotion in Text, pp. 45–53 (2010)
21. Lee-Thorp, J., Ainslie, J., Eckstein, I., Ontanon, S.: FNet: mixing tokens with fourier transforms. arXiv preprint arXiv:2105.03824 (2021)
22. Li, X., Song, K., Feng, S., Wang, D., Zhang, Y.: A co-attention neural network model for emotion cause analysis with emotional context awareness. In: Proceedings of the 2018 Conference on Empirical Methods in Natural Language Processing, pp. 4752–4757. Association for Computational Linguistics, Brussels, Belgium, October–November 2018. https://doi.org/10.18653/v1/D18-1506, https://aclanthology.org/D18-1506
23. Liu, Y., et al.: Roberta: a robustly optimized BERT pretraining approach. CoRR abs/1907.11692 (2019). http://arxiv.org/abs/1907.11692
24. Mikolov, T., Sutskever, I., Chen, K., Corrado, G.S., Dean, J.: Distributed representations of words and phrases and their compositionality. In: Advances in Neural Information Processing Systems, vol. 26 (2013)
25. Organization, W.H., et al.: Depression: a global crisis. World Mental Health Day, October 10 2012. World Federation for Mental Health, Occoquan, VA, USA (2012)
26. Pantic, M., Sebe, N., Cohn, J.F., Huang, T.S.: Affective multimodal human-computer interaction. In: Proceedings of the 13th ACM International Conference on Multimedia, Singapore, 6–11 November 2005, pp. 669–676. ACM (2005). https://doi.org/10.1145/1101149.1101299
27. Peng, Y., Chen, Q., Lu, Z.: An empirical study of multi-task learning on BERT for biomedical text mining. In: Proceedings of the 19th SIGBioMed Workshop on Biomedical Language Processing, BioNLP 2020, Online, 9 July 2020, pp. 205–214. Association for Computational Linguistics (2020). https://doi.org/10.18653/v1/2020.bionlp-1.22, https://doi.org/10.18653/v1/2020.bionlp-1.22
28. Poria, S., et al.: Recognizing emotion cause in conversations. Cogn. Comput. **13**(5), 1317–1332 (2021)

29. Singh, A., Hingane, S., Wani, S., Modi, A.: An end-to-end network for emotion-cause pair extraction. In: Proceedings of the Eleventh Workshop on Computational Approaches to Subjectivity, Sentiment and Social Media Analysis, pp. 84–91 (2021)

30. Spitzer, R.L., Cohen, J., Fleiss, J.L., Endicott, J.: Quantification of agreement in psychiatric diagnosis: a new approach. Arch. Gen. Psychiatry **17**(1), 83–87 (1967)

31. Uban, A.S., Chulvi, B., Rosso, P.: Understanding patterns of anorexia manifestations in social media data with deep learning. In: Proceedings of the Seventh Workshop on Computational Linguistics and Clinical Psychology: Improving Access, pp. 224–236. Association for Computational Linguistics, Online, June 2021. https://doi.org/10.18653/v1/2021.clpsych-1.24, https://aclanthology.org/2021.clpsych-1.24

32. Vaswani, A., et al.: Attention is all you need. In: Advances in Neural Information Processing Systems 30: Annual Conference on Neural Information Processing Systems 2017(December), pp. 4–9, 2017. Long Beach, CA, USA, pp. 5998–6008 (2017). https://proceedings.neurips.cc/paper/2017/hash/3f5ee243547dee91fbd053c1c4a845aa-Abstract.html

33. Yüksel, A., Bahadir-Yilmaz, E.: Relationship between depression, anxiety, cognitive distortions, and psychological well-being among nursing students. Perspect. Psychiatr. Care **55**(4), 690–696 (2019)

Domain-Aligned Data Augmentation for Low-Resource and Imbalanced Text Classification

Nikolaos Stylianou$^{(\boxtimes)}$ ⓘ, Despoina Chatzakou ⓘ, Theodora Tsikrika ⓘ, Stefanos Vrochidis ⓘ, and Ioannis Kompatsiaris ⓘ

Information Technologies Institute, Centre for Research and Technology Hellas, Thessaloniki, Greece
{nstylia,dchatzakou,theodora.tsikrika,stefanos,ikom}@iti.gr

Abstract. Data Augmentation approaches often use Language Models, pretrained on large quantities of unlabeled generic data, to conditionally generate examples. However, the generated data can be of subpar quality and struggle to maintain the same characteristics as the original dataset. To this end, we propose a Data Augmentation method for low-resource and imbalanced datasets, by aligning Language Models to in-domain data prior to generating synthetic examples. In particular, we propose the alignment of existing generic models in task-specific unlabeled data, in order to create better synthetic examples and boost performance in Text Classification tasks. We evaluate our approach on three diverse and well-known Language Models, four datasets, and two settings (i.e. imbalance and low-resource) in which Data Augmentation is usually deployed, and study the correlation between the amount of data required for alignment, model size, and its effects in downstream in-domain and out-of-domain tasks. Our results showcase that in-domain alignment helps create better examples and increase the performance in Text Classification. Furthermore, we find a positive connection between the number of training parameters in Language Models, the volume of fine-tuning data, and their effects in downstream tasks.

Keywords: Natural Language Processing · Data Augmentation · Low-resource data · Imbalanced data · Text Classification

1 Introduction

Modern Deep Learning applications typically require a very large amount of labeled training data [30] to operate in a satisfactory manner. While some Large Language Models [6,29,39] have been capable of achieving state-of-the-art performance with only a handful of labeled examples, Few-Shot learning, in which a small number of examples are provided as contextualized prompts, remains a challenging task [5]. Data Augmentation acts as a countermeasure to the lack of sufficiently labeled training data, serving as an effective strategy in artificially increasing the size of a training dataset.

J. Kamps et al. (Eds.): ECIR 2023, LNCS 13981, pp. 172–187, 2023.
https://doi.org/10.1007/978-3-031-28238-6_12

In addition, commonly, label distribution in the real world is rarely balanced, especially in the case of user generated content, a fact which is often reflected in the training data and thus resulting in poor performance of the trained models [14]. Solving class imbalance in the training data is usually dealt with over or under sampling techniques from the dominant or subservient class, respectively [8,25]. However, these approaches can lead to overfitting and loss of valuable information, respectively.

With Data Augmentation, the goal is to generate examples for specific classes such that the training dataset is increased. While traditional Natural Language Processing (NLP) Data Augmentation techniques [35,38] often struggle to maintain the correct label for the created examples, modern approaches utilizing Language Models (LMs) have made significant strides in this area [1,19,37]. Effectively, having the ability to automatically generate class-specific examples can significantly contribute to both low-resource scenarios where labeled data are scarce, as well as in tasks with label imbalance. What is more, as we avoid over and under sampling examples (e.g. by randomly creating copies or deleting examples from the minority and majority class, respectively) we minimize the risk of overfitting and information loss, respectively.

With LM-based Data Augmentation approaches, the use of LMs pretrained on generic domains often results in generated examples that do not reflect a task's specific characteristics. This paper explores the effects of in-domain alignment for LM-based Data Augmentation, i.e. fine-tuning pretrained LMs on domain specific data given a certain task. We evaluate our approach in both extremely low-resource and highly imbalanced settings for Text Classification. Specifically, we expand the work of [19], which establishes a method for utilizing popular and state-of-the-art LMs to create synthetic data.

Concretely, in this work, we fine-tune three diverse LMs (GPT2, BERT, and BART) on in-domain data to align the pretrained models with the domain's characteristics. This alignment is performed with three increasing sizes of in-domain data to examine the correlation between different LMs and in-domain data volume. We also investigate the effectiveness of Data Augmentation on datasets with user generated texts of differing lengths, namely the SST-2, SNIPS, and TREC datasets used in previous works [1,19] and also the Rotten Tomatoes (RT) dataset [26], which has significantly longer texts compared to the moderately sized texts of the other three datasets. Moreover, to further assess the robustness and validity of the in-domain alignment for Data Augmentation, we simulate extreme label imbalance for the in-domain corpora by under-sampling a single class of the training set. By artificially doubling and balancing the minority class example with domain-aligned LMs, we showcase the performance gains among aligned LM Data Augmentation, non-aligned LM Data Augmentation and no augmentation. Overall, we highlight the effects of in-domain alignment with a BERT-based classifier [12] for both in- and out-of-domain corpora.

To sum up, our contributions are as follows: (a) we showcase that in-domain alignment can help produce better results in low-resource and imbalanced Text Classification, (b) we investigate the volume of data required for alignment in different models and settings, and (c) we show that minority class Data Augmentation following in-domain alignment can greatly improve the performance in imbalanced Text Classification tasks.

2   Related Work

Data Augmentation (DA) has taken many forms, from simple alterations to the initial content [13,35] to automatically generating artificial (synthetic) examples [1,19,27].

Early approaches to DA focus on artificially altering the text through a series of transformations. EDA [35] proposes the replacement of certain words, word swapping, and random insertion and deletion of words to alter the original content, while ADEA [17] adds artificial noise by inserting strings of punctuation marks. Similarly, [38] uses interpolation and n-gram smoothing as a means of introducing noise, while [32] proposes the use of back-translation, i.e. translating a sample to another language and converting it back to the original language. SMERTI [13] presents a semantic approach towards text replacement with the focus being on maintaining the sentiment and fluency of the original text.

Utilizing the advantages of LM pretraining, a plethora of approaches have been designed based on LM capabilities [4]. In general, LM-based DA approaches exploit the training objectives and text prompting techniques to condition the predictions of different LMs so that they can create artificial examples. Towards this direction, CBERT [37] exploits the Masked Language Modeling (MLM) objective of BERT [12] and fine-tunes the model so that the masked tokens are conditioned by the provided label. LAMBADA [1] utilizes GPT2 [28], a generative LM, to create labeled examples after fine-tuning it on the training set via label prompting. These examples are first filtered by a baseline classifier to ensure that the predicted labels correspond to the one used to generate them, before being used in the final training set. Similarly, a per-instance prompted GPT2 model is presented in [3], to create similar texts which are then filtered based on embedding distances.

Based on these approaches, [19] proposes a unified framework in which different conditioning strategies are explored for BERT, GPT2, and BART [20] in a low-resource scenario. In comparison to past approaches, no filtering is performed due to the extremely low number of samples available for training. To counteract the disadvantages of the early approaches, such as being task-specific or having difficulty to create label-preserving examples, Polyjuice [36] aims to generate texts with specific perturbations and substitutions. This is achieved through a counterfactual sentence generation process rather than label conditioning used in all other LM-based methods.

Several methods have also been proposed to augment the data in latent space, rather than creating new examples. Specifically, Cutoff [31] creates noise in the latent space by zeroing rows or columns of the input vectors, hence removing tokens, features, or even spans of words, without the need to artificially create examples. Building upon this, HiddenCut [9] uses Cutoff layers within the Transformer architecture blocks [34] of the entire model to introduce noise between layers, similarly to how dropout works [2]. Similarly, Mixup [15] proposes a linear intepolation of textual samples from the same class to increase the input signals received by the model with the same number of available data. Lastly, CoDa [27] combines classic approaches, such as back-translation, with novel approaches, such as Cutoff and Mixup, as well as adversarial training to create better models. A recent survey on DA methods for text classification analyzes the aforementioned approaches in detail [3].

Previous approaches either fine-tune the respective LMs on the training data or generate latent examples which are hard to benchmark and tied to datasets. Consequently, we opt to focus on methods which can be evaluated with more means than only through the performance changes in downstream tasks. In particular, prior to fine-tuning on the training data, we first fine-tune the LMs on unlabeled in-domain data, to align the general language models with our effective domain such that the artificially generated examples better match the domain's characteristics and can therefore lead to better overall performance.

3   Domain-Aligned Data Augmentation

Our proposed method aims to utilize well-established pretrained LMs and a collection of task-specific unlabeled data, to generate synthetic examples in order to increase dataset size and eliminate imbalance. By initially fine-tuning the pretrained LMs on in-domain data, we aim to generate better examples for the task in hand and increase performance.

To that end, we build on top of [19] that uses three language models, BERT, GPT2, and BART, which are autoencoding (AE), autoregressive (AR) and sequence-to-sequence (seq2seq) LMs, respectively. These models are aligned with the domain-specific data following their original training objectives (Sect. 3.1), while the conditional generation is informed by the findings of [19] (Sect. 3.2).

Problem Formulation. Let $D_{Train} = \{x_i, y_i\}_n^1$ be a dataset for a task T containing n examples, where $x_i = \{w_j\}_m^1$ is an example in the dataset containing m words and y_i is the associated label of this example. And let $D_{Domain} = \{x_k\}_v^1$ be a dataset of v unlabeled examples, which can be easily acquired to match with task T. We assume that $v >> n$ with v being able to scale by acquiring more unlabeled available data. Also, let G be a LM pretrained on generic data. This work proposes $G_{Aligned}$ being the G fine-tuned on D_{Domain} and $G_{AlignedTuned}$ being the fine-tuned $G_{Aligned}$ LM on D_{Train}, while previous work has focused on G_{Tuned}, i.e., pretrained LMs fine-tuned only on D_{Train}. Finally, $D_{Synthetic}$ is the product of generating and adding s examples to D_{Train} using $G_{AlignedTuned}$ from a dataset $D_{Select} \subseteq D_{Train}$ containing only class examples under a threshold d from D_{Train}.

Our goal is to train a task specific model M such that $\text{Score}(M_{Synthetic}) > \text{Score}(M_{Train})$ trained on $D_{Synthetic}$ and D_{Train} respectively and Score is a task appropriate metric (e.g. Accuracy, F1-score, etc.). The process followed to achieve this is described in Algorithm 1.

Algorithm 1. Domain Aligned Data Augmentation

Input: Training dataset D_{Train}
   In-domain unlabeled data D_{Domain}
   Pretrained model G
   d: threshold number of examples per class
   s: number of examples to be generated
1: Fine-tune G using D_{Domain} to obtain $G_{Aligned}$
2: Fine-tune $G_{Aligned}$ using D_{Train} to obtain $G_{AlignedTuned}$
3: $D_{Synthetic} \leftarrow \{\}$
4: $D_{Select} \leftarrow$ ExampleSelector(D_{Train}, d)
5: **foreach** $\{x_i, y_i\} \in D_{Select}$ **do**
6:    Synthesize s examples $\{\hat{x}_i, \hat{y}_i\}_s^1$ using $G_{AlignedTuned}$
7:    $D_{Synthetic} \leftarrow D_{Synthetic} \cup \{\hat{x}_i, \hat{y}_i\}_s^1$
8: **end foreach**
9: $D_{Task} \leftarrow D_{Train} \cup D_{Synthetic}$

3.1   In-Domain Alignment

For the LM alignment, we fine-tune the pretrained LMs on D_{Domain} using different training objectives to obtain $G_{Aligned}$. Specifically, we tune BERT using only the MLM objective and discard the Next Sentence Prediction objective described in [12], as it has shown to not contribute towards better performance in downstream tasks [22]. GPT2 is tuned following the original objective, in an autoregressive setting [28]. Lastly, BART uses the denoising objective described in [20], following the same corruption strategies. For brevity, we do not describe these objectives and methods in detail and refer readers to the cited works.

3.2   Conditional Generation

Using the previously acquired $G_{Aligned}$, we further fine-tune the LMs on D_{Select} of each task so that we obtain $G_{AlignedTuned}$ which will be capable to conditionally generate new instances. Fine-tuning strategies for conditional generation are also different for each individual model. For BERT, we follow the CBERT approach [37], in which the model is first fine-tuned with a MLM objective with the class as a single token sentence followed by a separator token and the original text. After the model is tuned, random tokens from the original text are masked and the model predicts replacements, altering the original text. For GPT2, we follow [1] in which the class is also prepended to the original text followed by a separator token and the model is trained autoregressively. Lastly, BART operates similarly to CBERT with the label prepended to the start of the original text, followed by a separator token [19]. For BART, we present results in two masking strategies, word and span, due to performance variance based on masking [19].

   Readers are encouraged to follow the original works for details on the conditional generation process for each LM architecture. Conditional Generation is dynamically used for both low-resource and imbalanced data through the example selection threshold d. In the case of low-resource data, all labels are used simultaneously to obtain $G_{AlignedTuned}$ (i.e. $D_{Select} = D_{Train}$), while for imbalanced data the model is trained only on the minority classes samples and produces only examples for those classes (i.e. $D_{Select} \subseteq D_{Train}$).

3.3 Baseline Classifier

The task-specific model (M) is a BERT-based classifier, with a dropout layer and a feed-forward layer with Softmax activation. Specifically, for each input sequence, the latent representation of the *[CLS]* special token, which acts as a sentence representation, is forwarded through the added layers to get the final label prediction. We opt to use this as our baseline to closely match with previous works [1,19,37]. All results presented in Sect. 5 are the effects of training this model, with the same configuration, on different datasets.

4 Experimental Setting

4.1 Datasets

We evaluate the proposed approach using four classification datasets, two of which belong to the same domain, namely Movie Reviews are examined in this work, and two are out-of-domain, along with a single in-domain dataset which we consider as unlabeled for domain alignment. The datasets used are:

- **SST-2** [33] a binary sentiment classification dataset (positive and negative) on movie reviews.
- **RT** [26] a binary sentiment classification dataset (positive and negative) on long movie reviews.
- **SNIPS** [11] an intent classification dataset, identifying seven distinct intents (PlayMusic, GetWeather, RateBook, SearchScreeningEvent, SearchCreative-Work, AddToPlaylist, BookRestaurant).
- **TREC** [21] a question classification dataset identifying six question types (Description, Entity, Abbreviation, Human, Location, Numeric).

Taking into account the classification datasets chosen for our experiments, we use the **IMDB** dataset [24], i.e. a binary sentiment (positive and negative) classification dataset on movie reviews, as a Domain alignment corpus. As a result, we consider SST-2 and RT as in-domain datasets and SNIPS and TREC as out-of-domain datasets. We opted for standardized and simple datasets for our experiments so that they are easy to replicate and focused on our objective, rather than introducing multiple levels of complexity. Similarly, for SST-2, SNIPS and TREC we use the same dataset versions as previous works [19,37], while for RT we use the published version of the dataset.

Detailed statistics of the datasets used in this work are presented in Table 1. Importantly, we note that all the datasets used in previous studies (i.e., SST-2, SNIPS, and TREC) contain small examples, with a maximum of 52 words

Table 1. Data statistics for datasets used in downstream tasks and alignment.

Data	No. Examples (train/dev/test)	Example length (max/min/mean)
SST-2	6228/692/1821	52/2/19.28
RT	1200/400/400	2737/17/765.75
SNIPS	13084/700/700	35/2/9.09
TREC	4906/546/500	37/3/10.21
IMDB	75000/0/25000	2470/10/233.77

Table 2. Number of instances of the minority class after simulated imbalance.

Dataset		10%	30%	50%
SST-2	POS	324	972	1622
	NEG	298	896	1494
RT	POS	62	184	310
	NEG	67	189	315

per example across all datasets, while the RT and IMDB datasets we introduce contain longer sequences, with RT being used for evaluation and IMDB for in-domain alignment. Finally, the IMDB training set contains 50000 unlabeled examples, with the number of labeled examples being equal between training and testing.

4.2 In-Domain Data Volumes

LM fine-tuning aims to align the model while avoiding catastrophic forgetting. To fine-tune the three diverse LMs, the volume of data required and its effects in downstream tasks varies. As such, we split the IMDB train set in three volumes, *Small*, *Medium*, and *Large* of 18750, 37500, and 75000 examples, respectively, representing the 25%, 50%, and 100% of the training data. The volumes contain equal parts of positive, negative, and unlabeled examples.

4.3 Low Resource and Imbalance Experiments

Simulating Low Resource. To align with previous work, we perform experiments on both in-domain and out-of-domain datasets. In particular, similar to [19], we create folds containing 10 examples of each class for all datasets. Specifically, for SST-2, SNIPS, and TREC we create 15 folds by randomly sampling examples, while for RT we use the provided 10 folds of the original dataset and sub-sample the examples in those. For each fold, we double the training size by creating 1 synthetic example ($s = 1$) for every original example (Algorithm 1).

Simulating Imbalance. To evaluate the effects of our approach in different imbalanced scenarios, we perform experiments only on the in-domain datasets, given that this work focuses on in-domain alignment. In particular, to investigate the effects on various degrees of imbalance, we create three imbalanced datasets per class by sub-sampling each class of the training set for SST-2 and RT in 10%, 30% and 50% of all the class examples, while keeping the full number of examples of the opposite class. For each sub-sampling class, we create 5 folds through random example selection, to account with variance in the results. We evaluate both doubling the minority class examples ($s = 1$) as well as balancing the datasets ($s = 9$ for 10%, $s = 3$ for 30% and $s = 1$ for 50%), for each original minority example. Table 2 reports the number of instances of the minority class for each imbalance scenario.

4.4   Experimental Setup

We use three pretrained LMs for in-domain alignment and conditional generation. Their configuration during in-domain alignment is described in Table 3. All models were trained until convergence to the *Small*, *Medium*, and *Large* volumes of in-domain data.

Table 3. Language Model configuration for In-Domain Alignment.

Parameters	GPT2	BERT	BART
Common Name	"base"	"base-uncased"	"base"
Training Parameters	117M	110M	140M
Hidden Size	768	768	768
No Encoder/Decoder layers	12/0	0/12	6/6
No Attention Heads	12	12	12
Learning Rate	5e-6	5e-6	5e-6
Batch Size	64	64	32

For Conditional Generation in both Low-resource and Imbalance Setting, the models use the following configurations. GPT2 is trained for 25 epochs with a batch size of 32 and learning rate of 4e-5. Nucleus sampling is used for text generation with `top_p` 0.9, `top_k` 0 and `temperature` 0 [16]. BERT is trained for 10 epochs with a batch size of 8 and a learning rate of 4e-5. The masking probability is set to 15% and the maximum number of masked tokens is 256 per sequence. BART is trained for 30 epochs with a batch size of 12 and a learning rate of 1e-5. Token masking rate in both word and span masking is 40%. The number of examples generated depends on the experiment setting (Sect. 4.3) and the threshold (d) is set so that it selects all classes in low-resource setting and only the minority class in imbalanced setting.

Lastly, our Baseline classifier is using the `bert-base-uncased` configuration (Table 3), trained for 8 epochs with a batch size of 8 and a learning rate of 4e-5. In all experiments we use the Adam optimizer [18]. The codebase to reproduce all the experiments is available on Github[1].

---

[1] https://github.com/M4D-MKLab-ITI/Domain-aligned-Data-Augmentation.

5   Results

We consider two sets of results, for *low-resource setting* and for *imbalance setting*. In each setting, we compare our baseline classifier's performance trained on data with no augmentation, augmented using pretrained LMs (*Tuned*) [19], and augmented using domain-aligned LMs (*AlignedTuned*); the latter two double the minority class size, i.e., consider $s = 1$. In the imbalance setting, *Balanced* is further used to denote models which generate enough examples to balance the class instances, namely 9, 3, and 1 examples per minority class example, for 10%, 30%, and 50% imbalance, respectively (see also Sect. 4.3). For brevity, we use subscripts T, AT and B for *Tuned, AlignedTuned* and *Balanced* respectively.

All results presented are product of re-implementation (including the results from the *Tuned* methods) due to the degree of randomness in the example selection process for the fold generation of the SST2, SNIPS, and TREC datasets.

Low-resource setting. In this setting, Table 4 presents the results with no augmentation, the results with augmentation using pretrained LMs (*Tuned*), and the results with augmentation using domain-aligned LMs (*AlignedTuned*) by first listing per LM the best result among the three volume sizes, followed by the results per LM for each volume size.

For in-domain evaluation, we notice that, on SST-2 and RT, the classifier achieves overall better performance when trained on examples created from domain-aligned LMs. Comparing the best performing aligned model with their non-aligned counterpart, statistical significant performance increase ($p < 0.05$) is achieved in all experiments on the RT dataset[2] and we see a noticeable but not significant increase on SST-2 ($p \approx 0.20$). As an exception, the aligned BART with span masking on the SST-2 dataset exhibits a loss in overall Accuracy. This comes in line with the findings of [19], where different BART masking strategies perform better in different datasets. In out-of-domain evaluation, on TREC and SNIPS, the results reveal increase in Accuracy, but no statistically important, when the examples are generated from GPT2 and CBERT, while aligned BART generated examples generally appear to have an opposite effect.

Examining the performance of the domain-aligned LMs further, with respect to the volume sizes, we notice that in-domain performance tends to increase with volume size, before it slightly drops. However, even the degraded scores are overall better than the ones achieved by the non-aligned LM generated examples. In out-of-domain datasets, we notice a performance drop, proportional to the level of alignment of the LMs with the task in hand; exceptions rise in the form of CBERT and GPT2 where we notice an increase in mean Accuracy when aligned on *Small* and *Large* volumes respectively.

Imbalance Setting. For imbalanced evaluation, we test three different levels of imbalance with F1-score using, for the synthetic data generation, the best aligned LMs in terms of data volume from each LM type (see Table 4 for the best performing *AlignedTuned* LMs). We opt to use F1-score, as Accuracy is plagued with majority class bias and hence inherently flawed in this setting

---

[2] BARTword Aligned has a $p < 0.06$ due to high STD.

Table 4. Low-resource classifier performance with in-domain aligned models (T: *Tuned*, AT: *AlignedTuned*). Bold scores are the best score *per model per dataset*. Underlined scores indicate improvement over lower volume alignment and non-augmented scores *per model per dataset*. Bold & underlined scores are best scores *per dataset*.

Model	SST-2	RT	TREC	SNIPS
NoAug	52.817±4.174	55.5 ± 6.5	43.506 ± 11.364	85.714 ± 2.794
GPT2$_T$	57.250 ± 5.998	57.5 ± 7.158	55.146 ± 9.912	85.714 ± 2.794
CBERT$_T$	59.549 ± 5.706	61.5 ± 9.233	57.146 ± 8.554	87.171 ± 3.452
BARTword$_T$	59.205 ± 4.168	60.5 ± 7.566	58.786 ± 6.193	85.476 ± 3.198
BARTspan$_T$	59.769 ± 3.976	61.5 ± 7.762	57.386 ± 8.599	87.074 ± 2.835
GPT2$_{AT}$	**58.290** ± 5.071	**59.0** ± 6.244	**57.426** ± 4.768	**87.019** ± 3.163
CBERT$_{AT}$	**60.505** ± 6.137	**65.5** ± 7.889	**59.986** ± 8.107	**87.847** ± 2.128
BARTword$_{AT}$	**60.464** ± 4.628	**64.0** ± 12.806	56.040 ± 8.208	**85.876** ± 2.719
BARTspan$_{AT}$	58.528 ± 4.198	**65.0** ± 7.416	55.200 ± 8.304	85.477 ± 2.680
		Small Volume		
GPT2$_{AT}$	<u>57.744</u> ± 5.262	<u>58.0</u> ± 6.403	54.666 ± 8.514	<u>85.809</u> ± 2.885
CBERT$_{AT}$	**60.505** ± 6.137	<u>63.0</u> ± 9.797	**59.986** ± 8.107	**87.847** ± 2.128
BARTword$_{AT}$	58.586 ± 5.085	**64.0** ± 12.806	56.040 ± 8.208	**85.876** ± 3.676
BARTspan$_{AT}$	58.528 ± 4.198	<u>64.0</u> ± 9.695	55.200 ± 8.304	85.477 ± 2.680
		Medium Volume		
GPT2$_{AT}$	**58.290** ± 5.071	**59.0** ± 6.244	55.133 ± 7.838	<u>86.828</u> ± 2.872
CBERT$_{AT}$	60.505 ± 6.137	63.0 ± 9.797	59.986 ± 8.107	87.847 ± 2.128
BARTword$_{AT}$	**60.464** ± 4.628	61.5 ± 11.191	55.093 ± 10.711	85.847 ± 2.719
BARTspan$_{AT}$	58.455 ± 4.493	63.5 ± 7.762	50.386 ± 10.832	84.057 ± 3.553
		Large Volume		
GPT2$_{AT}$	57.426 ± 4.768	59.0 ± 8.306	**57.426** ± 4.768	**87.019** ± 3.163
CBERT$_{AT}$	58.056 ± 5.765	**65.5** ± 7.889	58.866 ± 8.161	87.647 ± 2.360
BARTword$_{AT}$	60.464 ± 4.628	61.5 ± 11.191	53.306 ± 10.421	84.561 ± 3.465
BARTspan$_{AT}$	58.191 ± 4.034	**65.0** ± 7.416	49.933 ± 8.806	85.028 ± 2.585

[7,23]. We compare the classifier's performance with examples generated from the aligned LMs to that with examples generated from pretrained LMs, as well as without any augmentation.

Starting with SST-2 (Table 5), we note that the aligned models performed better than their non-aligned counterparts in all settings, with better F1-score and lower standard deviation among folds. However, not all improvements are statistically important, with all GPT2 models failing to improve over baseline results. In addition, performance tends to degrade when artificially balancing the instances with all models.

Interestingly, statistical significant improvements are mostly noted when the Negative class is the subservient one, which maintains lower scores in all settings

Table 5. F1-score score of imbalanced setting with artificial imbalance in Positive (Pos) or Negative (Neg) label for the SST-2 dataset (T: *Tuned*, TB: *TunedBalanced*, AT: *AlignedTuned* and ATB: *AlignedTunedBalanced*). Bold scores are the best score *per model per dataset*. Bold & underlined scores are best scores per dataset. Statistical significant improvement ($p < 0.05$) over non-alinged counterparts is shown with *.

	SST-2					
Model	Pos 10%	Neg 10%	Pos 30%	Neg 30%	Pos 50%	Neg 50%
NoAug	82.65 ± 0	33.30 ± 0	89.24 ± 0	86.63 ± 0	90.14 ± 0	89.97 ± 0
GPT2$_T$	77.64 ± 3.5	56.41 ± 12.1	88.06 ± 1.0	83.06 ± 1.9	89.36 ± 1.1	87.61 ± 5.0
GPT2$_{AT}$	**78.81** ± 2.8*	**58.63** ± 5.1*	**88.56** ± 0.5	**83.53** ± 1.6	**89.57** ± 0.5	**88.37** ± 0.2*
GPT2$_{TB}$	57.82 ± 1.0	54.72 ± 1.3	75.13 ± 2.4	72.58 ± 2.2	89.36 ± 1.1	87.61 ± 5.0
GPT2$_{ATB}$	58.24 ± 1.3	55.03 ± 1.1	76.26 ± 2.1*	73.31 ± 1.9	**89.57** ± 0.5	**88.37** ± 0.2*
CBERT$_T$	87.48 ± 0.6	77.23 ± 1.6	90.5 ± 0.6	88.32 ± 0.7	91.03 ± 0.2	90.29 ± 0.6
CBERT$_{AT}$	<u>**87.99**</u> ± 0.7	<u>**78.64**</u> ± 0.9*	<u>**90.52**</u> ± 0.2	<u>**88.72**</u> ± 0.7*	<u>**91.4**</u> ± 0.2	<u>**90.58**</u> ± 0.4*
CBERT$_{TB}$	84.64 ± 1.2	76.34 ± 2.7	90.11 ± 0.4	87.68 ± 0.6	91.03 ± 0.2	90.29 ± 0.6
CBERT$_{ATB}$	85.12 ± 0.6	77.95 ± 1.2*	90.26 ± 0.4	88.70 ± 0.3*	<u>**91.4**</u> ± 0.2	<u>**90.58**</u> ± 0.4*
BART$_T$	82.68 ± 3.0	77.63 ± 2.4	89.79 ± 0.5	86.77 ± 1.2	90.79 ± 0.4	89.22 ± 0.4
BART$_{AT}$	**84.56** ± 1.4*	**78.10** ± 1.3*	**90.10** ± 0.3*	87.11 ± 0.1*	**91.12** ± 0.3	**89.47** ± 0.5
BART$_{TB}$	83.5 ± 0.7	77.14 ± 1.2	89.78 ± 0.8	88.40 ± 0.7	90.79 ± 0.4	89.22 ± 0.4
BART$_{ATB}$	84.32 ± 1.0	77.70 ± 1.0	89.88 ± 0.2	**88.60** ± 0.6	**91.12** ± 0.3	**89.47** ± 0.5

than with a Positive subservient class. These results verify an inherit difficulty in predicting negative examples due to high average mutual information between class examples, pointing to high similarity between same class texts, the use of sarcasm which portraits them as positive texts, and noise due to mislabeled examples [10].

In RT (Table 6), the performance advantage between Positive and Negative subservient class datasets is more equally divided, depending on the augmentation approach. In addition, in this longer text dataset the alignment gains are more pronounced with almost all aligned models achieving statistically significant improvement over their non aligned counterparts. More importantly, when the subservient class becomes balanced, we notice dramatic improvement in performance compared to just doubling the minority class examples.

Specifically, we note that in all the 10% imbalance cases on RT, the augmentation proved ineffective when doubling the subservient class examples and the classifier failed to predict the minority class. In comparison, the balanced counterparts almost double the performance in the 10% and 30% imbalance settings. This jump in performance is attributed both to the extra examples and the quality of the generated examples, which are longer and allow for more diversity.

Overall, balancing the minority class showcases very different behavior in SST-2 and RT, with the first degrading and the second improving. This is attributed to both the size of the datasets, where the subservient class at 50% imbalance of RT has the same number of instances as the subservient class at

Table 6. F1-score of imbalanced setting with artificial imbalance in Positive (Pos) or Negative (Neg) label for the RT dataset (T: *Tuned*, TB: *TunedBalanced*, AT: *Aligned-Tuned* and ATB: *AlignedTunedBalanced*). Bold scores are the best score *per model per dataset*. Bold & underlined scores are best scores per dataset. Statistical significant improvement ($p < 0.05$) over non-alinged counterparts is shown with *.

	RT					
Model	Pos 10%	Neg 10%	Pos 30%	Neg 30%	Pos 50%	Neg 50%
NoAug	33.33 ± 0	33.33 ± 0	73.29 ± 2.2	33.33 ± 0	83.85 ± 0	64.06 ± 0
GPT2$_T$	33.33 ± 0	33.33 ± 0	34.52 ± 1.8	39.67 ± 14.5	80.51 ± 3.6	71.21 ± 16.23
GPT2$_{AT}$	33.33 ± 0	33.33 ± 0	$42.00 \pm 8.3^*$	$45.12 \pm 12.6^*$	**81.31 ± 5.2**	**$78.56 \pm 6.5^*$**
GPT2$_{TB}$	59.95 ± 5.6	62.10 ± 5.4	80.06 ± 2.9	70.03 ± 13.8	80.51 ± 3.6	71.21 ± 16.23
GPT2$_{ATB}$	**$61.21 \pm 11.1^*$**	**$63.25 \pm 3.5^*$**	**81.11 ± 1.1**	**$78.09 \pm 3.7^*$**	**81.31 ± 5.2**	**$78.56 \pm 6.5^*$**
CBERT$_T$	33.33 ± 0	33.33 ± 0	70.74 ± 20.0	57.04 ± 20.9	72.17 ± 19.5	76.22 ± 19.4
CBERT$_{AT}$	33.33 ± 0	33.33 ± 0	$75.60 \pm 11.8^*$	$58.92 \pm 20.9^*$	**$84.41 \pm 0.9^*$**	**$82.35 \pm 3.5^*$**
CBERT$_{TB}$	74.92 ± 4.5	65.59 ± 6.4	82.39 ± 2.2	73.39 ± 3.3	72.17 ± 19.5	76.22 ± 19.4
CBERT$_{ATB}$	**$76.19 \pm 3.0^*$**	**$68.69 \pm 4.2^*$**	**$\underline{84.18} \pm 1.2^*$**	**$74.38 \pm 2.9^*$**	**$\underline{84.41} \pm 0.9^*$**	**$82.35 \pm 3.5^*$**
BART$_T$	33.33 ± 0	33.33 ± 0	39.28 ± 11.9	50.38 ± 21.1	37.38 ± 20.0	83.37 ± 2.5
BART$_{AT}$	33.33 ± 0	33.33 ± 0	39.50 ± 11.8	$57.47 \pm 18.8^*$	**$83.39 \pm 0.5^*$**	**$\underline{84.47} \pm 2.4^*$**
BART$_{TB}$	61.82 ± 13.0	55.69 ± 6.9	80.60 ± 3.2	73.84 ± 20.3	37.38 ± 20.0	83.37 ± 2.5
BART$_{ATB}$	**$62.32 \pm 9.5^*$**	**$66.11 \pm 6.2^*$**	**81.72 ± 3.3**	**$\underline{81.74} \pm 4.4^*$**	**$83.39 \pm 0.5^*$**	**$\underline{84.47} \pm 2.4^*$**

10% imbalance of SST-2, and their characteristics, i.e., short and long texts that grant different levels of generation freedom to the models. We intrinsically notice less diverse generated examples for SST-2 balancing, which led to overfitting issues of the characterstics of the generated examples. In RT, this phenomenon is universally less pronounced, leading to the increase in performance.

6 Discussion

Overall, we show that in-domain alignment can have a positive effect in example generation when the labeled set of training data for a certain classification task is either very small or characterized by imbalance.

In particular, we observe that in low-resource settings, using only a small amount of labeled data, we can generate synthetic examples for all classes, boosting the performance of the text classifier. Furthermore, by only creating synthetic examples of the minority classes in imbalanced scenarios, we significantly help improve the performance, especially in the case of RT. Importantly, the experimental results showcase even when balancing the dataset, synthetic examples are not a replacement for real examples, but they can lead to significantly better performance. We further found that generating more than one synthetic example from each training example, longer sequences allowed for higher degree of freedom to the model, which translated into improved performance. Short texts on the other hand exhibited the opposite effect, with repeated synthetic examples.

184 N. Stylianou et al.

In addition, by studying the effects of unlabeled data volume in the downstream tasks, we notice that different LM architectures operate best under different data volumes. A correlation between the number of training parameters of the LM and the volume of data exists, with CBERT operating better in the *Small* and *Medium* ranges, while also having the best lowest parameter count, GPT2 performing best in the *Medium* range, and BART performing best in the *Medium* and *Large* ranges.

This fluctuation in volume sizes of the same LM architecture is attributed to the different characteristics of the in-domain datasets. RT is characterized by considerably larger sequences than SST-2 and we dynamically select the number of tokens to be replaced or generated depending on the architecture. Hence, the longer the sequence, the more predictions are required by the LM and the harder it becomes to generate quality examples.

In terms of out-of-domain evaluation, performance generally degrades proportionally to the level of LM alignment. While this is expected, it is important to highlight it, as it limits the usability of the aligned LMs in other tasks. Overall, GPT2 appears to handle best out-of-domain tasks, which can be attributed to its autoregressive nature, while both other models replace only parts of the original examples through masking predictions.

However, in imbalanced settings and especially on the RT dataset we find that model performance varies depending on the class examples to be generated and the quantity of available training data. With positive examples as the minority class, performance is overall better in all models, with the exception of GPT2 which excels in generating negative examples on the RT dataset. As the same behavior is exhibited in both aligned and non-aligned models, it cannot be attributed to the alignment dataset, but can be attributed to the models' characteristics.

Overall, the aligned models improved over their non-aligned counter parts in both low-resource and imbalanced settings with mask-based augmentation models (i.e., BERT and BART) performed better than their generative counterpart (i.e., GPT2). However, examples generated from GPT2 were more diverse and different generation strategies can significantly impact performance.

7   Conclusions and Future Work

Current Data Augmentation approaches focus on either very specific augmentation techniques which are hard to transfer to other tasks or generic approaches to filter large quantities of automatically created examples. We propose the use of in-domain alignment for LM-based Data Augmentation in low-resource and imbalance Text Classification tasks.

By aligning the model, better synthetic examples are generated that can boost the performance of the in-domain tasks. We also find a positive correlation between volume of unlabeled data for in-domain alignment and downstream performance, as well as identify performance degrading point which can inform future applications.

While our approach creates better examples, generating a plethora of examples from a single example is non-trivial, as evident by our experimental results when balancing imbalanced datasets. Improving the text generation process so that it better scales to the generation of more examples, while remaining invariant to text characteristics, remains a future work. Besides the creation of better examples from LMs, our approach can be bootstrapped to semi-supervised or active learning approaches, following other works, that can help filter out generated examples and increase the performance further.

Acknowledgements. This project has received funding from the European Union's Horizon 2020 research and innovation programme under grant agreements No 101021797 (STARLIGHT) and No 833464 (CREST).

References

1. Anaby-Tavor, A., et al.: Do not have enough data? Deep learning to the rescue! In: Proceedings of the AAAI Conference on Artificial Intelligence, vol. 34, pp. 7383–7390 (2020)
2. Baldi, P., Sadowski, P.J.: Understanding dropout. In: Advances in Neural Information Processing Systems, vol. 26 (2013)
3. Bayer, M., Kaufhold, M.A., Buchhold, B., Keller, M., Dallmeyer, J., Reuter, C.: Data augmentation in natural language processing: a novel text generation approach for long and short text classifiers. Int. J. Mach. Learn. Cybern., 1–16 (2022)
4. Bayer, M., Kaufhold, M.A., Reuter, C.: A survey on data augmentation for text classification. ACM Comput. Surv. **55**(7), 1–39 (2022)
5. Bommasani, R., et al.: On the opportunities and risks of foundation models. arXiv preprint arXiv:2108.07258 (2021)
6. Brown, T., et al.: Language models are few-shot learners. Adv. Neural. Inf. Process. Syst. **33**, 1877–1901 (2020)
7. Brzezinski, D., Stefanowski, J., Susmaga, R., Szczech, I.: On the dynamics of classification measures for imbalanced and streaming data. IEEE Trans. Neural Netw. Learn. Syst. **31**(8), 2868–2878 (2019)
8. Chawla, N.V., Bowyer, K.W., Hall, L.O., Kegelmeyer, W.P.: Smote: synthetic minority over-sampling technique. J. Artif. Intell. Res. **16**, 321–357 (2002)
9. Chen, J., Shen, D., Chen, W., Yang, D.: HiddenCut: simple data augmentation for natural language understanding with better generalizability. In: Proceedings of the 59th Annual Meeting of the Association for Computational Linguistics and the 11th International Joint Conference on Natural Language Processing (Volume 1: Long Papers), pp. 4380–4390. Association for Computational Linguistics, August 2021. https://doi.org/10.18653/v1/2021.acl-long.338
10. Collins, E., Rozanov, N., Zhang, B.: Evolutionary data measures: understanding the difficulty of text classification tasks. In: Proceedings of the 22nd Conference on Computational Natural Language Learning, pp. 380–391 (2018)
11. Coucke, A., et al.: Snips voice platform: an embedded spoken language understanding system for private-by-design voice interfaces. arXiv preprint arXiv:1805.10190 (2018)

12. Devlin, J., Chang, M.W., Lee, K., Toutanova, K.: BERT: pre-training of deep bidirectional transformers for language understanding. In: Proceedings of the 2019 Conference of the North American Chapter of the Association for Computational Linguistics: Human Language Technologies, Volume 1 (Long and Short Papers), pp. 4171–4186 (2019)

13. Feng, S.Y., Li, A.W., Hoey, J.: Keep calm and switch on! Preserving sentiment and fluency in semantic text exchange. In: Proceedings of the 2019 Conference on Empirical Methods in Natural Language Processing and the 9th International Joint Conference on Natural Language Processing (EMNLP-IJCNLP), pp. 2701–2711. Association for Computational Linguistics, Hong Kong, November 2019. https://doi.org/10.18653/v1/D19-1272

14. Fernández, A., García, S., Galar, M., Prati, R.C., Krawczyk, B., Herrera, F.: Learning from Imbalanced Data Sets. Springer, Cham (2018). https://doi.org/10.1007/978-3-319-98074-4

15. Guo, H., Mao, Y., Zhang, R.: Augmenting data with Mixup for sentence classification: an empirical study. arXiv preprint arXiv:1905.08941 (2019)

16. Holtzman, A., Buys, J., Du, L., Forbes, M., Choi, Y.: The curious case of neural text degeneration. In: International Conference on Learning Representations (2020)

17. Karimi, A., Rossi, L., Prati, A.: AEDA: An easier data augmentation technique for text classification. In: Findings of the Association for Computational Linguistics: EMNLP 2021, pp. 2748–2754. Association for Computational Linguistics, Punta Cana, November 2021

18. Kingma, D.P., Ba, J.: Adam: a method for stochastic optimization. arXiv preprint arXiv:1412.6980 (2014)

19. Kumar, V., Choudhary, A., Cho, E.: Data augmentation using pre-trained transformer models. In: Proceedings of the 2nd Workshop on Life-long Learning for Spoken Language Systems, pp. 18–26. Association for Computational Linguistics, Suzhou, December 2020

20. Lewis, M., et al.: BART: denoising sequence-to-sequence pre-training for natural language generation, translation, and comprehension. In: Proceedings of the 58th Annual Meeting of the Association for Computational Linguistics, pp. 7871–7880 (2020)

21. Li, X., Roth, D.: Learning question classifiers. In: COLING 2002: The 19th International Conference on Computational Linguistics (2002)

22. Liu, Y., et al.: RoBERTa: a robustly optimized BERT pretraining approach. arXiv preprint arXiv:1907.11692 (2019)

23. Luque, A., Carrasco, A., Martín, A., de Las Heras, A.: The impact of class imbalance in classification performance metrics based on the binary confusion matrix. Pattern Recogn. **91**, 216–231 (2019)

24. Maas, A.L., Daly, R.E., Pham, P.T., Huang, D., Ng, A.Y., Potts, C.: Learning word vectors for sentiment analysis. In: Proceedings of the 49th Annual Meeting of the Association for Computational Linguistics: Human Language Technologies, pp. 142–150. Association for Computational Linguistics, Portland, June 2011

25. Mani, I., Zhang, I.: KNN approach to unbalanced data distributions: a case study involving information extraction. In: Proceedings of Workshop on Learning From Imbalanced Datasets, vol. 126, pp. 1–7. ICML (2003)

26. Pang, B., Lee, L.: A sentimental education: Sentiment analysis using subjectivity summarization based on minimum cuts. In: Proceedings of the 42nd Annual Meeting of the Association for Computational Linguistics (ACL 2004), pp. 271–278 (2004)

27. Qu, Y., Shen, D., Shen, Y., Sajeev, S., Chen, W., Han, J.: CoDA: contrast-enhanced and diversity-promoting data augmentation for natural language understanding. In: International Conference on Learning Representations (2020)

28. Radford, A., Wu, J., Child, R., Luan, D., Amodei, D., Sutskever, I., et al.: Language models are unsupervised multitask learners. OpenAI blog **1**(8), 9 (2019)

29. Rae, J.W., et al.: Scaling language models: methods, analysis & insights from training gopher. arXiv preprint arXiv:2112.11446 (2021)

30. Rosenfeld, J.S.: Scaling laws for deep learning. arXiv preprint arXiv:2108.07686 (2021)

31. Shen, D., Zheng, M., Shen, Y., Qu, Y., Chen, W.: A simple but tough-to-beat data augmentation approach for natural language understanding and generation. arXiv preprint arXiv:2009.13818 (2020)

32. Shleifer, S.: Low resource text classification with ULMFit and backtranslation. arXiv preprint arXiv:1903.09244 (2019)

33. Socher, R., et al.: Recursive deep models for semantic compositionality over a sentiment treebank. In: Proceedings of the 2013 Conference on Empirical Methods in Natural Language Processing, pp. 1631–1642 (2013)

34. Vaswani, A., et al.: Attention is all you need. In: Advances in Neural Information Processing Systems, vol. 30 (2017)

35. Wei, J., Zou, K.: EDA: easy data augmentation techniques for boosting performance on text classification tasks. In: Proceedings of the 2019 Conference on Empirical Methods in Natural Language Processing and the 9th International Joint Conference on Natural Language Processing (EMNLP-IJCNLP), pp. 6382–6388. Association for Computational Linguistics, Hong Kong, November 2019. https://doi.org/10.18653/v1/D19-1670

36. Wu, T., Ribeiro, M.T., Heer, J., Weld, D.S.: Polyjuice: generating counterfactuals for explaining, evaluating, and improving models. In: Proceedings of the 59th Annual Meeting of the Association for Computational Linguistics and the 11th International Joint Conference on Natural Language Processing (Volume 1: Long Papers), pp. 6707–6723 (2021)

37. Wu, X., Lv, S., Zang, L., Han, J., Hu, S.: Conditional BERT contextual augmentation. In: Rodrigues, J.M.F., et al. (eds.) ICCS 2019. LNCS, vol. 11539, pp. 84–95. Springer, Cham (2019). https://doi.org/10.1007/978-3-030-22747-0_7

38. Xie, Z., et al.: Data noising as smoothing in neural network language models. In: 5th International Conference on Learning Representations, ICLR 2017 (2017)

39. Yang, Z., Dai, Z., Yang, Y., Carbonell, J., Salakhutdinov, R.R., Le, Q.V.: XLNet: generalized autoregressive pretraining for language understanding. In: Wallach, H., Larochelle, H., Beygelzimer, A., d' Alché-Buc, F., Fox, E., Garnett, R. (eds.) Advances in Neural Information Processing Systems, vol. 32. Curran Associates, Inc. (2019)

Privacy-Preserving Fair Item Ranking

Jia Ao Sun[1,2(✉)], Sikha Pentyala[1,3], Martine De Cock[3,4], and Golnoosh Farnadi[1,2,5]

[1] Mila - Quebec AI Institute, Montréal, QC, Canada
{sunjiaao,farnadig}@mila.quebec
[2] Université de Montréal, Montréal, QC, Canada
[3] University of Washington, Tacoma, WA, USA
{sikha,mdecock}@uw.edu
[4] Ghent University, Ghent, Belgium
[5] HEC Montréal, Montréal, QC, Canada

Abstract. Users worldwide access massive amounts of curated data in the form of rankings on a daily basis. The societal impact of this ease of access has been studied and work has been done to propose and enforce various notions of fairness in rankings. Current computational methods for fair item ranking rely on disclosing user data to a centralized server, which gives rise to privacy concerns for the users. This work is the first to advance research at the conjunction of producer (item) fairness and consumer (user) privacy in rankings by exploring the incorporation of privacy-preserving techniques; specifically, differential privacy and secure multi-party computation. Our work extends the equity of amortized attention ranking mechanism to be privacy-preserving, and we evaluate its effects with respect to privacy, fairness, and ranking quality. Our results using real-world datasets show that we are able to effectively preserve the privacy of users and mitigate unfairness of items without making additional sacrifices to the quality of rankings in comparison to the ranking mechanism in the clear.

Keywords: Ranking · Privacy · Fairness

1 Introduction

Information systems, such as those used for search retrieval and recommendation, have become a ubiquitous part of our daily lives. These systems provide users with curated information such as content produced in social media or web pages, or organize the information in a way that is most relevant to users to aid their decision-making on what to buy, what to watch, or who to hire. These results are often outputted in the form of ranked lists. To maximize the utility of the system for each user, the ranked list is ordered by decreasing relevance (based on some score or rating) to the user. Users are then susceptible to paying most of their attention to the highest-ranked items in the ranked list, causing *position bias* [12,13]. For a single ranked list (generated for a single user or a single query), the

J. Kamps et al. (Eds.): ECIR 2023, LNCS 13981, pp. 188–203, 2023.
https://doi.org/10.1007/978-3-031-28238-6_13

attention given to each item would decrease at a faster rate than relevance, *i.e.*, lower-ranked items become disadvantaged by receiving disproportionately less attention relative to their relevance. When a sequence of ranked lists is generated in this manner, higher-ranked items reap increasingly disproportionately more attention relative to their relevance. This results in economic and social impacts on major stakeholders–the item providers, *a.k.a.* producers–of the information systems, leading to economic disparity, bias toward underrepresented producers, and unhealthy markets [14,17].

Methods for mitigating unfairness in rankings have been proposed in the literature [18,19,25,30,31]. In our work, we focus on post-processing techniques that reorder ranked lists to distribute exposure fairly [3,24]. These methods assume that the rankings or preferences of the users are available to a central entity that can then apply post-processing techniques to achieve fairness for the items. Centralizing user data can lead to privacy leakage or even intentional privacy violations, such as when the data is routinely sold when companies undergo bankruptcy [4]. The growing awareness of the need for more stringent user privacy protections, as well as the requirement to comply with regulations such as the GDPR[1], is prompting the increased use of privacy-enhancing technologies (PETs) in recommender systems [1,6,11,28,29]. To the best of our current knowledge, no works address the challenge of achieving fairness for producers (*a.k.a.* items) while preserving the privacy of consumers (*a.k.a.* users) in ranking systems. In this paper, we explore this under-researched area in the literature.

We address the problem where each user (client) u_l has a list ρ_l of items $d_1^l, d_2^l, \ldots, d_i^l, \ldots, d_n^l$ that was generated in a privacy-preserving manner and ranked according to relevance. During a post-processing phase for bias mitigation, we wish to alter every user's ranking ρ_l to optimize for equity of amortized attention [3] without requiring any user disclosure of ρ_l to a centralized entity. One could achieve this goal with secure multi-party computation (MPC), *i.e.*, by having each client encrypt their data to send to a set of MPC servers, having the MPC servers perform all computations needed for reranking over the encrypted data, and then return the results to the client to decrypt. While this approach would preserve end-to-end privacy and produce the same rankings as the bias mitigation method from Biega et al. [3] yields in the clear (*i.e.*, without any encryption), the computational and communication overhead would be prohibitively large for practical applications. We therefore propose a much more scalable approach by combining MPC to preserve input privacy with differential privacy (DP) to preserve output privacy. In our solution, MPC servers store intermediate results and then perturb them with DP noise, which clients subsequently use to perform local computations. We demonstrate the applicability of our proposed solution by experimenting on three real-world datasets. We empirically show that fairness for items can be improved while ensuring the privacy of users, without making additional sacrifices to the ranking quality.

---

[1] European General Data Protection Regulation https://gdpr-info.eu/.

2   Preliminaries

2.1   Fairness in Ranking

To maximize utility, ranking algorithms generate lists of items sorted by their predicted level of relevance to a query or user. The most relevant items are positioned at the top of the list and receive the most exposure. Fairness in rankings is concerned with distributing exposure to the ranked items in order to mitigate the consequences of position bias. Many fairness notions have been proposed in this context [5,22,24,30,31]. In this paper, we consider the fairness notion based on the attention received by items, which is dependent on the items' exposure, as proposed by Biega et al. [3].

Equity of Amortized Attention (EOAA). Biega et al. introduced the fairness notion of *equity of amortized attention* to achieve *amortized individual fairness* for a set of items $\mathcal{D} = \{d_1, d_2, \ldots, d_i, \ldots, d_n\}$ appearing in a sequence of relevance-based rankings $\rho_1, \rho_2, \ldots, \rho_l, \ldots, \rho_L$ for users $u_1, u_2, \ldots, u_l, \ldots, u_L$, respectively [3]. The position of item d_i in ranking ρ_l influences the amount of attention that d_i receives. EOAA is achieved if each item d_i receives cumulative attention A_i proportional to its cumulative relevance R_i, when amortized over a sequence of rankings. Biega et al. define a measure for this notion of fairness by taking the sum of the absolute differences between A_i and R_i for $i = 1 \ldots n$ as shown in Eq. 1:

$$unfairness(\rho_1, ..., \rho_L) = \sum_{i=1}^{n} |A_i - R_i| = \sum_{i=1}^{n} \left| \sum_{l=1}^{L} a_i^l - \sum_{l=1}^{L} r_i^l \right|. \qquad (1)$$

For each item d_i, r_i^l is its relevance score for user u_l, and a_i^l is the amount of attention it receives in ranking ρ_l. Biega et al. proposed to improve the EOAA of a sequence of relevance-based rankings $\rho_1, \rho_2, \ldots, \rho_l, \ldots, \rho_L$ by reranking each of the rankings sequentially to produce $\rho_1^*, \rho_2^*, \ldots, \rho_l^*, \ldots, \rho_L^*$. To rerank ρ_l, taking into account the previously computed rerankings $\rho_1^*, \rho_2^*, \ldots, \rho_{l-1}^*$, the following post-processing linear program (ILP) is solved:

$$\text{Minimize} \sum_{i=1}^{n} \sum_{j=1}^{n} \left| A_i^{l-1} + \hat{w}_j - (R_i^{l-1} + \hat{r}_i^l) \right| \cdot X_{i,j} \qquad (2)$$

$$\text{Subject to} \sum_{j=1}^{k} \sum_{i=1}^{n} \frac{2^{\hat{r}_i^l} - 1}{\log_2(j+1)} X_{i,j} \geq \theta \cdot DCG(\rho_l)@k \qquad (3)$$

$$X_{i,j} \in \{0,1\}, \forall_{i,j} \text{ and } \sum_i X_{i,j} = 1, \forall_j \text{ and } \sum_j X_{i,j} = 1, \forall_i. \qquad (4)$$

This ILP solves for n^2 decision variables $X_{i,j}$ that represent the reranking ρ_l^* of the items in ranking ρ_l. A_i^{l-1} and R_i^{l-1} are the accumulated attention and relevance of item d_i, respectively, over rerankings $\rho_1^*, \rho_2^*, \ldots, \rho_{l-1}^*$. $\hat{w}_j$ is the normalized attention weight of placing an item at position j, calculated as $\hat{w}_j = w_j / \sum_{t=1}^{n} w_t$, based on the geometric attention model $w_j = 0.5(0.5)^{j-1}$ assumed in [3]. $\hat{r}_i^l$ is the normalized relevance score of r_i^l, calculated as $\hat{r}_i^l = \frac{r_i^l - r_{\min}}{r_{\max} - r_{\min}} / \sum_{t=1}^{n} \frac{r_t^l - r_{\min}}{r_{\max} - r_{\min}}$. $r_{\max}$ and $r_{\min}$ are the maximum and minimum relevance scores, respectively, that a user can have for an item. The coefficient $\left| A_i^{l-1} + w_j - (R_i^{l-1} + r_i^l) \right|$ is the unfairness measure of placing an item d_i at position j in the current l^{th} reranking. The constraint in Eq. 3 bounds the quality of the first k items in the reranking in terms of its discounted cumulative gain (DCG), such that it is no lower than $0 \leq \theta \leq 1$ times the DCG of the top-k items of the original relevance-based ranking. The constraints in Eq. 4 specify that the n^2 decision variables $X_{i,j}$ are binary, and that there is a single 1 per j for all i's, and a single 1 per i for all j's. $X_{i,j} = 1$ indicates item d_i is placed in position j, and $X_{i,j} = 0$ otherwise.

The quality of reranking ρ_l^* is measured in terms of its divergence from the original relevance-based ranking ρ_l. This is quantified as $NDCG(\rho_l, \rho_l^*)$ in Eq. 5, where $DCG(\rho_l^*)$ is normalized by $DCG(\rho_l)$.

$$NDCG(\rho_l, \rho_l^*) = \frac{DCG(\rho_l^*)}{DCG(\rho_l)} \tag{5}$$

The maximum NDCG value is 1, and occurs when either $\rho_l = \rho_l^*$, or if items of equal relevance scores are shuffled with each other.

2.2   Privacy-Enhancing Technologies (PETs)

Differential Privacy (DP). DP provides formal guarantees that the result of computations on a dataset D is negligibly affected by the participation of a single user, thereby offering privacy through plausible deniability [9]. Formally, a randomized algorithm $\mathcal{F}$ provides ϵ-DP if for all pairs of neighboring datasets D and D' (i.e. datasets that differ in one entity), and for all subsets S of $\mathcal{F}$'s range

$$P(\mathcal{F}(D) \in S) \leq e^{\epsilon} \cdot P(\mathcal{F}(D') \in S), \tag{6}$$

where ϵ is the privacy budget or privacy loss. The smaller the value of ϵ, the stronger the privacy guarantees. An ϵ-DP algorithm $\mathcal{F}$ is usually created out of an algorithm $\mathcal{F}^*$ by adding noise that is proportional to the *sensitivity* of $\mathcal{F}^*$, in which the sensitivity measures the maximum impact a change in the underlying dataset can have on the output of $\mathcal{F}^*$. In our paper, $\mathcal{F}^*$ performs aggregation of the relevance scores and the attention weights of items across many users. The traditional DP paradigm–global DP–assumes a central curator who collects data from the users and injects controlled noise either to the inputs, outputs, or both when revealing the computed aggregation. Although this model provides *output privacy*, it requires users to disclose their private data with a central entity. To remove this need for sensitive user information disclosure, in this paper we emulate the central curator from the global DP paradigm with a set of untrusted servers running MPC protocols (see below).

Secure Multi-Party Computation (MPC). MPC [7] is an umbrella term for cryptographic approaches that enable computations over encrypted data. We follow the MPC as a service paradigm in which each data holder encrypts each value x from its input data by converting it into so-called secret shares and subsequently distributes these shares among a set of MPC servers. While the original value x can be trivially reconstructed when all shares are combined, none of the MPC servers by themselves learns anything about the value of x. Next, the MPC servers jointly execute protocols to perform computations over the secret shared values to obtain a secret sharing of the desired output value (in our paper, an aggregated unfairness measure perturbed with noise to provide DP). MPC is concerned with the protocol execution coming under attack by an adversary, which may corrupt one or more of the parties to learn private information or cause the result of the computation to be incorrect. The MPC protocols that we use in this paper are designed to prevent such attacks from being successful, and they are mathematically proven to guarantee *input privacy* and correctness.

Threat Model. An adversary can corrupt any number of parties. In a *dishonest-majority* setting, half or more of the parties may be corrupt, while in an *honest-majority* setting, more than half of the parties are honest (not corrupted). Furthermore, the adversary can be a *semi-honest* or a *malicious* adversary. While a party corrupted by a semi-honest or 'passive' adversary follows the protocol instructions correctly but tries to obtain additional information, parties corrupted by malicious or 'active' adversaries can deviate from the protocol instructions. The protocols we use in this paper are sufficiently generic to be used in dishonest-majority as well as honest-majority settings, with passive or active adversaries. This is achieved by changing the underlying MPC scheme to align with the desired security setting.

MPC computations are commonly done on integers modulo q, *i.e.*, in the ring $\mathbb{Z}_q$. For instance, in a well-known dishonest majority 2-party (2PC) computation setting with passive adversaries, a data holder converts its input data into $x = x_0 + x_1 \mod q$ and sends x_0 to MPC server S_0, and x_1 to S_1. We use $[\![x]\!] = (x_0, x_1)$ as a shorthand for a secret sharing of x throughout the paper. If the servers have received secret shares y_0 and y_1 of a value y from another data holder, then the servers can compute a secret sharing of $x + y$ as $[\![x + y]\!] = (x_0 + y_0, x_1 + y_1)$ without learning the values of x, y, or their sum. Besides addition, we use an MPC protocol π_{LAP} that enables the MPC servers to sample secret sharings of noise drawn from a Laplace distribution. π_{LAP} is substantially more complex and involves communication between the MPC servers in addition to operations on their own local shares. See [20] for details.

3   Related Work

3.1   Fairness in Rankings

Most work in fairness in rankings has been studied in the context of fairly distributing exposure to the elements of the ranked list. The elements of the ranked list could be people or items (*e.g.*, content, products, places). Much of the work involving exposure has centered around group fairness, where exposure should ideally be distributed equally among different groups defined by their protected attributes (such as gender or race). For example, Yang and Stoyanovich [30] proposed extending the traditional fairness concept of statistical parity to the ranking system, where being a member of a protected group does not influence a person's position in a ranking. Zehlike and Castillo [31] proposed a fair top-k ranking algorithm following the fairness notion of affirmative action, where a minimum number of protected group members are guaranteed in every top-k ranking (top 10, top 20, etc.). Celis et al. [5] proposed an algorithm addressing the constrained ranking maximization problem, where there is a limit to the number of sensitive items per protected group in the ranking, and no one group dominates. Sapiezynski et al. [22] also used statistical parity, but used a geometric distribution to model user attention as an analogue to exposure. Singh and Joachims [24] introduced fairness of exposure in rankings and proposed to use linear programming to optimize for the maximum utility in a ranking, subject to group fairness constraints. However, limited studies have been done on individual item fairness in rankings. One such work is Biega et al.'s [3] proposal of equity of attention, which aims to achieve amortized fairness over a sequence of rankings by distributing attention proportional to item relevance. Our work extends Biega et al.'s [3] reranking approach by implementing privacy-preserving measures at various stages of the ranking mechanism.

3.2   Privacy-Preserving Ranking Systems

Various PETs have been used in previous works on privacy-preserving learning-to-rank (LTR) systems. Dehghani et al. [8] used mimic learning, where only a model trained on the sensitive data is shared, and not the data itself. Furthermore, Laplace noise is used during aggregation as part of the DP guarantee. Yang et al. [29] used the information-theoretic privacy approach, which involves obfuscating data in accordance with a data distortion budget. Kharitonov [16] used a federated learning setup with evolution strategies optimization, and then incorporated local DP. Wang et al. [27] extended Kharitonov's work to larger datasets and found a substantial loss in utility compared to other non-private online learning-to-rank systems. Wang et al. [26] then used a federated learning setup and local DP, similar to Kharitonov [16], but incorporated a pairwise differentiable gradient descent (PDGD) optimization approach instead. Ge et al. [11] incorporated the Paillier homomorphic encryption algorithm in their PrivItem2Vec model. Among these previous works in privacy-preserving ranking systems, none have exploited the use of MPC together with DP.

3.3 Fair and Privacy-Preserving Ranking Systems

It is evident that both fairness in rankings and privacy-preserving methods in ranking systems have been well-studied; however, there is a dearth of research at the intersection of fairness and privacy in ranking systems. Resheff et al. [21] used privacy-adversarial training in their recommender system to obtain user representations that obfuscate users' sensitive attributes. This approach focuses on improving group fairness and preserves some user privacy by way of preventing implicit private information attacks. Sato [23] proposed a local ranking system framework that is independent from the centralized recommender system, where users can post-process the rankings they receive by themselves by setting their own fairness constraints to their preferences. Their privacy-preserving method is in each user developing their own recommender system. However, this is very computationally expensive and many users' devices do not have the processing power to maintain these systems. Both of these works address privacy and fairness with respect to the users' protected attributes, yet to the best of our knowledge, there has been no work so far that addresses privacy and fairness with respect to the amount of attention items receive in relation to their relevance.

4 Methodology

4.1 Problem Description

We consider a regression-style recommendation model $\mathcal{M}$ that is trained in a privacy-preserving manner (such as [16,28]), *i.e.*, the model does not leak any information about the training data. $\mathcal{M}$ is deployed at each user u_l to predict their relevance scores $\mathbf{r}^l = (r_1^l, r_2^l, \ldots, r_i^l, \ldots, r_n^l)$ on a global set of items $\mathcal{D}$, where $r_i^l = \mathcal{M}(u_l, d_i)$. In this process, the users' raw data–such as their preferences, demographics, embeddings, etc.–are not disclosed to anyone. User u_l's relevance-based ranking ρ_l is its list of items sorted in decreasing order of their relevance scores.

The post-processing technique described in Sect. 2.1 assumes that a central server S accesses the relevance-based rankings $\rho_1, \rho_2, \ldots, \rho_l, \ldots, \rho_L$ of all users and reranks them into $\rho_1^*, \rho_2^*, \ldots, \rho_l^*, \ldots, \rho_L^*$ to achieve *equity of amortized attention* without losing the ranking utility beyond the set threshold. This setup, which we refer to as the *centralized setup*, causes leakage of sensitive user information to the central server S, including:

(P_1) the preference of the user for all items in the form of relevance scores,
(P_2) the top-k items that the user is most likely to be interested in, and
(P_3) the order of the top-k items the user is most likely to be interested in.

Below we describe how we adapt the above post-processing method to address privacy issues (P_1)–(P_3) in order to achieve individual fairness for the items and preserve the privacy of the users.

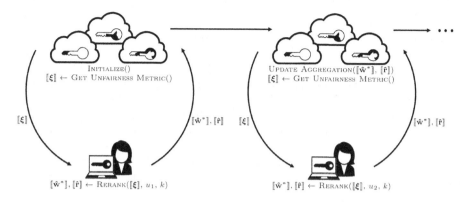

Fig. 1. Flow diagram of privacy-preserving fair item ranking algorithm[4]

4.2 Proposed Method

The key observation underlying our method is that each user u_l has all the information needed to solve the ILP to rerank their original ranking ρ_l into ρ_l^*, except for the values of A_i^{l-1} and R_i^{l-1} in Eq. 2. A_i^{l-1} and R_i^{l-1} are the accumulated attention and relevance of item d_i, respectively, over rerankings $\rho_1^*, \rho_2^*, \ldots, \rho_{l-1}^*$, i.e., the values of A_i^{l-1} and R_i^{l-1} depend on sensitive information from users $u_1, \ldots, u_{l-1}$ which we neither want to disclose to user u_l nor to a central server (as is the case in the centralized setup). In our solution, we therefore maintain encrypted versions of A_i^{l-1} and R_i^{l-1} which initially are 0 (see Procedure INITIALIZE in Algorithm 1), and are updated in a privacy-preserving manner each time a user completes their ILP computation. More precisely, no entity knows by itself at any point what the current values of A_i^{l-1} and R_i^{l-1} are. Instead, these values are split into secret shares and distributed over MPC servers who can jointly perform operations to update the shares, without ever learning the true values of the inputs or the results of the computations. Figure 1 illustrates the high-level flow of our solution, and Algorithm 1 presents the pseudo-code, which we explain in more detail below. We use $\mathbf{A}^{l-1}$ and $\mathbf{R}^{l-1}$ to denote the vectors $[A_1^{l-1}, \ldots, A_n^{l-1}]$ and $[R_1^{l-1}, \ldots, R_n^{l-1}]$, respectively.

The users update their rankings in sequence $l = 1 \ldots L$. When the sequence reaches user u_l, u_l first requests the current vector of differences $\mathbf{A}^{l-1} - \mathbf{R}^{l-1}$ from the MPC servers, prompting the MPC servers to compute a secret sharing $[\![\mathbf{A}^{l-1} - \mathbf{R}^{l-1}]\!]$ from their local secret shares of $\mathbf{A}^{l-1}$ and $\mathbf{R}^{l-1}$ (Procedure GET UNFAIRNESS METRIC). In principle, each MPC server could send their secret shares of $\mathbf{A}^{l-1} - \mathbf{R}^{l-1}$ to user u_l, which u_l could combine to construct the value of $\mathbf{A}^{l-1} - \mathbf{R}^{l-1}$. However, although $\mathbf{A}^{l-1} - \mathbf{R}^{l-1}$ consists of aggregated information only, this value could still leak information about the previous users $u_1, \ldots, u_{l-1}$ to user u_l, especially if u_l is one of the first to rerank. In our solution, this privacy loss is mitigated by having the MPC servers perturb the value at each index of $[\![\mathbf{A}^{l-1} - \mathbf{R}^{l-1}]\!]$ with Laplace noise *before* sending it to user u_l (Procedure GET UNFAIRNESS METRIC). We denote this perturbed secret sharing as $[\![\boldsymbol{\xi}]\!]$. The DP guarantee that the MPC servers provide in this manner is that the probability of returning any specific value of $\boldsymbol{\xi}$ is very similar to the probability of returning that value if the data of a previous user u_i ($i = 1 \ldots l - 1$) would have been left out of the computation of $[\![\mathbf{A}^{l-1} - \mathbf{R}^{l-1}]\!]$ (see Eq. 6). This entails that the value of $\boldsymbol{\xi}$ returned to user u_l does not leak information about the users who computed their rerankings before it was u_l's turn. To this end, the MPC servers generate secret shares of Laplace noise using the MPC-protocol π_{LAP} for secure sampling from a Laplace distribution as described by Pentyala et al. [20] and add these secret shares to $[\![\mathbf{A}^{l-1} - \mathbf{R}^{l-1}]\!]$, effectively emulating the global DP paradigm without having to rely on a central trusted aggregator. We provide more information about the scale parameter b for the Laplace noise below.

Algorithm 1. Privacy-Preserving Fair Item Ranking

Achieving EOAA privately over L users
$n \leftarrow$ number of items per ranking
$L \leftarrow$ number of rankings to rerank
$k \leftarrow$ number of items in quality constraint Eq. 3
$\mathbf{w} \leftarrow$ attention weights
$\hat{\mathbf{w}} \leftarrow$ NORMALIZE($\mathbf{w}$)
INITIALIZE()
for each u_l, where $l = 1 \ldots L$ do
    $[\![\boldsymbol{\xi}]\!] \leftarrow$ GET UNFAIRNESS METRIC()
    $[\![\hat{\mathbf{w}}^*]\!], [\![\hat{\mathbf{r}}]\!] \leftarrow$ RERANK($[\![\boldsymbol{\xi}]\!], u_l, k$)
    UPDATE AGGREGATION($[\![\hat{\mathbf{w}}^*]\!], [\![\hat{\mathbf{r}}]\!]$)
end for

User u_l subroutine
procedure RERANK($[\![\boldsymbol{\xi}]\!], u_l, k$)
    $\boldsymbol{\xi} \leftarrow (\epsilon/L) \sum [\![\boldsymbol{\xi}]\!]$
    $\mathbf{r} \leftarrow \mathcal{M}(u_l, \mathcal{D})$
    $\hat{\mathbf{r}} \leftarrow$ NORMALIZE($\mathbf{r}$)
    $\mathbf{s} \leftarrow$ ILP($\boldsymbol{\xi}, \hat{\mathbf{r}}, k$)
    $\hat{\mathbf{w}}^* \leftarrow \hat{\mathbf{w}}[\mathbf{s}]$
    Return $[\![\hat{\mathbf{w}}^*]\!], [\![\hat{\mathbf{r}}]\!]$
end procedure

MPC servers' subroutines
procedure INITIALIZE
    $\Delta f \leftarrow$ sensitivity calculated from Eq. 11
    $\epsilon \leftarrow$ privacy budget
    $[\![\mathbf{A}]\!] \leftarrow [\![0]\!]$
    $[\![\mathbf{R}]\!] \leftarrow [\![0]\!]$
end procedure

procedure GET UNFAIRNESS METRIC
    //MPC protocol for global DP
    $b \leftarrow \Delta f/(\epsilon/(n \cdot L))$
    $[\![\boldsymbol{\xi}]\!] \leftarrow [\![\mathbf{A} - \mathbf{R}]\!] + \pi_{\mathsf{LAP}}(b)$
    Return $[\![\boldsymbol{\xi}]\!]$
end procedure

procedure UPDATE AGGREGATION($[\![\hat{\mathbf{w}}^*]\!], [\![\hat{\mathbf{r}}]\!]$)
    //MPC protocol to perform aggregation
    $[\![\mathbf{A}]\!] \leftarrow [\![\mathbf{A}]\!] + [\![\hat{\mathbf{w}}^*]\!]$
    $[\![\mathbf{R}]\!] \leftarrow [\![\mathbf{R}]\!] + [\![\hat{\mathbf{r}}]\!]$
end procedure

Once user u_l has received secret shares of $\boldsymbol{\xi}$ from each MPC server, u_l can sum up the secret shares to get $\mathbf{A}^{l-1} - \mathbf{R}^{l-1} + \pi_{\mathsf{LAP}}(b)$, using this as a proxy for $\mathbf{A}^{l-1} - \mathbf{R}^{l-1}$, and then proceed to solve the ILP program in Sec. 2.1 (Procedure RERANK) for $X_{i,j}$ ($i = 1 \ldots n$ and $j = 1 \ldots n$). In our implementation, we scaled $\mathbf{A}^{l-1} - \mathbf{R}^{l-1} + \pi_{\mathsf{LAP}}(b)$ by a positive factor ϵ/L so that the solution of the ILP is easier to compute. We note that scaling by a positive factor does not affect the outcome in $X_{i,j}$. $X_{i,j}$ can then translate to a vector $\mathbf{s}$ of indices to reorder the normalized attention weights $\hat{\mathbf{w}}$. $\hat{\mathbf{w}}^*$ is the vector of attention weights distributed to each item at each index in the order of $\mathbf{s}$. Thus, $\hat{\mathbf{w}}^*$ and $\hat{\mathbf{r}}$ make up the reranking ρ_l^*. User u_l subsequently encrypts the values in these vectors by splitting them into secret shares $[\![\hat{\mathbf{w}}^*]\!]$ and $[\![\hat{\mathbf{r}}]\!]$, and distributes the shares among the MPC servers. This enables the MPC servers to update their secret shares of the aggregated values to $[\![\mathbf{A}^l]\!]$ and $[\![\mathbf{R}^l]\!]$ (Procedure UPDATE AGGREGATIONS), which they will need when the next user, u_{l+1}, makes a request. The whole process is repeated until all users have completed their reranking.

Scale Parameter b. To make our algorithm ϵ-DP, the MPC servers answer each query by adding noise to the true aggregate. To this end, the MPC servers sample noise from a Laplace distribution with mean 0 and scale $b = \Delta f / \epsilon'$ in which Δf denotes the sensitivity and ϵ' the privacy budget per query. Appropriate values for these parameters are described next. In Algorithm 1, each user u_l ($l = 1 \ldots L$) queries the MPC servers for aggregated information $A_i^{l-1} - R_i^{l-1}$ about each item d_i ($i = 1 \ldots n$) through the Procedure GET UNFAIRNESS METRIC. The total number of queries to be answered by the MPC servers is in other words $n \cdot L$. These queries are executed against overlapping datasets, as u_l queries the aggregated information of users $u_1, \ldots, u_{l-1}$; u_{l+1} queries the aggregated information of users $u_1, \ldots, u_l$, etc. Given a total privacy budget ϵ, we therefore allocate $\epsilon' = \epsilon / (n \cdot L)$ per query.

Δf is the sensitivity computed for the aggregate $A_i^{l-1} - R_i^{l-1}$, and is given by Eq. 11 that computes the maximum value that can be contributed to this aggregate by any single user.

$$\hat{r}_{\min} = \frac{r_{\min} - r_{\min}}{r_{\max} - r_{\min}} \bigg/ \left(\frac{r_{\min} - r_{\min}}{r_{\max} - r_{\min}} + (n-1) \left(\frac{r_{\max} - r_{\min}}{r_{\max} - r_{\min}} \right) \right) = 0 \qquad (7)$$

$$\hat{r}_{\max} = \frac{r_{\max} - r_{\min}}{r_{\max} - r_{\min}} \bigg/ \left(\frac{r_{\max} - r_{\min}}{r_{\max} - r_{\min}} + (n-1) \left(\frac{r_{\min} - r_{\min}}{r_{\max} - r_{\min}} \right) \right) = 1 \qquad (8)$$

$$\hat{w}_{\min} = \frac{w_n}{\sum_{j=1}^{n} w_j} \qquad (9)$$

$$\hat{w}_{\max} = \frac{w_1}{\sum_{j=1}^{n} w_j} \qquad (10)$$

$$\Delta f = \max(|\hat{w}_{\max} - \hat{r}_{\min}|, |\hat{w}_{\min} - \hat{r}_{\max}|) \qquad (11)$$

Equations 9, 10 are based on the normalized geometric attention model and Eqs. 7, 8 are based on the range of the normalized relevance scores that the MPC servers may receive. Computation of Δf and b is independent of the users' data and can be precomputed by one of the MPC servers in the clear, *i.e.*, without encryption (see Procedure INITIALIZE).

5   Results

5.1   Datasets

We use three recommender system datasets in our experiments: Amazon Digital Music,[5] Book Crossing,[6] and MovieLens-1M.[7] Each dataset contains information

---

[5] http://jmcauley.ucsd.edu/data/amazon/links.html.
[6] http://www2.informatik.uni-freiburg.de/~cziegler/BX/.
[7] https://grouplens.org/datasets/movielens/1m/.

Table 1. Statistics of datasets used to train each SVD model

Dataset	# Users	# Items	# Ratings	Rating levels
Amazon digital music	$478,235$	$266,414$	$836,006$	$1,2,...,5$
Book crossing	$77,805$	$185,973$	$433,671$	$1,2,...,10$
MovieLens 1M	$6,040$	$3,706$	$1,000,209$	$1,2,...,5$

about each user and item, and ratings that users gave to items. We detail the number of users, items, ratings, and the range of possible ratings of each dataset in Table 1.

5.2   Experimental Setup

We trained a singular value decomposition (SVD) model[8] for each dataset and predicted relevance scores for every user-item pair. We assume that these models were trained in a privacy-preserving manner. For our experiments, we select $n = 100$ items to rerank for all datasets, and rerank $L = 3000$ users' rankings for the Amazon Digital Music and Book Crossing datasets, and all $L = 6040$ users' rankings for the MovieLens-1M dataset. We use Gurobi[9] to solve the ILP in Eq. 2–4. We set $k = 100$, the quality loss constraint to $\theta = 0.8$, and calculate the sensitivity based on Eq. 11 to be $\Delta f = 1$ for all datasets. We perform an empirical analysis for privacy budget $\epsilon \in \{0.5, 1, 10, 100, 1000, 10000, 100000\}$.

All MPC computations are done in the MP-SPDZ framework [15] and performed over a ring $\mathbb{Z}_q$ with $q = 2^{64}$. We perform experiments in a dishonest majority security setting with 2 computing parties (2PC) and passive adversaries.

We evaluate our approach in terms of unfairness (Eq. 1) and utility (Eq. 5) and study the privacy-fairness-utility trade-offs.

5.3   Discussion

Fairness Vs. Privacy Trade-offs. We demonstrate the cost on item fairness when preserving the privacy of users in Fig. 2 (left column; Figs. 2a, 2c, and 2e). 'Centralized setup (no fairness)' are the unfairness measures without any bias mitigation, and 'Centralized setup (w/fairness)' are the unfairness measures when applying the post-processing technique from Sect. 2.1 in a centralized setup without any privacy protection. We ideally need privacy-preserving techniques that result in the unfairness metrics close to the 'Centralized setup (w/fairness)'. Our results show that our method can still improve fairness even with the addition of our privacy-preserving techniques. Specifically, our method preserves users' privacy at every step of the process, both when users transfer $[\![\hat{\mathbf{w}}^*]\!]$ and $[\![\hat{\mathbf{r}}]\!]$ to the servers, and when the servers transfer $[\![\boldsymbol{\xi}]\!]$ to the users.

---

[8] https://surpriselib.com/.

[9] https://www.gurobi.com/.

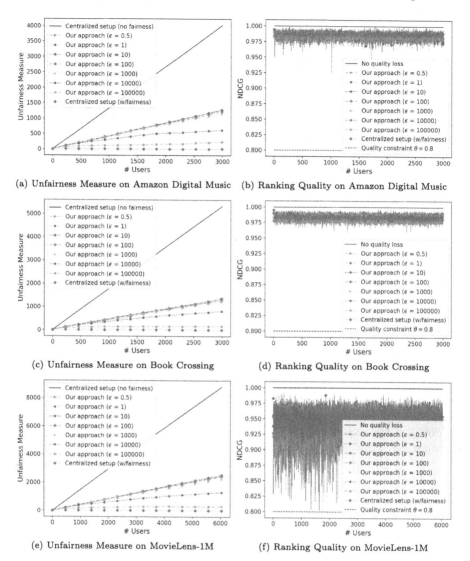

Fig. 2. Model performance on each dataset

We observe a trade-off between privacy and unfairness in our results, where a decrease in the privacy budget (*i.e.*, more privacy) imposes a higher cost to the fairness, which is in line with the literature. This is because the amount of noise added perturbs the values of the unfairness measure, which consequently affects how well the ILP can compute rerankings when compared to the centralized setup without differential privacy. We note that our solution is able to preserve input privacy even with higher ϵ.

Utility Vs. Privacy Trade-offs. We show the impact on reranking quality when introducing privacy for the users in Fig. 2 (right column; Figs. 2b, 2d, and 2f). $NDCG = 1$ represents the upper boundary of NDCG and indicates no change in the ordering of the relevance scores of the items compared to the original ranking. The dotted line at $NDCG = 0.8$ represents the quality constraint θ set in the ILP. Our results show that the quality of the rerankings are always maintained in the set boundary, $0.8 \leq NDCG \leq 1$, irrespective of the amount of noise added to preserve privacy. Our study shows that using our approach, the privacy of users can be preserved without losing utility beyond the threshold θ initially set in the ILP.

The resulting privacy-fairness trade-offs stem from preserving the output privacy. The utility-fairness trade-offs are due to the bias mitigation techniques and with and without privacy.

Runtime All experiments were performed on a 2.6 GHz 6-Core Intel Core i7 with 16 GB RAM. It takes about 0.67 s (averaged over $L = 6,040$ users) to rerank a user's ranking for $n = 100$ in a centralized setup. The additional cost in runtime due to added privacy is less than 5 s per client for a 2PC passive security setting with mixed circuits [10]. These runtimes vary across different security settings. With a 3PC passive security setting [2], this additional runtime can be reduced to less than 1 s. We believe that this increased cost in runtime to rerank a user's ranking is worth the gain in privacy. We note that the MPC schemes are normally divided into two phases: the offline phase and the online phase, and we have reported runtimes for both. The offline phase performs computations independent of the data and thus can be carried out prior to the users' rerankings. By doing so, the responsiveness of our approach can be further improved to make our approach feasible in practice and near real-time.

6   Conclusion and Future Work

We presented the novel idea of promoting producer (item) fairness while preserving the privacy of the consumers (users) in a recommendation ecosystem using post-processing techniques. We proposed an approach in which users work in tandem with secure multi-party computation (MPC) servers to rerank items, taking into account both relevance scores and attention weights. The MPC servers receive user data in encrypted form only (secret shares), and perform all computations over this data while it stays encrypted. Furthermore, whenever the MPC servers need to release aggregated information to a user, they perturb it with noise to provide differential privacy (DP) guarantees, thereby avoiding leakage of the data of any user to any other user.

We demonstrated that the incorporation of our privacy-preserving approach results in unfairness mitigation without additional cost to utility, through comparison to the centralized approach in which all users disclose their data to a central server. Our approach can be extended to other bias mitigation techniques and various other notions of fairness in rankings. We believe our work promotes research possibilities at the intersection of privacy and fairness in recommender systems, while also encouraging development of techniques for end-to-end privacy-preserving and fairness promoting pipelines for both producers and consumers in multi-stakeholder recommender systems.

Acknowledgements. This project was partially funded by the Canadian Institute for Advanced Research (CIFAR).

References

1. Ammad-Ud-Din, M., et al.: Federated collaborative filtering for privacy-preserving personalized recommendation system. arXiv preprint arXiv:1901.09888 (2019)
2. Araki, T., Furukawa, J., Lindell, Y., Nof, A., Ohara, K.: High-throughput semi-honest secure three-party computation with an honest majority. In: ACM SIGSAC Conference on Computer and Communications Security, pp. 805–817 (2016)
3. Biega, A.J., Gummadi, K.P., Weikum, G.: Equity of attention: amortizing individual fairness in rankings. In: 41st International ACM SIGIR Conference on Research & Development in Information Retrieval, pp. 405–414 (2018)
4. Canny, J.: Collaborative filtering with privacy. In: IEEE Symposium on Security and Privacy, pp. 45–57 (2002)
5. Celis, L.E., Straszak, D., Vishnoi, N.K.: Ranking with fairness constraints. In: 45th International Colloquium on Automata, Languages, and Programming (ICALP 2018). Leibniz International Proceedings in Informatics (LIPIcs), vol. 107, pp. 28:1–28:15 (2018)
6. Chai, D., Wang, L., Chen, K., Yang, Q.: Secure federated matrix factorization. IEEE Intell. Syst. **36**(5), 11–20 (2020)
7. Cramer, R., Damgard, I., Nielsen, J.: Secure Multiparty Computation and Secret Sharing. Cambridge University Press Print, New York (2015)
8. Dehghani, M., Azarbonyad, H., Kamps, J., de Rijke, M.: Share your model instead of your data: Privacy preserving mimic learning for ranking. arXiv preprint arXiv:1707.07605 (2017)
9. Dwork, C., Roth, A.: The algorithmic foundations of differential privacy. Found. Trends Theor. Comput. Sci. **9**(3–4), 211–407 (2014)
10. Escudero, D., Ghosh, S., Keller, M., Rachuri, R., Scholl, P.: Improved primitives for MPC over mixed arithmetic-binary circuits. In: Micciancio, D., Ristenpart, T. (eds.) CRYPTO 2020. LNCS, vol. 12171, pp. 823–852. Springer, Cham (2020). https://doi.org/10.1007/978-3-030-56880-1_29
11. Ge, Z., Liu, X., Li, Q., Li, Y., Guo, D.: PrivItem2Vec: a privacy-preserving algorithm for top-N recommendation. Int. J. Distrib. Sensor Netw. **17**(12) (2021)
12. Joachims, T., Granka, L., Pan, B., Hembrooke, H., Radlinski, F., Gay, G.: Evaluating the accuracy of implicit feedback from clicks and query reformulations in web search. ACM Trans. Inf. Syst. (TOIS) **25**(2), 7-es (2007)
13. Joachims, T., Radlinski, F.: Search engines that learn from implicit feedback. Computer **40**(8), 34–40 (2007)
14. Kay, M., Matuszek, C., Munson, S.A.: Unequal representation and gender stereotypes in image search results for occupations. In: Proceedings of the 33rd Annual ACM Conference on Human Factors in Computing Systems, pp. 3819–3828 (2015)

15. Keller, M.: MP-SPDZ: a versatile framework for multi-party computation. In: Proceedings of the 2020 ACM SIGSAC Conference on Computer and Communications Security, pp. 1575–1590 (2020)
16. Kharitonov, E.: Federated online learning to rank with evolution strategies. In: Proceedings of the 12th ACM International Conference on Web Search and Data Mining, pp. 249–257 (2019)
17. Liu, Y., Ge, K., Zhang, X., Lin, L.: Real-time attention based look-alike model for recommender system. In: Proceedings of the 25th ACM SIGKDD International Conference on Knowledge Discovery & Data Mining, pp. 2765–2773 (2019)
18. Mehrotra, R., McInerney, J., Bouchard, H., Lalmas, M., Diaz, F.: Towards a fair marketplace: counterfactual evaluation of the trade-off between relevance, fairness & satisfaction in recommendation systems. In: Proceedings of the 27th ACM International Conference on Information and Knowledge Management, pp. 2243–2251 (2018)
19. Morik, M., Singh, A., Hong, J., Joachims, T.: Controlling fairness and bias in dynamic learning-to-rank. In: Proceedings of the 43rd International ACM SIGIR Conference on Research & Development in Information Retrieval, pp. 429–438 (2020)
20. Pentyala, S., Neophytou, N., Nascimento, A., De Cock, M., Farnadi, G.: Privacy-preserving group fairness in cross-device federated learning. In: Proceedings of NeurIPS workshop on Algorithmic Fairness through the Lens of Causality and Privacy, Proceedings of Machine Learning Research (2023)
21. Resheff, Y.S., Elazar, Y., Shahar, M., Shalom, O.S.: Privacy and fairness in recommender systems via adversarial training of user representations. arXiv preprint arXiv:1807.03521 (2018)
22. Sapiezynski, P., Zeng, W., E Robertson, R., Mislove, A., Wilson, C.: Quantifying the impact of user attention on fair group representation in ranked lists. In: Companion Proceedings of the 2019 World Wide Web Conference, pp. 553–562 (2019)
23. Sato, R.: Private recommender systems: how can users build their own fair recommender systems without log data? In: Proceedings of the 2022 SIAM International Conference on Data Mining (SDM), pp. 549–557. SIAM (2022)
24. Singh, A., Joachims, T.: Fairness of exposure in rankings. In: Proceedings of the 24th ACM SIGKDD International Conference on Knowledge Discovery & Data Mining, pp. 2219–2228 (2018)
25. Singh, A., Joachims, T.: Policy learning for fairness in ranking. Adv. Neural Inf. Process. Syst. **32** (2019)
26. Wang, S., Liu, B., Zhuang, S., Zuccon, G.: Effective and privacy-preserving federated online learning to rank. In: Proceedings of the 2021 ACM SIGIR International Conference on Theory of Information Retrieval, pp. 3–12 (2021)
27. Wang, S., Zhuang, S., Zuccon, G.: Federated online learning to rank with evolution strategies: a reproducibility study. In: Hiemstra, D., Moens, M.-F., Mothe, J., Perego, R., Potthast, M., Sebastiani, F. (eds.) ECIR 2021. LNCS, vol. 12657, pp. 134–149. Springer, Cham (2021). https://doi.org/10.1007/978-3-030-72240-1_10
28. Wu, C., Wu, F., Cao, Y., Huang, Y., Xie, X.: FedGNN: federated graph neural network for privacy-preserving recommendation. arXiv preprint arXiv:2102.04925 (2021)
29. Yang, D., Qu, B., Cudré-Mauroux, P.: Privacy-preserving social media data publishing for personalized ranking-based recommendation. IEEE Trans. Knowl. Data Eng. **31**(3), 507–520 (2018)

30. Yang, K., Stoyanovich, J.: Measuring fairness in ranked outputs. In: Proceedings of the 29th International Conference on Scientific and Statistical Database Management, pp. 1–6 (2017)
31. Zehlike, M., Bonchi, F., Castillo, C., Hajian, S., Megahed, M., Baeza-Yates, R.: Fa* ir: a fair top-k ranking algorithm. In: Proceedings of the 2017 ACM on Conference on Information and Knowledge Management, pp. 1569–1578 (2017)

Multimodal Geolocation Estimation of News Photos

Golsa Tahmasebzadeh[1,2]([✉]) [iD], Sherzod Hakimov[3] [iD], Ralph Ewerth[1,2] [iD], and Eric Müller-Budack[1,2] [iD]

[1] TIB–Leibniz Information Centre for Science and Technology, Hannover, Germany
{golsa.tahmasebzadeh,ralph.ewerth,eric.mueller}@tib.eu
[2] L3S Research Center, Leibniz University Hannover, Hannover, Germany
[3] Computational Linguistics, University of Potsdam, Potsdam, Germany
sherzod.hakimov@uni-potsdam.de

Abstract. The widespread growth of multimodal news requires sophisticated approaches to interpret content and relations of different modalities. Images are of utmost importance since they represent a visual gist of the whole news article. For example, it is essential to identify the locations of natural disasters for crisis management or to analyze political or social events across the world. In some cases, verifying the location(s) claimed in a news article might help human assessors or fact-checking efforts to detect misinformation, i.e., fake news. Existing methods for geolocation estimation typically consider only a single modality, e.g., images or text. However, news images can lack sufficient geographical cues to estimate their locations, and the text can refer to various possible locations. In this paper, we propose a novel multimodal approach to predict the geolocation of news photos. To enable this approach, we introduce a novel dataset called Multimodal Geolocation Estimation of News Photos (*MMG-NewsPhoto*). *MMG-NewsPhoto* is, so far, the largest dataset for the given task and contains more than half a million news texts with the corresponding image, out of which 3000 photos were manually labeled for the photo geolocation based on information from the image-text pairs. For a fair comparison, we optimize and assess state-of-the-art methods using the new benchmark dataset. Experimental results show the superiority of the multimodal models compared to the unimodal approaches.

Keywords: Multimodal photo geolocalization · News analytics · Information retrieval

1 Introduction

Multimedia data have been growing exponentially on the Web and social media in the last decade. To convey information more efficiently, many news articles appear in a multimodal format, i.e., using both image and text. However, along with the proliferation of news articles, fake news has gathered momentum. Hence,

J. Kamps et al. (Eds.): ECIR 2023, LNCS 13981, pp. 204–220, 2023.
https://doi.org/10.1007/978-3-031-28238-6_14

Photo by Terra Eclipse (CC BY 2.0)

Photo by Aron Urb (CC BY 2.0)

Boehner says Obama pushing U.S. toward the "fiscal cliff"
[...] **John Boehner** accused President **Barack Obama** [...] **Boehner** and the **House of Representatives** leadership [...] to give **Obama** power to raise U.S. [...] **U.S.** credit [...]

a) GT: Washington, D.C., U.S, N.Amerika

Adapting for the Future at European University Conference
[...] **Brussels** to discuss [...] European higher education [...] **Ms. Wilson** said [...] **European** University Association conference in **Brussels** [...] **European** universities [...]

b) GT: Brussels, Belgium, Europe

Fig. 1. Samples from the *MMG-NewsPhoto* dataset. GT: Ground Truth location. Photos are replaced with similar ones due to license restrictions.

it is essential to organize, analyze, and contextualize the image content. The estimation of the geolocation of news images is an important aspect for various real-world applications. Example applications are news retrieval [1], image verification [10], and misinformation detection in news [34].

Most previous approaches for geolocation estimation of photos depend solely on visual information [16,17,25], and only a few methods process more than one modality [20,21]. Existing image-based methods are mainly focused on specific environments such as cities [5,17] or landmarks [2,7,42]. However, the image-based methods are unable to represent news-related geographic features such as *public personality* (Fig. 1a) or an *event* (Fig. 1b). Most of the multimodal approaches are based on the Yahoo Flickr Creative Commons 100 Million (YFCC100M) dataset [36], and depend on the tags provided with the images. However, they do not make use of rich textual information provided in news body that indicates possible photo locations (Fig. 1b). A multimodal dataset of news articles is *BreakingNews* [31], where the geolocation labels are provided by the news feed primarily taken from the RDF (Resource Description Framework) Site Summary (RSS) or, if not available, inferred using heuristics such as the publisher location or the story text [31]. However, the extracted geolocations can be inaccurate or even wrong. Another drawback of the *BreakingNews* dataset is that the labels of the test split are derived in the same way. Overall, there is a considerable need for a multimodal dataset of news articles that provides geolocation labels specifically for images, as well as multimodal solutions for geolocation estimation of news photos.

In this paper, we define the task of photo geolocalization as a multimodal problem. We propose a multimodal approach that considers visual and textual information from the news photo and body text that integrates hierarchical information of different granularities (spatial resolutions). The main contributions are summarized as follows: (1) We introduce the *MMG-NewsPhoto* (Multimodal Geolocation Estimation of News Photos) dataset that contains more than half a million news articles. The proposed dataset covers more than 14,000 cities and 241 countries across all continents within multiple news domains such as *Health,*

Business, *Society*, and *Politics*; (2) We provide extensive annotation guidelines and define news-specific visual concepts that represent the photo geolocation; (3) We propose a multimodal approach that leverages state-of-the-art visual and textual features for multimodal photo geolocalization; (4) We evaluate our proposed method on two datasets, including *MMG-NewsPhoto* and compare it against state-of-the-art methods, including some baseline re-implementations. The source code, dataset, and annotation guidelines are publicly available[1].

The remainder of the paper is structured as follows. Section 2 describes the related work. In Sect. 3, the acquisition of the dataset is explained. The proposed model for multimodal geolocation estimation is presented in Sect. 4, while the experimental setup and results are discussed in Sect. 5. Section 6 concludes the paper and outlines future directions.

2   Related Work

There are two main criteria to classify the approaches for geolocation estimation of photos: the environment target and the data type, i.e., images and multimodal data [9]. In this section, we briefly review related work on photo geolocation estimation and primarily focus on multimodal approaches, existing datasets, and their drawbacks.

Image-Based Approaches. Many existing methods based on image geolocalization focus on urban [5,17] and natural environments, such as mountains [4,37]. Some attempts estimate photo location at global scale without any prior assumptions about the environment. Most of them treat geolocation estimation as a classification problem [25,32,35,43]. Improvements were made, for example, by exploiting a retrieval approach and a large geo-tagged image database [40], using overlapping sets of visually similar cells [32], incorporating a hierarchical cell structure as well as environmental scene context [25], or leveraging the advantages of contrastive learning [19]. However, while these approaches achieve promising results solely based on visual information, news provides textual information that can further increase the performance, particularly in the absence of distinct geographical cues (Fig. 1b).

Multimodal Approaches. There are only few methods [11,20,21,31,33] that address geolocation estimation as a multimodal problem most of which rely on constructing large-scale geographical language models by generating a probabilistic model based on mentions of textual tags across the globe [20,21,33]. Crandall et al. [11] combine image content and textual metadata at two granularity levels, at a city level ($\approx$100 km) and landmark level ($\approx$100 m). Trevisiol et al. [38] process the textual information of a set of videos to determine their geo-relevance and to find frequent matching items. In case of lack of such information, they resort to visual features. Later, a multimodal approach was proposed by

---

[1] Source code & dataset: https://github.com/TIBHannover/mmg-newsphoto.

Ramisa et al. [31] where they combine visual features with text using the nearest neighbor method and Support Vector Regression (SVR).

Multimodal Datasets. Most multimodal approaches are based on the *YFCC 100M* dataset [36] or the *MediaEval Placing Task* benchmark datasets [23] including images, videos, and metadata. Another dataset proposed by Uzkent et al. [39] contains images and text from Wikipedia combined with satellite images. More recently, a dataset called *Multiple Languages and Modalities* (MLM) [1] has been introduced, which includes images along with multilingual texts from *Wikidata* [41]. Unlike the previous datasets, the *BreakingNews* introduced by Ramisa et al. [31] contains multimodal news articles and is the most relevant for our work. It includes image, text, caption, and metadata (such as geo-coordinates and popularity) and covers various domains such as *Sports*, *Politics*, and *Health*. The provided geolocation labels for both training and evaluation are extracted from the RSS, publisher, or news text. But as discussed in Sect. 1, these automatically derived locations can be inaccurate or even wrong. Instead, we provide high-quality manually annotated photo geolocations for fair and reliable evaluation. In addition, our proposed *MMG-NewsPhoto* dataset includes more than half a million samples (*BreakingNews* only includes around 60,000 samples with geolocations) from 241 countries and more than 14,000 cities across continents.

3 MMG-NewsPhoto Dataset

In this section, we explain the dataset creation (Sect. 3.1) and annotation process (Sect. 3.2) of the proposed *MMG-NewsPhoto* dataset for multimodal geolocation estimation of news images.

3.1 Dataset Creation

Datasets. We use the collection of articles provided by the *Good News* [6] and *CC-News* [24] datasets. *Good News* [6] is an image captioning dataset comprising 466,000 image-caption pairs. Based on web links to the news articles, we extract all articles with a body text, title, image link(s) with corresponding caption(s), and domain label(s). *CC-News* [24] includes 44 million documents written in English extracted from around 30,000 unique news sources. We sort the sources based on the number of news articles and scrape news documents from the top-20 sources in the same way as mentioned above. Finally, we download all the images and discard the ones with corrupted or inaccessible images. As a result, we end up with circa 10 million data samples, including body text, and at least one image-caption pair per sample acquired from both news sources.

Initial Removal. We remove redundant documents (except one) based on the cosine similarity (normalized to [0, 1]) of the body texts using *TF-IDF* (Term Frequency; Inverse Document Frequency) above a threshold of 0.5. Next, we

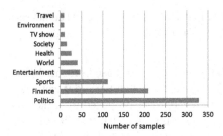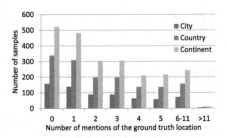

Fig. 2. Left: Test data distribution among domains. Right: Frequency of ground truth location mentions in the body text for the test split.

manually group the domain labels into ten categories such as *Health*, *Business*, and *Politics* (see full list in Fig. 2, left). Some domains such as *Art* and *Technology* include various invalid images for the task, i.e., ads or stock photos. We discard these types of images as they typically lack geographic content or do not correspond to the locations mentioned in the body text of news.

Location Linking. We assume that locations mentioned in a caption are good candidates for photo geolocation. We apply named entity recognition and disambiguation to extract all locations in the captions. Following related work [27], we use *spaCy* [15] to extract the named entities and use *Wikifier* [8] to link them to *Wikidata* entities. We only keep entities of type *Location* with valid geocoordinates (latitude, longitude) extracted from the *Wikidata* Property *P625*.

Photo Location Assignment. The location entities extracted from the captions do not always indicate the photo locations and can, for example, also refer to entity attributes, e.g., *"U.S. President Biden"*. Thus, captions are tokenized to extract certain prepositions, e.g., "across", "along", and "in", which combined with a location mention, are more likely to refer to the photo location. We keep samples for which the distance of one of 37 prepositions (See footnote 1) to the *claimed photo location* is at most two tokens. Furthermore, samples with more than one unique location are removed, resulting in exactly one *claimed photo location*.

Location Enrichment. We apply reverse-geocoding to map around 50,000 fine-grained locations (i.e., city, road, building, etc.) extracted from the captions to cities using *Nominatim* [29]. Next, we extract associated country (*Property P17*), continent (*Property P30*) and geo-coordinates (*Property P625*) from *Wikidata*.

Data Sampling. For manual annotation, 3000 samples are selected to construct the test dataset. To avoid bias, the samples are selected (1) from all domains, (2) from all continents, (3) from highly populated cities (minimum population of

500,000) and medium populated cities (population: 20,000 to 500,000), (4) with at least three unique locations mentioned in text, and (5) with different number of mentions of the ground-truth location in the body text. The latter ensures that simple cases with frequent mentions of the ground truth and complex cases, i.e., many locations mentioned in the text with somewhat equal frequencies, are included. For simple cases, a textual approach that leverages the frequency of named entities can already achieve high performance without even considering the image. Based on complex cases, we can analyze the direct impact of the image for multimodal geolocation estimation. The statistics for the test split are visualized in Fig. 2, right. From the remaining samples, 10% are randomly chosen for validation, and the rest is used for training.

3.2 Data Annotation Process and Guidelines

We give an in-depth explanation of the guidelines used for the manual annotation of the test split, which is aimed at making the assessment fair and transparent. The exact guidelines used during annotation are provided on our *GitHub* page (See footnote 1).

Geo-Representative Concepts. For photo geolocation estimation, a *geographically representative image* depicts concepts that help to identify its location. We group *geo-representative concepts* into two types: *strong* and *weak concepts*. A *strong concept* is a unique identity of a location, e.g., the appearance of the *Eiffel Tower* in an image that can unambiguously be assigned to the city *Paris*, country *France*, and continent *Europe*. A *weak concept*, on the other hand, provides clues for one or even a few specific locations but without sufficient evidence on its own. For example, a certain *President* is an identity of a country but can travel to different locations. Only multiple *weak concepts*, all of which correspond to the same location, in an image can lead to the identification of the geolocation of news photos. For instance, multiple *car plates* or *groups of people* can represent the corresponding country. As shown in Table 1, we define *strong* or *weak* visual concepts based on the following eight categories: *building, clothing, event, group of people, natural scenery, object, public personality* and *scene text*.

Annotation Questions (Q). Given an image-caption pair and the linked location of the caption, we ask each annotator the following questions:
Q1: *Is it a valid sample?* To determine whether a sample is valid for the identification of the photo geolocation, an annotator selects *"no"* if (1) the image is an ad, a stock photo, a web page, a map, or a data visualization, (2) the linked location is wrong, not a location, or not the *claimed photo location* (see paragraph *Photo Location Assignment*) of the caption. Otherwise, *"yes"* is chosen.
Q2: *Which weak* and *strong concepts* are shown in the image? The annotator selects the strong or weak concepts (Table 1) depicted in the image.

Table 1. Strong and weak visual concepts used in the annotation process.

Strong geo-representative concepts

Category	City	Country	Continent
Building	Buildings, landmarks	–	–
Clothing	–	Public service uniforms	–
Event	Social movements, sports competitions	Social movements, sports competitions, natural disasters, country elections, wars	Sports competitions, natural disasters
Group of people	–	–	–
Natural scenery	City-specific natural landmarks	Country-specific natural landmarks	Continent-specific natural landmarks
Object	Logos of events, organizations, etc.	Public service vehicles	–
Public personality	–	–	–
Scene text	Street signs with mentions of cities	Country names in signs	–

Weak geo-representative concepts

Category	City	Country	Continent
Building	–	Buildings with specific architectures	–
Clothing	Uniforms of sport clubs	Uniforms of soldiers, cultural costumes, national sport team uniforms	–
Event	–	–	–
Group of people	–	Residents of a country, common activity	–
Natural scenery	–	–	Land forms, flora, fauna
Object	–	Personal cars and/or car plates, flag, logo	–
Public personality	–	Politicians, athletes, celebrities	–
Scene text	–	Text in specific language	–

Q3: *Is the linked city (Q3.1), country (Q3.2), continent (Q3.3) shown in the image?* These questions are asked to obtain the ground-truth location at various granularities. A user selects *"yes"* if (1) at least one *strong concept* is visible, (2) a single *weak concept* occurs in high frequency (e.g., multiple *car plates*), (3) a combination of at least two distinct *weak concepts* is shown, or (4) a single *weak concept* with valid proof (e.g., a Web page that proves the location) is provided. Otherwise, *"no"* is selected. If *"yes"* is given as an answer, a confidence level is selected: *"very confident"*, *"confident"*, and *"not confident"*.
Q4: *What is the environmental setting of the image?* The user selects one of the following categories: *"indoor"*, *"outdoor urban"*, *"outdoor nature"* to indicate the environment in which an image was taken.

Q5: *Is it a closeup?* Since locations are usually difficult to predict for closeups, we asked the annotators to identify whether the image shows a closeup or not. **Q6: *Did you need external resources for Q3?*** The final question determines whether or not the annotator needed external resources to decide on Q3. If "Yes" is selected, we asked the annotators to provide the links.

Annotator Training. We employed four graduate students with computer science backgrounds who were paid 10 EUR per hour (slightly above the minimum wage in Germany in early 2022) for annotations. Furthermore, three experts (doctoral and postdoctoral researchers) with a research focus on computer vision and multimodal analytics provided annotations. All annotators were trained based on the annotation guidelines (See footnote 1). We performed two dry runs using 100 samples and discussed the results to refine the guidelines.

Annotation Process. The annotation task was performed in two steps as follows. (1) All annotators were asked to validate the 3000 samples according to Q1. Using majority voting, 1700 valid samples were obtained. (2) For each *valid* sample, Q2 to Q6 were annotated by three annotators, and majority voting was applied to select samples where two users agreed on the answer per question. Based on selected answers for Q3.1 to Q3.3, we obtained the final annotations. For all questions, the answer should be *"yes"*, with a confidence level of either *"very confident"* or *"confident"*. Samples, where at least two annotators selected the confidence level *"not confident"* were re-annotated by an expert. As a result, we obtained final annotations for Q3.1, Q3.2, and Q3.3, where the answers correspond to the granularity of the geolocation of images. These granularities are turned into three variants of the test data: Test_{city}, $\text{Test}_{country}$, $\text{Test}_{continent}$. Please note that finer granularity samples are subsets of coarser granularities.

Annotation Study Findings. Krippendorff's alpha [22] was used to calculate inter-annotator agreements for Q3. The agreements are 0.41 for *city*, 0.41 for *country*, and 0.51 for the *continent*, which we consider low to moderate. Responses to Q4 and Q5 indicated that 40.2% of the images are close-ups and 37.7% are indoor images, both of which typically depict few weak geo-representative concepts and are challenging for the photo-geolocation task. For 49.7% of the samples, annotators needed external resources (Q6) to decide whether the image showed the linked location. Overall, these numbers demonstrate the difficulty of the task for humans and explain the moderate inter-coder agreement for Q3.

Dataset Statistics. The *MMG-NewsPhoto* dataset includes 554,768 training, 60,893 validation, and 2259 test samples (sum for all granularities). The dataset contains 14,331 cities, 241 countries, and 6 continents. Table 2 shows data distribution among continents and top-10 countries. Since 1700 test samples and thus about 57% of the test samples are valid, we assume that train and validation sets contain a similar proportion of valid samples.

Table 2. Data distribution for continents (top) and top-10 countries (bottom).

	Europe	N.America	Asia	Oceania	Africa	S.America	Total
Train	190,064	188,175	121,045	20,468	21,096	13,920	554,768
Validation	21,041	20,675	13,120	2147	2,331	1,579	60,893
Test$_{city}$	196	189	215	13	27	20	660
Test$_{country}$	235	212	274	13	35	25	794
Test$_{continent}$	235	215	278	13	37	27	805
Total	211,769	209,466	134,932	22,654	22,526	15,573	617,920

	U.S.	U.K.	India	China	Australia	France	Japan	Germany	Spain	Russia
Train	173,584	82,917	27,435	18,390	17,018	16,347	15,669	14,477	13,702	9,330
Validation	19,076	9,253	3,024	2,007	1,805	1,766	1,732	1,569	1,459	1,055
Test$_{country}$	190	82	121	11	11	8	17	24	11	15

4   Multimodal Photo Geolocation Estimation

We define multimodal geolocation estimation of news photos as a classification task, where the photo location is predicted based on the visual content and contextual information from the accompanied body text. The number of $|\mathbb{C}_g|$ locations available in the dataset for a granularity g (e.g., city, country, or continent) are considered as target classes. The $|\mathbb{C}_g|$-dimensional one-hot encoded vector $\mathbf{y}_g = \langle y_1, y_2, \ldots, y_{|\mathbb{C}_g|} \rangle \in \{0, 1\}^{|\mathbb{C}_g|}$ represents the ground-truth location. In the remainder of this section, we define the features incorporated from state-of-the-art approaches and describe the multimodal architecture and loss function.

Textual Features. The pre-trained language model BERT (Bidirectional Encoder Representations from Transformers) [12] is employed to extract two distinct types of textual features, each with 768 dimensions, from the body text of the news article. (1) We average the embeddings extracted with BERT of each sentence to create a single vector B-Bd $\in \mathbb{R}^{768}$ to encode the global contextual information. (2) To create an entity-centric embedding, denoted as B-Et $\in \mathbb{R}^{768}$, we follow related work [27] and combine *spaCy* [15] and *Wikifier* [8] to link location, person, and event entities to *Wikidata*. The BERT embeddings for these entities are extracted based on their *Wikidata* label. Finally, we compute the average of the entity vectors taking into account multiple mentions of the same entity, as they may be more important for the geolocation of the photo.

Visual Features. To represent the *geo-representative visual concepts*, we rely on CLIP (Contrastive Language-Image Pretraining) [30]. We use ViT-B/32 image encoder to extract visual features with 512 dimensions denoted as CLIP$_i \in \mathbb{R}^{512}$.

Network Architecture. In our proposed model architecture, we aim to combine textual and visual features to predict photo geolocations on various granularities, i.e., city, country, and continent levels. Since the feature dimension of visual and textual features differ, we first encode each feature vector using

l_e fully-connected (FC) layers with n_e neurons each. Next, we concatenate these embeddings and feed them into l_o output FC-layers. In the hidden output layers, we use n_o neurons, and in the last output layer, the number of neurons corresponds to the number of locations $|\mathbb{C}_g|$ for a given granularity g. To leverage the hierarchical information, we employ individual classifiers for each granularity in city, country, and continent level to output probabilities $\hat{\mathbf{y}}_g \in \mathbb{R}^{|\mathbb{C}_g|}$ of size $|\mathbb{C}_{city}| = 14,331$, $|\mathbb{C}_{country}| = 241$, and $|\mathbb{C}_{continent}| = 6$. Please note that we use the *Rectified Linear Unit (ReLU)* activation function [28] for all layers except the last output layer that uses a *softmax*. More details are provided on GitHub (See footnote 1).

Loss Function. To aggregate the granularity classifiers and highlight the hierarchical attribution, we build a multi-task learning loss function as follows:

$$\mathcal{L} = \sum_g \lambda_g \mathcal{L}_g(\mathbf{y}_g, \hat{\mathbf{y}}_g), \text{ with } g \in \{\text{city, country, continent}\}, \qquad (1)$$

$$\mathcal{L}_g(\mathbf{y}_g, \hat{\mathbf{y}}_g) = -\mathbf{y}_g \log \hat{\mathbf{y}}_g - (1 - \mathbf{y}_g) \log(1 - \hat{\mathbf{y}}_g), \qquad (2)$$

where λ_g are the relative weights learned during training for the different granularities, considering the difference in magnitude between losses by consolidating the log standard deviation. The cross-entropy loss $\mathcal{L}_g$ for a single granularity $g \in \{\text{city, country, continent}\}$ is defined according to Eq. (2).

5 Experimental Setup and Results

This section presents the experimental setup, comparison of different architectures on the proposed *MMG-NewsPhoto* dataset as well as on *BreakingNews* [31].

5.1 Experimental Setup

Evaluation Metrics. We use the Great Circle Distance (GCD) between the geocoordinates of the predicted and ground-truth location at several tolerable error radii [13]. These values are 25, 200, and 2500 km for city, country, and continent, respectively. Furthermore, we measure the Accuracy@k that indicates whether the ground-truth location is within the top-k model predictions.

Hyperparameter Settings. To extract textual features, we limit the text to 500 tokens. We set the number of FC-layers to $l_e = 2$, $l_o = 2$ and choose $n_e = 1024$, $n_o = 512$ neurons (Sect. 4). While *single-task learning* model variants (denoted with stl) are optimized using a single granularity g, the remaining models use the multi-task loss presented in Eq. (1) to learn from hierarchical geographical information. We use the *Adam* optimizer [18], a learning rate of e-5, batch size of 256, weight decay of 0 for optimization, ReLU activation $max(0, x)$ [28], and norm [3] with a clamp $min = 1 \times 10^{-12}$. Before each layer, a dropout with a ratio of 0.1 is applied. We train all the models for 100 epochs and clip gradients with a max norm of 5. The model with the lowest loss on the validation set is used for evaluation.

Table 3. Fraction of samples [%] localized within a GCD of at most 25 km (**CI:** city level), 200 km (**CR:** country level), and 2500 km (**CT:** continent level) on *MMG-NewsPhoto*.

Approach	CI	CR	CT
base(M, f^*) [25]	10.3	20.2	40.9
CLIP$_i$	**30.6**	**65.5**	**78.3**
T-Freq	12.6	31.5	49.9
B-Bd	31.5	73.4	85.6
B-Et	31.4	73.7	83.5
B-Bd $\oplus$ B-Et	**32.1**	**74.7**	**84.6**
T-base(M, f^*)	31.2	58.8	70.7
VT$_{CM}$ [26]	22.3	50.1	60.1
CLIP$_i$ $\oplus$ B-Bd $\oplus$ B-Et	**43.0**	**76.7**	**83.4**

Table 4. Mean and median GCD divided by 1000 km on city level for the BreakingNews test set. Models trained on *MMG-NewsPhoto* are evaluated in a zero-shot setting on *BreakingNews* and MMG → BN means that the model is finetuned on *BreakingNews*.

Approach	Training	Mean	Median
CLIP$_i$	MMG	3.67	1.37
CLIP$_i$	MMG → BN	3.22	0.92
B-Bd $\oplus$ B-Et	MMG	2.26	**0.47**
B-Bd $\oplus$ B-Et	MMG → BN	**2.25**	0.51
CLIP$_i$ $\oplus$ B-Bd $\oplus$ B-Et	MMG	2.70	0.63
CLIP$_i$ $\oplus$ B-Bd $\oplus$ B-Et	MMG → BN	2.38	**0.50**
Places [31]	BN	3.40	**0.68**
W2V matrix [31]	BN	1.92	0.90
VGG19 + Places + W2V matrix [31]	BN	**1.91**	0.88

Baselines. We compare our models to the following baselines. Note that we did not fine-tune these models and used their official models or implementations.

base(M, f^*) [25] is a state-of-the-art model for photo geolocation estimation model based on ResNet-101 [14] pre-trained on a subset of *YFCC100M* [36].

T-base(M, f^*) is an extension of *base*(M, f^*) where its predictions are reduced to mentioned locations in the news body to include textual information.

T-Freq is based on language models for geo-tagging text [20,23,33]. We employ a statistical model based on frequency of entities per city using the train set. More details are provided in the supplemental material on *GitHub* (See footnote 1). The predicted location per sample is the one with the highest probability.

VT$_{CM}$ is based on the cross-modal entity consistency of image and text [26] based on persons, locations, and events. To get predictions per test image, we sort *Cross-modal Location Similarity (CMLS)* values and get the top k locations.

5.2 Results on MMG-NewsPhoto

Comparison of the Unimodal Models. As Table 3 shows, regarding the visual models, CLIP$_i$ noticeably outperforms the baseline *base*(M, f^*) [25]. Regarding the textual models, the B-Bd $\oplus$ B-Et surpasses the individual features. It indicates that both the contextual information and named entities and their frequencies play a vital role in the geolocation estimation of a news photo. Table 5 reports the results for Accuracy@k and shows that the CLIP$_i$ visual model is superior at the country and continent levels, but in the city-level CLIP$_i$ (stl) is slightly better. Among the textual models, the B-Bd $\oplus$ B-Et out-

Table 5. Accuracy@k (A@k) for different test sets (number of samples in brackets) of *MMG-NewsPhoto*. Approaches denoted with (stl) are trained on the respective test granularity g and do not use the multi-task loss in Equation (1).

Approach	Modality	Test$_{city}$ (660)				Test$_{country}$ (794)				Test$_{continent}$ (805)	
		A@1	A@2	A@5	A@10	A@1	A@2	A@5	A@10	A@1	A@2
$base(M, f^*)$ [25]	Visual	8.3	11.2	15.9	19.8	12.7	16.9	23.2	30.1	51.1	73.7
CLIP$_i$ (stl)	Visual	29.1	38.9	48.5	57.4	61.5	70.3	81.0	88.0	77.4	89.2
CLIP$_i$	Visual	27.9	37.7	48.5	58.0	61.5	70.9	80.4	85.0	78.1	90.8
T-$Freq$	Textual	10.5	14.1	19.2	24.5	31.1	38.4	48.0	54.3	55.8	70.8
B-Bd	Textual	27.9	38.2	49.2	60.2	69.5	76.2	84.3	88.7	85.0	92.8
B-Et	Textual	28.2	40.5	52.3	62.7	70.3	79.5	86.8	89.9	83.0	92.8
B-Bd ⊕ B-Et (stl)	Textual	28.9	39.2	50.9	62.9	70.4	78.0	86.0	91.1	83.6	92.8
B-Bd ⊕ B-Et	Textual	28.6	40.2	52.9	62.0	70.8	78.6	87.3	91.2	84.1	92.0
T-$base(M, f^*)$	Multimodal	27.1	36.5	43.3	44.2	62.8	74.2	79.5	80.1	75.4	86.0
VT_{CM} [26]	Multimodal	11.4	20.3	36.1	42.6	40.4	63.5	84.0	87.7	53.8	81.4
CLIP$_i$ ⊕ B-Bd ⊕ B-Et (stl)	Multimodal	37.9	50.9	62.7	71.2	73.6	82.2	89.5	92.2	81.9	90.3
CLIP$_i$ ⊕ B-Bd ⊕ B-Et	Multimodal	39.5	52.1	64.5	72.7	73.3	81.1	90.1	92.6	82.9	92.7

performs the rest at the country and continent levels, but it is not significantly better than B-Bd ⊕ B-Et (stl) in the city level.

Comparison of the Multimodal Models. As presented in Table 3, the combination of the best unimodal features, CLIP$_i$ ⊕ B-Bd ⊕ B-Et significantly outperforms all the other models in all granularity levels. Regarding Accuracy@k, Table 5 confirms the same results. For the multi-task setting, it was effective in all the granularities. In conclusion, the hierarchical information propagated from the larger granularity levels not only improves the performance in the smaller granularities, such as city but also in the country and the continent levels.

Comparison of Different Domains. Fig. 3, right presents the Accuracy@1 per domain for different models. As shown, the multimodal model outperforms in most of the domains. In domains like *Finance, Health*, and *Sports*, the visual model outperforms the textual model. In *TV show* and *World*, adding visual information does not help, and in the *Health* domain, additional textual information does not impact the performance.

Comparison of Different Concepts. Fig. 3, left shows the Accuracy@1 per concept (see Table 1). As presented, the proposed multimodal model outperforms the rest in all the concepts except *public personality* and *group of people*. Also, it is observed that, based on the multimodal model, the concept *event* results in the lowest, and *scene text* results in the highest performance.

Qualitative Results. Figure 4 illustrates the results of different models. As expected, the visual model fails when there are only weak geo-representative concepts (Fig. 4a). However, it succeeds when: (1) there is a strong concept

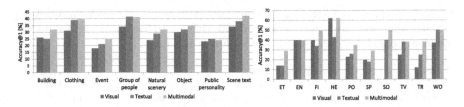

Fig. 3. Accuracy@1 [%] of the best performing visual, textual and multimodal models per concept (left) and per domain (right). ET: Entertainment, EN: Environment, FI: Finance, HE: Health, PO: Politics, SP: Sports, SO: Society, TR: Travel, TV: TV show, WO: World.

Fig. 4. Sample outputs from the *MMG-NewsPhoto* dataset with the predicted locations using best-performing textual, visual and multimodal models. Predictions written in bold are correct and correspond to the ground-truth (GT) locations. Images are replaced with similar ones due to license restrictions.

(such as a landmark in Fig. 4 b), or (2) a weak concept occurs in high frequency, e.g., *soldier* in Fig. 4d. The textual model fails when: (1) no relevant location is mentioned (Fig. 4b), (2) various irrelevant entities are mentioned, e.g., *U.S.* in Fig. 4d. As expected textual model succeeds if there are many relevant entities to the location (Fig. 4a, c). When the text mentions many topics irrelevant to the image, the multimodal model fails (Fig. 4d). Conversely, the multimodal model succeeds in either of the following conditions: (1) the text provides rich information (both in terms of entities and content) such as Fig. 4a, c, or (2) the image illustrates strong visual concepts, such as Fig. 4b.

5.3   Results on *BreakingNews*

Although the image locations provided by *BreakingNews* [31] can be inaccurate (discussed in Sect. 1), we perform experiments on the dataset for comparison. *BreakingNews* includes 33,376, 11,209, and 10,580 samples for train, validation, and testing. Ramisa et al. [31] treat the task as a regression problem where their models output the geo-coordinates. In our case, we handle the problem as a classification task to predict a specific city, country, or continent. Thus, we mapped the geo-coordinates to the closest city, country and continent classes in

MMG-NewsPhoto based on GCD. Table 4 presents the comparison of the proposed models with *BreakingNews* (abbreviated with BN) [31] approaches. The comparison is based on the Mean and Median GCD values [31]. We evaluate our approach in two settings. In the zero-shot setting, the model was trained on *MMG-NewsPhoto* and tested on *BreakingNews* without further optimization. In the second configuration, the best model on *MMG-NewsPhoto* is both fine-tuned and tested on *BreakingNews*. The B-Bd ⊕ B-Et model has the lowest Median value (470 km) in the zero-shot setting and outperforms VGG19 + Places + W2V matrix [31] (880 km). In general, the comparison confirms the feasibility of applying the proposed models to unseen examples. In the second setting (MMG → BN), $CLIP_i$ ⊕ B-Bd ⊕ B-Et outperforms all the *BreakingNews* baselines by 180 to 380 km of the median value. As observed, our models perform better using the median metric, i.e., our models are better for the majority of samples.

6   Conclusions and Future Work

This paper proposes a novel multimodal approach for geolocation estimation of news photos that integrates the hierarchical information of different granularities (spatial resolutions). For this purpose, we have introduced a novel dataset called *MMG-NewsPhoto* that contains more than half a million image-text pairs for more than 14,000 cities and 241 countries. We manually annotated 3000 samples for the evaluation to acquire different data variants at the granularity levels of city, country, and continent. We have compared our approach with several state-of-the-art approaches and baselines. Experiments showed that the combination of textual and visual features outperforms the compared models that rely only on features from a single modality. In future work, visual concepts (e.g., car plates, events, etc.), including scene text (e.g., on buildings, street signs, etc.), could be extracted for an improved geolocalization. Furthermore, the impact of photo geolocation estimation on tasks such as fake news detection, news recommendation, and cross-modal retrieval could be investigated.

Acknowledgements. This work was partially funded by the EU Horizon 2020 research and innovation program under the Marie Skłodowska-Curie grant agreement no. 812997 (CLEOPATRA ITN), and by the Ministry of Lower Saxony for Science and Culture (Responsible AI in digital society, project no. 51171145).

References

1. Armitage, J., Kacupaj, E., Tahmasebzadeh, G., Swati, Maleshkova, M., Ewerth, R., Lehmann, J.: MLM: a benchmark dataset for multitask learning with multiple languages and modalities. In: International Conference on Information and Knowledge Management, CIKM, pp. 2967–2974. ACM (2020). https://doi.org/10.1145/3340531.3412783
2. Avrithis, Y., Kalantidis, Y., Tolias, G., Spyrou, E.: Retrieving landmark and non-landmark images from community photo collections. In: International Conference on Multimedia, MM, pp. 153–162. ACM (2010). https://doi.org/10.1145/1873951.1873973

3. Ba, L.J., Kiros, J.R., Hinton, G.E.: Layer normalization. CoRR (2016). http://arxiv.org/abs/1607.06450
4. Baatz, G., Saurer, O., Köser, K., Pollefeys, M.: Large scale visual geo-localization of images in mountainous terrain. In: Fitzgibbon, A., Lazebnik, S., Perona, P., Sato, Y., Schmid, C. (eds.) ECCV 2012. LNCS, vol. 7573, pp. 517–530. Springer, Heidelberg (2012). https://doi.org/10.1007/978-3-642-33709-3_37
5. Berton, G.M., Masone, C., Caputo, B.: Rethinking visual geo-localization for large-scale applications. In: Conference on Computer Vision and Pattern Recognition, CVPR, pp. 4868–4878. IEEE (2022). https://doi.org/10.1109/CVPR52688.2022.00483
6. Biten, A.F., Gómez, L., Rusiñol, M., Karatzas, D.: Good news, everyone! context driven entity-aware captioning for news images. In: Conference on Computer Vision and Pattern Recognition, CVPR, pp. 12466–12475. Computer Vision Foundation/IEEE (2019). http://openaccess.thecvf.com/content_CVPR_2019/html/Biten_Good_News_Everyone_Context_Driven_Entity-Aware_Captioning_for_News_Images_CVPR_2019_paper.html
7. Boiarov, A., Tyantov, E.: Large scale landmark recognition via deep metric learning. In: Zhu, W., et al. (eds.) Proceedings of the 28th ACM International Conference on Information and Knowledge Management, CIKM, pp. 169–178. ACM (2019). https://doi.org/10.1145/3357384.3357956
8. Brank, J., Leban, G., Grobelnik, M.: Semantic annotation of documents based on wikipedia concepts. Informatica (Slovenia) (2018). http://www.informatica.si/index.php/informatica/article/view/2228
9. Brejcha, J., Čadík, M.: State-of-the-art in visual geo-localization. Pattern Anal. Appl. 20(3), 613–637 (2017). https://doi.org/10.1007/s10044-017-0611-1
10. Cheng, J., Wu, Y., AbdAlmageed, W., Natarajan, P.: QATM: quality-aware template matching for deep learning. In: Conference on Computer Vision and Pattern Recognition, CVPR. pp. 11553–11562. Computer Vision Foundation/IEEE (2019). http://openaccess.thecvf.com/content_CVPR_2019/html/Cheng_QATM_Quality-Aware_Template_Matching_for_Deep_Learning_CVPR_2019_paper.html
11. Crandall, D.J., Backstrom, L., Huttenlocher, D.P., Kleinberg, J.M.: Mapping the world's photos. In: International Conference on World Wide Web, WWW, pp. 761–770. ACM (2009). https://doi.org/10.1145/1526709.1526812
12. Devlin, J., Chang, M., Lee, K., Toutanova, K.: BERT: pre-training of deep bidirectional transformers for language understanding. In: Conference of the North American Chapter of the Association for Computational Linguistics: Human Language Technologies, NAACL-HLT, pp. 4171–4186. Association for Computational Linguistics (2019). https://doi.org/10.18653/v1/n19-1423
13. Hays, J., Efros, A.A.: IM2GPS: estimating geographic information from a single image. In: Conference on Computer Vision and Pattern Recognition, CVPR. IEEE Computer Society (2008)
14. He, K., Zhang, X., Ren, S., Sun, J.: Deep residual learning for image recognition. In: Conference on Computer Vision and Pattern Recognition, CVPR, pp. 770–778. IEEE Computer Society (2016). https://doi.org/10.1109/CVPR.2016.90
15. Honnibal, M., Montani, I.: spaCy 2: Natural language understanding with bloom embeddings, convolutional neural networks and incremental parsing (2017). https://spacy.io
16. Izbicki, M., Papalexakis, E.E., Tsotras, V.J.: Exploiting the earth's spherical geometry to geolocate images. In: European Conference on Machine Learning and Knowledge Discovery in Databases, ECML PKDD, pp. 3–19. Springer (2019). https://doi.org/10.1007/978-3-030-46147-8_1

17. Kim, H.J., Dunn, E., Frahm, J.: Learned contextual feature reweighting for image geo-localization. In: Conference on Computer Vision and Pattern Recognition, CVPR, pp. 3251–3260. IEEE Computer Society (2017). https://doi.org/10.1109/CVPR.2017.346

18. Kingma, D.P., Ba, J.: Adam: A method for stochastic optimization. In: International Conference on Learning Representations, ICLR (2015). http://arxiv.org/abs/1412.6980

19. Kordopatis-Zilos, G., Galopoulos, P., Papadopoulos, S., Kompatsiaris, I.: Leveraging efficientnet and contrastive learning for accurate global-scale location estimation. In: International Conference on Multimedia Retrieval, ICMR, pp. 155–163. ACM (2021). https://doi.org/10.1145/3460426.3463644

20. Kordopatis-Zilos, G., Papadopoulos, S., Kompatsiaris, I.: Geotagging text content with language models and feature mining. Proc. IEEE, 1971–1986 (2017). https://doi.org/10.1109/JPROC.2017.2688799

21. Kordopatis-Zilos, G., Popescu, A., Papadopoulos, S., Kompatsiaris, Y.: Placing images with refined language models and similarity search with pca-reduced VGG features. In: MediaEval 2016 Workshop. CEUR-WS.org (2016). http://ceur-ws.org/Vol-1739/MediaEval_2016_paper_13.pdf

22. Krippendorff, K.: Computing krippendorff's alpha-reliability (2011). https://repository.upenn.edu/asc_papers/43

23. Larson, M.A., Soleymani, M., Gravier, G., Ionescu, B., Jones, G.J.F.: The benchmarking initiative for multimedia evaluation: Mediaeval 2016. IEEE MultiMedia, 93–96 (2017). https://doi.org/10.1109/MMUL.2017.9

24. Mackenzie, J.M., Benham, R., Petri, M., Trippas, J.R., Culpepper, J.S., Moffat, A.: CC-News-En: A large english news corpus. In: International Conference on Information and Knowledge Management, CIKM, pp. 3077–3084. ACM (2020). https://doi.org/10.1145/3340531.3412762

25. Müller-Budack, E., Pustu-Iren, K., Ewerth, R.: Geolocation estimation of photos using a hierarchical model and scene classification. In: Ferrari, V., Hebert, M., Sminchisescu, C., Weiss, Y. (eds.) ECCV 2018. LNCS, vol. 11216, pp. 575–592. Springer, Cham (2018). https://doi.org/10.1007/978-3-030-01258-8_35

26. Müller-Budack, E., Theiner, J., Diering, S., Idahl, M., Ewerth, R.: Multimodal analytics for real-world news using measures of cross-modal entity consistency. In: International Conference on Multimedia Retrieval, ICMR, pp. 16–25. ACM (2020). https://doi.org/10.1145/3372278.3390670

27. Müller-Budack, E., Theiner, J., Diering, S., Idahl, M., Hakimov, S., Ewerth, R.: Multimodal news analytics using measures of cross-modal entity and context consistency. Int. J. Multimed. Inf. Retrieval 10(2), 111–125 (2021). https://doi.org/10.1007/s13735-021-00207-4

28. Nair, V., Hinton, G.E.: Rectified linear units improve restricted Boltzmann machines. In: Fürnkranz, J., Joachims, T. (eds.) International Conference on Machine Learning (ICML), pp. 807–814. Omnipress (2010). https://icml.cc/Conferences/2010/papers/432.pdf

29. Nominatim. https://nominatim.org/release-docs/latest/api/Reverse/. Accessed 19 May 2022

30. Radford, A., et al.: Learning transferable visual models from natural language supervision. In: International Conference on Machine Learning, ICML, pp. 8748–8763. PMLR (2021). http://proceedings.mlr.press/v139/radford21a.html

31. Ramisa, A., Yan, F., Moreno-Noguer, F., Mikolajczyk, K.: Breakingnews: Article annotation by image and text processing. IEEE Trans. Pattern Anal. Mach. Intell., 1072–1085 (2018). https://doi.org/10.1109/TPAMI.2017.2721945

32. Seo, P.H., Weyand, T., Sim, J., Han, B.: CPlaNet: enhancing image geolocalization by combinatorial partitioning of maps. In: Ferrari, V., Hebert, M., Sminchisescu, C., Weiss, Y. (eds.) ECCV 2018. LNCS, vol. 11214, pp. 544–560. Springer, Cham (2018). https://doi.org/10.1007/978-3-030-01249-6_33

33. Serdyukov, P., Murdock, V., van Zwol, R.: Placing flickr photos on a map. In: SIGIR Conference on Research and Development in Information Retrieval, SIGIR, pp. 484–491. ACM (2009). https://doi.org/10.1145/1571941.1572025

34. Singhal, S., Shah, R.R., Chakraborty, T., Kumaraguru, P., Satoh, S.: Spotfake: a multi-modal framework for fake news detection. In: IEEE International Conference on Multimedia Big Data, BigMM, pp. 39–47. IEEE (2019). https://doi.org/10.1109/BigMM.2019.00-44

35. Theiner, J., Müller-Budack, E., Ewerth, R.: Interpretable semantic photo geolocation. In: Winter Conference on Applications of Computer Vision, WACV, pp. 1474–1484. IEEE (2022). https://doi.org/10.1109/WACV51458.2022.00154

36. Thomee, B., et al.: The new data and new challenges in multimedia research. CoRR (2015). http://arxiv.org/abs/1503.01817

37. Tomesek, J., Cadík, M., Brejcha, J.: Crosslocate: cross-modal large-scale visual geo-localization in natural environments using rendered modalities. In: Winter Conference on Applications of Computer Vision, WACV, pp. 2193–2202. IEEE (2022). https://doi.org/10.1109/WACV51458.2022.00225

38. Trevisiol, M., Jégou, H., Delhumeau, J., Gravier, G.: Retrieving geo-location of videos with a divide & conquer hierarchical multimodal approach. In: International Conference on Multimedia Retrieval, ICMR, pp. 1–8. ACM (2013). https://doi.org/10.1145/2461466.2461468

39. Uzkent, B., et al.: Learning to interpret satellite images using Wikipedia. In: International Joint Conference on Artificial Intelligence, IJCAI, pp. 3620–3626. ijcai.org (2019). https://doi.org/10.24963/ijcai.2019/502

40. Vo, N.N., Jacobs, N., Hays, J.: Revisiting IM2GPS in the deep learning era. In: International Conference on Computer Vision, ICCV, pp. 2640–2649. IEEE Computer Society (2017)

41. Vrandecic, D., Krötzsch, M.: Wikidata: a free collaborative knowledgebase. Commun. ACM, 78–85 (2014). https://doi.org/10.1145/2629489

42. Weyand, T., Araujo, A., Cao, B., Sim, J.: Google landmarks dataset v2 - a large-scale benchmark for instance-level recognition and retrieval. In: Conference on Computer Vision and Pattern Recognition, CVPR, pp. 2572–2581. IEEE (2020). https://doi.org/10.1109/CVPR42600.2020.00265

43. Weyand, T., Kostrikov, I., Philbin, J.: PlaNet - photo geolocation with convolutional neural networks. In: Leibe, B., Matas, J., Sebe, N., Welling, M. (eds.) ECCV 2016. LNCS, vol. 9912, pp. 37–55. Springer, Cham (2016). https://doi.org/10.1007/978-3-319-46484-8_3

Topics in Contextualised Attention Embeddings

Mozhgan Talebpour[1(✉)], Alba García Seco de Herrera[1], and Shoaib Jameel[2]

[1] School of Computer Science and Electronic Engineering, University of Essex,
Colchester, UK
{mozhgan.talebpour,alba.garcia}@essex.ac.uk
[2] Electronics and Computer Science, University of Southampton, Southampton, UK
M.S.Jameel@southampton.ac.uk

Abstract. Contextualised word vectors obtained via pre-trained language models encode a variety of knowledge that has already been exploited in applications. Complementary to these language models are probabilistic topic models that learn thematic patterns from the text. Recent work has demonstrated that conducting clustering on the word-level contextual representations from a language model emulates word clusters that are discovered in latent topics of words from Latent Dirichlet Allocation. The important question is how such topical word clusters are automatically formed, through clustering, in the language model when it has not been explicitly designed to model latent topics. To address this question, we design different probe experiments. Using BERT and DistilBERT, we find that the attention framework plays a key role in modelling such word topic clusters. We strongly believe that our work paves way for further research into the relationships between probabilistic topic models and pre-trained language models.

1 Introduction

Pre-trained language models (PLMs), e.g., ELMo [35], Generative Pre-trained Transformer (GPT) [37], PaLM [11], and Bidirectional Encoder Representations from Transformers (BERT) [14] are pre-trained using large amounts of text data [24], for instance, BERT has been pre-trained on the BookCorpus and Wikipedia collections. During the domain-independent pre-training process, these models encode a variety of latent information, for instance, semantic and syntactic properties [57], as a result, these models can make reliable predictions even under a zero-shot setting in different applications [20,41,43]. While the pre-training process is computationally [51] and financially expensive [47], these models can be cheaply fine-tuned to reliably handle different downstream tasks such as document classification [1] and information retrieval [50,61], a process that is commonly referred to as transfer learning [32]. For instance, BERT has shown strong performance in natural language understanding [63], text summarisation [25], document classification [10] and other Natural Language Processing (NLP) downstream applications [43].

J. Kamps et al. (Eds.): ECIR 2023, LNCS 13981, pp. 221–238, 2023.
https://doi.org/10.1007/978-3-031-28238-6_15

Another class of models that continues to dominate the text mining landscape are probabilistic topic models (PTMs) [7,8]. These models are probabilistic approaches toward determining dominant topics in a text corpus in a completely unsupervised way. A latent topic is described as a probability distribution of words. Latent Dirichlet Allocation (LDA) [8] is a popular model for discovering topics. In LDA, the model learns to represent a document as a mixture of latent topics and each topic is represented by a mixture of words. When LDA is viewed as a matrix factorisation model, given a term document co-occurrence matrix as input and the number of topics, the model factorises the matrix into two low-dimensional matrices that are word topic and document topic representations. The word topic matrix captures the importance of words in the vocabulary of each topic whereas the document topic matrix captures the topic distribution in every document. While LDA has been a popular model that is based on Bayesian learning, a class of linear algebra-based model called Non-negative Matrix Factorisation (NMF) [26,56] has become equally popular to learn topics [33].

In [43], the authors dissected BERT to understand the property of every layer. They find that lower layers, i.e., layer 1 or 2 capture the linear word order, while the BERT's middle layers learn the syntactic information reliably and the higher layers capture the contextualised information. The authors in [49] and [45] showed that BERT word embedding clustering via simple algorithms such as k-means results in word clusters as if they are learned by a topic model. The authors conducted a series of qualitative probe experiments to find out that most of the word clusters of BERT resemble what is often discovered by the LDA model. While these studies make relevant observations, what is not well studied is how the topic information is encoded at the time of pre-training given that BERT or any other contextual language model is not designed to model topical word clusters. In this work, by conducting different probe experiments, we answer how BERT and DistilBERT [44] can capture clusters of words that resemble what is learnt by topic models. We find that it is the attention [4,9] mechanism in these language models that plays a key role in modelling what resembles word topics as discovered by the topic model.

2   Related Work

The main goal of PLMs [31] is to simulate human language understanding by finding the most probable words sequence and patterns. The traditional language model used probability distribution to predict the next word, but they were not very scalable such as those based on unigram, bigram or trigram language models [36]. The recently developed PLMs are trained using large amounts of text data where some of them exploit a strategy called masked language modelling in a self-supervised way. Once these models have been trained, they have been applied in a wide variety of applications. The key advantage of PLMs is that they can be applied on different downstream tasks [15] reliably.

BERT has been developed with stacked transformers [52] layers where each layer captures different properties in text data, e.g., some layers are ideal to

[CLS]	football		57	38	65	70
the	is	0	app	game	time	movie
player	played	1	apps	player	day	film
plays	in	2	developer	video game	work	show
football	a	3	application	gaming	hour	story
[SEP]	stadium	4	user	developer	week	episode
	[SEP]					

(a) Attention mechanism in BERT via visualisation in Layer 12. We observe that words that are central to the context are assigned high attention weights.

(b) Words, ordered by decreasing probability, obtained from the LDA model.

Fig. 1. Word importance visualization in BERT and LDA.

capture semantic information [48,53]. Transformers consist of encoder-decoder structures. The encoder transforms the sequence of input tokens into a high-level dimension. Decoder predicts input data from encoder [18]. However, in BERT only the encoder part of transformers has been used. There is an important concept in BERT called attention that assigns weights to different input features given their importance in the underlying task. One example is: given the text about *cats*, the model will pay more attention, via attention weights, to words such as *fur*, *eyes*, etc. BERT's attention has also been studied in [12] where the authors find that different attention heads focus on different aspects of language, e.g., they find that heads direct objects of verbs, determiners of nouns, objects of prepositions, and objects of possessive pronouns with far greater accuracy. While they have studied the syntactic and semantic information encoded in different attention heads, they have not separately probed latent topics as learned by the topic models such as LDA and NMF. In Fig. 1a, we depict how attention works obtained via a popular visualisation tool[1]. We input two sentences in sequence, where the first sentence "The player plays football." is followed by the second sentence "Football is played in a stadium.", and both describe the sport *football*. The visualisation tool depicts the case when we select the token "football" in the first sentence and how other semantically related tokens such as "football", "stadium", and "played" are highlighted with high attention weights.

Topic modelling is a machine learning technique that automatically discovers hidden topics in unlabelled data. A topic is defined as a probability distribution of words. While these topic models have been inspired by the latent concept-based models such as Latent Semantic Analysis (LSA) [13] and Probabilistic Latent Semantic Analysis (pLSA) [21], Latent Dirichlet Allocation (LDA) [8] has been widely applied to discover latent topics because it addressed some of the fundamental challenges in LSA and pLSA such as scaling on large datasets and overfitting. While in [27], the authors demonstrated that static word embeddings are related to SVD, which is the core algorithm used in LSA, what we demon-

---

[1] https://github.com/jessevig/bertviz.

strate here is that models such as PLMs implicitly learn latent topic information as encoded by the PTMs.

LDA has been trained considering the exchangeability [17] assumption meaning that word order does not matter in a document. These models describe documents as a mixture of words and each document comprises a mixture of topics defined by the user. Note that BERT does not model document-level information; there are extensions such as Sentence-BERT (SBERT) [40] to model documents.

In Fig. 1b, we depict a typical output obtained from LDA using a freely available online topic modelling visualisation tool[2]. We can observe from this output that there are five top-ranked probability words in some topics that are indexed by topic labels as discrete numbers. From topic index 57, we can infer that the topic describes computer or mobile applications and their development. Topic number 38 describes video gaming.

BERT has demonstrated state-of-the-art results in many NLP downstream tasks, such as natural language inference and information retrieval. Some previous studies have emphasised the importance of contextual information as an additional feature of topic modelling. In [3], for example, the importance of sentence contextual representation and neural topic model was investigated. In SBERT [40], embedding representation was used as the input to the prodLDA [46] neural topic model. If an input document length exceeded the SBERT predefined length, the rest of the document would be omitted. Despite this limitation, the model produced a higher coherence score when compared to Bag-of-Words (BoW) representation embedding. Some other studies have focused on how, and if, adding topic modelling information to a BERT model can lead to an improvement in its performance. In a study conducted by Peinelt et al., [34], they have used topic modelling to improve the BERT performance of semantic similarity domain applications like question answering. They have used BERT-base final layer's *[CLS]* token embedding as the corresponding embedding of an input document. Wang et al., [55] have argued that BERT contextual embedding can be improved by adding topical information to it. In their study, BERT embedding was derived from topics in the corpus. The findings of this research suggest that a word vector representation is equal to the weighted average of different topical vectors. If a topic has high importance in a corpus, words that are related to that topic gain higher importance.

In another related research conducted by [23], topical text classification was applied to a scientific domain dataset. The authors compared the findings of their research with SciBERT [2], which is a pre-trained language model based on BERT, but on scientific documents. Concatenation of BERT embedding and document topic vector was used as an input to a two-layer feed-forward neural network. In a recent study, [49], the role of BERT embedding was examined from a different perspective. This research argued that clustering token-level BERT embedding shares many similarities with topic modelling. The authors used different PLMs such as BERT, GPT-2 [38] and RoBERTa's [28] last three layers of embedding. This work found that except RoBERTa, BERT and GPT-2

---

[2] https://pyldavis.readthedocs.io/en/latest/index.html.

word-level clustering resulted in clusters that resemble close to those obtained using the LDA model. While LDA learns topics as a probability distribution of words, the word clusters obtained by clustering token-level embeddings in PLM cannot be confused with a probability distribution of words. What the authors showed is that there are some similarities between the word clusters of a PTM when compared with the clusters obtained from a PLM.

While the works mentioned above demonstrate important relationships between PTMs and PLMs, what is currently lacking is a further understanding of how latent topics are encoded in the PLM vectors and what component helps in encoding this information. There are works mentioned above that have trained latent topics with pre-trained language models in a unified way. The question is whether it is needed to learn latent topics with pre-trained language models again. While these works have shown quantitative improvements, it is unclear how latent topics are helping them improve upon the results.

3 Probe Tasks

The problem that we intend to study in this paper is whether latent topic information is automatically encoded in contextualised word embeddings. While it is not explicitly evident that latent topic information is encoded, we must design probe tasks. Our key goal is thus to understand how PLMs such as BERT and DistilBERT can discover word clusters that are often discovered by PTMs when they are not specifically designed to model such information. To this end, we first chose to study in more detail the role that attention heads play in the PLM model. It is because just as in a topic model, words that are central to the document's global context are assigned a high probability and words that are central to a topic are assigned a high probability. For instance, if the document is about "sports", words such as "football", "goal", and "player" will have a high probability in that document. Similarly, these words will occur with a high probability in the topic that is about sports. The attention mechanism too shows similar behaviour in the document where words that are central within the given contextual window are assigned high attention weights. Attention weight specifies the importance of a particular word when it is accompanied by other words [12] in a certain pre-defined contextual window.

We consider BERT-base uncased and DistilBERT-based uncased models as our PLMs because of their popularity and computational ease. We also know that the LDA model outputs word and document topic representations [8]. Given the number of factors or latent dimensions, NMF factorises the co-occurrence matrix into two low-dimensional matrices where one matrix encodes word clusters and the other matrix encodes document clusters. Since language models capture word-level patterns, we thus choose word topics in LDA and NMF. Since both LDA and NMF can explicitly be assigned to soft clusters based on their probability values, in the case of the attention representations, we must cluster them using a soft clustering algorithm. This would help us produce word clusters with cluster assignments.

There are other components that we could also study such as the role played when different transformers layers when stacked together. However, previous studies have already found out that the different layers capture different properties of text data, e.g., in BERT, lower layers capture linear word order, middle layers capture syntactic information whereas higher layers capture semantic information. None of these studies has found that word clusters resembling latent topics are also modelled by one of these layers after thorough experimentation. As a result, given their findings, we focus on the attention heads in PLMs first.

In BERT-base, there are 12 layers, each layer containing 12 attention heads. The attention head computes attention weights between all pairs of word combinations in an input sentence. Attention weight can be interpreted as an important criterion when considering two words simultaneously. For example (weather, sunny) pair's attention weight is higher than the (weather, desk) pair. It is because when BERT is trained on billions of tokens (weather, sunny) combinations occurred more frequently than other words such as "desk". Similarly, in the LDA model, if words such as "sports" and "football" occur, they will be assigned a high probability value in the word topic. DistilBERT is also based on the BERT-base model but is much lighter weight with respect to its parameters. It has been obtained after a process known as knowledge distillation [19,29] where the original bigger model known as the teacher was used to train the lighter-weight compressed student model to mimic its behaviour. It was found that in the case of DistilBERT, it retained most of BERT's advantages with a much-reduced parameter set.

Using two publicly available benchmark datasets, we conduct two different probe tasks to demonstrate the generalisability of our findings. In the first probe task, we conduct word-level clustering on the representations obtained from PTM and PLM models and compute the coherence measure which has been popularly used in topic models to evaluate the quality of the topics. In the case of the language models, we extract attention weights from each layer of the model and we obtain the word-attention representations for every word in the vocabulary. We then cluster these attention vectors using a clustering algorithm where the attention vectors are used as features. Through this attention clustering, we expect that words that are semantically related are clustered in one cluster. The motivation is that if the word clusters contain thematically related words, the clusters will demonstrate a high coherence measure. While there have been debates around the usefulness of coherence measure [22], in our study, we use the same measure to compare all models quantitatively.

We intend to probe if there is a comparable coherence performance between a layer of PLM and the word-topic representations obtained from PTM. By comparable, we mean whether the coherence results are numerically close to each other. If the coherence results are comparable, we can expect that in terms of the thematic modelling of words, the language model and the topic models are learning semantically related content. While the coherence probe task might not completely be relied upon, we design an additional probe task to find out the word overlaps between the word clusters obtained from the PLMs and PTMs.

Our motivation is that if the coherence value between the clusters is high then there must be a reliable overlap between the words in the clusters. Since the higher layers, 10, 11 and 12 in the case of the BERT-base model capture semantic information more than the lower layers, we expect that the clusters of words in the high layers of the language model will show higher commonality with those clusters that are learnt by the PLMs.

3.1 Experimental Settings

Datasets: We have used 20 NewsGroups (20NG) and IMDB datasets which are two popular datasets commonly used in the text mining community. The 20NG dataset contains about 18,000 documents in 20 news categories after removing duplicate and empty instances. IMDB dataset contains 50,000 movie reviews that have been labelled as positive or negative. The 20NG dataset contains several long documents whereas IMDB contains relatively short documents with relatively more noisy text.

Text Preprocessing: In the case of the PTMs, we have followed a common pre-processing strategy such as the removal of the stop words, the removal of punctuation, and non-ASCII characters. Through our experiments, we have found that if we do not remove stopwords from text, they tend to dominate most of the topics including increasing the dimensionality of the semantic space resulting in high space and time complexities. While some workaround have been proposed to model natural language using PTMs such as using asymmetric priors, they can be computationally intensive on large datasets [54]. In the case of the PLMs, we let the default pre-processor handle pre-processing, for instance, the BERT-base model has the WordPiece tokenizer. Using NLTK [5], we conducted sentence segmentation.

PLM Attention Weights: For every word in the vocabulary, we obtain the word attention weights from the BERT-base uncased and DistilBERT models. As BERT uses wordPiece tokenisation, if tokenised sentence length is more than 512 tokens, the input sentence is split which is common in the literature. Attention weights of all tokens in a sentence would be stored. If a word appears in different sentences, the average of all words' attention is used which is also commonly done including taking their average embedding of their word pieces [59]. We have obtained attention weights from every layer of BERT. BERT attention weight has been defined as an average of all attention heads in each layer.

We have obtained attention weights from the vanilla BERT-base model. Besides that, we have also obtained attention weights from the fine-tuned version of the BERT model to gauge the role fine-tuning might play in the process. Fine-tuning was done on the text classification task using labels associated with labelled instances in the 20NG and IMDB datasets. Through cross-validation in the fine-tuning process, we present the results of the best-performing model on the test set with the ideal model parameters obtained via a 30% held-out dataset. We have followed the same configuration with the vanilla DistillBERT-base model.

Topic Modelling: We have used the Latent Dirichlet Allocation (LDA) model implemented in Gensim [39] to discover latent topics in our datasets. In the 20NG dataset, we have varied the number of topics from 20 to 200. In the IMDB dataset, we varied the number of topics from 2 to 30 which gave us better results. We have used the NMF model implemented in Gensim. According to [58], larger datasets tend to have more topics than smaller ones. As a result, we have chosen different topic pools in different datasets. We have not chosen the number of topics to be equal to the dimensionality of the word vectors obtained from PTMs because PTMs tend to encode a variety of information in their vectors, e.g., syntactic and semantic information. Besides that, having many topics larger than what we have chosen above tends to result in sub-optimal latent topics leading to the deterioration of performance.

Clustering: We have used the soft Gaussian Mixture Models (GMM) [6] clustering algorithm on the embeddings obtained from PLMs. LDA is already a soft clustering model where probability values are used to assign soft clusters to instances [60]. In LDA, we can automatically obtain the word-topic assignments based on the probability values of words in each topic which is also true for clusters obtained via the GMM model. We used GMM because its implementation is widely and freely available in different software libraries.

Evaluation: In topic modelling, coherence measure has been widely used to evaluate the quality of the latent topics [30]. Coherence score "c-v" has been used in our setting which is available in the Gensim library. This measure has been adapted from the work of Roder et al. [42]. We use coherence to measure the semantic relatedness of tokens in the word clusters obtained from both PLM and PTM models. We also use the number of word overlaps between the top-k words in clusters obtained from the two models to gauge the word overlaps among the clusters. We set $k = 20$ which gives a reliable trade-off between selecting the most thematically related top-k words and not choosing (general or noisy) words with low probability estimated in the word clusters. To compute the word overlap values, for every topic in PTM and every layer's word cluster in PLMs, we computed the overlap between the top-k words, followed by computing the "mode" value. While there are metrics such as entropy and exclusivity [49], we will use these metrics in the extended version of this paper.

4   Discussion

We have computed cluster coherence values on two different datasets. Given two clusters with their respective coherence values. If one cluster's coherence value is higher than the other, the one with the higher coherence values is regarded as a coherent cluster, for instance, in the case of text, the tokens in the coherent clusters tend to be semantically associated with each other. In both LDA and NMF models, we have varied the number of topics to demonstrate the impact of topic clusters. In Table 1 we present the topic coherence results in the 20 Newsgroups and IMDB datasets for the LDA and NMF models. We observe that in the LDA model when the number of topics is 20, we obtain the best

Table 1. Coherence results for LDA and NMF models.

20 Newsgroups				IMDB			
LDA		NMF		LDA		NMF	
# topics	c-v	# topics	c-v	# topics	c-v	# topics	c-v
20	**0.518**	20	0.478	2	0.363	2	0.276
30	0.487	30	**0.504**	5	0.364	5	0.275
50	0.504	50	0.484	10	0.370	10	0.299
100	0.474	100	0.453	20	**0.461**	20	**0.300**
150	0.470	150	0.455	30	0.437	30	0.299
200	0.473	200	0.474				

coherence value of 0.518 in the 20NG dataset. In the case of the NMF model, the best coherence value is when the number of factors is 30 with 0.504 in the 20NG dataset. In the IMDB dataset, we also obtain the best coherence value when the number of topics is 20 in the LDA model with a value of 0.461 and the NMF model gives us the value 0.300 when the number of factors is 20.

PLM & 20NG Dataset: In the case of the vanilla BERT-base model in Table 2 (left), i.e., 20NG dataset, we notice that when the number of soft attention clusters is 50 there is some comparable performance with the coherence results. Precisely, we read from the table that for VB50 the coherence value is 0.503 in layer 11. This coherence value is numerically close to 0.518 when the number of topics is 20, and in the case of the NMF model, it is approximately equal to 0.504 when the number of factors is 30. This suggests that both LDA and vanilla BERT-base attention word clusters are semantically coherent when the number of soft clusters is 50. We also notice that the contextual layers are mainly playing a key role in modelling such semantically close words, i.e., layer 11. When we refer to the word overlaps in Table 4, we notice that the top 20 word overlaps are also consistent with the BERT-base model in layers 7, 8, 9 and 11. It means that out of 20 words, there are 17 overlapping words.

Upon comparing the results of the fine-tuned version of the BERT-base model where the fine-tuning was done on the classification task, we notice that soft clusters 50 and 100 in Table 2 lead to comparable coherence performances obtained by the LDA and NMF models in Table 1. Precisely, we read from the table that when the number of clusters is 50 and 100, we obtain the coherence value of 0.508 and 0.503, respectively that again are numerically comparable to 0.518 in the coherence table for LDA and 0.504 for the NMF model, i.e., Table 1. While it would be ideal to have these coherence results be equal, such results are difficult to obtain considering noise in the data and the randomness involved when initiating the training process of these semantic models. What is interesting in the case of the fine-tuned version of the BERT-base model is that two layers show comparable coherence performances and both these layers learn contextual information.

Table 2. 20NG GMM (left) and IMDB GMM (right) clustering on BERT-base attention weights on the left and the right. The values depict coherence results. VB refers to the vanilla BERT-base model and FT refers to the fine-tuned version. The number followed by VB and FT refers to the number of clusters specified in the GMM model.

Layer	VB30	VB50	VB100	VB150	VB200	FT30	FT50	FT100	FT150	FT200
1	0.360	0.502	0.489	0.481	0.343	0.333	0.477	0.502	0.466	0.330
2	0.346	0.480	0.463	0.464	0.334	0.327	0.479	0.480	0.462	0.333
3	0.329	0.450	0.453	0.448	0.323	0.315	0.466	0.450	0.459	0.324
4	0.328	0.466	0.461	0.452	0.332	0.324	0.466	0.466	0.461	0.332
5	0.33	0.460	0.448	0.449	0.324	0.325	0.458	0.460	0.453	0.329
6	0.33	0.459	0.455	0.451	0.318	0.347	0.466	0.459	0.460	0.337
7	0.337	0.478	0.471	0.454	0.325	0.346	0.495	0.478	0.479	0.347
8	0.347	0.469	0.468	0.470	0.336	0.353	**0.508**	0.469	**0.496**	0.359
9	0.346	0.486	0.474	0.471	0.344	**0.370**	**0.508**	0.486	0.494	0.360
10	0.373	0.480	0.483	0.476	0.360	0.368	0.502	0.480	0.494	0.358
11	0.369	**0.503**	0.489	0.481	0.360	0.357	0.483	**0.503**	0.489	**0.361**
12	**0.373**	0.502	**0.489**	**0.484**	**0.363**	0.364	0.485	0.502	0.480	0.355

Layer	VB2	VB5	VB10	VB20	VB30	FT2	FT5	FT10	FT20	FT30
1	0.411	0.390	0.365	0.358	0.355	0.455	0.442	0.372	0.348	0.333
2	0.473	0.447	0.374	0.347	0.352	0.469	**0.459**	0.420	0.391	0.384
3	0.480	0.414	0.418	0.385	0.366	0.501	0.452	0.415	**0.412**	**0.404**
4	**0.586**	0.478	0.444	0.421	0.404	0.457	0.388	0.386	0.383	0.380
5	0.583	**0.490**	0.439	0.426	0.422	0.410	0.383	0.350	0.359	0.356
6	0.563	0.477	**0.471**	0.429	0.405	0.452	0.438	0.396	0.374	0.357
7	0.546	0.485	0.431	0.425	0.416	0.489	0.403	0.399	0.374	0.366
8	0.510	0.438	0.428	0.414	0.415	**0.514**	0.427	0.395	0.397	0.369
9	0.452	0.410	0.393	0.383	0.373	0.438	0.425	0.366	0.377	0.374
10	0.476	0.430	0.381	0.351	0.349	0.426	0.417	0.369	0.361	0.349
11	0.425	0.429	0.400	0.398	0.385	0.430	0.391	0.346	0.354	0.347
12	0.523	0.454	0.446	**0.439**	**0.431**	0.469	0.433	**0.440**	0.402	0.385

Table 3. 20NG GMM (left) and IMDB GMM (right) clustering on DistilBERT attention weights. The values depict coherence results. VD refers to the vanilla DistilBERT model and FT refers to the fine-tuned version. The number followed by VD and FT refers to the number of clusters specified to the GMM model.

Layer	VD30	VD50	VD100	VD150	VD200	FT30	FT50	FT100	FT150	FT200
1	0.504	0.497	0.518	0.515	0.521	0.502	0.508	0.511	0.517	0.513
2	0.509	0.509	0.510	0.514	0.515	0.511	0.518	0.510	0.507	0.508
3	0.514	0.514	0.508	0.508	0.510	0.507	0.503	0.503	0.499	0.507
4	0.516	0.509	0.516	0.517	0.513	0.502	0.502	0.504	0.503	0.505
5	0.548	0.550	0.551	0.544	0.549	0.544	0.550	0.546	0.543	0.544
6	**0.593**	**0.572**	**0.572**	**0.573**	**0.568**	**0.573**	**0.576**	**0.573**	**0.571**	**0.571**

Layer	VD2	VD5	VD10	VD20	VD30	FT2	FT5	FT10	FT20	FT30
1	0.231	0.258	0.249	0.250	0.251	0.219	0.224	0.226	0.248	0.238
2	0.166	0.211	0.212	0.224	0.230	0.255	0.225	0.231	0.232	0.238
3	0.231	0.244	0.235	0.251	0.244	0.253	0.225	0.230	0.235	0.234
4	0.151	0.228	0.204	0.228	0.234	0.170	0.223	0.218	0.219	0.217
5	0.317	**0.347**	0.307	0.270	0.264	0.252	0.268	**0.274**	**0.272**	0.261
6	**0.334**	0.327	**0.325**	**0.317**	**0.312**	**0.288**	**0.270**	0.274	0.254	**0.264**

When we look at the topic associated with "computing technology" in the 20NG dataset, we noticed that words such as "organisation", "com", and "nntp" were among the overlapping words which suggest that both BERT and LDA learn thematically the same words. While it can be argued that even simple clustering algorithms such as k-means might generate clusters that are coherent and with high-overlapping words, we have found out that k-means does not lead to coherent clusters and the word overlap count was also very low, for instance, in most cases we found the word overlap values to be sometimes 1, and most often, 0.

In the case of the vanilla DistilBERT model presented in Table 3 (left), we notice that the higher layers demonstrate the highest soft cluster coherence results. What we notice is that the contextual layers show a higher degree of cluster coherence comparable to performance with the LDA model than with the NMF model in Table 1, for instance, the vanilla DistilBERT version with 200 soft clusters shows a relatively comparable performance when compared with the LDA model in Table 1. It can be argued that in terms of the absolute numbers the results in Table 3 are much higher than in Table 1 when we only look at the highest DistilBERT layers values. One of the reasons is that different pre-processing strategies have been chosen in both models. However, this was unavoidable because including stop words in the PTM models would result in noisy topics. Note that other layers such as Layer 4, soft cluster 30, in the case of the vanilla DistilBERT model compare well with the LDA coherence results. Layer 4 in the case of the DistilBERT model compares reliably with the soft cluster 30 when we consider the NMF model.

PLM & IMDB Dataset: In the IMDB dataset, Table 1 presents the ideal coherence value when the number of topics/factors is 20 for the LDA and the NMF models. For the LDA model, the coherence value is 0.461 and for the NMF model, the coherence value is 0.300. Referring to Table 2 (right), we see that the comparable LDA value is obtained in layer 6 in the vanilla BERT-base version when the number of soft clusters is 10. In the fine-tuned version, we see the comparable value in layer 8 when compared to the LDA model and when the number of soft clusters is 150. If we consider topic 30 in Table 1, we notice two comparable values in Table 2 in layer 12 which is a layer that captures contextual information more than any other layer when the vanilla soft clusters are 20 and 30.

In Table 4, most word overlaps occur in layers 5, 9, 11, and 12 and these results are consistent with the 20NG results where higher contextual layers have the maximum word overlap. We also notice that layers 6 and above have the most ideal coherence values indicating that if the clusters are coherent, they also have maximum word overlaps. It means that these clusters share common words. In DistilBERT, in Tables 3 and 4 we see that the NMF model tends to show comparable coherence values in the higher layers. In Table 4, we observe that the word overlaps are fairly uniformly distributed across layers. While the lower layers have shown to have maximum overlaps, we can notice that the upper layers too have a word similar overlaps. However, their coherence values are not comparable. It is because IMDB instances are short noisy sentences where the model seems to be performing not very reliably unlike the 20NG dataset. What is also noticeable from the results is that the fine-tuned version of the DistilBERT model does not show comparable coherence performance when compared with the NMF model. This could suggest that classification fine-tuning helps DistilBERT lose the latent topic information.

In summary: 1) the attention mechanism is an important component in the PLMs that help capture some patterns that are also captured by PTMs. 2) there is correspondence between the coherence results obtained from PLMs and PTMs because in most cases we obtain comparable coherence performance. 3) in PLMs, there are high word overlaps in the contextualised layers and clusters of words obtained from PTMs. 4) in most cases, it is the contextualised layer that captures the most commonality with PTMs.

One of the limitations of our work is that it does not experiment with other language models very different from BERT such as XLNet [62] and GPT-3 [16] to ascertain that similar conclusions could be also derived from them. However, what is important to note is that our conclusions point toward the importance of the attention mechanism rather than the way pre-training is done or the size of the dataset that has been used to pre-train the model, or the model design. We also have to verify whether the results are generalizable to even larger models such as BERT-large which requires much more computational resources to conduct this study.

We show another finding through Fig. 2 where we demonstrate the importance of the attention mechanism and how topic weights (probabilities) and attention weights tend to focus on the same words in a given context. To generate the figure, we have taken an example from the IMDB dataset. In the

Table 4. BERT (left) and DistilBERT (right) attention word overlap with LDA.

Layer	20NG	IMDB	Layer	20NG	IMDB
1	16	14	1	**12**	**10**
2	16	13	2	**12**	**10**
3	16	12	3	9	**10**
4	16	14	4	**12**	**10**
5	16	**17**	5	**12**	**10**
6	16	14	6	**12**	9
7	**17**	12			
8	**17**	12			
9	**17**	**17**			
10	16	11			
11	**17**	**17**			
12	16	**17**			

	words	BERT-based	DistilBERT-based	LDA	NMF
0	This	0.000000	0.000000	0.000000	0.000000
1	movie	0.071700	0.091100	0.041000	0.000000
2	is	0.000000	0.000000	0.000000	0.000000
3	terrible	0.111100	0.086300	0.002000	0.001000
4	but	0.000000	0.000000	0.000000	0.000000
5	it	0.000000	0.000000	0.000000	0.000000
6	has	0.000000	0.000000	0.000000	0.000000
7	some	0.000000	0.000000	0.000000	0.000000
8	good	0.102200	0.071100	0.010000	0.101000
9	effects	0.111700	0.097400	0.002000	0.003000

Fig. 2. Illustrating attention using a sentence from the IMDB dataset as an example. We have presented these results from the BERT-base layer 11 and DistilBERT-based layer 5. The number of topics/factors in the case of PTM is 20. The figure is used to demonstrate that these models tend to focus on relevant tokens within their context and assign lower weights to general tokens such as stopwords.

BERT-base model, layer 11 is examined because it is the contextual layer and has the highest word overlaps in Table 4. In the case of the DistilBERT-base model, we have selected layer 5 given that it is one of the contextual layers and has one of the highest word overlaps in Table 4. We have selected the number of topics as 20 and the number of NMF factors as 20 which is based on the results obtained in Table 1. What we observe from the figure all the models tend to focus on the relevant keywords in the context, for instance, we observe that PLMs focus on the words such as "good", "effects", "terrible", "movie" that are relevant to the movie and the PTMs too tend to focus on the same tokens in this context. What we learn from the figure is that PTMs and PLMs, while they are different, both tend to focus on the relevant words in a given contextual window. This figure helps us to draw some relationships between the attention weights and the topic probabilities in that they focus on the important words only. We also notice that common words such as stopwords are given less weightage by the models.

While the authors in [49] have found out that the word clusters obtained from some PLMs tend to cluster the contextualised word vectors that resemble what is learned by a topic model, our result suggests that it is the attention mechanism that is playing a key role in obtaining such results which is the key contribution of our work. It can also be argued that the contextualised token embeddings obtained from a PLM model can lead to almost similar conclusions, in this work, we wanted to explicitly study the role of the attention weights.

5   Conclusions

Topic modelling has remained a dominant modelling paradigm in the last decade with several topic models developed in the literature [64]. Topic models were not only modelled using Bayesian statistics but also linear algebra-based such as the NMF model. While both these models are formulated differently, they both tend to exhibit similar clustering properties. With the development of PLMs, these models have now taken over the landscape in text mining and NLP because they have outperformed existing baselines. Recent research points out that word-level clustering on BERT embeddings results in word clusters that share a close relationship with those discovered using topic models. As a result, this motivated us to study the reason which component in the language model helps capture such topic information when the model has not been explicitly designed to model latent word topics. Through probe tasks, we find that it is the attention mechanism that plays a key role in modelling word patterns that resemble something that is also discovered using topic models. We strongly believe that our work helps add further insight into the relationships between topic models and PLMs including the role that is played by the attention mechanism in the language model. In the future, we will conduct a thorough theoretical analysis to find out the key theoretical similarities between a topic model and a PLM. We will also study how different PLMs other than those that are based on BERT encode latent topics using attention weights.

Our results are not only applicable to NLP and document modelling fields in general, but the results are also relevant to information retrieval. For instance, in an information retrieval setting, we can only use features obtained from PLMs to retrieve relevant documents without having to worry about latent topics features that would potentially increase the number of features that might even degrade the performance of an information retrieval engine. Besides that, we may be injecting more redundant features into the information retrieval model. Topic models have been shown to improve information retrieval results and PLMs have been shown to demonstrate even better results. This could be because PLMs already have encoded a variety of features in their rich vector space that includes latent topics. As a result, the improvement that we see also comes from topics implicitly encoded in the PLM attention vectors. We thus believe that our paper will have a significant impact in the information retrieval field too.

References

1. Adhikari, A., Ram, A., Tang, R., Lin, J.: DocBERT: BERT for document classification. arXiv preprint arXiv:1904.08398 (2019)
2. Beltagy, I., Lo, K., Cohan, A.: SciBERT: a pretrained language model for scientific text. arXiv (2019)
3. Bianchi, F., Terragni, S., Hovy, D.: Pre-training is a hot topic: contextualized document embeddings improve topic coherence. arXiv (2020)
4. Bibal, A., et al.: Is attention explanation? an introduction to the debate. In: Proceedings of the 60th Annual Meeting of the Association for Computational Linguistics (Volume 1: Long Papers), pp. 3889–3900. Association for Computational Linguistics, Dublin, May 2022. https://doi.org/10.18653/v1/2022.acl-long.269. https://aclanthology.org/2022.acl-long.269
5. Bird, S., Klein, E., Loper, E.: Natural Language Processing with Python: Analyzing Text with the Natural Language Toolkit. O'Reilly Media, Inc. (2009)
6. Bishop, C.M.: Pattern recognition. Mach. Learn. **128**(9) (2006)
7. Blei, D.M.: Probabilistic topic models. Commun. ACM **55**(4), 77–84 (2012)

8. Blei, D.M., Ng, A.Y., Jordan, M.I.: Latent Dirichlet allocation. JMLR **3**, 993–1022 (2003)
9. Brunner, G., Liu, Y., Pascual, D., Richter, O., Ciaramita, M., Wattenhofer, R.: On identifiability in transformers. arXiv preprint arXiv:1908.04211 (2019)
10. Chalkidis, I., Fergadiotis, M., Malakasiotis, P., Androutsopoulos, I.: Large-scale multi-label text classification on EU legislation. arXiv (2019)
11. Chowdhery, A., et al.: Palm: scaling language modeling with pathways. arXiv preprint arXiv:2204.02311 (2022)
12. Clark, K., Khandelwal, U., Levy, O., Manning, C.D.: What does BERT look at? An analysis of BERT'S attention. arXiv (2019)
13. Deerwester, S., Dumais, S.T., Furnas, G.W., Landauer, T.K., Harshman, R.: Indexing by latent semantic analysis. J. Am. Soc. Inf. Sci. **41**(6), 391–407 (1990)
14. Devlin, J., Chang, M.W., Lee, K., Toutanova, K.: BERT: pre-training of deep bidirectional transformers for language understanding. arXiv (2018)
15. Edunov, S., Baevski, A., Auli, M.: Pre-trained language model representations for language generation. arXiv (2019)
16. Floridi, L., Chiriatti, M.: GPT-3: its nature, scope, limits, and consequences. Mind. Mach. **30**(4), 681–694 (2020)
17. Foti, N.J., Williamson, S.A.: A survey of non-exchangeable priors for Bayesian nonparametric models. IEEE Trans. Pattern Anal. Mach. Intell. **37**(2), 359–371 (2013)
18. Futami, H., Inaguma, H., Ueno, S., Mimura, M., Sakai, S., Kawahara, T.: Distilling the knowledge of BERT for sequence-to-sequence ASR. arXiv (2020)
19. Hahn, S., Choi, H.: Self-knowledge distillation in natural language processing. arXiv preprint arXiv:1908.01851 (2019)
20. Heo, S.H., Lee, W., Lee, J.H.: mcBERT: momentum contrastive learning with BERT for zero-shot slot filling. arXiv preprint arXiv:2203.12940 (2022)
21. Hofmann, T.: Probabilistic latent semantic analysis. arXiv (2013)
22. Hoyle, A., Goel, P., Hian-Cheong, A., Peskov, D., Boyd-Graber, J., Resnik, P.: Is automated topic model evaluation broken? The incoherence of coherence. In: Advances in Neural Information Processing Systems, vol. 34, pp. 2018–2033 (2021)
23. Iida, F., Pfeifer, R., Steels, L., Kuniyoshi, Y.: Lecture notes in artificial intelligence (subseries of lecture notes in computer science): Preface. AI 3139 (2004)
24. Lai, Y.A., Lalwani, G., Zhang, Y.: Context analysis for pre-trained masked language models. In: EMNLP, pp. 3789–3804 (2020)
25. Lamsiyah, S., Mahdaouy, A.E., Ouatik, S.E.A., Espinasse, B.: Unsupervised extractive multi-document summarization method based on transfer learning from BERT multi-task fine-tuning. JIS (2021)
26. Lee, D.D., Seung, H.S.: Learning the parts of objects by non-negative matrix factorization. Nature **401**(6755), 788–791 (1999)
27. Levy, O., Goldberg, Y.: Neural word embedding as implicit matrix factorization. In: Advances in Neural Information Processing Systems, vol. 27 (2014)
28. Liu, Y., et al.: RoBERTa: a robustly optimized BERT pretraining approach. arXiv (2019)
29. Lopes, R.G., Fenu, S., Starner, T.: Data-free knowledge distillation for deep neural networks. arXiv preprint arXiv:1710.07535 (2017)
30. Mimno, D., Wallach, H., Talley, E., Leenders, M., McCallum, A.: Optimizing semantic coherence in topic models. In: Proceedings of the 2011 Conference on Empirical Methods in Natural Language Processing, pp. 262–272 (2011)
31. Min, B., et al.: Recent advances in natural language processing via large pre-trained language models: a survey. arXiv preprint arXiv:2111.01243 (2021)

32. Mozafari, M., Farahbakhsh, R., Crespi, N.: A BERT-based transfer learning app-roach for hate speech detection in online social media. In: Cherifi, H., Gaito, S., Mendes, J.F., Moro, E., Rocha, L.M. (eds.) COMPLEX NETWORKS 2019. SCI, vol. 881, pp. 928–940. Springer, Cham (2020). https://doi.org/10.1007/978-3-030-36687-2_77

33. de Paulo Faleiros, T., de Andrade Lopes, A.: On the equivalence between algo-rithms for non-negative matrix factorization and latent Dirichlet allocation. In: ESANN (2016)

34. Peinelt, N., Nguyen, D., Liakata, M.: tBERT: topic models and BERT joining forces for semantic similarity detection. In: ACL, pp. 7047–7055 (2020)

35. Peters, M.E., et al.: Deep contextualized word representations. CoRR abs/1802.05365 (2018)

36. Ponte, J.M., Croft, W.B.: A language modeling approach to information retrieval. In: ACM SIGIR Forum, vol. 51, pp. 202–208. ACM New York (2017)

37. Radford, A., Narasimhan, K., Salimans, T., Sutskever, I.: Improving language understanding by generative pre-training (2018)

38. Radford, A., et al.: Language models are unsupervised multitask learners. OpenAI Blog 1(8), 9 (2019)

39. Rehurek, R., Sojka, P.: Gensim-python framework for vector space modelling. NLP Centre, Faculty of Informatics, Masaryk University, Brno, Czech Republic 3(2) (2011)

40. Reimers, N., Gurevych, I.: Sentence-BERT: sentence embeddings using Siamese BERT-networks. arXiv (2019)

41. Ristoski, P., Lin, Z., Zhou, Q.: KG-ZESHEL: knowledge graph-enhanced zero-shot entity linking. In: Proceedings of the 11th on Knowledge Capture Conference, pp. 49–56 (2021)

42. Röder, M., Both, A., Hinneburg, A.: Exploring the space of topic coherence mea-sures. In: Proceedings of the eighth ACM International Conference on Web Search and Data Mining, pp. 399–408 (2015)

43. Rogers, A., Kovaleva, O., Rumshisky, A.: A primer in BERTology: what we know about how BERT works. TACL 8, 842–866 (2020)

44. Sanh, V., Debut, L., Chaumond, J., Wolf, T.: DistilBERT, a distilled version of BERT: smaller, faster, cheaper and lighter. arXiv preprint arXiv:1910.01108 (2019)

45. Sia, S., Dalmia, A., Mielke, S.J.: Tired of topic models? Clusters of pretrained word embeddings make for fast and good topics too! arXiv (2020)

46. Srivastava, A., Sutton, C.: Autoencoding variational inference for topic models. arXiv (2017)

47. Strubell, E., Ganesh, A., McCallum, A.: Energy and policy considerations for deep learning in NLP. arXiv preprint arXiv:1906.02243 (2019)

48. Tenney, I., Das, D., Pavlick, E.: BERT rediscovers the classical NLP pipeline. arXiv preprint arXiv:1905.05950 (2019)

49. Thompson, L., Mimno, D.: Topic modeling with contextualized word representa-tion clusters. arXiv (2020)

50. Trabelsi, M., Chen, Z., Davison, B.D., Heflin, J.: Neural ranking models for doc-ument retrieval. Inf. Retrieval J. 24(6), 400–444 (2021). https://doi.org/10.1007/s10791-021-09398-0

51. Turc, I., Chang, M.W., Lee, K., Toutanova, K.: Well-read students learn better: on the importance of pre-training compact models. arXiv preprint arXiv:1908.08962 (2019)

52. Vaswani, A., et al.: Attention is all you need. In: NIPS, vol. 30 (2017)

53. Voita, E., Sennrich, R., Titov, I.: The bottom-up evolution of representations in the transformer: a study with machine translation and language modeling objectives. arXiv preprint arXiv:1909.01380 (2019)
54. Wallach, H., Mimno, D., McCallum, A.: Rethinking LDA: why priors matter. In: Advances in Neural Information Processing Systems, vol. 22 (2009)
55. Wang, Y., Bouraoui, Z., Anke, L.E., Schockaert, S.: Deriving word vectors from contextualized language models using topic-aware mention selection. arXiv (2021)
56. Wang, Y.X., Zhang, Y.J.: Nonnegative matrix factorization: a comprehensive review. IEEE Trans. Knowl. Data Eng. **25**(6), 1336–1353 (2012)
57. Warstadt, A., et al.: Investigating BERT'S knowledge of language: five analysis methods with NPIS. arXiv preprint arXiv:1909.02597 (2019)
58. Wei, X., Croft, W.B.: LDA-based document models for ad-hoc retrieval. In: SIGIR, pp. 178–185 (2006)
59. Wu, Y., et al.: Google's neural machine translation system: bridging the gap between human and machine translation. arXiv (2016)
60. Xu, W., Liu, X., Gong, Y.: Document clustering based on non-negative matrix factorization. In: Proceedings of the 26th Annual International ACM SIGIR Conference on Research and Development in Information Retrieval, pp. 267–273 (2003)
61. Yang, W., Zhang, H., Lin, J.: Simple applications of BERT for ad hoc document retrieval. arXiv preprint arXiv:1903.10972 (2019)
62. Yang, Z., Dai, Z., Yang, Y., Carbonell, J., Salakhutdinov, R.R., Le, Q.V.: XLNet: generalized autoregressive pretraining for language understanding. In: Advances in Neural Information Processing Systems, vol. 32 (2019)
63. Zhang, Z., et al.: Semantics-aware BERT for language understanding. In: AAAI, vol. 34, pp. 9628–9635 (2020)
64. Zhao, H., Phung, D., Huynh, V., Jin, Y., Du, L., Buntine, W.: Topic modelling meets deep neural networks: a survey. arXiv preprint arXiv:2103.00498 (2021)

New Metrics to Encourage Innovation and Diversity in Information Retrieval Approaches

Mehmet Deniz Türkmen[1]($\boxtimes$), Matthew Lease[2], and Mucahid Kutlu[1]

[1] Department of Computer Engineering, TOBB University of Economics and Technology, Ankara, Turkey
{m.turkmen,m.kutlu}@etu.edu.tr
[2] School of Information, University of Texas at Austin, Austin, TX, USA
ml@utexas.edu

Abstract. In evaluation campaigns, participants often explore variations of popular, state-of-the-art baselines as a low-risk strategy to achieve competitive results. While effective, this can lead to local "hill climbing" rather than a more radical and innovative departure from standard methods. Moreover, if many participants build on similar baselines, the overall diversity of approaches considered may be limited. In this work, we propose a new class of IR evaluation metrics intended to promote greater diversity of approaches in evaluation campaigns. Whereas traditional IR metrics focus on user experience, our two "innovation" metrics instead reward exploration of more divergent, higher-risk strategies finding relevant documents missed by other systems. Experiments on four TREC collections show that our metrics do change system rankings by rewarding systems that find such rare, relevant documents. This result is further supported by a controlled, synthetic data experiment, and a qualitative analysis. In addition, we show that our metrics achieve higher evaluation stability and discriminative power than the standard metrics we modify. To support reproducibility, we share our source code.

Keywords: Evaluation · Metrics · Information retrieval

1 Introduction

Researchers must balance risk vs. reward in prioritizing methods to investigate. Higher-risk methods offer the potential for a larger impact, but with a greater chance of sub-baseline performance. In contrast, lower-risk methods are more likely to yield improvement but may be incremental. A popular strategy to straddle such risk is to investigate variants of popular state-of-the-art models (e.g., use of pre-trained language models, such as GPT-3 [1]). While this represents a low-risk strategy to achieve competitive results, it can lead to local "hill climbing" rather than exploring higher-risk, more radical departures from current state-of-the-art methods. Moreover, if many researchers build on similar baselines, this can limit the overall diversity of approaches being explored in the field.

J. Kamps et al. (Eds.): ECIR 2023, LNCS 13981, pp. 239–254, 2023.
https://doi.org/10.1007/978-3-031-28238-6_16

In this work, we investigate a novel class of "innovation" evaluation metrics that seek to promote greater diversity among participant methods in evaluation campaigns. Such community benchmarking and evaluation campaigns play an important role in assessing the current state-of-the-art and promoting continuing advancements. For participants, evaluation campaigns provide a valuable testing ground for novel methods, and evaluation metrics chosen by a campaign can galvanize community attention on particular aspects of system performance. Evaluation campaign metrics thus help to steer a field.

Whereas traditional IR metrics focus on ranking quality for the user, our innovation metrics instead reward exploration of more divergent, higher-risk ranking methods that find relevant documents missed by most other systems. The key intuition is that a system finding relevant documents missed by other systems must differ in approach. Specifically, we modify standard Precision@K and Average Precision metrics to reward retrieval of such "rare" relevant documents missed by other systems. A simple mixture-weight parameter controls the relative weight placed on such rarity, and setting this to zero reverts to the original metric. As such, evaluation campaigns adopting our metrics could easily control the extent to which they want to reward diversity of approaches vs. more standard user-oriented performance measures.

Experiments over four TREC collections show that our proposed metrics do yield different rankings of systems compared to the existing metrics. In particular, we observe a steady decrease in rank correlation with official system rankings as greater weight is placed on finding rare, relevant documents. This means that if our metrics were adopted in practice, participants would be incentivized to retrieve more diverse relevant documents, with the potential to spur further innovation in the field. Additional results show that our metrics provide higher discriminative power and evaluation stability than the standard Precision@K and Average Precision metrics that we modify.

Contributions. 1) We propose a novel class of "innovation" metrics to stimulate greater diversity of document ranking approaches for evaluation campaigns. Future work is expected to expand and improve upon our initial metrics. 2) We propose new generalizations of classic P@K and AP metrics via a simple user-specified mixture weight. This allows weighting document rarity or trivial reversion to the standard metric. 3) Results over four TREC collections show our metrics change system rankings, as well as providing higher discriminative power and evaluation stability than the standard metrics we modify. 4) We share our source code to support reproducibility and follow-on work[1].

Our article is organized as follows. Section 2 describes our proposed metrics. Section 3 then presents an initial, controlled study using synthetic data to show how retrieving rare vs. common documents affects system rankings. Next, Sect. 4 presents our main results with TREC collections, including a qualitative analysis in Sect. 4.6. We then present discussion and limitations in Sect. 5. Section 6 discusses related work, and we conclude in Sect. 7.

---

[1] https://github.com/mdenizturkmen/ecir2023.

2 Proposed IR Metrics

Retrieval of *rare* documents (that few or no other systems retrieve) indicates that a system's ranking algorithm diverges from that of other systems. In this section, we introduce our two "innovation" metrics that seek to promote exploration of different approaches by rewarding retrieval of such rare, relevant documents. Specifically, we adapt Precision@K (P@K) (Sect. 2.1) and Average Precision (AP) metrics (Sect. 2.2), introducing a linear interpolation parameter α that balances the original metric vs. innovation by varying the weight placed on document rarity. In both cases, setting $\alpha = 0$ reverts to the original metric.

2.1 Rareness-Based Precision@K ($P@K_{Rareness}$)

We define our rareness-based precision-at-k as follows:

$$P@K_{Rareness} = \frac{1}{k} \sum_{i=1}^{k} Rel(d_i) \left(1 + \alpha R(d_i)\right) \tag{1}$$

where k is the rank cut-off value, $Rel(d_i)$ is a binary indicator function for whether d_i is relevant or not, $R(d_i)$ quantifies document rarity, and α is the aforementioned linear interpolation parameter. As noted earlier, setting $\alpha = 0$ reverts to the standard P@K formula. In the other direction, larger α values provide greater rewards for the retrieval of rare documents. Like the original P@K, only relevant documents contribute to the score (i.e., when $Rel(d_i) = 1$), so document rarity is immaterial when $Rel(d_i) = 0$. We define rarity $R(d)$ by:

$$R(d) = 1 - \frac{S_d}{S} \tag{2}$$

where S is the total number of systems and $S_d \geq 1$ is the number of those that retrieve document d. Rareness is bounded by $R(d) \in [0, \frac{(S-1)}{S}]$, minimized when a document is retrieved by all systems (i.e., $S_d = S$) and maximized when only one system retrieves d (i.e., $S_d = 1$). Therefore, as the number of systems S increases, retrieving rare documents becomes more valuable.

While α can be at any value, we recommend setting $\alpha \in [0, 1]$ yielding bounds of $P@K_{Rareness} \in [0, 2)$. The lower bound of $P@K_{Rareness} = 0$ occurs when all documents are non-relevant. The upper-bound is reached when $\alpha = 1$ and all retrieved documents are relevant and have maximal rarity $R(d) = \frac{(S-1)}{S}$, thus $P@K_{Rareness} = 2\frac{(S-1)}{S} < 2$.

2.2 Rareness Based Average Precision ($AP_{Rareness}$)

Assuming N_R relevant documents for a given topic, we define $AP_{Rareness}$ as:

$$AP_{Rareness} = \frac{1}{N_R} \sum_{i=1}^{k} Rel(d_i) P@K_{Rareness}(i) \tag{3}$$

When $\alpha = 0$, $P@K_{Rareness} = P@K$, and thus $AP_{Rareness} = AP$. $AP_{Rareness}$ directly inherits $P@K_{Rareness}$'s same lower-bound and upper-bound of $[0, 2)$.

3   Experiment with Synthetic Data

We first present a controlled, synthetic data experiment to explore the behavior
of $P@K_{Rareness}$ for varying $\alpha \in [0, 1]$ and numbers of D relevant documents
retrieved. We contrast the evaluation of two hypothetical systems: S_{rare} vs.
S_{common}, on a single topic (#1127540) from the Deep Learning Track 2020
(DLT20) [9], as if our hypothetical systems had participated with other real
participants. While S_{rare} always retrieves simulated relevant documents found
by no other system, S_{common} retrieves the most common, real relevant documents
first. We include all official runs from DLT20's document ranking task.

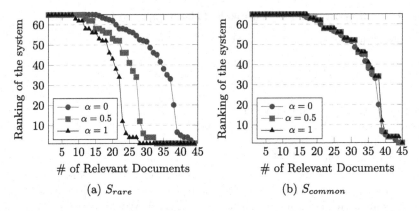

Fig. 1. Ranking of hypothetical systems, S_{rare} and S_{common}, for topic 1127540 of Deep
Learning Track 2020 based on $P@100_{Rareness}$. Experiments vary rarity weight α as well
as D, the number of relevant documents retrieved.

Figure 1 shows the $P@100_{Rareness}$ ranking of S_{rare} vs. S_{common}. Note that
a lower rank indicates a better system, with the best system being ranked first
(i.e., having rank 1). First, recall that when $\alpha = 0$, $P@K_{Rareness} = P@k$. In
this case, both S_{rare} and S_{common} are seen to exhibit the same $P@K_{Rareness}$
curve, as expected, since no weight is placed on rarity. Second, we see that
S_{common}'s ranking is largely unaffected by α since it always retrieves common
(i.e., non-rare) relevant documents. In contrast, the ranking of S_{rare} noticeably
changes across different α values. For example, it requires 28, 33, and 44 relevant
documents to be ranked first when α is set to 1, 0.5, and 0, respectively.

Overall, the results above validate our expectations regarding the behavior
of $P@k_{Rareness}$ under controlled conditions. It reverts toward standard $P@k$ at
$\alpha = 0$, and results place greater emphasis on rarity as we move toward $\alpha = 1$.

4   Experiments with Real Data

In this section, we first describe our experimental setup (Sect. 4.1). Next, we
compare our modified metrics vs. their original counterparts in terms of system
rankings (Sect. 4.2), discriminative power (Sect. 4.3), and evaluation stability

(Sect. 4.4). We also assess how our metrics are affected by the number of systems (Sect. 4.5). Furthermore, we conduct qualitative analysis to better understand the nature of rarely-retrieved documents (Sect. 4.6).

4.1 Experimental Setup

We use trec_eval[2] for calculations of classical evaluation metrics. We set the cut-off threshold to 100 for all metrics we use including ours. We use four different TREC collections, including TREC-5 [11], TREC-8 [12], Web Track 2014 (WT14) [7], and Deep Learning Track 2020 (DLT20) [9]. We carry out our experiments using all official runs from ad-hoc search tasks of TREC-5, TREC-8, and WT14, and the document ranking task of DLT20.

4.2 System Rankings

We compare system rankings for $P@100_{Rareness}$ and $AP_{Rareness}$ against rankings based on P@100 and AP, respectively, in order to observe the impact of rewarding rarity. We report Kendall's τ rank correlation. Experiments with τ_{AP} [31] yielded similar results and so are omitted.

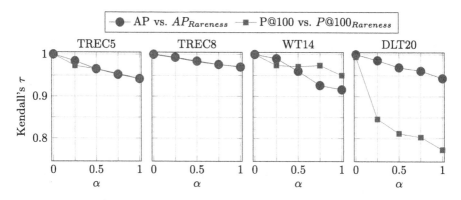

Fig. 2. Kendall's τ correlation between system rankings based on $P@100_{Rareness}$ vs. $P@100$ and system rankings based on AP vs. $AP_{Rareness}$.

Figure 2 shows Kendall's τ scores on four test collections for varying α. As expected, when $\alpha=0$, our modified metrics revert to their unmodified forms, thus yielding perfect $\tau = 1$ rank correlation. We observe steady trends of decreasing rank correlation with increasing α. While Kendall's τ scores for comparisons against AP and P@K metrics are similar in TREC-5 and TREC-8, they diverge in WT14 and DLT20. For instance, when we compare P@100 vs. $P@100_{Rareness}$ in DLT20, Kendall's τ is lower than 0.9 (a traditionally-accepted threshold for acceptable correlation [29]) for $\alpha > 0$. However, we do not observe this when we

---

[2] https://trec.nist.gov/trec_eval/.

Table 1. Discriminative power of metrics for 95% and 99% significance thresholds. The highest score for each collection and significance threshold is written in **bold**. Note that the total number of system pairs are 1830, 8256, 406, 2016 for TREC-5, TREC-8, WT14, DTL20, respectively.

Metric		TREC-5		TREC-8		WT14		DLT20	
	α	95%	99%	95%	99%	95%	99%	95%	99%
P@100	0.0	598	406	3666	2973	218	174	61	15
$P@100_{Rareness}$	0.5	680	457	3778	3050	213	168	173	24
$P@100_{Rareness}$	1.0	728	476	3825	3079	213	168	237	82
AP	0.0	541	334	3731	2976	211	169	320	169
$AP_{Rareness}$	0.5	632	404	3915	3209	216	168	352	209
$AP_{Rareness}$	1.0	701	467	4048	3363	214	170	376	238

compare $AP_{Rareness}$ vs. AP. This suggests that DLT20 systems retrieve many rare, relevant documents at low ranks, causing large changes in system rankings when we use $P@100_{Rareness}$. Smaller changes occur with $AP_{Rareness}$ as the impact of documents is diminished due to their low ranks.

4.3   Discriminative Power

Discriminative power indicates how well a metric can tell systems apart. Zhou et al. [34] measure discriminative power by counting the number of significantly different system pairs. We apply this same method to measure the discriminative power of our proposed metrics, using Tukey's HSD test as the statistical hypothesis test. Table 1 shows the number of significantly different pairs for baseline and our proposed metrics when we use 95% and 99% significance thresholds.

We observe that our metrics have higher discriminative power than baselines. Increasing α tends to increase discriminative power across test collections.

4.4   Stability

If an evaluation methodology is reliable, the measured performance of systems should be stable, i.e., should not change dramatically under different conditions. In order to measure the stability of metrics, we adopt Buckley and Voorhees [3]'s approach. We first sample T topics and calculate system scores on the sampled topic set only. Next, we compare each pair of systems to see which performs better. After repeating this process R times, we assess the stability of the comparison over the R trials. For example, imagine one system outperforms another in 700/1000 trials, yielding a stability of 0.7 for that pair. We take the average stability scores of all pairs as the overall metric stability. In our experiments, we arbitrarily set T to the half of the topic set in each collection (i.e., 22 ($= \lfloor 45/2 \rfloor$) for DLT20 and 25 ($= 50/2$) for the others). We set the number of trials $R = 1000$ but observed that the results largely converged

Table 2. Metric stability scores. Our metrics are most stable across collections.

Metric	α	TREC-5	TREC-8	WT14	DLT20
P@100	0.0	0.532	0.545	0.635	0.071
$P@100_{Rareness}$	0.5	0.642	0.628	0.716	0.128
$P@100_{Rareness}$	1.0	0.709	0.684	0.768	0.179
AP	0.0	0.513	0.580	0.433	0.425
$AP_{Rareness}$	0.5	0.578	0.623	0.517	0.444
$AP_{Rareness}$	1.0	0.633	0.656	0.585	0.466

after 100 trials. Results for baselines vs. proposed metrics are shown in Table 2. $P@100_{Rareness}$ and $AP_{Rareness}$ yield a higher stability score in all cases vs. their classic counterparts.

4.5   Impact of Number of Systems

As retrieval-rarity of documents depends on the participating systems, system scores and rankings might change when we use a different set of systems to calculate the rarity scores of documents. To test how scores of systems change as the systems to be evaluated vary, we conduct the experiment described in Algorithm 1. In particular, we first rank all systems [Line 1]. Then we randomly pick N number of systems [Line 5] and rank them [Line 6]. Subsequently, we get how these N systems are ranked initially (i.e., when all systems are used) [Line 7] and calculate the τ score between these two rankings [Line 8]. We repeat this process 1000 times [Lines 3–9] and calculate the average τ score [Line 10]. Table 3 shows the results for $N = 2^j, j \in [1-6]$ in TREC-8. We observe that correlation scores are generally very high, suggesting that rankings of systems are stable even though we use different sets of systems.

Algorithm 1. Experiment to Analyze Impact of Using N Participants
---
Input: $P \leftarrow$ The whole participant list
$\qquad N \leftarrow$ The number of selected systems
1: $R_o \leftarrow$ rank systems in P
2: $\tau_N \leftarrow 0$
3: $trials \leftarrow 1000$
4: **for all** $trials$ **do**
5: $\quad p_N \leftarrow$ randomly sample N systems from P
6: $\quad r_N \leftarrow$ rank systems in p_N
7: $\quad R_N \leftarrow$ filter systems $\in p_N$ from R_o
8: $\quad \tau_N \leftarrow \tau_N + \tau\_correlation(R_N, r_N)$
9: **end for**
10: $\tau_N \leftarrow \tau_N \ / \ trials$

Table 3. Impact of number of systems based on the experimental setup explained in Algorithm 1. We use TREC-8 for this experiment.

Metrics	N = 2	N = 4	N = 8	N = 16	N = 32	N = 64
$P@100_{Rareness}(\alpha = 1)$	0.970	0.976	0.983	0.987	0.992	0.995
$AP_{Rareness}(\alpha = 1)$	0.970	0.984	0.985	0.989	0.993	0.996

4.6 Qualitative Analysis

To better understand the nature of rarely-retrieved documents, we conducted the following qualitative analysis. We randomly selected six TREC-8 topics, computed the rarity $R(d)$ of each relevant document d, and then selected five documents with varying rarity scores. We manually analyzed how document relevance changes depending on rarity. In general, while commonly retrieved documents appear focused on the search topic, rarely retrieved documents differ in focus but still contain relevant passages.

Table 4 presents manually analyzed documents for topic 431, whose narrative states the information need: "latest developments in robotic technology". The relevant document FBIS4-44815 with minimal rarity is entitled, "Germany: Automation, Robotics Seen as Keys in Industrial", which seems directly relevant to the information need. In contrast, relevant document FBIS3-38782 ($R(d) = 0.81$) only indirectly mentions that a robot can be used for underwater photography, with the title "BND Warns Against Nuclear Terrorists".

If rarely retrieved documents are less relevant, why reward their retrieval? First, while the observation above may hold when all systems are roughly comparable, this is not always true. For example, manual runs have long been advocated in evaluation campaigns because they tend to differ markedly from automated runs and find relevant documents that other systems miss. In general, we cannot tell whether outlier systems are brilliant or remedial without human labels [27]. Second, our goal in this work is to encourage systems that diverge from the pack, with the hope that such divergence will correspond to improvement. The nature of research is that some amount of failure often precedes success, and that making larger departures is important to create a potential for larger improvements. Third, even if we assume a user-centered view, finding additional, less relevant documents can still be important in various cases: when there are few relevant documents, in a "total recall" task setting [24] or pooling [28], or as input to a rank fusion ensemble model [19]. We discuss these further in the next section.

5   Discussion and Limitations

In this section, we discuss various aspects and limitations of our work: motivation and concept of "innovation" metrics (Sect. 5.1), proposed methods (Sect. 5.2), our experimental design and findings (Sect. 5.3), and potential impacts and directions for future work (Sect. 5.4).

Table 4. Analyzed documents for topic 431. The relevant content column corresponds to sentences that might fulfill the information need. If there are multiple useful sentences in a document, the most informative one is selected.

Document ID	Rareness	Relevant Content
FBIS3-38782	0.81	One of the trapeze-like wings had broken off, the nose was missing, and in the dull gray water of Lake Constance even a diving robot of the "Sear Rover" type could send only diffuse video pictures from 159 m below the surface of the lake
LA020889-0003	0.62	A Japanese robot named Wabot II tickles a keyboard to produce original music as part of an exhibit at the Chicago Museum of Science and Industry
LA102589-0109	0.41	The trucks, equipped with robotic arms that hoist and empty containers, will collect trash every week and recyclable items twice monthly
LA092189-0061	0.22	Industrially they use robots for welding, painting or picking and placing items, for example
FBIS4-44815	0.08	New applications for service robots are opening up also in medicine and rehabilitation, in care for the aged and handicapped, in bureaus and logistics, in municipal activities, in households, in hobbies and recreation

5.1   Concept and Motivation

We envision potential benefit from stimulating greater diversity in document ranking methods. In regard to evaluation campaigns, we suggest the field would benefit if participants built upon a wider range of existing methods and/or investigated more radical departures from those methods. While today's evaluation campaigns are already healthy and vibrant, we believe it could be fruitful: 1) to reflect on, assess, and discuss as a community the ways in which we might further strengthen evaluation campaigns; 2) to focus on the diversity of approaches and innovation in particular, and how to promote higher-risk research with potential for greater gains; and 3) to operationalize metrics by which we might measure and optimize for such innovation in evaluation campaigns. Potential counterarguments could be that: a) campaign steering committees are already doing (1) and don't need larger community engagement in it; b) innovation is a complex construct that is best left to organic processes rather than trying to "force" it through explicit optimization; and c) one can argue that research construed as incremental is actually instrumental (i.e., small steps and minor variants can add up over time to large advances). Such discussion and debate seem healthy for a community, regardless of the outcome.

One controversial aspect of our work is the proposal of IR evaluation metrics that explicitly seek to optimize something other than retrieval quality for the user. In particular, the metrics we propose reward systems for retrieving relevant documents missed by other systems, but there is no obvious reason a user would prefer such rare relevant documents over common ones. In fact, less retrieved documents may tend to be less relevant on average and thus aptly lower-ranked (Sect. 4.6). In fact, prior work in meta-ranking (aka rank fusion) has exploited the number of systems that retrieve a given document as a useful feature in estimating document relevance [19]. However, our goal of promoting greater community diversity of ranking methods is not a user-oriented metric, but a field-oriented metric. Moreover, in seeking to promote higher-risk research, we may need to explore a variety of methods yielding sub-par results for the user before we discover a novel method that does provide a transformative advance. For example, years of research on (then) sub-par neural networks was necessary before yielding today's state-of-the-art deep learning methods [20].

Rewarding retrieval of rare relevant documents also has the potential to improve meta-ranking (aka rank fusion or ensemble ranking) and pooling [28]. For instance, ensemble models benefit from a diverse set of input systems that complement each other's shortcomings. Thus, including input systems that find unique relevant documents could boost ensemble performance. Pooling similarly benefits from the diversity of participating systems so that the pool finds as many relevant documents as possible. This helps to ensure that the pool is reusable for future systems using innovative approaches. Our metrics could thus encourage more diverse systems to improve meta-ranking and pooling. In the other direction, recall measures for those tasks might also be repurposed to measure and promote overall diversity and innovation of ranking approaches.

5.2 Proposed Metrics

The two specific innovation metrics we propose have a variety of limitations and represent only the tip of the iceberg of better innovation metrics. We expect future work will propose better metrics that surpass ours.

As noted above, the notion of innovation is a complex construct. Our metrics that reward retrieval of relevant documents missed by other systems are clearly crude metrics for quantifying such a complex construct. To the best of our knowledge, ours is the first metric for measuring and promoting such innovation, but the first effort seldom represents the only or best way. More sophisticated future work by others could model this construct with greater detail and fidelity.

While we have suggested combining $Rel(d_i)$ and $Rel(d) \cdot R(d)$ together into a single mixture for simplicity, an evaluation campaign could also use these as separate and complementary official metrics, akin to evaluating precision vs. recall separately rather than fusing them together into a single f-measure metric. On the other hand, our mixture approach can also be seen as an easy way to generalize existing metrics to consider additional aspects of performance. Because our modified metrics revert to their standard counterpart metrics when $\alpha = 0$, generalization allows use in that original, more restricted setting while also permitting greater flexibility in incorporating additional factors when $\alpha > 0$. While we focus on generalizing existing metrics to include consideration of document rarity, other researchers might incorporate other aspects of system performance into traditional metrics using similar linear mixtures.

At a more mundane level, because our metrics are bounded by $[0, 2)$, it may be useful to renormalize them to a more standard $[0, 1]$ range. While this might be done to values post hoc, hindsight instead suggests two minor revisions to formulas for future use. First, re-define rarity as $R'(d) = 1 - \frac{S_d - 1}{S - 1} \in [0, 1]$ for $S, S_d >= 1$, maximized when $S_d = S$. Second, re-define $P'@K_{rareness}$ as:

$$P'@K_{rareness} = \frac{1}{k} \sum_{i=1}^{k} \left[(1 - \alpha)Rel(d_i) + \alpha\, Rel(d_i)\, R(d_i) \right] \qquad (4)$$

where we now constrain $\alpha \in [0, 1]$ as a probability. This mixture model formulation directly bounds $P'@K_{rareness} \in [0, 1]$.

Our metrics assume linearity in: 1) how we quantify rarity $R(d)$; and 2) the mixture model between the classic metric and rarity. If we consider IR's rich history exploring many variant functions for inverse-document frequency (IDF) to weight rare terms [26], one could imagine similarly exploring many other weighting functions for rarity. Regarding the mixture model, while we have assumed a fixed α across topics, future work might also investigate a hyperparameter approach (akin to Dirichlet smoothing [33]) to intelligently vary α per topic in relation to per topic factors, such as the number of relevant documents.

Yet another idea would be to incorporate document importance alongside rarity in the reward metric for innovation. Intuitively, finding a relevant document that other systems miss is more important when there are few relevant documents in total. As an example, assume for some topic that a given relevant document is only retrieved by a single system. If there are only two relevant documents in total, finding that second relevant document may be vital to satisfying a user's information need. On the other hand, if there were 100 relevant documents, finding the 100th document may provide minimal further value. This would suggest extending the metric to consider the number of relevant documents for each topic.

Finally, our use of P@K and AP assumes binary relevance judgments. Future work could extend innovation metrics to graded relevance judgments.

5.3   Experimental Design and Findings

While we evaluated over four test collections to assess generality, we did not explore the properties of these test collections in detail, or how those varying properties could impact our findings. In addition, expanding our coverage to further test collections could further assess the robustness of findings. Finally, it could be useful to conduct a qualitative inspection of the meta-data descriptions of the best-performing systems (submitted by participants along with their TREC runs) in order to assess the correlation between system descriptions vs. which systems perform best when scored by our innovation metrics.

5.4   Expected Use and Impact

Imagine our metrics were adopted by an evaluation campaign and one or more participating systems sought to optimize them. Beyond the broad goal of promoting higher-risk research and accelerating field innovation, this would be expected to specifically lead to more diverse document rankings. Assuming a fixed evaluation budget (i.e., the number of documents that human judges will review), less overlap across document rankings would mean that we could only pool to a lower depth for the same cost. However, whether this would lead to a more or less complete document pool remains an open, empirical question, likely dependent on the setting of α used. For evaluation campaigns that permit participants to submit multiple runs and distinguish an "official" run (contributing to pooling) vs. additional runs (scored by the official run pool), whether official vs. additional runs would be used to set S_d would also impact subsequent findings.

A well-known issue in IR is the reusability of pools. A very different system might find relevant documents all other systems missed, but if it did not participate in the pool, it would be penalized in evaluation rather than rewarded. Similarly, when we quantify rarity $R(d)$ based on participating systems, there are questions of reusability for future systems evaluating on an existing pool. Moreover, we would expect that a system optimizing for such rarity would be even more likely to run into this problem in practice. Another common distinction made is between methods to create reusable test collections (e.g., pooling) vs. methods to efficiently rank a current set of systems (e.g., StatAP [21] and MTC [4]). Similarly, our rarity metrics will return different scores depending on the other participating systems in the pool. A limitation of our work is that we only rank systems participating in a shared-task, leaving study of reusability for future work.

6   Related Work

To the best of our knowledge, no existing IR evaluation metrics consider the innovativeness of systems. While we frame this *wrt.* rarely-retrieved documents, prior work has usefully designed metrics to evaluate systems reliably with missing judgments, such as Bpref [2] and infAP [30]. These metrics aim to predict the performance of systems with incomplete judgments. In contrast, our focus is to promote innovation in document ranking methods.

To handle missing judgments, a number of studies have explored how to select documents to be judged such as Move-To-Front [8] and MaxMean [16]. These studies aim to maximize the number of relevant documents because unjudged documents are assumed to be non-relevant. As a document is more likely to be relevant if retrieved by many systems, commonly-retrieved documents are more likely to be judged than rarely retrieved ones. However, in contrast to these document selection methods, we assign more weight to rarely-retrieved ones.

In modifying P@K and AP, we have followed standard practice in aggregating scores over topics using a simple arithmetic average. However, various other aggregate statistics have been proposed. Robertson [23] asserts that the impact of hard topic scores is diminished on the overall score with the arithmetic mean. He thus recommends geometric mean instead. Ravana and Moffat [22] show that geometric mean average precision (GMAP) is better at handling variability in topic difficulty than arithmetic mean average precision (MAP). Mizzaro [17] proposes normalized mean average precision (NMAP), which takes into account topic difficulty. He defines topic difficulty as 1-(average AP score). Unlike these studies, we focus on retrieval difficulty at the document level. In addition, prior studies on topic difficulty work on how to aggregate traditional IR metrics.

As noted earlier, while we assumed binary relevance judgments and modify only P@K and AP metrics, many other metrics exist, beyond binary relevance, that could be extended to innovation. Prominent examples include normalized discounted cumulative gain (nDCG) [14], and rank biased precision (RBP) [18], which assume that users will examine documents in the retrieval order and might stop examining whenever their information need is satisfied. Such rank-based metrics ascribe more weight to documents at higher ranks. Other important evaluation metrics include miss (i.e., the fraction of non-retrieved documents that are relevant) [13], fallout [15], expected reciprocal rank [5], weighted reciprocal rank [10], and O-measure [25].

Prior work has also proposed metrics rewarding the diversity within a single document ranking in relation to novelty and coverage of different topic facets. For instance, Zhai et al. [32] propose three metrics – subtopic recall metric (S-recall), subtopic precision (S-precision) and weighted subtopic precision (WS-precision) – that consider redundancy in ranked lists. Clarke et al. [6] extend nDCG by rewarding novelty and covering multiple topic aspects. In contrast, we quantify diversity across systems rather than within a single ranked list. In particular, we reward systems for retrieving relevant documents that other systems miss.

7   Conclusion

We propose a new class of IR evaluation metrics designed to promote exploration of higher-risk, more radical departures from current state-of-the-art methods. These "innovation metrics" reward retrieval of relevant documents missed by other systems. The key intuition is that finding relevant documents missed by other systems suggests a markedly different approach. More specifically, we generalize classic Precision@K and Average Precision metrics via a simple mixture-weight parameter controlling the relative reward for finding relevant documents other systems miss. Setting this to zero reverts to the original metric.

Experiments over four TREC collections show that our proposed metrics yield different system rankings compared to the existing metrics. In particular, we observe a steady decrease in rank correlation with official system rankings as reward increases for finding rare, relevant documents. These results are further supported by a controlled, synthetic data experiment, as well as qualitative analysis. Collectively, results suggest that if our metrics were adopted in practice, participants would be incentivized to retrieve more diverse relevant documents, with the potential to spur further innovation in the field. Finally, we also show that our metrics provide higher discriminative power and evaluation stability than the standard Precision@K and Average Precision metrics that we modify.

To the best of our knowledge, ours is the first proposal of IR evaluation metrics designed to explicitly measure and promote innovation in ranking methods. That said, the first attempt at any endeavor is seldom the only or best way to accomplish it. Our two proposed metrics have a variety of limitations and represent only the tip of the iceberg for imagining this new class of innovation metrics. Consequently, we expect future metrics will be proposed that surpass ours in better modeling the complex construct of innovation, and in doing so, will further advance the cause of promoting innovation in ranking methods.

Acknowledgments. We thank the reviewers for their valuable feedback. This research was supported in part by the Scientific and Technological Research Council of Turkey (TUBITAK) ARDEB 3501 (Grant No 120E514) and by Good Systems (https://goodsystems.utexas.edu), a UT Austin Grand Challenge to develop responsible AI technologies. Our opinions are our own.

References

1. Brown, T., et al.: Language models are few-shot learners. In: Advances in Neural Information Processing Systems, vol. 33, pp. 1877–1901 (2020)
2. Buckley, C., Voorhees, E.M.: Retrieval evaluation with incomplete information. In: Proceedings of the 27th Annual International ACM SIGIR Conference on Research and Development in Information Retrieval, pp. 25–32 (2004)
3. Buckley, C., Voorhees, E.M.: Evaluating evaluation measure stability. In: ACM SIGIR Forum, vol. 51, pp. 235–242. ACM New York (2017)
4. Carterette, B., Allan, J., Sitaraman, R.: Minimal test collections for retrieval evaluation. In: Proceedings of the 29th Annual International ACM SIGIR Conference on Research and Development in Information Retrieval, pp. 268–275 (2006)
5. Chapelle, O., Metlzer, D., Zhang, Y., Grinspan, P.: Expected reciprocal rank for graded relevance. In: Proceedings of the 18th ACM Conference on Information and Knowledge Management, pp. 621–630 (2009)
6. Clarke, C.L., et al.: Novelty and diversity in information retrieval evaluation. In: Proceedings of the 31st Annual International ACM SIGIR Conference on Research and Development in Information Retrieval, pp. 659–666 (2008)
7. Collins-Thompson, K., Macdonald, C., Bennett, P., Diaz, F., Voorhees, E.M.: Trec 2014 web track overview. Technical report, MICHIGAN UNIV ANN ARBOR (2015)
8. Cormack, G.V., Palmer, C.R., Clarke, C.L.: Efficient construction of large test collections. In: Proceedings of the 21st Annual International ACM SIGIR Conference on Research and Development in Information Retrieval, pp. 282–289 (1998)
9. Craswell, N., Mitra, B., Yilmaz, E., Campos, D.: Overview of the TREC 2020 deep learning track. arXiv preprint arXiv:2102.07662 (2021)
10. Eguchi, K., Oyama, K., Ishida, E., Kando, N., Kuriyama, K.: Overview of the web retrieval task at the third NTCIR workshop. In: NTCIR. Citeseer (2002)
11. Harman, D., Voorhees, E.: Overview of the fifth text retrieval conference (TREC-5). In: Harman, D., Voorhees, E. (eds.) Information Technology: The Fifth Text REtrieval Conference (TREC-5), National Institute of Standards and Technology Special Publication, pp. 500–238 (1996)

12. Hawking, D., Voorhees, E., Craswell, N., Bailey, P., et al.: Overview of the TREC-8 web track. In: TREC (1999)
13. Heine, M.: Information-retrieval from classical databases from a signal-detection standpoint-a review. Inf. Technol. Res. Dev. Appl. **3**(2), 95–112 (1984)
14. Järvelin, K., Kekäläinen, J.: IR evaluation methods for retrieving highly relevant documents. In: ACM SIGIR Forum, vol. 51, pp. 243–250. ACM, New York (2017)
15. Kraft, D.H., Bookstein, A.: Evaluation of information retrieval systems: a decision theory approach. J. Am. Soc. Inf. Sci. **29**(1), 31–40 (1978)
16. Losada, D.E., Parapar, J., Barreiro, A.: Multi-armed bandits for adjudicating documents in pooling-based evaluation of information retrieval systems. Inf. Process. Manag. **53**(5), 1005–1025 (2017)
17. Mizzaro, S.: The good, the bad, the difficult, and the easy: something wrong with information retrieval evaluation? In: Macdonald, C., Ounis, I., Plachouras, V., Ruthven, I., White, R.W. (eds.) ECIR 2008. LNCS, vol. 4956, pp. 642–646. Springer, Heidelberg (2008). https://doi.org/10.1007/978-3-540-78646-7_71
18. Moffat, A., Zobel, J.: Rank-biased precision for measurement of retrieval effectiveness. ACM Trans. Inf. Syst. (TOIS) **27**(1), 1–27 (2008)
19. Nuray, R., Can, F.: Automatic ranking of information retrieval systems using data fusion. Inf. Process. Manag. **42**(3), 595–614 (2006)
20. Onal, K.D., et al.: Neural information retrieval: at the end of the early years. Inf. Retrieval J. **21**(2), 111–182 (2018)
21. Pavlu, V., Aslam, J.: A practical sampling strategy for efficient retrieval evaluation. Northeastern University, College of Computer and Information Science (2007)
22. Ravana, S.D., Moffat, A.: Exploring evaluation metrics: Gmap versus map. In: Proceedings of the 31st Annual International ACM SIGIR Conference on Research and Development in Information Retrieval, pp. 687–688 (2008)
23. Robertson, S.: On GMAP: and other transformations. In: Proceedings of the 15th ACM International Conference on Information and Knowledge Management, pp. 78–83 (2006)
24. Roegiest, A., Cormack, G.V., Clarke, C.L., Grossman, M.R.: TREC 2015 total recall track overview. In: TREC (2015)
25. Sakai, T.: On the task of finding one highly relevant document with high precision. Inf. Media Technol. **1**(2), 1025–1039 (2006)
26. Salton, G., Buckley, C.: Term-weighting approaches in automatic text retrieval. Inf. Process. Manag. **24**(5), 513–523 (1988)
27. Soboroff, I., Nicholas, C., Cahan, P.: Ranking retrieval systems without relevance judgments. In: Proceedings of the 24th Annual International ACM SIGIR Conference on Research and Development in Information Retrieval, pp. 66–73 (2001)
28. Spark-Jones, K.: Report on the need for and provision of an 'ideal' information retrieval test collection. Computer Laboratory (1975)
29. Voorhees, E.M.: Variations in relevance judgments and the measurement of retrieval effectiveness. Inf. Process. Manag. **36**(5), 697–716 (2000)
30. Yilmaz, E., Aslam, J.A.: Estimating average precision with incomplete and imperfect judgments. In: Proceedings of the 15th ACM International Conference on Information and Knowledge Management, pp. 102–111 (2006)
31. Yilmaz, E., Aslam, J.A., Robertson, S.: A new rank correlation coefficient for information retrieval. In: Proceedings of the 31st Annual International ACM SIGIR Conference on Research and Development in Information Retrieval, pp. 587–594 (2008)

32. Zhai, C., Cohen, W.W., Lafferty, J.: Beyond independent relevance: methods and evaluation metrics for subtopic retrieval. In: ACM SIGIR Forum, vol. 49, pp. 2–9. ACM New York (2015)
33. Zhai, C., Lafferty, J.: A study of smoothing methods for language models applied to ad hoc information retrieval. In: ACM SIGIR Forum, vol. 51, pp. 268–276. ACM, New York (2017)
34. Zhou, K., Lalmas, M., Sakai, T., Cummins, R., Jose, J.M.: On the reliability and intuitiveness of aggregated search metrics. In: Proceedings of the 22nd ACM International Conference on Information & Knowledge Management, pp. 689–698 (2013)

Probing BERT for Ranking Abilities

Jonas Wallat[1]($\boxtimes$) (iD), Fabian Beringer[1], Abhijit Anand[1] (iD),
and Avishek Anand[1,2] (iD)

[1] L3S Research Center, Hannover, Germany
{jonas.wallat,fabian.beringer,abhijit.anand}@l3s.de
[2] TU Delft, Delft, Netherlands
avishek.anand@tudelft.nl

Abstract. Contextual models like BERT are highly effective in numerous text-ranking tasks. However, it is still unclear as to whether contextual models understand well-established notions of relevance that are central to IR. In this paper, we use *probing*, a recent approach used to analyze language models, to investigate the ranking abilities of BERT-based rankers. Most of the probing literature has focussed on linguistic and knowledge-aware capabilities of models or axiomatic analysis of ranking models. In this paper, we fill an important gap in the information retrieval literature by conducting a layer-wise probing analysis using four probes based on lexical matching, semantic similarity as well as linguistic properties like coreference resolution and named entity recognition. Our experiments show an interesting trend that BERT-rankers better encode ranking abilities at intermediate layers. Based on our observations, we train a ranking model by augmenting the ranking data with the probe data to show initial yet consistent performance improvements (The code is available at https://github.com/yolomeus/probing-search/).

1 Introduction

Large contextual models such as BERT [14] have delivered impressive and robust performance gains in many NLP and IR tasks. However, these over-parameterized contextual models are still used as functional black boxes with little understanding of what the contextual embedding spaces actually encode. Towards this, *probing* was introduced as a procedure to investigate whether specific linguistic properties or factual information are present in contextual text representations [6], which enable large contextual models to perform well on language tasks. Probes offer insight into otherwise functionally opaque contextual models. Most of the effort in designing probes is to ground the behavior of large contextual models in well-understood linguistic properties and world knowledge. For example, a *part-of-speech* (POS) probe investigates to what degree contextual representations encode POS information in their representations. This innate ability to encode POS is typically investigated by learning a lightweight classifier, called a *probe*, to predict the POS property from the embeddings. The performance of a probe measures the quality of the *contextual representations.*

© The Author(s), under exclusive license to Springer Nature Switzerland AG 2023
J. Kamps et al. (Eds.): ECIR 2023, LNCS 13981, pp. 255–273, 2023.
https://doi.org/10.1007/978-3-031-28238-6_17

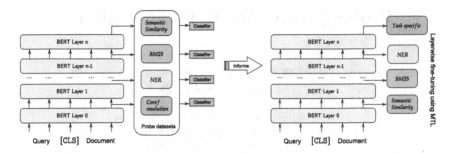

Fig. 1. Procedural overview: in the first set of experiments, we probe for different abilities of neural ranking models (e.g., BM25, semantic similarity). We then utilize the information where the model best captures these properties to give additional training signals to that specific layer during multi-task learning.

Consequently, various task-specific probing tasks have been developed to investigate contextual embeddings for linguistic and factual knowledge [6,36,48,55].

This paper focuses on large contextual models that have been applied with major success in information retrieval tasks. However, there is limited work on probing for IR and, particularly, to *text ranking tasks*. Until now, most studies focused on probing for linguistic [23,48] or factual knowledge [35,36] of pre-trained models, e.g., finding that BERT's layers and their abilities coincide with the classical NLP pipeline [47] or that dependency parse trees can be decoded from BERT's embeddings [23]. There has also been work on investigating the evolution of higher-level factual and linguistic knowledge through the layers of large contextual models [47,52]. Most of the existing work in explaining the behavior of contextual ranking models is through IR axioms [7,41,51]. Although axioms are well-established, formal descriptions of *what makes a good text ranker*, they have limited modeling of semantic similarity and have been shown to have limited applicability to explain neural rankers [7,51].

1.1 Research Questions

We aim to fill the gap of characterizing the performance of neural rankers in terms of IR abilities by proposing probing methods. Through probing, we try to understand the behavior of ranking models by grounding it on well-understood IR properties and best practices for text ranking – *matching, semantic similarity,* in conjugation with essential linguistic properties of *named entity recognition,* and *coreference resolution.* We answer the following research questions:

RQ 1. What abilities do neural rankers acquire to perform the ranking task?

RQ 2. Can we apply the knowledge to build better ranking models?

1.2 Summary of Contributions

First, we construct probing datasets for the probing tasks of lexical matching, semantical similarity, named entity recognition (NER), and coreference resolution from the MS MARCO dataset [49]. Next, we measure – (a) the degree to which a ranking model understands the IR property (*accuracy*) and, (b) the degree to which the property is extractable from the ranking model (*minimum description length*). Figure 1 depicts and overview of our experiments. We conduct extensive experiments using multiple probes over multiple layers of a BERT-fine-tuned ranking model. Other than existing works that predominantly report only probing results, we operationalize our findings by constructing multitask learning-based ranking models using auxiliary tasks based on probes.

Results. Our probing study shows that ranking models prioritize lexical and semantical similarity and coreference information over NER abilities. Moreover, we usually find intermediary layers (4–6) to best capture these concepts. We also find that training ranking models in a multi-task learning setup (i.e., ranking and the aforementioned ranking subtasks) can be beneficial - especially when we use our probing results to inform on which layer to train the subtasks.

2   Related Work

Probing large and overparameterized contextual models was introduced by Conneau et al. [11] in the NLP community to improve their interpretability. This work aims to probe neural rankers to understand their IR abilities as a step towards explainable IR [4]. Specifically, probing is a posthoc interpretability approach that, instead of optimizing fidelity [40,45,46], tries to ground the knowledge or abilities stored in the parametric memory of neural rankers.

2.1   Probing for Linguistic Properties

Tenney et al. [48] proposed aggregating individual word embeddings to move from word-level probing to subsentences, allowing to probe for coreference and other semantic, long-range concepts. Consequently, many works used this methodology. Zhao et al. [56] investigated how contextualized BERT embeddings are. Tenney et al. [47] probed BERT and found early layers to focus on lower-level concepts, such as *syntax*, and more-involved higher layers on concepts such as *semantics*. Subsequent work improving the probing paradigm either by contextualizing the probing results with suitable baselines [21,54], introducing control tasks [22], or characterizing embedding vs classifier performance [37,50]. For detailed overview of the probing literature until 2019, we refer to the review by Belinkov and Glass [6]. We include many of the best practices in the literature in our work. Many works have investigated task-specific probing [2,52]. Most related to our work is Wallat et al. [52], who also perform a layer-wise probing to check the retention of factual knowledge in BERT. Their layer-wise analysis suggests that most factual knowledge resides in the later layers of the models, with the ranking model outperforming other fine-tuned models in knowledge retention. We instead probe for ranking abilities.

2.2   Probing in IR

In the context of IR, MacAvaney et al. [31] study the ranking models using a large set of diagnostic probes such as *term-frequency*. They also study the effects of shuffling word orders or paraphrasing on the ranking performance. Fan et al. [16] show that the ranking model improved in capturing *synonym detection* information while sacrificing the ability to identify named entities. While both of these works investigated the abilities of IR models, they focus on only the final representation of that is derived from the last layer of the model. We believe that investigating the flow of information through the intermediate layers can yield additional insights. Furthermore, both [16,31] do not contextualize the probing results with standard probing baselines like control tasks, or a measure of *ease of extraction* (e.g., MDL) as recommended in the probing literature [5].

2.3   Axiomatic Interpretability

Similar to probing, neural rankers have also been diagnosed or interpreted using IR axioms [7,41,51]. These works either directly rank documents according to specific axioms such as "if document A contains more query terms than document B, then A should be ranked higher" [20], check whether rankers conform with axioms using diagnostic datasets [41], or try to explain neural rankers with these axioms [7,51]. However, most of these approaches have reported limited success. Völske et al. [51] find that axiomatic explanations frequently fail if models are not confident in their decision and that the existing axioms are insufficient in explaining the complex decisions of ranking models [7]. By investigating the acquired abilities of ranking models, we position our work between the existing high-level investigation into factual knowledge containment [52] and explaining model decisions by shallow features (i.e., axioms) [51].

2.4   Understanding Relevance Factors Without Probing

Apart from probing, the attention patterns of ranking models have been under investigation, finding that redundant attention often focuses on tokens with a high document frequency (e.g., punctuation) [53] and that the attention captures inverse-document frequency information [9]. Furthermore, Qiao et al. [38] investigate the attention and term-matching behavior of BERT and find that it focuses more on query tokens that appear in the document, suggesting attention and lexical matching being deciding factors for BERT's performance gains. Rau and Kamps [39] study the role of NLP abilities in the effectiveness of neural ranking models. By constructing inputs without word order information, they find that while word order seems highly relevant for BERT's pretraining, it is not necessary for relevance estimation.

2.5   Data Augmentation in IR

In the second part of our paper, we use additional training signals from our probe tasks to train ranking models. While there is existing work that utilized information such as BM25 to train rankers either with weak supervision [13] or by data augmentation [3,44], our work is, to the best of our knowledge, the first to operationalize probing results to build more effective ranking models. We specifically probe the representations of the common early interaction BERT ranker as proposed by Nogueira et al. [33], which applies a linear layer to the [CLS] token in order to estimate relevance. Besides early interaction methods, there have been recent works on late interaction models [18,26,29], where independent document and query representations only interact in the last layer.

3 Probing BERT Ranking Models

The ability of a text ranker to effectively rank documents given an underspecified query is based on many well-understood principles in IR like term matching, document frequency, and length normalization, among others [32]. In this work, we are interested in BERT rankers, but our analysis can naturally be extended to other overparameterized contextual rankers with multiple transformer layers.

3.1 Problem Statement

Given a trained (or fine-tuned) text ranking model $\mathcal{M}$, we are interested in measuring the degree to which output representations of $\mathcal{M}$ satisfy or adhere to well-understood ranker properties. For each ranking property i, a *probing dataset* P_i is constructed. To measure if a property i is well-captured in $\mathcal{M}$, our objective is to train a probing classifier or simple a *probe* g_i given the output representation/s or embedding from $\mathcal{M}$ to generalize on the probing dataset P_i.

3.2 Layerwise Probing

We conduct probing analysis on multiple layers of $\mathcal{M}$ to assess the evolution of ranking properties across layers of the ranking model. For each of our ranking subtasks (Sect. 3.4) and each layer of the model, we train a simple MLP classifier over the model's output representations or embeddings.

We follow the probing paradigm that is based on the general assumption that an *above-chance performance* on the probing tasks indicates the presence of task knowledge in the embeddings. These probing performances need to be put into context by how hard the task is (e.g., by comparing performance with suitable baselines [54]) and how much of the performance can actually be attributed to the classifier [37,50]. Towards addressing these concerns, we first carefully select random and pre-trained baselines to compare against (refer Sect. 4.2) and secondly use the *minimum description length* (MDL) to measure attributability. Next, we detail our probing setup with MDL.

3.3 Probing with Minimum Description Length

By applying the information-theoretic concept of minimum description length to the probing paradigm, Voita and Titov [50] address the question: *how well the model encodes certain information?* If the embedding encodes a concept such as named entities more efficiently, it can describe this information more precisely. In that case, the minimum description length will be shorter than in embeddings that do not capture named entity information.

To compute MDL, we use the online code definition [42]. For this, the dataset $D = \{(x_i, y_i)\}_{i=1}^{N}$ is divided into timesteps $1 = t_0 < t_1 < \ldots < t_S = N$. After encoding block t_0 with a uniform code, for each following timestep, a probing model p_{θ_i} is trained on the samples $(1, \ldots, t_i)$ and used to predict over data points $(t_i + 1, \ldots, t_{i+1})$. The full MDL is then computed as a sum over the codelengths of each p_{θ_i} and the uniform encoding of the first block:

$$L(y_{1:n}|x_{1:n}) = t_1 \log_2 C - \sum_{i=1}^{S-1} \log_2 p_{\theta_i}(y_{t_i+1:t_{i+1}}|x_{t_i+1:t_{i+1}}) \tag{1}$$

where C is the number of target classes. Following Voita and Titov [50], we choose timesteps at 0.1, 0.2, 0.4, 0.8, 1.6, 3.2, 6.25, 12.5, 25, 50 and 100% of the dataset.

Similarly to Fayyaz et al. [17], we reformulate MDL to *compression*. For this, MDL is scaled in relation to the codelength of a uniform encoding:

$$\texttt{compression} = \frac{N \log_2(C)}{\text{MDL}} \tag{2}$$

where N is the number of targets, and C is the number of target classes. Since MDL depends on the total number of targets, a relative measure, like compression, is more practical for comparing tasks. Furthermore, both accuracy and compression are to be maximized, while MDL is to be minimized.

MDL is only defined for classification tasks as it requires the number of target classes. Therefore, we reformulate regression tasks to classification tasks by binning target scores into $k = 10$ equally sized class bins.

3.4   Probing Tasks

For a selection of principled ranking abilities, we utilize well-known abilities of ranking models from the information retrieval (IR) literature: Arguably, one of the most fundamental ranking subtasks is a model's ability to match text, which has been widely used either for classical ranking models [43] or to inform the pre-finetuning of neural rankers [27]. Furthermore, we probe for the ranking model's ability to match according to the semantic meaning [30]. Given that a large part of queries focus on entities and that named entity recognition (NER) can have a positive impact on IR [25], we include NER as one of our tasks. Lastly, we include coreference resolution, which is not canonically associated with principled ranking. With the established importance of entity recognition, we wonder how well ranking models can perform the matching of entity surface forms between queries and documents.

For our experiments, we compile a list of abilities that neural ranking models might employ for predicting document relevance. We choose our tasks as follows:

BM25 Prediction. The BM25 algorithm [43] uses lexical matching to estimate relevance and is widely used in ranking. We ask whether neural rankers encode the necessary information to perform well at measuring lexical similarity. The BM25 formula includes inverted document frequencies of the terms; therefore, to accurately predict BM25, the ranker needs to implicitly learn term distributions in the dataset. We use query document pairs from the MS MARCO test set and predicted BM25 scores as labels to create the probing dataset.

Semantic Similarity. Like lexical matching, it seems very probable that part of the ranking model's performance can be attributed to semantic matching. We test whether semantic similarity information resides in the embeddings of our rankers. Similar to existing work in axiomatic IR [51], we estimate the semantic similarity between query and document pairs by the cosine similarity between the average GloVe [34] query and document embeddings (after stop-word removal).

Named Entity Recognition. Since user queries usually ask for some information about entities, we test the models' ability to identify entities. To do so, we use the Spacy [24] named entity recognizer and tag all named entities in MS MARCO query-document pairs.

Coreference Resolution. Queries are often underspecified [10]. We, therefore, include the probing task of coreference resolution between entity mentions in the query and surface form occurrences in the document into our suite of tasks. Given a query "trump birthplace", the task is to match an entity from the query ("trump") to surface forms in a document (e.g., "Donald Trump", "the former president"). To find coreference pairs, we use Huggingface's neuralcoref[1].

4   Experimental Setup

4.1   Datasets

MS MARCO: We use the TREC Deep Learning track (2019) dataset (TREC-DL) for evaluation. Our models are evaluated on the TREC-DL test split which contains 200 queries. For creating training and development splits we use MS MARCO, containing 532k queries. To retrieve documents from the corpus of $\sim$ 8.8mio passages, we use BM25.

Probing: Since our (contextual) ranking models are trained on MS MARCO, we explicitly use the MS MARCO test set to create our probing datasets. For this, we uniformly sample 60k query-passage pairs, where 40k are used for training, and 10k for validation and testing, respectively.

4.2   Models

We conduct our probing experiments on BERT [14], using three different base models throughout our experiments:

---

[1] https://github.com/huggingface/neuralcoref.

1. **BERT-BASE-UNCASED** - the publicly available[2] pre-trained BERT model consisting of 12-layer, 768 dimensions, 12-heads, 110M parameters. The length of the input is restricted to 512 tokens.
2. **BERT-MSM-PASSAGE** - bert-base model, fine-tuned on MS MARCO for the TREC-DL 2019 *passage* level ranking task [12]
3. **BERT-MSM-DOC** - the bert-base model, fine-tuned on MS MARCO for the TREC-DL 2019 *document* level ranking task.

The ranking models were trained with a similar setup as Nogueira et al. [33] for up to 20 epochs on using the binary cross-entropy objective.

4.3   Training Probe Models

For all tasks, we train a 2-layer MLP probe model with self-attention pooling (similar to [48]) for up to a maximum of 50 epochs and perform early stopping after 10 epochs if no improvement in validation loss has been measured. As an optimization algorithm, we use Adam [28] with a batch size of 32 and clip gradients with an L2-norm greater than 5. We start with a learning rate of 1e–4 and half it at the end of an epoch if the validation loss does not improve.

5   Results

To establish which ranking ability is learned by fine-tuning on ranking datasets (**RQ 1**), we compare the performance of a fine-tuned passage ranking (BERT-MSM-PASSAGE) and a document ranking model (BERT-MSM-DOC) to two baselines: 1) a pretrained model without fine-tuning, and 2) model with random weight initialization. For a pre-trained model, we use a BERT model (BERT-BASE-UNCASED). Furthermore, we use BERT input embeddings with random weight initialization as a source of random embeddings [54].

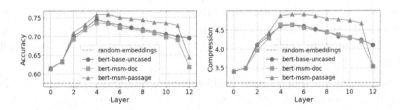

Fig. 2. Probing results over the layers for the BM25 task.

---

[2] https://huggingface.co/bert-base-uncased.

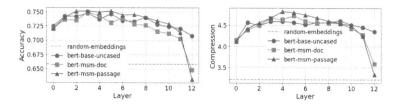

Fig. 3. Probing results over the layers for the semantic similarity task.

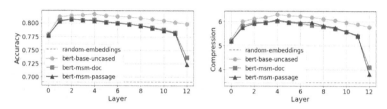

Fig. 4. Probing results over the layers for the NER task.

5.1 Matching Ability of Ranking Models

Figure 2 presents the degree to which fine-tuned BERT models have learned to predict BM25 or, in other words, exhibit the ability to perform term matching. The plot on the left shows *task accuracy* and the plot on the right shows *compression* over the layers (metric introduced in Sect. 3.2). First of all, and expectedly, we can see that all three models capture more BM25 information than random embeddings to a large degree. While the accuracy seems to differ only slightly, we can observe that the compression of BERT-MSM-PASSAGE is markedly higher than for the other two models. A higher compression score means the BM25 information is more easily decodable from the ranking models' embeddings. By probing all layers of our models, we can also understand in which layer the matching ability is best captured. It is evident that the BM25 knowledge increases until layer 5 or 6 and then slowly decreases until layer 11. In layer 12, the performance decreases starkly - a result that is in line with multiple works finding that the last layer is the most task-specific and therefore performs worse in other tasks ([2,52] inter alia). Additionally, recent work by Ghasemi et al. [19] suggests that BERT rankers do not fully rely on lexical matching, which is also indicated by BM25 knowledge decreasing in the later layers.

5.2 Ability to Capture Semantic Similarity

The probing results for semantic similarity are shown in Fig. 3. Again, we can observe similar trends. Semantic similarity appears to be best captured in layer 4 (compared to layers 5 or 6 for BM25). Like with BM25, we can see the ranking models' compressions to be slightly improved over the pre-trained model – suggesting that training the models on ranking emphasizes understanding and capturing semantic similarity.

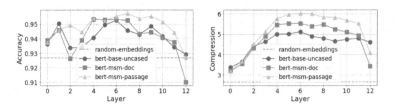

Fig. 5. Probing results over the layers for the coreference resolution task.

5.3 Other Abilities

Figures 4 and 5 show the probing performance on NER and coreference resolution, respectively. Interestingly, we find that the ranking models do not spot entities better than a pre-trained model (Fig. 4). Although the results suggest that the identification of entities is not a priority, matching surface forms of entities is better encoded after fine-tuning on the ranking task.

5.4 Insights and Summary

Our first insight is that, compared to BERT-MSM-DOC, BERT-MSM-PASSAGE shows a better accuracy-compression trade-off in all the auxiliary tasks considered. In other words, not only does BERT-MSM-PASSAGE exhibit primitive ranking abilities, but these abilities are easily extractable for text ranking tasks. Second, all considered auxiliary tasks are best encoded at intermediary layers and slowly decrease towards the final layer. This shows that deep contextual models used as rankers extract features that are in some sense compositional in nature, with lower-level abilities being exhibited in the lower layers. We believe that the abilities we deal with are intermediate abilities. Existing layerwise studies have shown that ranking models exhibit higher-level abilities in the last few layers [52]. Finally, we observe that BM25, semantic similarity, and coreference resolution are better encoded in ranking models. NER, on the other hand, seems to be deprioritized by the re-ranking models in our study, confirming earlier results [16].

6 Can the Probing Results Be Used for Building Better Rankers?

Until now, we have established that fine-tuned ranking models exhibit basic linguistic and information retrieval abilities. To answer **RQ 2**, we operationalize our findings. Towards this, along with the ranking training set, we construct three task datasets (BM25, NER, semantic similarity). As in this setting, we aim for learning ranking on MS MARCO, we only use queries from the train set to prevent test overlap. For each task, we sample $100k$ queries and, using BM25, retrieve 10 documents each. This results in 1 million samples per task which is approximately the size of our pointwise MS MARCO training set.

We employ a multi-task learning (MTL) setup where we train the ranking task together with individual ranking subtasks (Sect. 3.4). To support the model's learning process, we directly funnel the subtask signal into the model at the corresponding layer where it was best captured (as identified during the probing experiments) and supply the ranking signal in the last layer.

6.1 MTL Training

Multi-task learning is an approach of training multiple tasks in parallel with shared representations to share knowledge across tasks [8]. This has been shown to improve generalization. To train on multiple tasks simultaneously, we uniformly draw samples from the pool of both datasets until the batch size is reached. We then pass the resulting mixed batch through the language model and retrieve the intermediate output representations at each layer. For simplicity, average pooling over the sequence dimension is performed at the desired layers, and a task-specific 2-layer MLP is applied, which takes the following form:

$$\text{MLP}(x) = W_1 \sigma(W_0 x + b_0) + b_1 \tag{3}$$

with $W_0 \in \mathbb{R}^{m \times n}$, $W_1 \in \mathbb{R}^{n \times k}$ and $b_0 \in \mathbb{R}^n$, $b_1 \in \mathbb{R}^k$ as learnable parameters and σ as the RELU activation. Analogously to our probing experiments, we cast regression to classification tasks by binning the targets into $k = 10$ categories. For our loss function, we use the simple scaling scheme proposed in [1]

$$\mathcal{L}(y_i, \hat{y}_i) = \frac{\text{CE}(y_i, \hat{y}_i)}{\log k_i} \tag{4}$$

where y_i and $\hat{y}_i$ are target and prediction for datapoint i respectively, CE is the cross-entropy loss and k_i denotes the number of target classes for point i, e.g. for a binary target $k_i = 2$. For experiments with the pairwise objective, we similarly use margin loss with $\lambda = 0.2$.

Table 1. Effect of different loss objectives on ranking with BM25 as auxiliary task on the TREC-DL 2019 dataset. pt and pr refer to the pointwise and pairwise training objectives. * marks a significant improvement (p-value < 0.1).

Model	Layer	MAP	MRR	nDCG@10	nDCG@20	P@10	P@20
Ranking (pt-baseline)	12	0.436	0.926	0.678	**0.653**	0.784	**0.685**
Ranking + BM25 (pt)	5	**0.437**	**0.947**	**0.682**	0.652	**0.791**	0.680
Ranking (pr-baseline)	12	0.433	**0.965**	0.681	0.652	0.772	0.670
Ranking + BM25 (pr)	5	**0.452***	**0.965**	**0.685**	**0.673***	**0.786**	**0.708***

Table 2. Effect of layers on MTL performance on the TREC-DL 2019 and 2020 dataset. While we train the ranking task (pointwise loss) always on the final layer, we experiment with different layers for the auxiliary tasks (BM25, named entity recognition, semantic similarity). Bold values indicate the best performance out of all configurations of that specific model (e.g., for all Ranking+BM25 models). */** mark a significant improvement (p-value $< 0.1/0.05$ respectively).

Model	Layer	TREC 19			TREC 20		
		MAP	MRR	nDCG@10	MAP	MRR	nDCG@10
Ranking (baseline)	12	0.436	0.926	0.678	0.446	0.875	0.674
Ranking + BM25	5	0.437	0.947	0.682	0.454	0.900	0.680
	6	**0.439**	**0.953***	**0.690**	**0.460****	**0.932***	**0.689**
	12	0.420	0.912	0.659	0.450	0.927	0.668
Ranking + NER	4	**0.447***	**0.950**	0.685	**0.466****	**0.922****	**0.705****
	5	0.444	0.934	0.680	0.451	0.859	0.679
	12	**0.447**	0.944	**0.688**	0.464**	0.912	0.691
Ranking + Sem	1	0.436	**0.934***	**0.682**	0.451	0.910	**0.687**
	4	**0.440**	0.928	**0.682**	0.453	0.897	0.677
	12	0.436	0.928	0.669	**0.458***	**0.913**	0.683

6.2   MTL Results

First, we train both ranking-only and MTL (Ranking + BM25) models in pairwise and pointwise fashion. Table 1 presents these results.

The experiment suggests that the multi-task training setup with training BM25 on layer 5, as well as ranking on layer 12, improves the overall task performance. While there is an improvement in the pointwise training, we observe larger improvements in the pairwise setting.

Insight. Combining the ranking task with auxiliary tasks can improve the overall ranking performance.

6.3   Effect of MTL Layers on Performance

Next, we investigate if selecting the layer with the best probing performance does hold a benefit over choosing the last layer in our MTL setup.

Table 2 presents the results of multi-task training setups with the ranking task on layer 12 and auxiliary tasks on varying layers. Given significantly higher training times in the pairwise setting, we trained these models with a pointwise objective. It is evident that for the BM25 task, there is a benefit to selecting the MTL layer according to the probing results. Using the 12th layer for training both ranking and BM25 leads to a degradation in ranking performance (compared to the baseline model). Adding semantic-similarity based data augmentation, however, yields no clear trend on the TREC-DL 2019 and 2020 datasets. We hypothesize that BERT embeddings and the self-attention mechanism are sufficient in estimating query document similarity for the re-ranking task. Also, the construction of gold labels by using GloVe embeddings might not capture semantic similarity as it is used by BERT. For NER, we see all chosen layers to be beneficial. This might be the result of ranking models dropping NER to some capacity (see Fig. 4) and directly forcing the model to include NER information being helpful for the ranking task and specifically the entity-driven MS MARCO dataset. The probing study results suggest that NER information is not prioritized while acquiring the ability to rank passages.

Insight. Choosing the MTL layer according to the probing results can outperform choosing the last layer.

6.4   Threats to Validity

The general shortcoming of probing studies is that a high probing accuracy is not a causal reason for applicability during inference [5,15,50]. Secondly, the decrease in probing task performance over the later layers suggests that the model prioritizes other, potentially compositional, information over our considered IR abilities. At this point, we do not fully understand what information is *used* for relevance estimation. The MTL experiments are a first step towards applying the information gathered from probing studies and are able to show some statistically significant improvements using a very simple MTL setup. The question of how much performance improvement is possible by augmenting additional training signal at intermediary layers will require additional research on the optimal *location*, *tasks*, as well as the right *amount* of training signal to be supplied.

7   Discussion and Conclusion

In this paper, we study the abilities acquired by neural ranking models. To do so, we construct probing datasets from MS MARCO and study how well ranking models encode lexical and semantic similarity, named entity recognition, and coreference resolution. We find ranking models to better encode lexical and semantic similarity as well as coreference resolution. Unlike previous work, which only investigated the final layer, we find these abilities to be best captured at an intermediary layer and to drop toward the final layer, posing the question of what information ranking models utilize for relevance estimation. We later use this information on which layers best encode the tasks to inform our multi-task learning setup. Our experiments show that training the ranking task on the final, and the auxiliary task (e.g., lexical similarity) on the layer with the best probing performance can outperform training both tasks on the final layer. More work, exceeding our naive MTL setup, has to be done to see how much improvement really is possible. Nevertheless, we see potential in adding ranking subtasks to the training setup for improving generalization and data efficiency. To the best of our knowledge, this is the first work to show that the probing results are not purely informational and can be used to improve the model-building process.

Acknowledgements. This research was (partially) funded by the Federal Ministry of Education and Research (BMBF), Germany under the project LeibnizKILabor with grant No. 01DD20003.

References

1. Aghajanyan, A., Gupta, A., Shrivastava, A., Chen, X., Zettlemoyer, L., Gupta, S.: Muppet: Massive multi-task representations with pre-finetuning. In: Moens, M., Huang, X., Specia, L., Yih, S.W. (eds.) Proceedings of the 2021 Conference on Empirical Methods in Natural Language Processing, EMNLP 2021, Virtual Event / Punta Cana, Dominican Republic, 7–11 Nov 2021, pp. 5799–5811. Association for Computational Linguistics (2021). https://doi.org/10.18653/v1/2021.emnlp-main.468

2. van Aken, B., Winter, B., Löser, A., Gers, F.A.: How does BERT answer questions?: A layer-wise analysis of transformer representations. In: Zhu, W., et al. (eds.) Proceedings of the 28th ACM International Conference on Information and Knowledge Management, CIKM 2019, Beijing, China,3–7 Nov 2019, pp. 1823–1832. ACM (2019). https://doi.org/10.1145/3357384.3358028

3. Anand, A., Leonhardt, J., Rudra, K., Anand, A.: Supervised contrastive learning approach for contextual ranking. In: Crestani, F., Pasi, G., Gaussier, É. (eds.) ICTIR 2022: The 2022 ACM SIGIR International Conference on the Theory of Information Retrieval, Madrid, Spain, 11–12 July 2022, pp. 61–71. ACM (2022). https://doi.org/10.1145/3539813.3545139

4. Anand, A., Lyu, L., Idahl, M., Wang, Y., Wallat, J., Zhang, Z.: Explainable information retrieval: a survey. CoRR abs/2211.02405 (2022). https://arxiv.org/abs/2211.02405

5. Belinkov, Y.: Probing classifiers: promises, shortcomings, and advances. Comput. Linguist. **48**(1), 207–219 (2022). https://doi.org/10.1162/coli_a_00422

6. Belinkov, Y., Glass, J.: Analysis methods in neural language processing: a survey. Trans. Assoc. Comput. Linguist. **7**, 49–72 (2019). https://doi.org/10.1162/tacl_a_00254. https://www.aclweb.org/anthology/Q19-1004

7. Câmara, A., Hauff, C.: Diagnosing bert with retrieval heuristics. In: Jose, J.M., et al. (eds.) ECIR 2020. LNCS, vol. 12035, pp. 605–618. Springer, Cham (2020). https://doi.org/10.1007/978-3-030-45439-5_40

8. Caruana, R.: Multitask learning. Mach. Learn. **28**(1), 41–75 (1997). https://doi.org/10.1023/A:1007379606734

9. Choi, J., Jung, E., Lim, S., Rhee, W.: Finding inverse document frequency information in BERT. CoRR abs/2202.12191 (2022). https://arxiv.org/abs/2202.12191

10. Clarke, C.L.A., Kolla, M., Vechtomova, O.: An effectiveness measure for ambiguous and underspecified queries. In: Azzopardi, L., et al. (eds.) ICTIR 2009. LNCS, vol. 5766, pp. 188–199. Springer, Heidelberg (2009). https://doi.org/10.1007/978-3-642-04417-5_17

11. Conneau, A., Kruszewski, G., Lample, G., Barrault, L., Baroni, M.: What you can cram into a single $&!#* vector: probing sentence embeddings for linguistic properties. In: Gurevych, I., Miyao, Y. (eds.) Proceedings of the 56th Annual Meeting of the Association for Computational Linguistics, ACL 2018, Melbourne, Australia, 15–20 July 2018, Volume 1: Long Papers, pp. 2126–2136. Association for Computational Linguistics (2018). https://doi.org/10.18653/v1/P18-1198. https://www.aclweb.org/anthology/P18-1198/

12. Craswell, N., Mitra, B., Yilmaz, E., Campos, D., Voorhees, E.M.: Overview of the TREC 2019 deep learning track. CoRR abs/2003.07820 (2020). https://arxiv.org/abs/2003.07820

13. Dehghani, M., Zamani, H., Severyn, A., Kamps, J., Croft, W.B.: Neural ranking models with weak supervision. In: Kando, N., Sakai, T., Joho, H., Li, H., de Vries, A.P., White, R.W. (eds.) Proceedings of the 40th International ACM SIGIR Conference on Research and Development in Information Retrieval, Shinjuku, Tokyo, Japan, 7–11 August 2017, pp. 65–74. ACM (2017). https://doi.org/10.1145/3077136.3080832

14. Devlin, J., Chang, M.W., Lee, K., Toutanova, K.: BERT: pre-training of deep bidirectional transformers for language understanding. In: Proceedings of the 2019 Conference of the North American Chapter of the Association for Computational Linguistics: Human Language Technologies, Volume 1 (Long and Short Papers), pp. 4171–4186. Association for Computational Linguistics, Minneapolis, Minnesota (2019). https://doi.org/10.18653/v1/N19-1423. https://www.aclweb.org/anthology/N19-1423

15. Elazar, Y., Ravfogel, S., Jacovi, A., Goldberg, Y.: Amnesic probing: behavioral explanation with amnesic counterfactuals. Trans. Assoc. Comput. Linguistics 9, 160–175 (2021). https://doi.org/10.1162/tacl_a_00359. https://doi.org/10.1162/tacl_a_00359

16. Fan, Y., Guo, J., Ma, X., Zhang, R., Lan, Y., Cheng, X.: A linguistic study on relevance modeling in information retrieval. In: Leskovec, J., Grobelnik, M., Najork, M., Tang, J., Zia, L. (eds.) WWW 2021: The Web Conference 2021, Virtual Event / Ljubljana, Slovenia, 19–23 Apr 2021, pp. 1053–1064. ACM / IW3C2 (2021). https://doi.org/10.1145/3442381.3450009

17. Fayyaz, M., Aghazadeh, E., Modarressi, A., Mohebbi, H., Pilehvar, M.T.: Not all models localize linguistic knowledge in the same place: a layer-wise probing on bertoids' representations. In: Bastings, J., et al. (eds.) Proceedings of the Fourth BlackboxNLP Workshop on Analyzing and Interpreting Neural Networks for NLP, BlackboxNLP@EMNLP 2021, Punta Cana, Dominican Republic, 11 Nov 2021, pp. 375–388. Association for Computational Linguistics (2021). https://aclanthology.org/2021.blackboxnlp-1.29

18. Formal, T., Piwowarski, B., Clinchant, S.: SPLADE: sparse lexical and expansion model for first stage ranking. In: Diaz, F., Shah, C., Suel, T., Castells, P., Jones, R., Sakai, T. (eds.) SIGIR 2021: The 44th International ACM SIGIR Conference on Research and Development in Information Retrieval, Virtual Event, Canada, 11–15 July 2021, pp. 2288–2292. ACM (2021). https://doi.org/10.1145/3404835.3463098

19. Ghasemi, N., Hiemstra, D.: BERT meets cranfield: uncovering the properties of full ranking on fully labeled data. In: Proceedings of the 16th Conference of the European Chapter of the Association for Computational Linguistics: Student Research Workshop, pp. 58–64. Association for Computational Linguistics, Online (2021). https://aclanthology.org/2021.eacl-srw.9

20. Hagen, M., Völske, M., Göring, S., Stein, B.: Axiomatic result re-ranking. In: Mukhopadhyay, S., et al.(eds.) Proceedings of the 25th ACM International Conference on Information and Knowledge Management, CIKM 2016, Indianapolis, IN, USA, 24–28 October 2016, pp. 721–730. ACM (2016). https://doi.org/10.1145/2983323.2983704

21. Hewitt, J., Ethayarajh, K., Liang, P., Manning, C.D.: Conditional probing: measuring usable information beyond a baseline. In: Moens, M., Huang, X., Specia, L., Yih, S.W. (eds.) Proceedings of the 2021 Conference on Empirical Methods in Natural Language Processing, EMNLP 2021, Virtual Event / Punta Cana, Dominican Republic, 7–11 November 2021, pp. 1626–1639. Association for Computational Linguistics (2021). https://doi.org/10.18653/v1/2021.emnlp-main.122

22. Hewitt, J., Liang, P.: Designing and interpreting probes with control tasks. In: Proceedings of the 2019 Conference on Empirical Methods in Natural Language Processing and the 9th International Joint Conference on Natural Language Processing (EMNLP-IJCNLP), pp. 2733–2743. Association for Computational Linguistics, Hong Kong, China (2019). https://doi.org/10.18653/v1/D19-1275. https://aclanthology.org/D19-1275

23. Hewitt, J., Manning, C.D.: A structural probe for finding syntax in word representations. In: Proceedings of the 2019 Conference of the North American Chapter of the Association for Computational Linguistics: Human Language Technologies, Volume 1 (Long and Short Papers), pp. 4129–4138. Association for Computational Linguistics, Minneapolis, Minnesota (2019). https://doi.org/10.18653/v1/N19-1419. https://www.aclweb.org/anthology/N19-1419

24. Honnibal, M., Montani, I.: Natural language understanding with bloom embeddings, convolutional neural networks and incremental parsing. Unpublished Software Application. https://spacy.io (2017)

25. Khalid, M.A., Jijkoun, V., de Rijke, M.: The impact of named entity normalization on information retrieval for question answering. In: Macdonald, C., Ounis, I., Plachouras, V., Ruthven, I., White, R.W. (eds.) ECIR 2008. LNCS, vol. 4956, pp. 705–710. Springer, Heidelberg (2008). https://doi.org/10.1007/978-3-540-78646-7_83

26. Khattab, O., Zaharia, M.: Colbert: efficient and effective passage search via contextualized late interaction over BERT. In: Huang, J., et al. (eds.) Proceedings of the 43rd International ACM SIGIR conference on research and development in Information Retrieval, SIGIR 2020, Virtual Event, China, 25–30 July 2020, pp. 39–48. ACM (2020). https://doi.org/10.1145/3397271.3401075

27. Kim, M., Ko, Y.: Multitask fine-tuning for passage re-ranking using BM25 and pseudo relevance feedback. IEEE Access 10, 54254–54262 (2022). https://doi.org/10.1109/ACCESS.2022.3176894

28. Kingma, D.P., Ba, J.: Adam: a method for stochastic optimization. In: Bengio, Y., LeCun, Y. (eds.) 3rd International Conference on Learning Representations, ICLR 2015, San Diego, CA, USA, 7–9 May 2015, Conference Track Proceedings (2015). https://arxiv.org/abs/1412.6980

29. Leonhardt, J., Rudra, K., Khosla, M., Anand, A., Anand, A.: Efficient neural ranking using forward indexes. In: Laforest, F., et al. (eds.) WWW 2022: The ACM Web Conference 2022, Virtual Event, Lyon, France, 25–29 April 2022, pp. 266–276. ACM (2022). https://doi.org/10.1145/3485447.3511955

30. Li, H., Xu, J.: Semantic matching in search. Found. Trends Inf. Retr. 7(5), 343–469 (2014). https://doi.org/10.1561/1500000035

31. MacAvaney, S., Feldman, S., Goharian, N., Downey, D., Cohan, A.: ABNIRML: analyzing the behavior of neural IR models. Trans. Assoc. Comput. Linguistics 10, 224–239 (2022). https://doi.org/10.1162/tacl_a_00457

32. Manning, C.D., Raghavan, P., Schütze, H.: Introduction to information retrieval. Cambridge University Press (2008). https://doi.org/10.1017/CBO9780511809071. https://nlp.stanford.edu/IR-book/pdf/irbookprint.pdf

33. Nogueira, R.F., Cho, K.: Passage re-ranking with BERT. CoRR abs/1901.04085 (2019). https://arxiv.org/abs/1901.04085
34. Pennington, J., Socher, R., Manning, C.: GloVe: global vectors for word representation. In: Proceedings of the 2014 Conference on Empirical Methods in Natural Language Processing (EMNLP), pp. 1532–1543. Association for Computational Linguistics, Doha, Qatar (2014). https://doi.org/10.3115/v1/D14-1162. https://aclanthology.org/D14-1162
35. Petroni, F., et al.: KILT: a benchmark for knowledge intensive language tasks. In: Proceedings of the 2021 Conference of the North American Chapter of the Association for Computational Linguistics: Human Language Technologies, pp. 2523–2544. Association for Computational Linguistics, Online (2021). https://doi.org/10.18653/v1/2021.naacl-main.200. https://aclanthology.org/2021.naacl-main.200
36. Petroni, F., et al.: Language models as knowledge bases? In: Proceedings of the 2019 Conference on Empirical Methods in Natural Language Processing and the 9th International Joint Conference on Natural Language Processing (EMNLP-IJCNLP), pp. 2463–2473. Association for Computational Linguistics, Hong Kong, China (2019). https://doi.org/10.18653/v1/D19-1250. https://aclanthology.org/D19-1250
37. Pimentel, T., Saphra, N., Williams, A., Cotterell, R.: Pareto probing: trading off accuracy for complexity. In: Proceedings of the 2020 Conference on Empirical Methods in Natural Language Processing (EMNLP), pp. 3138–3153. Association for Computational Linguistics, Online (2020). https://doi.org/10.18653/v1/2020.emnlp-main.254. https://www.aclweb.org/anthology/2020.emnlp-main.254
38. Qiao, Y., Xiong, C., Liu, Z., Liu, Z.: Understanding the behaviors of BERT in ranking. CoRR abs/1904.07531 (2019). https://arxiv.org/abs/1904.07531
39. Rau, D., Kamps, J.: The role of complex NLP in transformers for text ranking. In: Crestani, F., Pasi, G., Gaussier, É. (eds.) ICTIR 2022: The 2022 ACM SIGIR International Conference on the Theory of Information Retrieval, Madrid, Spain, 11–12 July 2022, pp. 153–160. ACM (2022). https://doi.org/10.1145/3539813.3545144
40. Rennings, D., Lyu, L., Anand, A.: Listwise explanations for ranking models using multiple explainers. In: Advances in Information Retrieval - 45th European Conference on IR Research, ECIR 2023, Dublin, Ireland, Proceedings, Part I. Lecture Notes in Computer Science, Springer (2023)
41. Rennings, D., Moraes, F., Hauff, C.: An axiomatic approach to diagnosing neural IR models. In: Azzopardi, L., Stein, B., Fuhr, N., Mayr, P., Hauff, C., Hiemstra, D. (eds.) ECIR 2019. LNCS, vol. 11437, pp. 489–503. Springer, Cham (2019). https://doi.org/10.1007/978-3-030-15712-8_32
42. Rissanen, J.: Universal coding, information, prediction, and estimation. IEEE Trans. Inf. Theory 30(4), 629–636 (1984). https://doi.org/10.1109/TIT.1984.1056936
43. Robertson, S.E., Walker, S., Jones, S., Hancock-Beaulieu, M., Gatford, M.: Okapi at TREC-3. In: Harman, D.K. (ed.) Proceedings of The Third Text REtrieval Conference, TREC 1994, Gaithersburg, Maryland, USA, 2–4 Nov 1994. NIST Special Publication, vol. 500–225, pp. 109–126. National Institute of Standards and Technology (NIST) (1994). https://trec.nist.gov/pubs/trec3/papers/city.ps.gz
44. Rudra, K., Anand, A.: Distant supervision in BERT-based Adhoc document retrieval. In: d'Aquin, M., Dietze, S., Hauff, C., Curry, E., Cudré-Mauroux, P. (eds.) CIKM 2020: The 29th ACM International Conference on Information and Knowledge Management, Virtual Event, Ireland, 19–23 October 2020, pp. 2197–2200. ACM (2020). https://doi.org/10.1145/3340531.3412124

45. Singh, J., Anand, A.: EXS: explainable search using local model agnostic interpretability. In: Culpepper, J.S., Moffat, A., Bennett, P.N., Lerman, K. (eds.) Proceedings of the Twelfth ACM International Conference on Web Search and Data Mining, WSDM 2019, Melbourne, VIC, Australia, 11–15 Feb 2019, pp. 770–773. ACM (2019). https://doi.org/10.1145/3289600.3290620

46. Singh, J., Anand, A.: Model agnostic interpretability of rankers via intent modelling. In: Hildebrandt, M., Castillo, C., Celis, L.E., Ruggieri, S., Taylor, L., Zanfir-Fortuna, G. (eds.) FAT* 2020: Conference on Fairness, Accountability, and Transparency, Barcelona, Spain, 27–30 Jan 2020, pp. 618–628. ACM (2020). https://doi.org/10.1145/3351095.3375234

47. Tenney, I., Das, D., Pavlick, E.: BERT rediscovers the classical NLP pipeline. In: Proceedings of the 57th Annual Meeting of the Association for Computational Linguistics, pp. 4593–4601. Association for Computational Linguistics, Florence, Italy (2019). https://doi.org/10.18653/v1/P19-1452. https://www.aclweb.org/anthology/P19-1452

48. Tenney, I., et al.: What do you learn from context? probing for sentence structure in contextualized word representations. In: 7th International Conference on Learning Representations, ICLR 2019, New Orleans, LA, USA, 6–9 May 2019. OpenReview.net (2019). https://openreview.net/forum?id=SJzSgnRcKX

49. Thorne, J., Vlachos, A., Christodoulopoulos, C., Mittal, A.: FEVER: a large-scale dataset for fact extraction and Verification. In: Proceedings of the 2018 Conference of the North American Chapter of the Association for Computational Linguistics: Human Language Technologies, Volume 1 (Long Papers), pp. 809–819. Association for Computational Linguistics, New Orleans, Louisiana (2018). https://doi.org/10.18653/v1/N18-1074. https://aclanthology.org/N18-1074

50. Voita, E., Titov, I.: Information-theoretic probing with minimum description length. In: Proceedings of the 2020 Conference on Empirical Methods in Natural Language Processing (EMNLP), pp. 183–196. Association for Computational Linguistics, Online (2020). https://doi.org/10.18653/v1/2020.emnlp-main.14. https://www.aclweb.org/anthology/2020.emnlp-main.14

51. Völske, M., et al.: Towards axiomatic explanations for neural ranking models. In: Hasibi, F., Fang, Y., Aizawa, A. (eds.) ICTIR 2021: The 2021 ACM SIGIR International Conference on the Theory of Information Retrieval, Virtual Event, Canada, 11 July 2021, pp. 13–22. ACM (2021). https://doi.org/10.1145/3471158.3472256

52. Wallat, J., Singh, J., Anand, A.: BERTnesia: investigating the capture and forgetting of knowledge in BERT. In: Proceedings of the Third BlackboxNLP Workshop on Analyzing and Interpreting Neural Networks for NLP, pp. 174–183. Association for Computational Linguistics, Online (2020). https://doi.org/10.18653/v1/2020.blackboxnlp-1.17. https://aclanthology.org/2020.blackboxnlp-1.17

53. Zhan, J., Mao, J., Liu, Y., Zhang, M., Ma, S.: An analysis of BERT in document ranking. In: Huang, J., et al. (eds.) Proceedings of the 43rd International ACM SIGIR conference on research and development in Information Retrieval, SIGIR 2020, Virtual Event, China, 25–30 July 2020, pp. 1941–1944. ACM (2020). https://doi.org/10.1145/3397271.3401325

54. Zhang, K., Bowman, S.: Language modeling teaches you more than translation does: Lessons learned through auxiliary syntactic task analysis. In: Proceedings of the 2018 EMNLP Workshop BlackboxNLP: Analyzing and Interpreting Neural Networks for NLP, pp. 359–361. Association for Computational Linguistics, Brussels, Belgium (2018). https://doi.org/10.18653/v1/W18-5448. https://www.aclweb.org/anthology/W18-5448

55. Zhang, X., Ramachandran, D., Tenney, I., Elazar, Y., Roth, D.: Do language embeddings capture scales? In: Proceedings of the Third BlackboxNLP Workshop on Analyzing and Interpreting Neural Networks for NLP, pp. 292–299. Association for Computational Linguistics, Online (2020). https://doi.org/10.18653/v1/2020.blackboxnlp-1.27. https://aclanthology.org/2020.blackboxnlp-1.27

56. Zhao, M., Dufter, P., Yaghoobzadeh, Y., Schütze, H.: Quantifying the contextualization of word representations with semantic class probing. In: Findings of the Association for Computational Linguistics: EMNLP 2020, pp. 1219–1234. Association for Computational Linguistics, Online (2020). https://doi.org/10.18653/v1/2020.findings-emnlp.109. https://www.aclweb.org/anthology/2020.findings-emnlp.109

Clustering of Bandit with Frequency-Dependent Information Sharing

Shen Yang[1,2], Qifeng Zhou[1,2]([⊠]) [iD], and Qing Wang[3]

[1] Department of Automation, Xiamen University, Xiamen, China
yangshen@stu.xmu.edu.cn, zhouqf@xmu.edu.cn
[2] Xiamen Key Laboratory of Big Data Intelligent Analysis and Decision-Making,
Xiamen, China
[3] Intelligent IT Operations, IBM T.J. Watson Research Center, New York, NY, USA
qing.wang1@ibm.com

Abstract. In today's business marketplace, the great demand for developing intelligent interactive recommendation systems is growing rapidly, which sequentially suggest users proper items by accurately predicting their preferences, while receiving up-to-date feedback to promote the overall performance. Multi-armed bandit, which has been widely applied to various online systems, is quite capable of delivering such efficient recommendation services. To further enhance online recommendations, many works have introduced clustering techniques to fully utilize users' information. These works consider symmetric relations between users, i.e., users in one cluster share equal weights. However, in practice, users usually have different interaction frequency (i.e., activeness) in one cluster, and their collaborative relations are unsymmetrical. This brings a challenge for bandit clustering since inactive users lack the capability of leveraging these interaction information to mitigate the cold-start problem, and further affect active ones belonging to one cluster. In this work, we explore user activeness and propose a frequency-dependent clustering of bandit model to deal with the aforementioned challenge. The model learns representation of each user's cluster by sharing collaborative information weighed based on user activeness, i.e., inactive users can utilize the collaborative information from active ones in the same cluster to optimize the cold start process. Extensive studies have been carefully conducted on both synthetic data and two real-world datasets indicating the efficiency and effectiveness of our proposed model.

Keywords: Interactive recommendation system · Weighed clustering · Multi-armed bandit model

1 Introduction

Personalized recommendation system (RecSys), aiming to provide users valuable information and services accurately from massive amounts of data, plays a

© The Author(s), under exclusive license to Springer Nature Switzerland AG 2023
J. Kamps et al. (Eds.): ECIR 2023, LNCS 13981, pp. 274–287, 2023.
https://doi.org/10.1007/978-3-031-28238-6_18

significant role in information services of modern society. Specifically, the main focus of RecSys is to be capable of recommending items to satisfy users' requirements, so as to maximize some metrics (e.g., exposure, user satisfaction or product profit) [5]. For this reason, traditional recommendation algorithms [1] like collaborative filtering-based recommendation [10,14] and content-based recommendation [19], have emerged, which predict items that users are interested in based on historical data, and the predictions are at an individual level.

The offline training and online testing pattern of traditional recommendation algorithms results in their inappropriateness to provide high-quality online recommendations under the rapidly changing scenes of information services, e.g., cold start problems [15]. The situation of online recommendation is similar to the multi-armed bandit problem [2,18], that is, when new items or users are introduced into the system, RecSys needs to strike a balance between satisfying user interests and exploring new items to maximize long-term benefits. As a kind of reinforcement learning algorithm, bandit model can effectively solve the exploitation and exploration problem of cold start in online recommendations, and have become a popular recommendation algorithm, especially contextual bandit models [21]. LinUCB proposed in [12] is one classic contextual bandit model. With the development of social networks, researchers found that the implementation of clustering and collaborative filtering among users can make fuller use of feedback to obtain better recommendations [3,7,14,22]. The main idea of these works is to discover the underlying user clusters and share information within it for more accurate recommendations. However, current clustering-based bandit models suppose symmetric relations exist between the same-cluster users and share equal weights as well, resulting in the same collaborative effect among users [13]. To put it simply, inactive users who have low interaction frequency can pose negative impact on the collaborative effect of the same-cluster users, leading to the uncertainty of recommendation. Therefore, it is valuable for inactive users to learn more information from active ones, while active users filter out "harmful" information from inactive ones in the same cluster.

In this paper, we study the asymmetric relations between users in a cluster and propose a new frequency-dependent clustering bandit model (FreqCB) to solve the aforementioned issues. We use the information of user activeness (i.e., interaction) for recommendation in the bandit setting, and the information sharing process is carried out through user activeness. In this way, inactive users can be hardly affected active ones from the same cluster by avoiding the collaborative effect, while inactive users can leverage active users' information to obtain a recommendation that fits their cluster's interests. Extensive studies have been conducted on both synthetic data and real-world datasets to indicate the effectiveness of FreqCB.

Our main contributions are as follows:

- We develop a method for capturing the influence of user activeness to optimize (e.g., speed up) the cold start process and get better recommendations.
- We propose a frequency-dependent clustering bandit model under the framework of upper-confidence bound. The model learns representation of each user's cluster by sharing collaborative information weighed based on user activeness.

- We demonstrate the effectiveness of our proposed model in extensive studies conducted on both synthetic data and real-world datasets.

2 Related Work

In this section, we highlight our studies and compare them with existing works related to our approach, including clustering-based bandit models and collaborative filtering-based bandit models.

Researchers have found that dividing users into different clusters and customizing the bandits for each cluster can enhance the bandit models' performance. The technique of online clustering of bandit is first studied in [7], which is inspired by [4] encoding social relationships in a graph for clustering bandit and other earlier works. [4] proposes GOB.Lin which runs a bandit model on a network and makes the nodes of the graph share information with its neighbor nodes by introducing the graph's Laplacian matrix. But GOB.Lin has its limitations: it is the solution only for networked bandit problems and the clusters can't be updated adaptively with the varying of user's parameter vectors. [7] follows up and designs CLUB, an adaptive clustering of bandit strategy on graph. CLUB deletes edges between users whose interests are no longer similar to each other and forms a new cluster on the new graph after edges deleting. SCLUB in [13] is a generalized version of CLUB. SCLUB proposes new splitting, merging, and updating methods on clustering to identify the underlying clusters faster and more accurate. Besides, [17] develops DYNUCB, utilizing k-means clustering method [16] to dynamically cluster all users based on the similarity of their parameter vectors. But the number of underlying clusters is unknown and needs to be specified as a hyper parameter.

Based on [7,9] proposes distributed confidence ball algorithms for solving linear bandit problems in peer to peer networks with limited communication capabilities. To address the challenge of uncertain estimation of user interests and user clustering, [22] proposes ClexB for the online recommendation, which is a contextual bandit policy that incorporates knowledge sharing via adaptive clustering and learning user interest via exploring clustering.

In these mentioned works, users are not grouped in a context-dependent approach. [14] proposes COFIBA which takes advantage of clustering and collaborative filtering from both user and item sides. [6] presents CAB based on the linear contextual bandit framework and implements the context-dependent feedback sharing mechanism over users in a flexible manner. LOCB [3] starts with a set of seeds (e.g., users), then updates recursively the seeds set and neighbors of each seed after pulling module.

These algorithms employ clustering or collaborative filtering on either user or item side to share information in/across clusters, which does not consider the asymmetric relationship between users. However, in a bandit-setting recommendation system with collaborative filtering, it's a universal case: active users provide more feedback to the recommendation system, whereas inactive/new users can hardly have affect on it. Hence, more noise would be brought in when using the equal weights of active and inactive users to represent a cluster. Our paper proposes FreqCB, a frequency-dependent clustering bandit model incorporating user activeness to represent the clusters.

3 Problem Formulation

The set of N users is represented by $U = \{u_1, u_2, \cdots, u_N\}$. Learning procedure goes as a sequence: At each round $t = [1, 2, \ldots, T]$, the learning agent receives a user $i_t \in U$ with a set of context vectors $H_{i_t} = \{x_1, x_2, \cdots, x_K\}$, where $\| x \| \leq 1, x \in \mathbb{R}^d$ for all x in H_{i_t}. H_{i_t} consists of the vectors of candidate arms (e.g., items) to be recommended for user i_t. The learning agent then selects one arm a_t, whose feature vector is $\bar{x} \in H_{i_t}$ to recommend for user i_t according to the historical feedback $\mathcal{Q}(t) = \{i_\tau, a_\tau, r_\tau\}_{\tau=1}^t$, and observes payoff $r_t \in \mathbb{R}$, which is a function of both user i_t and the recommended arm a_t. After receiving feedback, the agent updates its parameters and continue to recommend. The objective of the agent is to select a proper policy $\pi = \{a_1, a_2, \cdots, a_T\}$ to minimize the cumulative regret [11], which is defined below,

$$Regret_\pi = \sum_{t=1}^{T} (r_t^* - r_t), \tag{1}$$

where r_t^* denotes payoff of the optimal choice at time t. This goal is a long-term cumulative reward that can be regarded as the target of the multi-armed bandit problem, where each item denotes each arm. To accomplish this goal, assuming r_t is a linear function of a_t's feature, which has been successfully used in bandit problems. The function is defined as follows:

$$r_t = \omega_{i_t}^{\mathrm{T}} \bar{x} + \epsilon_{a_t} \tag{2}$$

where ω_{i_t} is a d-dimension parameter vector which indicates the preference of user i_t, $\bar{x}$ is the feature vector of arm a_t, and ϵ_{a_t} is Gaussian noise.

LinUCB, as a classic solution for multi-armed bandit problem, holds a linear combination assumption between the user's feature vector and context vector, and pulls the arm $\bar{x}$ with the largest score, which is defined below:

$$\bar{x} = \arg\max_{x_k \in H} (\omega^{\mathrm{T}} x_k + \alpha \sqrt{x_k^{\mathrm{T}} M^{-1} x_k}), \tag{3}$$

where H is the set of context vectors, x_k denotes context vector of the k-th arm in candidate arms, ω is the parameters of current user, $M \in \mathbb{R}^{d \times d}$ is the correlation matrix of current user which contains information of arms that the user pulls before, α is a hyper parameter to combine the expectation (the former term) and standard deviation (the latter term) of reward. The deviation term aims to balance the trade-off between exploration and exploitation, that is, the larger deviation, the more exploration. ω is estimated by ridge regression as follows:

$$\omega = M^{-1} b, \tag{4}$$

where $b \in \mathbb{R}^d$ is the corresponding response vector (e.g., the corresponding d click/no-click user feedback).

Different from standard bandit models, we focus on clusters among users and arms that valid information can be shared for better recommendation. Existing related works only consider symmetrical relations between users in one cluster, but asymmetric relations are more practical in reality.

4   Solution and Algorithms

In this section, we propose a frequency-dependent clustering bandit algorithm, FreqCB (as shown in Algorithm 1) to explore the effectiveness of fully utilizing users' activeness to improve the performance of online recommendation. In FreqCB, we treat the users' interaction frequency as their activeness, which makes sense practically.

FreqCB maintains vector ω, vector b and correlation matrix M for each user and each cluster at timestamp t. We set three hyperparameters α, β, and γ. β, γ are used in the procedure of splitting and merging user clusters, while α is for exploring the upper-confidence bound in the reward prediction. The interact frequency of each user is set to 1 initially for computational feasibility. The initial cluster is set as the whole users with index 1. This algorithm proceeds in stages. The g_{th} stage contains 2^g timestamps totally. In each stage, the algorithm can pay more attention on exploring inaccurate user clusters and leave accurate clusters unaffected. Every user i is set unserved in the beginning of the g_{th} stage: $ch_i = 0$; if user i is served, $ch_i = 1$. Note that the subscript t denotes the timestamp. At t timestamp, we get the user $i_t \in U$ to be served and the context vectors $H_t = \{x_1, x_2, \cdots, x_K\}$ of items available for recommendation.

The cluster $c_j (i_t \in c_j)$ can be found and the algorithm computes the bandit parameters of user cluster c_j by $\omega_{c_j} = M_{c_j}^{-1} b_{c_j}$ and recommend a_t, whose feature vector is $\bar{x}$, for user i_t according to the standard LinUCB form in Eq. 3:

$$\bar{x} = \arg\max_{x_k \in H_{i_t}} (\omega_{c_j}^{\mathrm{T}} x_k + \alpha \sqrt{x_k^{\mathrm{T}} M_{c_j}^{-1} x_k}). \tag{5}$$

After receiving the feedback r_t given by user i_t, the algorithm updates the information of user i_t and cluster c_j (see Algorithm 2). Note that in the information updating procedure of cluster c_j, we introduce the user activeness for the evolution of M_{c_j} and b_{c_j}. To share information between similar users, FreqCB aggregates the information of users in the same cluster c_j weighted by their interaction frequency (activeness) to capture more accurate information of cluster c_j:

$$M_{c_j} = I + \sum_{i \in c_j} \hat{f}_i (M_i - I), b_{c_j} = \sum_{i \in c_j} \hat{f}_i b_i, \tag{6}$$

where $\hat{f}_i$ is the normalized frequency relating to cluster c_j:

$$\hat{f}_i = \frac{f_i}{\sum_{q \in c_j} f_q}, i \in c_j. \tag{7}$$

Algorithm 1. FreqCB

1: Input: exploration hyperparameters α, β, γ.
2: Initialize $b_i = 0 \in \mathbb{R}^d$, $M_i = I \in \mathbb{R}^{d \times d}$ and $f_i = 1$ for each $i \in U$.
3: Initialize the set of cluster indexes $S = \{1\}$.
4: Initialize the first cluster $c_1 = U, b_{c_1} = 0 \in \mathbb{R}^d, M_{c_1} = I \in \mathbb{R}^{d \times d}$.
5: **for** stage $g = 1, 2, \cdots$ **do**
6:     Set $ch_i = 0$ for user i in each cluster.
7:     Set $\omega_{c_j} = M_{c_j}{}^{-1} b_{c_j}$ for each cluster c_j, j is the cluster index in S.
8:     **for** $\tau = 1, 2, \cdots, 2^g$ **do**
9:         Compute current timestamp: $t = 2^g + \tau - 2$.
10:        Get user $i_t \in U$ to be served.
11:        Get get context vectors $H_{i_t} = \{x_1, \cdots, x_K\}$ for candidates to be recommended.
12:        Get the index j of cluster c_j that user i_t belongs to.
13:        Compute the coefficient of cluster c_j: $\omega_{c_j} = M_{c_j}{}^{-1} b_{c_j}$.
14:        Recommend item a_t corresponding to $\bar{x} = \arg\max_{x_k \in H_{i_t}} (\omega_{c_j}{}^{\mathrm{T}} x_k + \alpha \sqrt{x_k^{\mathrm{T}} M_{c_j}{}^{-1} x_k})$.
15:        Receive the feedback r_t given by user i_t.
16:        Update the information of user i_t and cluster c_j by calling UPDATE.
17:        Split user i_t from cluster c_j and form a new cluster $c_{j'}$ by calling SPLIT.
18:        Set $ch_{i_t} = 1$.
19:        Merge similar clusters by calling MERGE.
20:    **end for**
21: **end for**

Algorithm 2. UPDATE

1: $M_{i_t} = M_{i_t} + \bar{x}\bar{x}^{\mathrm{T}}$, $b_{i_t} = b_{i_t} + r_t \bar{x}$, $f_{i_t} = f_{i_t} + 1$
2: $\omega_{i_t} = M_{i_t}^{-1} b_{i_t}$
3: Obtain $\hat{f}_{i_t}$ by normalizing the frequency of users in c_j.
4: $M_{c_j} = I + \sum_{i \in c_j} \hat{f}_i (M_i - I)$, $b_{c_j} = \sum_{i \in c_j} \hat{f}_i b_i$, $F_{c_j} = F_{c_j} + 1$
5: $\omega_{c_j} = M_{c_j}{}^{-1} b_{c_j}$
6: Compute $p_{c_j} = \frac{F_{c_j}}{\|c_j\| t}$
7: Compute $p_{i_t} = \frac{f_{i_t}}{t}$

By Eq. 6, the algorithm could assist inactive users who have few interaction records leverage from active ones for effective clustering analysis, and could prevent cluster representations from being devastated by inaccurate representations of inactive users.

If user i_t is not consistent with the current cluster c_j, FreqCB splits user i_t out (see Algorithm 3). Different with SCLUB [13], FreqCB introduces user activeness in the evolution of M_{c_j} and b_{c_j}:

$$M_{c_j} = M_{c_j} - \hat{f}_{i_t}(M_{i_t} - I), b_{c_j} = b_{c_j} - \hat{f}_{i_t}\bar{x}. \tag{8}$$

Algorithm 3. SPLIT

1: Set $Q(F) = \sqrt{\frac{1+\ln(1+F)}{1+F}}$

2: **if** $\|\omega_{i_t} - \omega_{c_j}\| > \beta(Q(F_{i_t}) + Q(F_{c_j}))$ or $\|p_{i_t} - p_{c_j}\| > \gamma Q(t)$ **then**

3:    Set $j' = \max S + 1$.

4:    Obtain $\hat{f}_{i_t}$ by normalizing the frequency of users in c_j.

5:    $M_{c_j} = M_{c_j} - \hat{f}_{i_t}(M_{i_t} - I)$, $b_{c_j} = b_{c_j} - \hat{f}_{i_t}b_{i_t}$, $F_{c_j} = F_{c_j} - f_{i_t}$

6:    $c_j = c_j - \{i_t\}$

7:    $M_{c_{j'}} = M_{i_t}$, $b_{c_{j'}} = b_{i_t}$, $F_{c_{j'}} = f_{i_t}$, $c_{j'} = \{i_t\}$

8: **end if**

Algorithm 4. MERGE

1: **for** any two served clusters j_1 and j_2 **do**

2:    **if** $\|\omega_{c_{j_1}} - \omega_{c_{j_2}}\| < \frac{\beta}{2}(Q(F_{c_{j_1}}) + Q(F_{c_{j_2}}))$ and $\|p_{c_{j_1}} - p_{c_{j_2}}\| < \gamma Q(t)$ **then**

3:      $c_{j_1} = c_{j_1} \cup c_{j_2}$

4:      $F_{c_{j_1}} = F_{c_{j_1}} + F_{c_{j_2}}$

5:      Obtain $\hat{f}_i$ by normalizing the frequency of users in c_{j_1}.

6:      $M_{c_{j_1}} = I + \sum_{i \in c_{j_1}} \hat{f}_i(M_i - I)$

7:      $b_{c_{j_1}} = \sum_{i \in c_{j_1}} \hat{f}_i b_i$

8:      Delete c_{j_2}

9:    **end if**

10: **end for**

Based on Eq. 8, the information of user i_t can be deleted from cluster c_j and other users' information in cluster c_j remains. Then two served clusters will be merged if they are consistent at some level (see Algorithm 4). Similarly, according to Eq. 6, FreqCB introduces user activeness in the evolution of new clusters. When this stage ends, FreqCB continues to the next stage until it reaches the final timestamp. β and γ are hyper parameters to control the gap between user and cluster, and the gap between different clusters.

To generate the candidate items for H_{i_t}, we randomly select $K - 1$ items from all items and randomly select one item from nonzero payoff items of user i_t to form a candidate item set of size K.

Complexity of Implementation. Assuming a total of T rounds, K items in H_{i_t}, and d dimensions for vectors x_k, we can analysis the complexity of FreqCB by each round. Each recommendation takes $O(Kd^2)$ time, where estimated vector for the cluster that the current user belongs to need to be computed as well as predicted scores for items in H_{i_t}. After receiving feedback, users frequency is calculated, and both user and cluster representation are updated. Note that the number of users in the cluster is at most n. Thus, the time complexity for the update is $O(nd^2 + nd)$. Each split check and each merge check take $O(d)$, respectively.

In the rounds where all current clusters are accurate, FreqCB does not continue to split and merge after one split check and m merge check. Therefore, it costs $O((K+n)d^2+(n+m)d)$ under this circumstance. In the rounds where cluster structure is being exploring, each split takes $O(n+d^2)$ time and each merge costs $O(d^3+nd)$. Note that the number of merge times is at most $n(n-1)/2$, so the time complexity in this situation is $O(n^2d^3 + (K+n+1)d^2 + (n^3+n)d+n)$. And the time complexity of FreqCB for T rounds (the worst case) is

$$O(T(n^2d^3 + (K + n + 1)d^2 + (n^3 + n)d + n)).$$

In addition, the time complexity for LinUCB is $O(TKd^2)$. Since SCLUB allows split and merge for cluster exploration, its time complexity (in expectation) for T rounds is:

$$O(TKd^2 + Tmd + (\frac{nd^3}{\delta_1} + \frac{n^2d^3}{\delta_2})\ln{(T)}),$$

where δ_1 and δ_2 are simplified parameters in [13].

5   Experiments

In this section, we evaluate FreqCB[1] on both synthetic data and two real-world datasets comparing with strong baselines.

5.1   Compared Algorithms

- **Random.** It randomly picks one of all K items in every timestep, without learning user parameters.
- **ϵ-greedy.** It randomly selects one of all K items with probability ϵ, otherwise selects the item with the highest empirical mean with probability $1 - \epsilon$.
- **Lin-One.** It runs a single instance of LinUCB to serve all users, which means all users belong to the same cluster.
- **Lin-Ind.** It runs an independent instance of LinUCB per user to make a recommendation in a fully personalized style, which means each user forms one cluster by himself.
- **CAB.** It is based on the linear contextual bandit framework and incorporates collaborative filtering. Users having similar behavior form one cluster in a context dependent way, then CAB selects arms by using the information of the whole cluster.
- **SCLUB.** It incorporates collaborative filtering by graph-based clustering. User clusters are regarded as graphs and they are maintained by splitting, merging, and updating. Users in the same cluster share collaborative information with others.

---

[1] Our code and data is available at https://github.com/holywoodys/FreqCB.

Algorithm 5. Evaluation

1: Inputs: $T > 0$
2: Initial: $G_0 = 0$, a zero total reward at $t = 0$
3: **for** $t = 1, 2, \cdots, T$ **do**
4:    Get event feedback $e_t = \{i_t, a_t, r_t\}$
5:    $\mathcal{Q}(t) = \mathcal{Q}(t-1) \cup e_t$
6:    $G_t = G_{t-1} + r_t$
7:    Current CTR $= G_t/t$
8: **end for**
9: Outputs: final CTR $= G_T/T$

5.2   Experiments on Synthetic Data

Dataset Description. The synthetic data contains 100 users split into ten clusters, and 1,000 articles which are grouped into ten clusters as well. The size of each user cluster is set as 10. For article feature vectors, we allocate different numbers of 0 randomly to the first five dimensions as a seed vector of a group, and the values of other non-zero dimensions are generated from Gaussian distribution. Then, their sixth dimension is set to 1 and the vectors are transformed into unit vectors. In the sequence of recommendation, the size of candidate item set H_{i_t} is 25 ($K = 25$), and the served user i_t is generated uniformly at random over 100 users.

Evaluation Method. The synthetic data includes statistic information of user-item interactions. To mimic the real-world dataset, We allocate z items to each user and take these as items they interact with. In every synthetic data, z is fixed for every user for simplicity. We hold the setting that if a user has interacted with the recommended item, the policy gets 1 payoff. We evaluate algorithm performance by Click-Through Rate (CTR). Note that whether the cold start process is accelerated can be judged by final CTR. If the algorithm learns the user's interest faster, it can achieve more accurate recommendations in the follow-up, thereby increasing the final CTR. The concrete evaluation method is described in Algorithm 5. In our experiments, without loss of generality, hyper parameter γ and β are fixed. We test different α, a common exploration parameter that all baselines share.

Experimental Results. We run each algorithm in seven different parameter settings with $T = 50,000$, where α is set as $\{0.001, 0.005, 0.01, 0.07, 0.1, 0.5, 1\}$. Each setting is tested on five runs and we record the average CTRs, standard deviation, max and min CTR. The best results for each algorithm are shown in Table 1. We can observe that FreqCB ($\alpha = 0.07$) achieves the best CTR over all results. Specifically, the best result of FreqCB improves over SCLUB's best result by 14.4%. The reason is that FreqCB enables active users to filter out the "harmful" information from inactive users, and inactive users to learn more from active users. Therefore, the cluster representation ω_j can be improved.

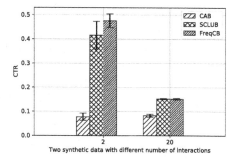

Fig. 1. CTR on different synthetic data. The abscissa represents the numbers of interactions. To mimic real-world data, in our settings, numbers of interactions z denote the numbers of items that each user interacts with.

To study the influence of users' activeness, we test clustering bandit model on different synthetic data and report their best results in Fig. 1. The horizontal axis denotes user activeness which reflects the interaction frequency in the synthetic data. Figure 1 shows that when the user activeness is at a low level, our model outperforms other clustering bandit models, and when the user's activeness becomes higher, the performance of our model tends to be consistent with SCLUB.

Table 1. CTR on synthetic data.

Algorithm	Mean	std	max	min
Random	0.04234	0.000504	0.04282	0.04160
Lin-one ($\alpha = 0.1$)	0.30435	0.012441	0.31776	0.28692
Lin-ind ($\alpha = 0.01$)	0.14651	0.016498	0.16880	0.12972
CAB ($\alpha = 0.001$)	0.07792	0.014791	0.09840	0.06388
SCLUB ($\alpha = 0.01$)	0.41366	0.060439	0.49294	0.35398
FreqCB ($\alpha = 0.07$)	**0.47581**	0.028275	0.50606	0.42456

5.3   Experiments on Real-world Dataset

Dataset Description. The following experiments are conducted on two real-world datasets:

(1) **LastFM** dataset is extracted from the music streaming service Last.fm. This dataset contains a social network with 1,892 users, 12,717 bi-directional user friend relations, and 17,632 artists. There are 11,946 tags used to tag artists by users. We preprocess tags by breaking down them into single words, then remove special symbols and convert duplicate tags into one, e.g., we consider the tag "metal metal metal" as "metal". The dataset contains the tag assignments of artists provided by each particular user, so we collect all tags that an artist has

been tagged to create a TF-IDF context vector $x \in \mathbb{R}^d$, which uniquely represents the current artist, and utilize Principal Component Analysis for dimension reduction. Besides, the dataset contains information about artists that users have listened to, so we utilize this information to create binary payoffs: if a user listened to an artist at least once the payoff is 1, otherwise the payoff is 0. The way of generating recommendation candidates is mentioned in Sect. 4.

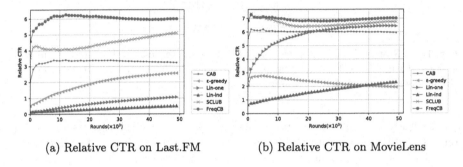

(a) Relative CTR on Last.FM   (b) Relative CTR on MovieLens

Fig. 2. The relative CTR of the proposed model and baselines on the two datasets, Last.FM and MovieLens.

(2) **25 m MovieLens** dataset is an extension of [8], which contains 25 million ratings and more than 1 million tag applications for 62,423 movies by 162,541 users. For simplicity, we choose randomly 1,000 users from the users whose rating numbers ≥ 100 for experiments. To avoid the prior information for bandit models, we use the same method as LastFM dataset does to get movie features. The binary payoff is given by 5-star rating: if the rating is more than 3, the payoff is 1, otherwise the payoff is 0.

Experimental Results. In our experiments, the interact frequency of each user is set to 1 initially. When a user is served, his interact frequency updates according to Algorithm 2. We test two datasets with $T = 50,000$ samples and record the CTRs per thousand. For all algorithms, the exploration parameter α is set as $\{0.1, 0.3, 0.5, 0.7, 1.0\}$, respectively. The evaluation policy is described in Algorithm 5. Each dataset has been run five times and we compute the average relative CTR [20] for comparison. Experimental results are shown in Fig. 2.

For both two real-world datasets, our model FreqCB outperforms SCLUB as well as other strong baselines. The final CTR of FreqCB improves over SCLUB by 17.45% and 3.94% on LastFM and MovieLens, respectively.

For LastFM, in the beginning, Lin-one and Lin-ind perform worse because the users' feedback is not enough to estimate users' parameters accurately. As recommendation proceeds, Lin-one outperforms Lin-ind in the long run due to its better collaborative effect over all users. Compared to MovieLens dataset, the performance of baselines without applying collaborative personalization (e.g., Lin-one and Lin-ind) is relatively poor. This can be attributed to the generation way of LastFM dataset [6]. The interactions between users and music are few and lots of songs to be played are generated by the music streaming service of Last.fm. Hence, the nonzero payoff music may not be the one the user likes actually and the collaborative effect over all users is weak.

For MovieLens dataset, users' interactions are frequent, so the models incorporating collaboration between users (e.g., Lin-one, CAB, SCLUB, FreqCB) achieve better performance than others (e.g., ϵ-greedy, Lin-ind).

Figure 3 provides the visualization of cluster representations derived from SCLUB and FreqCB on LastFM dataset, respectively. The given results are under each model's best parameter setting. SCLUB acquires 1,891 clusters and obtains 0.201 CTR, while FreqCB achieves 1,411 clusters and obtains 0.261 CTR. From the visualization of two models, we find that SCLUB cannot effectively merge some similar clusters, especially the clusters shown at the bottom left of Fig. 3a. Figure 3b exhibits discernible clustering effect, which qualitatively indicate that FreqCB is capable of leveraging asymmetric collaborative signals to have better cluster representations.

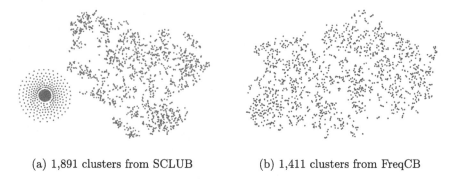

(a) 1,891 clusters from SCLUB   (b) 1,411 clusters from FreqCB

Fig. 3. Visualization of the learned t-SNE transformed cluster representations derived from SCLUB and FreqCB. Each point represents a user cluster from Last.FM dataset.

6 Conclusion

In this paper, we propose a frequency-dependent clustering bandit model for personalized online recommendations. Our model is capable of achieving collaborative filtering effectively by considering user activeness. The experimental results on both synthetic data and real-world datasets indicate that our model has promising results compared with other state-of-the-art models when considering the frequency of user interactions with the system.

There are some remained directions worthwhile to be explored. One future direction is to extend our solution to some other clustering of bandit strategies, where loosening the hard assumption that one user only belongs to one group is a prospective way to improve the quality of recommendation. Another interesting direction is to apply both users and items clustering for better interactive collaboration.

Acknowledgement. This work is partially supported by China Natural Science Foundation under grant (No. 62171391) and the Natural Science Foundation of Fujian Province of China under grant (No. 2020J01053).

References

1. Adomavicius, G., Tuzhilin, A.: Toward the next generation of recommender systems: a survey of the state-of-the-art and possible extensions. IEEE Trans. Knowl. Data Eng. **17**(6), 734–749 (2005). https://doi.org/10.1109/TKDE.2005.99
2. Auer, P., Cesa-Bianchi, N., Freund, Y., Schapire, R.E.: The nonstochastic multi-armed bandit problem. SIAM J. Comput. **32**(1), 48–77 (2002). https://doi.org/10.1137/S0097539701398375
3. Ban, Y., He, J.: Local clustering in contextual multi-armed bandits. In: Proceedings of the Web Conference 2021, pp. 2335–2346, WWW 2021. Association for Computing Machinery, New York, NY, USA (2021). https://doi.org/10.1145/3442381.3450058
4. Cesa-Bianchi, N., Gentile, C., Zappella, G.: A gang of bandits (2013)
5. Gangan, E., Kudus, M., Ilyushin, E.: Survey of multiarmed bandit algorithms applied to recommendation systems. Int. J. Open Inf. Technol. **9**, 12–27 (2021)
6. Gentile, C., Li, S., Kar, P., Karatzoglou, A., Zappella, G., Etrue, E.: On context-dependent clustering of bandits. In: Precup, D., Teh, Y.W. (eds.) Proceedings of the 34th International Conference on Machine Learning. Proceedings of Machine Learning Research, vol. 70, pp. 1253–1262. PMLR, 6–11 August 2017. https://proceedings.mlr.press/v70/gentile17a.html
7. Gentile, C., Li, S., Zappella, G.: Online clustering of bandits. In: Xing, E.P., Jebara, T. (eds.) Proceedings of the 31st International Conference on Machine Learning. Proceedings of Machine Learning Research, vol. 32, pp. 757–765. PMLR, Beijing, China, 22–24 June 2014. https://proceedings.mlr.press/v32/gentile14.html
8. Harper, F.M., Konstan, J.A.: The MovieLens datasets: history and context. ACM Trans. Interact. Intell. Syst. **5**(4), 1–19 (2015). https://doi.org/10.1145/2827872
9. Korda, N., Szorenyi, B., Li, S.: Distributed clustering of linear bandits in peer to peer networks. In: Balcan, M.F., Weinberger, K.Q. (eds.) Proceedings of The 33rd International Conference on Machine Learning. Proceedings of Machine Learning Research, vol. 48, pp. 1301–1309. PMLR, New York, New York, USA, 20–22 June 2016. https://proceedings.mlr.press/v48/korda16.html
10. Koren, Y., Bell, R., Volinsky, C.: Matrix factorization techniques for recommender systems. Computer **42**(8), 30–37 (2009). https://doi.org/10.1109/MC.2009.263
11. Lattimore, T., Szepesvári, C.: Bandit Algorithms. Cambridge University Press, Cambridge (2020). https://doi.org/10.1017/9781108571401
12. Li, L., Chu, W., Langford, J., Schapire, R.E.: A contextual-bandit approach to personalized news article recommendation. In: Proceedings of the 19th International Conference on World Wide Web, WWW 2010, pp. 661–670. Association for Computing Machinery, New York, NY, USA (2010). https://doi.org/10.1145/1772690.1772758, https://doi.org/10.1145/1772690.1772758
13. Li, S., Chen, W., Li, S., Leung, K.S.: Improved algorithm on online clustering of bandits (2019)
14. Li, S., Karatzoglou, A., Gentile, C.: Collaborative filtering bandits. In: Proceedings of the 39th International ACM SIGIR Conference on Research and Development in Information Retrieval, SIGIR 2016, pp. 539–548. Association for Computing Machinery, New York, NY, USA (2016). https://doi.org/10.1145/2911451.2911548

15. Lika, B., Kolomvatsos, K., Hadjiefthymiades, S.: Facing the cold start problem in recommender systems. Expert Syst. Appl. **41**(4, Part 2), 2065–2073 (2014). https://doi.org/10.1016/j.eswa.2013.09.005, https://www.sciencedirect.com/science/article/pii/S0957417413007240

16. MacQueen, J.: Some methods for classification and analysis of multivariate observations. In: Proceedings of the Fifth Berkeley Symposium on Mathematical Statistics and Probability, vol. 1, pp. 281–297, Oakland, CA, USA (1967)

17. Nguyen, T.T., Lauw, H.W.: Dynamic clustering of contextual multi-armed bandits. In: Proceedings of the 23rd ACM International Conference on Conference on Information and Knowledge Management, CIKM 2014, pp. 1959–1962. Association for Computing Machinery, New York, NY, USA (2014). https://doi.org/10.1145/2661829.2662063

18. Pandey, S., Chakrabarti, D., Agarwal, D.: Multi-armed bandit problems with dependent arms. In: Proceedings of the 24th International Conference on Machine Learning, ICML 2007, pp. 721–728. Association for Computing Machinery, New York, NY, USA (2007). https://doi.org/10.1145/1273496.1273587

19. Pazzani, M.J., Billsus, D.: Content-based recommendation systems. In: Brusilovsky, P., Kobsa, A., Nejdl, W. (eds.) The Adaptive Web. LNCS, vol. 4321, pp. 325–341. Springer, Heidelberg (2007). https://doi.org/10.1007/978-3-540-72079-9_10

20. Wang, Q., et al.: Online interactive collaborative filtering using multi-armed bandit with dependent arms. IEEE Trans. Knowl. Data Eng. **31**(8), 1569–1580 (2019). https://doi.org/10.1109/TKDE.2018.2866041

21. Xu, X., Dong, F., Li, Y., He, S., Li, X.: Contextual-bandit based personalized recommendation with time-varying user interests. Proc. AAAI Conf. Artif. Intell. **34**(04), 6518–6525 (2020). https://doi.org/10.1609/aaai.v34i04.6125, https://ojs.aaai.org/index.php/AAAI/article/view/6125

22. Yang, L., Liu, B., Lin, L., Xia, F., Chen, K., Yang, Q.: Exploring clustering of bandits for online recommendation system. In: Fourteenth ACM Conference on Recommender Systems, RecSys 2020, pp. 120–129. Association for Computing Machinery, New York, NY, USA (2020). https://doi.org/10.1145/3383313.3412250

Graph Contrastive Learning with Positional Representation for Recommendation

Zixuan Yi[(✉)], Iadh Ounis, and Craig Macdonald

University of Glasgow, Glasgow, UK
z.yi.1@research.gla.ac.uk,
{iadh.ounis,craig.macdonald}@glasgow.gla.ac.uk

Abstract. Recently, graph neural networks have become the state-of-the-art in collaborative filtering, since the interactions between users and items essentially have a graph structure. However, a major issue with the user-item interaction graph in recommendation is the absence of the positional information of users/items, which limits the expressive power of graph recommenders in distinguishing the users/items with the same neighbours after propagating several graph convolution layers. Such a phenomenon further induces the well-known over-smoothing problem. We hypothesise that we can obtain a more expressive graph recommender through graph positional encoding (e.g., Laplacian eigenvector) thereby also alleviating the over-smoothing problem. Hence, we propose a novel model named Positional Graph Contrastive Learning (PGCL) for top-K recommendation, which aims to explicitly enhance graph representation learning with graph positional encoding in a contrastive learning manner. We show that concatenating the learned graph positional encoding and the pre-existing users/items' features in each feature propagation layer can achieve significant effectiveness gains. To further have sufficient representation learning from the graph positional encoding, we use contrastive learning to jointly learn the correlation between the pre-exiting users/items' features and the positional information. Our extensive experiments conducted on three benchmark datasets demonstrate the superiority of our proposed PGCL model over existing state-of-the-art graph-based recommendation approaches in terms of both effectiveness and alleviating the over-smoothing problem.

1 Introduction

Personalised recommendation is a widely used technology to improve the quality of information services, which aims to predict a group of items that users might intend to purchase according to their preferences. The effective personalisation of the recommendation results, typically rely on rich available data, in particular the historical user-item interactions [11]. Recent advances in Graph Neural Networks (GNNs) provided a strong and fundamental opportunity to develop effective personalised recommendations [29]. Specifically, GNNs adopt embedding propagation to aggregate neighbourhood embeddings iteratively through

J. Kamps et al. (Eds.): ECIR 2023, LNCS 13981, pp. 288–303, 2023.
https://doi.org/10.1007/978-3-031-28238-6_19

connectivities on a bipartite user-item graph. By stacking the multiple propagation layers, each node on the graph can access high-order neighbours' information through the message passing scheme [31], rather than only modelling the direct interactions between users and items. With their advantages in handling structural data and exploring structural information on a graph, graph recommenders have attained a state-of-the-art recommendation performance [6].

Despite the success of the graph recommenders, the current graph feature propagation function only repeatedly aggregates neighbourhood embeddings that are adjacent to the target node. As a consequence, the conventional message passing scheme of the graph recommenders typically fails to differentiate two users with the same interacted items and all user representations converge to a constant after propagating several graph convolution layers [2]. This problem can be further amplified after stacking multiple layers [10], leading to the well-known over-smoothing problem [17]. Indeed, this limitation is now well understood in the context of the equivalence of GNNs with the Weisfeiler-Leman (WL) test [30] for graph isomorphism [22,34], which further confirms the limited expressive power of the current graph recommenders. Consequently, there is a stronger motivation for proposing a new graph recommender that is more expressive in distinguishing the users/items with the same neighbours after graph convolution and hence to further amplify the difference between the users/items further apart. Indeed, many approaches have been proposed to alleviate the limited expressive power of the GNNs, to some extent, by considering the positional encoding (PE) information of nodes for enriching the nodes' features [4,14,25]. Graph positional encoding approaches [3,4,37] typically consider a global positioning or a unique representation of the users/items in the graph, which can encode a graph-based distance between the users/items. To leverage the advantage of positional encoding, in this paper, we also use a graph-specific learned positional encoding as a unique ID for each user/item and inject these positional encodings into each feature propagation layer to improve the expressive power of graph recommenders.

Inspired by recent studies [15,32,33,39], which have shown the superior ability of Contrastive Learning (CL) to construct supervised signals from correlations within raw data, we also investigate in this paper the possibility of leveraging CL to explore the correlations among learned graph positional encodings and address the limited expressive power problem in graph recommenders. A typical approach [32,40] to apply CL to recommendations on graphs is to first augment the user-item bipartite graph with noise or structure perturbations, and then to maximise the agreement of the augmented user/item embeddings via a graph encoder. To address the limited expressive power of graph recommenders, we propose a novel recommendation model named Positional Graph Contrastive Learning (PGCL) for top-K recommendation, which aims to use existing graph positional encoding methods to improve the expressive power of graph recommenders and further enhances the integrated user/item representations through a noise-based augmentation method. To be more specific, PGCL provides additional positional information to existing graph recommenders by injecting the learned graph positional encoding into each feature propagation layer. Furthermore, in order to prevent distorting the users' intents, we apply a noise-based augmentation technique to such position-enhanced user/item embeddings – this aims to maintain the users' intent unchanged while adding distance properties to the learned user/item representations. To summarise, in this paper, we argue that graph recommenders enriched with our proposed graph positional encoding can effectively improve their expressive power while alleviating the over-smoothing problem. Indeed, we show that with the integration of contrastive learning, our PGCL model enforces the divergence of the learned user/item representations resulting in an improved recommendation performance.

Our contributions in this paper are as follows: (1) We propose a personalised graph-based recommendation model for top-K recommendation, which leverages the learned graph positional encoding to facilitate a new message passing scheme for existing graph recommenders; (2) We apply noise-based augmentation on position-enhanced user/item embeddings and examine the impact of the resulting PGCL model using different ranking metrics; (3) We conduct extensive experiments on three benchmark datasets and demonstrate the effectiveness of PGCL in comparison to the existing state-of-the-art graph recommenders; (4) By comparing with the existing baselines, we show that PGCL is more expressive by stacking multiple layers and can alleviate the over-smoothing problem by reducing the over-smoothness of user/item embeddings.

2 Related Work

In this section, we discuss the related methods and techniques to our conducted study, namely graph-based recommendation, graph positional encoding and graph contrastive learning for recommendation.

Graph-Based Recommendation: Graph-based recommenders [10,20,29] typically exploit the message passing scheme in the user-item graph by propagating information from local neighbours and integrating the collaborative signals into a user/item representation. However, the existing approaches (e.g., Light-GCN [10]) follow the original message passing scheme, which is known to suffer from over-smoothing due to its repeated aggregation of local information. As a result, the existing graph recommenders only propagate homogeneous features (e.g., IDs) from the original neighbours, which are not expressive enough to distinguish the users/items with the same neighbours after stacking several graph convolution layers. Unlike prior works, we leverage the positional representation of each user/item that relies on positional features (e.g., Laplacian eigenvectors) and inject the learned positional encodings into each feature propagation layer of the existing graph recommenders so as to enhance their expressive power.

Graph Positional Encoding: The notion of positional encodings (PEs) in graphs is not a trivial concept, as there exists no canonical way of ordering nodes [14]. Various studies [3,5,8,14,18,28,36] have exploited positional encodings on graphs to improve the expressiveness of GNNs. Many earlier studies [21,23] used index positional encoding to enhance conventional GNNs in terms of their associated model expressiveness. For example, GRP [23] devised a positional encoding by assigning to each node an identifier that depends on the index ordering. This approach can be computationally expensive as it needs to account for all $n!$ node permutations to guarantee a higher expressiveness. Therefore, some prior studies (e.g., [16,37]) have applied a more efficient distance positional encoding to enhance the model expressiveness of GNNs. For example, P-GNN [37] enhanced the model expressiveness by projecting the distances between a target node and randomly sampled nodes into a position-aware embedding. However, a large number of sampled nodes will include most of the nodes on the graph, thus leading to insufficient positional embeddings. DEGNN [16] modeled a distance positional encoding by capturing distances between nodes using landing probabilities of random walks. However, this approach cannot scale to large-scale graphs because of the cost of computing the power matrices. Alternatively, Laplacian eigenvectors [1] have been shown to be good candidates for graph positional encoding, since Laplacian eigenvectors form a meaningful local coordinate system while preserving the global graph structure. In particular, we can pre-compute the Laplacian eigenvectors/eigenvalues and provide a unique ID for each node, which solves the scalability issue on the user-item graph in a recommender system and further enhances the pre-existing node features by merging Laplacian eigenvectors/eigenvalues. Another alternative approach used by APPNP [13] provided an improved graph feature propagation scheme with Personalised PageRank [9], which particularly addresses the over-smoothing problem in a random walk manner. In this paper, we leverage the Laplacian eigenvectors and the random walk operator to define a new relative positional encoding in recommender systems. Unlike the above prior works, we allocate the learned positional encodings to a separate message passing function to generate the user/item positional embedding from the neighbours' positional information.

Graph Contrastive Learning for Recommendation: Recently, graph-based recommendation approaches [19,32,35,38] have benefited from contrastive learning, because its ability to extract contrastive signals from the raw data is well-aligned with the recommender systems' needs for more collaborative filtering signals. SGL [32] adopted different augmentation operators such as edge dropout and node dropout, which aim to capture the essential information of the original user-item bipartite. The authors of SimGCL [40] claimed that graph augmentations highly distort the original graph and applied a more effective noise-based augmentation on a user/item representation level. As discussed above, we aim to inject the learned positional encodings into the node features to enhance the final user/item embeddings. Hence, applying a representation-level augmentation is more reasonable than perturbing the graph structure. To the best of our knowledge, our proposed PGCL model is the first graph-based recommendation model to enhance user/item representations by contrasting augmented user/item embeddings with a learned graph positional encoding.

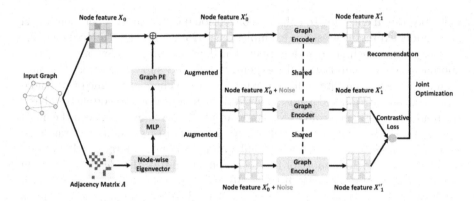

Fig. 1. The architecture of our PGCL model

3   Model Architecture

In this section, we first present the personalised recommendation task in Sect. 3.1. We describe our proposed PGCL model, the architecture of which is illustrated in Fig. 1. In Sect. 3.2, we define the graph positional encoding. Next, we illustrate the motivation of decoupling the positional encoding from the conventional message passing function in Sect. 3.3. In Sect. 3.4, we apply contrastive learning for effectively learning the positional encoding and optimise our PGCL model jointly with a pairwise ranking loss.

3.1   Preliminaries

In this paper, we focus on addressing the ranking-based recommendation task. Conceptually, we consider a recommender system with a user set $\mathcal{U}$ and an item set $\mathcal{I}$. In order to facilitate the description of graph recommenders, we use $\mathcal{G} = (\mathcal{V}, \mathcal{E})$ to denote an interaction graph, where the node set $\mathcal{V} = \mathcal{U} \cup \mathcal{I}$ includes all users and items. $\mathcal{E}$ is the set of edges. e_u denotes the user feature for user u and e_i denotes the item feature for item i. In addition, p_u and p_i denote the positional feature of the user and item, respectively. The layers are indexed by ℓ, where $\ell = 0$ denotes the input layer. For a given user u or item i, there is a positional feature p_u or p_i on an interaction graph $\mathcal{G}$. We aim to estimate the users' preferences through a graph encoder f, which can recommend the top-K items for a target user u.

3.2   Definition of Initial Graph Positional Encoding

As mentioned Sect. 2, we aim to use the Laplacian eigenvectors and the random walk operator to define the positional encoding (PE) in recommendation. In this section, we define the initial positional encoding of a given user u or item i.

Laplacian PE: Laplacian PE (LapPE) is a spectral technique that embeds graphs into an Euclidean space, and is defined via the factorisation of the graph's Laplacian $\Delta = \mathrm{I} - D^{-1/2}AD^{-1/2} = U^T \Lambda U$, where I is the identity matrix, A is the adjacency matrix, D is the degree matrix, and matrices Λ and U correspond to the Laplacian eigenvalues and Laplacian eigenvectors of a graph, respectively. In this work, we consider the Laplacian eigenvector as the initial graph positional encoding, which is defined as follows:

$$p_i^{\text{LapPE}} = [U_{i1}, U_{i2}, \cdots, U_{ik}] \in \mathbb{R}^k \tag{1}$$

As a consequence, LapPE is expected to provide a unique ID for each user/item representation and is distance-sensitive w.r.t. the Euclidean norm.

Random Walk PE: Apart from LapPE, we also investigate the random walk-based method [5] to generate the graph positional encoding. Hence, we use Random Walk PE (RWPE), which is a method based on the random walk diffusion process. Formally, RWPE is defined with k-steps of random walks as follows:

$$p_i^{\text{RWPE}} = \left[\mathrm{RW}_i, \mathrm{RW}_i^2, \cdots, \mathrm{RW}_i^k \right] \in \mathbb{R}^k \tag{2}$$

where $\mathrm{RW} = AD^{-1}$ is the random walk operator. As such, RWPE provides a unique node representation under the condition that each user/item has a unique k-hop topological neighbourhood [16] for a sufficiently large k.

Finally, the initial graph PE of the network is obtained by projecting LapPE or RWPE into a d-dimensional feature vector with a Multi-Layer Perceptron (MLP) network:

$$p_i^{\ell=0} = \mathrm{MLP}\left(p_i^{\text{PE}}\right) = W^0 p_i^{\text{PE}} + b^0 \in \mathbb{R}^d, \tag{3}$$

where $W^0 \in \mathbb{R}^{d \times k}$ and $b^0 \in \mathbb{R}^d$ are the learned parameters of the MLP network. As illustrated in Fig. 1, we leverage LapPE or RWPE to generate the initial positional encoding through a MLP network.

3.3   Feature Propagation with Learned Positional Encoding

Figure 1 illustrates how PGCL concatenates the graph PE and the pre-existing node feature X' which is generated by user/item IDs. As discussed in Sect. 2, we aim to decouple the graph PE from the conventional message passing function. Hence, we propose a message passing function for the graph PEs. The layer update equations is defined as follows:

$$p_u^{\ell+1} = \mathrm{Tanh}\left(\mathrm{AGG}\left(p_u^\ell, \{p_i^\ell\}_{i \in \mathcal{N}_u}\right)\right), \tag{4}$$

where Tanh is the activation function, and AGG is an aggregation function that combines the positional information of the adjacent item nodes. Since the graph positional encoding is interpreted as a unique positional ID of a given node (see Sect. 2), we expect to use the message passing scheme to exploit high-order positional information of the positional features through the aggregation operation. Next, we aim to integrate the graph PE p_i into user representations. Analogously, we can obtain the updated positional embeddings of the items.

As illustrated in Fig. 1, we concatenate the graph PE – which is generated by Eq. (4) – with the existing node feature X, similar to the Transformers [27] network structure:

$$e_u^{\ell+1} = \text{AGG}\left(\begin{bmatrix} e_u^\ell \\ p_u^\ell \end{bmatrix}, \left\{ \begin{bmatrix} e_i^\ell \\ p_i^\ell \end{bmatrix} \right\}_{i \in \mathcal{N}_u}\right) \tag{5}$$

where e_u and e_i denote the representations of user u and item i, respectively, $\mathcal{N}_u$ is the neighbourhood of the user u, and p_u & p_i denote the position representations of user u and item i, respectively. We use LightGCN [10] to aggregate the concatenated result of the pre-exiting item feature e_i and with the item positional feature p_i. Hence, the feature propagation equation is defined as follows:

$$\mathbf{e}_u^{\ell+1} = \sum_{i \in \mathcal{N}_u} \frac{1}{\sqrt{|\mathcal{N}_u|}\sqrt{|\mathcal{N}_i|}} \begin{bmatrix} e_i^\ell \\ p_i^\ell \end{bmatrix}, \tag{6}$$

$$\text{with } p_i^\ell = \text{Tanh}\left(W_1 p_i^{\ell-1} + \sum_{i \in \mathcal{N}_i} \frac{1}{\sqrt{|\mathcal{N}_u||\mathcal{N}_i|}} \left(W_1 p_u^{\ell-1} + W_2 \left(p_u^{\ell-1} \odot p_i^{\ell-1}\right)\right)\right) \tag{7}$$

where W_1 and W_2 are trainable weight matrices. The main difference of our feature propagation layer with the standard graph recommenders is a separated message passing function for the graph PE, which injects the graph PE into each propagation layer. For this reason, we expect PGCL to provide less over-smoothed user/item embeddings, thereby alleviating the over-smoothing problem of the existing graph recommenders.

3.4   Self-augmented Learning

As discussed in Sect. 2, since a graph perturbation has the possibility to distort the user-item bipartite graph, applying a representation-level augmentation on the learned graph PE is more rational than perturbing the graph structure. Following Yu et al. [40], we apply a noise-based augmentation on the representation level for both the integrated user and item embeddings. For example, given a user embedding e_u, which integrates its graph positional feature p_u, we can generate an augmented user representation by adding a noise vector Δ_u as follows:

$$\mathbf{e}_u' = \mathbf{e}_u + \Delta_u', \mathbf{e}_u'' = \mathbf{e}_u + \Delta_u'', \, e_u \in \mathbb{R}^d \tag{8}$$

$$\text{with } \Delta_u = e_x \odot \text{sign}\,(\mathbf{e}_u) \odot \epsilon, \, e_x \in \mathbb{R}^d \sim U(0,1) \tag{9}$$

where $\mathbf{e}_u'$ and $\mathbf{e}_u''$ are two augmented user representations, e_x is a vector that is generated by random numbers from a uniform distribution, and ϵ is a hyper-parameter to control the strength of the user representation perturbation with a range in $[0,1]$. Goodfellow et al. [7] have also shown that a linear pertur-bation in high-dimensional spaces can generate sufficient samples. As such, in addition to applying the noise-based augmentation, we also aim to enforce that the integrated user/item representations further spread out in the entire embed-ding space so as to fully exploit the expressive power of the embedding space. In the graph contrastive recommendation scenario [32,40], the target is to gen-erate a better user/item representation via data augmentations. Hence we use InfoNCE [24], to maximise the agreement of two augmented representations:

$$\mathcal{L}_{cl}^{user} = -\log \frac{\exp\left(\mathbf{e}_u'^{\top}\mathbf{e}_u''/\tau\right)}{\sum_{i=1}^{n}\exp\left(\mathbf{e}_u'^{\top}\mathbf{e}_n/\tau\right)} \tag{10}$$

where $\mathbf{e}_n$ is the embedding of a different user, and τ is a hyper-parameter that adjusts the dynamic range of the resulting loss value. Analogously, we can calcu-late the contrastive loss of a target item $\mathcal{L}_{cl}^{item}$. Therefore, we obtain a combined contrastive loss that acts as an auxiliary loss for top-K recommendation tasks as follows: $\mathcal{L}_{cl} = \mathcal{L}_{cl}^{user} + \mathcal{L}_{cl}^{item}$. To better mine the user/item representations in recommendation, we adopt a multi-task training strategy to jointly optimise the widely used pair-wise ranking objective, namely Bayesian Personalised Ranking (BPR) [26], and the contrastive learning objective $\mathcal{L}_{cl}$:

$$\mathcal{L} = \lambda_1\mathcal{L}_{cl} + \sum_{(u,i,j)\in D_s} \ln \sigma(y_{ui} - \mathbf{e}_u^{\top}\mathbf{e}_i) + \lambda_2 \|\Theta\|_2^2 \tag{11}$$

where the second term is the BPR loss, $\mathbf{e}_u$ is the user embedding, $\mathbf{e}_i$ denotes the positive item embedding and y_{ui} is the ground truth value, which indi-cates whether the paired user and item have interacted, $D_s = \{(u,i,j)|(u,i) \in R^+, (u,j) \in R^-\}$ is the set of the training data, R^+ indicates the observed interactions and R^- indicates the unobserved interactions; $\sigma(\cdot)$ is the sigmoid function, Θ is the set of model parameters in the BPR loss, while λ_1 and λ_2 are hyper-parameters to control the strengths of the contrastive learning and L_2 regularisation, respectively. Through propagating the integrated user/item rep-resentations in multiple feature propagation layers with Eq. (6), we obtain mul-tiple user/item embeddings from each layer, then we concatenate each user/item embedding e_u^ℓ, so that the final embedding collectively contains information from each layer. Hence, we can estimate the relevant score between a user and item by minimising the multi-task learning loss in Eq. (11).

4   Experiments

We now examine the performance of PGCL through experiments on three real-world datasets, in comparison to four existing state-of-the-art graph recommen-dation models. To demonstrate the effectiveness of PGCL, we conduct experi-ments to answer the following three research questions:

RQ1: How does the PGCL model perform in top-K recommendation compared with existing baselines?

Table 1. Statistics of the used datasets.

Dataset	Users	Items	Interactions	Density
Gowalla	39,657	31,211	1,072,325	0.087%
Yelp2018	28,361	43,142	1,481,472	0.121%
Amazon-Kindle	116,417	72,439	1,643,646	0.019%

RQ2: How do different positional encodings and augmentation methods impact the recommendation performance?
RQ3: Is our PGCL model more expressive than LightGCN thereby alleviating the over-smoothing problem compared to the baselines based on LightGCN?

4.1 Datasets and Evaluation Protocol

We evaluate our PGCL model using three real-world datasets, namely *Yelp2018*[1], *Gowalla*[2] and *Amazon-Kindle*[3]. Table 1 shows the statistics of these datasets. Following He et al. [10] and Wang et al. [29], we randomly split the above datasets into training, validation, and testing sets with a 7:1:2 ratio. We use two commonly used evaluation metrics: Recall@K and NDCG@K to evaluate the performance of top-K recommendation. We follow [40] in setting K = 20 and report the average performance achieved for all users in the testing set. We use the Adam [12] optimiser in both our PGCL model and the four baseline models. We apply early-stopping during training, terminating the training when the validation loss does not decrease for 50 epochs. To determine the hyper-parameters in both PGCL and the baseline models, we apply a grid search on the validation set. Specifically, we tune our PGCL model by varying the learning rate in $\{10^{-2}, 10^{-3}, 10^{-4}\}$. The learning rates of the baseline models are also tuned according to the suggested ranges in [10], for a fair comparison. Similarly, we also tune each of λ_1, λ_2 and ϵ within the range of $\{0, 0.1, 0.2, ..., 1.0\}$. A detailed analysis of the models' performance with different layer settings is shown in Sect. 4.5.

4.2 Baselines

We compare the effectiveness of PGCL[4] with four existing strong baselines. In the following, we briefly describe these baselines: **(1) NGCF** [29] is a classical GNN-based model that first captures the high-order connectivity information in the embedding function by stacking multiple embedding propagation layers. **(2) LightGCN** [10] is another GNN-based model that has evolved from NGCF. It simplifies the design in the feature propagation by removing the non-linear activation and the transformation matrices. This approach has been widely used as a

---

[1] https://www.yelp.com/dataset.
[2] https://snap.stanford.edu/data/loc-gowalla.html.
[3] https://jmcauley.ucsd.edu/data/amazon/.
[4] Source code is available at: https://github.com/zxy-ml84/PGCL.

strong graph recommender for top-K recommendation [6]. **(3) SGL** [32] leverages contrastive learning for GNN-based models. With LightGCN as the encoder of the users/items, SGL adopts different augmentation operators such as edge dropout and node dropout, on the pre-existing features of the users/items. This approach can implicitly identify the important nodes from different augmentations [41]. **(4) SimGCL** [40] is effective in improving LightGCN with different augmentations, which is similar to SGL. It removes the dropout-based augmentations from SGL and devises a noise-based augmentation on the user/item representation level with an increased recommendation performance. In addition, to examine the effectiveness of the Laplacian positional encoding (see Eq. (1)), we compare PGCL to a variant called **PGCL**$_{w/o\,CL}$. Different from PGCL, PGCL$_{w/o\,CL}$ only concatenates the positional encoding (from Eq. (7)) with the pre-existing users/items' features from LightGCN, without applying contrastive learning (Eq. (10)).

4.3   Performance Comparison with Baselines (RQ1)

Table 2 compares our proposed PGCL model with four used baselines. We particularly compare PGCL to the strongest baseline, whose performance is highlighted with an underline in the table. From the table, we observe that for all three datasets, PGCL outperforms all the baseline models on all metrics, and statistically significantly in most cases according to the paired t-test with Holm-Bonferroni correction. This result demonstrates the rationality and effectiveness of injecting graph PE to the graph feature propagation layer and incorporating the augmented positional and pre-existing node features (i.e. IDs). For a given GNN-based method (NGCF, LightGCN, PGCL$_{w/o\,CL}$, PGCL), we evaluate the usefulness of leveraging the graph PE in enriching the user/item representations. Comparing NGCF, LightGCN and PGCL$_{w/o\,CL}$, we observe that PGCL$_{w/o\,CL}$ performs generally better than both NGCF and LightGCN on all three used datasets. This result demonstrates the benefit of injecting the learned positional encoding into the pre-exiting users/items' features to estimate the users' preferences. On the other hand, as can be observed in Table 2, PGCL$_{w/o\,CL}$ performs worse than PGCL on all three used datasets. This result illustrates the importance of contrastive learning in providing additional supervised signals during training. For the contrastive learning method, we also evaluate the usefulness of different augmentations by comparing our PGCL model with SGL and SimGCL. Table 2 shows that the noise-based methods (SimGCL, PGCL) markedly outperform the dropout-based method (SGL) on both the Yelp2018 and Amazon-Kindle datasets. This result shows the marginal effect of graph perturbation and the effectiveness of using noise-based augmentation. Moreover, Table 2 also shows that PGCL outperforms SimGCL by a large margin on all metrics (significantly on Gowalla), which demonstrates that graph PE can enrich the user/item representations as an additional feature. Hence, in answer to RQ1, we conclude that our proposed PGCL model can effectively leverage both the graph positional feature and the augmented user/item representations, thereby enhancing the existing graph recommender models with significant performance improvements.

Table 2. Experimental results for PGCL in comparison to other baselines. The best performance is highlighted in bold and the second best result is highlighted with underline. * denotes a significant difference compared to the result of PGCL using the paired t-test with the Holm-Bonferroni correction for $p < 0.01$.

Dataset	Yelp2018		Gowalla		Amazon-Kindle	
Methods	Recall@20	NDCG@20	Recall@20	NDCG@20	Recall@20	NDCG@20
NGCF	0.0502*	0.0417*	0.0889*	0.0592*	0.1893*	0.1285*
LightGCN	0.0542*	0.0437*	0.0996*	0.0635*	0.2230*	0.1644*
SGL	0.0563*	0.0449*	<u>0.1071</u>	<u>0.0671</u>	0.2331*	0.1726*
SimGCL	<u>0.0577</u>	<u>0.0466</u>	0.1068*	0.0664*	<u>0.2425</u>	<u>0.1801</u>
PGCL$_{w/oCL}$	0.0581*	0.0477*	0.1059*	0.0675*	0.2420*	0.1803*
PGCL	**0.0608**	**0.0501**	**0.1122**	**0.0699**	**0.2572**	**0.1934**
%Improv.	5.37%	7.51%	4.76%	4.17%	6.06%	7.38%

4.4 Ablation Study (RQ2)

To investigate the impact of each component of our PGCL model and different graph positional encodings (PE), Table 3 shows how the performance of PGCL changes when we start with LightGCN as the basic graph encoder and apply contrastive positional encoding on top of it so as to conclude on the effectiveness of graph PE and contrastive learning. Table 3 shows that the PGCL$_{LapPE}$ variant, which uses the Laplacian eigenvalue and a representation level augmentation achieves the best performance on all datasets. This promising result is due to the addition of the unique learned positional features of the users/items and an effective representation learning on the users/items' embeddings. Specifically, we observe from Table 3 that both LightGCN$_{RWPE}$ and LightGCN$_{LapPE}$ achieve an effectiveness gain compared with LightGCN. This result demonstrates the effectiveness of graph PE. One possible reason is that the graph PE denotes a unique positional information to the user/item embedding in each feature propagation layer. For the PGCL$_{RWPE}$ and the LightGCN$_{RWPE}$ variants, which use a random walk operator in Table 3, there is a performance reduction compared with PGCL$_{LapPE}$ and LightGCN$_{LapPE}$, which indicates that a global ID (LapPE) is more beneficial than a local ID (RWPE) for the user-item interaction data in recommender systems. Moreover, we also observe that there is an effectiveness improvement from LightGCN$_{LapPE}$ to PGCL$_{LapPE}$. This suggests that both the PE and the pre-existing users/items' features provide an additional supervised signal through contrastive learning. Hence, in answer to RQ2, we conclude that PGCL successfully leverages graph positional encodings to learn effective user/item representations in a contrastive learning scheme.

4.5 The Over-Smoothing Problem (RQ3)

After showing that PGCL is effective in improving LightGCN, we now study the characteristics of graph PE in terms of their usefulness against over-smoothing. In this section, we investigate the over-smoothing problem by comparing PGCL

Table 3. PGCL performance in terms of Recall@20 and NDCG@20 on the used datasets. * denotes a significant difference compared to the result of PGCL using the paired t-test with the Holm-Bonferroni correction for $p < 0.01$.

Dataset	Yelp2018		Gowalla		Amazon-Kindle	
Methods	Recall@20	NDCG@20	Recall@20	NDCG@20	Recall@20	NDCG@20
LightGCN	0.0542*	0.0437*	0.0996*	0.0635*	0.2230*	0.1644*
LightGCN$_{RWPE}$	0.0556*	0.0458*	0.1014*	0.0631*	0.2303*	0.1713*
LightGCN$_{LapPE}$	0.0581*	0.0477*	0.1059*	0.0675*	0.2420*	0.1803*
PGCL$_{RWPE}$	0.0583*	0.0486*	0.1068*	0.0674*	0.2451*	0.1815*
PGCL$_{LapPE}$	**0.0608**	**0.0501**	**0.1122**	**0.0699**	**0.2572**	**0.1934**

Table 4. Performance comparison between PGCL and LightGCN at different layers. The peak performance for each method is highlighted in bold.

Dataset		Yelp2018		Gowalla		Amazon-Kindle	
Layers	Methods	Recall@20	NDCG@20	Recall@20	NDCG@20	Recall@20	NDCG@20
1 Layer	LightGCN	0.0531	0.0433	0.0982	0.0622	0.2214	0.1635
	PGCL	0.0570 (+7.3%)	0.0477 (+10.2%)	0.109 (+11.0%)	0.0677 (+8.8%)	0.2553 (+15.3%)	0.1917 (+17.2%)
2 Layers	LightGCN	0.0519	0.0421	0.0993	0.0630	0.2225	0.1641
	PGCL	0.0582 (+12.1%)	0.0493 (+17.1%)	0.1106 (+11.4%)	0.0686 (+8.9%)	0.2561 (+15.1%)	0.1921 (+17.1%)
3 Layers	LightGCN	0.0536	0.0435	**0.0996**	**0.0635**	**0.2230**	**0.1644**
	PGCL	0.0580 (+8.2%)	0.0488 (+12.2%)	0.1112 (+11.6%)	0.0692 (+9.0%)	0.2565 (+15.0%)	0.1927 (+17.2%)
4 Layers	LightGCN	**0.0542**	**0.0437**	0.0991	0.0632	0.2224	0.1640
	PGCL	0.0595 (+9.8%)	0.0496 (+13.5%)	0.1115 (+12.5%)	0.0697 (+10.3%)	**0.2572 (+15.6%)**	**0.1934 (+17.9%)**
5 Layers	LightGCN	0.0538	0.0427	0.0987	0.0630	0.2217	0.1637
	PGCL	**0.0608 (+13.0%)**	**0.0501 (+14.6%)**	**0.1122 (+13.7%)**	**0.0699 (+11.0%)**	0.2562 (+15.6%)	0.1925 (+17.6%)

and LightGCN with different layer settings in Table 4. As shown in Table 4, both PGCL and LightGCN reach their best effectiveness within 5 graph layers. In addition, all PGCL variants are effective in improving LightGCN under different layer settings on all used datasets. The largest improvements are observed on the Amazon-Kindle dataset where PGCL can remarkably improve LightGCN by 15.6% on Recall and 17.9% on NDCG with a 4-layer setting. Specifically, PGCL continues to reach a higher recommendation performance on the Gowalla and Amazon-Kindle datasets with more layers while LightGCN already reaches its peak performance at 3-layer. This result indicates that injecting the learned graph positional encoding (PE) can benefit the general message passing scheme by encoding the graph PE as additional features and can improve the expressive power of LightGCN with an increased models' depth. On the other hand, as can be seen in Table 4, PGCL does not reach its peak performance at the highest layer on the Amazon-Kindle dataset. This observation indicates that our PGCL model tends to suffer from over-smoothing when using a higher number of layers. We leave the investigation of the over-smoothing problem as a future work direction.

To further examine the effectiveness of the graph PE, we conduct a further analysis on the over-smoothness values for both the 2-layer PGCL and all 2-layer baselines. We use the over-smoothness of second-order embedding to evaluate the PGCL's capability of alleviating the over-smoothing problem.

Following He et al. [10], we calculate the users' over-smoothness that have an overlap on the interacted items. In particular, as in [10], we use a smoothness metric to evaluate the over-smoothness of the users/items. A higher value indicates less over-smoothing. Similarly we can also obtain the over-smoothness for the item embeddings. Table 5 shows the over-smoothness values of PGCL and the various used baseline models. The results show that our PGCL model obtains the largest O-Smoothness$_u$ and O-Smoothness$_i$ values, which indicate that more effective user/item embeddings are generated with the learned graph PE. Comparing LightGCN and PGCL$_{w/o CL}$, we note that the graph positional encoding exhibits a large gain on O-Smoothness$_u$ and O-Smoothness$_i$ while improving the recommendation performance at the same time. We also compare the impact of using the noise-based augmentation in addressing the over-smoothness problem. According to the results in Table 5, PGCL outperforms PGCL$_{w/o CL}$ both in over-smoothness and recommendation performance by a large margin, which demonstrates the effectiveness of mining augmented user/item embeddings through contrastive learning. Hence, in answer to RQ3, we conclude that PGCL successfully alleviates the over-smoothing problem by injecting the learned graph positional encoding to each feature propagation layer. This further shows that a graph positional encoding learned with a separate message passing function can lead to a more expressive graph recommender.

Table 5. Over-smoothness comparison of the 2-layer user/item embeddings between PGCL and the baselines. O-Smoothness$_u$ and O-Smoothness$_i$ represent the over-smoothness of users/items, respectively. A higher over-smoothness value indicates less over-smoothing (i.e. a higher value is better).

Dataset	Yelp2018			Gowalla		
Methods	O-Smoothness$_u$↑	O-Smoothness$_i$↑	Recall@20↑	O-Smoothness$_u$↑	O-Smoothness$_i$↑	Recall@20↑
LightGCN	10747.4	8318.5	0.0542	14634.6	6314.2	0.0996
SimGCL	12187.5	9932.3	0.0577	15043.1	6939.1	0.1068
PGCL$_{w/o CL}$	13317.8	10177.4	0.0581	15257.4	7192.5	0.1063
PGCL	13978.4	11748.1	**0.0608**	16462.3	7936.8	**0.1122**

5   Conclusions

In this work, we proposed the PGCL model to tackle the over-smoothing problem of graph recommenders by leveraging graph positional encoding. Specifically, we used Laplacian eigenvector as graph positional encoding to endow the user/item embedding in each feature propagation layer. In particular, we updated the learned graph positional encoding with a separated message passing function and merged it with the pre-existing users/items' features. We further encoded users/items' preferences by contrasting the augmented user/item representations with the learned graph positional encodings. Our results on three benchmark datasets showed that PGCL effectively leverages graph positional encoding along with the commonly-used graph recommenders and provides a significant improvement in comparison with the existing baselines. Moreover, we conducted an ablation study to investigate the effect of using different positional encodings for our PGCL model and concluded that the Laplacian eigenvector is more beneficial for the user-item interaction data. Furthermore, we showed that PGCL is more expressive because it can stack more layers with an improved recommendation performance while reducing the over-smoothness of user/item embeddings compared to the baselines.

References

1. Belkin, M., Niyogi, P.: Laplacian eigenmaps for dimensionality reduction and data representation. Neural Comput. **15**(6), 1373–1396 (2003)
2. Chen, M., Wei, Z., Huang, Z., Ding, B., Li, Y.: Simple and deep graph convolutional networks. In: Proceedings of the 37th International Conference on Machine Learning (2020)
3. Dwivedi, V.P., Bresson, X.: A generalization of transformer networks to graphs. arXiv preprint arXiv:2012.09699 (2020)
4. Dwivedi, V.P., Joshi, C.K., Laurent, T., Bengio, Y., Bresson, X.: Benchmarking graph neural networks. arXiv preprint arXiv:2003.00982 (2020)
5. Dwivedi, V.P., Luu, A.T., Laurent, T., Bengio, Y., Bresson, X.: Graph neural networks with learnable structural and positional representations. In: Proceedings of the 10th International Conference on Learning Representations (2022)
6. Gao, C., Wang, X., He, X., Li, Y.: Graph neural networks for recommender system. In: Proceedings of the 15th ACM International Conference on Web Search and Data Mining (2022)
7. Goodfellow, I.J., Shlens, J., Szegedy, C.: Explaining and harnessing adversarial examples. In: Proceedings of the 4th International Conference on Learning Representations (2015)
8. Grégoire, M., Dexiong, C., Margot, S., Julien, M.: GraphiT: encoding graph structure in transformers. arXiv preprint arXiv:2106.05667 (2021)
9. Haveliwala, T.H.: Topic-sensitive PageRank. In: Proceedings of the 11th International Conference on World Wide Web (2002)
10. He, X., Deng, K., Wang, X., Li, Y., Zhang, Y., Wang, M.: LightGCN: simplifying and powering graph convolution network for recommendation. In: Proceedings of the 43rd International ACM SIGIR Conference on Research and Development in Information Retrieval (2020)
11. KG, S., Sadasivam, G.S.: A survey on personalized recommendation techniques. Int. J. Recent Innov. Trends Comput. Commun. **2**(6), 1385–1395 (2014)
12. Kingma, D.P., Ba, J.: Adam: a method for stochastic optimization. In: Proceedings of the 3rd International Conference on Learning Representations (2014)
13. Klicpera, J., Bojchevski, A., Günnemann, S.: Predict then propagate: graph neural networks meet personalized PageRank. arXiv preprint arXiv:1810.05997 (2018)
14. Kreuzer, D., Beaini, D., Hamilton, W., Létourneau, V., Tossou, P.: Rethinking graph transformers with spectral attention. In: Advances in Neural Information Processing Systems, vol. 34 (2021)
15. Lee, D., Kang, S., Ju, H., Park, C., Yu, H.: Bootstrapping user and item representations for one-class collaborative filtering. In: Proceedings of the 44th International ACM SIGIR Conference on Research and Development in Information Retrieval (2021)

16. Li, P., Wang, Y., Wang, H., Leskovec, J.: Distance encoding-design provably more powerful GNNs for structural representation learning. arXiv preprint arXiv:2009.00142 (2020)
17. Li, Q., Han, Z., Wu, X.M.: Deeper insights into graph convolutional networks for semi-supervised learning. In: Proceedings of the 32th AAAI Conference on Artificial Intelligence (2018)
18. Lim, D., et al.: Sign and basis invariant networks for spectral graph representation learning. arXiv preprint arXiv:2202.13013 (2022)
19. Liu, S., Ounis, I., Macdonald, C.: An MLP-based algorithm for efficient contrastive graph recommendations. In: Proceedings of the 45th International ACM SIGIR Conference on Research and Development in Information Retrieval (2022)
20. Liu, S., Ounis, I., Macdonald, C., Meng, Z.: A heterogeneous graph neural model for cold-start recommendation. In: Proceedings of the 43rd International ACM SIGIR Conference on Research and Development in Information Retrieval (2020)
21. Loukas, A.: What graph neural networks cannot learn: depth vs width. arXiv preprint arXiv:1907.03199 (2019)
22. Morris, C., et al.: Weisfeiler and Leman go neural: Higher-order graph neural networks. In: Proceedings of the 33th AAAI Conference on Artificial Intelligence (2019)
23. Murphy, R., Srinivasan, B., Rao, V., Ribeiro, B.: Relational pooling for graph representations. In: Proceedings of the 36th International Conference on Machine Learning (2019)
24. van den Oord, A., Li, Y., Vinyals, O.: Representation learning with contrastive predictive coding. arXiv preprint arXiv:1807.03748 (2018)
25. Rampášek, L., Galkin, M., Dwivedi, V.P., Luu, A.T., Wolf, G., Beaini, D.: Recipe for a general, powerful, scalable graph transformer. In: Proceedings of the 36th Conference on Neural Information Processing Systems (2022)
26. Rendle, S., Freudenthaler, C., Gantner, Z., Schmidt-Thieme, L.: BPR: Bayesian personalized ranking from implicit feedback. In: Proceedings of the 25th Conference on Uncertainty in Artificial Intelligence (2009)
27. Vaswani, A., et al.: Attention is all you need. In: Advances in Neural Information Processing Systems, vol. 30 (2017)
28. Wang, X., Ounis, I., Macdonald, C.: Leveraging review properties for effective recommendation. In: Proceedings of the 30th International Conference on World Wide Web (2021)
29. Wang, X., He, X., Wang, M., Feng, F., Chua, T.S.: Neural graph collaborative filtering. In: Proceedings of the 42nd International ACM SIGIR Conference on Research and Development in Information Retrieval (2019)
30. Weisfeiler, B., Leman, A.: The reduction of a graph to canonical form and the algebra which appears therein. NTI, Series 2(9), 12–16 (1968)
31. Welling, M., Kipf, T.N.: Semi-supervised classification with graph convolutional networks. In: Proceedings of the 5th International Conference on Learning Representations (2016)
32. Wu, J., et al.: Self-supervised graph learning for recommendation. In: Proceedings of the 44th International ACM SIGIR Conference on Research and Development in Information Retrieval (2021)
33. Xie, X., et al.: Contrastive learning for sequential recommendation. In: Proceedings of the 38th IEEE International Conference on Data Engineering (2022)
34. Xu, K., Hu, W., Leskovec, J., Jegelka, S.: How powerful are graph neural networks? In: Proceedings of the 6th International Conference on Learning Representations (2018)

35. Yi, Z., Wang, X., Ounis, I., Macdonald, C.: Multi-modal graph contrastive learning for micro-video recommendation. In: Proceedings of the 45th International ACM SIGIR Conference on Research and Development in Information Retrieval (2022)
36. Ying, C., et al.: Do transformers really perform badly for graph representation? In: Advances in Neural Information Processing Systems, vol. 34 (2021)
37. You, J., Ying, R., Leskovec, J.: Position-aware graph neural networks. In: Proceedings of the 36th International Conference on Machine Learning (2019)
38. Yu, J., Yin, H., Li, J., Gao, M., Huang, Z., Cui, L.: Enhance social recommendation with adversarial graph convolutional networks. IEEE Trans. Knowl. Data Eng. **34**, 3727–3739 (2020)
39. Yu, J., Yin, H., Li, J., Wang, Q., Hung, N.Q.V., Zhang, X.: Self-supervised multi-channel hypergraph convolutional network for social recommendation. In: Proceedings of the 30th International Conference on World Wide Web (2021)
40. Yu, J., Yin, H., Xia, X., Chen, T., Cui, L., Nguyen, Q.V.H.: Are graph augmentations necessary? Simple graph contrastive learning for recommendation. In: Proceedings of the 45th International ACM SIGIR Conference on Research and Development in Information Retrieval (2022)
41. Yu, J., Yin, H., Xia, X., Chen, T., Li, J., Huang, Z.: Self-supervised learning for recommender systems: a survey. In: Proceedings of the 45th International ACM SIGIR Conference on Research and Development in Information Retrieval (2022)

Domain Adaptation for Anomaly Detection on Heterogeneous Graphs in E-Commerce

Li Zheng[1,2], Zhao Li[3,4(✉)], Jun Gao[1,2(✉)], Zhenpeng Li[5], Jia Wu[6], and Chuan Zhou[7]

[1] The Key Laboratory of High Confidence Software Technologies, Ministry of Education, Beijing, China
[2] School of Computer Science, Peking University, Beijing, China
{greezheng,gaojun}@pku.edu.cn
[3] Zhejiang University, Hangzhou, China
lzjoey@gmail.com
[4] Hangzhou Link2Do Technology, Hangzhou, China
[5] Alibaba Group, Hangzhou, China
zhen.lzp@alibaba-inc.com
[6] School of Computing, Macquarie University, Sydney, Australia
jia.wu@mq.edu.cn
[7] Academy of Mathematics and System Science, Chinese Academy of Sciences, Beijing, China
zhouchuan@amss.ac.cn

Abstract. Anomaly detection models have been the indispensable infrastructure of e-commerce platforms. However, existing anomaly detection models on e-commerce platforms face the challenges of "cold-start" and heterogeneous graphs which contain multiple types of nodes and edges. The scarcity of labeled anomalous training samples on heterogeneous graphs hinders the training of reliable models for anomaly detection. Although recent work has made great efforts on using domain adaptation to share knowledge between similar domains, none of them considers the problem of domain adaptation between heterogeneous graphs. To this end, we propose a **Domain Adaptation** method for heterogeneous **GRaph Anomaly Detection** in **E**-commerce (**DAGrade**). Specifically, DAGrade is designed as a domain adaptation approach to transfer our knowledge of anomalous patterns from label-rich source domains to target domains without labels. We apply a heterogeneous graph attention neural network to model complex heterogeneous graphs collected from e-commerce platforms and use an adversarial training strategy to ensure that the generated node vectors of each domain lay in the common vector space. Experiments on real-life datasets show that our method is capable of transferring knowledge across different domains and achieves satisfactory results for online deployment.

Keywords: Domain adaptation · Anomaly detection · Heterogeneous graph · E-commerce

© The Author(s), under exclusive license to Springer Nature Switzerland AG 2023
J. Kamps et al. (Eds.): ECIR 2023, LNCS 13981, pp. 304–318, 2023.
https://doi.org/10.1007/978-3-031-28238-6_20

1   Introduction

Due to the widespread use of smart mobile devices, billions of users have engaged in online shopping. E-commerce platforms such as *Taobao*[1] and *Lazada*[2] have become an essential part of our modern lives. However, fraud behavior poses a severe threat to these platforms [5], and anomaly detection models for anti-fraud play an important role in maintaining a satisfactory user experience on these platforms [1,2].

Anomaly detection on e-commerce platforms faces the challenges of "cold-start" [21,23] and heterogeneous graphs which contain multiple types of nodes and edges. The scarcity of labeled anomalous training samples on heterogeneous graphs hinders the training of reliable models for anomaly detection [22]. As shown in Fig. 1, we wish to build a model for detecting anomalous behaviors on *Lazada* in Southeast Asia. Although *Lazada* provides a huge-sized heterogeneous graph with rich data of users' purchase transactions, we are still unable to build a workable anomaly detection model because of the insufficiency of labeled anomalous training examples. On the other hand, for the platform of *Taobao* we have already accumulated plenty of anomalous training examples which can potentially help the construction of the *Lazada* model, because anomalous users are likely to share similar patterns between China and Southeast Asia.

Recent work has made great efforts on using domain adaptation to share knowledge between two similar domains. For example, the works [3,11,17] use labeled normal and anomalous data in both source and target domains to capture anomaly patterns, yet labeled target domain data are not available during training. The work [10] only needs labeled normal data for the target domain, which is still difficult to meet in our scenario, due to the high cost of manual labeling. In most cases, data in the target domain is completely unlabeled, which requires training the data in an unsupervised way. Although the above work can address the knowledge sharing problem between similar domains, they are incapable of handling knowledge and patterns hidden behind heterogeneous graphs in both source and target domains.

From the view of graph data analysis, representation learning of heterogeneous graphs which embeds heterogeneous graphs into a low-dimensional vector space becomes popular due to the development of graph neural networks (GNNs) which can automatically propagate and aggregate structure and content information between neighboring nodes [4,7,16]. Typically, HeterGNN [32] and HGAT [29] combine GNNs with the idea of metapath2vec [8], which enables GNNs to capture the heterogeneous structure information with attention [28]. A recent work HGT [13], by using a Transformer-like framework [27], models the meta-type triplets on different edges of a heterogeneous graph, so that nodes can aggregate information from neighbors with proper weight values.

Indeed, none of the existing work considers the problem of domain adaptation between heterogeneous graphs. Therefore, we propose a **D**omain

---

[1] https://www.taobao.com.
[2] https://www.lazada.com.

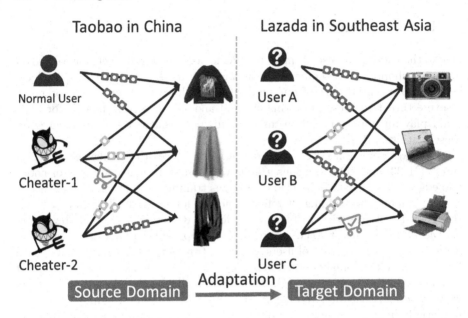

Fig. 1. An illustration of domain adaptation between e-commerce platforms of *Taobao* in China and *Lazada* in Southeast Asia. In the source domain of *Taobao*, we have already known some anomalous patterns extracted from *Taobao*'s heterogeneous transaction network, e.g., malicious users recommend/buy a cheating product of poor quality while giving negative comments to its competitive products of high quality. On the other hand, the target domain of *Lazada* lacks labeled data necessary to independently develop a solid model. By transferring the knowledge of anomalous patterns from *Taobao* to *Lazada*, we can discover more anomalous behaviors.

Adaptation method for heterogeneous **GR**aph **A**nomaly **D**etection in **E**-commerce (**DAGrade**). Specifically, we adopt the domain adaptation method, using the common vector space as a bridge to transfer knowledge from the source domain to the target one, with the help of training signals from anomaly classifier, domain discriminator, and reconstruction error. To ensure that the anomaly classifier trained with the help of labeled data from the source domain can be used directly on the data from the target domain, an adversarial training method is utilized to make the generated node vectors of each domain reside in the same vector space. In addition, we apply HGT [13] to learn messages between different types of nodes. The contributions are summarized as follows:

- We first study the problem of domain adaptation for anomaly detection on heterogeneous graphs.
- We propose a new anomaly detection model **DAGrade** for discovering anomalous edges from heterogeneous graphs. DAGrade is a domain adaptation method that can transfer the patterns of anomalous nodes from source domains to target domains. An attentional heterogeneous graph neural network is used to model heterogeneous graphs. Moreover, we use adversarial

learning to ensure that the nodes of both source and target domains are mapped to the same vector space.

– Experiments on real-life *Taobao* and *Lazada* datasets show that the proposed method DAGrade is capable of transferring knowledge of anomaly detection across different domains and achieves satisfactory results.

2   Related Work

Domain Adaptation for Anomaly Detection. The existing domain adaptation methods for anomaly detection are usually supervised or semi-supervised, which require not only the labeled data from the source domain but also all or part of the labeled data from the target domain. [11] and [3] take the imbalance of samples into account in anomaly detection, but they need normal and abnormal samples to train. The work [17] trains the latent domain vectors using fully supervised data across domains, which enables the approach to infer anomaly detectors using latent domain information. [31] learns domain-invariant representation through cross-domain encoders and adversarial generators, so that the anomaly classifier can be directly used in the target domain.

There are also several studies that approach the issue as a one-class classification task and simply use regular data from the source and target domains during training, such as [10,14,30].

Heterogeneous Graph Representation. Heterogeneous graph representation embeds the heterogeneous graph into a low-dimensional vector space and uses node embeddings to provide support for downstream tasks. Heterogeneous graph representation is an extension of homogeneous graph representation that places more emphasis on the different kinds of nodes and edges that make up the graph. According to the technique used, existing methods can be divided into path-based and GNN-based methods.

Inspired by the skip-gram method [19], PTE [25] divides the heterogeneous graph into several homogeneous graphs, on each of which PTE performs the skip-gram framework to keep similar nodes closer in vector space. To capture the graph heterogeneity around nodes, a random walk strategy based on meta-path is introduced in metapath2vec [8] and its extensions [6,9,20], which pays more attention to the interaction between specific types of nodes.

Graph neural network has been widely studied and applied for the representation of heterogeneous graphs after the convolution operation was introduced into the homogeneous graph by GCN [16], which flexibly combines structural and content information. RGCN [24] extends GCN by learning a specific weight matrix for each type of edge. HeterGNN [32] uses type-specific RNNs to capture the information of different types of neighbor nodes. Inspired by [8], HGAT [29] augments the graph by sampling meta-paths and learn weight matrices.

More recently, HGT [13] parameterizes each meta-type triple on the edge and uses Transformer to propagate information to nodes and neighbors in a self-attention architecture.

3   Preliminaries

3.1   Heterogeneous Graphs

A heterogeneous graph can be denoted as $\mathcal{G} = \{\mathcal{V}, \mathcal{E}\}$, where $\mathcal{V}$ and $\mathcal{E}$ are a set of nodes and edges respectively. Let $n = |\mathcal{V}|$ and $m = |\mathcal{E}|$. An edge $e = (u, v) \in \mathcal{E}$ denotes a link between nodes u and v. We use an adjacency matrix $A \in \mathbb{R}^{n \times n}$ to denote all the edges in $\mathcal{E}$, where $\forall (u, v) \in \mathcal{E}, A_{u,v} = 1$.

Type Mapping. Nodes in $\mathcal{V}$ have different types. For example, in an academic graph, there are three types of nodes, i.e. authors, papers and conferences. We use $\mathcal{T}_v$ to represent a set of node types. Similarly, edges also have different types, and we use $\mathcal{T}_e$ to denote a set of edge types. For a node v and an edge e, we use the mapping functions as follows,

$$\varphi(v) : V \to \mathcal{T}_v, \tag{1}$$
$$\psi(e) : E \to \mathcal{T}_e. \tag{2}$$

Meta-type Triplets. Each edge is associated with two nodes, and the relationship of an edge depends on type of the edge and types of the two associated nodes. A meta-type triplet of an edge $e = (u, v)$ can be denoted as

$$\langle \varphi(u), \psi(e), \varphi(v) \rangle. \tag{3}$$

Features. For each node $v \in \mathcal{V}$, v is associated with a feature vector of $\mathbf{x}_v$ of $k_{\varphi(v)}$ dimensions. Note that $k_{\varphi(v)}$ are different due to different types of nodes. For convenience, we denote the feature matrix of all nodes as X.

3.2   Data Distributions in Domains

Given two heterogeneous graphs $\mathcal{G}_s = \{\mathcal{V}_s, \mathcal{E}_s\}$ and $\mathcal{G}_t = \{\mathcal{V}_t, \mathcal{E}_t\}$. Graph $\mathcal{G}_s$ is collected from source domain, while graph $\mathcal{G}_t$ from target domain. For a domain graph, the node features X, adjacency matrix A, and edge labels Y are under a domain-specific distribution p, i.e.,

$$\{X_s, A_s, Y_s\} \sim p_s, \tag{4}$$
$$\{X_t, A_t, Y_t\} \sim p_t, \tag{5}$$

where p_s and p_t are distributions of the source and target domains, respectively. During training, source data $\{X_s, A_s, Y_s\}$ and target data $\{X_t, A_t\}$ are available, but the edge labels Y_t of the target domain is unknown.

3.3 Problem Definition

The goal of DAGrade is to detect anomalies in the edge set $\mathcal{E}_t$ of the target graph. Specifically, for an edge $e \in \mathcal{E}_t$, we wish to estimate a score $\mathcal{C}(e)$, *i.e.*, the anomalous probability, of edge e. Generally, we do not have labeled data at the target domain, and wish to borrow the knowledge from the source domain. We use $\{X_s, A_s, Y_s\}$ from the source domain and $\{X_t, A_t\}$ from the target domain to train the model. When testing, we compare the edge label Y_t of the target domain with the anomalous score $\mathcal{C}(e)$.

4 DAGrade Method

Figure 2 shows the framework of our DAGrade method. The core idea of DAGrade is to embed the nodes from source and target domains into a vector space by using a heterogeneous attentional graph encoder as described in Sect. 4.1, and a domain discriminator (Sect. 4.2) is used to ensure that the vectors from different domains follow the similar distribution. As demonstrated in Sect. 4.3, we train an anomaly classifier using labeled edges in the source domain. In Sect. 4.4, the reconstruction error is used as the training signal of nodes embedding in the target domain. We introduce loss functions in Sect. 4.5.

4.1 Heterogeneous Graph Attentional Encoder

An essential component is to map heterogeneous graphs to low-dimensional feature vectors. In this work, we use a heterogeneous attentional encoder based on heterogeneous graph transformer (HGT) [13], where the graph mapping is built upon four steps, i.e., feature extraction, heterogeneous attention, heterogeneous message passing, and heterogeneous aggregation.

Feature Exaction. The dimensions of features may be different for different types of nodes. For a node u, the dimension of its features is $k_{\varphi(u)}$, and its initial embedding is obtained as below:

$$\mathbf{h}_u^{(0)} = \mathbf{W}_{\varphi(u)}\mathbf{x}_u + \mathbf{b}_{\varphi(u)} \tag{6}$$

where d is the dimension of hidden states, and $\mathbf{W}_{\varphi(u)} \in \mathbb{R}^{k_{\varphi(u)} \times d}$ and $\mathbf{b}_{\varphi(u)}$ are the weights and bias of the linear transformation for node type $\varphi(u)$ at the initialization stage.

Heterogeneous Attention. For an edge $e = (u, v) \in \mathcal{E}$ of $\mathcal{G}$, we calculate the importance of node u to node v with respect to edge e as follows.

$$\mathbf{Att}(u, e, v) = \underset{\forall u \in N(v)}{\mathrm{Softmax}} \left(\underset{i \in [1,h]}{\|} \mathrm{ATT\text{-}head}^i(u, e, v) \right) \tag{7}$$

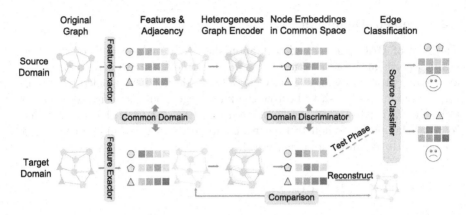

Fig. 2. DAGrade framework

where h is the number of heads. Each ATT-head takes the form:

$$\text{ATT-head}^i(u, e, v) = K^i(u) W_{\psi(e)}^{ATT} Q^i(v)^T \tag{8}$$

$$K^i(u) = \text{K-Linear}_{\varphi(u)}^i\left(\mathbf{h}_u^{(l-1)}\right) \tag{9}$$

$$Q^i(v) = \text{Q-Linear}_{\varphi(v)}^i\left(\mathbf{h}_v^{(l-1)}\right) \tag{10}$$

where the linear projections $\text{K-Linear}_{\varphi(u)}^i : \mathbb{R}^d \to \mathbb{R}^{\frac{d}{h}}$ and $\text{Q-Linear}_{\varphi(v)}^i : \mathbb{R}^d \to \mathbb{R}^{\frac{d}{h}}$ are for key and query vectors, respectively. To calculate the similarity between node u and v, we take the bilinear form in Eq. (8).

Considering that different types of edges also have an impact on node similarity, we use edge type-specific weights $W_{\psi(e)}^{ATT} \in \mathbf{R}^{\frac{d}{h} \times \frac{d}{h}}$, where $\psi(e)$ represent the type of edge e.

For the i-th ATT-head, we project the hidden states of node u and node v from $(l-1)$-th layer to the i-th Key-vector $K^i(u)$ and $Query$-vector $Q^i(v)$ with the linear projections $\text{K-Linear}_{\varphi(u)}^i : \mathbb{R}^d \to \mathbb{R}^{\frac{d}{h}}$ and $\text{Q-Linear}_{\varphi(v)}^i : \mathbb{R}^d \to \mathbb{R}^{\frac{d}{h}}$. Note that the subscript of each linear projection is different, which means that each type of node has a unique projection method.

Heterogeneous Message Passing. For node v, messages passed from its neighbors along edges are naturally affected by the types of nodes u and the types of edges. Along the edge $e = (u, v)$, the message passed from the neighbor node u to node v can be expressed as:

$$\mathbf{Mes}(u, e, v) = \underset{i \in [1, h]}{||} \text{MES-head}^i(u, e, v), \tag{11}$$

where h is the number of heads and each MES-head takes the form:

$$\text{MES-head}(u, e, v) = \text{V-Linear}_{\varphi(u)}^i\left(\mathbf{h}_u^{(l-1)}\right) W_{\psi(e)}^{MES} \tag{12}$$

where V-Linear$^i_{\varphi(u)} : \mathbb{R}^d \to \mathbb{R}^{\frac{d}{h}}$ is the linear projection for value vectors. For the i-th MES-head, we project the hidden states of the neighbor u from $(l-1)$-th layer to the i-th $Value$-vector $Q^i(u)$ with the linear projection V-Linear$^i_{\varphi(u)} : \mathbb{R}^d \to \mathbb{R}^{\frac{d}{h}}$.

Similar to the attention stage, we set edge type-specific weights $W^{MES}_{\psi(e)} \in \mathbf{R}^{\frac{d}{h} \times \frac{d}{h}}$, where $\psi(e)$ represent the type of edge e.

Heterogeneous Aggregation. With message passed and heterogeneous multi-head attention received we can simply calculate the updated hidden states of node v at l-th layer:

$$\tilde{\mathbf{h}}^{(l)}_v = \underset{\forall u \in N(v)}{\oplus} \Big(\mathbf{Att}(u,e,v) \cdot \mathbf{Mes}(u,e,v) \Big) \tag{13}$$

Through Eq. (13), we aggregate the messages of all neighboring nodes to node v according to their respective importance. In order to prevent over-smoothing with the increase of depth, we adopt the idea of identity mapping from residual network by adding self-loop to the hidden state of node v:

$$\mathbf{h}^{(l)}_v = \text{A-Linear}_{\varphi(v)}\big(\sigma(\tilde{\mathbf{h}}^{(l)}_v) \big) + \mathbf{h}^{(l-1)}_v, \tag{14}$$

where the linear projection A-Linear$^i_{\varphi(v)} : \mathbb{R}^d \to \mathbb{R}^d$ is specific for node type $\varphi(v)$, and $\sigma(\cdot)$ is the sigmoid activation function.

HAE. We abbreviate above four steps as **HAE** (heterogeneous attentional encoder), *i.e.*, for a node u,

$$\mathbf{h}_u = \mathbf{HAE}(\mathbf{x}_u; \Theta_H), \tag{15}$$

where Θ_H represents the trainable weight parameters in the linear projections that appears in the above four steps. After performing HGT on source domain and target domain respectively, we get the representations:

$$H_s = \mathbf{HAE}(X_s; \Theta_s), \tag{16}$$
$$H_t = \mathbf{HAE}(X_t; \Theta_t), \tag{17}$$

where Θ_s and Θ_t denote **HAE** trainable parameters of source domain and target domain, respectively.

4.2   Domain Discriminator

The anomaly classifier in the source domain can be reused in the target domain as we need to project both the source domain and the target domain into the same feature space in order to transfer the knowledge obtained from the source domain to the target domain.

For a chosen representation $\mathbf{h}_u = \mathbf{HAE}(\mathbf{x}_u; \Theta_H)$ and its domain label y_u, we use a domain discriminator to get the predicted domain probability:

$$\hat{y}_u = \mathbf{D}_d\big(\mathbf{HAE}(\mathbf{x}_u; \Theta_H); \Theta_d\big) \in (0, 1), \tag{18}$$

where $\hat{y}_u \in (0, 1)$ represents probability that the node embedding belongs to the target domain. In this paper, a lower $\hat{y}_u$ indicates more chances that u belongs to the source domain. The ground-truth domain label is denoted as $d_u \in \{d_s, d_t\}$ for node u.

For a node, the cross-entropy is used to measure the loss of its domain discriminating:

$$L_d(\mathbf{x}_u, y_u; \Theta_H, \Theta_d) = -\mathbb{I}_{d_u = d_t} log(\hat{y}_u) - \big(1 - \mathbb{I}_{d_u = d_t}\big) log(1 - \hat{y}_u) \tag{19}$$

The total loss of the domain discriminator is

$$\mathcal{L}_d = \frac{1}{n_s} \sum_{u \in \mathcal{V}_s} L_d(\mathbf{x}_u, y_u; \Theta_s, \Theta_d) + \frac{1}{n_t} \sum_{v \in \mathcal{V}_t} L_d(\mathbf{x}_v, y_v; \Theta_t, \Theta_d) \tag{20}$$

4.3   Anomaly Classifier for Source Domain

We train an anomaly classifier on data in source domain with edge labels. For a training sample edge $e = (u, v)$ and its data $(\mathbf{x}_u, \mathbf{x}_v, y_e)$, we obtain the representation $\mathbf{h}_u$ and $\mathbf{h}_v$ of two nodes from $\mathbf{HAE}(\cdot; \Theta_s)$. The anomaly classifier outputs a prediction for e:

$$\hat{y}_e = \mathcal{C}(e) = \mathbf{C}(\mathbf{h}_u, \mathbf{h}_v; \Theta_c) \in (0, 1) \tag{21}$$

where Θ_c is the trainable parameters of the classifier. We call the Eq. (21) as the score function of our method. We use the cross-entropy to measure the loss in predicting the anomaly score of e:

$$L_c(\mathbf{x}_u, \mathbf{x}_v, y_e; \Theta_s, \Theta_c) = -y_e log(\hat{y}_e) - (1 - y_e) log(1 - \hat{y}_e) \tag{22}$$

The total loss of the anomaly classifier on source domain training data is:

$$\mathcal{L}_c = \frac{1}{m_s} \sum_{e \in \mathcal{E}_s} \mathbb{E}_{(\mathbf{x}_u, \mathbf{x}_v, y_e) \sim p_s} L_c(\mathbf{x}_u, \mathbf{x}_v, y_e; \Theta_s, \Theta_c) \tag{23}$$

4.4   Reconstruction in Target Domain

Since the edge labels of the target domain are not visible during the training phase, we can not directly use the cross-entropy loss to train a classifier in a supervised way like the source domain. We take the difference between the original graph and the reconstructed graph as the training signal for the target domain.

After we get node embeddings, the reconstruction error [15] on an edge $e = (u, v)$ is:

$$L_r(\mathbf{x}_u, \mathbf{x}_v; \Theta_t) = -log[p(\mathbf{A}_{u,v} = 1 | \mathbf{x}_u, \mathbf{x}_v)] \tag{24}$$

The total reconstruction loss of the entire graph of target domain is:

$$\mathcal{L}_r = \mathbb{E}_{H_t \sim (X_t, \mathbf{A})} L_r(\mathbf{x}_u, \mathbf{x}_v; \Theta_t) = ||\mathbf{A} - \hat{\mathbf{A}}||_2^2 \tag{25}$$

where $\hat{\mathbf{A}} = \sigma(H_t \cdot H_t^T)$ is the inner-production of representation H_t and $\sigma(\cdot)$ is the sigmoid function.

4.5 Loss Functions

For domain discriminator $\mathbf{D}_d$, the objective is to distinguish the source node embeddings from the target node embeddings, while for **HAE**, its objective is to fool $\mathbf{D}_d$. For anomaly classifier, the objective is to reduce the difference between Y_s and $\hat{Y}_s$. Reducing the reconstruction error is helpful to the training of target domain encoders. Using an adversarial leaning configuration [12], the training objective:

$$\mathcal{L}_1 = \mathcal{L}_c - \lambda_d \mathcal{L}_d + \lambda_r \mathcal{L}_r, \tag{26}$$
$$\mathcal{L}_2 = \mathcal{L}_c + \lambda_d \mathcal{L}_d + \lambda_r \mathcal{L}_r \tag{27}$$

where loss weights λ_d and λ_r are the hyper-parameters for loss functions $\mathcal{L}_d$ and $\mathcal{L}_r$, respectively. The training of the method can be written as a joint optimization with respect to $\mathcal{L}_1$ and $\mathcal{L}_2$:

$$\min_{\Theta_H, \Theta_c} \mathcal{L}_1 \tag{28}$$

$$\min_{\Theta_H, \Theta_c, \Theta_d} \mathcal{L}_2 \tag{29}$$

5 Experiments

5.1 Experimental Setup

Datasets. The details of the two datasets are listed in Table 1. The Taobao dataset is exacted from *Taobao*, which is the largest e-commerce site in China, generating billions of online transactions on a daily basis. The Lazada dataset is exacted from *Lazada*, which is one of the largest e-commerce sites in Southeast Asia and targets at users from Indonesia, Malaysia, the Philippines, and Thailand.

Both datasets contain a huge number of transaction records between users and products on the e-commerce sites, which are naturally mixed with potential anomalies. Each node represents a user or a product of the site, and each edge indicates an interaction between a user and a product on the site. Each node

314 L. Zheng et al.

Table 1. Statistics of datasets

Dataset	$|\mathcal{V}|$	$|\mathcal{E}|$	$|\mathcal{T}_v|$	$|\mathcal{T}_e|$	Ano. rate
Taobao	92,317	291,828	2	2	6.43%
Lazada	26,150	82,764	2	2	19.34%

corresponds to a feature vector, and the vector dimension depends on the type of the node. The features of nodes are obtained by summarizing their historical behavior records on the website in a month.

The experimental goal of this paper is to detect anomalous edges in the target domain without labeled data. We choose AUC (the area under the ROC curve) as the metric to compare the performance of our method with baselines.

Baselines. We compare our method DAGrade with the following baselines:

- IF (Isolation Forest) [18] is an unsupervised anomaly detection method which finds sparsely distributed outliers by recursively segmenting the dataset.
- DS (DeepSphere) [26] encodes nodes by an auto-encoder and finds a hypersphere with the minimized radius as the borderline for both normal and anomalous data.
- DNN (Deep Neural Network) means to use a multi-layer perception to encode nodes.
- GCN (Graph Convolutional Network) [16] combines structural and content information by propagating and aggregating messages between nodes.
- HGT (Heterogeneous Graph Transformer) [13] models the meta-type triples on different types of edges for a heterogeneous graph with a self-attention transformer framework.

Implementation Details. Our method and all baselines are implemented with Python-3.7 and TensorFlow-1.14. All the experiments are conducted on an Ubuntu Server 18.04LTS machine with a NVIDIA RTX2080Ti GPU (11GB memory), 32-core Intel Xeon CPU (2.3 GHz), and 256 GB of RAM. To be fair, we use 16 as the model dimension of hidden states in neural networks for all the baselines, and the layer number of the network architecture is 2. Multi-layer perceptions are used as anomaly classifiers of source domain, and the depth of MLP is set to 2. The dropout rate is set to 0.2. Adam is chosen as the optimizer for all methods with the same learning rate.

5.2 Experimental Results

We conduct experiments on two domain adaptation situations: Taobao to Lazada, and Lazada to Taobao. The hyper-parameters $\{\lambda_d, \lambda_r\}$ for losses are set to $\{2.0, 1.0\}$ and $\{3.0, 2.0\}$ for the above situations, respectively. The number of attention heads h is set to 4 for DAGrade.

We adopt two training strategies for baselines and use prefix *DA-* and *U-* to represent that the method is trained with/without a domain discriminator.

Table 2. AUC results for the target domain.

Methods	Taobao → Lazada	Lazada → Taobao
IF	0.481	0.518
DS	0.528	0.545
U-GCN	0.496 ± 0.066	0.742 ± 0.026
U-HGT	0.633 ± 0.029	0.784 ± 0.028
DA-DNN	0.524 ± 0.053	0.723 ± 0.091
DA-GCN	0.550 ± 0.040	0.764 ± 0.039
DAGrade	**0.681 ± 0.009**	**0.831 ± 0.018**
-w/o $\mathcal{L}_r$	0.660 ± 0.012	0.822 ± 0.022
-w/o **HAE**	0.594 ± 0.025	0.768 ± 0.032

Considering the fluctuation of the results caused by the random initialization of the parameters, we repeat the corresponding experiments five times and calculate the average and standard deviation.

The experimental results are summarized in Table 2. We can see that the proposed method DAGrade outperforms all the baselines significantly and consistently on these domain settings. In particular, DAGrade achieves relative performance gains over DA- baselines by 12–30% in terms of AUC for Lazada→Taobao. The comparison with the U-HGT results illustrates the improvement brought by domain adaptation. The gap between our results and baseline DA-GCN shows the benefit from modeling of heterogeneous graphs.

5.3 Ablation Study

We perform an ablation study for our method DAGrade. We conduct experiments to study the influence of reconstruction loss and the heterogeneous attentional encoder, respectively.

The experimental results are summarized in Table 2. As shown in the table, without the heterogeneous attentional encoder, the AUC of the method decreases significantly, which demonstrates the benefit of modeling heterogeneous information. Removing reconstruction loss $\mathcal{L}_r$ also resulted in a small decrease of AUC, suggesting that there exists some domain-specific knowledge in the target domain.

5.4 Parameter Study

We study the effects of hyper-parameters on DAGrade, including the loss weight λ_d for domain discriminator loss and the loss weight λ_r for reconstruction error in target domain. The range of λ_d is $\{0.5, 1.0, 2.0, 3.0, 4.0, 5.0\}$. The range of λ_r is $\{0.2, 0.5, 1.0, 2.0, 3.0, 4.0, 5.0, 10\}$. Other parameters are set to optimum.

We choose Lazada dataset as the source domain and Taobao dataset as the target domain. We evaluate the classification result for target domain in terms of AUC metric.

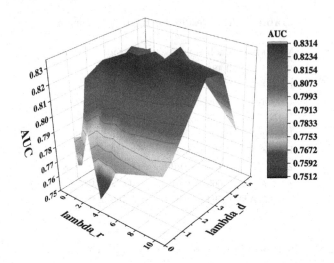

Fig. 3. AUC results on Lazada→Taobao with respect to different parameters.

As shown in Fig. 3, AUC increases significantly when λ_d increases from 0.5 to 2.0, and reaches its peak among 2.0 and 3.0. With the increase of λ_r, AUC is gradually improved, and reaches the maximum when $\lambda_r = 2.0$. After λ_d and λ_r reach the optimization, AUC gradually decreases as they further increase. AUC is relatively stable with the change of λ_r, but it is very sensitive to λ_d. These results are due to the fact that λ_d is the loss weight of the domain discriminator, which is an important module for DAGrade to ensure that data from different domains are projected into the same vector space.

6   Conclusions

We propose a novel anomaly detection method for analyzing heterogeneous graphs on e-commerce platforms. Based on an attentional heterogeneous graph neural network model, the knowledge of anomaly detection is transferred from the source domain to a new target domain via a domain adaptation approach. Our method achieves satisfactory performance in experiments on examples of *Taobao* and *Lazada* networks, and the results show the effectiveness of our method on domain adaptation for anomaly detection on heterogeneous graphs in e-commerce.

Acknowledgment. This work was partially supported by NSFC under Grant No. 62272008 and 61832001.

References

1. Abdallah, A., Maarof, M.A., Zainal, A.: Fraud detection system: a survey. J. Netw. Comput. Appl. **68**, 90–113 (2016)
2. Ahmed, M., Mahmood, A.N., Islam, M.R.: A survey of anomaly detection techniques in financial domain. Futur. Gener. Comput. Syst. **55**, 278–288 (2016)
3. Al-Stouhi, S., Reddy, C.K.: Transfer learning for class imbalance problems with inadequate data. Knowl. Inf. Syst. **48**(1), 201–228 (2016)
4. Bruna, J., Zaremba, W., Szlam, A., LeCun, Y.: Spectral networks and locally connected networks on graphs. arXiv preprint arXiv:1312.6203 (2013)
5. Cai, Y., Zhu, D.: Fraud detections for online businesses: a perspective from blockchain technology. Financial Innov. **2**(1), 1–10 (2016). https://doi.org/10.1186/s40854-016-0039-4
6. Cen, Y., Zou, X., Zhang, J., Yang, H., Zhou, J., Tang, J.: Representation learning for attributed multiplex heterogeneous network. In: Proceedings of the 25th ACM SIGKDD International Conference on Knowledge Discovery & Data Mining, pp. 1358–1368 (2019)
7. Defferrard, M., Bresson, X., Vandergheynst, P.: Convolutional neural networks on graphs with fast localized spectral filtering. In: Advances in Neural Information Processing Systems, pp. 3844–3852 (2016)
8. Dong, Y., Chawla, N.V., Swami, A.: metapath2vec: scalable representation learning for heterogeneous networks. In: Proceedings of the 23rd ACM SIGKDD International Conference on Knowledge Discovery and Data Mining, pp. 135–144 (2017)
9. Fu, T.Y., Lee, W.C., Lei, Z.: HIN2Vec: explore meta-paths in heterogeneous information networks for representation learning. In: Proceedings of the 2017 ACM on Conference on Information and Knowledge Management, pp. 1797–1806 (2017)
10. Fujita, H., Matsukawa, T., Suzuki, E.: One-class selective transfer machine for personalized anomalous facial expression detection. In: VISIGRAPP (5: VISAPP), pp. 274–283 (2018)
11. Ge, L., Gao, J., Ngo, H., Li, K., Zhang, A.: On handling negative transfer and imbalanced distributions in multiple source transfer learning. Stat. Anal. Data Mining ASA Data Sci. J. **7**(4), 254–271 (2014)
12. Goodfellow, I., et al.: Generative adversarial nets. In: Advances in Neural Information Processing Systems, vol. 27 (2014)
13. Hu, Z., Dong, Y., Wang, K., Sun, Y.: Heterogeneous graph transformer. In: Proceedings of the Web Conference 2020, pp. 2704–2710 (2020)
14. Idé, T., Phan, D.T., Kalagnanam, J.: Multi-task multi-modal models for collective anomaly detection. In: 2017 IEEE International Conference on Data Mining (ICDM), pp. 177–186. IEEE (2017)
15. Kipf, T.N., Welling, M.: Variational graph auto-encoders. arXiv preprint arXiv:1611.07308 (2016)
16. Kipf, T.N., Welling, M.: Semi-supervised classification with graph convolutional networks. In: International Conference on Learning Representations (ICLR) (2017)
17. Kumagai, A., Iwata, T., Fujiwara, Y.: Transfer anomaly detection by inferring latent domain representations. In: Advances in Neural Information Processing Systems, pp. 2471–2481 (2019)
18. Liu, F.T., Ting, K.M., Zhou, Z.H.: Isolation forest. In: 2008 Eighth IEEE International Conference on Data Mining, pp. 413–422. IEEE (2008)
19. Mikolov, T., Sutskever, I., Chen, K., Corrado, G.S., Dean, J.: Distributed representations of words and phrases and their compositionality. Adv. Neural. Inf. Process. Syst. **26**, 3111–3119 (2013)

20. Park, C., Kim, D., Zhu, Q., Han, J., Yu, H.: Task-guided pair embedding in heterogeneous network. In: Proceedings of the 28th ACM International Conference on Information and Knowledge Management, pp. 489–498 (2019)
21. Patcha, A., Park, J.M.: An overview of anomaly detection techniques: Existing solutions and latest technological trends. Comput. Netw. **51**(12), 3448–3470 (2007)
22. Robertson, W.K., Maggi, F., Kruegel, C., Vigna, G., et al.: Effective anomaly detection with scarce training data. In: NDSS. Citeseer (2010)
23. Roth, K., Pemula, L., Zepeda, J., Schölkopf, B., Brox, T., Gehler, P.: Towards total recall in industrial anomaly detection. arXiv preprint arXiv:2106.08265 (2021)
24. Schlichtkrull, M., Kipf, T.N., Bloem, P., van den Berg, R., Titov, I., Welling, M.: Modeling relational data with graph convolutional networks. In: Gangemi, A., et al. (eds.) ESWC 2018. LNCS, vol. 10843, pp. 593–607. Springer, Cham (2018). https://doi.org/10.1007/978-3-319-93417-4_38
25. Tang, J., Qu, M., Mei, Q.: PTE: predictive text embedding through large-scale heterogeneous text networks. In: Proceedings of the 21th ACM SIGKDD International Conference on Knowledge Discovery and Data Mining, pp. 1165–1174 (2015)
26. Teng, X., Yan, M., Ertugrul, A.M., Lin, Y.R.: Deep into hypersphere: robust and unsupervised anomaly discovery in dynamic networks. In: Proceedings of the Twenty-Seventh International Joint Conference on Artificial Intelligence (2018)
27. Vaswani, A., et al.: Attention is all you need. In: Advances in Neural Information Processing Systems, pp. 5998–6008 (2017)
28. Veličković, P., Cucurull, G., Casanova, A., Romero, A., Liò, P., Bengio, Y.: Graph attention networks. In: International Conference on Learning Representations (2018). https://openreview.net/forum?id=rJXMpikCZ
29. Wang, X., et al.: Heterogeneous graph attention network. In: The World Wide Web Conference, pp. 2022–2032 (2019)
30. Yamaguchi, M., Koizumi, Y., Harada, N.: AdaFlow: domain-adaptive density estimator with application to anomaly detection and unpaired cross-domain translation. In: ICASSP 2019–2019 IEEE International Conference on Acoustics, Speech and Signal Processing (ICASSP), pp. 3647–3651. IEEE (2019)
31. Yang, Z., Bozchalooi, I.S., Darve, E.: Anomaly detection with domain adaptation. arXiv preprint arXiv:2006.03689 (2020)
32. Zhang, C., Song, D., Huang, C., Swami, A., Chawla, N.V.: Heterogeneous graph neural network. In: Proceedings of the 25th ACM SIGKDD International Conference on Knowledge Discovery & Data Mining, pp. 793–803 (2019)

Short Papers

Improving Neural Topic Models with Wasserstein Knowledge Distillation

Suman Adhya[ID] and Debarshi Kumar Sanyal[✉][ID]

Indian Association for the Cultivation of Science, Jadavpur 700032, India
adhyasuman30@gmail.com, debarshisanyal@gmail.com

Abstract. Topic modeling is a dominant method for exploring document collections on the web and in digital libraries. Recent approaches to topic modeling use pretrained contextualized language models and variational autoencoders. However, large neural topic models have a considerable memory footprint. In this paper, we propose a knowledge distillation framework to compress a contextualized topic model without loss in topic quality. In particular, the proposed distillation objective is to minimize the cross-entropy of the soft labels produced by the teacher and the student models, as well as to minimize the squared 2-Wasserstein distance between the latent distributions learned by the two models. Experiments on two publicly available datasets show that the student trained with knowledge distillation achieves topic coherence much higher than that of the original student model, and even surpasses the teacher while containing far fewer parameters than the teacher. The distilled model also outperforms several other competitive topic models on topic coherence.

Keywords: Topic modeling · Knowledge distillation · Wasserstein distance · Contextualized topic model · Variational autoencoder

1 Introduction

Topic modeling has come up as an important technique to analyze large document corpora and extract their themes automatically [1,26,30]. Therefore, they are frequently used to obtain an overview of the topics in document archives and web search results, match queries and documents, and diversify search results [11,28]. While latent Dirichlet allocation (LDA) [5] is the classical topic modeling algorithm, recent approaches exploit deep neural networks, specifically, variational autoencoders (VAEs) [13]. ProdLDA [24] is a well-known VAE-based topic model that uses a product of experts and a Laplace approximation to the Dirichlet prior. Bianchi et al. [3] recently proposed *CombinedTM*, a contextualized topic model that feeds into the VAE of ProdLDA a distributed representation of the document built with a pre-trained language model (PLM) like sentence-BERT (SBERT) [22] along with a bag-of-words (BoW) representation of the document. It achieves state-of-the-art topic coherence on many benchmark data sets. Given a VAE-based topic model pre-trained on a corpus, one can pass a document from the corpus through the VAE encoder and recover its topics. A remarkable

J. Kamps et al. (Eds.): ECIR 2023, LNCS 13981, pp. 321–330, 2023.
https://doi.org/10.1007/978-3-031-28238-6_21

feature of contextualized topic models is that, if the PLM is multilingual and the input to the encoder solely consists of contextualized representations from the PLM, it is possible to train the model in one language and test it in another, making it a zero-shot topic model, also called *ZeroShotTM* [4]. Increasing the network complexity like the depth or width of the neural networks in the VAE might improve the coherence of the generated topics but produces a larger memory footprint, thereby making it difficult to store and use the topic models on resource-constrained devices. Using only contextualized embeddings in the input would also reduce the model size but could hit the topic quality as well.

In this paper, we investigate if a VAE-based topic model can be compressed without compromising topic coherence. For this purpose, we use knowledge distillation (KD), which involves a teacher model to improve the performance of a smaller student model [12]. While KD has been used for classification tasks in image [10] and text processing [17], this paper tackles an unsupervised learning problem for a generative model. Specifically, we distill knowledge from a CombinedTM teacher to a smaller ZeroShotTM student. In standard KD [12], the aim is to minimize the cross-entropy between the soft labels produced by the student and the teacher models along with the Kullback-Leibler (KL) divergence between their respective output distributions. But even if the two distributions have very little dissimilarity with each other, the KL-divergence may reach a very high value, and if the two distributions are not overlapping at all, it explodes to infinity [19]. To avoid these issues, we choose 2-Wasserstein distance [18] instead of KL-divergence in distillation loss. Our distillation process minimizes the cross-entropy between the soft labels produced by the teacher and the student, *and* the square of the 2-Wasserstein distance between the latent distributions learned by the two models. Wasserstein distance arises in the theory of optimal transport and measures how 'close' two distributions are [9,21,27]. Unlike the KL divergence, if the Wasserstein between two distributions is high, this actually represents that the underlying distributions are very different from each other.

In summary, our contributions are: **(1)** We propose a 2-Wasserstein distance-based knowledge distillation framework for neural topic models. We call our method *Wasserstein knowledge distillation*. To the best of our knowledge, this is the first work on inter-VAE knowledge distillation for topic modeling. **(2)** Experiments on two public datasets show that in terms of topic coherence, the distilled model significantly outperforms the student and even scores better than the teacher. The distilled model also beats several strong baselines on topic coherence. This demonstrates the efficacy of our approach. We have made our code publicly available[1].

2   Background on Wasserstein Distance

Let $(\mathcal{X}, d)$ be a complete separable metric space with metric d and equipped with a Borel σ-algebra. Let $\mathcal{P}(\mathcal{X})$ denote the space of all probability measures defined on $\mathcal{X}$ with finite p-th moment for $p \geq 1$. If $\mathbb{P}_1, \mathbb{P}_2 \in \mathcal{P}(\mathcal{X})$, then $\Pi(\mathbb{P}_1, \mathbb{P}_2)$

---

[1] https://github.com/AdhyaSuman/CTMKD.

is defined to be the set of measures $\pi \in \mathcal{P}(\mathcal{X}^2)$ having $\mathbb{P}_1$ and $\mathbb{P}_2$ as marginals. The pth Wasserstein distance between the two probability measures $\mathbb{P}_1$ and $\mathbb{P}_2$ in $\mathcal{P}(\mathcal{X})$ is defined as

$$W_p(\mathbb{P}_1, \mathbb{P}_2) = \left(\inf_{\pi \in \Pi(\mathbb{P}_1, \mathbb{P}_2)} \int_{\mathcal{X}^2} d(x,y)^p \, \mathrm{d} \, \pi(x,y) \right)^{1/p} \tag{1}$$

$W_p(\mathbb{P}_1, \mathbb{P}_2)$ is intuitively the minimum 'cost' of transforming $\mathbb{P}_1$ to $\mathbb{P}_2$ (or vice versa) [27]. Consider $\mathcal{X} = \mathbb{R}^n$ with d as the Euclidean norm. Suppose $\mathbb{P}_1 = \mathcal{N}(\mu_1, \Sigma_1)$, and $\mathbb{P}_2 = \mathcal{N}(\mu_2, \Sigma_2)$ are normal distributions with means $\mu_1, \mu_2 \in \mathbb{R}^n$ and symmetric positive semi-definite covariance matrices $\Sigma_1, \Sigma_2 \in \mathbb{R}^{n \times n}$. From [18], the squared 2-Wasserstein distance between $\mathbb{P}_1$ and $\mathbb{P}_2$ is given by:

$$W_2(\mathbb{P}_1, \mathbb{P}_1)^2 = \|\mu_1 - \mu_2\|_2^2 + \text{trace} \left(\Sigma_1 + \Sigma_2 - 2(\Sigma_2^{1/2} \Sigma_1 \Sigma_2^{1/2})^{1/2} \right) \tag{2}$$

Wasserstein distance has been used to train various machine learning models including classifiers [7], Boltzmann machines [16], and generative adversarial networks [2], where it is found to be a better loss metric than KL-divergence.

3 Proposed Framework for Knowledge Distillation

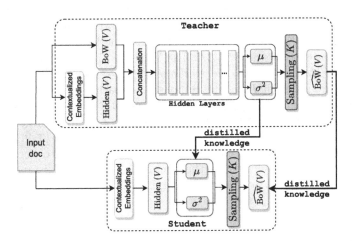

Fig. 1. Framework for knowledge distillation from CombinedTM to ZeroShotTM.

Our framework for KD is shown in Fig. 1. The teacher and the student models are both VAEs. The teacher T is a CombinedTM [3] that takes as input x a document encoded as the concatenation of the document's normalized BoW representation $x_{\text{BoW}} \in \mathbb{R}^V$, where V is the vocabulary size, and its contextualized embedding x_{ctx} scaled to dimension V by a linear layer. The student is a ZeroShotTM [4]. While the student's encoder takes only the document's contextualized representation, its decoder still needs the BoW vector during training, but it is not necessary when we use only its trained encoder to infer the topics for a given document. The teacher's encoder is a multi-layer feed-forward neural network (FFNN) while we make the student's encoder an FFNN with one hidden layer.

A VAE-based topic model works as follows [24]. Suppose it has to learn K topics from a corpus. The VAE encoder having weights W learns the approximate posterior distribution $q_W(z|x)$ represented by mean $\mu \in \mathbb{R}^K$ and variance $\sigma^2 \in \mathbb{R}^K$ for an input instance x. The decoder samples a vector $z \sim q_W(z|x)$ using the reparameterization trick [13], and produces the document-topic vector $\theta = \texttt{softmax}(z)$, which is passed through a shallow FFNN with weight matrix $\beta_{K \times V}$ to learn a distribution $p_\beta(x|z)$. The VAE is trained by backpropagation to minimize the following loss $\mathcal{L}_{\text{VAE}}$:

$$\mathcal{L}_{\text{VAE}} = \mathcal{L}_{\text{NLL}} + \mathcal{L}_{\text{KL}} \equiv -\mathbb{E}_{z \sim q_W(z|x)}\big[\log p_\beta(x|z)\big] + D_{\text{KL}}\big(q_W(z|x) \,\|\, p(z)\big) \quad (3)$$

where $\mathcal{L}_{\text{NLL}}$ is the expected negative log-likelihood of the reconstructed BoW, and $\mathcal{L}_{\text{KL}}$ is a regularizer measuring the KL-divergence of the encoder's output $q_W(z|x)$ from the assumed prior $p(z)$ of the latent distribution.

Now suppose that the teacher has been already trained on a dataset to learn K topics, and that, after training, the weights of its encoder and decoder are W_T^* and β_T^*, respectively. We will use this frozen teacher model to train the student *with KD* to learn K topics from the same dataset and the same vocabulary. We denote this KD-trained student by S'. Let the weights in its encoder and decoder be $W_{S'}$ and $\beta_{S'}$, respectively, at the start of some iteration during the training of S'. Given an input instance x, the student's loss function has two components: *(i) Loss associated with student VAE:* The VAE loss $\mathcal{L}_{\text{VAE}}$ is given by Eq. (3). *(ii) Loss associated with knowledge distillation:* While training S', every instance x is passed through both T and S'. Suppose the teacher's encoder outputs the K-variate Gaussian $\mathcal{N}(z|\mu_T, \sigma_T^2)$ while the student's encoder outputs the K-variate Gaussian $\mathcal{N}(z|\mu_{S'}, \sigma_{S'}^2)$. Note that instead of a full covariance matrix, a diagonal covariance matrix (encoded as a vector) is learned [3,4]. Let $\Sigma_T = \texttt{diag}(\sigma_T)$ and $\Sigma_{S'} = \texttt{diag}(\sigma_{S'})$, which are easily observed to be symmetric positive semi-definite. We calculate the squared 2-Wasserstein distance between the distributions learned by T and S' using Eq. (2):

$$\mathcal{L}_{\text{KD-2W}} = \|\mu_T - \mu_{S'}\|_2^2 + \text{trace}\left(\Sigma_T + \Sigma_{S'} - 2(\Sigma_{S'}^{1/2} \Sigma_T \Sigma_{S'}^{1/2})^{1/2}\right) \quad (4)$$

We propose to minimize $\mathcal{L}_{\text{KD-2W}}$ so that the distribution learned by the student is pulled close to that of the teacher. The decoder of the teacher and that of the student produce unnormalized logits $u_T = \beta_T^\top \theta$ and $u_{S'} = \beta_{S'}^\top \theta$, respectively. We compute the cross-entropy loss $\mathcal{L}_{\text{KD-CE}}$ between the soft labels $\texttt{softmax}(u_T/t)$ and $\texttt{softmax}(u_{S'}/t)$ where t is the softmax temperature (hyperparameter) [12]. In addition to identifying the most probable class, the soft labels formed by a higher softmax temperature ($t > 1$) capture the correlation between the labels, which is desired in the distillation framework. The total loss due to KD is

$$\mathcal{L}_{\text{KD}} = \mathcal{L}_{\text{KD-2W}} + t^2 \mathcal{L}_{\text{KD-CE}} \quad (5)$$

Finally, with $\alpha \in [0,1]$ as a hyperparameter, the total loss for the student S' is

$$\mathcal{L}_{S'} = (1 - \alpha)\mathcal{L}_{\text{VAE}} + \alpha\mathcal{L}_{\text{KD}} \qquad (6)$$

4   Experimental Setup

We have performed all experiments in OCTIS [25], which is an integrated framework for topic modeling. We use the following datasets from OCTIS: **20NG**, which contains $16,309$ newsgroup documents on 20 different subjects [25], and **M10** comprising 8355 scientific publications from 10 distinct research areas [20]. For each dataset, the vocabulary contains the 2K most common words in the corpus. We represent each topic by its top-10 words. We use **Normalized Pointwise Mutual Information (NPMI)** [15] and **Coherence Value (CV)** [14,23] to measure topic coherence. NPMI of a topic is high if the words in the topic tend to co-occur. CV is calculated using an indirect cosine measure along with the NPMI score over a boolean sliding window. Higher values of NPMI and CV are better.

The experiments are done for topic counts $K \in \{20, 50, 100\}$ on the 20NG dataset and for topic counts $K \in \{10, 20, 50, 100\}$ on the M10 dataset, where 20 and 10 are the golden number of categories for 20NG and M10, respectively. We denote the teacher (CombinedTM) by **T**, the student (ZeroShotTM) by **S**, and the distilled student model (ZeroShotTM) by **SKD**. The encoder in **T** uses 768-dimensional contextualized sentence embeddings (SBERT) from paraphrase-distilroberta-base-v2. The encoders in **S** and **SKD** use 384-dimensional SBERT embeddings from all-MiniLM-L6-v2 model.

Using the Bayesian optimization framework of OCTIS, we have calculated the optimal number of hidden layers H in the teacher's encoder (which takes as input the concatenation of a document's contextualized and BoW representations) from the set $\{1, 2, \ldots, 10\}$ that maximizes the NPMI for the teacher. As shown in Table 1, on 20NG dataset, we found $H = 1$ for topic count $K \in \{20, 50\}$ and $H = 5$ for $K = 100$; on M10, we observed $H = 4$ for $K = 10$, $H = 5$ for $K = 20$, $H = 2$ for $K = 50$, and $H = 3$ for $K = 100$. Each hidden layer of the teacher contains 100 neurons.

Table 1. The optimal number of hidden layers H in the encoder of the teacher **T** for each dataset and different topic counts K.

Dataset	K	H	Dataset	K	H
	20	1		10	4
20NG	50	1	M10	20	5
	100	5		50	2
				100	3

We have tuned the hyperparameters $\alpha \in [0,1]$ and $t \in \{1, 2, \ldots, 5\}$ for **SKD** in OCTIS. For performance analysis, we compare these models with **ProdLDA** [24], **NeuralLDA** [24], **Embedded Topic Model (ETM)** [6] and **LDA** [5], already implemented in OCTIS. We use the default parameters unless otherwise mentioned. All models are trained for 100 epochs with a batch size of 64. Each reported performance score is the median over 5 runs (except for **T**, where we use a single run as it must be frozen for KD).

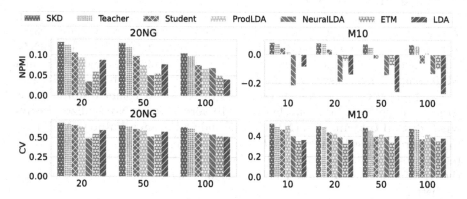

Fig. 2. Coherence scores (**NPMI** and **CV**) for different topic models on two datasets: **20NG** and **M10**. The X-axis is marked with the topic counts used for each dataset.

5   Results

Models **S** and **SKD** contain the same number of parameters, which is smaller than that of **T**. The sizes of all the models depend on the SBERT dimension, the number and size of hidden layers, the number of topics, and the vocabulary size. For example, for 20 topics in 20NG, **T** takes 6.14 MB while **SKD** 2.74 MB (for parameters and buffers) – a reduction in model size by 55.4%. In general, the compression ranged from 37.6% to 56.3%.

Figure 2 shows the coherence scores for each topic model for all topic settings and datasets. **SKD** achieves the highest NPMI and CV scores. Among **T**, **S**, and **SKD**, we find **SKD** performs much better than **S** and even modestly better than **T**. On 20NG, the NPMI scores of (**T**, **S**, **SKD**) are $(0.125, 0.106, 0.132)$ for $K = 20$, $(0.121, 0.098, 0.130)$ for $K = 50$, and $(0.098, 0.076, 0.105)$ for $K = 100$, so the maximum gain of **SKD** over **S** is 38.2% and that over **T** is 7.4%. Similarly on M10, the NPMI scores are $(0.073, 0.046, 0.084)$ for $K = 10$, $(0.076, 0.037, 0.08)$ for $K = 20$, $(0.053, -0.027, 0.073)$ for $K = 50$, and $(0.059, -0.06, 0.07)$ for $K = 100$. Thus, on M10, **SKD** improves NPMI of **S** by over 100% for $K \in \{50, 100\}$, and that of **T** by at most 37.7%. Student outperforming the teacher is surprising but has been reported earlier for *supervised* tasks [8, 29].

When we deleted any one of the two loss terms from $\mathcal{L}_{KD}$ in Eq. (5), NPMI and CV of **SKD** dropped (see Table 2). Thus, although the simpler model and weaker SBERT lower the student's performance, the knowledge distilled from the teacher's encoder and decoder vastly improves it.

Table 2. Ablation study for the distillation loss term defined in Eq. (5). For each metric, the median over five independent runs for each topic count is mentioned.

KD-loss ($\mathcal{L}_{KD}$)	20NG						M10							
	NPMI			CV			NPMI				CV			
	20	50	100	20	50	100	10	20	50	100	10	20	50	100
$\mathcal{L}_{KD\text{-}2W} + \mathcal{L}_{KD\text{-}CE}$	**0.132**	**0.130**	**0.105**	**0.687**	**0.657**	**0.638**	**0.084**	**0.080**	**0.073**	**0.070**	**0.522**	**0.499**	**0.485**	**0.475**
$\mathcal{L}_{KD\text{-}2W}$	0.109	0.114	0.089	0.659	0.638	0.615	0.051	0.049	0.037	0.043	0.498	0.479	0.459	0.452
$\mathcal{L}_{KD\text{-}CE}$	0.110	0.105	0.083	0.653	0.629	0.588	0.042	0.052	0.016	0.023	0.485	0.464	0.425	0.425

The higher performance of the contextualized topic models over other topic models agrees with similar results in [3,4]. In Table 3, we compare qualitatively some aligned topics learned by **T**, **S**, and **SKD** from the 20NG corpus. For the first three topics, **SKD** displays more word overlap than **S** with the corresponding topics from **T**, showing that **T** and **SKD** learn similar topic-word distributions. Interestingly, the fourth topic from **SKD** contains more healthcare-related words than the fourth topic from **T** although the latter is also primarily on healthcare; this shows that **SKD** can produce more coherent topics than **T**.

Table 3. Some selected topics output when **T**, **S**, and **SKD** models are run on the 20NG corpus for 20 topics. If a word in a topic from **S** or **SKD** is shared with the corresponding topic in **T**, then it is in **bold** otherwise it is in *italic*.

Model	ID	Topics
T	0	gun, law, firearm, crime, weapon, assault, amendment, state, police, permit
	11	russian, turkish, people, village, genocide, armenian, muslim, population, greek, army
	17	oil, engine, ride, front, road, chain, bike, motorcycle, water, gas
	3	health, make, president, patient, medical, people, doctor, disease, work, year
S	0	**law**, *people*, **state**, *government*, **gun**, **amendment**, *constitution*, **firearm**, **crime**, *privacy*
	1	**armenian**, **village**, *soldier*, *soviet*, **muslim**, *troop*, **turkish**, **russian**, **genocide**, *land*
	17	**engine**, *car*, *mile*, **ride**, **bike**, **oil**, **front**, *wheel*, **motorcycle**, *tire*
	7	**medical**, **disease**, *study*, *treatment*, **doctor**, **patient**, **health**, *food*, *risk*, *percent*
SKD	0	**gun**, **law**, **weapon**, **firearm**, **amendment**, **crime**, *bill*, **assault**, *constitution*, **police**
	11	**turkish**, **genocide**, **armenian**, **russian**, **village**, **population**, *israeli*, *war*, *attack*, **muslim**
	17	**ride**, **engine**, *car*, **bike**, **motorcycle**, **front**, **oil**, *motor*, **road**, *seat*
	3	**health**, **medical**, **doctor**, **disease**, **patient**, *insurance*, **treatment**, *drug*, *care*, *risk*

6   Conclusion

We have proposed a 2-Wasserstein loss-based knowledge distillation framework to compress a contextualized topic model. Experiments on two datasets show that the pruned topic model produces topics with coherence better than that of the topics produced by the student and even the larger teacher model. This is a new method for neural topic distillation. In the future, we would like to study it analytically and apply it to distill knowledge across other neural topic models.

References

1. Adhya, S., Sanyal, D.K.: What does the Indian Parliament discuss? An exploratory analysis of the question hour in the Lok Sabha. In: Proceedings of the LREC 2022 Workshop on Natural Language Processing for Political Sciences, Marseille, France, pp. 72–78. European Language Resources Association, June 2022. https://aclanthology.org/2022.politicalnlp-1.10

2. Arjovsky, M., Chintala, S., Bottou, L.: Wasserstein generative adversarial networks. In: Proceedings of the International Conference on Machine Learning, pp. 214–223. PMLR (2017). https://proceedings.mlr.press/v70/arjovsky17a.html

3. Bianchi, F., Terragni, S., Hovy, D.: Pre-training is a hot topic: contextualized document embeddings improve topic coherence. In: Proceedings of the 59th Annual Meeting of the Association for Computational Linguistics and the 11th International Joint Conference on Natural Language Processing (Volume 2: Short Papers), pp. 759–766 (2021). https://aclanthology.org/2021.acl-short.96/

4. Bianchi, F., Terragni, S., Hovy, D., Nozza, D., Fersini, E.: Cross-lingual contextualized topic models with zero-shot learning. In: The 16th Conference of the European Chapter of the Association for Computational Linguistics. Association for Computational Linguistics (2021). https://aclanthology.org/2021.eacl-main.143/

5. Blei, D.M., Ng, A.Y., Jordan, M.I.: Latent Dirichlet allocation. J. Mach. Learn. Res. **3**, 993–1022 (2003). https://www.jmlr.org/papers/volume3/blei03a/blei03a.pdf

6. Dieng, A.B., Ruiz, F.J., Blei, D.M.: Topic modeling in embedding spaces. Trans. Assoc. Comput. Linguist. **8**, 439–453 (2020). https://doi.org/10.1162/tacl_a_00325

7. Frogner, C., Zhang, C., Mobahi, H., Araya-Polo, M., Poggio, T.: Learning with a Wasserstein loss. In: Proceedings of the 28th International Conference on Neural Information Processing Systems, NIPS 2015, vol. 2, pp. 2053–2061. MIT Press, Cambridge (2015). https://papers.neurips.cc/paper/5679-learning-with-a-wasserstein-loss.pdf

8. Furlanello, T., Lipton, Z., Tschannen, M., Itti, L., Anandkumar, A.: Born again neural networks. In: Proceedings of the International Conference on Machine Learning, pp. 1607–1616. PMLR (2018). https://proceedings.mlr.press/v80/furlanello18a.html

9. Gao, R., Kleywegt, A.: Distributionally robust stochastic optimization with Wasserstein distance. Math. Oper. Res. (2022). https://doi.org/10.1287/moor.2022.1275

10. Gou, J., Yu, B., Maybank, S.J., Tao, D.: Knowledge distillation: a survey. Int. J. Comput. Vis. **129**(6), 1789–1819 (2021). https://doi.org/10.1007/s11263-021-01453-z

11. Guo, J., Cai, Y., Fan, Y., Sun, F., Zhang, R., Cheng, X.: Semantic models for the first-stage retrieval: a comprehensive review. ACM Trans. Inf. Syst. (TOIS) **40**(4), 1–42 (2022). https://doi.org/10.1145/3486250

12. Hinton, G., Vinyals, O., Dean, J.: Distilling the knowledge in a neural network, **2**(7). arXiv preprint arXiv:1503.02531 (2015)

13. Kingma, D.P., Welling, M.: Auto-encoding variational Bayes. In: Proceedings of the International Conference on Learning Representations (2014). https://arxiv.org/abs/1312.6114

14. Krasnashchok, K., Jouili, S.: Improving topic quality by promoting named entities in topic modeling. In: Proceedings of the 56th Annual Meeting of the Association for Computational Linguistics (Volume 2: Short Papers), pp. 247–253 (2018). https://aclanthology.org/P18-2040/

15. Lau, J.H., Newman, D., Baldwin, T.: Machine reading tea leaves: automatically evaluating topic coherence and topic model quality. In: Proceedings of the 14th Conference of the European Chapter of the Association for Computational Linguistics, Gothenburg, Sweden, pp. 530–539. Association for Computational Linguistics, April 2014. https://aclanthology.org/E14-1056
16. Montavon, G., Müller, K.R., Cuturi, M.: Wasserstein training of restricted Boltzmann machines. In: Advances in Neural Information Processing Systems, vol. 29 (2016). https://papers.nips.cc/paper/6248-wasserstein-training-of-restricted-boltzmann-machines
17. Nityasya, M.N., Wibowo, H.A., Chevi, R., Prasojo, R.E., Aji, A.F.: Which student is best? A comprehensive knowledge distillation exam for task-specific BERT models. arXiv preprint arXiv:2201.00558 (2022)
18. Olkin, I., Pukelsheim, F.: The distance between two random vectors with given dispersion matrices. Linear Algebra Appl. **48**, 257–263 (1982). https://doi.org/10.1016/0024-3795(82)90112-4
19. Ozair, S., Lynch, C., Bengio, Y., van den Oord, A., Levine, S., Sermanet, P.: Wasserstein dependency measure for representation learning. In: Wallach, H., Larochelle, H., Beygelzimer, A., d'Alché-Buc, F., Fox, E., Garnett, R. (eds.) Advances in Neural Information Processing Systems, vol. 32. Curran Associates, Inc. (2019). https://proceedings.neurips.cc/paper/2019/file/f9209b7866c9f69823201c1732cc8645-Paper.pdf
20. Pan, S., Wu, J., Zhu, X., Zhang, C., Wang, Y.: Tri-party deep network representation. In: Proceedings of the Twenty-Fifth International Joint Conference on Artificial Intelligence, IJCAI 2016, pp. 1895–1901. AAAI Press (2016). https://www.ijcai.org/Proceedings/16/Papers/271.pdf
21. Panaretos, V.M., Zemel, Y.: Statistical aspects of Wasserstein distances. Ann. Rev. Stat. Appl. **6**, 405–431 (2019). https://doi.org/10.1146/annurev-statistics-030718-104938
22. Reimers, N., et al.: Sentence-BERT: sentence embeddings using Siamese BERT-networks. In: Proceedings of the 2019 Conference on Empirical Methods in Natural Language Processing and the 9th International Joint Conference on Natural Language Processing (EMNLP-IJCNLP), pp. 3982–3992. Association for Computational Linguistics (2019). https://aclanthology.org/D19-1410
23. Röder, M., Both, A., Hinneburg, A.: Exploring the space of topic coherence measures. In: Proceedings of the Eighth ACM International Conference on Web Search and Data Mining, pp. 399–408 (2015). https://doi.org/10.1145/2684822.2685324
24. Srivastava, A., Sutton, C.: Autoencoding variational inference for topic models. In: International Conference on Learning Representations (2017). https://openreview.net/forum?id=BybtVK9lg
25. Terragni, S., Fersini, E., Galuzzi, B.G., Tropeano, P., Candelieri, A.: OCTIS: comparing and optimizing topic models is simple! In: Proceedings of the 16th Conference of the European Chapter of the Association for Computational Linguistics: System Demonstrations, pp. 263–270 (2021). https://aclanthology.org/2021.eacl-demos.31/
26. Vayansky, I., Kumar, S.A.: A review of topic modeling methods. Inf. Syst. **94**, 101582 (2020). https://doi.org/10.1016/j.is.2020.101582
27. Villani, C.: Topics in Optimal Transportation, vol. 58. American Mathematical Society (2021). https://www.math.ucla.edu/~wgangbo/Cedric-Villani.pdf

28. Zhai, C., Geigle, C.: A tutorial on probabilistic topic models for text data retrieval and analysis. In: The 41st International ACM SIGIR Conference on Research & Development in Information Retrieval, pp. 1395–1398 (2018). https://doi.org/10. 1145/3209978.3210189

29. Zhang, L., Song, J., Gao, A., Chen, J., Bao, C., Ma, K.: Be your own teacher: improve the performance of convolutional neural networks via self distillation. In: Proceedings of the IEEE/CVF International Conference on Computer Vision, pp. 3713–3722 (2019). https://doi.ieeecomputersociety.org/10.1109/ICCV.2019.00381

30. Zhang, Y., Jiang, T., Yang, T., Li, X., Wang, S.: HTKG: deep keyphrase generation with neural hierarchical topic guidance. In: Proceedings of the 45th International ACM SIGIR Conference on Research and Development in Information Retrieval, SIGIR 2022, pp. 1044–1054. Association for Computing Machinery, New York (2022). https://doi.org/10.1145/3477495.3531990

Towards Effective Paraphrasing
for Information Disguise

Anmol Agarwal[1]([✉]) [iD], Shrey Gupta[1] [iD], Vamshi Bonagiri[1] [iD], Manas Gaur[2] [iD],
Joseph Reagle[3] [iD], and Ponnurangam Kumaraguru[1] [iD]

[1] International Institute of Information Technology, Hyderabad, India
{anmol.agarwal,shrey.gupta}@students.iiit.ac.in,
vamshi.b@research.iiit.ac.in, pk.guru@iiit.ac.in
[2] University of Maryland, Baltimore County, USA
manas@umbc.edu
[3] Northeastern University, Boston, USA
joseph@reagle.org

Abstract. Information Disguise (*ID*), a part of computational ethics in Natural Language Processing (*NLP*), is concerned with best practices of textual paraphrasing to prevent the non-consensual use of authors' posts on the Internet. Research on ID becomes important when authors' written online communication pertains to sensitive domains, e.g., mental health. Over time, researchers have utilized AI-based automated word spinners (e.g., SpinRewriter, WordAI) for paraphrasing content. However, these tools fail to satisfy the purpose of ID as their paraphrased content still leads to the source when queried on search engines. There is limited prior work on judging the effectiveness of paraphrasing methods for ID on search engines or their proxies, neural retriever (*NeurIR*) models. We propose a framework where, for a given sentence from an author's post, we perform iterative perturbation on the sentence in the direction of paraphrasing with an attempt to confuse the search mechanism of a NeurIR system when the sentence is queried on it. Our experiments involve the subreddit "r/AmItheAsshole" as the source of public content and Dense Passage Retriever as a NeurIR system-based proxy for search engines. Our work introduces a novel method of phrase-importance rankings using perplexity scores and involves multi-level phrase substitutions via beam search. Our multi-phrase substitution scheme succeeds in disguising sentences 82% of the time and hence takes an essential step towards enabling researchers to disguise sensitive content effectively before making it public. We also release the code of our approach. (https://github.com/idecir/idecir-Towards-Effective-Paraphrasing-for-Information-Disguise)

Keywords: Neural information retrieval · Adversarial retrieval · Paraphrasing · Information disguise · Computational ethics

---

A. Agarwal and S. Gupta—Authors contributed equally.
M Gaur—Research with KAI[2] Lab @ UMBC.

1 Introduction

When a researcher quotes, verbatim, from an online post about a sensitive topic (e.g., politics, mental health, drug use), this could bring additional, unwanted scrutiny to the author of that post [1]. The supportive role of Reddit also allows a swarm of tracking technologies to pick the authors of the posts as subjects without consent, which can lead others to authors' profiles and posting history, using which other aspects of personal identity might be inferred [9].

Consequently, some researchers alter verbatim phrases, so their sources are not easily locatable via search services (e.g., Google Search). Researchers leverage traditional (e.g., summarization) or AI-based paraphrasing (e.g., Quillbot [6]) methods to disguise the content. These strategies are inspired by Bruckman et al.'s two most prominent methods of disguise: (a) Verbatim Quoting and (b) Paraphrasing [4]. Until recently, there have been no quantitative tests of the efficacy of such disguise methods and no description of how to do it well. In 2022, in an analysis of 19 Reddit research reports which had claimed to have heavily disguised the content, it was found that 11 out of the 19 reports failed to disguise their sources sufficiently; that is, one or more of their sources could be located via search services [20]. A complementary report [21] tested the efficacy of both human and automated paraphrasing techniques (i.e., Spin Rewriter and WordAI). The report's authors concluded that while word spinners (typically used for generating plagiarism and content farms) could improve the practice of ethical disguise, the research community needed openly specified techniques whose (non) locatability and fidelity to meaning and fluency were well understood.

In this work, we examine the ID problem through the purview of Black Box Adversarial NLP (e.g., perturbations) and NeurIR. So far, most of the methods in Adversarial NLP focus on downstream tasks such as classification and use labeled datasets to train their adversarial paraphrasing model in a supervised fashion [10]. In the context of ID, an effective paraphrase of a sentence should make a semantic NeurIR under-rank the source of the sentence. Our proposed method is entirely *unsupervised* as we use only the document ranks returned by the retriever to guide our model. Our research is not directed toward plagiarism. Instead, it focuses on preventing authors' from being a target of non-consensual experiments because of their online content.

We make the following contributions to the current research: (a) We devise a computational method based on expert rules [21] that attack phrases using BERT and counter-fitting vectors. (b) We automate the method to prioritize attack locations using perplexity metric to determine phrase importance ranking in Sect. 3.1. (c) We define a novel adaptation of beam search to make multi-level and multi-word perturbations for dynamic paraphrasing in Sect. 3.2. (d) We analyze the success of our proposed approach in Sect. 4 as a trade-off between locatability[1] and semantic similarity. In addition, experts in journalism and communication studies validate our insights. We use *Universal Sentence Encoder (USE)* [5] semantic similarity metric to ensure that the meaning is preserved

---

[1] A source document has high locatability if a system engine retrieves it in the top-K results when queried with one of the sentences within the document.

after paraphrasing. Due to the API limitations on Google Search, we test our approach on a NeurIR system - *Dense Passage Retriever (DPR)* [13].

2 Related Work

Paraphrasing is a well-studied problem in NLP literature [30] and so are the limitations of deep language models in NLP [23]. However, we study paraphrasing from the perspective of ID, a requirement for ethical research in NLP. Prior works in Adversarial NLP, such as the work by Alzantot et al., focused on word-level perturbations in sentences that fool a sentiment classifier [3]. Following it, Jia et al. [11] proposed a family of functions to induce robustness in NLP models working on sentiment classification (*SC*) and natural language inference (*NLI*). Jin et al. [12] introduced TextFooler, capable of generating paraphrased text, successfully confusing models for SC and NLI. Experiments with diversity-aware paraphrasing metrics like Jeffrey's divergence, Word Mover's Distance (*WMD*), etc., did not yield sentences that would make the author's identity on search engines non-locatable [27,29]. Most prior work on paraphrasing methods are specific to classification [7,8,15,16,22,28] and NLI [3,17] tasks. For a (query, document) pair, the prior work on adversarial retrieval [19,25,26] focuses mainly on causing highly relevant/non-relevant documents to be demoted/promoted in the rankings by making minimal changes in the document text. Our work is different in that (i) the document store for the retriever is fixed; (ii) we try to perturb the text in queries to demote the rank of the source post; (iii) we aim to use the paraphrased queries to recreate the paraphrased version of the document which can be made public by researchers. Hence, we investigate adversarial retrieval from the Information Disguise perspective and develop a method that performs retriever-guided paraphrasing of content for applications requiring ID.

3 Methodology

Dataset: We collected 2000 posts from *r/AmItheAsshole*. Given a post, we split its content into *chunks* (*documents*) with sentence boundaries preserved and include each chunk in the document store for DPR. We extracted 1748 one-line sentences (averaging 23 words/sentence) across these posts, which caused at least one of the documents extracted from the source post to be ranked within the top 2 when queried on DPR.

Problem Formulation: Given a sentence s_t derived from post P, let $\mathcal{R}(s_t, P)$ denote the numerically lowest rank among the ranks of documents derived from post P when s_t is queried on DPR. We aim to generate s_p, i.e., a paraphrase of s_t, to maximize $\mathcal{R}(s_p, P)$ for making the post P non-locatable, under the constraints that $Sim(s_t, s_p) \geq \epsilon$, where ϵ is a chosen semantic similarity threshold.

System Architecture: Let $s_t = \langle w_1, .., w_L \rangle$ be a sentence with L tokens. As pointed out in [28], if we intend to perturb n disjoint substrings (phrases) of a string, with each substring having m potential replacements, the search space of potential paraphrases of s_t will have $(m + 1)^n - 1$ possibilities. To reduce this complexity, we first discuss our method to paraphrase with single-phrase perturbations and then extend it for multi-phrase perturbations using beam search over the search space.

3.1 Level One Attack with Single Phrase Perturbation

Identifying Attackable Phrases of the Sentence: For s_t, there are $\frac{L*(L-1)}{2} +$ L candidate phrases of the form: $ph \equiv \langle w_l,.., w_r \rangle$ $(l \leq r)$ which can be replaced. To consider attacking only those substrings with proper independent meaning, we consider only those substrings which are present as a node in the constituency-based parse tree of the query s_t obtained using the Berkeley Neural Parser [14].

Ranking Attackable Parse Tree Nodes Based on Perplexity: Let the set of nodes in constituency-based parse tree T for s_t be H. Let N_T^{str} *be the substring present at node* N_T of the parse tree. We define the score $PLL(N_T)$ for the node as follows: **(Step 1)** Mask N_T^{str} within s_t and use BERT masked language model to find the most likely substitution Z. **(Step 2)** Replace mask with Z resulting in new sentence $S = \langle w_1,..,w_{l-1},Z,w_{r+1},..,w_L \rangle$. **(Step 3)** $PLL(N_T)$ = Pseudo log-likelihood of the sentence S obtained from BERT by iteratively masking every word in the sentence and then summing the log probabilities, as also done in [2,24].

Let $AN(s_t) = \{N_{T1},.., N_{TP}\}$ be the top P nodes when ranked on basis of highest $PLL(N_T)$ scores. Since $PLL(s_t)$ is constant across all parse tree nodes, the difference $(PLL(N_T) - PLL(s_t))$ and hence, $PLL(N_T)$ helps us capture the peculiarity of phrase N_T^{str} and hence, its contribution in making the source document d locatable when queried using s_t (see Fig. 1).

Generating Suggestions for Attacking at a Parse Tree Node: For each parse tree node N_T in $AN(s_t)$, we generate candidate perturbations in 2 ways: **(a)** Bert-masking based candidates $B_{cand}(N_T)$: Generated by masking N_T^{str} within s_t and using BERT to generate 10 replacements [8], **(b)** Synonym-based candidates $CF_{cand}(N_T)$: For nodes containing a single token $(l=r)$, we replace N_T^{str} i.e., $\langle w_{l=r} \rangle$ with the 10 nearest neighbours in counter-fitting word embedding space [18], as also done in [28], for producing synonym-based replacements. This leads to the set of candidate perturbations when attacking via the parse tree node N_T to be $Sug_{cand}(N_T) = B_{cand}(N_T) \cup CF_{cand}(N_T)$. As a result, the set of the candidate perturbations (i.e., $CP(s_t)$) derived for query s_t will be the union of candidate perturbations across top-ranked P parse tree nodes in the constituency-based parse tree of s_t: $CP(s_t) = \bigcup_{N_T \in AN(s_t)} Sug_{cand}(N_T)$.

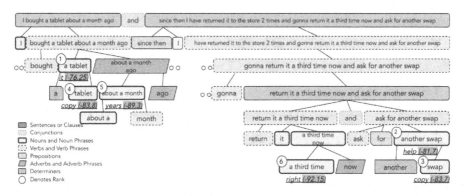

Fig. 1. Constituency-based parse tree of s_t = "I bought a tablet about a month ago and since then I have returned it to the store 2 times and gonna return it a third time now and ask for another swap." The highest ranked node has a PLL score of -76 after replacing *"a tablet"* with *"it"*. The second highest ranked node has a PLL score of -81.7 after replacing *"another swap"* with *"help"*. (The ethical import of this sample sentence is minimal; it was taken from an innocuous submission (i.e., dings on a tablet) without any identifying information; it was quickly severed from its author (deleted) and consequently does not appear to be indexed by Pushshift or Google. This is applicable for both Fig. 1 and Fig. 2)

3.2   Augmenting the Level One Attack to Multiple Levels

The state-space for multiphrase perturbation can be considered a tree where the node at level *num* includes *num* phrase substitutions on s_t and is obtained using the method described in Sect. 3.1 *num* times sequentially. As discussed, the size of this search space is vast. Hence, we use *beam search* where the number of nodes expanded at each level in the search tree is restricted to beam width k. This selection of k nodes is achieved based on a heuristic function that scores each node based on its potential to have a quality solution in its subtree (see Fig. 2).

Algorithm 1 Multilevel Perturbation for Paraphrasing

Input: s_t: Original sentence
    ϵ: Semantic similarity threshold
    MaxLevels: Number of levels in beam search tree
    $f(\cdot)$: Scoring function for ranking candidates for next beam level
    $SimScore(\cdot,\cdot)$: Semantic Similarity between 2 sentences
    MaxBeamSize: Maximum number of nodes in a beam level
    α: Parameter for differently weighing locatability and semantic similarity
    $\geqslant$: Comparator function

1: CurrentNodesInBeam $\leftarrow$ $List[BeamNode(s_t)]$
2: NextLevelCandidates $\leftarrow$ $priority\_queue(\geqslant)$
3: **for** $i \leftarrow 1$ to MaxLevels :
4:     **for** $\mathcal{N} \in$ CurrentNodesInBeam :
5:       $\mathcal{N}$.MakeParseTree()
6:       $\mathcal{N}$.FilterAlreadyAttackedLocations()
7:       top-p $\leftarrow \mathcal{N}$.RankAndFetchAttackLocations()
8:       Candidates $\leftarrow \bigcup_{N_T \in \text{top-p}} Sug_{cand}(N_T)$
9:       Candidates $\leftarrow FilterAndKeep(x \in$ Candidates s.t $Sim(x, s_t) \geqslant \epsilon)$
10:      UpdateBestResults()
11:      **for** $C \in$ Candidates :
12:        NextLevelCandidates.$UniquePush(C)$
13:      **end for**
14:     **end for**
15:     CurrentNodesInBeam $= List[BeamNode(x)]$ for x in NextLevelCandidates]
16:     NextLevelCandidates.clear()
17: **end for**

Algorithm Explanations: (**1**) *FilterAlreadyAttackedLocations()*: Removes those nodes in the parse tree whose phrase has already been replaced once in an attack on one of the previous beam levels. (**2**) *RankAndFetchAttackLocations()*: Ranks the remaining parse tree nodes based on perplexity scores defined in

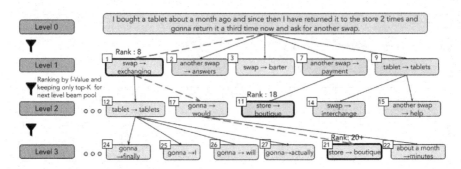

Fig. 2. Beam Search Tree for a sample query s_t (only 5 nodes shown at each level for clarity). The best attack at levels 1 and 2, corresponding to node IDs 1 and 11, pushed $\mathcal{R}(s_p, P)$ from 1 to 8 and 18, respectively. The best attack at level 3 (shown by the dashed path): "I bought a tablet about a month ago, and since then, I have returned it to the *boutique* 2 times, and *would* return it a third time now and ask for another *exchanging*.": succeeded in displacing all documents from the source post outside top 20 while maintaining high semantic similarity of 0.93 with the original query.

Sect. 3.1. **(3)** *f(s)*: For a candidate s, the heuristic score to estimate its potential is calculated as $(1-\alpha)*Sim(s, s_t)+\alpha*(R(s,P)-1)/20$. **(4)** *UniquePush()*: The priority queue order is determined by operator $\geq$ on $f(s)$ for each candidate s. The size is restricted to "beam width", and s is not pushed if another element in the queue already has the same text as s.

4 Evaluation

We measure the success of our attack on s_t, i.e., query Q based on whether all documents extracted from source post P are absent from the top-K (K=1, 5, 10, and 20) retrieved documents by DPR when s_t is queried. To measure the overall effectiveness of our perturbation mechanism, we define the *Hit-Rate@K (HR)* metric given by: $HR@K = \frac{(\Sigma_i^{N_q} checkTopK(Q_i))}{N_q}$, where $checkTopK(Q_i)$ returns 1 or 0 depending on whether any of the extracted documents from the target post for query Q_i is in the top-K retrieved documents by DPR and N_q is the total number of queries, i.e., 1748.

For single-level perturbations, as shown in Table 1, we compare the performance of the attack schemes using BERT and counter-fitting vectors by varying the semantic similarity threshold (ϵ) and the number of parse tree nodes to be attacked ($P = 1, 5, All/*$), after ranking based on perplexity scores. We see that for lower values of ϵ, BERT substitutions are effective at reducing $HR@K$, and preserving the grammatical structure, but the replacements do not take semantics of the replaced phrase into account. Therefore, we filter out suggestions using a high semantic similarity threshold ($\epsilon = 0.95$). On the other hand, attacking using counter-fitting vectors replaces words with close synonyms and preserve meaning, so it performs slightly better than BERT on high thresholds, fooling the retriever 23% of the time when K = 5, P = 5, and $\epsilon = 0.95$ compared to BERT which succeeds only 15% of the time.

Table 1. Hit-Rate@K (HR) scores for single-level perturbation approach when attacking using (1) Bert-masking based candidates only, (2) Synonym-based candidates only. Scores are shown across varying values of ϵ (minimum semantic similarity required between the paraphrase and original sentence) and P (the number of parse tree nodes to attack after perplexity-based ranking).

HR	Bert-masking based candidates only									Synonym-based candidates only								
	$\epsilon = 0$			$\epsilon = 0.5$			$\epsilon = 0.95$			$\epsilon = 0$			$\epsilon = 0.5$			$\epsilon = 0.95$		
	P=1	P=5	P=*	P=1	P=5	P=*	P=1	P=5	P=*	P=1	P=5	P=*	P=1	P=5	P=*	P=1	P=5	P=*
K = 1	0.24	0.07	0.04	0.35	0.09	0.05	0.86	0.56	0.36	0.39	0.19	0.14	0.39	0.20	0.14	0.70	0.40	0.31
K = 5	0.48	0.25	0.17	0.59	0.29	0.20	0.96	0.85	0.72	0.70	0.47	0.39	0.70	0.53	0.39	0.91	0.77	0.63
K = 10	0.56	0.34	0.26	0.67	0.39	0.29	0.97	0.92	0.83	0.79	0.60	0.52	0.79	0.60	0.52	0.96	0.85	0.76
K = 20	0.64	0.44	0.35	0.75	0.50	0.40	0.99	0.96	0.92	0.86	0.71	0.64	0.86	0.71	0.64	0.97	0.92	0.87

For multi-level attacks using beam search, we use a combination of BERT and counter-fitting based attack strategy owing to the insights obtained from single-level attack experiments. In Table 2, we report the $HR@K$ value for $MaxLevels = \{1, 2, 3\}$ i.e., when perturbing $MaxLevels$ disjoint phrases within the sentence. We see that only 29% ($1 - HR@20$ at Level 1) of the attacks work in the first level. The lowest hit rate is obtained at level 3, where attacking just via 4 parse tree nodes is enough for our attack to succeed 82% of the time in sending all documents from the source post outside top 20 despite being constrained under $\epsilon = 0.8$. This success rate is higher than the single-level perturbation success rate: 65% even when $\epsilon = 0$ and P=*.

Table 2. Hit-Rate@K for Beam Search where: MaxLevels = 3, MaxBeamSize = 10, $\alpha = 0.8$, $\epsilon = 0.8$, and P = 4.

HR@K	Level 1	Level 2	Level 3
K = 1	0.18	0.06	**0.04**
K = 5	0.46	0.17	**0.10**
K = 10	0.60	0.24	**0.13**
K = 20	0.71	0.34	**0.18**

5 Conclusion

We introduce a novel black-box framework for effectively paraphrasing text for Information Disguise. Our method uses an unsupervised approach for paraphrasing, where we rank potential attack areas via perplexity scores and generate perturbations using BERT and counter-fitting word vectors. We expand our approach into a multi-phrase substitution setting enabled via beam search. We succeeded in effectively disguising 82% of the queries by displacing their sources outside a rank of the top 20 when queried on DPR while maintaining high semantic similarity. Our approach can be used to effectively disguise an entire post by concatenating the perturbed versions of the individual sentences within the post from neural retrievers. However, currently, our approach does not take the grammatical quality of the paraphrased sentences into account. Due to the large number of requests made to the retriever to disguise a sentence, our approach is unlikely to work on actual search engines due to API limits. Achieving comparable results as ours while not exceeding the API limits is an interesting problem we wish to solve in our future work.

338   A. Agarwal et al.

References

1. Adams, N.N.: 'Scraping' reddit posts for academic research? addressing some blurred lines of consent in growing internet-based research trend during the time of covid-19. Int. J. Soc. Res. Methodol., 1–16 (2022). https://doi.org/10.1080/13645579.2022.2111816
2. Alikaniotis, D., Raheja, V.: The unreasonable effectiveness of transformer language models in grammatical error correction. In: Proceedings of the Fourteenth Workshop on Innovative Use of NLP for Building Educational Applications, pp. 127–133. Association for Computational Linguistics, Florence, August 2019. https://doi.org/10.18653/v1/W19-4412. https://aclanthology.org/W19-4412
3. Alzantot, M., Sharma, Y., Elgohary, A., Ho, B.J., Srivastava, M., Chang, K.W.: Generating natural language adversarial examples. In: Proceedings of the 2018 Conference on Empirical Methods in Natural Language Processing. pp. 2890–2896. Association for Computational Linguistics, Brussels, October–November 2018. https://doi.org/10.18653/v1/D18-1316. https://aclanthology.org/D18-1316
4. Bruckman, A.: Studying the amateur artist: a perspective on disguising data collected in human subjects research on the Internet. Ethics Inf. Technol. 4(3), 217–231 (2002)
5. Cer, D., et al.: Universal sentence encoder, March 2018
6. Fitria, T.N.: Quillbot as an online tool: Students' alternative in paraphrasing and rewriting of english writing. Englisia: J. Lang. Educ. Humanities 9(1), 183–196 (2021)
7. Gao, J., Lanchantin, J., Soffa, M.L., Qi, Y.: Black-box generation of adversarial text sequences to evade deep learning classifiers. In: 2018 IEEE Security and Privacy Workshops (SPW), pp. 50–56 (2018)
8. Garg, S., Ramakrishnan, G.: BAE: BERT-based adversarial examples for text classification. In: Proceedings of the 2020 Conference on Empirical Methods in Natural Language Processing (EMNLP), pp. 6174–6181. Association for Computational Linguistics, Online, November 2020. https://doi.org/10.18653/v1/2020.emnlp-main.498. https://aclanthology.org/2020.emnlp-main.498
9. HRW: "how dare they peep into my private life?" October 2022. https://www.hrw.org/report/2022/05/25/how-dare-they-peep-my-private-life/childrens-rights-violations-governments
10. Iyyer, M., Wieting, J., Gimpel, K., Zettlemoyer, L.: Adversarial example generation with syntactically controlled paraphrase networks. In: Proceedings of the 2018 Conference of the North American Chapter of the Association for Computational Linguistics: Human Language Technologies, Volume 1 (Long Papers), pp. 1875–1885. Association for Computational Linguistics, New Orleans, June 2018. https://doi.org/10.18653/v1/N18-1170. https://aclanthology.org/N18-1170
11. Jia, R., Raghunathan, A., Göksel, K., Liang, P.: Certified robustness to adversarial word substitutions. In: Proceedings of the 2019 Conference on Empirical Methods in Natural Language Processing and the 9th International Joint Conference on Natural Language Processing (EMNLP-IJCNLP), pp. 4129–4142. Association for Computational Linguistics, Hong Kong, November 2019. https://doi.org/10.18653/v1/D19-1423. https://aclanthology.org/D19-1423
12. Jin, D., Jin, Z., Zhou, J.T., Szolovits, P.: Is bert really robust? a strong baseline for natural language attack on text classification and entailment. In: Proceedings of the AAAI Conference on Artificial Intelligence, vol. 34, pp. 8018–8025 (2020)

13. Karpukhin, V., et al.: Dense passage retrieval for open-domain question answering. In: Proceedings of the 2020 Conference on Empirical Methods in Natural Language Processing (EMNLP), pp. 6769–6781. Association for Computational Linguistics, Online, November 2020. https://doi.org/10.18653/v1/2020.emnlp-main.550. https://aclanthology.org/2020.emnlp-main.550

14. Kitaev, N., Cao, S., Klein, D.: Multilingual constituency parsing with self-attention and pre-training. In: Proceedings of the 57th Annual Meeting of the Association for Computational Linguistics, pp. 3499–3505 (2019)

15. Li, J., Ji, S., Du, T., Li, B., Wang, T.: Textbugger: generating adversarial text against real-world applications. In: 26th Annual Network and Distributed System Security Symposium, NDSS 2019, San Diego, California, USA, 24–27 February 2019. The Internet Society (2019). https://www.ndss-symposium.org/ndss-paper/textbugger-generating-adversarial-text-against-real-world-applications/

16. Li, L., Ma, R., Guo, Q., Xue, X., Qiu, X.: BERT-ATTACK: adversarial attack against BERT using BERT. In: Proceedings of the 2020 Conference on Empirical Methods in Natural Language Processing (EMNLP), pp. 6193–6202. Association for Computational Linguistics, Online, November 2020. https://doi.org/10.18653/v1/2020.emnlp-main.500. https://aclanthology.org/2020.emnlp-main.500

17. Minervini, P., Riedel, S.: Adversarially regularising neural nli models to integrate logical background knowledge. In: Conference on Computational Natural Language Learning (2018)

18. Mrkšić, N., et al.: Counter-fitting word vectors to linguistic constraints. In: Proceedings of the 2016 Conference of the North American Chapter of the Association for Computational Linguistics: Human Language Technologies, pp. 142–148. Association for Computational Linguistics, San Diego, June 2016. https://doi.org/10.18653/v1/N16-1018. https://aclanthology.org/N16-1018

19. Raval, N., Verma, M.: One word at a time: adversarial attacks on retrieval models. arXiv preprint arXiv:2008.02197 (2020)

20. Reagle, J.: Disguising Reddit sources and the efficacy of ethical research. Ethics Inf. Technol. 24(3), September 2022

21. Reagle, J., Gaur, M.: Spinning words as disguise: shady services for ethical research? First Monday, January 2022

22. Ren, S., Deng, Y., He, K., Che, W.: Generating natural language adversarial examples through probability weighted word saliency. In: Annual Meeting of the Association for Computational Linguistics (2019)

23. Ribeiro, M.T., Wu, T., Guestrin, C., Singh, S.: Beyond accuracy: Behavioral testing of NLP models with CheckList. In: Proceedings of the 58th Annual Meeting of the Association for Computational Linguistics. pp. 4902–4912. Association for Computational Linguistics, Online, July 2020. https://doi.org/10.18653/v1/2020.acl-main.442. https://aclanthology.org/2020.acl-main.442

24. Salazar, J., Liang, D., Nguyen, T.Q., Kirchhoff, K.: Masked language model scoring. In: Proceedings of the 58th Annual Meeting of the Association for Computational Linguistics, pp. 2699–2712. Association for Computational Linguistics, Online, July 2020. https://doi.org/10.18653/v1/2020.acl-main.240. https://aclanthology.org/2020.acl-main.240

25. Wang, Y., Lyu, L., Anand, A.: Bert rankers are brittle: a study using adversarial document perturbations. In: Proceedings of the 2022 ACM SIGIR International Conference on Theory of Information Retrieval, ICTIR 2022, pp. 115–120. Association for Computing Machinery, New York (2022). https://doi.org/10.1145/3539813.3545122. https://doi.org/10.1145/3539813.3545122

26. Wu, C., Zhang, R., Guo, J., de Rijke, M., Fan, Y., Cheng, X.: Prada: practical black-box adversarial attacks against neural ranking models. ACM Trans. Inf. Syst., December 2022. https://doi.org/10.1145/3576923. https://doi.org/10.1145/3576923

27. Xu, Q., Zhang, J., Qu, L., Xie, L., Nock, R.: D-page: diverse paraphrase generation. CoRR abs/1808.04364 (2018). https://arxiv.org/abs/1808.04364

28. Yoo, J.Y., Qi, Y.: Towards improving adversarial training of NLP models. In: Findings of the Association for Computational Linguistics: EMNLP 2021, pp. 945–956. Association for Computational Linguistics, Punta Cana, Dominican Republic, November 2021. https://doi.org/10.18653/v1/2021.findings-emnlp.81. https://aclanthology.org/2021.findings-emnlp.81

29. Zhao, W., Peyrard, M., Liu, F., Gao, Y., Meyer, C.M., Eger, S.: MoverScore: Text generation evaluating with contextualized embeddings and earth mover distance. In: Proceedings of the 2019 Conference on Empirical Methods in Natural Language Processing and the 9th International Joint Conference on Natural Language Processing (EMNLP-IJCNLP), pp. 563–578. Association for Computational Linguistics, Hong Kong, November 2019. https://doi.org/10.18653/v1/D19-1053. https://aclanthology.org/D19-1053

30. Zhou, J., Bhat, S.: Paraphrase generation: a survey of the state of the art. In: Proceedings of the 2021 Conference on Empirical Methods in Natural Language Processing, pp. 5075–5086 (2021)

Generating Topic Pages for Scientific Concepts Using Scientific Publications

Hosein Azarbonyad[(✉)], Zubair Afzal, and George Tsatsaronis

Elsevier, Amsterdam, The Netherlands
{h.azarbonyad,zubair.afzal,g.tsatsaronis}@elsevier.com

Abstract. In this paper, we describe *Topic Pages*, an inventory of scientific concepts and information around them extracted from a large collection of scientific books and journals. The main aim of *Topic Pages* is to provide all the necessary information to the readers to understand scientific concepts they come across while reading scholarly content in any scientific domain. *Topic Pages* are a collection of automatically generated information pages using NLP and ML, each corresponding to a scientific concept. Each page contains three pieces of information: a definition, related concepts, and the most relevant snippets, all extracted from scientific peer-reviewed publications. In this paper, we discuss the details of different components to extract each of these elements. The collection of pages in production contains over $360,000$ *Topic Pages* across 20 different scientific domains with an average of 23 million unique visits per month, constituting it a popular source for scientific information.

Keywords: Scientific document processing · Definition extraction · Multi-document summarization

1 Introduction

Technical terminology is an important piece of scientific publications [6,7]. Scientists and researchers use technical terminology and concepts to convey information concisely. As a result, there is an overwhelming and growing number of scientific concepts in any scientific domain, adding to the difficulties scientists have to catch up with the ever-growing list of technical concepts and new content. Knowledge sources such as *Wikipedia* can provide useful information on technical and scientific concepts to a large extent, however, due to their *"wisdom-of-crowds"* creation method there are many omissions and errors, and they may not always be a trustworthy source to understand and refer to a scientific concept. Our *Topic Pages*[1] proposition creates a knowledge source in a *"wisdom-of-experts"* fashion, as the information on scientific concepts is extracted from iconic scientific books in the domain, or from high-impact peer-reviewed scientific publications on the topic (Fig. 1).

Each *Topic Page* is centered around one scientific concept and contains a definition for the concept, a set of related concepts, and a set of relevant snippets

---

[1] https://www.elsevier.com/solutions/sciencedirect/topics.

J. Kamps et al. (Eds.): ECIR 2023, LNCS 13981, pp. 341–349, 2023.
https://doi.org/10.1007/978-3-031-28238-6_23

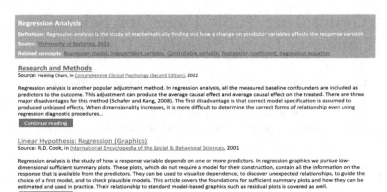

Fig. 1. An example *Topic Page* presenting the concept "regression analysis"', with a definition, related concepts, and a set of relevant snippets extracted from articles and books.

all extracted from scientific peer-reviewed articles and books. The definition comprises one sentence extracted from books and journals that provides a brief, yet concise, description of the concept. Snippets are text excerpts from books or journals, relevant to the concept, and provide contextual information about the concept. Related concepts are a set of most relevant concepts to the given concept that can help users to explore the relevant terminology around their concept of interest.

The collection contains over 360,000 *Topic Pages* in 20 different scientific domains. These topic pages are hyperlinked from publications in ScienceDirect[2], which is one of the largest scientific publication search engines and databases containing over 18 million full-text articles, helping users to navigate to the corresponding *Topic Page* when they encounter an unfamiliar scientific concept in an article with just one click. There are over 5.8 million articles that provide hyperlinks that we have created from scientific articles to topic pages. *Topic Pages* attract over 23 million unique visits per month.

In the remainder of the paper, we briefly review related work in Sect. 2, we describe the technical pipeline for generating *Topic Pages* in Sect. 3, we evaluate empirically the most challenging module of the pipeline, which is the definition extraction, in Sect. 4 and we conclude in Sect. 5 by arraying some limitations of the current technical solution and provide pointers to future work.

2   Related Work

To the best of our knowledge, there is no similar solution to the one introduced in this paper for automatically generating topic pages for scientific concepts. Most of the related work falls under the definition extraction task, and this is where we put the focus in this section. Early work on definition extraction task was focused on rule-based and pattern-matching approaches [3,8,20],

---

[2] https://www.sciencedirect.com/.

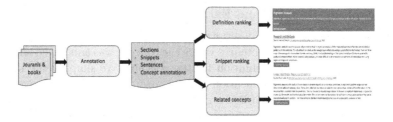

Fig. 2. An overview of the topic pages generation pipeline including all essential components.

often resulting in low recall given their limited coverage. Supervised models have also been proposed and shown to be more effective than the rule-based methods for this task [6,9,14–16]. These models use statistical information regarding concepts, as well as structural information of the sentences such as part of speech (*POS*) tags to distinguish definitional from non-definitional sentences. More recent work for definition extraction focused on using neural models for the task [2,5,7,11,13,18,19]. Notably, *LSTM* [11] and a combination of *CNN* and *LSTM* [2] have been used to learn the structure of definitional sentences. In our work, we also introduce and use a combined *LSTM+CNN* model but, different from [2], we capture both semantic (learned from the sentence itself) and structural information within sentences (learned from POS tags). A joint model that encodes sentences and their structure has been used before in [7], but, unlike the task tackled in that work, our definition extraction component assumes that the term is known and tries to detect whether the given candidate sentence is a good definition for the term or not.

3   Topic Pages Pipeline

There are four main components for generating *Topic Pages*, as shown in Fig. 2: an annotation module, a definition ranking module, a snippet ranking module, and a related concept extraction module.

3.1   Article Annotation

The annotation module receives content in *XML* format, finds concepts' mentions in articles and books, and then feeds the sentences and snippets mentioning a concept into the subsequent components. Each section in the article is considered a snippet. After we perform sentence splitting, we annotate concepts in sentences by using a simple dictionary look-up against the *Omniscience* taxonomy [12] which is a taxonomy of scientific concepts. If an abbreviation for the concept is proposed in the text, such as "Machine Learning (ML)", then the abbreviation (ML) is also added as an alias for the concept and is looked in the article. We use the Schwartz and Hearst method [17] to detect such abbreviations. If multiple concepts partially share some span (of an annotation), we annotate the span with the longest concept and ignore the short annotation.

3.2   Definition Ranking and Extraction

Definitions provide a concise description of the concept. For each concept, and per domain, we rank all the sentences where the concept was annotated and select the top-ranked one as the definition for the concept. We simplify the machine learning task to binary classification where, given a concept and a candidate sentence, the model predicts if it is a good definition for the concept or not. For a target concept, candidate sentences are ranked based on the score the classifier assigns to them and the top-ranked sentence is used as the definition. We use two different models for the definition classification task: an *LSTM+CNN* and a *SciBERT* model.

LSTM+CNN Model. Previous work [2,11] used *LSTM* [4] and *CNN* [10] models to classify sentences in the definition classification task. We use a combined approach that uses two *LSTMs* and two *CNNs*: one *LSTM* gets the actual sentence as the input and captures the sequential patterns of terms, and the other *LSTM* gets the *POS* tags of the words in the sentence as the input and captures the sequential patterns of syntax in the sentence. One *CNN* gets the actual sentence as the input captures the spatial distribution of terms, and the other *CNN* gets the *POS* tags of the words in the sentence as the input and captures the spatial distribution of grammatical elements. We concatenate the representations learned by each of these models and feed it to a feed-forward *MLP* layer which does the classification, using cross-entropy loss for training.

SciBERT Model. We use the *SciBERT* model [1] which is trained on scientific articles. As input, we feed the concept and the candidate sentence separated with a special token ([SEP]) to the *SciBERT* model and get the representation of the [CLS] token. This representation is then fed to a simple feed-forward layer which does the classification, using cross-entropy loss for optimization.

3.3   Snippet Ranking

For a given concept, all snippets annotated with the concept are collected and ranked by a snippet ranking method. The top 10 snippets are used for generating the *Topic Page* for the concept. We use a lexical matching model that scores snippets using a simple location-aware term frequency score as follows:

$$F(c, s) = \frac{tf(c, s)}{|s|} * (1 - \frac{l_1(c)}{|s|}) \tag{1}$$

where c and s are a concept and a snippet respectively, $tf(c, s)$ is the frequency of c in s, $|s|$ is the length of s, and $l_1(c)$ is the location of the first occurrence of c in s. Hence, the earlier the concept is mentioned in a snippet, the higher the score the snippet would receive.

3.4   Related Concept Extraction

To find the most relevant concepts to a given concept, we retrieve all co-occurring concepts in snippets. Concepts are then ranked based on the number of their co-occurrence with the target concept and the top 5 concepts are selected as the related concepts to the target concept.

Table 1. Performance of different definition classification models on the WCL dataset in terms of macro-averaged precision, recall, and F1.

Model	Precision	Recall	F1
Jin et al. [6]	0.92	0.79	0.85
Li et al. [11]	0.90	0.92	0.91
Navigli and Velardi [14]	**0.99**	0.61	0.85
LSTM+CNN	0.94	0.91	0.92
SciBERT	0.94	**0.93**	**0.93**

4 Results

In this section, we describe the scientific content collection and the used taxonomy (*Omniscience*) that are the basis of the *Topic Pages*. We further discuss the results of the different definition ranking models on two datasets and provide some statistics of the generated *Topic Pages* and usage statistics over time. We leave a large-scale evaluation of the snippet ranking and the related concept extraction modules to future work.

4.1 Datasets and Baselines

We use a collection of articles published in over $2,700$ journals as well as the content of $43,000$ books to generate the *Topic Pages*. This collection contains over 18 million articles and book chapters in *XML* format. All journals and books belong to different scientific domains. We use the *OmniScience* taxonomy to build the *Topic Pages*, which contain over 700K concepts for the 20 domains.

To evaluate the performance of the definition ranking module, we use the *WCL* dataset [14] which contains $4,619$ sentences labeled either as definitional ("good") or non-definitional ("bad") sentences regarding a concept. We follow the same setup as [11] for training and evaluating models on this dataset. We additionally use a proprietary dataset containing $43,368$ sentences extracted from articles and books distributed across 8 different domains and labeled by subject matter experts for the definition evaluation task as either "good" or "bad" definitions regarding a concept. We compare the performance of several models including the *LSTM+CNN*, *SciBERT*, Navigli and Velardi [14], Li et al. [11], and Jin et al. [6] on the *WCL* dataset. We further evaluate the performance of the best-performing models on the proprietary dataset. For the *LSTM+CNN* model, the batch size is set to 32, the number of hidden layers of the *LSTM* model is set to 128, and word embeddings are initiated with *GloVe* and fine-tuned during training. The *MLP* module has a hidden layer with 256 dimensions trained for 10 epochs. The *SciBERT* model is trained for 8 epochs with a batch size of 16. We perform 10-fold cross-validation and report the average performance.

4.2 Results of the Definition Extraction Models

Table 1 shows the performance of different models on the *WCL* dataset. This dataset is extracted from Wikipedia and most of the Wikipedia-based definitions follow a similar structure, making them easy to classify. The *SciBERT* model achieves the best *F1* score on this dataset. Navigli and Velardi [14] have higher precision than all models but a very low recall compared to *SciBERT*. The higher performance of the *SciBERT* model compared to the *LSTM+CNN* model shows that *SciBERT* can learn both sequential and spatial distribution of words in definitional sentences as well as the structural information within such sentences.

We further evaluate the performance of the top-performing models (*SciBERT* and *LSTM+CNN*) on the proprietary dataset which is much larger than the *WCL* dataset; results are shown in Table 3. This dataset contains definitions from various sources. Unlike Wikipedia-based definitions, definitions extracted from different books and journals do not follow a similar structure which makes the classification task more difficult, hence the lower performance of the two models compared to the *WCL* dataset. The *SciBERT* model outperforms the *LSTM+CNN* model on this dataset as well across all domains. This again confirms the ability of the *SciBERT* model in modeling semantics and the structure of definitions. Moreover, *SciBERT* has consistently higher performance than the *LSTM+CNN* on all individual domains except *Social Sciences*. As *SciBERT* is pre-trained on publications in the biomedical and computer science domains the low performance of this model on domains such as *Social Sciences* may be attributed to this fact. On the other hand, as the results show, *SciBERT* performs better on domains such as *Chemistry* and *Material Sciences* as such domains are closer to its trained domains.

Table 2. Example of errors (false positives) of the SciBERT-based models.

Concept	Definition	Error source
Association List	An association list is simply a list of name value pairs	Too generic
Hierarchical DB	In a hierarchical DB relationships are defined by storage structure	Too generic
Habilitation	The acquisition of abilities not possessed previously	Too specific
Sample Space	the set of all possible outcomes in a probability model	Partially good

Other than the domain difference, the additional errors should be attributed to the inherent difficulty of the task. Based on our analysis, the biggest sources of errors are the false positives which are mainly caused by generic, specific, or partially good definitions. Table 2 shows examples of definitions wrongly labeled by the *SciBERT* model and the possible explanation for the errors. Generic definitions are good definitions but they cover a very broad aspect of the concept. Specific definitions are also good definitions but they contain unnecessary additional information. Partially good definitions cover only some essential aspects of the concept. All these cases are labeled as "bad definitions" by the subject-matter experts but detected as "good definitions" by the model. To handle such

Table 3. Performance of the *LSTM+CNN* and *SciBERT* models on five domains.

Domain	SciBERT			LSTM+CNN		
	Precision	Recall	F1	Precision	Recall	F1
Chemistry	0.78	0.80	0.79	0.69	0.68	0.68
Earth Sciences	0.80	0.84	0.82	0.66	0.64	0.65
Material Sciences	0.80	0.88	0.83	0.50	0.49	0.49
Computer Science	0.56	0.60	0.58	0.43	0.48	0.45
Social Sciences	0.39	0.43	0.41	0.38	0.46	0.42
All domains	**0.79**	**0.78**	**0.78**	0.70	0.69	0.69

cases, the model should have an understanding of the generality or specificity of the concept which can be quite challenging to model.

The *Topic Pages* product contains over 363,000 topic pages in 20 different scientific domains. Topic pages have over 23 million visits per month making them one of the popular knowledge bases among researchers and students. There are about 63,000 concepts without a definition on *Topic Pages* mostly due to the bad performance of the current production model (*LSTM+CNN*) in some domains.

5 Conclusions and Discussion

In this paper, we introduced *Topic Pages*, a publicly available knowledge base for scientific concepts with their definitions, most relevant concepts, and snippets providing more context around them. We described all the major components combined to build this resource. The pipeline for generating *Topic Pages* can be used on top of any document collection as well as a taxonomy to build a similar resource in any domain. With over 363,000 topic pages in 20 different scientific domains, and more than 23 million unique visitors per month, *Topic Pages* are one of the popular knowledge bases among researchers and students. We described all major components of the pipeline for extracting different pieces of information necessary to generate the pages. In this work, we mainly focused on building a high-performance definition extraction model. To this end, we used an *LSTM+CNN* and a *SciBERT* model. Empirical evaluation shows that both models can outperform existing models for the definition classification and extraction task. However, the *SciBERT* model still needs to be improved for domains such as *Social Sciences*. The biggest drawback of using *SciBERT* for such domains is that this model is pre-trained on mostly biomedical articles and, therefore, it cannot model all other domains as well. As a future work, we would like to exploit the concepts and their definitions extracted from Wikipedia as well as expand our dataset to further fine-tune the *SciBERT* model for such domains. As another future work, we are going to use the click-through data we have collected as a proxy to train supervised models for related concept extraction and snippet ranking components.

References

1. Beltagy, I., Lo, K., Cohan, A.: SciBERT: a pretrained language model for scientific text. In: Proceedings of the 2019 Conference on Empirical Methods in Natural Language Processing and the 9th International Joint Conference on Natural Language Processing (EMNLP-IJCNLP), pp. 3615–3620 (2019)
2. Espinosa-Anke, L., Schockaert, S.: Syntactically aware neural architectures for definition extraction. In: Proceedings of the 2018 Conference of the North American Chapter of the Association for Computational Linguistics (NAACL), pp. 378–385 (2018)
3. Hearst, M.A.: Automatic acquisition of hyponyms from large text corpora. In: COLING, pp. 539–545 (1992)
4. Hochreiter, S., Schmidhuber, J.: Long short-term memory. Neural Comput. **9**(8), 1735–1780 (1997)
5. Jain, A., Gupta, N., Mujumdar, S., Mehta, S., Madhok, R.: Content driven enrichment of formal text using concept definitions and applications. In: Proceedings of the 29th on Hypertext and Social Media, pp. 96–100 (2018)
6. Jin, Y., Kan, M.Y., Ng, J.P., He, X.: Mining scientific terms and their definitions: a study of the ACL anthology. In: Proceedings of the 2013 Conference on Empirical Methods in Natural Language Processing (EMNLP), pp. 780–790 (2013)
7. Kang, D., Head, A., Sidhu, R., Lo, K., Weld, D.S., Hearst, M.A.: Document-level definition detection in scholarly documents: existing models, error analyses, and future directions. In: Proceedings of the First Workshop on Scholarly Document Processing, pp. 196–206 (2020)
8. Klavans, J.L., Muresan, S.: A method for automatically building and evaluating dictionary resources. In: Proceedings of the Third International Conference on Language Resources and Evaluation (LREC), pp. 231–234 (2002)
9. Kobyliński, Ł., Przepiórkowski, A.: Definition extraction with balanced random forests. In: International Conference on Natural Language Processing, pp. 237–247 (2008)
10. LeCun, Y., Bottou, L., Bengio, Y., Haffner, P.: Gradient-based learning applied to document recognition. Proc. IEEE **86**(11), 2278–2324 (1998)
11. Li, S., Xu, B., Chung, T.L.: Definition extraction with LSTM recurrent neural networks. In: Chinese Computational Linguistics and Natural Language Processing Based on Naturally Annotated Big Data, pp. 177–189 (2016)
12. Malaisé, V., Otten, A., Coupet, P.: Omniscience and extensions-lessons learned from designing a multi-domain, multi-use case knowledge representation system. In: European Knowledge Acquisition Workshop, pp. 228–242 (2018)
13. Murthy, S.K., et a.: Accord: a multi-document approach to generating diverse descriptions of scientific concepts. arXiv preprint arXiv:2205.06982 (2022)
14. Navigli, R., Velardi, P.: Learning word-class lattices for definition and hypernym extraction. In: Proceedings of the 48th Annual Meeting of the Association for Computational Linguistics (ACL), pp. 1318–1327 (2010)
15. Reiplinger, M., Schäfer, U., Wolska, M.: Extracting glossary sentences from scholarly articles: a comparative evaluation of pattern bootstrapping and deep analysis. In: ACL-2012 special workshop on rediscovering 50 years of discoveries, pp. 55–65 (2012)
16. Roig Mirapeix, M., Espinosa Anke, L., Camacho-Collados, J.: Definition extraction feature analysis: from canonical to naturally-occurring definitions. In: Proceedings of the Workshop on the Cognitive Aspects of the Lexicon, pp. 81–91 (2020)

17. Schwartz, A.S., Hearst, M.A.: A simple algorithm for identifying abbreviation definitions in biomedical text. In: Biocomputing 2003, pp. 451–462 (2002)
18. Veyseh, A., Dernoncourt, F., Dou, D., Nguyen, T.: A joint model for definition extraction with syntactic connection and semantic consistency. In: AAAI, pp. 9098–9105 (2020)
19. Veyseh, A.P.B., Dernoncourt, F., Tran, Q.H., Nguyen, T.H.: What does this acronym mean? Introducing a new dataset for acronym identification and disambiguation. In: COLING, pp. 3285–3301 (2020)
20. Westerhout, E.: Definition extraction using linguistic and structural features. In: Proceedings of the 1st Workshop on Definition Extraction, pp. 61–67 (2009)

De-biasing Relevance Judgements for Fair Ranking

Amin Bigdeli[1]([✉]), Negar Arabzadeh[2], Shirin Seyedsalehi[1], Bhaskar Mitra[3], Morteza Zihayat[1], and Ebrahim Bagheri[1]

[1] Toronto Metropolitan University, Toronto, Canada
{abigdeli,shirin.seyedsalehi,mzihayat,bagheri}@torontomu.ca
[2] University of Waterloo, Waterloo, Canada
narabzad@uwaterloo.ca
[3] Microsoft Research, Montreal, Canada
bmitra@microsoft.com

Abstract. The objective of this paper is to show that it is possible to significantly reduce stereotypical gender biases in neural rankers without modifying the ranking loss function, which is the current approach in the literature. We systematically de-bias gold standard relevance judgement datasets with a set of balanced and well-matched query pairs. Such a de-biasing process will expose neural rankers to comparable queries from across gender identities that have associated relevant documents with compatible degrees of gender bias. Therefore, neural rankers will learn not to associate varying degrees of bias to queries from certain gender identities. Our experiments show that our approach is able to (1) systematically reduces gender biases associated with different gender identities, and (2) at the same time maintain the same level of retrieval effectiveness.

1 Introduction

There have been both qualitative and quantitative studies that have effectively shown that stereotypical biases are prevalent in various natural language processing and Information Retrieval (IR) techniques, models and datasets [1,2,6,8,9,11, 17,23,24]. Given these tools are often deployed at scale, such biases have the potential to directly impact the lives of many people. More specifically within the context of IR, biased retrieval methods can exacerbate biases by exposing users to a set of biased documents in response to user queries. In order to systematically address such biases, various researchers have proposed methods that can help measure and/or mitigate systematic biases, such as gender biases, in IR methods [7,20,21]. For instance, Rekabsaz et al. [21] compared different neural ranking models and found that the ranked list of documents returned by neural ranking models are more inclined towards the male gender compared to traditional retrieval methods such as BM25. Building on this work, Rekabsaz et al. [20] later proposed the AdvBert model, which is a Bert re-ranker, which leverages adversarial training to de-bias the output encoder of the Bert model from gender inclination. The authors reported that AdvBert increases the fairness level of the ranked list of documents.

J. Kamps et al. (Eds.): ECIR 2023, LNCS 13981, pp. 350–358, 2023.
https://doi.org/10.1007/978-3-031-28238-6_24

However, this comes at the cost of reduced retrieval effectiveness. Recently, some studies have investigated reducing stereotypical biases while maintaining retrieval effectiveness [3,4,22]. For instance, the authors in [22] propose a neural ranking model that incorporates a notion of gender bias in the loss function to reduce gender bias exposure in the retrieved documents. Furthermore, in [3], the authors propose a bias-aware negative sampling training strategy that represents those documents that are not only irrelevant but also biased towards a particular gender as negative samples to the model. As a result, the model learns the concept of relevance and avoids gender biases.

Our work in this paper builds on the foundations of the earlier work [3,4,11, 20,22] and attempts to address biases exposed by IR methods while maintaining their effectiveness. While existing studies have shown that it is possible to reduce gender biases among the retrieved list of documents associated with gender neutral queries, none of them investigated psychological biases that exist among the retrieved documents of gender affiliated queries. Our work is inspired by the observations made by [5] that shows gold standard relevance judgement collections such as MS MARCO may include systematic psychological biases. We propose a methodical approach for augmenting relevance judgement datasets with automatically generated pairs of query-documents that can systematically reduce biases when used to train neural rankers. We show that neural rankers trained on our proposed de-biased relevance judgement datasets exhibit significantly lower biases while maintaining comparable levels of retrieval effectiveness.

In summary, our work delivers the following main contributions. First, we propose a systematic approach for automatically building query-document pairs that can be used for training neural rankers. Second, we show how combining our proposed query-document pairs with existing gold standard relevance judgement datasets can lead to the training of less biased neural rankers that have competitive effectiveness. We conduct our experiments on the MS MARCO passage collection and use three widely adopted psychological and stereotypical gender bias measurement methods to show that decrease in bias happens effectively regardless of how gender biases are measured. We also report the effectiveness of our approach on the MS MARCO passage retrieval task.

2   Proposed Approach

In this work, our hypothesis is that a neural ranker trained on a balanced relevance judgment dataset has a lower likelihood of exhibiting biased retrieval performance. We aim to augment relevance judgement datasets with query and document pairs that have controlled and matched degrees of bias and hence allow the retrieval method to learn a balanced measure of relevance without being inclined towards certain stereotypical biases. We hypothesize that the augmentation of an existing relevance judgement dataset with our proposed dataset leads to a consistent reduction in bias while maintaining comparable effectiveness. Developing such a balanced dataset cannot be accomplished through crowdsourcing due to several reasons: (1) the collection of a large number of judged queries by human participants is very expensive; and (2) given gender biases may be unconsciously embedded in the labelers' beliefs, it is still possible that the newly collected data still suffer from such biases. Therefore, we propose an automated method to generate pairs of query and relevant documents that have controlled degrees of bias.

2.1 Query-Relevant Document Pairs Generation

Given our objective, we need to first automatically generate a set of queries and their associated relevant documents that would be then further filtered to ensure a balanced representation of bias. To this end, we adopt a translation approach that translates a document into a query representation. We train a transformer model based on existing query-relevant document pairs that are already available in the relevance judgement dataset. Thus, the transformer learns to generate queries for an input document. The details of the transformer is provided in the experimental setup section. With the transformer, we are able to generate queries for each document in any given document corpus; producing a set of query and relevant document pairs.

Given the generated pairs of query and relevant documents, we need to selectively choose comparable queries from different genders. We recognize that gender identities go beyond a binary framework, and in practice requires a careful treatment of a spectrum of gender identities, however given available datasets consist predominantly of binary gender queries, we build two classes of queries affiliated with the male and female genders. The idea is that there should always be a corresponding query in one gender that matches a query in the other where the documents associated with these queries show comparable degrees of stereotypical bias. Such an approach would develop a relevance judgement dataset that controls for bias across different gender-affiliated queries. We determine the gender of each query using the proposed model in [5] which is fine-tuned over the manually classified gendered queries released by [21]. We assume that the gender of the document associated with each query to be the same as the gender of the query. As such, we produce a large number of query and relevant document pairs that have been predictively labeled with gender affiliation information.

2.2 Balancing Biases on Query-Document Pairs

With the generated query-document pairs, we aim to perform a controlled matching process with balanced representation of queries and documents from each gender affiliation such that the matched queries exhibit the same degrees of bias regardless of the gender of the query or document. To this end, we assume, as suggested in the literature [12,18], that each document can be characterized through a set of psychological processes such as affective, cognitive, and perceptual processes, to name a few. Such psychological characteristics of a document can be captured through the widely-adopted Linguistic Inquiry and Word Count (LIWC) toolkit [19]. Let us assume that each document can be characterized by a set of n different psychological characteristics, namely $P_1, P_2, ...P_n$. Let $\phi(d)$ be the document psychological characteristic representation for document d, based on its psychological characteristics as $\phi(d) = [P_1(d), P_2(d), ..., P_n(d)]$

where $\phi(d)$ is an n-dimensional vector whose individual elements quantify the different psychological characteristics observed in document d. We benefit from this document representation to perform the matching process between queries affiliated with different gender identities. Consider D_f and D_m to be the set of relevant documents affiliated with female and male queries, respectively. The degree of similarity of a document in $d_i \in D_f$ and another $d_j \in D_m$ is computed as the cosine similarity of their representations $\phi(d_i)$ and $\phi(d_j)$.

In order to build comparable pairs of queries from across gender identities, for each query in one gender identity, we identify a matching query from the other gender identity such that their associated relevant documents' psychological characteristics are most similar to one another. This will produce a collection of pairs of queries from different gender identities that are associated with relevant documents that have similar psychological characteristics. The benefit of this is that given the queries from each gender identity are paired through a matching process, the degree of bias exposed to each gender-affiliated query is no different than the other and hence bias is controlled across the two classes. We propose that the augmentation of existing relevance judgment datasets such as MS MARCO with our proposed matched query-document pairs has the potential to systematically control the stereotypical gender biases.

3   Experiments

Passage Collection. We employed the MS MARCO passage collection dataset that consists of 8,841,822 passages [14].

Query Sets. For the purpose of measuring psychological characteristics, we use the set of gendered queries introduced in [5]. We also employ two different query sets that consist of neutral queries. The first query set is a human-annotated dataset, which consists of 1,765 neutral queries [21]. The other dataset [20] consists of 215 queries in which the queries are neutral in nature, but the retrieved documents exhibit biases.

Bias Measurement. We adopt two strategies to measure gender biases and refer to these as *proxy measures of bias* because while they have been used in the most recent papers on gender bias in IR, they have not yet been empirically or theoretically shown to be the best or at least reliable measures of bias. The **first approach** relies on measuring differences observed across pairs of gender-affiliated queries. We measure the degree of bias based on the metric proposed in [5] which measures bias as the degree to which male and female affiliations are observed within a document based on psychometric properties offered in LIWC [19]. We measure the difference between the psychological characteristics of queries affiliated with different gender identities as a sign of bias towards a certain gender identity. The **second approach** is based on the bias measurement strategy proposed in [21]. The authors propose two metrics based on (1) presence (Boolean) and (2) term frequency of gendered terms for measuring gender bias within a document. They further expand their proposed metrics over the retrieved list of documents for the queries in the dataset by proposing the

Average Ranking Bias (ARaB) metric, which calculates the degree to which a ranked list of documents are biased towards a specific gender.

3.1 Experimental Setup

To generate query and relevant document pairs, we fine-tuned a T5 transformer model based on the query-document pairs of the MSMARCO training set as suggested in [16]. Using this transformer, we generate queries for each document. Furthermore, to estimate the gender affiliation of queries, we adopt the BERT model released in [5].

As a result of generating queries based on the T5 transformer and estimating query gender affiliations using the fine-tuned BERT, we produce 298,389 female, 460,776 male, and 8,056,297 neutral queries. Each of these queries are associated with one relevant judgment document used to generate the query. Furthermore, for each document in this collection, we produce $\phi(d)$ based on LIWC psychological characteristics, namely affective processes, cognitive processes, drives, and personal concerns, and their subprocesses, which constitute a total of 22 subprocesses. Inspired by [10], we augment the small training set of MS MARCO with data from our generated query-document pairs using different ratios with 10% increments.

Based on the de-biased datasets, we leverage the BERT transformer model for passage ranking introduced by Nogueira et al. [15] and train BERT-base-uncased on the original dataset, i.e., the small training set of MS MARCO, as well as the newly developed de-biased datasets. We use OpenMatch [13] to fine-tuning for the ranking task with batch size of 64, learning rate of 2e-5, and epoch of 1. We also set the max document length and max query length to 150 and 20, respectively. We publicly release our code, models, results and datasets for general use.[1] We note that while the query-document pairs are included in our dataset, the predicted gender affiliation of the queries are hidden as these were solely predicted based on a fine-tuned BERT model and may not be reflective of true gender affiliations.

Table 1. MRR on original and de-biased datasets. * indicates statistically significant decrease in effectiveness. (two-tailed paired t-test 95% confidence).

Training set	Ratio	MRR@10	Reduction (%)
Original	–	0.3080	–
De-biased	0.05	0.3100	0.65%
	0.15	0.3039	−1.33%
	0.25	0.3002	−2.53%
	0.35	0.2905	−5.68%*

---

[1] https://github.com/aminbigdeli/balanced-relevance-judgment-collection.

Table 2. Impact of training on de-biased dataset on the difference in psychological characteristics of gender-affiliated queries.

Training dataset		Affective processes	Cognitive processes	Drives	Personal concerns
Original dataset	Female queries	0.0315	0.0725	0.0545	0.0600
	Male Queries	0.0290	0.0521	0.0641	0.0829
	Difference	0.0025	0.0204	0.0095	0.0229
De-biased dataset	Female queries	0.0304	0.0730	0.0536	0.0546
	Male queries	0.0288	0.0563	0.0624	0.0747
	Difference	0.0016	0.0167	0.0088	0.0201
Reduction (%)		36.00%	18.13%	7.37%	12.22%

3.2 Results and Findings

Impact on Retrieval Effectiveness. The objective of our work has been to reduce proxy measures of bias while maintaining retrieval effectiveness. As such, we investigate how the same model [15] performs when trained on different training datasets including the MS MARCO small training set and the de-biased datasets with different ratios. We measure retrieval effectiveness based on the 6,980 queries of the small dev set of MS MARCO collection based on the standard leaderboard metric, i.e., MRR@10.

Table 1 shows the results of the model when trained based on different augmentation ratios. We increased the augmentation ratio until the retrieval effectiveness of the model dropped significantly below the performance of the model that was trained on MS MARCO dataset without augmentation.

Table 3. The impact of training BERT-base-uncased on the de-biased dataset on proxy measures of gender bias based on different neutral query sets. Reduction (%) values are computed based on actual metric values, while the metric values are rounded to three decimal points.

Query set	Training set	TF ARaB		Boolean ARaB		LIWC	
		Value	Reduction	Value	Reduction	Value	Reduction
QS1	Original	0.072	–	0.059	–	0.011	–
	De-biased	0.059	18.05%	0.049	16.95%	0.011	5.98%
QS2	Original	0.029	–	0.017	–	0.006	–
	De-biased	0.019	34.48%	0.011	35.29%	0.005	16.67%

It is expected that as the number of synthetically generated data pairs used to augment the original dataset increases, the effectiveness of the model drops gradually. This is because the query-document pairs are included in our synthetic dataset such that they would balance the degrees of bias and since they are synthetic query-document pairs, they are not as effective for training the model to learn query-document relevance. On the other hand, the expectation would be that a larger ratio of synthetic data would lead to drop in bias. As such, the preference would be to include as much synthetically generated data as possible to reduce bias. As shown in Table 1, we increase the ratio until the time when the decrease in performance becomes, statistically speaking, significantly lower than the model trained on the original set. This happens when the ratio is set to 35%. Therefore, we employ the ratio of 25% in the rest of our experiment.

Impact on Proxy Measures of Bias. We investigate the impact of our approach on the reduction of the proxy measures of bias.

Bias Observed on Gender Affiliated Queries: We adopt the gendered queries released in [5] and calculate the psychological characteristics observed in the ranked list of each of the models trained on original MS MARCO in comparison to the one trained on the 25% de-biased dataset.

As shown in Table 2, the model trained on the de-biased dataset substantially reduces the differences found on the expression of psychological characteristics between the queries in the different gender identities. A higher reduction in the differences between the gender identities, in Table 2, is a positive indication of the success to bridge the gap between the representation of psychological characteristics in documents retrieved in relation to the gendered queries.

Bias Observed on Neutral Queries: We adopt two different query sets: (**QS1**) 1,765 neutral queries from [21], and (**QS2**) 215 queries from [20]. We investigate if the `BERT-base-uncased` model trained on the de-biased dataset is able to reduce the gender biases among the ranked documents for neutral queries. We report the level of gender bias among the top-10 ranked list of documents for neutral queries using both classes of ARaB metric proposed in [21] in Table 3. As shown in the table, gender inclination among the ranked list of documents for neutral queries decreases significantly when they are retrieved by the model trained on the de-biased dataset in terms of both Boolean and Term Frequency ARaB measures. To validate our findings, we also report the difference between the male and female affiliation for the top-10 ranked list of documents to measure the degree of gender inclination in Table 3. As shown, the reduction of bias associated with gender affiliation computed by LIWC is consistent with both of the ARaB measures and can be observed over all of the datasets. According

Table 4. BERT-Tiny trained on our de-biased dataset vs ADVBERT-Tiny. Reduction (%) values are computed based on actual metric values, while the metric values are rounded to three decimal points and reported in this table.

Query set	Training set	Utility	TF ARaB		Boolean ARaB		LIWC	
		MRR@10	Value	Reduction (%)	Value	Reduction (%)	Value	Reduction (%)
QS1 (Rekabsaz et. al 2020)	Original	0.219	0.076	–	0.063	–	0.012	–
	De-biased	0.199	0.047	38.15%	0.042	32.06%	0.010	10.74%
	ADVBERT	0.189	0.064	15.78%	0.058	7.30%	0.009	24.79%
QS2 (Rekabsaz et. al 2021)	Original	0.175	0.005	–	0.006	–	0.005	–
	De-biased	0.163	0.001	79.19%	0.000	97.01%	0.005	11.11%
	ADVBERT	0.149	0.009	−85.98%	0.007	−16.67%	0.005	14.81%

to Table 3, the proposed de-biased dataset for training neural ranking models can reduce gender inclination among the retrieved list of documents for neutral queries.

Comparative Analysis. We compare our work with a recent method proposed by [20], known as ADVBERT. As suggested in their paper, we adopt the BERT-Tiny model and train it based on the method proposed by the authors over the original MS MARCO dataset. We additionally, train the same BERT-Tiny model without adversarial training on the original MS MARCO dataset as well as our proposed de-biased dataset. We report the performance of these models on the two sets of neutral queries. Table 4 shows the results in terms of ARaB and LIWC for the three models and across two query sets. Our approach shows superior retrieval effectiveness compared to ADVBERT. This speaks to the objective of our work to maintain effectiveness while addressing bias. In terms of the proxy measures of bias and specifically when considering the ARaB metrics, our proposed approach shows consistent superior performance over ADVBERT in both query sets and variations of the ARaB metric. On the other hand, when comparing the degree of bias reduction based on the LIWC-based gender affiliation, we find that ADVBERT has a higher degree of bias reduction but this has come at the cost of effectiveness.

4 Concluding Remarks

We proposed an approach to generate matched query-document pairs across gender identities for systematically reducing stereotypical biases that are learnt by neural rankers. Our approach distinguishes itself from existing methods in that (1) it systematically reduces gender biases, and also (2) maintains comparable levels of retrieval effectiveness.

References

1. Baeza-Yates, R.: Bias on the web. Commun. ACM **61**(6), 54–61 (2018)
2. Baeza-Yates, R.: Bias in search and recommender systems. In: Fourteenth ACM Conference on Recommender Systems, p. 2 (2020)
3. Bigdeli, A., Arabzadeh, N., Seyedsalehi, S., Zihayat, M., Bagheri, E.: A light-weight strategy for restraining gender biases in neural rankers. In: Hagen, M., et al. (eds.) ECIR 2022. LNCS, vol. 13186, pp. 47–55. Springer, Cham (2022). https://doi.org/10.1007/978-3-030-99739-7_6
4. Bigdeli, A., Arabzadeh, N., Seyersalehi, S., Zihayat, M., Bagheri, E.: On the orthogonality of bias and utility in ad hoc retrieval. In: Proceedings of the 44rd International ACM SIGIR Conference on Research and Development in Information Retrieval (2021)
5. Bigdeli, A., Arabzadeh, N., Zihayat, M., Bagheri, E.: Exploring gender biases in information retrieval relevance judgement datasets. In: Hiemstra, D., Moens, M.-F., Mothe, J., Perego, R., Potthast, M., Sebastiani, F. (eds.) ECIR 2021. LNCS, vol. 12657, pp. 216–224. Springer, Cham (2021). https://doi.org/10.1007/978-3-030-72240-1_18

6. Caliskan, A., Bryson, J.J., Narayanan, A.: Semantics derived automatically from language corpora contain human-like biases. Science **356**(6334), 183–186 (2017)

7. Fabris, A., Purpura, A., Silvello, G., Susto, G.A.: Gender stereotype reinforcement: measuring the gender bias conveyed by ranking algorithms. Inf. Process. Manag. **57**(6), 102377 (2020)

8. Font, J.E., Costa-Jussa, M.R.: Equalizing gender biases in neural machine translation with word embeddings techniques. arXiv preprint arXiv:1901.03116 (2019)

9. Gerritse, E.J., Hasibi, F., de Vries, A.P.: Bias in conversational search: the double-edged sword of the personalized knowledge graph. In: Proceedings of the 2020 ACM SIGIR on International Conference on Theory of Information Retrieval, pp. 133–136 (2020)

10. Ju, J.H., Yang, J.H., Wang, C.J.: Text-to-text multi-view learning for passage re-ranking. arXiv preprint arXiv:2104.14133 (2021)

11. Klasnja, A., Arabzadeh, N., Mehrvarz, M., Bagheri, E.: On the characteristics of ranking-based gender bias measures. In: 14th ACM Web Science Conference 2022, pp. 245–249 (2022)

12. Li, J., Ott, M., Cardie, C., Hovy, E.: Towards a general rule for identifying deceptive opinion spam. In: Proceedings of the 52nd Annual Meeting of the Association for Computational Linguistics (Volume 1: Long Papers), pp. 1566–1576 (2014)

13. Liu, Z., Zhang, K., Xiong, C., Liu, Z.: OpenMatch: an open-source package for information retrieval. arXiv e-prints pp. arXiv-2102 (2021)

14. Nguyen, T., et al.: MS MARCO: a human generated machine reading comprehension dataset. In: CoCo@ NIPS (2016)

15. Nogueira, R., Cho, K.: Passage re-ranking with BERT. arXiv preprint arXiv:1901.04085 (2019)

16. Nogueira, R., Lin, J., Epistemic, A.: From doc2query to docTTTTTquery. Online preprint (2019)

17. Olteanu, A., et al.: FACTS-IR: fairness, accountability, confidentiality, transparency, and safety in information retrieval. In: ACM SIGIR Forum, vol. 53, pp. 20–43. ACM, New York (2021)

18. Ott, M., Choi, Y., Cardie, C., Hancock, J.T.: Finding deceptive opinion spam by any stretch of the imagination. arXiv preprint arXiv:1107.4557 (2011)

19. Pennebaker, J.W., Francis, M.E., Booth, R.J.: Linguistic inquiry and word count: LIWC 2001. **71**(2001) (2001)

20. Rekabsaz, N., Kopeinik, S., Schedl, M.: Societal biases in retrieved contents: measurement framework and adversarial mitigation for BERT rankers. arXiv preprint arXiv:2104.13640 (2021)

21. Rekabsaz, N., Schedl, M.: Do neural ranking models intensify gender bias? In: Proceedings of the 43rd International ACM SIGIR Conference on Research and Development in Information Retrieval, pp. 2065–2068 (2020)

22. Seyedsalehi, S., Bigdeli, A., Arabzadeh, N., Mitra, B., Zihayat, M., Bagheri, E.: Bias-aware fair neural ranking for addressing stereotypical gender biases. In: EDBT, pp. 2–435 (2022)

23. Seyedsalehi, S., Bigdeli, A., Arabzadeh, N., Zihayat, M., Bagheri, E.: Addressing gender-related performance disparities in neural rankers. In: Proceedings of the 45th International ACM SIGIR Conference on Research and Development in Information Retrieval, pp. 2484–2488 (2022)

24. Sun, T., et al.: Mitigating gender bias in natural language processing: literature review. arXiv preprint arXiv:1906.08976 (2019)

A Study of Term-Topic Embeddings
for Ranking

Lila Boualili[1,2(✉)] and Andrew Yates[1,3]

[1] Max Planck Institute for Informatics, Saarbrücken, Germany
[2] IRIT, University of Toulouse III, Toulouse, France
`lila.boualili@irit.fr`
[3] University of Amsterdam, Amsterdam, Netherlands
`a.c.yates@uva.nl`

Abstract. Contextualized representations from transformer models have significantly improved the performance of neural ranking models. Late interactions popularized by ColBERT and recently compressed with clustering in ColBERTv2 deliver state-of-the-art quality on many benchmarks. ColBERTv2 uses centroids along with occurrence-specific delta vectors to approximate contextualized embeddings without reducing ranking effectiveness. Analysis of this work suggests that these centroids are "term-topic embeddings". We examine whether term-topic embeddings can be created in a differentiable end-to-end way, finding that this is a viable strategy for removing the separate clustering step. We investigate the importance of local context for contextualizing these term-topic embeddings, analogous to refining centroids with delta vectors. We find this end-to-end approach is sufficient for matching the effectiveness of the original contextualized embeddings.

1 Introduction

Contextualized representations from transformer models like BERT [2] have become the default representations in neural information retrieval (IR), achieving substantial gains in text ranking [7]. Notably, late interactions introduced in ColBERT [6] and subsequent variants [5,13,15] are state-of-the-art. In this paradigm, queries and documents are encoded into token-level vectors, and relevance is computed based on all-to-all soft matching between the query and document token vectors. Consequently, this approach greatly increases the space footprint required since all token vectors are indexed.

To reduce the storage footprint of late interactions, ColBERTv2 [13] uses residual compression. The authors study the semantic space produced by ColBERT and find that token representations localize in a small number of regions corresponding to the contextual topics of a token (e.g., *tornado*-blizzard vs. *tornado*-hurricane). Hence, this semantic space can be summarized, with high precision, by a set of centroids (e.g., hundreds of thousands), obtained through k-means clustering along with minor refinements at the dimension level. That is, each vector in a document is encoded by combining its nearest centroid with a quantized residual vector to account for the difference (delta) between the centroid and the original token vector. These vectors are "term-topic embeddings" characterizing a term by its topical context. Later work leverages these centroids to speed up the search latency of late interaction [12].

ColBERTv2's centroid-based approach to compression indicates that late-interaction vectors can be effectively encoded with coarse-grained term-topic embeddings (centroids) along with fine-grained refinements (residual vectors). This decomposition of contextualized token representations achieves its goal of reducing index size, but the multi-step process makes the approach difficult to study. The process is not differentiable due to the k-means clustering step, and the residual vectors lack a clear conceptual purpose.

In this work, we study how term-topic embeddings (centroids) can be learned end-to-end and whether local context is sufficient to refine these embeddings (residual) to better approximate ColBERT's original embeddings. We refer to ColBERT as the Oracle, given our goal is to approximate the contextualized token embeddings it produces. By creating term-topic embeddings end-to-end, we simplify the training process and enable analysis, which we leverage to study the role of local context for refinement.

Our results indicate that term-topic embeddings can be learned as part of the training process, rather than in a separate clustering step. In fact, term-topic embeddings alone provide up to 97% of the ranking quality achieved by ColBERT's original contextualized embeddings. Further integration of information from the local context provides the refinements necessary to match the ranking quality of ColBERT's original contextualized embeddings.

2   Analysis Framework

In order to analyze the importance of local context for contextualizing term-topic embeddings, we devise two end-to-end differentiable probing modules for ColBERT late-interaction vectors: (1) a Term Topic Module (TTM) to explore whether term-topic embeddings can be effectively learned end-to-end, and (2) a Local Context Module (LCM) to investigate the importance of surrounding tokens for contextualizing term-topic embeddings produced by the TTM.[1]

2.1   Term Topic Module (TTM)

Instead of clustering the contextualized representations produced by ColBERT for each target corpus, we explore whether term-topic embeddings can be learned and generalized across corpora. To do so, the TTM learns to decompose the semantic space produced by ColBERT, at the *token level*, into a small set of K static sub-embeddings per token. These sub-embeddings are intended to capture different topics closely tied to the token (e.g., *right* answer vs. *right* hand), analogously to ColBERTv2's centroids.

Given a contextualized representation c_i of a token produced by the ColBERT Oracle, TTM generates an approximation by combining the sub-embeddings of

---

[1] https://github.com/BOUALILILila/Term-Topic-Embeddings.

this token $s_i^{1:K}$ with a weighted average. Each weight reflects the importance of each sub-embedding to the contextualized token vector c_i using multi-headed attention weights [14]. The Oracle's embedding c_i is used for the query vector and the sub-embeddings are used for the key vectors. The attention weights are then max-pooled over the heads to output a single attention weight per token sub-embedding. Finally, the token sub-embeddings are combined using a weighted average of their corresponding attention weights. TTM sees each token occurrence as a combination (superposition) of the token's different meanings (sub-embeddings) weighted according to their contribution (importance) to the occurrence-specific context.

2.2   Local Context Module (LCM)

Not all fine-grained topic information may accurately be captured by a fixed set of term-topic embeddings due to the long tail. We devise the LCM to provide an additional mechanism for refining the topic. LCM uses a simplified contextualization layer to inject local context information into TTM-produced embeddings.

Formally, given a token t_i and its TTM embedding e_i, LCM applies multi-headed attention on the context window of size ws around t_i. The contextualized Oracle representation c_i of the central token t_i is used for the query vector, while the embeddings $e_{i-ws:i+ws}$ of the context tokens are used for the key and value vectors. The central token's refined representation is obtained by concatenating the results from all attention heads and projecting through the output layer. Finally, we formulate the output of LCM as a gating function,[2] and split the input token embedding into two independent parts $(e_i^1, e_i^2) \in \mathbb{R}^{D/2}$:

$$LCM(e_i) = (e_i^1 \oplus WCA(e_i^2)) \cdot W^{Gate} \tag{1}$$

where $\oplus$ denotes concatenation, WCA is the windowed cross-attention, and $W^{Gate} \in \mathbb{R}^{D \times D}$ is a feed-forward layer. Intuitively, the LCM can be viewed as using the Oracle's embeddings to determine the relevance of nearby TTM embeddings and then incorporating them.

2.3   Supervision

In an initial phase, we pre-train our probing modules from scratch using distillation from the contextualized representation of a distilBERT-based ColBERT Oracle [4]. The purpose of this pre-training is to build generalizable static sub-embeddings that can be easily transferred across corpora. To do so, we take single text sequences extracted from the TREC-CAR collection [3]. This collection is drawn from Wikipedia pages offering a diversity of contents ideal for learning generalizable static sub-embeddings. We optimize our module parameters to minimize the Mean Squared Error (MSE) loss between randomly sampled token representations S_s and the equivalent contextualized representations S_t produced by the Oracle.

---

[2] Our approach is inspired by work on representation-independent gated MLPs [8].

After pre-training, we freeze the sub-embedding and fine-tune the rest of our module parameters for the ranking task. We adopt ColBERT's *late-interaction* mechanism to compute relevance scores using the representations produced by our modules. We fine-tune our modules, analogously to how the Oracle was fine-tuned, with triples containing a query, a relevant passage, and a less relevant passage from the MS MARCO [10] train set. We use the *Margin-MSE* loss [4] to mimic the pairwise differences in passage scores of a set of cross-encoder teachers:

$$\mathcal{L}_{rank}(M_s) = MSE((M_s^+ - M_s^-), (M_t^+ - M_t^-)) \tag{2}$$

where the set of teachers M_t provides the relevance score for both passages w.r.t query for our student model M_s.

3   Experiments

We conduct experiments to investigate (1) whether term-topic embeddings can be learned end-to-end and (2) the impact of local contextualization on these embeddings. We report the official metrics on four ranking datasets. **MS MARCO** comprises $6,980$ sparsely judged development queries (Dev) [10]. We use Anserini's [16] implementation of BM25 with default parameters to retrieve the top-1000 candidate passages for reranking. We also consider the densely-judged query sets of 43 and 54 queries from the **TREC Deep Learning (DL)** passage reranking tracks of 2019 (DL'19) [1] and 2020 (DL'20) [1]. We rerank the official organizers' BM25 runs. We additionally include experiments on the **DL-Hard** passage benchmark [9], focusing on 50 challenging and complex queries partially from DL'19 and '20, by reranking the authors' BM25 run baseline.

Given a query and a list of passage candidates, our task is to rerank the passages according to their relevance to the query using late interactions. We consider reranking as an efficient and fair setting where the candidate documents always remain the same.

Table 1. TTM reranking effectiveness with variable number of token subembeddings K. Our module's best results are in **bold**.

	K	#Sub-embeddings	MRR@10
BM25	–	–	0.184
ColBERT	–	–	0.342
TTM	1	$30,522$	0.218
	5	$152,610$	0.317
	10	$305,220$	0.330
	15	$457,830$	**0.332**

3.1   Learned Term-Topic Embeddings for Late-Interactions

First, we study how our TTM-produced term-topic embeddings compare to the original contextualized representations produced by ColBERT. We empirically

Table 2. Reranking effectiveness of TTM-LCM with different context window lengths (ws). Our module's best results are in **bold**.

	ws	Dev MRR@10	DL'19 nDCG@10	DL'20 nDCG@10	DL-Hard nDCG@10
BM25	–	0.184	0.506	0.480	0.304
TREC-Best (no ensembles)	all	–	0.731	0.746	0.408
ColBERT	all	0.342	0.713	0.699	0.394
SRM	0	0.330	0.707	0.682	0.382
TTM-LCM	1	**0.343**	0.721	0.721	0.369
	2	0.341	0.723	0.717	0.406
	3	0.341	0.715	0.713	0.407
	4	0.342	0.719	0.723	0.387
	5	0.342	**0.728**	**0.727**	**0.409**
	all	0.337	0.717	0.707	0.382

investigate the optimal number of sub-embeddings required to represent the semantic space of a token with high precision.

Table 1 reports the ranking performance of term-topic embeddings with a varying number K of sub-embeddings per token ranging from a single static embedding per token, up to $K = 15$ sub-embeddings on MS MARCO Dev. As the results show, learning term-topic embeddings in an end-to-end way via TTM can hold up to 97% of the Oracle's performance while combining only a small number of static sub-embeddings summarizing token semantics.

Compared to the bag-of-words BM25 retriever, even using a single static embedding ($K = 1$) to represent all token occurrences is more effective. Increasing the number of sub-embeddings leads to better performance up to $K = 10$, where effectiveness stabilizes. Using $K = 15$ brings no significant gains over $K = 10$ while requiring 50% more embedding parameters, hence we use $K = 10$ for the rest of our analysis. Nonetheless, further refinement is required to match the Oracle's performance.

3.2 Local Contextualization for Refining Term-Topic Embeddings

We next analyze whether the local context of tokens can sufficiently refine TTM embeddings to incorporate fine-grained topical variations due to context. Table 2 reports the results of contextualizing term-topic embeddings via LCM with different context window lengths. Interestingly, by considering only the direct neighbors of a token (ws = 1) in LCM, the refined representations already match ColBERT's effectiveness on MS MARCO Dev. Considering the sub-word tokenization employed, we can hypothesize that with ws = 1 the LCM is possibly aggregating full-word representations from its sub-tokens. Further enlarging the context window leads to comparable performance with the Oracle. On DL'19 and '20, we notice slight improvements over the Oracle effectiveness with minor variations due to the window size. When we focus on the challenging queries in DL-Hard, results suggest that longer context windows can be beneficial and ws

= 1 is less effective for these queries. Interestingly, with a local context window of ws = 5, our approach matches the performance of the best run on DL-Hard. Using a window large enough to encompass all sequence tokens reduces performance, which may be due to the fact that tokens are unordered within the window (i.e., the LCM itself has no notion of token position).

LCM is a simpler implementation of a transformer using a very restrained local context in a single attention layer. Nevertheless, it is sufficiently informative to refine term-topic embeddings and match the Oracle's global contextualization. This could motivate lightweight local attention mechanisms for neural ranking models.

Table 3. Sample query-passage token matches from MS MARCO passage

Query	Module(s)	Top matching sampled tokens				
Pain in **right** arm	ColBERT	right (14.8)	Left (11.0)	West (8.5)	Upper (8.3)	Straight (8.3)
	TTM	Right (10.7)	Left (7.4)	Rights (6.6)	North (6.1)	West (5.6)
	TTM-LCM	Right (12.8)	Left (9.4)	West (7.4)	Straight (7.2)	Wrong (7.2)
Right to own arms	ColBERT	Right (14.1)	Rights (11.7)	Freedom (8.7)	Power (8.5)	Free (8.5)
	TTM	Right (9.8)	Rights (8.6)	Free (5.5)	Liberty (5.5)	Freedom (5.5)
	TTM-LCM	Right (11.4)	Rights (10.2)	Freedom (7.9)	Liberty (7.6)	Freedoms (7.5)
Operating **system**	ColBERT	System (15.2)	Systems (13.5)	pc (10.5)	Computer (10.4)	Server (10.1)
	TTM	System (10.7)	Systems (9.4)	Computer (6.9)	Software (6.7)	Unix (6.7)
	TTM-LCM	System (12.4)	Systems (12.1)	Unix (9.2)	Linux (9.0)	Software (9.0)
Nervous **system**	ColBERT	System (15.2)	Systems (13.4)	Nervous (9.4)	Brain (9.3)	Tract (9.2)
	TTM	System (9.6)	Systems (8.7)	Computer (6.4)	Unix (6.2)	Linux (6.1)
	TTM-LCM	System (13.5)	Systems (11.5)	Nervous (9.0)	Peripheral (9.0)	Central (8.7)

3.3 Case Study

Table 3 shows the impact of local contextualization on token similarity across different sampled contexts. We collect documents returned for both queries and then consider tokens within those documents. We report dot-product scores between query-passage tokens using TTM with $K = 10$ and TTM-LCM with ws = 1.

The first query searches for "right" in the sense of direction. Using term-topic embeddings (TTM) matches tokens related to direction like "left", but it also matches "rights" in the sense of legal rights. Adding local contextualization (LCM) increases the similarity to tokens like "right" or "west" and removes the strong matching to unrelated senses ("rights"). The second query uses "right" in the sense of legal rights. Both TTM and TTM-LCM are able to distinguish the correct sense of the term and behave closely to the Oracle. On the other hand, the queries related to "system" show the importance of local context to determine the correct meaning induced by the surrounding tokens of non-polysemous terms. TTM matches both "operating system" and "nervous system" to computer-related systems; this bias could be induced by the training data containing a significant number of occurrences of "system" in computer-oriented contexts. Here, LCM makes a drastic improvement by matching terms related to nerves (medical topic), demonstrating the importance of the local context.

3.4   Out of Domain Generalizability

We verify the zero-shot generalization capabilities of our approach, and more specifically, the generalizability of our sub-embeddings to the large TripClick benchmark [11], which is an out-of-domain collection of click log data from the medical-domain search engine *Trip Database*. It contains $1.5M$ passages, and $3,525$ test queries distributed into three query sets with $1,175$ queries each, namely Head, Torso, and Tail queries, grouped by their frequency.

We use the fine-tuned TTM-LCM on MS MARCO, with K = 10 and ws = 1, without further fine-tuning on TripClick, which is used as a held-out test set. We rerank the top 200 candidate passages retrieved by the BM25 implementation in Anserini [16] with default parameters.

Table 4. Ranking effectiveness of TTM-LCM (K = 10 and ws = 1) on TripClick. Best results are indicated in **bold**.

	TripClick head		TripClick torso		TripClick tail	
	MRR@10	nDCG@10	MRR@10	nDCG@10	MRR@10	nDCG@10
BM25	0.301	0.149	0.305	0.224	0.263	0.285
ColBERT	0.480	0.164	0.395	0.233	0.326	0.271
TTM-LCM	**0.510**	**0.169**	**0.400**	**0.240**	**0.329**	**0.283**

We report, in Table 4, the zero-shot performance of the TTM-LCM combination, with K = 10 and ws = 1, compared to the Oracle and the BM25 retriever. The results show that TTM-LCM exhibit the same, and even slightly better, zero-shot performance as the Oracle across the three different query sets, notably on torso and tail queries which are rare queries. This suggests that simpler local contextualization of learned term-topic embeddings can generalize as well as the transformer-contextualization process in the Oracle.

In the end, our approach cannot only yield match the Oracle's performance on in-domain MS MARCO passage ranking benchmarks but can also exhibit on-par zero-shot performance with ColBERT's contextualization on out-of-domain collections.

4   Conclusion

Contextualized representations have been widely adopted for their soft-matching effectiveness in the context of ranking. We presented in this study a framework for analyzing term-topic embeddings and the impact of local contextualization for ranking. By using an end-to-end differentiable module, we demonstrated that learning term-topic embeddings summarizing the semantics of a token at a high level using a small set of static embeddings, is a viable alternative to a separate clustering step. We also find that a restrained context window is informative enough to contextualize term-topic embeddings and match the representation quality of ColBERT's contextualized representations in the context of ranking.

References

1. Craswell, N., Mitra, B., Yilmaz, E., Campos, D., Voorhees, E.M.: Overview of the TREC 2019 deep learning track. arXiv preprint arXiv:2003.07820 (2020)
2. Devlin, J., Chang, M., Lee, K., Toutanova, K.: BERT: pre-training of deep bidirectional transformers for language understanding. In: Proceeding of the 2019 NAACL-HLT Conference, vol. 1. ACL, June 2019
3. Dietz, L., Verma, M., Radlinski, F., Craswell, N.: TREC complex answer retrieval overview. In: TREC (2017)
4. Hofstätter, S., Althammer, S., Schröder, M., Sertkan, M., Hanbury, A.: Improving efficient neural ranking models with cross-architecture knowledge distillation. arXiv preprint arXiv:2010.02666 (2020)
5. Hofstätter, S., Khattab, O., Althammer, S., Sertkan, M., Hanbury, A.: Introducing neural bag of whole-words with colBERTer: contextualized late interactions using enhanced reduction. arXiv preprint arXiv:2203.13088 (2022)
6. Khattab, O., Zaharia, M.: ColBERT: efficient and effective passage search via contextualized late interaction over BERT, pp. 39–48. Association for Computing Machinery, New York (2020)
7. Lin, J., Nogueira, R., Yates, A.: Pretrained Transformers for Text Ranking: BERT and Beyond. Synthesis Lectures on Human Language Technologies, vol. 14, no. 4, pp. 1–325 (2021)
8. Liu, H., Dai, Z., So, D., Le, Q.V.: Pay attention to MLPS. In: Advances in Neural Information Processing Systems, vol. 34, pp. 9204–9215 (2021)
9. Mackie, I., Dalton, J., Yates, A.: How deep is your learning: the DL-HARD annotated deep learning dataset. In: Proceedings of the 44th International ACM SIGIR Conference on Research and Development in Information Retrieval, SIGIR 2021, pp. 2335–2341. Association for Computing Machinery, New York (2021)
10. Nguyen, T., et al.: MS MARCO: a human generated machine reading comprehension dataset. In: CoCo@ NIPs (2016)
11. Rekabsaz, N., Lesota, O., Schedl, M., Brassey, J., Eickhoff, C.: TripClick: the log files of a large health web search engine. In: Proceedings of the 44th International ACM SIGIR Conference on Research and Development in Information Retrieval, SIGIR 2021, pp. 2507–2513. Association for Computing Machinery, New York (2021)
12. Santhanam, K., Khattab, O., Potts, C., Zaharia, M.: PLAID: an efficient engine for late interaction retrieval. In: Hasan, M.A., Xiong, L. (eds.) Proceedings of the 31st ACM International Conference on Information & Knowledge Management, Atlanta, GA, USA, 17–21 October 2022, pp. 1747–1756. ACM (2022)
13. Santhanam, K., Khattab, O., Saad-Falcon, J., Potts, C., Zaharia, M.: ColBERTv2: effective and efficient retrieval via lightweight late interaction. arXiv preprint arXiv:2112.01488 (2021)
14. Vaswani, A., et al.: Attention is all you need. In: Advances in Neural Information Processing Systems, pp. 5998–6008 (2017)
15. Wang, X., Macdonald, C., Tonellotto, N., Ounis, I.: Pseudo-relevance feedback for multiple representation dense retrieval. In: Proceedings of the 2021 ACM SIGIR International Conference on Theory of Information Retrieval, ICTIR 2021, pp. 297–306. Association for Computing Machinery, New York (2021)
16. Yang, P., Fang, H., Lin, J.: Anserini: enabling the use of Lucene for information retrieval research. In: Proceedings of the 40th International ACM SIGIR Conference on Research and Development in Information Retrieval, pp. 1253–1256 (2017)

Topic Refinement in Multi-level Hate Speech Detection

Tom Bourgeade[1]([✉])(iD), Patricia Chiril[3](iD), Farah Benamara[1,2](iD),
and Véronique Moriceau[1](iD)

[1] IRIT, Université de Toulouse, CNRS, Toulouse INP, UT3, Toulouse, France
{tom.bourgeade,farah.benamara,veronique.moriceau}@irit.fr
[2] IPAL, CNRS-NUS-ASTAR, Singapore, Singapore
[3] University of Chicago, Chicago, IL, USA
pchiril@uchicago.edu

Abstract. Hate speech detection is quite a hot topic in NLP and various annotated datasets have been proposed, most of them using binary generic (hateful vs.
non-hateful) or finer-grained specific (sexism/racism/etc.) annotations, to account
for particular manifestations of hate. We explore in this paper how to transfer
knowledge across both different manifestations, and different granularity or levels of hate speech annotations from existing datasets, relying for the first time on
a multilevel learning approach which we can use to refine generically labelled
instances with specific hate speech labels. We experiment with an easily extensible Text-to-Text approach, based on the T5 architecture, as well as a combination
of transfer and multitask learning. Our results are encouraging and constitute a
first step towards automatic annotation of hate speech datasets, for which only
some or no fine-grained annotations are available.

1 Motivation

Hate Speech (HS hereafter) has become a widespread phenomenon on social media
platforms like Twitter, and automated detection systems are thus required to deal with it.
In spite of no universally accepted definition of HS, these messages may express threats,
harassment, intimidation or *"disparage a person or a group on the basis of some characteristic such as race, color, ethnicity, gender, sexual orientation, nationality, religion,
or other characteristic"* [26]. HS may have different topical focuses: misogyny, sexism,
racism, xenophobia, etc. Which can be referred to as *hate speech topics*. For each HS
topic, hateful content is directed towards specific *targets* that represent the community
(individuals or groups) receiving the hatred.[1] HS is thus, by definition, *target-oriented*,
and it involves different ways of linguistically expressing hateful content such as references to racial or sexist stereotypes, the use of negative and positive emotions, swearing
terms, etc., all of which have to be considered if one is to train effective automated HS
detection systems.

---

[1] For example, black people and white people represent possible targets when the topical focus
is *racism* [31], while women are the targets when the topical focus is *misogyny* or *sexism* [22].
Warning: *This paper includes tweets that may contain instances of vulgarity, degrading terms
and/or hate speech.*

J. Kamps et al. (Eds.): ECIR 2023, LNCS 13981, pp. 367–376, 2023.
https://doi.org/10.1007/978-3-031-28238-6_26

Indeed, such systems would be invaluable for a variety of applications, from automated content classification and moderation, to (potentially malicious) community detection and analysis on social media [9].

To that end, various datasets of human-annotated tweets have been proposed, most often using binary *generic* (e.g., HS/not HS), or multi-label *specific* schemas (e.g., racism/sexism/neither). Unfortunately, due (in great parts) to the lack of clear consensus on these HS annotation schemas [21], gathering enough data to train models that generalize these concepts effectively is difficult. Various approaches have been proposed to palliate these issues: for example, transfer learning has been successfully used in a variety of NLP settings, in particular thanks to the Transformer architecture [33], which allows to leverage large quantities of unannotated text, by fine-tuning pre-trained models such as BERT [7] on tasks for which annotated data is more sparse, such as HS detection [1,17,24,25].

A complementary type of approach is Multi-Task Learning (MTL) [5,18,23], in which one can leverage different tasks and datasets by jointly training a single architecture on multiple objectives at once, sharing all (or parts) of its parameters between them. [32] were the first to showcase how MTL might be used to generalize HS detection models across a variety of datasets, and later on, [16].

Recently, [4] experimented with transferring specific manifestations of hate across HS topics on a varied set of such datasets, showing that MTL could be used to jointly predict both the hatefulness and the topical focus of specific HS instances.

These studies, however, usually consider generic and specific HS datasets as independent (train on one set and test on another) without accounting for common properties shared between both different manifestations of hate, as well as different levels or granularity of annotation. We take here a different perspective and investigate, to our knowledge for the first time, HS detection in a Multi-Level scenario, by answering the following question: *Could instances of generic HS be refined with specific labels, using a model jointly trained on these two levels of annotations?* To this end, we propose:

1. **An easily extensible multitask and multilevel setup designed for HS topic refinement of generic HS instances**, based on the T5 architecture [29], which can be used to generate new specific HS labels (see Fig. 1).
2. **A qualitative and error analyses of the refined labels produced by this approach**, applied to two popular generic HS datasets from the literature.

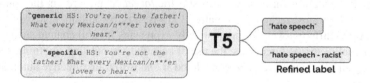

Fig. 1. Illustration of our topic refinement approach based on the T5 architecture

2   Datasets

As our main objective is investigating the problem of *transferring knowledge from different datasets, with different annotation granularity and different topical focuses*, we leverage six manually annotated HS corpora from previous studies. We selected these datasets as they are freely available to the research community. Among them, two are generic (Davidson [6] and Founta [14][2]), and four are specific about four different HS topics: *misogyny* (the Automatic Misogyny Identification (AMI) dataset collection from both IberEval [11] and Evalita [10]), *misogyny and xenophobia* (the HatEval dataset [2]), and *racism* and *sexism* (the Waseem dataset [34]). Each of these HS topics targets either gender (sexism and misogyny) and/or ethnicity, religion or race (xenophobia and racism). In Table 1 we summarize the corpora used in this study.

For the purpose of our experiments, we performed some simplifying split and merge operations on their classes, and their associated labels. For all datasets, we considered the respective 'negative' (i.e., not HS) classes to be equivalent, and used the unified negative-class label "nothing". In addition, as we are using both generic and specific HS datasets, we merged positives instances from generic datasets in a unified generic class labelled "HS". The Offensive and Abusive instances were removed from these datasets, as these concepts often co-exist with HS, but without a clear distinction [21, 27].

For the specific HS corpora, we made the simplification of merging the classes related to sexism and misogyny into the single unified label "HS-sexist". Similarly, we merged racism and xenophobia into the unified label "HS-racist". These labels are designed with T5's *text-to-text* nature in mind (cf. next section): the generic HS label overlaps part of the specific ones, thus a "misprediction" (or more accurately, a partial prediction in this multi-level scenario setup) at training time should only incur a partial error signal (e.g. predicting only "hate speech" in the *specific HS* task, the correct label being "hate speech - racist", incurs less error than predicting "nothing") (see Table 1).

As noted by a number of previous works [12,13,20,21], these types of merging of classes/labels may not be desirable, as each dataset has its own annotation schema. However, as the goal of this work is to explore the viability of HS topic refinement with currently available datasets, we chose to use this simplified annotation schema, and thus consider this added source of label noise to be part of the experimental setting. Addressing these issues, by expanding or reworking this set of labels will likely be explored in future work.

3   Experiments and Cross-Dataset Evaluation

3.1   Models

We rely primarily on a T5 (**T**ext-**t**o-**T**ext **T**ransfer **T**ransformer) architecture [29]. We also experiment with a RoBERTa [19] model, which we use here in an MTL architecture, as a point of comparison for evaluating the performances of these two models across datasets, outside of label refinement (see Sect. 4).

---

[2] At the moment of collecting the data, from the original dataset (http://ow.ly/BqCf30jqffN) we were able to retrieve only 44,898 tweets. See [20] for more details.

Table 1. General overview of the datasets used in this study.

Dataset	Original classes and sizes (with our T5 labels in bold)	T5 task prefix
Davidson	**HS**: Hate (1,430); **nothing**: Neither (4,160)	generic HS
Founta	**HS**: Hate (1,996); **nothing**: Normal (37,889)	generic HS
Waseem	**HS-racist**: Racism (1,957); **HS-sexist**: Sexism (3,216); **nothing**: None (11,315)	specific HS
HatEval	**HS-racist**: Immigrant (2,617); **HS-sexist**: Women (2,845); **nothing**: Not HS (7,509)	specific HS
Evalita	**HS-sexist**: Misogyny (2,245); **nothing**: Not Misogyny (2,755)	specific HS
IberEval	**HS-sexist**: Misogyny (1,851); **nothing**: Not Misogyny (2,126)	specific HS

T5 proposes a way to unify text generation and classification tasks in NLP, by reframing all of them as *text-to-text* problems. This allows the model to both better leverage its pre-training on large quantities of unsupervised text data, but also greatly simplifies MTL setups. Indeed, instead of requiring additional per-task label-space projection layers, the same fine-tuned weights can be used to perform each desired task, which can be indicated to the model by prepending input instances with some task-specific prefix text. MTL with RoBERTa, on the other hand, is traditionally performed by constructing some kind of projection layer (or layers) for each task in the training set, each with their separate target label-space.

We also experimented with BERT-like models which are domain-adapted for HS and toxic language detection, such as fBERT [30], HateBERT [3], or ToxDectRoBERTa [36], but they yielded similar cross-dataset performances, and so to conserve space, we do not present these results.

3.2   Experiments and Results

For the T5 model, we initially experimented with different prefixes and task labels configurations, but settled on "`generic HS:`" and "`specific HS:`", for the generic and specific HS datasets, respectively. In this setup, the model is fine-tuned without task or dataset specific information added, but rather, only the level of HS classification available and/or requested (*is HS present or not?* vs. *which specific topic of HS?*). We refer to this particular configuration using unified prefixes as T5-Refine.

To ascertain how well this configuration is able to learn both of these tasks, we perform a comparative evaluation of performance across datasets alongside other configurations, similar but not intended for topic refinement. As such, we also trained our models with MTL architectures as follows.

RoBERTa-MTL: This is a RoBERTa-base classifier, in the "classic" MTL configuration with one dedicated classification layer per task/dataset (a simple linear projection of the [CLS] token; see [7] or [19] for more details), on the same set of multi-level datasets. (output labels: HS/nothing for Davidson & Founta; HS-sexist/ nothing for Evalita & IberEval; HS-racist/HS-sexist/nothing for Waseem & HatEval);

T5-MTL: This is a fine-tuned T5-base model with task-specific prefixes (the names of the corresponding datasets) (output labels: HS/HS-racist/HS-sexist/nothing for all datasets), used here as an intermediate point of comparison between the previous two models (i.e., RoBERTa-MTL and T5-Refine).

Table 2. Comparative evaluation of our models across generic vs. specific HS datasets.

Test sets	Generic			Specific			All		
Model	P	R	$F1$	P	R	$F1$	P	R	$F1$
RoBERTa-MTL	65.14	71.09	67.23	80.84	81.15	77.68	73.49	76.44	72.78
T5-MTL	65.91	64.07	64.83	78.56	75.95	75.79	72.64	70.39	70.66
T5-Refine	63.00	65.06	63.92	79.32	73.59	73.62	71.68	69.60	69.08

We trained T5-Refine on all the training datasets combined (with generic/specific HS task prefixes) while RoBERTa-MTL and T5-MTL models were trained in a multi-task fashion (one head/task prefix per dataset) on the train set of each dataset. Experiments were performed with the AllenNLP [15] and Huggingface Transformers library [35]. Models were trained for a maximum of 12 epochs, with early stopping (patience 4 on validation loss), a batch size of 6, and gradient accumulation of 12. For T5 (RoBERTa) we use the AdaFactor (AdamW) optimizer with a learning rate =1e-3 (1e-5), determined by manual hyperparameter fine-tuning.

Table 2 presents the aggregated averaged results in terms of F-score ($F1$), precision (P), and recall (R) for the three models when tested on: all generic HS test sets (Davidson and Founta), all specific HS (Waseem, HatEval, Evalita, and IberEval) test sets, and all 6 combined test sets.

Table 3 present a more detailed view of these results, in terms of macro F1-scores only (for conciseness): for clarity, the multi-topic datasets (HatEval and Waseem) have been split into single-topic subsets (HatEval sexist/Waseem sexist and HatEval racist/Waseem racist). Then, for each dataset, "HS" and "not HS" correspond to each respective (sub)set's relevant binarized HS positive and negative classes (HS[-sexist/-racist]/nothing), alongside the Macro Averaged F1-scores. As can be observed, our HS topic refinement model, T5-Refine, despite training under the most difficult configuration (unified label-space and topic-level merged task prefixes), does not showcase significantly degraded cross-dataset performance, compared to the more task dedicated models.

4   Hate Speech Topic Refinement

Using the trained T5-Refine model, we can thus request it to produce specific HS labels for instances of generic HS datasets, here, Davidson and Founta, by simply switching to the specific task prefix at inference time. Table 4 presents a few illustrative examples, of what we consider to be successfully refined labels (examples #1–4), as well as errors (examples #5–9).

To judge the quality of these newly produced labels, we sample 600 instances (200 from each of: [gold = HS | predicted = nothing]; [gold = <any> | predicted = HS - sexist]; [gold = <any> | predicted = HS - racist], where <any> stands for all the possible gold labels) for each of the two generic HS datasets, and compare the predicted labels with the dataset's gold labels, but also with our own human re-annotation[3]

---

[3] Performed by a computational scientist and two of the authors of this paper.

Table 3. Detailed evaluation results per-dataset (F1-scores).

	Generic		Specifc (Gender)				Specifc (Race)	
Label	Davidson	Founta	Evalita	IberEval	HatEval sexist	Waseem sexist	HatEval racist	Waseem racist
RoBERTa-MTL								
HS	87.53	30.64	83.11	88.01	67.74	78.09	62.64	80.53
not HS	96.51	96.85	86.18	92.07	53.24	93.31	34.80	96.36
Macro	**92.02**	**63.75**	**84.65**	**90.04**	**60.49**	**85.70**	48.72	**88.44**
T5-MTL								
HS	93.82	25.64	64.97	91.41	63.73	68.25	59.90	96.42
not HS	80.45	97.75	71.22	84.97	51.66	92.62	63.96	79.08
Macro	87.13	61.70	68.09	88.19	57.70	80.44	**61.93**	87.75
T5-Refine								
HS	79.47	24.31	73.65	91.49	42.38	72.07	38.51	74.42
not HS	93.68	97.16	79.65	85.22	63.49	92.25	58.39	94.38
Macro	86.58	60.74	76.65	88.35	52.93	82.16	48.45	84.40

of those same instances. For both datasets, after manually re-annotating with specific HS labels, the final label was assigned according to a majority vote (at least two annotators always ended up agreeing, so no adjudication was necessary).[4] For Founta, the re-annotations process shows that in ~19% of the cases the instances gold-labelled as "HS" belong to a type of abusive language different from the ones investigated in this paper (e.g., offensive language, reporting/denunciation of hate speech, homophobia, islamohobia, etc.), which were re-annotated as out-of-scope. We obtain similar findings for Davidson, though at a larger scale (~57%). After discarding the instances re-annotated as out-of-scope, we obtained a "soft" agreement (coercing HS - racist and HS - sexist labels as equivalent to the generic HS gold label) with the gold labels of 25% for Founta, and 70% for Davidson. In contrast, the refined HS labels exactly match the human re-annotations in 52% of the in-scope instances for Davidson, and in 44% for Founta. While not perfect, overall, the annotators agree almost twice as often with the model-refined labels than with the gold labels for Founta. For Davidson this agreement instead decreases by 18%.

Qualitatively, we believe the main cause of mis-refinement stems from the significant number of merely offensive or abusive instances having been misannotated as hateful in model's training data, when they should be distinct according to datasets' annotation schemes (see last example of Table 4), which is a known problem in HS detection [12,28]. For example, in Davidson, all the instances containing the substring "b*tch" are gold-labelled as HS, regardless of context of use. After re-annotating, 19% were found to be actually HS - sexist, and 78% out-of-scope (more than 70% offensive). Similarly, the substring "f*g" was gold-labelled as HS, with 88% re-annotated as out-of-scope (mostly offensive, with less than 18% found to be homophobic). This is likely the cause of a number of false positive refined labels, which we

---

[4] Fleiss' kappas for the three-way re-annotation: 0.59 for Davidson and 0.62 Founta.

Table 4. Examples of refined labels obtained from our approach.

#	Dataset	Instance	Gold Label	Refined Label
1	Davidson	*Our people. Now is the time for the Aryan race 2 stand up and say"no more". Before the mongerls turn the world into a ghetto slum.*	HS	HS-racist
2	Davidson	*RT @USER: It's unattractive when girls act ghetto*	nothing	HS-sexist
3	Founta	*US attack/siege caused "1/3 #Yemeni #children acutely malnourished"- Says @USER #EndYemenSiege [URL]*	HS	nothing
4	Founta	*@USER @USER Don't think the world is as ignorant as you.Just because you think a certain law doesn't exist,doesn't make it true,you look foolish.*	HS	nothing
5	Founta	*Islamic State says U.S. 'being run by an idiot' [URL]*	HS	HS-racist
6	Founta	*I just watched a video with a crowd of white ppl shouting n**ga & goin crazy to songs about black men killing each other & it made me so sad*	HS	HS-racist
7	Davidson	*@USER: Lowkey called that faggot a faggot.*	HS	HS-sexist
8	Davidson	*Happpppppy Birthdayyyy lol. Niggahs is really 21 in this bitch . [URL]*	HS	HS-sexist
9	Davidson	*#SomethingIGetAlot Are you... asian? black? Hawaiian? gay? retarded? drunk?*	HS	HS-sexist

argue should not be annotated/refined as HS: for example, reporting of HS, either correctly (#3) or incorrectly refined (#5–6), or offensive language (#8).

Due to our limited unified specific HS labels, the model also struggles with instances containing neither sexist or racist HS (example #7), or those containing multiple simultaneous HS topics (#9): in both cases, a potential solution could be to add training datasets which are annotated for more varied and/or multiple targets per instance, such as [8] for example. Despite those issues, the model was still successful at producing a number of coherent refined labels (examples #1–2), or even "corrected" negative labels for some instances (examples #3–4).

5   Conclusion and Perspectives

In this paper, we show that multilevel and multitask learning for the purpose of topic refinement in HS appears to be a viable way to palliate the relative lack of specific HS annotated data. We experimented with a T5 architecture which presents a number of advantages for future improvements: namely, it is significantly easier to extend after-the-fact, as new tasks and datasets may be further fine-tuned on, without having to modify the model's architecture to accommodate new labels or levels of annotation. This may enable taking into account other topics of HS, such as homophobia, ableism, etc., which may be present in smaller quantities in generic HS datasets, through the use of Few-Shot learning, for example.

Acknowledgments. We would like to thank Walid Younes for helping in the re-annotation process. This work has been carried out in the framework of the STERHEOTYPES project funded by the Compagnia San Paolo 'Challenges for Europe'. The research of Farah Benamara is also partially supported by DesCartes: The National Research Foundation, Prime Minister's Office, Singapore under its Campus for Research Excellence and Technological Enterprise (CREATE) program.

References

1. Alonso, P., Saini, R., Kovács, G.: Hate speech detection using transformer ensembles on the HASOC dataset. In: Karpov, A., Potapova, R. (eds.) SPECOM 2020. LNCS (LNAI), vol. 12335, pp. 13–21. Springer, Cham (2020). https://doi.org/10.1007/978-3-030-60276-5_2
2. Basile, V., et al.: SemEval-2019 Task 5: multilingual detection of hate speech against immigrants and women in twitter. In: Proceedings of the 13th International Workshop on Semantic Evaluation. Minneapolis, Minnesota, USA, pp. 54–63. Association for Computational Linguistics (Jun 2019). https://doi.org/10.18653/v1/S19-2007, https://aclanthology.org/S19-2007
3. Caselli, T., Basile, V., Mitrović, J., Granitzer, M.: HateBERT: retraining BERT for abusive language detection in english. In: Proceedings of the 5th Workshop on Online Abuse and Harms (WOAH 2021), pp. 17–25. Association for Computational Linguistics, Online (Aug 2021). https://doi.org/10.18653/v1/2021.woah-1.3, https://aclanthology.org/2021.woah-1.3
4. Chiril, P., Pamungkas, E.W., Benamara, F., Moriceau, V., Patti, V.: Emotionally informed hate speech detection: a multi-target perspective. Cogn. Comput. **14**(1), 322–352 (2021). https://doi.org/10.1007/s12559-021-09862-5
5. Collobert, R., Weston, J.: A unified architecture for natural language processing: deep neural networks with multitask learning. In: Proceedings of the 25th International Conference on Machine Learning, pp. 160–167. ICML 2008, Association for Computing Machinery, New York, NY, USA (Jul 2008). https://doi.org/10.1145/1390156.1390177, https://doi.org/10.1145/1390156.1390177
6. Davidson, T., Warmsley, D., Macy, M., Weber, I.: Automated hate speech detection and the problem of offensive language. In: Proceedings of the International AAAI Conference on Web and Social Media. vol. 11(1), pp. 512–515 (May 2017), https://ojs.aaai.org/index.php/ICWSM/article/view/14955
7. Devlin, J., Chang, M.W., Lee, K., Toutanova, K.: BERT: pre-training of deep bidirectional transformers for language understanding. In: Proceedings of the 2019 Conference of the North American Chapter of the Association for Computational Linguistics: Human Language Technologies, vol. 1 (Long and Short Papers), pp. 4171–4186. Association for Computational Linguistics, Minneapolis, Minnesota (Jun 2019). https://doi.org/10.18653/v1/N19-1423, https://www.aclweb.org/anthology/N19-1423
8. ElSherief, M., et al.: Latent hatred: a benchmark for understanding implicit hate speech. In: Proceedings of the 2021 Conference on Empirical Methods in Natural Language Processing, pp. 345–363. Association for Computational Linguistics, Online and Punta Cana, Dominican Republic (Nov 2021). https://doi.org/10.18653/v1/2021.emnlp-main.29, https://aclanthology.org/2021.emnlp-main.29
9. Evkoski, B., Pelicon, A., Mozetič, I., Ljubešić, N., Novak, P.K.: Retweet communities reveal the main sources of hate speech. PLOS ONE **17**(3), e0265602 (2022). https://doi.org/10.1371/journal.pone.0265602, https://journals.plos.org/plosone/article?id=10.1371/journal.pone.0265602
10. Fersini, E., Nozza, D., Rosso, P.: Overview of the evalita 2018 task on automatic misogyny identification (AMI). In: Caselli, T., Novielli, N., Patti, V., Rosso, P. (eds.) Proceedings of the Sixth Evaluation Campaign of Natural Language Processing and Speech Tools for Italian. Final Workshop (EVALITA 2018) co-located with the Fifth Italian Conference on Computational Linguistics (CLiC-it 2018), Turin, Italy, 12–13 Dec 2018. CEUR Workshop Proceedings, vol. 2263. CEUR-WS.org (2018). http://ceur-ws.org/Vol-2263/paper009.pdf

11. Fersini, E., Rosso, P., Anzovino, M.: Overview of the task on automatic misogyny identification at IberEval 2018. In: Rosso, P., Gonzalo, J., Martínez, R., Montalvo, S., de Albornoz, J.C. (eds.) Proceedings of the Third Workshop on Evaluation of Human Language Technologies for Iberian Languages (IberEval 2018) co-located with 34th Conference of the Spanish Society for Natural Language Processing (SEPLN 2018), Sevilla, Spain, 18 Sep 2018. CEUR Workshop Proceedings, vol. 2150, pp. 214–228. CEUR-WS.org (2018). http://ceur-ws.org/Vol-2150/overview-AMI.pdf

12. Fortuna, P., Soler, J., Wanner, L.: Toxic, Hateful, Offensive or Abusive? What are we really classifying? An empirical analysis of hate speech datasets. In: Proceedings of the Twelfth Language Resources and Evaluation Conference. Marseille, France, pp. 6786–6794. European Language Resources Association (May 2020). https://aclanthology.org/2020.lrec-1.838

13. Fortuna, P., Soler-Company, J., Wanner, L.: How well do hate speech, toxicity, abusive and offensive language classification models generalize across datasets? Information Processing & Management **58**(3), 102524 (2021). https://doi.org/10.1016/j.ipm.2021.102524, https://www.sciencedirect.com/science/article/pii/S0306457321000339

14. Founta, A.M., et al.: Large scale crowdsourcing and characterization of twitter abusive behavior. In: Twelfth International AAAI Conference on Web and Social Media (Jun 2018). https://www.aaai.org/ocs/index.php/ICWSM/ICWSM18/paper/view/17909

15. Gardner, M., et al.: AllenNLP: a deep semantic natural language processing platform. In: Proceedings of Workshop for NLP Open Source Software (NLP-OSS). Melbourne, Australia, pp. 1–6. Association for Computational Linguistics (Jul 2018). https://doi.org/10.18653/v1/W18-2501, https://aclanthology.org/W18-2501

16. Kapil, P., Ekbal, A.: A deep neural network based multi-task learning approach to hate speech detection. Knowl. Based Syst. **210**, 106458 (2020). https://doi.org/10.1016/j.knosys.2020.106458, https://www.sciencedirect.com/science/article/pii/S0950705120305876

17. Kovács, G., Alonso, P., Saini, R.: Challenges of hate speech detection in social media. SN Comput. Sci. **2**(2), 1–15 (2021). https://doi.org/10.1007/s42979-021-00457-3

18. Liu, X., He, P., Chen, W., Gao, J.: Multi-task deep neural networks for natural language understanding. In: Proceedings of the 57th Annual Meeting of the Association for Computational Linguistics. Florence, Italy, pp. 4487–4496. Association for Computational Linguistics (Jul 2019). https://doi.org/10.18653/v1/P19-1441, https://www.aclweb.org/anthology/P19-1441

19. Liu, Y., et al.: RoBERTa: a robustly optimized BERT pretraining approach. arXiv:1907.11692 [cs] (Jul 2019)

20. Madukwe, K., Gao, X., Xue, B.. In Data We Trust: A Critical Analysis of Hate Speech Detection Datasets. In: Proceedings of the Fourth Workshop on Online Abuse and Harms. pp. 150–161. Association for Computational Linguistics, Online (Nov 2020). https://doi.org/10.18653/v1/2020.alw-1.18, https://aclanthology.org/2020.alw-1.18

21. Malmasi, S., Zampieri, M.: Challenges in discriminating profanity from hate speech. J. Exp. Theor. Artif. Intell. **30**(2), 187–202 (2018). https://doi.org/10.1080/0952813X.2017.1409284

22. Manne, K.: Down Girl: The Logic of Misogyny. Oxford University Press (2017)

23. Martínez Alonso, H., Plank, B.: When is multitask learning effective? Semantic sequence prediction under varying data conditions. In: Proceedings of the 15th Conference of the European Chapter of the Association for Computational Linguistics. Valencia, Spain. vol. 1, Long Papers, pp. 44–53. Association for Computational Linguistics (Apr 2017), https://aclanthology.org/E17-1005

24. Mathew, B., Saha, P., Yimam, S.M., Biemann, C., Goyal, P., Mukherjee, A.: HateXplain: a benchmark dataset for explainable hate speech detection. In: Proceedings of the AAAI Conference on Artificial Intelligence. vol. 35(17), pp. 14867–14875 (May 2021). https://ojs.aaai.org/index.php/AAAI/article/view/17745

25. Mutanga, R.T., Naicker, N., Olugbara, O.O.: Hate speech detection in twitter using transformer methods. Int. J. Adv. Comput. Sci. Appl. (IJACSA) **11**(9) (2020). https://doi.org/10.14569/IJACSA.2020.0110972, https://thesai.org/Publications/ViewPaper?Volume=11&Issue=9&Code=IJACSA&SerialNo=72

26. Nockleby, J.T.: Hate speech. In: L.W. Levy., K.L. Karst. (eds.), Encyclopedia of the American Constitution, 2nd edn. pp. 1277–1279 (2000)

27. Poletto, F., Basile, V., Sanguinetti, M., Bosco, C., Patti, V.: Resources and benchmark corpora for hate speech detection: a systematic review. Lang. Resour. Eval. **55**(2), 477–523 (2021)

28. Poletto, F., Basile, V., Sanguinetti, M., Bosco, C., Patti, V.: Resources and benchmark corpora for hate speech detection: a systematic review. Lang. Resour. Eval. **55**(2), 477–523 (2020). https://doi.org/10.1007/s10579-020-09502-8

29. Raffel, C., et al.: Exploring the limits of transfer learning with a unified text-to-text transformer. J. Mach. Learn. Res. **21**(140), 1–67 (2020). http://jmlr.org/papers/v21/20-074.html

30. Sarkar, D., Zampieri, M., Ranasinghe, T., Ororbia, A.: fBERT: a neural transformer for identifying offensive content. In: Findings of the Association for Computational Linguistics: EMNLP 2021. Punta Cana, Dominican Republic, pp. 1792–1798. Association for Computational Linguistics (Nov 2021). https://doi.org/10.18653/v1/2021.findings-emnlp.154, https://aclanthology.org/2021.findings-emnlp.154

31. Silva, L., Mondal, M., Correa, D., Benevenuto, F., Weber, I.: Analyzing the targets of hate in online social media. In: Proceedings of the 10th International Conference on Web and Social Media, ICWSM 2016, pp. 687–690. AAAI Press (2016). 10th International Conference on Web and Social Media, ICWSM 2016; Conference date: 17–05-2016 Through 20–05-2016

32. Talat, Z., Thorne, J., Bingel, J.: Bridging the Gaps: multi task learning for domain transfer of hate speech detection. In: Golbeck, J. (ed.) Online Harassment. HIS, pp. 29–55. Springer, Cham (2018). https://doi.org/10.1007/978-3-319-78583-7_3

33. Vaswani, A., et al.: Attention is All you Need. In: Advances in Neural Information Processing Systems. vol. 30. Curran Associates, Inc. (2017). https://papers.nips.cc/paper/2017/hash/3f5ee243547dee91fbd053c1c4a845aa-Abstract.html

34. Waseem, Z., Hovy, D.: Hateful symbols or hateful people? Predictive features for hate speech detection on twitter. In: Proceedings of the NAACL Student Research Workshop. San Diego, California, pp. 88–93. Association for Computational Linguistics (Jun 2016). https://doi.org/10.18653/v1/N16-2013, https://aclanthology.org/N16-2013

35. Wolf, T., et al.: Transformers: state-of-the-art natural language processing. In: Proceedings of the 2020 Conference on Empirical Methods in Natural Language Processing: System Demonstrations, pp. 38–45. Association for Computational Linguistics, Online (Oct 2020). https://doi.org/10.18653/v1/2020.emnlp-demos.6, https://aclanthology.org/2020.emnlp-demos.6

36. Zhou, X., Sap, M., Swayamdipta, S., Choi, Y., Smith, N.: Challenges in automated debiasing for toxic language detection. In: Proceedings of the 16th Conference of the European Chapter of the Association for Computational Linguistics: Main Volume. pp. 3143–3155. Association for Computational Linguistics, Online (Apr 2021). https://doi.org/10.18653/v1/2021.eacl-main.274, https://aclanthology.org/2021.eacl-main.274

Is Cross-Modal Information Retrieval Possible Without Training?

Hyunjin Choi[✉], Hyunjae Lee, Seongho Joe, and Youngjune Gwon

Samsung SDS, Seoul, Korea
{hjjin.choi,h8.lee}@samsung.com

Abstract. Encoded representations from a pretrained deep learning model (e.g., BERT text embeddings, penultimate CNN layer activations of an image) convey a rich set of features beneficial for information retrieval. Embeddings for a particular modality of data occupy a high-dimensional space of its own, but it can be semantically aligned to another by a simple mapping without training a deep neural net. In this paper, we take a simple mapping computed from the least squares and singular value decomposition (SVD) for a solution to the Procrustes problem to serve a means to cross-modal information retrieval. That is, given information in one modality such as text, the mapping helps us locate a semantically equivalent data item in another modality such as image. Using off-the-shelf pretrained deep learning models, we have experimented the aforementioned simple cross-modal mappings in tasks of text-to-image and image-to-text retrieval. Despite simplicity, our mappings perform reasonably well reaching the highest accuracy of 77% on recall@10, which is comparable to those requiring costly neural net training and fine-tuning. We have improved the simple mappings by contrastive learning on the pretrained models. Contrastive learning can be thought as properly biasing the pretrained encoders to enhance the cross-modal mapping quality. We have further improved the performance by multilayer perceptron with gating (gMLP), a simple neural architecture.

1 Introduction

Cross-modal information retrieval takes in one modality (or type) of data as a query to retrieve semantically related data of another type. There is a fundamental challenge in measuring the similarity between the query and the outcome having different modalities. Research in cross-modal retrieval has naturally focused on learning or training a joint subspace where different modalities of data can be compared directly.

Recently, pretraining deep learning models with large-scale data has proved effective for creating applications in computer vision and natural language processing (NLP). Available publicly, pretrained models are a powerful encoder of characteristic features for data onto an embedding space. Pretrained models are valid in a unimodal scenario, and it is difficult to purpose them for cross-modal

J. Kamps et al. (Eds.): ECIR 2023, LNCS 13981, pp. 377–385, 2023.
https://doi.org/10.1007/978-3-031-28238-6_27

(or multimodal) usage. Joint training of different data modalities in a large scale would be extremely difficult and costly (or it may not be feasible).

In this paper, we compute a simple mapping instead for cross-modal translation via the least squares and singular value decomposition (SVD). Embedding representations from a pretrained unimodal encoder are aligned *semantically* to embeddings from another encoder for different modality by the mapping. We have carried out an experimental validation of our approach for cross-modal tasks of text-to-image and image-to-text retrieval. Given a text query, the text-to-image mapping translates a text embedding onto the subspace for image embeddings where the translated text embedding can be directly compared to those of images, and vice versa for the image-to-text mapping.

We can improve the performance of our simple mappings by the choice of pretrained unimodal encoders used. There are off-the-shelf pretrained language models properly biased by contrastive learning with the Natural Language Inference (NLI) dataset. Usually, pretrained image models are already biased properly from training for object classification. An external component such as outer neural layers can further enhance the cross-modal performance. We demonstrate the improved performance by adding an outer multilayer perceptron with gating (gMLP) [10], a simple neural architecture.

Our contributions are as follows: i) encoded unimodal representations from off-the-shelf pretrained models can be aligned by a simple, training-free mapping for cross-modal information retrieval; ii) despite simplicity, our cross-modal mappings perform reasonably well reaching the highest accuracy of 77% on recall@10 comparable to deep neural nets with costly training and fine-tuning; iii) optionally, proper biases introduced by contrastive learning and outer neural architecture such as gated MLP can improve the cross-modal retrieval performance of the proposed mappings.

2   Related Work

Pretraining of deep learning models on large-scale data has flourished under unimodal assumptions. Recently, a self-supervised method makes automated training with unlimited data available on the Internet possible. In computer vision, the success of VGG [17] and ResNet [6] is immensely followed while BERT [3] and GPT [1] have achieved a similar success in NLP. Information retrieval can tremendously benefit from pretrained models although they are valid in a unimodal scenario only.

For the case of cross-modal retrieval, pairing up semantically equivalent data modalities can be considered. To learn a shared embedding subspace, one can explore the idea of jointly training semantically related data with different modalities [7,9,12,13,15,18,20,23]. Not necessarily for cross-modal information retrieval, these approaches have set language-vision benchmark tasks and achieved good downstream performances.

Attention mechanism used in masked language modeling can also be applied to a joint image-text encoder as in VL-BERT [18]. The recent explorations of learning image representations directly from semantic counterparts in natural language have partly inspired us. CLIP [14] and ALIGN [8] train images with relevant natural language captions to obtain rich vision-language representations for improving performance on diverse downstream tasks such as text-to-image matching and retrieval. They have demonstrated a simple contrastive learning setup capture better representations without heavily relying on labeled data or a sophisticated neural architecture. To achieve a good performance, however, substantive training effort is inevitable (in a scale of hundreds of high-spec GPUs and billions of training examples).

Instead of laborious and expensive training spent by recent related approaches, we have decided to experiment with simple mappings computable from the least squares and linear projections. The mappings translate a representation from pretrained models in one modality to another such that representations of different modalities can be directly compared for cross-modal retrieval. We take off-the-shelf BERT, RoBERTa, and ViT [4] for encoding text and images. The Transformer encoder [21] has originated for NLP, but attention mechanisms have grown their benefits in vision. Transformers are computationally more efficient than convolutional neural nets while providing an on-par or better results for vision tasks. ViT uses the pretrained weights with the JFT-300M dataset [19]. Because there is no inductive bias inherent in CNNs such as translation equivariance and locality, the CNN pretraining requires an enormous amount of data if trained from scratch.

3 Approach

Pretrained unimodal encoders take in text and image modalities of data as depicted in Fig. 1. Our goal is to embed one input modality and translate it onto an embedding subspace of the other through a cross-modal mapping. After the translation, embeddings are in the same semantic space and directly compared. That is, the similarity between data examples in different modalities can be examined by computing the inner product of their embedding vectors. In this section, we describe our approach and explain how to compute simple cross-modal mappings.

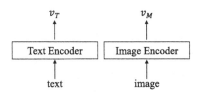

Fig. 1. Encoders for extracting vectors from text and image.

3.1   System of Least Squares via Normal Equations

Our first method is to learn the text-to-image mapping directly from paired image-text data (and vice versa for image-to-text mapping) via the least squares. Suppose text T and image M that are the source and the target of the mapping (or a linear projection) $\mathbf{\Phi}$. We seek the solution to the problem $\mathbf{V}_T \mathbf{\Phi} = \mathbf{V}_M$ with

$$
\mathbf{V}_T = \begin{bmatrix} - & \mathbf{v}_T^{(1)} & - \\ - & \mathbf{v}_T^{(2)} & - \\ & \vdots & \\ - & \mathbf{v}_T^{(n)} & - \end{bmatrix}, \; \mathbf{V}_M = \begin{bmatrix} - & \mathbf{v}_M^{(1)} & - \\ - & \mathbf{v}_M^{(2)} & - \\ & \vdots & \\ - & \mathbf{v}_M^{(n)} & - \end{bmatrix}, \; \mathbf{v}_T^{(i)} = \begin{bmatrix} t_1^{(i)} \\ t_2^{(i)} \\ \vdots \\ t_d^{(i)} \end{bmatrix}^\top, \; \mathbf{v}_M^{(i)} = \begin{bmatrix} m_1^{(i)} \\ m_2^{(i)} \\ \vdots \\ m_d^{(i)} \end{bmatrix}^\top
$$

where $\mathbf{V}_T$ and $\mathbf{V}_M$ are datasets that contain n embeddings for text T and image M with each $\mathbf{v} \in \mathbb{R}^d$. With $\mathbf{\Phi} = \begin{bmatrix} \phi^{(1)} \; \phi^{(2)} \ldots \phi^{(j)} \ldots \phi^{(d)} \end{bmatrix}$ whose element $\phi^{(j)} \in \mathbb{R}^d$ is a column vector, each $\mathbf{V}_T \phi^{(j)} = [m_j^{(1)} m_j^{(2)} \ldots m_j^{(n)}]$ gives a problem of the least squares. Since $k = 1, \ldots, d$, we have a system of d least-square problems that can be solved linear algebraically via the normal equation: $\mathbf{\Phi}^* = \left(\mathbf{V}_T^\top \mathbf{V}_T\right)^{-1} \mathbf{V}_T^\top \mathbf{V}_M$.

3.2   Solution to the Procrustes Problem

Given two data matrices, a source $\mathbf{V}_T$ and a target $\mathbf{V}_M$, the orthogonal Procrustes problem [16] describes approximation of a matrix searching for an orthogonal projection that most closely maps $\mathbf{V}_T$ to $\mathbf{V}_M$. Formally, we write

$$
\mathbf{\Psi}^* = \arg \min_{\mathbf{\Psi}} \| \mathbf{V}_T \mathbf{\Psi} - \mathbf{V}_M \|_F \quad \text{s.t. } \mathbf{\Psi}^\top \mathbf{\Psi} = \mathbf{I} \tag{1}
$$

The solution to Eq. (1) has the closed-form $\mathbf{\Psi}^* = \mathbf{X}\mathbf{Y}^\top$ with $\mathbf{X}\mathbf{\Sigma}\mathbf{Y}^\top = \text{SVD}(\mathbf{V}_M \mathbf{V}_T^\top)$, where SVD is the singular value decomposition. The Procrustes solution $\mathbf{\Psi}$ gives our second choice for the cross-modal mapping. Similarly, the Procrustes problem for image-to-text mapping can be set up and solved by SVD.

3.3   Optional Considerations

In this section, we describe optional considerations, perhaps with little training, that can improve the baseline cross-modal retrieval performance. Contrastive learning can be set up to minimize the distance between a pair of image and text examples of semantic equivalence (*i.e.*, the text description matches the image) [8,9,14,23]. Conversely, contrastive learning maximizes the distance of a non-matching pair. Because forming non-matching pairs (negatives) can be automated, contrastive learning is convenient to learn the joint representation. Also, it has demonstrated promising results on downstream vision-language tasks. Our approach is differentiated from others over the nonlinear architecture that follows the front-end bimodal (image-text) encoders. We explain our architectural components as follows.

gMLP Blocks. We train multi-layer perceptron (MLP) with gating (gMLP) [10], a nonlinear projection layer following the bimodal encoders. Stacked MLP layers without self-attention are used to capture semantic relationship of image and text data despite having a much lightened network. More importantly, gMLP has exactly the same input and output shape as BERT's and ViT's. This makes sense for using the feature vectors extracted by the bimodal encoders without modification. Our contrastive learning approach emphasizes the training on top of the bimodal encoder output.

Contrastive Learning Objective. From a pair of image and text examples, the bimodal encoders compute $v_T^{(i)}$ and $v_M^{(i)}$ as output to form $\mathcal{D} = \{(v_T^{(i)}, v_M^{(i)})\}_{i=1}^m$ that are applied to the gMLP nonlinear projection layers. We take the cross-entropy loss for contrastive learning with N in-batch negatives:

$$\ell_i = -\log \frac{e^{\text{sim}(h_i, h_i^+)/\tau}}{\sum_{j=1}^N e^{\text{sim}(h_i, h_j^+)/\tau}}, \tag{2}$$

where τ is a temperature hyperparameter, $\text{sim}(h_1, h_2)$ is the cosine similarity $\frac{h_1^\top h_2}{\|h_1\| \cdot \|h_2\|}$, and h_i and h_i^+ denote output vector of $v_T^{(i)}$ and $v_M^{(i)}$. Optionally, negative pairs can be automatically generated, and the loss function can trivially be modified.

4 Experiments

4.1 Setup

Dataset. To set up cross-modal information retrieval tasks (in both text-to-image and image-to-text directions), we use Flickr30k [22] that contains 31,000 images collected from Flickr, each of which is provided with five descriptive sentences by human annotators. A training set of 29,783 pairs and a test set of 1,000 pairs are used.

Evaluation. For evaluating the performance of cross-modal retrieval, we use the test partition from Flickr30k. We use the standard evaluation criteria used in most prior work on image-text retrieval task. We adopt recall@$x = 1, 5, 10, 20, 100$ as our evaluation metric.

Pretrained Encoders. We use the large BERT [3] and RoBERTa [11] models as our text encoders. Both are representative Transformer-based pretrained language models. We also use a large ViT pretrained on the ImageNet-21k dataset as our image encoder. ViT produces the image embeddings (a hidden dimension of 1,024 and a 32×32 patch size that produces 50 hidden vectors) that can be taken in as input to the gMLP layers without any modification.

Table 1. Cross-modal retrieval results on Flickr30K. For comparison, we add the results from CLIP [14]. ('+' means an enhanced encoder by biasing.)

Training	Encoder	Image to text					Text to image				
		R@1	R@5	R@10	R@20	R@100	R@1	R@5	R@10	R@20	R@100
No Training/GPU	BERT	11.5	37.6	50.5	63.3	86.8	15.5	39.4	53.4	67.0	89.4
	RoBERTa	18.1	44.4	58.6	70.1	90.1	17.1	42.5	56.3	70.3	91.5
	BERT+	29.4	**65.2**	75.9	85.5	**95.7**	20.5	47.5	62.0	74.1	**93.4**
	RoBERTa+	**31.9**	**65.2**	**77.2**	**85.8**	95.0	**22.6**	**51.5**	**64.4**	**75.5**	92.9
Outer Layer Training (1 1080ti * 0.3 h)	BERT	20.4	51.5	67.0	79.7	95.4	24.9	53.2	68.2	79.1	94.5
	RoBERTa	16.3	43.1	56.5	71.4	93.4	16.4	41.4	55.6	66.7	90.4
	BERT+	**37.5**	**71.6**	**81.5**	**87.8**	**97.0**	**33.4**	**65.4**	**77.6**	**84.9**	**96.4**
	RoBERTa+	14.9	44.7	58.0	72.4	93.2	16.2	42.0	54.6	66.5	91.3
Full Encoder Retraining (256 V100 * 12 days)	CLIP	88.0	98.7	99.4	–	–	68.7	90.6	95.2	–	–

Query text: A crowd of people stand in the background of a set of tables draped in white with various used toys set upon it.

Top 1   Top 2   Top 3   Top 4   Top 5

Query image

Top 1: A crowd of people stand in the background of a set of tables draped in white with various used toys set upon it.
Top 2: A group of people at a festival are blowing bubbles and enjoying the attractions.
Top 3: A large group of men , women , and children gather on a lawn for a concert.
Top 4: A musical group performs in front of a live audience, adorning colorful neon clothing and in front of a screen backdrop.
Top 5: A large display of artifacts are in a large room and around a very large display are six people with two people near wooden benches.

Fig. 2. Qualitative results of no training model. The dashed line shows the correct retrieval results.

4.2   Results

Simple Cross-Modal Mappings. Our baseline (i.e., no training) results are presented in Table 1. We choose higher number of two different linear mapping methods (least squares and SVD). This simple method reaches 58.6/56.3% recall@10. We find one of two linear mapping methods fail in some cases. We hypothesize this is caused by poor sentence embeddings of text encoders.

Using Enhanced Text Encoders via Biasing. The image encoder is originally pretrained to classify 1,000 object classes. On the contrary, the text encoder has not been trained to capture fine-grained sentence representations. Choi *et al.* [2] show sentence embeddings of BERT without additional learning give poor performance. We hypothesize that this will adversely affect the performance of cross-modal representation. To alleviate the lack of properly contextualized sentence embeddings, we adopt enhanced text encoder (noted BERT+ and RoBERTa+ in Table 1) from SimCSE [5] which uses self-supervised contrastive learning for text encoders. Using enhanced text encoder results in 77.2/64.4% recall@10 as reported in Table 1. Our qualitative results (as shown in Fig. 2) show cross-modal retrieval is possible without training.

Optional gMLP Outer Layer. Contrastive learning for the gMLP projection yields the best score of 81.5/77.6% recall@10. The training takes less than 20 min on a single NVIDIA 1080Ti graphics card.

5   Conclusion

With a plethora of large-scale pretrained deep learning models, we have posed an intriguing hypothesis for a light-weight, earth-saving approach to cross-modal information retrieval. Unimodal representations computed by a pretrained model form a high-dimensional embedding subspace of its own. Despite the mess, encoded representations from off-the-shelf pretrained models for different modalities of data can be semantically aligned without additional training. We have formulated classical problems to solve for a simple mapping, which is computable without training but by the least squares and SVD. We have experimented with publicly available pretrained models for text-to-image and image-to-text retrieval tasks. Our simple approach seems to have a good potential for improvement, particularly from the future enhancement of pretrained models. Optionally, if we allow little training to properly bias unimodal encoders and add outer gMLP layers, we can significantly improve the performance of cross-modal information retrieval.

References

1. Brown, T., et al.: Language models are few-shot learners. In: Advances in Neural Information Processing Systems, vol. 33, pp. 1877–1901 (2020)
2. Choi, H., Kim, J., Joe, S., Gwon, Y.: Evaluation of BERT and ALBERT sentence embedding performance on downstream NLP tasks. In: 2020 25th International Conference on Pattern Recognition (ICPR), pp. 5482–5487 (2021). https://doi.org/10.1109/ICPR48806.2021.9412102
3. Devlin, J., Chang, M., Lee, K., Toutanova, K.: BERT: pre-training of deep bidirectional transformers for language understanding. In: NAACL-HLT (1), pp. 4171–4186. Association for Computational Linguistics (2019)
4. Dosovitskiy, A., et al.: An image is worth 16 × 16 words: transformers for image recognition at scale. In: International Conference on Learning Representations (2021). https://openreview.net/forum?id=YicbFdNTTy
5. Gao, T., Yao, X., Chen, D.: SimCSE: simple contrastive learning of sentence embeddings. In: Proceedings of the 2021 Conference on Empirical Methods in Natural Language Processing, pp. 6894–6910 (2021)
6. He, K., Zhang, X., Ren, S., Sun, J.: Deep residual learning for image recognition. In: CVPR, pp. 770–778. IEEE Computer Society (2016)

7. Huang, Z., Zeng, Z., Liu, B., Fu, D., Fu, J.: Pixel-BERT: aligning image pixels with text by deep multi-modal transformers. CoRR abs/2004.00849 (2020)
8. Jia, C., et al.: Scaling up visual and vision-language representation learning with noisy text supervision. In: Meila, M., Zhang, T. (eds.) Proceedings of the 38th International Conference on Machine Learning, ICML 2021, Virtual Event. Proceedings of Machine Learning Research, 18–24 July 2021, vol. 139, pp. 4904–4916. PMLR (2021)
9. Li, X., et al.: OSCAR: object-semantics aligned pre-training for vision-language tasks. In: Vedaldi, A., Bischof, H., Brox, T., Frahm, J.-M. (eds.) ECCV 2020. LNCS, vol. 12375, pp. 121–137. Springer, Cham (2020). https://doi.org/10.1007/978-3-030-58577-8_8
10. Liu, H., Dai, Z., So, D., Le, Q.V.: Pay attention to MLPs. In: Thirty-Fifth Conference on Neural Information Processing Systems (2021). https://openreview.net/forum?id=KBnXrODoBW
11. Liu, Y., et al.: RoBERTa: a robustly optimized BERT pretraining approach. CoRR abs/1907.11692 (2019). http://arxiv.org/abs/1907.11692
12. Lu, J., Batra, D., Parikh, D., Lee, S.: ViLBERT: pretraining task-agnostic visiolinguistic representations for vision-and-language tasks. In: Advances in Neural Information Processing Systems, vol. 32. Curran Associates, Inc. (2019). https://proceedings.neurips.cc/paper/2019/hash/c74d97b01eae257e44aa9d5bade97baf-Abstract.html
13. Qi, D., Su, L., Song, J., Cui, E., Bharti, T., Sacheti, A.: ImageBERT: cross-modal pre-training with large-scale weak-supervised image-text data. CoRR abs/2001.07966 (2020)
14. Radford, A., et al.: Learning transferable visual models from natural language supervision. In: Meila, M., Zhang, T. (eds.) Proceedings of the 38th International Conference on Machine Learning, ICML 2021, Virtual Event. Proceedings of Machine Learning Research, 18–24 July 2021, vol. 139, pp. 8748–8763. PMLR (2021)
15. Sariyildiz, M.B., Perez, J., Larlus, D.: Learning visual representations with caption annotations. In: Vedaldi, A., Bischof, H., Brox, T., Frahm, J.-M. (eds.) ECCV 2020. LNCS, vol. 12353, pp. 153–170. Springer, Cham (2020). https://doi.org/10.1007/978-3-030-58598-3_10
16. Schönemann, P.: A generalized solution of the orthogonal procrustes problem. Psychometrika 31(1), 1–10 (1966). https://doi.org/10.1007/BF02289451
17. Simonyan, K., Zisserman, A.: Very deep convolutional networks for large-scale image recognition. In: Bengio, Y., LeCun, Y. (eds.) 3rd International Conference on Learning Representations, ICLR 2015, San Diego, CA, USA, 7–9 May 2015, Conference Track Proceedings (2015). http://arxiv.org/abs/1409.1556
18. Su, W., et al.: VL-BERT: pre-training of generic visual-linguistic representations. In: 8th International Conference on Learning Representations, ICLR 2020, Addis Ababa, Ethiopia, 26–30 April 2020. OpenReview.net (2020)
19. Sun, C., Shrivastava, A., Singh, S., Gupta, A.: Revisiting unreasonable effectiveness of data in deep learning era. CoRR abs/1707.02968 (2017). http://arxiv.org/abs/1707.02968
20. Tan, H., Bansal, M.: LXMERT: learning cross-modality encoder representations from transformers. In: Inui, K., Jiang, J., Ng, V., Wan, X. (eds.) Proceedings of the 2019 Conference on Empirical Methods in Natural Language Processing and the 9th International Joint Conference on Natural Language Processing, EMNLP-IJCNLP 2019, Hong Kong, China, 3–7 November 2019, pp. 5099–5110. Association for Computational Linguistics (2019)

21. Vaswani, A., et al.: Attention is all you need. In: Advances in Neural Information Processing Systems, vol. 30 (2017)
22. Young, P., Lai, A., Hodosh, M., Hockenmaier, J.: From image descriptions to visual denotations: new similarity metrics for semantic inference over event descriptions. Trans. Assoc. Comput. Linguist. **2**, 67–78 (2014)
23. Zhang, Y., Jiang, H., Miura, Y., Manning, C.D., Langlotz, C.P.: Contrastive learning of medical visual representations from paired images and text. CoRR abs/2010.00747 (2020)

Adversarial Adaptation for French Named Entity Recognition

Arjun Choudhry[1,2]([✉]) [iD], Inder Khatri[1] [iD], Pankaj Gupta[1] [iD], Aaryan Gupta[1] [iD], Maxime Nicol[2] [iD], Marie-Jean Meurs[2] [iD], and Dinesh Kumar Vishwakarma[1] [iD]

[1] Biometric Research Laboratory, Delhi Technological University, New Delhi, India
choudhry.arjun@gmail.com, dinesh@dtu.ac.in
[2] IKB Lab, Université du Québec à Montréal, Montréal, QC, Canada
nicol.maxime@courrier.uqam.ca, meurs.marie-jean@uqam.ca

Abstract. Named Entity Recognition (NER) is the task of identifying and classifying named entities in large-scale texts into predefined classes. NER in French and other relatively limited-resource languages cannot always benefit from approaches proposed for languages like English due to a dearth of large, robust datasets. In this paper, we present our work that aims to mitigate the effects of this dearth of large, labeled datasets. We propose a Transformer-based NER approach for French, using adversarial adaptation to similar domain or general corpora to improve feature extraction and enable better generalization. Our approach allows learning better features using large-scale unlabeled corpora from the same domain or mixed domains to introduce more variations during training and reduce overfitting. Experimental results on three labeled datasets show that our adaptation framework outperforms the corresponding non-adaptive models for various combinations of Transformer models, source datasets, and target corpora. We also show that adversarial adaptation to large-scale unlabeled corpora can help mitigate the performance dip incurred on using Transformer models pre-trained on smaller corpora.

Keywords: Named entity recognition · Adversarial adaptation · Transformer · Limited resource languages · Large-scale corpora

1 Introduction

Named Entity Recognition (NER) is the task of identifying and extracting specific entities from unstructured text, and labeling them into predefined classes. Over the years, NER models for high-resource languages, like English, have seen noticeable improvements in task performance owing to model architecture advancements and the availability of large, labeled datasets. In sharp contrast, languages like French still lack openly available, large-scale, labeled, robust

---

A. Choudhry and I. Khatri—Equal Contribution.

datasets free from biases, and have few general-domain language models, and barely any domain-specific ones. Creating large, robust NER datasets requires manual annotation and is prohibitively expensive, calling for less labeled data-reliant approaches, particularly for limited and low-resource languages.

Over the years, a variety of deep learning-based NER approaches have been proposed [22]. Some of the older approaches made use of external texts, or gazetteers, for the disambiguation of input words [21]. Several works have used combinations of Recurrent Neural Networks like LSTMs and GRUs with Conditional Random Fields to improve model performance for a variety of NER tasks [7–9].

Recent works have also incorporated the use of pre-trained contextualized language models for improved NER performance. Copara et $al.$ [2] proposed an ensemble-based NER framework for Biomedical NER in French, using various combinations of French language models. Liu et $al.$ [12] pre-trained a NER-BERT model on a large NER corpus to counter the underlying discrepancies between the language model and the NER dataset. They observed significantly better performance than the standard BERT [3] Transformer model across nine domains.

Researchers have also started incorporating various domain adaptation approaches for adapting NER models from high-resource domains to low-resource domains [13,25]. Peng et $al.$ [19] further proposed an entity-aware domain-adaptive framework using an attention layer to enable the improved transfer of features for models trained on one domain to another domain using adversarial learning. They observed noticeably better performance in cross-domain settings than without domain adaptation. However, due to the lack of robust and labeled datasets across multiple domains in limited-resource languages, these domain adaptive approaches are mostly restricted to high-resource languages. Wang et $al.$ [26] further introduced an adversarial perturbation approach for reduced overfitting, while also proposing the use of Gated-CNN to fuse the spatial information between adjacent words. However, they tested their approach on English datasets and noticed marginal gains over baseline approaches.

In this work, which follows our preliminary exploration [1], we incorporate the use of adversarial adaptation to improve the performance of NER models in in-domain settings for French. We propose a Transformer-based NER approach, which uses adversarial adaptation to counter the lack of large-scale labeled NER datasets in French. This helps us evaluate the use of adversarial adaptation in enabling a model to learn improved, generalized features by adapting them to large-scale unlabeled corpora that are readily available and easy to generate for even low-resource languages. We further train Transformer-based NER models on labeled source datasets and use larger corpora from similar or mixed domains as target sets for improved feature learning by the model. Our proposed approach helps outsource wider domain and general feature knowledge from easily available large unlabeled corpora. We limit the purview of our evaluation to the French language in this paper. However, our approach could further be applied to other limited and low-resource languages for improved NER performance, as well as for other downstream tasks. This paper is organized as follows: the proposed methodology is introduced in Sect. 2. Section 3 presents our experiments and discusses the obtained results while Sect. 4 concludes our findings.

2   Proposed Methodology

2.1   Datasets and Preprocessing

In this work, we use the WikiNER French [18], WikiNeural French [24], and Europeana French [17] datasets as the labeled source datasets used for the supervised training branch in our model. The Europeana dataset contains text extracted from historic European newspapers using Optical Character Recognition (OCR). However, the dataset contains some OCR errors, thus making it noisy. This can lead to lower performance by models trained on Europeana.

For the unlabeled target corpora, we use WikiNER and WikiNeural datasets, and the Leipzig Mixed French corpus[1]. We remove the labels for the former two datasets for use as large, unlabeled target corpora. These corpora enable us to evaluate the impact of adapting the models to similar-domain data, as well as more generalized corpora.

During preprocessing, we convert all NER tags to Inside-Outside-Beginning (IOB) [20] format and store the datasets in the CoNLL 2002 [23] NER format for easier data input during training.

2.2   Adversarial Adaptation to Similar Domain Corpus

Adversarial adaptation aids in selecting domain-invariant features which are transferable between source and target datasets [5]. Compared to other approaches to domain adaptation, adversarial adaptation incorporates a domain discriminator into the classification framework. A domain discriminator acts as a domain classifier and is trained on the features retrieved by the feature extractor layer in the framework. It is tasked with distinguishing between the features obtained from the source and target sets. With the help of a Gradient Reversal Layer (GRL) [4], the gradient flow of the domain discriminator is utilized to penalize the feature extractor for learning the domain-specific features, thus causing the feature extractor to learn domain-invariant features. GRL reverses the gradient direction and thus helps train different components of the neural network *adversarially*. This enables the feature extractor to yield features free from domain biases present in the source domain but not the target domain.

We propose the use of adversarial adaptation of NER models to large-scale, unlabeled corpora from similar or mixed domains. This helps us enable the model to extract relatively more generalizable features from the same domain as the source dataset, without altering the feature extractor significantly to learn overly generalized features that reduce in-domain performance. Adversarial adaptation to similar domain corpora thus helps reduces the risk of overfitting on the source

---

[1] https://wortschatz.uni-leipzig.de/en/download/French.

intricate training set-specific features, as it aids the feature extractor in extract-ing more generalizable features, indistinguishable from the features from the large-scale target corpora. This helps counter the dearth of large and robust training datasets in languages like French, which are readily available in English.

We evaluate our approach for three scenarios: source and target datasets are from the same domain; source and target datasets are from relatively different domains; and the source belongs to a certain domain while the target dataset is a mixed-domain, large-scale, general corpus. The latter scenario helps generalize the model further, reducing the domain-specific feature extraction in favor of features common to both the source dataset and the target corpora. Figure 1 graphically illustrates our proposed framework.

Our Transformer model is trained using two losses: the NER classifier loss L_{NER}, defined in Eq. 1, and the adversarial loss L_{adv}, defined in Eq. 2. The total loss is defined in Eq. 3. L_{NER} is the standard loss that penalizes the Trans-former model's token classification error, encouraging it to make more accurate entity predictions. It is further responsible for optimizing the NER Classifier's weights. L_{adv} is used adversarially, where the Transformer model is optimized in a way to maximize L_{adv}, while the domain classifier is optimized to minimize L_{adv}. To achieve this, we use a Gradient Reversal Layer between the Transformer model and the domain classifier. The Gradient Reversal Layer acts as an identity function during forward propagation, but during back-propagation, it multiplies its input by -1. This causes the back-propagated gradient to perform gradient ascent on the Transformer model with respect to the domain classifier's classi-fication loss, rather than gradient descent. We specifically use the adversarial domain classifier as a discriminator in our framework.

$$L_{NER} = \min_{\theta_f, \theta_n} \sum_{i=1}^{n_s} L_n^i \tag{1}$$

$$L_{adv} = \min_{\theta_d}(\max_{\theta_f}(\sum_{i=1}^{n_s} L_{ds}^i + \sum_{j=1}^{n_t} L_{dt}^j)) \tag{2}$$

$$L_{Total} = L_{NER} + \alpha(L_{adv}) \tag{3}$$

Here, n_s and n_t represent the number of samples in source and target sets respectively, θ_d, θ_n, and θ_f are the number of parameters for domain classifier, NER classifier, and Transformer model respectively, and L_{ds} and L_{dt} represent the Negative log-likelihood loss (NLLL) for the source and target respectively. We introduce another parameter α, which is the ratio between L_{NER} and L_{adv} in the total loss. This helps us with correctly penalizing the NER classifier and the domain classifier. We found the optimum value of α to be equal to 2, as this led to the best experimental results.

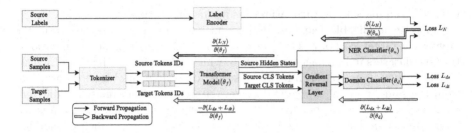

Fig. 1. Graphical representation of our adversarial adaptation framework for training NER models on source and target sets.

2.3 Language Models for NER

Recent NER research has incorporated large language models due to their contextual knowledge learned during pretraining [6,15,27]. We evaluate the efficacy of our proposed approach on three French language models: CamemBERT-base [16], CamemBERT-Wikipedia-4 GB (a variant of CamemBERT pre-trained on only 4 GB of Wikipedia French corpus), and FlauBERT-base [10]. FlauBERT is based on the BERT [3] architecture, while CamemBERT is based on the RoBERTa [11] architecture. FlauBERT is trained on nearly half the amount of training data (71 GB) as CamemBERT (138 GB). All three language models provide 768-dimensional word embeddings. Comparing CamemBERT-base and CamemBERT-Wiki-4 GB helps us analyze if we can replace language models pre-trained on large corpora with models pre-trained on smaller corpora adapted to unlabeled corpora during fine-tuning on a downstream task.

3   Experiments and Results

We evaluated the performance of our approach for various combinations of language models, source datasets, and target datasets. Each model was evaluated on the test subset (20% of the total data) of the source dataset. All domain-adaptive and baseline models were trained for up to 16 epochs. The training was stopped when peak validation accuracy for the in-domain validation set was achieved. We used a batch size of 16 for each experiment. We used the AdamW optimizer [14], an optimized version of Adam with weight decay with a learning rate of 0.00002 and a learning rate scheduler while training. We used a Transformer encoder-based layer with 512 units for feature extraction. Both the NER and domain classifier branches consisted of two dense layers each. We used a gradient reversal layer for the adversarial domain classification branch, using the last two hidden embeddings as input. These experiments are fully reproducible, and the systems are made available as open-source[2] under GNU GPL v3.0. Table 1 illustrates our results. Some prominent findings observed are:

---

[2] https://github.com/Arjun7m/AA_NER_Fr.

Table 1. Performance evaluation of our proposed adversarial adaptation approach to large-scale corpora for various combinations of models, source, and target sets. We observe noticeably improved performance for the adversarial adaptation models as compared to their corresponding non-adaptive models across almost all settings.

Model	Source	Target	Precision	Recall	F1-score
CamemBERT-Wiki-4GB	WikiNER	–	0.911	0.925	0.918
		WikiNeural	**0.966**	**0.963**	**0.969**
		Mixed-Fr	0.956	0.962	0.959
	WikiNeural	–	0.859	0.872	0.866
		WikiNER	**0.872**	**0.891**	**0.881**
		Mixed-Fr	0.870	0.879	0.875
	Europeana	–	0.728	0.642	0.682
		WikiNER	0.738	**0.691**	**0.714**
		Mixed-Fr	**0.774**	0.640	0.701
CamemBERT-base	WikiNER	–	0.960	0.968	0.964
		WikiNeural	**0.973**	0.976	**0.975**
		Mixed-Fr	0.972	**0.978**	0.974
	WikiNeural	–	0.943	0.950	0.946
		WikiNER	0.943	**0.953**	**0.948**
		Mixed-Fr	**0.946**	0.950	**0.948**
	Europeana	–	0.927	0.933	0.930
		WikiNER	0.911	0.927	0.920
		Mixed-Fr	**0.942**	**0.943**	**0.943**
FlauBERT-base	WikiNER	–	0.963	0.964	0.963
		WikiNeural	0.964	0.968	0.966
		Mixed-Fr	**0.974**	**0.972**	**0.973**
	WikiNeural	–	0.934	0.946	0.940
		WikiNER	0.935	**0.950**	**0.942**
		Mixed-Fr	**0.941**	0.943	**0.942**
	Europeana	–	0.835	0.863	0.849
		WikiNER	0.855	**0.865**	0.860
		Mixed-Fr	**0.882**	0.854	**0.867**

Models Trained Using Our Adversarial Adaptation Framework Consistently Outperformed Their Non-adaptive Counterparts Across all Metrics. We observed that the adversarial adaptation models showed significant performance improvements across Precision, Recall, and F1-score over their non-adaptive counterparts across almost all combinations of source datasets, target datasets, and language models. This is beneficial for low and limited-resource languages, where adversarial adaptation to unlabelled corpora can mitigate the need for creating robust labeled datasets for training NER models.

Adversarial Adaptation Can Help Alleviate Some of the Performance Loss Incurred on Using Smaller Models. On fine-tuning the CamemBERT-Wiki-4GB model using our adversarial approach, we observed performance similar to or close to the non-adapted CamemBERT-base model for select settings. In fact, CamemBERT-Wiki-4GB model, when trained on the WikiNER dataset and adapted to WikiNeural corpus, outperformed the unadapted CamemBERT-base model. Nearly every language model reported improved results with adversarial adaptation. Thus, the use of adversarial adaptation during fine-tuning can act as a substitute for using larger language models for downstream tasks, thus leading to reduced computational costs for pretraining as well as fine-tuning.

Adapting NER models to the same domain target corpus as the source dataset generally leads to slightly better performance than adapting to a mixed domain corpora. We observed that models adapted to a corpus from the same domain as the source dataset (like when WikiNER and WikiNeural are used as source and target datasets, or vice versa) showed similar to slightly better performance than the same models were adapted to general domain corpora. However, both of these cases were almost always better than the corresponding unadapted models.

Adapting NER models to a mixed-domain target corpus generally leads to better performance than adapting the models to a corpus from a different domain corpus. We observed that models, when adapted to a mixed-domain corpus (like in the case of Europeana to Mixed-Fr), showed noticeably better performance than the corresponding models adapted to a corpus from a slightly different domain (as in the case of Europeana to WikiNER). However, in most scenarios, both of these settings led to better performance than the unadapted setting.

4   Conclusion and Future Work

In this work, we proposed a Transformer-based Named Entity Recognition framework using adversarial adaptation to large-scale similar domain or mixed domain corpora for improved feature learning in French. We adapted our models using our framework to corpora from the same domain as the source dataset as well as mixed-domain corpora. We evaluated our approach on three French language models, three target datasets, and three large-scale corpora. We observed noticeably improved performance using our proposed approach, as compared to models trained without our approach, across almost all Transformers and datasets. We further observed that adapting a model to a large-scale corpus for a downstream task can help alleviate some of the performance loss incurred by using smaller language models pre-trained on less robust corpora. Our proposed framework can further be applied to other low and limited-resource languages for a variety of downstream tasks.

In the future, we intend to evaluate the efficacy of our approach for multi-lingual and cross-lingual NER, particularly for out-of-domain settings using a multi-lingual corpus as the target dataset for adaptation and multi-lingual transformer embeddings. This can help reduce the dependence on labeled datasets for each language for NER. We further aim to evaluate the impact of the underlying language script used as the target data, and how performance is affected upon using a target corpora composed of texts from languages from different scripts for improved generalization over the language-independent features.

Acknowledgments. This research was enabled by support provided by Calcul Québec, The Digital Research Alliance of Canada and MITACS.

References

1. Choudhry, A., et al.: Transformer-based named entity recognition for French using adversarial adaptation to similar domain corpora (2022). https://doi.org/10.48550/ARXIV.2212.03692. Accessed 11 Jan 2023
2. Copara, J., Knafou, J., Naderi, N., Moro, C., Ruch, P., Teodoro, D.: Contextualized French language models for biomedical named entity recognition. In: Actes de la 6e conférence conjointe Journées d'Études sur la Parole (JEP, 33e édition), Traitement Automatique des Langues Naturelles (TALN, 27e édition), Rencontre des Étudiants Chercheurs en Informatique pour le Traitement Automatique des Langues (RÉCITAL, 22e édition). Atelier DÉfi Fouille de Textes. pp. 36–48. ATALA et AFCP, Nancy, France (2020), https://aclanthology.org/2020.jeptalnrecital-deft.4. Accessed 11 Jan 2023
3. Devlin, J., Chang, M.W., Lee, K., Toutanova, K.: BERT: pre-training of deep bidirectional transformers for language understanding. In: Proceedings of the 2019 Conference of the North American Chapter of the Association for Computational Linguistics: Human Language Technologies, Volume 1 (Long and Short Papers). pp. 4171–4186. Association for Computational Linguistics, Minneapolis, Minnesota (2019). https://doi.org/10.18653/v1/N19-1423. Accessed 11 Jan 2023
4. Ganin, Y., Lempitsky, V.: Unsupervised Domain Adaptation by Backpropagation (2014). https://doi.org/10.48550/ARXIV.1409.7495. Accessed 11 Jan 2023
5. Ganin, Y., et al.: Domain-adversarial training of neural networks. J. Mach. Learn. Res. **17**(1), 2096–2030 (2016). http://jmlr.org/papers/v17/15-239.html. Accessed 11 Jan 2023
6. Gong, C., Tang, J., Zhou, S., Hao, Z., Wang, J.: Chinese named entity recognition with Bert. DEStech Trans. Comput. Sci. Eng. (2019). https://doi.org/10.12783/dtcse/cisnrc2019/33299. Accessed 11 Jan 2023
7. Gridach, M., Haddad, H.: Arabic named entity recognition: a bidirectional GRU-crf approach. In: Gelbukh, A. (ed.) CICLing 2017. LNCS, vol. 10761, pp. 264–275. Springer, Cham (2018). https://doi.org/10.1007/978-3-319-77113-7_21
8. Habibi, M., Weber, L., Neves, M., Wiegandt, D.L., Leser, U.: Deep learning with word embeddings improves biomedical named entity recognition. Bioinformatics **33**(14), 37–48 (2017). https://doi.org/10.1093/bioinformatics/btx228. Accessed 11 Jan 2023

9. Lample, G., Ballesteros, M., Subramanian, S., Kawakami, K., Dyer, C.: Neural architectures for named entity recognition. In: Proceedings of the 2016 Conference of the North American Chapter of the Association for Computational Linguistics: Human Language Technologies, pp. 260–270. Association for Computational Linguistics, San Diego, California (2016). https://doi.org/10.18653/v1/N16-1030. Accessed 11 Jan 2023

10. Le, H., et al.: FlauBERT: unsupervised language model pre-training for French. In: Proceedings of The 12th Language Resources and Evaluation Conference, pp. 2479–2490. European Language Resources Association, Marseille, France (2020). https://www.aclweb.org/anthology/2020.lrec-1.302. Accessed 11 Jan 2023

11. Liu, Y., et al.: RoBERTa: a Robustly optimized bert pretraining approach (2019). https://doi.org/10.48550/ARXIV.1907.11692. Accessed 11 Jan 2023

12. Liu, Z., Jiang, F., Hu, Y., Shi, C., Fung, P.: NER-BERT: a Pre-trained model for low-resource entity tagging (2021). https://doi.org/10.48550/ARXIV.2112.00405. Accessed 11 Jan 2023

13. Liu, Z., et al.: CrossNER: evaluating cross-domain named entity recognition. Proc. AAAI Conf. Artif. Intell. **35**(15), 13452–13460 (2021). https://doi.org/10.1609/aaai.v35i15.17587. Accessed 11 Jan 2023

14. Loshchilov, I., Hutter, F.: Decoupled weight decay regularization (2017). https://doi.org/10.48550/ARXIV.1711.05101. Accessed 11 Jan 2023

15. Lothritz, C., Allix, K., Veiber, L., Bissyandé, T.F., Klein, J.: Evaluating pre-trained transformer-based models on the task of fine-grained named entity recognition. In: Proceedings of the 28th International Conference on Computational Linguistics. pp. 3750–3760. International Committee on Computational Linguistics, Barcelona, Spain (Online) (2020). https://doi.org/10.18653/v1/2020.coling-main.334. Accessed 11 Jan 2023

16. Martin, L., et al.: CamemBERT: a tasty french language model. In: Proceedings of the 58th Annual Meeting of the Association for Computational Linguistics, pp. 7203–7219. Association for Computational Linguistics, Online (2020). https://doi.org/10.18653/v1/2020.acl-main.645. Accessed 11 Jan 2023

17. Neudecker, C.: An open corpus for named entity recognition in historic newspapers. In: Proceedings of the Tenth International Conference on Language Resources and Evaluation (LREC'16)., pp. 4348–4352. European Language Resources Association (ELRA), Portorož, Slovenia (2016), https://aclanthology.org/L16-1689. Accessed 11 Jan 2023

18. Nothman, J., Ringland, N., Radford, W., Murphy, T., Curran, J.R.: Learning multilingual named entity recognition from wikipedia. Artif. Intell. **194**, 151–175 (2013). https://doi.org/10.1016/j.artint.2012.03.006. Accessed 11 Jan 2023

19. Peng, Q., Zheng, C., Cai, Y., Wang, T., Xie, H., Li, Q.: An entity-aware adversarial domain adaptation network for cross-domain named entity recognition (Student Abstract). Proc. AAAI Conf. Artif. Intell.**35**(18), 15865–15866 (2021). https://doi.org/10.1609/aaai.v35i18.17929. Accessed 11 Jan 2023

20. Ramshaw, L.A., Marcus, M.P.: Text chunking using transformation-based learning (1995). https://doi.org/10.48550/ARXIV.CMP-LG/9505040. Accessed 11 Jan 2023

21. Ratinov, L., Roth, D.: Design challenges and misconceptions in named entity recognition. In: Proceedings of the Thirteenth Conference on Computational Natural Language Learning (CoNLL-2009), pp. 147–155. Association for Computational Linguistics, Boulder, Colorado (2009). https://aclanthology.org/W09-1119. Accessed 11 Jan 2023

22. Roy, A.: Recent trends in named entity recognition (NER) (2021). https://doi.org/ 10.48550/ARXIV.2101.11420. Accessed 11 Janu 2023
23. Sang, E.F.T.K.: Introduction to the CoNLL-2002 shared task: language-independent named entity recognition (2002). https://doi.org/10.48550/ARXIV. CS/0209010. Accessed 11 Jan 2023
24. Tedeschi, S., Maiorca, V., Campolungo, N., Cecconi, F., Navigli, R.: WikiNEuRal: combined neural and knowledge-based silver data creation for multilingual NER. In: Findings of the Association for Computational Linguistics: EMNLP 2021, pp. 2521–2533. Association for Computational Linguistics, Punta Cana, Dominican Republic (2021). https://doi.org/10.18653/v1/2021.findings-emnlp.215. Accessed 11 Jan 2023
25. Wang, J., Kulkarni, M., Preotiuc-Pietro, D.: Multi-domain named entity recognition with genre-aware and agnostic inference. In: Proceedings of the 58th Annual Meeting of the Association for Computational Linguistics, pp. 8476–8488. Association for Computational Linguistics, Online (2020). https://doi.org/10.18653/v1/ 2020.acl-main.750. Accessed 11 Jan 2023
26. Wang, J., Xu, W., Fu, X., Xu, G., Wu, Y.: ASTRAL: adversarial trained LSTM-CNN for named entity recognition. Knowl.-Bsed Syst. **197**, 105842 (2020). https:// doi.org/10.1016/j.knosys.2020.105842. Accessed 11 Jan 2023
27. Yan, R., Jiang, X., Dang, D.: Named entity recognition by using XLNet-BiLSTM-CRF. Neural Process. Lett. **53**(5), 3339–3356 (2021). https://doi.org/10.1007/ s11063-021-10547-1. Accessed 11 Jan 2023

Exploring Fake News Detection with Heterogeneous Social Media Context Graphs

Gregor Donabauer[✉] and Udo Kruschwitz

Information Science, University of Regensburg, Regensburg, Germany
{gregor.donabauer,udo.kruschwitz}@ur.de

Abstract. Fake news detection has become a research area that goes way beyond a purely academic interest as it has direct implications on our society as a whole. Recent advances have primarily focused on text-based approaches. However, it has become clear that to be effective one needs to incorporate additional, contextual information such as spreading behaviour of news articles and user interaction patterns on social media. We propose to construct heterogeneous social context graphs around news articles and reformulate the problem as a graph classification task. Exploring the incorporation of different types of information (to get an idea as to what level of social context is most effective) and using different graph neural network architectures indicates that this approach is highly effective with robust results on a common benchmark dataset.

Keywords: Fake news detection · Social media networks · Graph machine learning

1 Introduction

Detecting online information disorders is an important step to guarantee free expression and discourse, e.g. [7]. A traditional approach to address this problem is manual fact-checking by experts, which is highly labour-intensive, but more recently automated methods that make use of external knowledge bases and natural language processing (NLP) have been proposed [10,34], sometimes as a tool to support the fact-checker [22]. Recent studies have demonstrated the growing importance of social media as a source of information for many people [23,27]. In this context it has been pointed out that a range of information types extracted from social media and the relations between such entities are useful for detecting fake news, e.g., [21,28,32,33]. Relations between news items and related information signals have been modelled to extract additional features by applying graph neural networks (GNNs) on graph structures (as discussed below), but so far this has typically either been limited to individual social media features or in combination with features that are not taken from social media (which makes it hard to assess the relevance of *social context only*).

To link different types of information they can be modelled as heterogeneous graphs [23], and such networks have the capability of introducing multiple types

© The Author(s), under exclusive license to Springer Nature Switzerland AG 2023
J. Kamps et al. (Eds.): ECIR 2023, LNCS 13981, pp. 396–405, 2023.
https://doi.org/10.1007/978-3-031-28238-6_29

of relations around a news article within a *single data structure*, which allows to see each graph as an independent data point. However it is not clear if such data structures are superior compared to homogeneous graphs that model only one relation type, as spreading and interaction behaviour between true and fake news generally heavily differs on social media [31].

There is a substantial body of work that demonstrates the use of GNNs, but it remains an open question as to how suitable they are for the use-case of fake news detection with *social media context features only*. It is also unclear which social media signals are most effective here and if there are significant differences between GNN architectures. To investigate these questions, our contributions can be summarized as follows: **(1)** We introduce **HetSMCG**, a methodology to construct extensive **Het**erogeneous **S**ocial **M**edia **C**ontext **G**raphs around news articles fusing multiple types of information in a single graph and reformulate fake news detection as a *graph classification problem*; **(2)** We evaluate a range of experimental settings exploring information sources taken from *social media only* and different GNN architectures to understand how the task can effectively be solved; **(3)** We compare how performance between heterogeneous multi-relation graphs and homogeneous single-relation graphs differs; **(4)** We make all our code and in-depth experimental results available to the community to foster reproducibility and sharing of resources.[1]

2   Related Work

One way of classifying the different research directions at a high level is to distinguish *content-based* methods that focus on the stories themselves, *context-based* approaches tapping into, e.g., social media signals and *intervention-based* approaches [25]. We limit our focus on the context-based paradigm with graph-based representations in our brief contextualization within existing related work.

The majority of context-based approaches for fake news detection use graph structures to model relations, e.g. by creating user-article graphs and utilizing the neighborhood structure to embed news nodes that can then be classified [4]. However, the number of information types is limited to users and news items. In similar problem formulations, user-post interaction graphs are considered [20,24]. Mehta et al. [18] treat fake news detection as a node classification task based on a large graph of users, news and publishers (though they do not model differences in edge types and do not limit information to social media context). In follow-up work they also introduce link prediction in this graph structure to find previously unavailable connections [19]. Lu and Li [16] also consider user graphs to extract social interactions using GNNs. In combination with recurrent networks, that leverage retweet patterns, they classify related news articles. Similar approaches that use temporal retweet patterns have been proposed [8,29]. Formulating fake news detection as graph classification has been applied by Dou et al. [8]. They calculated user representations by leveraging timeline tweet embeddings and use the outcome as node features to model news propagation paths associated with

---

[1] https://github.com/doGregor/Graph-FakeNewsNet.

articles. Another propagation-based approach was introduced by Song et al. [29]. In their work graph-structured, temporally evolving retweet patterns have been used to classify whether an article is fake or not. Two other approaches very similar to our work construct large heterogeneous graphs of articles, users, social media features like tweets, and publishers – again not limited to social media context [15,23]. They are classifying news nodes as either true or fake. Thus, incoming news nodes need to be integrated into these grown data structures to produce accurate results. With our approach we aim at proposing a method where each news article can be seen as an independent data point (graph).

We conclude that graph structures are a powerful way to include different dimensions of information and thus to improve fake news detection. However, linking many different entity types (and not just one) in independent data structures has barely been investigated yet (and even less if we only consider information types taken from social media in a heterogeneous way). We aim at contributing to filling this gap with our research.

3   Methodology and Experiments

3.1   Dataset and Graph Construction Method

The only common benchmark dataset that provides rich social context information in network-like structures is FakeNewsNet [26] which is why we adopt it. FakeNewsNet consists of two different datasets, *Politifact* and *GossipCop*. Both datasets are from different domains (PolitiFact for political news and GossipCop for entertainment stories) which allows to distinguish between domain-specific performances. We also use a mix of the two subsets (full *FakeNewsNet*) to see how our approach performs if the news are not limited to a closed domain. Twitter's terms and condition state that tweets and other information cannot be publicly released directly.[2] We therefore recrawl all data (thereby acquiring a dataset that will slightly differ from what others have used so far).

We construct a heterogeneous social media context graph $\mathcal{G} = (\mathcal{V}, \mathcal{E})$ around each news article. Thus, each snapshot consists of a set of disjoint vertex sets $\mathcal{V} = \mathcal{V}_N \cup \mathcal{V}_T \cup \mathcal{V}_U$ where $\mathcal{V}_i \cap \mathcal{V}_j = \emptyset, \forall i \neq j$. $\mathcal{V}_N$ represents news articles and $|\mathcal{V}_N| = 1$ for each graph. Node features for this type of vertex are obtained using BERT-base document embeddings, i.e. $\forall v_n \in \mathcal{V}_N : v_n \in \mathbb{R}^{768}$. All embeddings are generated using flairNLP[3]. Set $\mathcal{V}_T$ includes all tweets that refer to the original news article, all tweets retweeting those posts and the latest timeline tweets of each user. Since in our experiments we are incrementally increasing the amount of information types in the heterogeneous graphs, the number of nodes in $\mathcal{V}_T$ varies depending on the experimental setup. Node features are a concatenation of textual BERT-base document embeddings, as well as retweet count and favorite count, i.e. $\forall v_t \in \mathcal{V}_T : v_t \in \mathbb{R}^{770}$. Finally, $\mathcal{V}_U$ is the set of user nodes where features are a concatenation of BERT-base profile description embeddings, follower count,

---

[2] https://developer.twitter.com/en/developer-terms/agreement-and-policy.

[3] https://github.com/flairNLP/flair.

friends count, favorites count and statuses count, i.e. $\forall v_u \in \mathcal{V}_U : v_u \in \mathbb{R}^{772}$. We also evaluate graphs with text embeddings only and leave out other features (e.g. number of favorite count) in that case. Figure 1 shows a sampe graph that includes all mentioned types of information with the news article node at the centre of the graph. Edges connecting the nodes are satisfying constraints according to the node types they link together. More specifically, we use at most three types of edges: tweets citing news articles $((v_t, \tau_{TN}, v_n) \in \mathcal{E} \rightarrow v_t \in \mathcal{V}_T, v_n \in \mathcal{V}_N)$, users posting tweets (which also applies to users posting retweets and users posting timeline tweets) $((v_u, \tau_{UT}, v_t) \in \mathcal{E} \rightarrow v_u \in \mathcal{V}_U, v_t \in \mathcal{V}_T)$ and tweets retweeting tweets $((v_t, \tau_{TT}, u_t) \in \mathcal{E} \rightarrow v_t \in \mathcal{V}_T, u_t \in \mathcal{V}_T)$. The graphs fuse all types of information (the news piece as well as a range of different social media context features) without any prior aggregation step.

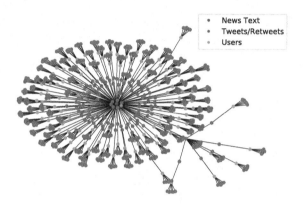

- News Text
- Tweets/Retweets
- Users

Fig. 1. Example of a heterogeneous social context graph constructed around a news article.

3.2 Experimental Setup

We construct graphs by incrementally increasing the amount of included data as follows: **(1)** We start with only considering tweets related to the news article; **(2)** We add user profiles of the people who posted the tweets to the graphs; **(3)** We include the five latest timeline posts of those users; **(4)** We add retweets related to the initial Twitter posts; **(5)** We combine all types of information mentioned so far. For simplicity we set the number of timeline tweets to five to keep the relation between number of all nodes and number of timeline tweet nodes in the graph manageable. The in-depth exploration of different settings is left as future work. To keep the structure of graphs consistent we exclude samples where news data are no longer available. Graphs with fewer than five vertices per node type are skipped to make sure that at least some social media features are available (other studies sometimes even set higher limits of, e.g., a minimum of 15 tweets [17]). Furthermore, we make all edges undirected to improve message passing during graph convolution. For all experiments we use 5-fold cross-validation and report the average scores as results. We keep the overall number of graphs and the cross-validation split consistent for all setups and datasets to obtain comparable results. The number of graphs per dataset amounts to 483 for PolitiFact (real: 235; fake: 248), 12,214 for GossipCop (real: 10,067; fake: 2,147) and thus 12,697 for the full FakeNewsNet dataset (real: 10,302; fake: 2,395) which consists of a mixture of the two subsets.

We are also flattening all graphs into non-heterogeneous structures to understand how such a setup performs compared to the one described so far. Since homogeneous graphs do not explicitly model disjoint sets of vertices and use a single adjacency matrix, we have to unify all feature dimensions and we do that by either pruning all nodes to a dimension of 768 (only text) or zero-padding all nodes to a dimension of 772. We classify each graph using two-layer vanilla graph neural networks and evaluate three different types of graph convolution: (1) GraphSAGE [11]; (2) Graph Attention Convolution (GAT) [30]; (3) Heterogeneous Graph Transformer (HGT) [13]. We first pass the graphs through the two graph convolutional layers. This operation is performed using separate weights with respect to every edge type. Next, we perform global mean pooling over all nodes of the same type, concatenate this information and feed it through a dropout layer before generating the prediction with a two-node linear layer. For details on the models we refer to our GitHub.

All models are trained with the same hyperparameter setting to obtain comparable results. We use 20 train epochs, a batch size of 16 and a learning-rate of $8e^{-5}$. As the number of real and fake graphs is unbalanced for GossipCop and the full FakeNewsNet dataset, we use class weights during experiments with those data. Our implementations are based on PyTorch Geometric[4] [9]. Training and testing are executed using a single NVIDIA GTX 1080 GPU with an overall graphical memory size of 11GB. Even though we use high-dimensional node features and a large number of graphs during our experiments, training and inference in the described setup only take about 15 min on a single GPU (with respect to the highest number of graphs).

4   Results

In line with common practice in NLP, we report precision, recall and macro F1 scores for all setups [14, Ch. 4]. Detailed results are reported on our GitHub. As previously mentioned, the values are the average results obtained by 5-fold cross validation. In general, excluding social media metrics as node features (e.g., retweet count and number of followers) yields better results than concatenating them with text embedding to initially represent the graph nodes. We observe the highest scores for Politifact with setup (5) (0.979 macro F1 and accuracy),

---

[4] https://github.com/pyg-team/pytorch_geometric.

for GossipCop with setup (3) (0.972 macro F1 and 0.983 accuracy) and the full FakeNewsNet dataset with setup (5) (0.966 macro F1 and 0.979 accuracy), only considering textual features. Interestingly, we get good results for Politifact from the point of adding retweets. Timeline tweets instead are not important. For GossipCop we observe very similar performance for all setups once including tweets and users. Here, the other features (retweets and timeline tweets) do not significantly change the model performance. As the full FakeNewsNet dataset mainly consists of GossipCop, we can observe very similar results here. The best performing GNN convolution type is throughout all setups HGT.

The results for the setups with homogeneous graphs are all significantly lower at $p < 0.05$ using paired t-test compared to their equivalent with heterogeneous graphs. This observation holds true for truncated and padded homogeneous graphs for all three datasets. We demonstrate how modelling disjoint sets of nodes and multiple types of edges improves the representation of social context and leads to overall significantly better fake news detection performance.

For comparison, we use two recently published approaches as strong baselines.[5] As we excluded some of the articles from our experiments (due to missing social context information), we rerun both baseline systems with the same data used in our experiments. The competitive baselines are: **(1) CMTR** [12]: BERT classified texts that were preprocessed using summarization techniques as well as additional social media features like comments. **(2) HetTransformer** [15]: node classification in a heterogeneous graph featuring user, post and news nodes using an encoder-decoder transformer model. We use our results obtained by the same setup (HGT and all types of information) for comparison against the baseline scores even if this setup does not always reach the highest performance in each setup (to avoid cherry-picking). Table 1 reports competitive and robust results on all datasets and a new benchmark performance on the Politifact subset (though this is not statistically significant). Rerunning HetTransformer we observe much lower results on Politifact compared to those reported in their paper. This might be caused by a greater influence of recrawling the data given this subset's overall small size.

5 Discussion

An interesting observation is that the results for the Politfact dataset improve when we include retweet data in our network structures. It has already been pointed out that fake news tend to have more retweets than real news [26], and the results provide some support for the utility of this feature. Moreover, it has also previously been reported that misinformation in general spreads more effectively due to its emotionalizing content [2]. Due to the small size of the dataset (real: 239; fake: 261), this could have a large influence on classification

---

[5] There are many other studies reporting results on FakeNewsNet or conceptually related problems that could have been adopted as baselines, e.g. [1,6,8,23,29]. We picked the two most recent and what we judged to be most competitive baselines, leaving rerunning of other baselines as future work.

Table 1. F1 and accuracy scores for *Politifact, GossipCop* and the full *FakeNewsNet* dataset. Significant differences to our approach (at $p < 0.05$ and Bonferroni correction) are marked with **.

Approach	Politifact		Gossipcop		FakeNewsNet	
	F1	ACC	F1	ACC	F1	ACC
CMTR [12]	0.965	0.965	0.854**	0.922**	0.859**	0.920**
HetTransformer [15]	0.900**	0.900**	**0.994**	**0.997**	**0.985**	**0.991**
HetSMCG (our approach)	**0.979**	**0.979**	0.969	0.982	0.966	0.979

performance. The much larger GossipCop and full FakeNewsNet datasets (real: 12,610; fake: 3,185) seems to be less influenced by such information type-specific characteristics. Since we assign retweets to the same node type as article-related tweets and user timeline tweets we address the problem of information type-specific performance gaps. An observation that applies to Politifact and the full FakeNewsNet is that including all types of information leads to the best performing setup. For Gossipcop retweets appear less relevant. However, using all available social media context results in competitive high-performance figures which is also supported by the results of the statistical significance tests. For our experiments we also find that there are only marginal differences between the different GNN architectures while HGT gives the highest performance.

We should also point out some limitations of our work. Even though we compare our results against strong, recently published baselines one needs to be careful when reporting and interpreting improvements over existing systems, e.g. [5]. Generalisability of insights is a related issue and results in NLP are sometimes overclaimed in the recent literature [3]. Providing access to all code and data in our Github repository helps addressing this concern (in addition to supporting reproducibility of results).

6   Conclusion

We have demonstrated how including a variety of social media context information can improve fake news detection. By modelling the problem as a graph classification task using heterogeneous graph data structures we achieve competitive results on a real-world dataset. Incorporating all the available contextual information turns out to be generally most effective for our approach.

Acknowledgments. This work was supported by the project *COURAGE: A Social Media Companion Safeguarding and Educating Students* funded by the Volkswagen Foundation, grant number 95564. We want to thank all reviewers for their insightful comments that helped us to improve our work.

References

1. Azri, A., Favre, C., Harbi, N., Darmont, J., Noûs, C.: MONITOR: a multimodal fusion framework to assess message veracity in social networks. In: Bellatreche, L., Dumas, M., Karras, P., Matulevičius, R. (eds.) Advances in Databases and Information Systems, pp. 73–87. Springer International Publishing, Cham (2021). https://doi.org/10.1007/978-3-030-82472-3_7
2. Bakir, V., McStay, A.: Fake news and the economy of emotions. Digital J. **6**(2), 154–175 (2018). https://doi.org/10.1080/21670811.2017.1345645
3. Bowman, S.: The dangers of underclaiming: reasons for caution when reporting how NLP systems fail. In: Proceedings of the 60th Annual Meeting of the Association for Computational Linguistics (Volume 1: Long Papers), pp. 7484–7499. Association for Computational Linguistics, Dublin, Ireland, May 2022. https://aclanthology.org/2022.acl-long.516
4. Chandra, S., Mishra, P., Yannakoudakis, H., Nimishakavi, M., Saeidi, M., Shutova, E.: Graph-based modeling of online communities for fake news detection. CoRR abs/2008.06274 (2020). https://arxiv.org/abs/2008.06274
5. Church, K.W., Kordoni, V.: Emerging trends: SOTA-chasing. Nat. Lang. Eng. **28**(2), 249–269 (2022). https://doi.org/10.1017/S1351324922000043
6. Das, S.D., Basak, A., Dutta, S.: A heuristic-driven uncertainty based ensemble framework for fake news detection in tweets and news articles. Neurocomputing. **491**, 607–620 (2022). https://doi.org/10.1016/j.neucom.2021.12.037, https://www.sciencedirect.com/science/article/pii/S0925231221018750
7. Dori-Hacohen, S., Sung, K., Chou, J., Lustig-Gonzalez, J.: Restoring healthy online discourse by detecting and reducing controversy, misinformation, and toxicity online. In: Proceedings of the 44th International ACM SIGIR Conference on Research and Development in Information Retrieval, pp. 2627–2628. Association for Computing Machinery, New York, NY, USA (2021). https://doi.org/10.1145/3404835.3464926
8. Dou, Y., Shu, K., Xia, C., Yu, P.S., Sun, L.: User preference-aware fake news detection. In: Proceedings of the 44th International ACM SIGIR Conference on Research and Development in Information Retrieval, pp. 2051–2055. Association for Computing Machinery, New York, NY, USA (2021). https://doi.org/10.1145/3404835.3462990
9. Fey, M., Lenssen, J.E.: Fast graph representation learning with PyTorch geometric. In: ICLR Workshop on Representation Learning on Graphs and Manifolds (2019)
10. Guo, Z., Schlichtkrull, M., Vlachos, A.: A survey on automated fact-checking. Trans. Assoc. Comput. Linguist. **10**, 178–206 (2022). https://doi.org/10.1162/tacl_00454
11. Hamilton, W.L., Ying, R., Leskovec, J.: Inductive representation learning on large graphs. In: Proceedings of the 31st International Conference on Neural Information Processing Systems, pp. 1025–1035. NIPS 2017, Curran Associates Inc., Red Hook, NY, USA (2017)
12. Hartl, P., Kruschwitz, U.: Applying automatic text summarization for fake news detection. In: Proceedings of the Language Resources and Evaluation Conference. pp. 2702–2713. European Language Resources Association, Marseille, France, June 2022
13. Hu, Z., Dong, Y., Wang, K., Sun, Y.: Heterogeneous graph transformer. In: Proceedings of The Web Conference 2020, pp. 2704–2710. WWW 2020, Association for Computing Machinery, New York, NY, USA (2020). https://doi.org/10.1145/3366423.3380027

14. Jurafsky, D., Martin, J.: Speech and language processing: an introduction to natural language processing, computational linguistics, and speech recognition (2023), 3rd edition (draft), 7 January 2023. https://web.stanford.edu/~jurafsky/slp3/

15. Li, T., Sun, Y., Hsu, S.l., Li, Y., Wong, R.C.W.: Fake news detection with heterogeneous transformer (2022). https://doi.org/10.48550/ARXIV.2205.03100, https://arxiv.org/abs/2205.03100

16. Lu, Y.J., Li, C.T.: GCAN: Graph-aware co-attention networks for explainable fake news detection on social media. In: Proceedings of the 58th Annual Meeting of the Association for Computational Linguistics, pp. 505–514. Association for Computational Linguistics, July 2020. https://doi.org/10.18653/v1/2020.acl-main.48, https://aclanthology.org/2020.acl-main.48

17. Lukasik, M., Cohn, T., Bontcheva, K.: Point process modelling of rumour dynamics in social media. In: Proceedings of the 53rd Annual Meeting of the Association for Computational Linguistics and the 7th International Joint Conference on Natural Language Processing (Volume 2: Short Papers), pp. 518–523. Association for Computational Linguistics, Beijing, China, July 2015. https://doi.org/10.3115/v1/P15-2085, https://aclanthology.org/P15-2085

18. Mehta, N., Goldwasser, D.: Tackling fake news detection by interactively learning representations using graph neural networks. In: Proceedings of the First Workshop on Interactive Learning for Natural Language Processing, pp. 46–53. Association for Computational Linguistics, August 2021. https://doi.org/10.18653/v1/2021.internlp-1.7, https://aclanthology.org/2021.internlp-1.7

19. Mehta, N., Pacheco, M., Goldwasser, D.: Tackling fake news detection by continually improving social context representations using graph neural networks. In: Proceedings of the 60th Annual Meeting of the Association for Computational Linguistics (Volume 1: Long Papers), pp. 1363–1380. Association for Computational Linguistics, Dublin, Ireland, May 2022. https://aclanthology.org/2022.acl-long.97

20. Min, E., et al.: Divide-and-conquer: post-user interaction network for fake news detection on social media. In: Proceedings of the ACM Web Conference 2022, pp. 1148–1158. WWW 2022, Association for Computing Machinery, New York, NY, USA (2022). https://doi.org/10.1145/3485447.3512163

21. Mosca, E., Wich, M., Groh, G.: Understanding and interpreting the impact of user context in hate speech detection. In: Proceedings of the Ninth International Workshop on Natural Language Processing for Social Media, pp. 91–102. Association for Computational Linguistics (2021). https://doi.org/10.18653/v1/2021.socialnlp-1.8, https://aclanthology.org/2021.socialnlp-1.8

22. Nakov, P., et al.: Automated fact-checking for assisting human fact-checkers. In: Zhou, Z.H. (ed.) Proceedings of the Thirtieth International Joint Conference on Artificial Intelligence, IJCAI-2021, pp. 4551–4558. International Joint Conferences on Artificial Intelligence Organization, August 2021. https://doi.org/10.24963/ijcai.2021/619

23. Nguyen, V.H., Sugiyama, K., Nakov, P., Kan, M.Y.: Fang: leveraging social context for fake news detection using graph representation. In: Proceedings of the 29th ACM International Conference on Information & Knowledge Management. p. 1165–1174. CIKM 2020, Association for Computing Machinery, New York, NY, USA (2020). https://doi.org/10.1145/3340531.3412046

24. Rode-Hasinger, S., Kruspe, A., Zhu, X.X.: True or false? Detecting false information on social media using graph neural networks. In: Proceedings of the Eighth Workshop on Noisy User-generated Text (W-NUT 2022), pp. 222–229. Association for Computational Linguistics, Gyeongju, Republic of Korea, October 2022. https://aclanthology.org/2022.wnut-1.24

25. Sharma, K., Qian, F., Jiang, H., Ruchansky, N., Zhang, M., Liu, Y.: Combating fake news: a survey on identification and mitigation techniques. ACM Trans. Intell. Syst. Technol. (TIST) **10**(3), 1–42 (2019)

26. Shu, K., Mahudeswaran, D., Wang, S., Lee, D., Liu, H.: Fakenewsnet: a data repository with news content, social context, and spatiotemporal information for studying fake news on social media. Big Data **8**(3), 171–188 (2020). https://doi.org/10.1089/big.2020.0062. pMID: 32491943

27. Shu, K., Sliva, A., Wang, S., Tang, J., Liu, H.: Fake news detection on social media: a data mining perspective. SIGKDD Explor. Newsl. **19**(1), 22–36 (2017). https://doi.org/10.1145/3137597.3137600

28. Shu, K., Wang, S., Liu, H.: Beyond news contents: the role of social context for fake news detection. In: Proceedings of the Twelfth ACM International Conference on Web Search and Data Mining, pp. 312–320. WSDM 2019, Association for Computing Machinery, New York, NY, USA (2019). https://doi.org/10.1145/3289600.3290994

29. Song, C., Shu, K., Wu, B.: Temporally evolving graph neural network for fake news detection. Inf. Process. Manage. **58**(6), 102712 (2021). https://doi.org/10.1016/j.ipm.2021.102712, https://www.sciencedirect.com/science/article/pii/S0306457321001965

30. Veličković, P., Cucurull, G., Casanova, A., Romero, A., Liò, P., Bengio, Y.: Graph attention networks. In: International Conference on Learning Representations (2018). https://openreview.net/forum?id=rJXMpikCZ

31. Vosoughi, S., Roy, D., Aral, S.: The spread of true and false news online. Science. **359**(6380), 1146–1151 (2018). https://doi.org/10.1126/science.aap9559, https://www.science.org/doi/abs/10.1126/science.aap9559

32. Weinzierl, M., Harabagiu, S.: Identifying the adoption or rejection of misinformation targeting COVID-19 vaccines in twitter discourse. In: Proceedings of the ACM Web Conference 2022, pp. 3196–3205. WWW 2022, Association for Computing Machinery, New York, NY, USA (2022). https://doi.org/10.1145/3485447.3512039

33. Yang, R., Wang, X., Jin, Y., Li, C., Lian, J., Xie, X.: Reinforcement subgraph reasoning for fake news detection. In: Proceedings of the 28th ACM SIGKDD Conference on Knowledge Discovery and Data Mining. p. 2253–2262. KDD 2022, Association for Computing Machinery, New York, NY, USA (2022). https://doi.org/10.1145/3534678.3539277

34. Zhou, X., Zafarani, R.: A survey of fake news: fundamental theories, detection methods, and opportunities. ACM Comput. Surv. **53**(5), 1–40 (2020). https://doi.org/10.1145/3395046

Justifying Multi-label Text Classifications
for Healthcare Applications

João Figueira[(✉)][iD], Gonçalo M. Correia[iD], Michalina Strzyz[iD],
and Afonso Mendes[iD]

Priberam Labs, Lisbon, Portugal
{joao.figueira,goncalo.correia,michalina.strzyz,amm}@priberam.pt

Abstract. The healthcare domain is a very active area of research for
Natural Language Processing (NLP). The classification of medical records
according to codes from the International Classification of Diseases (ICD)
is an essential task in healthcare. As a very sensitive application, the automatic classification of personal medical records cannot be immediately
trusted without human approval. As such, it is desirable for classification models to provide reasons for each decision, such that the medical
coder can validate model predictions without reading the entire document.
AttentionXML is a multi-label classification model that has shown high
applicability for this task and can provide attention distributions for each
predicted label. In practice, we have found that these distributions do not
always provide relevant spans of text. We propose a simple yet effective
modification to AttentionXML for finding spans of text that can better
aid the medical coders: splitting the BiLSTM of AttentionXML into a forward and a backward LSTM, creating two attention distributions that find
the leftmost and rightmost limits of the text spans. We also propose a novel
metric for the usefulness of our model's suggestions by computing the drop
in confidence from masking out the selected text spans. We show that our
model has a similar classification performance to AttentionXML while surpassing it in obtaining relevant text spans.

Keywords: Healthcare · Multi-label classification · Span extraction

1 Introduction

In the medical domain, deep learning models, despite showing great performance
in some tasks, are not being widely deployed due to a lack of transparency behind
their decisions [9], which is often required in healthcare applications. In practice, due to the sensitivity of such applications and the need for accountability,
automatic decisions require a final validation of a human practitioner who will
confirm or reject them, thus using them as a decision aid. Therefore, by providing justifications behind machine decisions, a machine learning model will endow
the practitioner with a quick way to confirm them.

The classification of medical documents regarding diagnosis and procedures
described within the record is one such case. Classifying these documents according to ICD codes [5] is of utmost importance in many healthcare facilities.

J. Kamps et al. (Eds.): ECIR 2023, LNCS 13981, pp. 406–413, 2023.
https://doi.org/10.1007/978-3-031-28238-6_30

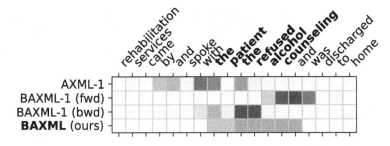

Fig. 1. Attention weights distribution for a segment of a MIMIC document, for the ICD9 label *305.00: nondependent alcohol abuse unspecified drinking behavior*. AXML-1 places attention on tokens close to the relevant "the patient refused alcohol counseling" span, but not on it. The backward and forward models place attention on the left and right of the relevant span, respectively. Our joining method is shown to be able to locate the relevant span.

The list of ICD codes present in a medical episode encodes the clinical process by providing a common ontology for recording, reporting, billing, and monitoring diseases in a standard and comparable way. Without a machine learning model, the *medical coder* (*i.e.* a doctor trained to identify which ICD codes are present in medical records) needs to go through extensive documents describing what the patient went through during their admission, which can be a time and cost-consuming task. By using a multi-label text classification model trained on identifying ICD codes present in the document, this process can be expedited. However, due to the sensitive nature of the task, the coder will still need to confirm each of the suggested labels. As such, the medical coder's job could be further expedited by having the model pinpoint the sections of the text most relevant to the validation of each predicted label.

AttentionXML (AXML) [22] is a neural network model that can be used for multi-label text classification that has been shown able to obtain promising results on clinical data [3]. In this task, a medical document is given as input for the model to predict the set of mentioned procedures and diagnoses, via their appropriate ICD codes [3,11,20]. Thanks to its attention module, AXML could, in theory, aid medical coders by revealing the highest attention weights of the model for each label to help them confirm or reject model suggestions, in a similar fashion as in Mullenbach et al. [11]. However, in our experiments, we have found that the attention weights often pinpointed unimportant sections of the input text, thus not providing useful help to the coders. In this work, we propose a **text span spotter** model that retrieves the relevant sections of the text for each label that AXML produces (Fig. 1). We describe how this model accurately selects these passages and how practitioners can use it to quickly approve or disapprove labels.

Our contributions are as follows: (i) propose a span identifier for AXML that accurately identifies token spans that explain a given label, with a small cost in performance; (ii) show the effectiveness of the added module in reasoning decisions compared to the original AXML model; (iii) propose a novel metric, *CoLoRWorM*, Confidence Loss on Relevant Word Masking, to measure the usefulness of a text span as a decision suggestion, based on the impact of masking that span.

2 Related Work

Our work is focused on extreme multi-label text classification (XMTC) applied to the medical domain. Particularly, we focus our research efforts on obtaining reasonings for label decisions out of AXML [22], a popular XMTC model. AXML is a label tree-based deep learning model that builds a probabilistic label tree (PLT) out of separate models, each using an attention neural module between the tokens in the text and the large label space. Previous work has applied AXML to healthcare datasets [3], but there has not been extensive research on using this XMTC model for relevant span extraction, nor investigating why the attention weights obtained by the vanilla model often do not provide useful information to practitioners.

Lately, explainability of deep neural network has attracted a lot of attention and several recent works propose methods to obtain explanations from neural network models, which can be particularly useful in the medical domain [12]. Some explainability methods are model-agnostic and *post hoc*, which obtain explanations by running another model on top of the original one [10,14]. While convenient, these methods do not always accurately explain the model's behaviour [1]. There have also been recent criticisms of obtaining *post hoc* explanations of black-box models, suggesting instead to make them interpretable from the start [15]. We propose to train an additional module, the spotter, along with the main AXML model in order to obtain better reasoning out of the attention weights. The proposed spotter is not explaining the label decision of AXML, it is locating what is likely the most relevant interpretable section of the text for the generation of that label because it shares its objective with AXML; however, the architecture was adapted to be able to extract this information. While there are some works that criticize and propose caution when assuming that attention weights correlate with explanations [6,17,21], in this work, we don't aim to explain the decisions of AXML, but rather to select relevant sections for a particular label as an aid to a future user of the software.

3 AttentionXML for Medical Documents

Architecture. AXML is a neural network model for extreme multi-label text classification that uses an attention mechanism to obtain weighted representations of each label based on the input documents. Particularly, AXML is composed of five layers: An **input word embedding layer**, initialized with pretrained embeddings; A **bi-directional long short-term memory (BiLSTM)**

layer, which is a recurrent neural network used to process sequence-like inputs such as text. It captures representations in both directions of the sequence using forward and backward LSTMs trained in the two directions of the sequence. The output of each LSTM is concatenated; A **label-to-token attention layer** [2], that computes the (linear) combination of context vectors generated by the BiL-STM with a learned query vector specific to each label. This generates a matrix of label-token weights that are used in a weighted sum of context vectors to create a label-specific document embedding; And a **fully-connected output layer**, where each label-specific document embedding is fed to two fully connected neural network layers. The last layer projects the representation into a single score value. The weights of these dense layers are shared between all labels.

Applicability for ICD Coding. Given this architecture, the neural component of AXML is lightweight, especially when compared to recent widely used transformer-based models in NLP research [8]. This is highly desirable in our particular application since the hardware infrastructure in most hospitals does not often allow for the use of heavy models. In its original formulation, AXML runs the above architecture several times and obtains the final scores by building a PLT with the scores of each network's output. Since the classification results did not differ much between this approach and using a single network, we use a single network without building a PLT (AXML-1 in [22]) in our experiments. This way, there is a single attention layer, which is beneficial for our approach. Additionally, ICD coding does not have a label space as extensive as other XMTC tasks, which was the main motivation for the PLT.

Retrieving Useful Text Sections for Each Label Decision. Given that AXML-1 has a single attention layer that provides token-label attention weights, it can be used as a native way to provide relevant tokens for the model decisions. We found it was often the case that the attention weights were selecting tokens irrelevant to the label decision, such as commas, that were close to but not containing the relevant text section. Given that the embeddings that are fed onto the attention layer are generated by the BiLSTM, i.e. they are contextual, it is possible for a single contextual text embedding to have accumulated sufficient information from its neighborhood for the current classification task. For this reason, the attention layer can attribute a high attention value to an arbitrary token in the neighborhood of the relevant section and still produce a document embedding with all the relevant information, which leads to the problem we were faced with.

Architecture of the *Spotter* Model. In order to obtain better text spans that help validate a machine's decisions, we propose a novel model architecture to be trained in tandem with AXML-1.

The *spotter* model consists of two separate models, each similar to AXML-1 in their architecture but having the BiLSTM replaced with forward or backward LSTM (thus a "broken" into its two components). We will call these models *BAXML-1-f*, and *BAXML-1-b*.

Algorithm 1. Peak joining

Input: Attention weights from the *spotter* model for some document-label pair, $f\_attentions$, and $b\_attentions$. Both are float lists with length L.
K, number of peaks to be extracted, and D, max distance between joined peaks.

Output: Joined attention weights for suggestion spans $j\_attentions$.

Initialize $j\_attentions = [0, 0, ...]$, size L
$f\_peaks = find\_peaks(f\_attentions, K)$
$b\_peaks = find\_peaks(b\_attentions, K)$
for b_peak **in** b_peaks **do**
    for f_peak **in** f_peaks **do**
        if $pos(f\_peak) \geq pos(b\_peak)$ **and** $pos(f\_peak) \leq pos(b\_peak) + D$ **then**
            $span\_attention = val(f\_peak)/2 + val(b\_peak)/2$
            $pointer = pos(b\_peak)$
            while $pointer \leq pos(f\_peak)$ **do**
                $j\_attentions[pointer] = max(span\_attention, j\_attentions[pointer])$
                $pointer = pointer + 1$

Since the LSTMs can accumulate relevant information onto arbitrary tokens in the neighborhood of the relevant sections, and that these tokens will later be favored by the attention layer, using separate LSTMs with different directions will restrict the positions at which these tokens will be found. With the forward LSTM of *BAXML-1-f*, we can guarantee that the relevant information can only be accumulated on a token at the end of, or to the right of, the relevant section. Similarly, with the backward LSTM of *BAXML-1-b*, we can guarantee that the relevant information can only be accumulated on a token at the start of, or to the left of, the relevant section. This way it is possible to find relevant spans delimited by the high attention tokens of these two models.

In Algorithm 1 we show the process for joining the attention outputs from the *spotter* model. The algorithm finds the top-K attention peaks from each model and identifies corresponding pairs that are not further apart than D tokens. The joined distribution populates the span between the position of these two peaks with the average attention value of both. These improved attention distributions are used to highlight relevant sections for the medical coders but are not used in the classification inference of the model. An example of the obtained token weight distributions is shown in Fig. 1.

4   Experiments and Results

To provide a more objective evaluation of our model, we propose a novel metric for the usefulness of selected relevant spans and tokens: *CoLoRWorM*, Confidence Loss on Relevant Word Masking. The metric is defined as:

$$C(x; e) = \frac{\sum_{l \in L_x} f(x, l) - f(mask(x, e(x, l), k), l)}{|L_x|}, \tag{1}$$

Table 1. *CoLoRWorM* results, masking 10, 20, and 30 tokens, evaluated on the MIMIC test set, using a AXML-1 model trained for ICD classification on the MIMIC train set as the *judge* model. Spotter model used parameters K = 30, and D = 10.

Span extraction system	CWM@10	CWM@20	CWM@30
Random (baseline)	−0.0001	0.0003	0.0021
AXML-1 attention weights	0.1261	0.1489	0.1597
BAXML-1 forward attention weights	0.1480	0.1857	0.2056
BAXML-1 backward attention weights	0.1451	0.1892	0.2156
Spotter joined attention weights	**0.1658**	**0.2325**	**0.2696**

Table 2. Multi-label ICD9 text classification Precision@K results. Measured on the MIMIC test set.

Model	P@1	P@3	P@5	P@8
AXML-1	91.60	85.87	80.46	72.43
AXML-1 (frozen embeddings)	91.47	85.50	79.52	71.27
BAXML-1-forward	91.66	85.48	79.98	72.06
BAXML-1-backward	91.42	85.51	80.26	72.44
BAXML-1-fwd + BAXML-1-bwd	**92.34**	**86.38**	**81.06**	**73.35**

where x is a document; L_x is the set of gold labels for that document; f is the *judge* classification model that outputs a confidence value for each document-label pair; e is the extraction model, that outputs a distribution of relevance weights over the tokens of the document; and $mask$ is the masking function, that will mask the top k tokens in the document that have the highest weights attributed by the extraction model.

Intuitively, if a selected token or span is very useful towards the decision regarding a specific label, masking these tokens should cause a significant loss in confidence for some arbitrary *judge* classification model. In Table 1, we show the *CoLoRWorM* results for all models. We can observe that our joining method outperforms using the attention weights of any model in isolation. Surprisingly, using either of the unidirectional LSTM models would already be an improvement over the BiLSTM of AXML-1.

Classification Performance and Ablation. We evaluate our models on a multi-label ICD9 classification task on the MIMIC-III dataset [7], which contains approximately 52k discharge summaries in English. Our experiments were performed on the MIMIC-full dataset, which contains labeled documents with frequent, few-shot, and zero-shot labels. Word embeddings were initialized with 300-dimensional English GloVe vectors [13].

Results are shown in Tables 2 and 3. All listed models were finetuned for this task. We evaluated models using the Precision@K metric along with F1, RP@K [4] and nDCG@K [16].

Table 3. Multi-label ICD9 text classification F1, RP@15, and nDCG@15 results. Measured on the MIMIC test set.

Model	F1	RP@15	nDCG@15
AXML-1	55.37	65.94	64.41
AXML-1 (frozen embeddings)	54.40	65.06	63.68
BAXML-1-forward	54.92	65.46	64.65
BAXML-1-backward	54.83	65.46	64.68
BAXML-1-fwd + BAXML-1-bwd	**55.82**	**66.50**	**65.60**

Using the separate LSTMs together (trained as two separate models, later joined by averaging their outputs) is shown to be the best-performing approach, eliminating the need for a normal AXML-1 model when using our span *spotter* system. This performance gain can possibly come from token embeddings being separately finetuned for the forward and backward LSTMs, and the joined models have overall more parameters than AXML-1. We also show a significant gain in finetuning the word embedding layer.

5 Conclusions

We have shown how the architecture of AXML can be modified for usage in a machine-assisted classification scenario for the medical domain. Our proposed model and algorithm should find similar usability in other domains where a similar level of accountability is desirable and model decisions should be similarly validated by a human user, expedited by the span extraction system.

We put forth a novel metric that helps evaluate the usability of extracted tokens and spans. We use it to show that our model is an improvement over AXML while slightly surpassing it in classification metrics.

For future work, we intend on comparing our extracted spans with the spans obtained from explainability systems, such as LIME [14], SHAP [10], Integrated Gradients [19], and Saliency [18], as well as evaluating it on other datasets.

Acknowledgments. We thank the anonymous reviewers for their feedback. We are grateful to our team at Priberam Labs for the insightful group discussion and feedback on the initial draft. This work was supported by the project IntelligentCare - Intelligent Multimorbidity Management System (Reference LISBOA-01-0247-FEDER-045948) co-financed by the ERDF (European Regional Development Fund) through the Lisbon Portugal Regional Operational Program - LISBOA 2020 and by the Portuguese Foundation for Science and Technology (FCT) under CMU Portugal Program.

References

1. Alvarez-Melis, D., Jaakkola, T.S.: On the robustness of interpretability methods. In: Proceedings of WHI (2018)

2. Bahdanau, D., Cho, K., Bengio, Y.: Neural machine translation by jointly learning to align and translate. In: Proceedings of the ICLR (2015)
3. Chalkidis, I., Fergadiotis, M., Kotitsas, S., Malakasiotis, P., Aletras, N., Androutsopoulos, I.: An empirical study on large-scale multi-label text classification including few and zero-shot labels. In: Proceedings of the EMNLP (2020)
4. Chalkidis, I., Fergadiotis, M., Malakasiotis, P., Androutsopoulos, I.: Large-scale multi-label text classification on EU legislation. In: Proceedings of the ACL (2019)
5. International Classification of Diseases (ICD). https://www.who.int/classifications/classification-of-diseases
6. Jain, S., Wallace, B.C.: Attention is not explanation. In: Proceedings of the NAACL (2019)
7. Johnson, A.E., et al.: MIMIC-III, a freely accessible critical care database. Sci. Data 3(1), 160035 (2016)
8. Lee, J., et al.: BioBERT: a pre-trained biomedical language representation model for biomedical text mining. Bioinformatics 36(4), 1234–1240 (2020)
9. Li, X., et al.: Interpretable deep learning: interpretation, interpretability, trustworthiness, and beyond. Knowl. Inf. Syst. 64(12), 3197–3234 (2022)
10. Lundberg, S., Lee, S.I.: A unified approach to interpreting model predictions. In: Proceedings of the NeurIPS (2017)
11. Mullenbach, J., Wiegreffe, S., Duke, J., Sun, J., Eisenstein, J.: Explainable prediction of medical codes from clinical text. In: Proceedings of the NAACL (2018)
12. Naylor, M., French, C., Terker, S., Kamath, U.: Quantifying explainability in NLP and analyzing algorithms for performance-explainability tradeoff. In: Proceedings of the IMLH (2021)
13. Pennington, J., Socher, R., Manning, C.: GloVe: global vectors for word representation. In: Proceedings of the EMNLP (2014)
14. Ribeiro, M.T., Singh, S., Guestrin, C.: "Why should i trust you?": Explaining the predictions of any classifier. In: Proceedings of the NAACL (2016)
15. Rudin, C.: Stop explaining black box machine learning models for high stakes decisions and use interpretable models instead. Nature Mach. Intell. 1(5), 206–215 (2019)
16. Schütze, H., Manning, C.D., Raghavan, P.: Introduction to information retrieval, vol. 39. Cambridge University Press, Cambridge (2008)
17. Serrano, S., Smith, N.A.: Is attention interpretable? In: Proceedings of the ACL (2019)
18. Simonyan, K., Vedaldi, A., Zisserman, A.: Deep inside convolutional networks: visualising image classification models and saliency maps. CoRR (2013)
19. Sundararajan, M., Taly, A., Yan, Q.: Axiomatic attribution for deep networks. In: Proceedings of the ICML (2017)
20. Vu, T., Nguyen, D.Q., Nguyen, A.: A label attention model for ICD coding from clinical text. In: Proceedings of the IJCAI (2020)
21. Wiegreffe, S., Pinter, Y.: Attention is not not explanation. In: Proceedings of the EMNLP (2019)
22. You, R., Zhang, Z., Wang, Z., Dai, S., Mamitsuka, H., Zhu, S.: AttentionXML: label tree-based attention-aware deep model for high-performance extreme multi-label text classification. In: Proceedings of the NeurIPS (2019)

Doc2Query--: When Less is More

Mitko Gospodinov, Sean MacAvaney[✉], and Craig Macdonald

University of Glasgow, Glasgow, UK
2024810G@student.gla.ac.uk,
{Sean.MacAvaney,Craig.Macdonald}@glasgow.ac.uk

Abstract. Doc2Query—the process of expanding the content of a document before indexing using a sequence-to-sequence model—has emerged as a prominent technique for improving the first-stage retrieval effectiveness of search engines. However, sequence-to-sequence models are known to be prone to "hallucinating" content that is not present in the source text. We argue that Doc2Query is indeed prone to hallucination, which ultimately harms retrieval effectiveness and inflates the index size. In this work, we explore techniques for filtering out these harmful queries prior to indexing. We find that using a relevance model to remove poor-quality queries can improve the retrieval effectiveness of Doc2Query by up to 16%, while simultaneously reducing mean query execution time by 30% and cutting the index size by 48%. We release the code, data, and a live demonstration to facilitate reproduction and further exploration (https://github.com/terrierteam/pyterrier_doc2query).

1 Introduction

Neural network models, particularly those based on contextualised language models, have been shown to improve search effectiveness [3]. While some approaches focus on re-ranking document sets from a first-stage retrieval function to improve precision [27], others aim to improve the first stage itself [4]. In this work, we focus on one of these first-stage approaches: Doc2Query [29]. This approach trains a sequence-to-sequence model (e.g., T5 [33]) to predict queries that may be relevant to a particular text. Then, when indexing, this model is used to *expand* the document by generating a collection of queries and appending them to the document. Though computationally expensive at index time [34], this approach has been shown to be remarkably effective even when retrieving using simple lexical models like BM25 [28]. Numerous works have shown that the approach can produce a high-quality pool of results that are effective for subsequent stages in the ranking pipeline [19,20,23,40].

However, sequence-to-sequence models are well-known to be prone to generate content that does not reflect the input text – a defect known in literature as "hallucination" [25]. We find that existing Doc2Query models are no exception. Figure 1 provides example generated queries from the state-of-the-art T5 Doc2Query model [28]. In this example, we see that many of the generated queries cannot actually be answered by the source passage (score ≤ 1).

© The Author(s), under exclusive license to Springer Nature Switzerland AG 2023
J. Kamps et al. (Eds.): ECIR 2023, LNCS 13981, pp. 414–422, 2023.
https://doi.org/10.1007/978-3-031-28238-6_31

Original Passage: Barley (Hordeum vulgare L.), a member of the grass family, is a major cereal grain. It was one of the first cultivated grains and is now grown widely. Barley grain is a staple in Tibetan cuisine and was eaten widely by peasants in Medieval Europe. Barley has also been used as animal fodder, as a source of fermentable material for beer and certain distilled beverages, and as a component of various health foods.	Generated Queries: (1) where does barley originate from · (2) what is the name of the cereal grain used in tibetan cooking? · (3) what is barley used for · (1) what is barley in food · (0) what is bare wheat · (3) what family of organisms is barley in · (1) why is barley important in tibetan diet · (3) what is barley · (2) where is barley grown · (1) where was barley first grown and eaten · (1) where was barley first used ...

Fig. 1. Example passage from MS MARCO and generated queries using the T5 Doc2Query model. The relevance of each query to the passage is scored by the authors on a scale of 0–3 using the TREC Deep Learning passage relevance criteria.

Based on this observation, we hypothesise that retrieval performance of Doc2Query would improve if hallucinated queries were removed. In this paper, we conduct experiments where we apply a new filtering phase that aims to remove poor queries prior to indexing. Given that this approach *removes* queries, we call the approach Doc2Query-- (Doc2Query-minus-minus). Rather than training a new model for this task, we identify that relevance models are already fit for this purpose: they estimate how relevant a passage is to a query. We therefore explore filtering strategies that make use of existing neural relevance models.

Through experimentation on the MS MARCO dataset, we find that our filtering approach can improve the retrieval effectiveness of indexes built using Doc2Query-- by up to 16%; less can indeed be more. Meanwhile, filtering naturally reduces the index size, lowering storage and query-time computational costs. Finally, we conduct an exploration of the index-time overheads introduced by the filtering process and conclude that the gains from filtering more than make up for the additional time spent generating more queries. The approach also has a positive impact on the environmental costs of applying Doc2Query; the same retrieval effectiveness can be achieved with only about a third of the computational cost when indexing. To facilitate last-metre, last-mile, and complete reproduction efforts [36], we release the code, indices, and filtering scores. (See footnote 1) In summary, we contribute a technique to improve the effectiveness and efficiency of Doc2Query by filtering out queries that do not reflect the original passage.

2   Related Work

The classical lexical mismatch problem is a key one in information retrieval - documents that do not contain the query terms may not be retrieved. In the literature, various approaches have addressed this: query reformulation – including stemming, query expansion models (e.g. Rocchio, Bo1 [1], RM3 [12]) – and document expansion [9,30,35]. Classically, query expansion models have been popular, as they avoid the costs associated with making additional processing for each document needed for document expansion. However, query expansion may result in reduced performance [11], as queries are typically short and the necessary evidence to understand the context of the user is limited.

The application of latent representations of queries and documents, such as using latent semantic indexing [8] allow retrieval to not be driven directly by lexical signals. More recently, transformer-based language models (such as BERT [6]) have resulted in representations of text where the contextualised meaning of words are accounted for. In particular, in dense retrieval, queries and documents are represented in embeddings spaces [14,37], often facilitated by Approximate Nearest Neighbour (ANN) data structures [13]. However, even when using ANN, retrieval can still be inefficient or insufficiently effective [15].

Others have explored approaches for augmenting lexical representations with additional terms that may be relevant. In this work, we explore Doc2Query [29], which uses a sequence-to-sequence model that maps a document to queries that it might be able to answer. By appending these *generated* queries to a document's content before indexing, the document is more likely to be retrieved for user queries when using a model like BM25. An alternative style of document expansion, proposed by MacAvaney et al. [19] and since used by several other models (e.g., [10,39,40]), uses the built-in Masked Language Modelling (MLM) mechanism. MLM expansion generates individual tokens to append to the document as a bag of words (rather than as a sequence). Although MLM expansion is also prone to hallucination,[1] the bag-of-words nature of MLM expansion means that individual expansion tokens may not have sufficient context to apply filtering effectively. We therefore focus only on sequence-style expansion and leave the exploration of MLM expansion for future work.

3   Doc2Query--

Doc2Query-- consists of two phases: a *generation* phrase and a *filtering* phase. In the generation phase, a Doc2Query model generates a set of n queries that each document might be able to answer. However, as shown in Fig. 1, not all of the queries are necessarily relevant to the document. To mitigate this problem, Doc2Query-- then proceeds to a filtering phase, which is responsible for eliminating the generated queries that are least relevant to the source document. Because hallucinated queries contain details not present in the original text (by definition), we argue that hallucinated queries are less useful for retrieval than non-hallucinated ones. Filtering is accomplished by retaining only the most relevant p proportion of generated queries over the entire corpus. The retained queries are then concatenated to their corresponding documents prior to indexing, as per the existing Doc2Query approach.

More formally, consider an expansion function e that maps a document to n queries: $e : \mathbf{D} \mapsto \mathbf{Q}^n$. In Doc2Query, each document in corpus $\mathcal{D}$ are concatenated with their expansion queries, forming a new corpus $\mathcal{D}' = \{\text{Concat}(d, e(d)) \mid d \in \mathcal{D}\}$, which is then indexed by a retrieval system. Doc2Query-- adds a filtering mechanism that uses a relevance model that maps a query and document to a real-valued relevance score $s : \mathbf{Q} \times \mathbf{D} \mapsto \mathbb{R}$ (with larger values indicating higher

---

[1] For instance, we find that SPLADE [10] generates the following seemingly-unrelated terms for the passage in Fig. 1 in the top 20 expansion terms: *reed, herb*, and *troy*.

relevance). The relevance scoring function is used to filter down the queries to those that meet a certain score threshold t as follows:

$$\mathcal{D}' = \Big\{ \text{Concat}\big(d, \{q \mid q \in e(d) \wedge s(q, d) \geq t\}\big) \mid d \in \mathcal{D} \Big\} \qquad (1)$$

The relevance threshold t is naturally dependent upon the relevance scoring function. It can be set empirically, chosen based on operational criteria (e.g., target index size), or (for a well-calibrated relevance scoring function) determined *a priori*. In this work, we combine the first two strategies: we pick t based on the distribution of relevance scores across all expansion queries. For instance, at $p = 0.3$ we only keep queries with relevance scores in the top 30%, which is $t = 3.215$ for the ELECTRA [31] scoring model on the MS MARCO dataset [26].

4 Experimental Setup

We conduct experiments to answer the following research questions:

RQ1 Does Doc2Query-- improve the effectiveness of document expansion?
RQ2 What are the trade-offs in terms of effectiveness, efficiency, and storage when using Doc2Query--?

Datasets and Measures. We conduct tests using the MS MARCO [26] v1 passage corpus. We use five test collections:[2] (1) the MS MARCO Dev (small) collection, consisting of 6,980 queries (1.1 qrels/query); (2) the Dev2 collection, consisting of 4,281 (1.1 qrels/query); (3) the MS MARCO Eval set, consisting of 6,837 queries (held-out leaderboard set); (4/5) the TREC DL'19/'20 collections, consisting of 43/54 queries (215/211 qrels/query). We evaluate using the official task evaluation measures: Reciprocal Rank at 10 (RR@10) for Dev/Dev2/Eval, nDCG@10 for DL'19/'20. We tune systems[3] on Dev, leaving the remaining collections as held-out test sets.

Models. We use the T5 Doc2Query model from Nogueira and Lin [28], making use of the inferred queries released by the authors (80 per passage). To the best of our knowledge, this is the highest-performing Doc2Query model available. We consider three neural relevance models for filtering: ELECTRA[4] [31], MonoT5[5] [32], and TCT-ColBERT[6] [16], covering two strong cross-encoder models and one strong bi-encoder model. We also explored filters that use the probabilities from the generation process itself but found them to be ineffective and therefore omit these results due to space constraints.

---

[2] ir-datasets [21] IDs: `msmarco-passage/dev/small`, `msmarco-passage/dev/2`, `msmarco-passage/eval/small`, `msmarco-passage/trec-dl-2019/judged`, `msmarco-passage/trec-dl-2020/judged`.

[3] BM25's $k1$, b, and whether to remove stopwords were tuned for all systems; the filtering percentage (p) was also tuned for filtered systems.

[4] `crystina-z/monoELECTRA_LCE_nneg31`.

[5] `castorini/monot5-base-msmarco`.

[6] `castorini/tct_colbert-v2-hnp-msmarco`.

Tools and Environment. We use the PyTerrier toolkit [22] with a PISA [17,24] index to conduct our experiments. We deploy PISA's Block-Max WAND [7] implementation for BM25 retrieval. Inference was conducted on an NVIDIA 3090 GPU. Evaluation was conducted using the ir-measures package [18].

Table 1. Effectiveness and efficiency measurements for Doc2Query-- and baselines. Significant differences between Doc2Query and their corresponding filtered versions for Dev, Dev2, DL'19 and DL'20 are indicated with * (paired t-test, $p < 0.05$). Values marked with † are taken from the corresponding submissions to the public leaderboard.

	RR@10			nDCG@10		ms/q	GB
System	Dev	Dev2	Eval	DL'19	DL'20	MRT	Index
BM25	0.185	0.182	†0.186	0.499	0.479	5	0.71
Doc2Query ($n = 40$)	0.277	0.265	†0.272	0.626	0.607	30	1.17
w/ELECTRA Filter (30%)	***0.316**	***0.310**	–	**0.667**	**0.611**	**23**	**0.89**
w/MonoT5 Filter (40%)	*0.308	*0.298	0.306	0.650	**0.611**	29	0.93
w/TCT Filter (50%)	*0.287	*0.280	–	0.640	0.599	30	0.94
Doc2Query ($n = 80$)	0.279	0.267	–	0.627	0.605	30	1.41
w/ELECTRA Filter (30%)	***0.323**	***0.316**	0.325	**0.670**	**0.614**	**23**	**0.95**
w/MonoT5 Filter (40%)	*0.311	*0.298	–	0.665	0.609	28	1.04
w/TCT Filter (50%)	*0.293	*0.283	–	0.642	0.588	28	1.05

5   Results

We first explore RQ1: whether relevance filtering can improve the retrieval of Doc2Query models. Table 1 compares the effectiveness of Doc2Query with various filters. We observe that all the filters significantly improve the retrieval effectiveness on the Dev and Dev2 datasets at both $n = 40$ and $n = 80$. We also observe a large boost in performance on the Eval dataset.[7] Though the differences in DL'19 and DL'20 appear to be considerable (e.g., 0.627 to 0.670), these differences are not statistically significant.

Digging a little deeper, Fig. 2 shows the retrieval effectiveness of Doc2Query with various numbers of generated queries (in dotted black) and the corresponding performance when filtering using the top-performing ELECTRA scorer (in solid blue). We observe that performing relevance filtering at each value of n

---

[7] Significance cannot be determined due to the held-out nature of the dataset. Further, due to restrictions on the number of submissions to the leaderboard, we only are able to submit two runs. The first aims to be a fair comparison with the existing Doc2Query Eval result, using the same number of generated queries and same base T5 model for scoring. The second is our overall best-performing setting, using the ELECTRA filter at $n = 80$ generated queries.

improves the retrieval effectiveness. For instance, keeping only 30% of expansion queries at $n = 80$, performance is increased from 0.279 to 0.323 – a 16% improvement.

In aggregate, results from Table 1 and Fig. 2 answer RQ1: Doc2Query-- filtering can significantly improve the retrieval effectiveness of Doc2Query across various scoring models, numbers of generated queries (n) and thresholds (p).

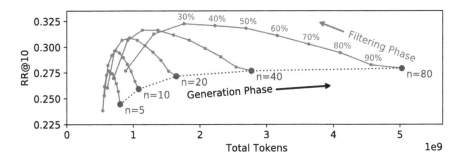

Fig. 2. Effectiveness (RR@10) on the Dev set, compared with the total number of indexed tokens. The generation phase is shown in dotted black (at various values of n), and the ELECTRA filtering phase is shown in solid blue (at various values of p). (Color figure online)

Next, we explore the trade-offs in terms of effectiveness, efficiency, and storage when using Doc2Query--. Table 1 includes the mean response time and index sizes for each of the settings. As expected, filtering reduces the index size since fewer terms are stored. For the best-performing setting ($n = 80$ with ELECTRA filter), this amounts to a 48% reduction in index size (1.41 GB down to 0.95 GB). Naturally, such a reduction has an impact on query processing time as well; it yields a 30% reduction in mean response time (30 ms down to 23 ms).

Doc2Query-- filtering adds substantial cost an indexing time, mostly due to scoring each of the generated queries. Table 2 reports the cost (in hours of GPU time) of the generation and filtering phases. We observe that ELECTRA filtering can yield up to a 78% increase in GPU time ($n = 10$). However, we find that the improved effectiveness makes up for this cost. To demonstrate this, we allocate the time spent filtering to generating additional queries for each passage. For instance, the 15 h spent scoring $n = 5$ queries could instead be spent generating 6 more queries per passage (for a total of $n = 11$). We find that when comparing against an unfiltered n that closely approximates the total time when filtering, the filtered results consistently yield significantly higher retrieval effectiveness. As the computational budget increases, so does the margin between Doc2Query and Doc2Query--, from 4% at 34 h up to 12% at 216 h.

Table 2. Retrieval effectiveness comparison for comparable indexing computational budgets (in hours of GPU time). Values of n without a filter are chosen to best approximate the total compute hours or the Dev effectiveness of the corresponding filtered version. Significant differences between in RR@10 performance are indicated with * (paired t-test, $p < 0.05$).

n	Filter	GPU hours Gen+Filt = Tot	RR@10 Dev	Dev2	Comment
5	ELECTRA	20 + 15 = 34	**0.273**	**0.270**	
11	*None*	34 + 0 = 34	*0.261	*0.256	*−4% Dev RR for sim. GPU hrs*
31	*None*	99 + 0 = 99	**0.273**	0.265	*×2.9 GPU hrs to match Dev RR*
10	ELECTRA	32 + 25 = 57	**0.292**	**0.292**	
18	*None*	59 + 0 = 59	*0.270	*0.260	*−8% Dev RR for sim. GPU hrs*
20	ELECTRA	66 + 47 = 113	**0.307**	**0.303**	
36	*None*	113 + 0 = 113	*0.275	*0.265	*−10% Dev RR for sim. GPU hrs*
40	ELECTRA	128 + 86 = 214	**0.316**	**0.310**	
68	*None*	216 + 0 = 216	*0.279	*0.267	*−12% Dev RR for sim. GPU hrs*

From the opposite perspective, Doc2Query consumes 2.9× or more GPU time than Doc2Query-- to achieve similar effectiveness ($n = 13$ with no filter vs. $n = 5$ with ELECTRA filter). Since the effectiveness of Doc2Query flattens out between $n = 40$ and $n = 80$ (as seen in Fig. 2), it likely requires a massive amount of additional compute to reach the effectiveness of Doc2Query-- at $n \geq 10$, if that effectiveness is achievable at all. These comparisons show that if a deployment is targeting a certain level of effectiveness (rather than a target compute budget), Doc2Query-- is also preferable to Doc2Query.

These results collectively answer RQ2: Doc2Query-- provides higher effectiveness at lower query-time costs, even when controlling for the additional compute required at index time.

6 Conclusions

This work demonstrated that there are untapped advantages in generating natural-language for document expansion. Specifically, we presented Doc2Query--, which is a new approach for improving the effectiveness and efficiency of the Doc2Query model by filtering out the least relevant queries. We observed that a 16% improvement in retrieval effectiveness can be achieved, while reducing the index size by 48% and mean query execution time by 30%.

The technique of filtering text generated from language models using relevance scoring is ripe for future work. For instance, relevance filtering could potentially apply to approaches that generate alternative forms of queries [38], training data [2], or natural language responses to queries [5]—all of which are potentially affected by hallucinated content. Furthermore, future work could explore approaches for relevance filtering over masked language modelling expansion [19], rather than sequence-to-sequence expansion.

Acknowledgements. Sean MacAvaney and Craig Macdonald acknowledge EPSRC grant EP/R018634/1: Closed-Loop Data Science for Complex, Computationally- & Data-Intensive Analytics.

References

1. Amati, G., Van Rijsbergen, C.J.: Probabilistic models of information retrieval based on measuring the divergence from randomness. ACM Trans. Inf. Syst. **20**(4), 357–389 (2002)
2. Bonifacio, L., Abonizio, H., Fadaee, M., Nogueira, R.: InPars: unsupervised dataset generation for information retrieval. In: Proceedings of SIGIR (2022)
3. Dai, Z., Callan, J.: Deeper text understanding for IR with contextual neural language modeling. In: Proceedings of SIGIR (2019)
4. Dai, Z., Callan, J.: Context-aware document term weighting for ad-hoc search. In: Proceedings of the Web Conference (2020)
5. Das, R., Dhuliawala, S., Zaheer, M., McCallum, A.: Multi-step retriever-reader interaction for scalable open-domain question answering. In: Proceedings of ICLR (2019)
6. Devlin, J., Chang, M.W., Lee, K., Toutanova, K.: BERT: pre-training of deep bidirectional transformers for language understanding. In: Proceedings of NAACL-HLT (2019)
7. Ding, S., Suel, T.: Faster top-k document retrieval using block-max indexes. In: Proceedings of SIGIR (2011)
8. Dumais, S.T., Furnas, G.W., Landauer, T.K., Deerwester, S., Harshman, R.: Using latent semantic analysis to improve access to textual information. In: Proceedings of SIGCHI CHI (1988)
9. Efron, M., Organisciak, P., Fenlon, K.: Improving retrieval of short texts through document expansion. In: Proceedings of SIGIR (2012)
10. Formal, T., Piwowarski, B., Clinchant, S.: SPLADE: sparse lexical and expansion model for first stage ranking. In: Proceedings of SIGIR (2021)
11. He, B., Ounis, I.: Studying query expansion effectiveness. In: Proceedings of ECIR (2009)
12. Jaleel, N.A., et al.: Umass at TREC 2004: novelty and HARD. In: TREC (2004)
13. Johnson, J., Douze, M., Jegou, H.: Billion-scale similarity search with GPUs. IEEE Trans. Big Data **7**(03), 535–547 (2021)
14. Khattab, O., Zaharia, M.: ColBERT: efficient and effective passage search via contextualized late interaction over BERT. In: Proceedings of SIGIR (2020)
15. Lin, J., Ma, X., Mackenzie, J., Mallia, A.: On the separation of logical and physical ranking models for text retrieval applications. In: Proceedings of DESIRES (2021)
16. Lin, S.C., Yang, J.H., Lin, J.: In-batch negatives for knowledge distillation with tightly-coupled teachers for dense retrieval. In: Proceedings of RepL4NLP (2021)
17. MacAvaney, S., Macdonald, C.: A python interface to PISA! In: Proceedings of SIGIR (2022)
18. MacAvaney, S., Macdonald, C., Ounis, I.: Streamlining evaluation with ir-measures. In: Proceedings of ECIR (2022)
19. MacAvaney, S., Nardini, F.M., Perego, R., Tonellotto, N., Goharian, N., Frieder, O.: Expansion via prediction of importance with contextualization. In: Proceedings of SIGIR (2020)
20. MacAvaney, S., Tonellotto, N., Macdonald, C.: Adaptive re-ranking with a corpus graph. In: Proceedings of CIKM (2022)

21. MacAvaney, S., Yates, A., Feldman, S., Downey, D., Cohan, A., Goharian, N.: Simplified data wrangling with ir_datasets. In: Proceedings of SIGIR (2021)
22. Macdonald, C., Tonellotto, N.: Declarative experimentation in information retrieval using PyTerrier. In: Proceedings of ICTIR (2020)
23. Mallia, A., Khattab, O., Suel, T., Tonellotto, N.: Learning passage impacts for inverted indexes. In: Proceedings of SIGIR (2021)
24. Mallia, A., Siedlaczek, M., Mackenzie, J., Suel, T.: PISA: performant indexes and search for academia. In: Proceedings of OSIRRC@SIGIR (2019)
25. Maynez, J., Narayan, S., Bohnet, B., McDonald, R.: On faithfulness and factuality in abstractive summarization. In: Proceedings of ACL (2020)
26. Nguyen, T., Rosenberg, M., Song, X., Gao, J., Tiwary, S., Majumder, R., Deng, L.: MS MARCO: a human generated machine reading comprehension dataset. In: Proceedings of CoCo@NIPS (2016)
27. Nogueira, R., Cho, K.: Passage re-ranking with BERT. ArXiv abs/1901.04085 (2019)
28. Nogueira, R., Lin, J.: From doc2query to docttttttquery (2019)
29. Nogueira, R., Yang, W., Lin, J.J., Cho, K.: Document expansion by query prediction. ArXiv abs/1904.08375 (2019)
30. Pickens, J., Cooper, M., Golovchinsky, G.: Reverted indexing for feedback and expansion. In: Proceedings of CIKM (2010)
31. Pradeep, R., Liu, Y., Zhang, X., Li, Y., Yates, A., Lin, J.: Squeezing water from a stone: a bag of tricks for further improving cross-encoder effectiveness for reranking. In: Proceedings of ECIR (2022)
32. Pradeep, R., Nogueira, R., Lin, J.: The expando-mono-duo design pattern for text ranking with pretrained sequence-to-sequence models. ArXiv abs/2101.05667 (2021)
33. Raffel, C., et al.: Exploring the limits of transfer learning with a unified text-to-text transformer. J. Mach. Learn. Res. **21**(140), 5485–5551 (2020)
34. Scells, H., Zhuang, S., Zuccon, G.: Reduce, reuse, recycle: green information retrieval research. In: Proceedings of SIGIR (2022)
35. Tao, T., Wang, X., Mei, Q., Zhai, C.: Language model information retrieval with document expansion. In: Proceedings of HLT-NAACL (2006)
36. Wang, X., MacAvaney, S., Macdonald, C., Ounis, I.: An inspection of the reproducibility and replicability of TCT-ColBERT. In: Proceedings of SIGIR (2022)
37. Xiong, L., Xiong, C., Li, Y., Tang, K.F., Liu, J., Bennett, P.N., Ahmed, J., Overwijk, A.: Approximate nearest neighbor negative contrastive learning for dense text retrieval. In: Proceedings of ICLR (2021)
38. Yu, S.Y., Liu, J., Yang, J., Xiong, C., Bennett, P.N., Gao, J., Liu, Z.: Few-shot generative conversational query rewriting. In: Proceedings of SIGIR (2020)
39. Zhao, T., Lu, X., Lee, K.: SPARTA: efficient open-domain question answering via sparse transformer matching retrieval. arXiv abs/2009.13013 (2020)
40. Zhuang, S., Zuccon, G.: TILDE: term independent likelihood model for passage re-ranking. In: Proceedings of SIGIR (2021)

Towards Quantifying the Privacy of Redacted Text

Vaibhav Gusain[ID] and Douglas Leith[(✉)][ID]

Trinity College Dublin, Dublin, Ireland
{gusainv,doug.leith}@tcd.ie

Abstract. In this paper we propose use of a k-anonymity-like approach for evaluating the privacy of redacted text. Given a piece of redacted text we use a state of the art transformer-based deep learning network to reconstruct the original text. This generates multiple full texts that are consistent with the redacted text, i.e. which are grammatical, have the same non-redacted words etc., and represents each of these using an embedding vector that captures sentence similarity. In this way we can estimate the number, diversity and quality of full text consistent with the redacted text and so evaluate privacy.

Keywords: Transformers · Text privacy · Data leaks · k-anonymity

1 Introduction

Redacting a piece of text involves replacing selected words with an uninformative mask symbol. Redaction is widely used, but is generally carried out manually and there has been little analysis of the degree of privacy obtained. Note that evaluating text privacy is generally not straightforward since even when a word is redacted it might still be possible to reliably estimate it from the surrounding text i.e. the context of the redacted word may be revealing.

Machine learning models for text embedding are often trained by masking out individual words in a piece of text and selecting a model that best reconstructs the missing text. The idea here is that similar words appear in a similar context. In particular, transformer-based neural networks such as BART [6] adopt this approach and achieve state of the performance in many natural language processing tasks.

Given a piece of redacted text, in this paper we apply transformer-based neural networks to try to reconstruct the original text. For example, when the text `he was stationed at singapore` is redacted to `he was stationed at <mask>` then the top 5 reconstructed text predictions by BART are shown in Table 1. It can be seen that the reconstructed text is grammatical, consistent with the redacted text (has the same non-redacted words etc.) and plausible even though in this example it does not correctly predict the missing word.

In this paper we study using such predicted reconstructions as the basis for a quantitative privacy metric for redacted text. This is motivated by the observation that the number of reconstructions that are estimated with high confidence

D. Leith—This work was supported by Science Foundation Ireland grant 16/IA/4610.

J. Kamps et al. (Eds.): ECIR 2023, LNCS 13981, pp. 423–429, 2023.
https://doi.org/10.1007/978-3-031-28238-6_32

Table 1. Left-hand table: Top 5 reconstructions by BART for the redacted sentence `he was stationed at <mask>`. The values shown in the second column are the corresponding confidence values output by BART. Right-hand table: top prediction by BART as the number of redacted words is increased.

he was stationed at <mask>	
he was stationed at the	0.62
he was stationed at:	0.58
he was stationed at Gettysburg	0.49
he was stationed at Ft.	0.48
he was stationed at Knox	0.47

Redacted sentence	BART top prediction
<mask> was <mask> at singapore	This article was originally published at singapore
<mask> <mask> <mask> at singapore	A look at singapore
<mask> <mask> <mask> <mask> singapore	Singapore singapore

can be expected to provide an approximate k-anonymity [9] measure i.e. a measure of "Hiding in the crowd" privacy since there are at-least K sentences that are plausibly consistent with the redacted text. Since the reconstructions are represented as embedded vectors that capture sentence similarity (similar sentences are represented by nearby vectors) then we can also estimate the diversity of the reconstructions.

This work reported here is just a first, exploratory step but we find that this general approach shows promise.

Rather than evaluating k-anonymity and text diversity, we begin by considering the text quality of the predictions since this turns out to be a useful predictor of privacy in coarse classification tasks such as sentiment analysis, news article categorisation and medical condition (e.g. has cancer or not). We find that there is a thresholding effect, whereby beyond a certain level of redaction the quality tends to drop sharply. By carrying out simulated attacks against the redacted text we find that the drop in BART prediction quality strongly correlates with a decrease in attack effectiveness. The proposed approach therefore has the potential to provide a practical, useful estimate of redacted text privacy.

1.1   Related Work

Text Redaction. Despite the widespread use of redaction, there has been very little work on quantifying the privacy of redacted text or on evaluating robustness to attacks that seek to generate privacy leaks. Instead most work to date has focused on identifying personal data with text so that it can be redacted. See, for example, [2] which considers discovery of names, home towns etc. in student discussion boards, and also the references therein. The closest work to the present paper is probably [1] which considers randomly redacting words to ensure a form of differential privacy and evaluates utility using a transformer neural net. However, there is no evaluation of the robustness of the redacted text to adversarial attacks (which is primarily what we use transformer neural nets for here) and the interpretation of differential privacy in the context of redacted text remains unclear (in [1] the surrounding context of a redacted word is ignored, yet will often have an important impact on the degree of privacy achieved).

Text Reconstruction. Predicting missing text has been the subject of a great deal work in recent years. The state of the art uses transformer-based neural net architectures, following the breakthrough performance achieved by BERT. BART [6] is a transformer-based neural net that targets reconstruction of text damaged by spelling mistakes, missing words etc. Roughly speaking it is an amalgamation of BERT and GPT2, consisting of a bidirectional encoder which is very similar to BERT and a left-to-right decoder which is very similar to GPT2. This design allows BART to even predict arbitrary length of text for a single mask token which cant be achieved with BERT.

2   Quantifying BART Text Quality

The right-hand table in Table 1 shows how the top predicted sentence reconstruction by BART varies as the number of redacted words is increased. It can be seen that by the time four out the five words in the sentence are redacted the BART prediction degrades and is no longer grammatical. In our experiments (see below), we find that this behaviour is a common feature of the BART reconstructions. Of course, it is quite reasonable behaviour since at this point there is so little information left in the redacted sentence that BART has few clues as to how it might be reconstructed. Equally, the point where this information loss occurs is obviously also of great interest from a privacy viewpoint.

Rather than considering just the top prediction by BART, we proceed by considering the top N predictions, typically with N=100. We then estimate the fraction of these predictions which are not grammatical, and investigate the use of this as a measure of privacy.

In general, it is not trivial to estimate whether a sentence is grammatical or not. Fortunately we do not need to solve the general problem but can instead exploit the fact that BART predictions tend to either be fairly grammatical or else are grossly non-grammatical e.g. with many repetitions of the same word (as can be seen in Table 1) and/or with many repetitions of punctuation and spurious characters. That is, the BART predictions tend to either be reasonable text or to be "gibberish".

To classify a sentence as gibberish or not, in our experiments we use Algorithm 1 although other choices are of course possible. Algorithm 1 combines a standard gibberish detector Nostrill [4] with a measure of the fraction of words from the original (non-redacted) sentence that overlap with the predicted sentence. Hyperparameter C controls the weight attached to each measure.

3   Experimental Measurements

3.1   Datasets Used

We evaluated performance on five datasets: four standard text classification datasets BBCnews [3], Amazon-Fine-food [8], AGnews [11], IMDB [7] plus the

Algorithm 1. Algorithm used to classify BART predictions as gibberish or not. Si is the actual input sentence without the mask, Sp is the BART prediction and C is a hyperparameter that checks the number of overlapping words between Si and Sp. It returns True if the prediction is estimated to be gibberish else it returns false

gibberish = use Nostrill to check if the Sp is gibberish or not.
if gibberish **then**
    return gibberish
else
    return customGibberish(Si,Sp,C)
end if
customGibberish(Si, Sp, C) :
$Si \leftarrow$ number of uniquewords in Si
$Sp \leftarrow$ number of uniquewords in Sp
$p \leftarrow$ number of common words in Si and Sp / length(Si)
gibber = p*100/ length(Si)
if gibber $<= C$ **then**
    return True
else
    return False
end if

Medal medical dataset [10]. BBCnews has fives classes (Business, Entertainment, Politics, Sport, Tech), Amazon-Fine-food has review stars and reviews with greater than 3 stars were assigned to one class and the rest to another class, AGnews has four classes (World, Sports, Business, Sci/Tech), IMDB has two classes (positive and negative sentiment), Medal has two classes (text specifically about cancer diseases, plus the rest). Each dataset was split 80:20 into a training and a test dataset, with the training dataset being available to the adversary but not the test dataset. The datasets are sampled so that they are balanced by category.

3.2   Threat Model

The attacker can observe redacted text, and a training data subset of each dataset. The redacted text is derived from held out data not available to the attacker. The aim of the attacker is to discover the category of the text e.g. for a movie review to discover the sentiment, for a news article to discover the news category.

3.3   Reconstruction Quality Metric

For each redacted sentence we take the top 100 reconstruction predictions from BART and apply Algorithm 1 to classify them as either gibberish or not, assigning value +1 for gibberish and 0 otherwise. We calculate the mean of these 100 values.

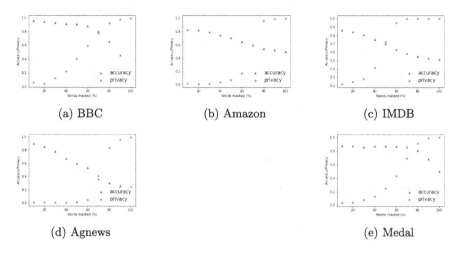

(a) BBC   (b) Amazon   (c) IMDB

(d) Agnews   (e) Medal

Fig. 1. Measured privacy metric and attack accuracy for each dataset as the fraction of redacted text is varied from 0 to 100%.

3.4 Privacy Attack Performance Metric

Using the training data for each dataset the adversary trains a classifier based on a TFIDF [5] vectoriser and a logistic regression model (for these datasets it is known that classifiers of this sort are able to achieve high accuracy). Given redacted text, the attacker then uses this classifier to estimate the category of the text. We evaluate the success of this reconstruction using the mean accuracy of these predictions i.e. the fraction of redacted sentences for which the category is correctly estimated. The test data is balanced, so accuracy is an informative performance measure.

3.5 Redaction Strategy

For each dataset we encode the words using a TFIDF vectoriser (discarding words with document frequency less than 10%). We then vary the level of redaction by replacing a random X percent of words by a mask token, varying X from 0 to 100%. Using TFIDF in this way avoid ineffectual masking of stop words and other uninformative words. Other redaction strategies are, of course, possible.

3.6 Additional Material

We will post our implementations and the associated data on github.

3.7 Results

Figure-1 shows the measured privacy metric and attack accuracy for each dataset as the fraction of redacted text is varied. It can be seen that as the percentage of masked words is increased the classification accuracy decreases while the privacy metric increases.

When less than around 20% of words are redacted, the privacy metric is close to zero for every dataset, indicating that BART consistently reconstructs grammatical sentences that are consistent with the redacted text. Analysis of the top 100 BART predictions (not included here) show little diversity in the sense that the sentence embedding vectors tend to cluster together. The attack accuracy is correspondingly also consistently high.

When greater than around 80% of words are redacted, then the privacy metric is close to 100% and the attack accuracy is approximately the reciprocal of the number of categories i.e. comparable with a random coin toss.

Between 20 and 80% redaction the privacy metric increases and the attack accuracy correspondingly decreases. By selecting a level of redaction that ensures the privacy metric is above a target threshold, e.g. 70%, then these measurements indicate that a good level of robustness against the reconstruction attack can be obtained across a wide range of datasets.

3.8   Discussion

Due to lack of space we do not include an evaluation of utility here, which can be expected to degrade as privacy increases. However, we note briefly that we have evaluated next word prediction performance for the Medal dataset vs privacy and find that the utility remains high even when redaction achieves a high level of resistance against estimation of medical condition.

We use attack accuracy as a proxy for privacy, since it is difficult to apply standard privacy metrics such as k-anonymity and differential privacy to natural language text data. However, initial results indicate that it may be possible to estimate a metric similar to k-anonymity by clustering the embedding vectors of the BART predictions and counting the number of distinct clusters. In the regime where BART predictions are grammatical (redaction level less than 20% in Fig. 1) these clusters reflect semantic diversity, whereas in the regime where BART predictions produce lower quality text the clusters tends to become less informative. However, we leave proper analysis of these aspects to future work.

Initial results also suggest that the nature of the privacy threat is relevant to the level of redaction needed. To prevent disclosure of broad textual aspects such as sentiment or new category our results show that a high level of redaction is necessary, but preventing disclosure of more fine-grained aspects might be achievable with lower levels of redaction. Again, we leave further study of this to future work.

References

1. Adelani, D.I., Davody, A., Kleinbauer, T., Klakow, D.: Privacy guarantees for de-identifying text transformations. arXiv preprint arXiv:2008.03101 (2020)
2. Bosch, N., Crues, R., Shaik, N., Paquette, L.: "hello,[redacted]": Protecting student privacy in analyses of online discussion forums. Grantee Submission (2020)

3. Greene, D., Cunningham, P.: Practical solutions to the problem of diagonal dominance in kernel document clustering. In: Proceedings of 23rd International Conference on Machine learning (ICML 2006), pp. 377–384. ACM Press (2006)

4. Hucka, M.: Nostril: A nonsense string evaluator written in python. J. Open Source Softw. **3**(25), 596 (2018). https://doi.org/10.21105/joss.00596

5. Jing, L.P., Huang, H.K., Shi, H.B.: Improved feature selection approach TFIDF in text mining. In: Proceedings of International Conference on Machine Learning and Cybernetics, vol. 2, pp. 944–946 (2002). https://doi.org/10.1109/ICMLC.2002.1174522

6. Lewis, M., et al.: Bart: denoising sequence-to-sequence pre-training for natural language generation, translation, and comprehension. arXiv preprint arXiv:1910.13461 (2019)

7. Maas, A.L., Daly, R.E., Pham, P.T., Huang, D., Ng, A.Y., Potts, C.: Learning word vectors for sentiment analysis. In: Proceedings of the 49th Annual Meeting of the Association for Computational Linguistics: Human Language Technologies, pp. 142–150. Association for Computational Linguistics, Portland, Oregon, USA, June 2011. http://www.aclweb.org/anthology/P11-1015

8. McAuley, J.J., Leskovec, J.: From amateurs to connoisseurs: modeling the evolution of user expertise through online reviews. In: Proceedings of the 22nd international conference on World Wide Web, pp. 897–908 (2013)

9. Samarati, P., Sweeney, L.: Protecting privacy when disclosing information: k-anonymity and its enforcement through generalization and suppression (1998)

10. Wen, Z., Lu, X.H., Reddy, S.: MeDAL: medical abbreviation disambiguation dataset for natural language understanding pretraining. In: Proceedings of the 3rd Clinical Natural Language Processing Workshop. Association for Computational Linguistics (2020). https://doi.org/10.18653/v1/2020.clinicalnlp-1.15

11. Zhang, X., Zhao, J.J., LeCun, Y.: Character-level convolutional networks for text classification. CoRR abs/1509.01626 (2015). arxiv.org:1509.01626

Detecting Stance of Authorities Towards Rumors in Arabic Tweets: A Preliminary Study

Fatima Haouari[(✉)] and Tamer Elsayed

Computer Science and Engineering Department, Qatar University, Doha, Qatar
{200159617,telsayed}@qu.edu.qa

Abstract. A myriad of studies addressed the problem of rumor verification in Twitter by either utilizing evidence from the propagation networks or external evidence from the Web. However, none of these studies exploited evidence from trusted authorities. In this paper, we define the task of detecting the stance of authorities towards rumors in tweets, i.e., whether a tweet from an authority agrees, disagrees, or is unrelated to the rumor. We believe the task is useful to augment the sources of evidence utilized by existing rumor verification systems. We construct and release the first Authority STance towards Rumors (AuSTR) dataset, where evidence is retrieved from authority timelines in Arabic Twitter. Due to the relatively limited size of our dataset, we study the usefulness of existing datasets for stance detection in our task. We show that existing datasets are somewhat useful for the task; however, they are clearly insufficient, which motivates the need to augment them with annotated data constituting stance of authorities from Twitter.

Keywords: Evidence · Claims · Social media

1 Introduction

Existing studies for rumor verification in social media exploited the propagation networks as a source of evidence, where they focused on the stance of replies [8,13,24,29,33,34], structure of replies [9,12,14,19,27,32], and profile features of retweeters [26]. Recently, Dougrez-Lewis et al. [17] proposed augmenting the propagation networks with evidence from the Web. To our knowledge, no previous research has investigated exploiting evidence for rumor verification in social media from the timelines of trusted authorities, where an authority is *an entity with the real knowledge or power to verify or deny a specific rumor* [11]. We believe that detecting stance of relevant authorities towards rumors can be a great asset to augment the sources of evidence utilized by existing rumor verification systems. It can also serve as a valuable tool for fact-checkers to automate their process of checking authority tweets to verify certain rumors. It is worth mentioning that stance of authorities can be just one (*but* important) source of evidence that compliment other sources and by itself may not (in some cases) be fully trusted to decide the veracity of rumors.

J. Kamps et al. (Eds.): ECIR 2023, LNCS 13981, pp. 430–438, 2023.
https://doi.org/10.1007/978-3-031-28238-6_33

In this paper, we conduct a preliminary study for detecting stance of authorities towards rumors spreading in Twitter in the Arab world. Exploiting sources of evidence for Arabic rumor verification in Twitter is still under-studied; existing studies exclusively focused on the tweet text for verification [2,5,18,20,28,31]. A notable exception is the work done by Haouari et al. [19] that utilized the replies, their structure, and repliers' profile features to verify Arabic COVID-19 rumors. Several studies addressed Arabic stance detection in Twitter; however, the target was a specific topic not rumors [6,15,22]. A few datasets for stance detection for Arabic claim verification were released recently, where the evidence is either news articles [3,10] or manually-crafted sentences [23]. However, there is no dataset where the rumors are tweets and the evidence is retrieved from authority timelines, neither in Arabic nor in other languages. To fill this gap, the contribution of our work is four-fold: (1) we define the task of detecting stance of authorities towards rumors in tweets, (2) we construct and release the first Authority STance for Rumors (AuSTR) dataset,[1] (3) we present the first study on the usefulness of existing stance detection datasets for our task, and (4) we perform a failure analysis to gain insights for the future work on the task. The research question we aim to address in this work is whether the existing datasets of Arabic stance detection for claim verification are useful for detecting the stance of authorities in Arabic tweets.

The remainder of this paper is organized as follows. We outline the construction methodology of AuSTR in Sect. 2. Our experimental setup is presented in Sect. 3. We discuss and analyze our results in Sect. 4. Finally, we conclude and suggest some future directions in Sect. 5.

2   Constructing AuSTR Dataset

To construct AuSTR where both the rumor and evidence are tweets, we exploit both fact-checking articles and variant authority Twitter accounts.

Exploiting Fact-Checking Articles. Fact-checkers who attempt to verify rumors usually provide in their fact-checking articles some examples of social media posts (e.g., tweets) propagating the specific rumors, and other posts from trusted authorities that constitute evidence to support their verification decisions. To construct AuSTR, we exploit both examples of tweets: stating rumors and showing evidence from authorities as provided by those fact-checkers. Specifically, we used AraFacts [4], a large dataset of Arabic rumors collected from 5 fact-checking websites. From those rumors, we selected only the ones that are expressed in tweets and have evidence in tweets as well.[2] We then extracted the rumor-evidence pairs as follows. For *true* and *false* rumors, we selected a single tweet example and all provided evidence tweets, which are then labeled as having *agree* and *disagree* stances respectively.[3] If the fact-checkers provided

---

[1] https://github.com/Fatima-Haouari/AuSTR.
[2] We contacted the authors of AraFacts to get this information as it was not released.
[3] We only kept evidence expressed in *text* rather than in image or video.

the authority account but stated no evidence was found to support or deny the rumor, we selected one or two tweets from the authority timeline posted soon before the rumor time, and assigned the *unrelated* label to the pairs.

Exploiting Authority Accounts. Given that fact-checkers focus more on *false* rumors than *true* ones, we ended up with only 4 *agree* pairs as opposed to 118 *disagree* pairs following the above step. To further expand our *agree* pairs, we did the reverse of the previous approach, where we collected the evidence first. Specifically, we started from a set of Twitter accounts of authorities (e.g., ministers, presidents, embassies, organization accounts, etc.) covering most of the Arab countries and multiple domains (e.g., politics, health, and sports), and selected recent tweets stating claims from their timelines. For each claim, we used Twitter search interface to look for tweets from regular users expressing it, but tried to avoid exact duplicates. Finally, to get closer to the real scenario, where percentage of *unrelated* tweets is usually higher than percentages of *agree* and *disagree* tweets in the authority timelines, we further expanded the *unrelated* pairs by selecting one or two *unrelated* recent tweets from the authority timeline posted before the rumor time for each *agree* and *disagree* pairs.

Overall, we end up with 409 pairs covering 171 unique claims, where 41 are *true* and 130 are *false*. Among those pairs, 118 are *disagree* (29%), 62 are *agree* (15%), and 229 are *unrelated* (56%).

3   Experimental Setup

Datasets. To study the usefulness of existing Arabic datasets that target stance for claim verification, we adopted the following ones for training:

1. **ANS** [23] of 3,786 **(claim, sentence)** pairs, where claims were extracted from news article titles from trusted sources, then annotators were asked to generate *true* and *false* sentences towards them by adopting paraphrasing and contradiction respectively. The sentences are annotated as either *agree*, *disagree*, or *other* towards the claims.
2. **ArabicFC** [10] of 3,042 **(claim, article)** pairs, where claims are extracted from a single fact-checking website verifying political claims about war in Syria, and articles collected by searching Google using the claim. The articles are annotated as either *agree*, *disagree*, *discuss*, or *unrelated* to the claim.
3. **AraStance** [3]: 3,676 **(claim, article)** pairs, where claims extracted from 3 Arabic fact-checking websites covering multiple domains and Arab countries. The articles were collected and annotated similar to ArabicFC.

To train our models, we considered only three labels, namely, *agree*, *disagree*, or *unrelated*. For ANS and AraStance, we used the same data splits provided by the authors; however, we split the ArabicFC into 70%, 10%, and 20% of the claims for training, development, and testing respectively[4]. When splitting data, we assigned all pairs having the same claim to the same split. Table 1 shows

---

[4] We release ArabicFC splits for reproducibility.

the size of different data splits of the three datasets. Due to the limited size of AuSTR, in this work, we opt to utilize it only as a *test set* while using the above datasets for training to show their usefulness in our task.

Table 1. Data splits of the Arabic stance datasets used for training.

Label	ANS			ArabicFC			AraStance		
	Train	Dev	Test	Train	Dev	Test	Train	Dev	Test
Agree	903	268	130	323	32	119	739	129	154
Disagree	1686	471	242	66	8	13	309	76	64
Unrelated	63	16	7	1464	198	410	1553	294	358
Total	2652	755	379	1853	238	542	2601	499	576

Stance Models. To train our stance models, we fine-tuned BERT [16] to classify whether the evidence sentence/article *agrees* with, *disagrees* with, or is *unrelated* to the claim. We feed BERT the claim text as sentence A, the evidence as sentence B (truncated if needed) separated by the [SEP] token. Finally, we use the contextual representation of the [CLS] token as input to a single classification layer with three output nodes, added on top of the BERT architecture to compute the probability for each class of stance.

Various Arabic BERT-based models were released recently [1,7,21,25,30]; we opted to choose ARBERT [1] as it was shown to achieve better performance on the stance datasets adopted in our work [3]. We adopted the authors' setup [3] by training the models for a maximum of 25 epochs, where early stopping was set to 5 and sequence length to 512. We trained 7 different models in an ablation study using different combinations of the stance datasets mentioned earlier.

4   Results and Discussion

The research question we address in this preliminary study is whether the existing stance detection datasets are useful or not in our task. To answer it, we use combinations of the existing datasets for training and AuSTR for testing. We also show how models trained on those combinations perform on their own corresponding in-domain test sets. While the results on the in-domain test sets are not comparable, since those test sets are different, they constitute an estimated upper bound performance. To evaluate the models, we report per-class F_1 and macro-F_1 scores. Table 2 presents the performance results of all experiments, which demonstrate several interesting observations.

First, we notice that almost all models (except a few) were able to achieve higher performance on their own in-domain test sets compared to AuSTR. This shows that domain adaptation was not very effective (thus in-domain data for our task is required for training the models).

Second, when using individual stance datasets for training, the model trained on AraStance clearly outperformed the others in all measures when tested on AuSTR. We note that ArabicFC is severely imbalanced, where the *disagree* class represents only 3.3% of the data, yielding a very poor performance on that class even when tested on its own in-domain test set. A similar conclusion was found by previous studies [3,10]. As for ANS, evidence is manually crafted, which is not as realistic as tweets from authorities. Alternatively, AraStance claims are extracted from three fact-checking websites,[5] covering multiple domains and Arab countries, similar to AuSTR, and the evidence is represented in articles written by journalists, not manually crafted.

Third, when tested on AuSTR, the model trained on all datasets combined exhibits the best performance on the *disagree* class; however its performance was severely degraded compared to the AraStance model on the *agree* class. This indeed needs further investigation.

Furthermore, we observe that AraStance achieved the highest $F_1(D)$ when used solely for training, and whenever combined with the other datasets. To investigate this, we manually examined a 10% random sample of *disagreeing* training articles. We found they have common words such as *rumors*, *not true*, *denied*, and *fake*; similar keywords appear in some *disagreeing* tweets of AuSTR.

Finally, we observe that there is a clear discrepancy in the performance across different classes. Considering the model trained on all datasets for example, $F_1(A)$ is 0.74 while $F_1(D)$ is 0.65. Moreover, it is clear that detecting the *disagree* stance is the most challenging subtask, which we expect to benefit from in-domain training. Overall, we believe training and testing on tweets is very different, as they are very short and informal, which needs special pre-processing.

Table 2. Performance on both the in-domain test sets and AuSTR, measured in per-class F_1 (A: Agree, D: Disagree, U: Unrelated) and macro-F_1. On AuSTR, bold and underlined values indicate best and second-best performance respectively.

Training set	Test on in-domain set				Test on AuSTR			
	$F_1(A)$	$F_1(D)$	$F_1(U)$	m-F_1	$F_1(A)$	$F_1(D)$	$F_1(U)$	m-F_1
ANS	0.824	0.901	0.923	0.882	0.653	0.578	0.709	0.647
ArabicFC	0.770	0.090	0.915	0.591	0.641	0.434	0.799	0.625
AraStance	0.898	0.833	0.95	0.894	**0.837**	0.613	<u>0.865</u>	**0.772**
ANS+ArabicFC	0.807	0.866	0.899	0.857	0.678	0.587	0.862	0.709
ANS+AraStance	0.893	0.909	0.955	0.919	0.743	0.629	0.847	0.740
ArabicFC+AraStance	0.765	0.555	0.897	0.739	<u>0.754</u>	<u>0.635</u>	0.862	0.750
All three datasets	0.778	0.742	0.889	0.803	0.741	**0.646**	**0.866**	<u>0.751</u>

Failure Analysis. We conducted a failure analysis on 17 examples from AuSTR that failed to be predicted correctly by *all* of our 7 trained models. We found

---

[5] Claims are collected from sources other than the ones we used to construct AuSTR.

that we can attribute the failures to two main reasons: (1) *Writing Style*, where the authority is denying a rumor about herself speaking in the first person. This constitutes 64.7% of the examined failures. We believe this is due to the fact that none of the stance datasets we used for training have evidence written by authorities themselves, as the source was either news articles written by journalists, or paraphrased or contradicted news headlines manually crafted by annotators. (2) *Indirect Disagreement/Agreement*, where the authority is indirectly denying/supporting the rumor. Examples of both types of failures are presented in Table 3. These findings motivate the need to augmenting existing stance datasets with rumor-evidence pairs from Twitter to further improve the performance of detecting the stance of authorities towards rumors from their tweets.

Table 3. Sample examples failed to be predicted correctly by <u>all</u> models. Failure types are writing style, indirect disagreement, and indirect agreement in order.

Rumor tweet [posting date]	[Stance] evidence tweet [posting date]
Mortada Mansour passed away recently of a heart attack. [29-10-2021]	**[Disagree]@Mortada5Mansour**: I am having my dinner now, and after a few minutes I will share a voice and video to reassure you, and I will reply to those who disturbed my family members in my village and caused the anxiety to all my fans. [29-10-2021]
Egypt does not give a vaccine to its citizens, the Gulf countries sponsor them: Saudi Arabia/Sultanate of Oman/Qatar refuses their intervention, so there is no other than Kuwait, the country of humanity that receives them and feeds them. What is the mysterious secret? Kuwait treats Egypt with special treatment. [07-05-2021]	**[Disagree]@mohpegypt**: Information about the #coronavirus vaccine. To book a vaccine, please visit the website http://egcovac.mohp.gov.eg or go to the nearest health unit (for citizens who have difficulty registering online). For more information, please call the hotline: 15335 #together_rest_assured. [10-05-2021]
Urgent The headquarters of the fourth channel was stormed by the militias of the Sadrist movement in the capital, Baghdad. [04-11-2022]	**[Agree]@MAKadhimi**:The attack on one of the Iraqi media outlets, and the threat to the lives of its employees, is a reprehensible act and represents the highest level of transgression against the law and freedom of the press and does not fall within the peaceful and legal practices and protests. We directed that the perpetrators be held accountable, and that protection be tightened on press institutions. [04-11-2022]

5 Conclusion and Future Work

In this paper, we defined the task of detecting stance of authorities towards rumors in tweets, and released the first dataset for the task targeting Arabic rumors. We studied the usefulness of existing Arabic datasets for stance detection for claim verification in our task. Based on our experiments and failure analysis, we found that although existing stance datasets showed to be somewhat useful for the task, they are obviously insufficient and there is a need to augment them with stance of authorities from Twitter data. In addition to expanding AuSTR to have sufficient training data for the task that can be used solely or to augment existing stance datasets, we plan to explore and contribute with stance models specific to the task.

Acknowledgments. The work of Fatima Haouari was supported by GSRA grant# GSRA6-1-0611-19074 from the Qatar National Research Fund. The work of Tamer Elsayed was made possible by NPRP grant# NPRP11S-1204-170060 from the Qatar National Research Fund (a member of Qatar Foundation). The statements made herein are solely the responsibility of the authors.

References

1. Abdul-Mageed, M., Elmadany, A., et al.: ARBERT & MARBERT: deep bidirectional transformers for Arabic. In: Proceedings of the 59th Annual Meeting of the Association for Computational Linguistics and the 11th International Joint Conference on Natural Language Processing (Volume 1: Long Papers), pp. 7088–7105 (2021)
2. Al-Yahya, M., Al-Khalifa, H., Al-Baity, H., AlSaeed, D., Essam, A.: Arabic fake news detection: comparative study of neural networks and transformer-based approaches. Complexity **2021**, 1–10 (2021)
3. Alhindi, T., Alabdulkarim, A., Alshehri, A., Abdul-Mageed, M., Nakov, P.: AraStance: a multi-country and multi-domain dataset of Arabic stance detection for fact checking. In: NLP4IF 2021, p. 57 (2021)
4. Ali, Z.S., Mansour, W., Elsayed, T., Al-Ali, A.: AraFacts: the first large Arabic dataset of naturally occurring claims. In: Proceedings of the Sixth Arabic Natural Language Processing Workshop, pp. 231–236 (2021)
5. Alqurashi, S., Hamoui, B., Alashaikh, A., Alhindi, A., Alanazi, E.: Eating garlic prevents COVID-19 infection: detecting misinformation on the Arabic content of Twitter. arXiv preprint arXiv:2101.05626 (2021)
6. Alqurashi, T.: Stance analysis of distance education in the Kingdom of Saudi Arabia during the COVID-19 pandemic using Arabic Twitter data. Sensors **22**(3), 1006 (2022)
7. Antoun, W., Baly, F., Hajj, H.: AraBERT: transformer-based model for Arabic language understanding. In: LREC 2020 Workshop Language Resources and Evaluation Conference, 11–16 May 2020, p. 9 (2020)
8. Bai, N., Meng, F., Rui, X., Wang, Z.: A multi-task attention tree neural net for stance classification and rumor veracity detection. Appl. Intell. 1–11 (2022)
9. Bai, N., Meng, F., Rui, X., Wang, Z.: Rumor detection based on a Source-Replies conversation Tree Convolutional Neural Net. Computing **104**(5), 1155–1171 (2022)

10. Baly, R., Mohtarami, M., Glass, J., Màrquez, L., Moschitti, A., Nakov, P.: Integrating stance detection and fact checking in a unified corpus. In: Proceedings of the 2018 Conference of the North American Chapter of the Association for Computational Linguistics: Human Language Technologies, vol. 2 (Short Papers), pp. 21–27. Association for Computational Linguistics, New Orleans, Louisiana, June 2018

11. Barrón-Cedeño, A., et al.: The CLEF-2023 CheckThat! Lab: checkworthiness, subjectivity, political bias, factuality, and authority of news articles and their sources. In: Proceedings of the 45th European Conference on Information Retrieval (ECIR 2023) (2023)

12. Bian, T., et al.: Rumor detection on social media with bi-directional graph convolutional networks. In: Proceedings of the AAAI Conference on Artificial Intelligence, pp. 549–556 (2020)

13. Chen, L., Wei, Z., Li, J., Zhou, B., Zhang, Q., Huang, X.J.: Modeling evolution of message interaction for rumor resolution. In: Proceedings of the 28th International Conference on Computational Linguistics, pp. 6377–6387 (2020)

14. Choi, J., Ko, T., Choi, Y., Byun, H., Kim, C.K.: Dynamic graph convolutional networks with attention mechanism for rumor detection on social media. Plos One **16**(8), e0256039 (2021)

15. Darwish, K., Magdy, W., Zanouda, T.: Improved stance prediction in a user similarity feature space. In: Proceedings of the 2017 IEEE/ACM International Conference on Advances in Social Networks Analysis and Mining 2017, pp. 145–148 (2017)

16. Devlin, J., Chang, M.W., Lee, K., Toutanova, K.: BERT: Pre-training of Deep Bidirectional Transformers for Language Understanding. arXiv preprint arXiv:1810.04805 (2018)

17. Dougrez-Lewis, J., Kochkina, E., Arana-Catania, M., Liakata, M., He, Y.: PHEMEPlus: enriching social media rumour verification with external evidence. In: Proceedings of the Fifth Fact Extraction and VERification Workshop (FEVER), pp. 49–58 (2022)

18. Elhadad, M.K., Li, K.F., Gebali, F.: COVID-19-FAKES: a Twitter (Arabic/English) dataset for detecting misleading information on COVID-19. In: Barolli, L., Li, K.F., Miwa, H. (eds.) INCoS 2020. AISC, vol. 1263, pp. 256–268. Springer, Cham (2021). https://doi.org/10.1007/978-3-030-57796-4_25

19. Haouari, F., Hasanain, M., Suwaileh, R., Elsayed, T.: ArCOV19-rumors: Arabic COVID-19 Twitter dataset for misinformation detection. In: Proceedings of the Sixth Arabic Natural Language Processing Workshop, pp. 72–81 (2021)

20. Hasanain, M., et al.: Overview of CheckThat! 2020 Arabic: automatic identification and verification of claims in social media. In: CLEF (2020)

21. Inoue, G., Alhafni, B., Baimukan, N., Bouamor, H., Habash, N.: The interplay of variant, size, and task type in Arabic pre-trained language models. In: Proceedings of the Sixth Arabic Natural Language Processing Workshop, pp. 92–104 (2021)

22. Jaziriyan, M.M., Akbari, A., Karbasi, H.: ExaASC: a general target-based stance detection corpus in Arabic language. In: 2021 11th International Conference on Computer Engineering and Knowledge (ICCKE), pp. 424–429. IEEE (2021)

23. Khouja, J.: Stance prediction and claim verification: an Arabic perspective. In: Proceedings of the Third Workshop on Fact Extraction and VERification (FEVER). Association for Computational Linguistics, Seattle, USA (2020)

24. Kumar, S., Carley, K.: Tree LSTMs with convolution units to predict stance and rumor veracity in social media conversations. In: Proceedings of the 57th Annual Meeting of the Association for Computational Linguistics. Association for Computational Linguistics, Florence, Italy, July 2019

25. Lan, W., Chen, Y., Xu, W., Ritter, A.: An empirical study of pre-trained trans-formers for Arabic information extraction. In: Proceedings of the 2020 Conference on Empirical Methods in Natural Language Processing (EMNLP), pp. 4727–4734. Association for Computational Linguistics, Online, November 2020
26. Liu, Y., Wu, Y.F.B.: Early detection of fake news on social media through prop-agation path classification with recurrent and convolutional networks. In: Thirty-Second AAAI Conference on Artificial Intelligence (2018)
27. Ma, J., Gao, W., Wong, K.F.: Rumor detection on Twitter with tree-structured recursive neural networks. In: Proceedings of the 56th Annual Meeting of the Asso-ciation for Computational Linguistics (Volume 1: Long Papers), pp. 1980–1989 (2018)
28. Mahlous, A.R., Al-Laith, A.: Fake news detection in Arabic tweets during the COVID-19 pandemic. Int. J. Adv. Comput. Sci. Appl. **12**(6), 778–788 (2021)
29. Roy, S., Bhanu, M., Saxena, S., Dandapat, S., Chandra, J.: gDART: improving rumor verification in social media with discrete attention representations. Inf. Pro-cess. Manage. **59**(3), 102927 (2022)
30. Safaya, A., Abdullatif, M., Yuret, D.: KUISAIL at SemEval-2020 task 12: BERT-CNN for offensive speech identification in social media. In: Proceedings of the Fourteenth Workshop on Semantic Evaluation, pp. 2054–2059. International Com-mittee for Computational Linguistics, Barcelona (Online), December 2020
31. Sawan, A., Thaher, T., Abu-el-rub, N.: Sentiment analysis model for fake news identification in Arabic tweets. In: 2021 IEEE 15th International Conference on Application of Information and Communication Technologies (AICT), pp. 1–6 (2021)
32. Song, C., Shu, K., Wu, B.: Temporally evolving graph neural network for fake news detection. Inf. Process. Manage. **58**(6), 102712 (2021)
33. Wu, L., Rao, Y., Jin, H., Nazir, A., Sun, L.: Different absorption from the same sharing: sifted multi-task learning for fake news detection. In: Proceedings of the 2019 Conference on Empirical Methods in Natural Language Processing and the 9th International Joint Conference on Natural Language Processing (EMNLP-IJCNLP). Association for Computational Linguistics, Hong Kong, China, Novem-ber 2019
34. Yu, J., Jiang, J., Khoo, L.M.S., Chieu, H.L., Xia, R.: Coupled hierarchical trans-former for stance-aware rumor verification in social media conversations. In: Pro-ceedings of the 2020 Conference on Empirical Methods in Natural Language Processing (EMNLP), pp. 1392–1401. Association for Computational Linguistics, Online, November 2020

Leveraging Comment Retrieval for Code Summarization

Shifu Hou[1], Lingwei Chen[2], Mingxuan Ju[1], and Yanfang Ye[1(✉)]

[1] University of Notre Dame, Notre Dame, IN 46556, USA
{shou,mju2,yye7}@nd.edu
[2] Wright State University, Dayton, OH 45435, USA
lingwei.chen@wright.edu

Abstract. Open-source code often suffers from mismatched or missing comments, leading to difficult code comprehension, and burdening software development and maintenance. In this paper, we design a novel code summarization model CodeFiD to address this laborious challenge. Inspired by retrieval-augmented methods for open-domain question answering, CodeFiD first retrieves a set of relevant comments from code collections for a given code, and then aggregates presentations of code and these comments to produce a natural language sentence that summarizes the code behaviors. Different from current code summarization works that focus on improving code representations, our model resorts to external knowledge to enhance code summarizing performance. Extensive experiments on public code collections demonstrate the effectiveness of CodeFiD by outperforming state-of-the-art counterparts across all programming languages.

Keywords: Code summarization · Comment retrieval · Heterogeneous graph neural network · Fusion-in-Decoder

1 Introduction

Software developers benefit from billions of lines of source code that reside in online repositories [13,30]. Due to social coding properties, code often suffers from comments being mismatched or missing [11,29]. This makes code comprehension more difficult, which could easily increase the burden of software development and maintenance [26]. Hence, correctly summarizing the code behaviors is important and useful. As it is very expensive to manually acquire high-quality summarization, automatic yet effective code summarization pipelines are needed to address this laborious challenge.

Automatic code summarization is a rapidly expanding research area. Retrieval approaches were first proposed as a practice to exploit code keywords and similarity [25,32], which are limited to code formulation and easily fail when identifiers and methods are poorly named. Inspired by natural machine translation (NMT) from natural language processing (NLP), sequence-to-sequence (seq2seq) models then came to the forefront that read in the code as a sequence of tokens and generate a natural language sentence as a sequence of words [8,19,27]. As source code written in formal programming languages is syntactically structured [2], seq2seq models have recently adapted

© The Author(s), under exclusive license to Springer Nature Switzerland AG 2023
J. Kamps et al. (Eds.): ECIR 2023, LNCS 13981, pp. 439–447, 2023.
https://doi.org/10.1007/978-3-031-28238-6_34

to more advanced graph-to-sequence (graph2seq) models. They leverage code structure and context through abstract syntax tree or constituency parsing tree [18], to boost the effectiveness of NMT techniques on code summarization [3,9,17,34].

Though the seq2seq and graph2seq models provide successful principles to solve the ambiguities and expressiveness in both source code and natural language descriptions, their inputs are inherently self-contained and struggle to leverage any external knowledge. In other words, while attending to depict source code and learn higher-level code representations for summarization, this line of research rarely takes advantage of any other relevant supplementary contexts. Hence our goal here is to investigate how much code summarization can benefit from retrieving external resources.

Retrieval-augmented pipelines from other fields such as open-domain question answering explore a retriever-reader framework, where a set of relevant passages are retrieved to enhance the knowledge coverage for question answering [14,15]. Inspired by their huge success, some recent works [22,32] start to shift such retrieval-augmented paradigms to extract different external resources for code summarization, which, however, either fail to capture useful connections between code snippets using traditional Dense Passage Retrieval (DPR) [15], or lead to unsatisfying performance improvement by introducing noisy information from external resources.

To address these limitations, in this paper, we propose a novel model that resorts to passage-like contexts from the collected data for code summarization. More specifically, the extracted supporting contexts refer to available comments paired with source code in the large training data collections. We argue that these text comments that are analogous to passages may contain "evidence" to the source code. To this end, on top of the state-of-the-art reader Fusion-in-Decoder [14], we design a retriever-reader framework for code summarization, called CodeFiD, which is shown in Fig. 1. In our Code-FiD, a retriever selects top k relevant comments for a given code using dense representations, where we deploy heterogeneous graph [7] and in-batch negatives training [15] to fully leverage cross-fertilization of source code and comments. Then an FiD reader takes the source code along with its retrieved comments as inputs and aggregates their presentations to produce the final code summary.

2 Notations and Problem Definition

Code Summarization. A given code is denoted as a token sequence $x = (x_1, x_2, \ldots, x_n)$. A code summarization model is based on encoder-decoder architecture [10], where the encoder maps the sequence of tokens to a sequence of representations $\mathbf{z} = (\mathbf{z}_1, \mathbf{z}_2, \ldots, \mathbf{z}_n)$; the decoder produces the output natural language sentence $y = (y_1, y_2, \ldots, y_m)$ by maximizing the conditional probability $p(y_1, y_2, \ldots, y_m \mid \mathbf{z})$, such that:

$$y^* = \underset{y}{\mathrm{argmax}} \sum_{t=1}^{m} \log p(y_t \mid y_{<t}, \mathbf{z}) \tag{1}$$

In this paper, instead of introducing syntactic structure to facilitate code representation learning, we rely on external knowledge with respect to relevant comments from collections to supplement code and boost its summarizing performance.

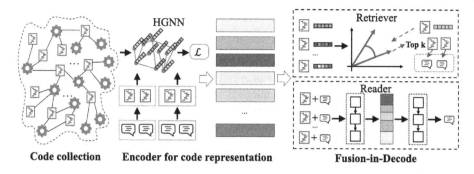

Fig. 1. In CodeFiD, a set of relevant comments are selected by HGNN-based retriever; then reader takes the code and retrieved comments to generate the summary.

Comment Retrieval. Given a code x and a large set of comments $\mathcal{C}$, comment retrieval is to compute the similarity between x and $\mathcal{C}$ using a similarity measuring function f in order to retrieve k ($k \geq 1$) comments $c_k \in \mathcal{C}$ of which representation vectors are the closest to the code vector:

$$c_k = \begin{cases} \text{argmax}_{c \in \mathcal{C}} \ f(\mathbf{x}, \mathbf{c}) & k = 1 \\ \text{argmax}_{c \in \mathcal{C}, f(\mathbf{x}, \mathbf{c}) < f(\mathbf{x}, \mathbf{c}_{k-1})} \ f(\mathbf{x}, \mathbf{c}) & k > 1 \end{cases} \quad (2)$$

where $\mathbf{x} \in \mathbb{R}^d$ and $\mathbf{c} \in \mathbb{R}^d$ are representation vectors for the code x and the comment c, respectively. In this paper, we define the similarity function f between x and c using dot product of their vectors, which has been widely used in retrieval research [15].

3 Proposed Model

3.1 Retriever

For the retrieval of supporting comments, a typical way is to train encoders to jointly embed code and comments into unified vector space by minimizing a ranking loss with positive and negative $\langle x, c \rangle$ pairs as training instances [4,27]. Two code examples are given as follows, where code 1's comment is "*Parses the kml file and updates Google transit feed object with the extracted information*", and code 2's comment is "*Parses the given kml dom tree and updates Google transit feed object*". Though two blocks of code are very different, their comments are close to each other, as code 1 invokes an API defined by code 2 with some sharing identifiers. However, joint embedding paradigm may not be able to effectively catch such connections between code snippets.

```
def Parse(self, filename, feed): #code 1
    dom = minidom.parse(filename)
    self.ParseDom(dom, feed)

def ParseDom(self, dom, feed): #code 2
    shape_num = 0
    for n in dom.getElementsByTagName('Placemark'):
        p = self.ParsePlacemark(n)
        if p.IsPoint():
```

```
        self.stopNameRe.search(p.name)
    elif p.IsLine():
        self.ConvertPlacemarkToShape(p, feed)
```

To solve this issue, here we design a new yet more structured retriever to fully lever-age cross-fertilization of code and comments: (1) the code data is first abstracted as a heterogeneous graph to model code interactions; (2) code representations are propagated and updated over this graph, which is completely guided by comment similarity, and then (3) top k pieces of relevant training code are selected for the given code using the learned representations, where k comments paired with these selected codes are finally retrieved to facilitate code summarization.

Encoder Using Heterogeneous Graph. As aforementioned, code pieces are related to each other through APIs and identifiers. Considering that code, APIs, and identifiers are of different types, we elaborate a heterogeneous graph (HG) [6,7,28,33] to represent the code data. To avoid introducing unexpected noises into graph, we intuitively extract those meaningful APIs and identifiers for HG construction. Specifically, the HG derived from code data collection is denoted as $G = (V, E, \mathbf{X})$, where V is node set, E is edge set to connect nodes when APIs/identifiers are included in code, and $\mathbf{X} \in \mathbb{R}^{|V| \times d}$ is node feature matrix initialized using pretrained CodeBERT [8]. Through HG, it is easy to identify the relationships between any code pairs. Afterwards, we feed the resulting HG into a heterogeneous graph neural network (HGNN) $g_\theta(\cdot)$ [31] to learn the higher-level code representations $\mathbf{Z} = g_\theta(\mathbf{X}) \in \mathbb{R}^{|V| \times d'}$ that take advantage of heterogeneous neighborhood aggregation and code interactions, where d' is the embedding size.

Training. Training HGNN encoder is a metric learning problem [16], such that the similarity between code representations can be a good ranking function. To achieve this goal, we need positive and negative code pairs to minimize the loss, which are unavailable explicitly at this stage. As we aim to back-propagate comment similarity to guide the updates on code representations, we design the following formulation: for each code x and its comment c, any code from the training data whose pairing comments are k nearest neighbors of c is considered as a positive of x, and any code from the remaining is a negative of x. In this way, HGNN encoder can create a vector space such that similar comment pairs will enforce smaller distance between their code representations, while dissimilar comments will lead to large code representation discrepancy. To enable this positive and negative formulation, all comments are first mapped to embedding space using pretrained BERT [5] before fed to nearest neighbor searching.

Let $\mathcal{X} = \{\langle x_i, x_i^+, x_{i,1}^-, \ldots, x_{i,m}^- \rangle\}_{i=1}^n$ be the training data that consists of n instances, where each instance includes one code snippet to summarize, one positive code snippet as well as m negative code snippets. We can thus optimize the HGNN encoder by minimizing the following loss:

$$\mathcal{L}(\boldsymbol{\theta}) = -\frac{1}{n} \sum_{x \in \mathcal{X}} \log \frac{e^{f(\mathbf{z}, \mathbf{z}^+)}}{e^{f(\mathbf{z}, \mathbf{z}^+)} + \sum_{j=1}^m e^{f(\mathbf{z}, \mathbf{z}_j^-)}} \tag{3}$$

Since we define $f(\cdot, \cdot)$ as dot product, we can use in-batch negatives [15] to reuse the computations and expedite the training in a more effective manner.

Table 1. The performance (BLEU-4) of different summarization pipelines on all datasets.

Model	ALL	Ruby	JavaScript	Go	Python	Java	PHP
DistillCodeT5	20.01	15.75	16.42	20.21	20.59	20.51	26.58
PolyglotCodeBERT	19.06	14.75	15.80	18.77	18.71	20.11	26.23
CoTexT	18.55	14.02	14.96	18.86	19.73	19.06	24.68
CodeBERT	17.83	12.16	14.90	18.07	19.06	17.65	25.16
Seq2Seq	14.32	9.64	10.21	13.98	15.93	15.09	21.08
RENCOS	20.44	15.95	16.77	21.26	20.90	20.30	27.48
REDCODER	21.36	16.27	17.93	21.62	21.01	**22.94**	28.42
CodeFiD (Ours)	**22.24**	**16.97**	**18.52**	**23.05**	**22.40**	22.14	**30.21**
w/ random retriver	17.95	11.61	13.26	16.88	18.85	18.66	23.95
w/ codebert retriver	21.02	16.01	17.21	22.17	21.55	21.51	28.75
w/o retriver	18.25	13.82	14.35	18.42	19.35	19.10	24.08

Comment Retrieval. To retrieve relevant comments for a given code, we proceed with two steps based on code vectors output by HGNN encoder: (1) select k code snippets whose representations are the closest to the given code in the same way defined in Sect. 2; and then (2) directly retrieve the pairing comments from these k code snippets as augmentation to support code summarization.

3.2 Reader

As this paper focuses on the investigation of the retrieval-augmented benefit for code summarization, we directly use FiD [14] to perform this task, which is based on a T5 model pretrained on unsupervised data [24]. More specifically, each retrieved comment is concatenated with the code, and then fed to the encoder independently from other comments to derive k different embedding outputs. These outputs are all concatenated to be processed by the decoder using attention mechanism to generate the final code summary. Similar to open domain question answering implemented in FiD, though it is simple, this reader yields two significant advantages [14]: (1) scalable to large number of comments, and (2) effective to learn from multiple comments.

4   Experiments

4.1   Experimental Setup

Data. We test our CodeFiD model on the CodeSearchNet dataset [12], which includes 908,224 training corpus, 44,689 validation corpus and 52,561 test corpus. This dataset has six programming languages, including Go, Java, JavaScript, PHP, Python and Ruby.

Implementation Details. We set for the number of retrieved comments per code as $k = 10$. We also evaluate its impact in Sect. 4.3. The parameter settings of HGNN and FiD are directly taken from [14,31]. All the experiments are performed under servers

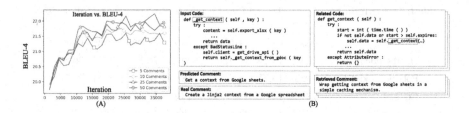

Fig. 2. Performance of CodeFiD regarding the number of retrieved comments and a case study.

equipped with one RTX A6000 48GB GPU. As for software, we use the public repository of FiD$_{base}$[1] for reader, and DGL[2] for HGNN-based retriever.

Evaluation Metrics. We use BLEU-4 score [21] to measure the quality of generated code summaries, which calculates the similarity (i.e., cumulative 4-gram precision) between the generated sequence and reference sequence.

4.2 Comparisons with Baselines

We evaluate our proposed model CodeFiD by comparisons with recent code summarization models, including DistillCodeT5 [20], PolyglotCodeBERT [1], CoTexT [23], CodeBERT [8], Seq2Seq [20], and two retrieval-augmented models RENCOS [32] and REDCODER [22]. The results are reported in Table 1. We can observe that using retrieval yields significant performance gains. Despite using T5 network as encoder and decoder, CodeFiD enables retrieved comments augmented to code input to outperform existing state-of-the-art models. The best performing baselines are DistillCodeT5 (non-retrieval) and REDCODER (retrieval-augmented), where CodeFiD delivers an average improvement of 2.23 BLEU-4 score from DistillCodeT5 and further 0.88 BLEU-4 score from REDCODER across all programming languages.

4.3 Impact of Number of Retrieved Comments

We conduct the sensitivity analysis of how different choices of number of retrieved comments k choices will affect the code summarization performance of CodeFiD. This evaluation is performed on single Python corpus. As illustrated in Fig. 2(A), when we enlarge k from 5 to 50, the performance difference is trivial at lower steps, while the BLEU-4 score tends to rise to a higher level for larger retrieved comment number, especially at latter epochs. Considering that larger k requires higher training computational budget, $k = 10$ seems a good trade-off between effectiveness and efficiency, whose average runtime for a batch (40 instances) costs 3.25 s.

---

[1] github.com/facebookresearch/FiD.
[2] www.dgl.ai.

4.4 Ablation Study

We also conduct ablation study to investigate the component contributions to CodeFiD performance. We formulate three alternative models, which are illustrated in Table 1. We can see HGNN retriever plays a crucial role in our model, which improves an average 3.99 BLUE score against the model without retriever. Random retriever underperforms by introducing irrelevant contexts that degrades code representations. CodeBERT retriever is promising, but fails to process the cases that rely on code interactions. Such a case is shown in Fig. 2(B), where CodeFiD benefits from structured retriever to locate the related code and retrieve its comment, which in turn provides the necessary evidence to produce the correct summary.

5   Conclusion

In this paper, we propose CodeFiD with a retriever-reader framework for code summarization. Specifically, our HGNN-based retriever selects a set of highly relevant comments, and then an FiD reader takes the source code along with its retrieved comments as inputs and aggregates their presentations to produce the final code summary. Extensive experiments on public code collections demonstrate the effectiveness of CodeFiD which outperforms state-of-the-art baselines. The improvement entailed by CodeFiD indicates that external knowledge, such as relevant comments from other code exploited in this paper, is beneficial for code summarization, which sheds light on a new direction for improving code summarization performance.

Acknowledgments. This work is partially supported by the NSF under grants IIS-2209814, IIS-2203262, IIS-2214376, IIS-2217239, OAC-2218762, CNS-2203261, CNS-2122631, CMMI-2146076, and the NIJ 2018-75-CX-0032. Any opinions, findings, and conclusions or recommendations expressed in this material are those of the authors and do not necessarily reflect the views of any funding agencies.

References

1. Ahmed, T., Devanbu, P.: Multilingual training for software engineering. arXiv preprint arXiv:2112.02043 (2021)
2. Allamanis, M., Barr, E.T., Devanbu, P., Sutton, C.: A survey of machine learning for big code and naturalness. ACM Comput. Surv. (CSUR) **51**(4), 1–37 (2018)
3. Alon, U., Brody, S., Levy, O., Yahav, E.: code2seq: generating sequences from structured representations of code. In: International Conference on Learning Representations (ICLR) (2019)
4. Chen, L., Hou, S., Ye, Y., Xu, S.: Attributed heterogeneous information network embedding for code retrieval. In: Heterogeneous Information Network Analysis and Applications (2021)
5. Devlin, J., Chang, M.W., Lee, K., Toutanova, K.: BERT: pre-training of deep bidirectional transformers for language understanding. In: Proceedings of NAACL-HLT (2019)
6. Fan, Y., Hou, S., Zhang, Y., Ye, Y., Abdulhayoglu, M.: Gotcha-sly malware! scorpion a metagraph2vec based malware detection system. In: Proceedings of the 24th ACM SIGKDD International Conference on Knowledge Discovery & Data Mining, pp. 253–262 (2018)

7. Fan, Y., Ju, M., Hou, S., Ye, Y., Wan, W., Wang, K., Mei, Y., Xiong, Q.: Heterogeneous temporal graph transformer: An intelligent system for evolving android malware detection. In: Proceedings of the 27th ACM SIGKDD Conference on Knowledge Discovery & Data Mining. pp. 2831–2839 (2022)
8. Feng, Z., et al.: CodeBERT: a pre-trained model for programming and natural languages. In: Findings of the Association for Computational Linguistics: EMNLP, pp. 1536–1547 (2020)
9. Hellendoorn, V.J., Sutton, C., Singh, R., Maniatis, P., Bieber, D.: Global relational models of source code. In: International conference on learning representations (2019)
10. Hou, S., Chen, L., Ye, Y.: Summarizing source code from structure and context. In: 2022 International Joint Conference on Neural Networks (IJCNN), pp. 1–8. IEEE (2022)
11. Hu, X., Li, G., Xia, X., Lo, D., Jin, Z.: Deep code comment generation. In: ICPC, pp. 200–210. IEEE (2018)
12. Husain, H., Wu, H.H., Gazit, T., Allamanis, M., Brockschmidt, M.: CodeSearchnet challenge: evaluating the state of semantic code search. arXiv preprint arXiv:1909.09436 (2019)
13. Iyer, S., Konstas, I., Cheung, A., Zettlemoyer, L.: Summarizing source code using a neural attention model. In: ACL, pp. 2073–2083 (2016)
14. Izacard, G., Grave, E.: Leveraging passage retrieval with generative models for open domain question answering. In: Proceedings of the 16th Conference of the European Chapter of the Association for Computational Linguistics, pp. 874–880 (2021)
15. Karpukhin, V., et al.: Dense passage retrieval for open-domain question answering. In: Proceedings of the 2020 Conference on Empirical Methods in Natural Language Processing (EMNLP), pp. 6769–6781 (2020)
16. Kulis, B., et al.: Metric learning: a survey. Found. Trends® Mach. Learn. 5(4), 287–364 (2013)
17. LeClair, A., Haque, S., Wu, L., McMillan, C.: Improved code summarization via a graph neural network. In: ICPC, pp. 184–195 (2020)
18. Ling, X., et al.: Deep graph matching and searching for semantic code retrieval. TKDD 15(5), 1–21 (2021)
19. Loyola, P., Marrese-Taylor, E., Matsuo, Y.: A neural architecture for generating natural language descriptions from source code changes. arXiv preprint arXiv:1704.04856 (2017)
20. Lu, S., et al.: CodeXglue: a machine learning benchmark dataset for code understanding and generation. arXiv preprint arXiv:2102.04664 (2021)
21. Papineni, K., Roukos, S., Ward, T., Zhu, W.J.: Bleu: a method for automatic evaluation of machine translation. In: Association for Computational Linguistics, pp. 311–318 (2002)
22. Parvez, M.R., Ahmad, W.U., Chakraborty, S., Ray, B., Chang, K.W.: Retrieval augmented code generation and summarization. arXiv preprint arXiv:2108.11601 (2021)
23. Phan, L., et al.: Cotext: Multi-task learning with code-text transformer. arXiv preprint arXiv:2105.08645 (2021)
24. Raffel, C., et al.: Exploring the limits of transfer learning with a unified text-to-text transformer. J. Mach. Learn. Res. 21(140), 1–67 (2020)
25. Rodeghero, P., McMillan, C., McBurney, P.W., Bosch, N., D'Mello, S.: Improving automated source code summarization via an eye-tracking study of programmers. In: ICSE, pp. 390–401 (2014)
26. Xia, X., Bao, L., Lo, D., Xing, Z., Hassan, A.E., Li, S.: Measuring program comprehension: a large-scale field study with professionals. IEEE Trans. Softw. Eng. 44(10), 951–976 (2017)
27. Yao, Z., Peddamail, J.R., Sun, H.: CoaCor: code annotation for code retrieval with reinforcement learning. In: The World Wide Web Conference, pp. 2203–2214 (2019)
28. Ye, Y., et al.: Out-of-sample node representation learning for heterogeneous graph in real-time android malware detection. In: 28th International Joint Conference on Artificial Intelligence (IJCAI) (2019)

29. Ye, Y., et al.: ICSD: an automatic system for insecure code snippet detection in stack over-flow over heterogeneous information network. In: Proceedings of the 34th Annual Computer Security Applications Conference, pp. 542–552 (2018)
30. Ye, Y., Li, T., Adjeroh, D., Iyengar, S.S.: A survey on malware detection using data mining techniques. ACM Comput. Surv. (CSUR) **50**(3), 1–40 (2017)
31. Zhang, C., Song, D., Huang, C., Swami, A., Chawla, N.V.: Heterogeneous graph neural network. In: Proceedings of the 25th ACM SIGKDD International Conference on Knowledge Discovery & Data Mining, pp. 793–803 (2019)
32. Zhang, J., Wang, X., Zhang, H., Sun, H., Liu, X.: Retrieval-based neural source code sum-marization. In: 2020 IEEE/ACM 42nd International Conference on Software Engineering (ICSE), pp. 1385–1397. IEEE (2020)
33. Zhao, J., Wang, X., Shi, C., Hu, B., Song, G., Ye, Y.: Heterogeneous graph structure learning for graph neural networks. In: Proceedings of the AAAI Conference on Artificial Intelli-gence, vol. 35, pp. 4697–4705 (2021)
34. Zügner, D., Kirschstein, T., Catasta, M., Leskovec, J., Günnemann, S.: Language-agnostic representation learning of source code from structure and context. arXiv preprint arXiv:2103.11318 (2021)

CPR: Cross-Domain Preference Ranking with User Transformation

Yu-Ting Huang[1], Hsien-Hao Chen[2], Tung-Lin Wu[3], Chia-Yu Yeh[3],
Jing-Kai Lou[4], Ming-Feng Tsai[2], and Chuan-Ju Wang[3]($\boxtimes$)

[1] Graduate Program of Data Science, National Taiwan University and Academia
Sinica, Taipei, Taiwan
r11946008@ntu.edu.tw
[2] National Chengchi University, Taipei, Taiwan
mftsai@nccu.edu.tw
[3] Academia Sinica, Taipei, Taiwan
{howard0100000,agiblida,cjwang}@citi.sinica.edu.tw
[4] KKStream Limited, Taipei, Taiwan
kaelou@kkstream.com

Abstract. Data sparsity is a well-known challenge in recommender systems. One way to alleviate this problem is to leverage knowledge from relevant domains. In this paper, we focus on an important real-world scenario in which some users overlap two different domains but items of the two domains are distinct. Although several studies leverage side information (e.g., user reviews) for cross-domain recommendation, side information is not always available or easy to obtain in practice. To this end, we propose cross-domain preference ranking (CPR) with a simple yet effective user transformation that leverages *only* user interactions with items in the source and target domains to transform the user representation. Given the proposed user transformation, CPR not only successfully enhances recommendation performance for users having interactions with target-domain items but also yields superior performance for cold-start users in comparison with state-of-the-art cross-domain recommendation approaches. Extensive experiments conducted on three pairs of cross-domain recommendation datasets demonstrate the effectiveness of the proposed method in comparison with existing cross-domain recommendation approaches. Our codes are available at https://github.com/cnclabs/codes.crossdomain.rec.

Keywords: Cross-domain recommendation · Cold-start problem ·
Collaborative filtering

1 Introduction

With the continued growth of online services, recommender systems have become ubiquitous, playing an essential role in providing quality information to users. In general, recommendation algorithms can be divided into content-based filtering and collaborative filtering, among which matrix factorization (MF) [9,14] and its derivative structures [5,20] have demonstrated their superior performance in some scenarios. However, in many real-world cases, existing approaches are unsatisfactory due to data sparsity and the cold-start problem.

One way to address these challenges is usually termed *cross-domain recommendation*, which relies on the concept of knowledge transfer [8,15,23]. Typically, given a set of user-item interactions from multiple domains, interactions from domains with richer information, the *source domains*, are used to improve the recommendation effectiveness of the domains with sparser information, the *target domains*. The primary assumption behind this kind of method is that knowledge discovered from the source domains is to some extent correlated to that in the target domains. Over the past few years, various cross-domain recommendation algorithms have been proposed [3,19,21], which transfer latent factors learned from the source domain into the target domain. However, these focus mainly on overlapping item scenarios, for instance, movie recommendation via knowledge transfer between the Movielens and Netflix datasets.

Nevertheless, the shared-user scenario is of vital importance, as the shared-user scenario reflects many online platforms that provide more than one service or type of content to users. For example, companies like Amazon and Apple provide streaming media and music services (e.g., Amazon Prime Video and Amazon Music). The "one user one account" tendency for different services makes it easier to leverage user-item interactions from different domains to improve recommendation performance. One reason why some studies pay little attention to this vital problem is the availability of such shared-user data: it is always privately owned by companies. Moreover, existing studies for such shared-user cross-domain recommendation usually leverage additional side information (e.g., text or metadata) to boost performance [11,19,21,22]; however, side information is not always available or easy to obtain in practice. As a result, several models, including that proposed in this paper, focus on purely leveraging user-item interactions to build the recommendation models [7,12,13,16]. However, these models do not satisfactorily address one of the most critical problems in cross-domain recommendation—the cold-start problem.

To this end, we propose a simple yet effective cross-domain preference ranking (CPR) algorithm with *user transformation*, inspired by TransE [1] and TransRec [4]. Specifically, CPR leverages user-item interactions in both the source and target domains to transform user representations for each user. Such a transformed representation consists of three components: (a) a pseudo user representation, (b) a user representation generated from the interactions with the source domain's items, and (c) a user representation generated from the interactions with the target domain's items. Note that the proposed user transformation addresses the cold-start problem in a simple yet elegant manner: at the inference stage, for a cold-start user (i.e., a user with item interactions in the source domain only), the representation of the user for recommending items in the target domain involves only component (b). We conduct extensive experiments on three pairs of publicly real-world cross-domain recommendation datasets, the results of which demonstrate that CPR not only successfully enhances the recommendation quality for users having interactions with target-domain items but also significantly outperforms existing methods for cold-start users.

2   Methodology

2.1   Problem Definition

In this work, without loss of generality, we consider the recommendation scenario involving two domains with disjoint item sets, namely, a source-domain item set and a target-domain item set(denoted as I^S and I^T, respectively); there exists a set of users having interactions with items from both domains, namely *shared users*. Formally, we denote the set of users having interactions with items in I^S (I^T) as U^S (U^T, respectively) and the shared users as $U^{\text{shared}} = U^S \cap U^T$ and $U^{\text{shared}} \neq \emptyset$.

Let $I = I^S \cup I^T$ and $U = U^S \cup U^T$. The goal of the proposed CPR approach is to learn the representation matrix $\Theta \in \mathbb{R}^{(|U|+|I|) \times d}$ mapping each user and item to a d-dimensional embedding vector. The learned embedding vectors enable us 1) to enhance the recommendation performance for users in U^{shared} by leveraging user-item interactions from the source domain and 2) to obtain satisfactory recommendation lists of items in the target domain for so-called cold-start users, i.e., users having interactions with items in the source domain only. Accordingly, in our later experiments, we separately evaluate the recommendation performance for the following three sets of users: 1) target users, U^T 2) shared users, U^{shared} and 3) cold-start users, $U^{\text{cold}} = U^S \setminus U^{\text{shared}}$.

2.2   Proposed CPR Approach

Given a user u, let I_u^S (I_u^T) denote the set of items in the source domain (target domain, respectively) that u has interacted with. The proposed CPR approach models the relations among I^S, U^S, I^T, U^T. Here, we use the concept of preference ranking [18] to describe such complex relations, for which the objective is to find an embedding matrix Θ that maximizes the posterior probability: $p(\Theta| >_{u,s,t}) \propto p(>_{u,s,t} |\Theta)p(\Theta)$, where $>_{u,s,t}$ indicates the preference structure between two items for the given u, s, t, where $u \in U^T$, $s \in I_u^S$, and $t \in I_u^T$.

To transfer knowledge from the source domain into the target domain, we bridge the non-overlapped I^S and I^T with the following user representation transformation: for each user $u \in U$, we have

$$\Theta_u = f(\Theta_u^{\text{pseudo}}, a_{I_u^S}, a_{I_u^T}), \tag{1}$$

in which Θ_u^{pseudo} denotes a learnable pseudo user representation for user u, $a_{I_u^S} = 1/|I_u^S| \sum_{i \in I_u^S} \Theta_i$, and $a_{I_u^T} = 1/|I_u^T| \sum_{i \in I_u^T} \Theta_i$, where Θ_u and Θ_i denote the representation for user $u \in U$ and item $i \in I$ respectively. Note that function f in Eq. (1) can be an arbitrary function such as summation, concatenation, or even neural networks with non-linear layers. For simplicity, in this paper, we choose summation, combining the three components as

$$\Theta_u = \Theta_u^{\text{pseudo}} + \boldsymbol{a}_{I_u^{\text{S}}} + \boldsymbol{a}_{I_u^{\text{T}}}. \tag{2}$$

With the transformation in Eq. (2), we formulate the maximum posterior estimator to derive our optimization criterion for CPR as

$$\text{CPR-OPT} := p(>_{u,s,t}|\Theta)p(\Theta)$$

$$= \ln \prod_{u \in U^{\text{T}}} \prod_{s \in I_u^{\text{S}}} \prod_{t \in I_u^{\text{T}}} \prod_{\substack{t^+ \in I_u^{\text{T}} \\ t^- \in I^{\text{T}} \setminus I_u^{\text{T}}}} p(t^+ >_{u,s,t} t^-|\Theta)p(\Theta)$$

$$= \sum_{u \in U^{\text{T}}} \sum_{s \in I_u^{\text{S}}} \sum_{t \in I_u^{\text{T}}} \sum_{\substack{t^+ \in I_u^{\text{T}} \\ t^- \in I^{\text{T}} \setminus I_u^{\text{T}}}} \ln \sigma \left(\langle \Theta_u, (\Theta_{t^+} - \Theta_{t^-}) \rangle \right) - \lambda ||\Theta||^2, \tag{3}$$

where $\sigma(\cdot)$ denotes the sigmoid function, $\langle \cdot, \cdot \rangle$ denotes the inner product for two vectors, and λ is a regularization parameter. Note that with Eq. (2), Eq. (3) can be decomposed into the following three components:

(a) $\langle \Theta_u^{\text{pseudo}}, (\Theta_{t^+} - \Theta_{t^-}) \rangle$, to model the item preference ranking between t^+ and t^- for user u;
(b) $\langle \boldsymbol{a}_{I_u^{\text{S}}}, (\Theta_{t^+} - \Theta_{t^-}) \rangle$, to model the item similarity to the items averaged from the source domain for user u regarding items t^+ and t^-;
(c) $\langle \boldsymbol{a}_{I_u^{\text{T}}}, (\Theta_{t^+} - \Theta_{t^-}) \rangle$, to model the item similarity to the items averaged from the target domain for user u regarding items t^+ and t^-.

For computational efficiency, in the training process, we follow the strategy used by Chiang et al. [2] to deal with the average representations $\boldsymbol{a}_{I_u^{\text{S}}}$ and $\boldsymbol{a}_{I_u^{\text{T}}}$ by sampling one item from each of the domains for optimization. Note that for users having interactions with I^{T} only (i.e., $u \in U^{\text{T}} \setminus U^{\text{shared}}$), we sample items only from I_u^{T} as $I_u^{\text{S}} = \emptyset$. The objective in Eq. (3) is then maximized by adopting asynchronous stochastic gradient ascent (ASGD) [17] to efficiently update the embedding matrix Θ in parallel.

As we mainly model preference ranking for I^{T} (see Θ_{t^+} and Θ_{t^-} in (a), (b), and (c) above), we sample only those users who have interactions with items in the target domain (i.e., $u \in U^{\text{T}}$). Therefore, for the cold-start users, U^{cold}, as $I_u^{\text{T}} = \emptyset$ and Θ_u^{pseudo} is a zero-valued vector (see Eq. (2)), at the inference stage, we adopt $\Theta_u = \boldsymbol{a}_{I_u^{\text{S}}}$ to calculate the inner product with item representations to obtain the recommendation.

3 Experiments

3.1 Dataset and Experimental Setup

We evaluated our method on three real-world cross-domain recommendation datasets. Specifically, each pair of cross-domain datasets comprises a pair of

Amazon review datasets from two different but relevant domains: (1) HK-CSJ: "Home and Kitchen" and "Clothing, Shoes, and Jewelry"; (2) MT-B: "Movies and TV" and "Books"; (3) SPO-CSJ: "Sports and Outdoors" and "Clothing, Shoes, and Jewelry". Note that we used the official 5-core datasets, in which all users and items have at least 5 reviews; also, we used only the data from the latest two years. Additionally, we chose the higher-density domain as the source domain, and the other as the target domain, where density is defined as ($\frac{\text{interactions}}{\text{users} \times \text{items}}$). For each dataset, A-B, we denote the source domain as A and the target domain as B. The details and statistics of the three dataset pairs are summarized in Table 1.

Table 1. Dataset statistics

	HK-CSJ		MT-B		SPO-CSJ	
Users	107,325	180,008	18,526	224,867	29,391	180,008
Shared-user ratio	25.4%	15.1%	19.7%	1.6%	21.1%	3.4%
Items	40,513	63,757	10,828	123,899	14,230	63,757
Interactions	825,814	1,500,124	188,926	3,399,620	223,550	1,500,124
Density	0.0190%	0.0131%	0.0942%	0.0122%	0.0535%	0.0131%

For each dataset, we first sorted each user's logs according to the timestamps and then adopted the commonly used leave-one-out strategy for evaluation [6].[1] Recall that as mentioned in Sect. 2.1, we evaluated the recommendation performance for three sets of users: 1) target users; U^{T}; 2) shared users, U^{shared}; and 3) cold-start users, U^{cold}. Due to the large amount of data, we evaluated users sampled from the designated scenario; specifically, for each scenario, we sampled 3,500 users for each dataset pair. Note that when assessing the performance on cold-start users, instead of sampling users from U^{cold}, we sampled 3,500 users from U^{shared} and removed all their interactions with items in the target domain for training.[2] We compare CPR with four baselines; (1) Bayesian personalized ranking (**BPR**) [18] and (2) **LightGCN** [5], which are classic and state-of-the-art single-domain recommendation algorithms, respectively, and cross-domain recommendation algorithms (3) **EMCDR** [13] and (4) **Bi-TGCF** [12].

We used LightFM [10] for the BPR implementation with the embedding dimension of 100, the learning rate of 0.025, and the L_2 regularizer with $\lambda = 0.0001$. For EMCDR, the initial embeddings were the above BPR embeddings trained by LightFM and we kept all other settings as stated in the EMCDR paper. For LightGCN[3] and Bi-TGCF[4], we used the source codes provided by

---

[1] For each user, we reserved the latest interaction as the test item and randomly sampled 99 negative items that the user did not interact with; we then evaluated how well the model ranked the test item against the negative ones.

[2] We did this because there was no target-domain ground truth for users in U^{cold}.

[3] https://github.com/gusye1234/LightGCN-PyTorch.

[4] https://github.com/sunshinelium/Bi-TGCF.

Table 2. Test users from target users

	HK–CSJ		MT–B		SPO–CSJ	
	HR@10	NDCG@10	HR@10	NDCG@10	HR@10	NDCG@10
BPR	0.4403	0.3080	0.5254	0.3324	0.4289	0.2905
BPR$^+$	0.3674	0.2381	0.5203	0.3316	0.4006	0.2660
LightGCN	0.5117	0.3945	0.8454	0.6736	0.5077	0.3824
LightGCN$^+$	†0.5377	†0.4070	†0.8594	†0.6820	0.5217	0.3877
EMCDR	0.4106	0.2775	0.5166	0.3266	0.4266	0.2888
Bi-TGCF	0.5369	0.3939	0.8391	0.6424	†0.5520	†0.4020
CPR	*0.5677	*0.4290	*0.8954	*0.7145	**0.5534**	**0.4183**
Improv	5.58%	5.42%	4.19%	4.76%	0.26%	4.05%

the authors; in particular, for LightGCN, we maintained all the authors' original settings and only increased the mini-batch size to 4096 for speeding up. For Bi-TGCF, we set the embedding propagation layer to {64, 64, 64}, the learning rate to 0.001, the mini-batch size to 65536, the negative sampling ratio to 4, and the message dropout ratio to 0.1. All other settings followed those in the original paper. For CPR, we set the embedding size as 100, the learning rate of 0.025, and L_2 regularizer with $\lambda = 0.0025$. We adopted the early stopping strategy with a maximum epoch number of 200 for all baselines and our model; the training procedure stops if the performance has not been improved for five consecutive epochs.

3.2   Experimental Results

Tables 2, 3 and 4 tabulate the results for the target user, shared user and, cold-start user scenarios, respectively. In the tables, the best performance is in bold-face; '$\dagger$' indicates the best performing method among all the baselines; '*' and 'Improv. (%)' denote statistical significance at $p < 0.05$ with a paired t-test and the percentage improvement of our model, respectively, with respect to the best performing baseline. Note that for the two single-domain baselines, BPR and LightGCN, the plus symbol (i.e., BPR$^+$ and LightGCN$^+$) denotes that we used the user-item interactions from both source and target domains to train the models, whereas those without the plus symbol indicate models trained on interactions from the target domain only (which is the conventional single-domain recommendation). From Tables 2 to 4, we offer two main observations from the experiments: (1) CPR outperforms all baselines for most cases; remarkably, it significantly outperforms the baselines for the cold-start user scenario except for the MT-B dataset, the shared-user ratio of which is the lowest among the three dataset pairs; (2) Bi-TGCF serves as a strong cross-domain baseline for most of the scenarios, but surprisingly, in the shared-user scenario, we observe that LightGCN, a single-domain model, exhibited better performance than Bi-TGCF.

Table 3. Test users from shared users

	HK–CSJ		MT–B		SPO–CSJ	
	HR@10	NDCG@10	HR@10	NDCG@10	HR@10	NDCG@10
BPR	0.2837	0.1750	0.1874	0.1143	0.2249	0.1330
BPR$^+$	0.2560	0.1405	0.1874	0.1140	0.2186	0.1208
LightGCN	0.3520	0.2450	†0.4263	†0.3216	0.3803	0.2640
LightGCN$^+$	†0.3714	†0.2508	0.4160	0.3128	0.3674	0.2566
EMCDR	0.2566	0.1434	0.2089	0.1250	0.1680	0.0861
Bi-TGCF	0.3583	0.2368	0.4174	0.2925	†0.3900	†0.2662
CPR	*0.3929	*0.2729	*0.4594	*0.3441	*0.4154	*0.2929
Improv	5.77%	8.81%	7.77%	7.00%	6.52%	10.01%

Table 4. Test users from cold-start users

	HK–CSJ		MT–B		SPO–CSJ	
	HR@10	NDCG@10	HR@10	NDCG@10	HR@10	NDCG@10
BPR$^+$	0.2417	0.1327	0.1351	0.0810	0.1806	0.0942
LightGCN$^+$	0.1380	0.0748	0.0580	0.0287	0.1386	0.0833
EMCDR	†0.2514	†0.1407	†0.2034	†0.1203	0.1466	0.0762
Bi-TGCF	0.2477	0.1370	0.1211	0.0686	†0.2569	†0.1548
CPR	*0.3160	*0.1899	*0.1760	*0.1014	*0.3371	*0.2100
Improv	25.68%	34.90%	−13.48%	−15.68%	31.26%	35.62%

4   Conclusion

We present CPR, a cross-domain recommendation approach that tackles the shared-user scenario for recommendation; the proposed method leverages *only* the interactions of users to items in both source and target domains to transform user representations. The proposed user transformation addresses the cold-start problem in a simple yet elegant manner. With the proposed user transformation, CPR effectively addresses the cold-start problem while simultaneously improving overall recommendation performance.

References

1. Bordes, A., Usunier, N., Garcia-Durán, A., Weston, J., Yakhnenko, O.: Translating embeddings for modeling multi-relational data. In: Proceedings of the 26th International Conference on Neural Information Processing Systems, vol. 2, pp. 2787–2795 (2013)
2. Chiang, W., Liu, X., Si, S., Li, Y., Bengio, S., Hsieh, C.: Cluster-GCN: an efficient algorithm for training deep and large graph convolutional networks. In: Proceedings of the 25th ACM SIGKDD International Conference on Knowledge Discovery & Data Mining, pp. 257–266 (2019)

3. Gao, C., et al.: Cross-domain recommendation without sharing user-relevant data. In: Proceedings of the 30th International Conference on World Wide Web, pp. 491–502 (2019)
4. He, R., Kang, W., McAuley, J.: Translation-based recommendation. In: Proceedings of the 11th ACM Conference on Recommender Systems, pp. 161–169 (2017)
5. He, X., Deng, K., Wang, X., Li, Y., Zhang, Y., Wang, M.: LightGCN: simplifying and powering graph convolution network for recommendation. In: Proceedings of the 43rd International ACM SIGIR Conference on Research and Development in Information Retrieval, pp. 639–648 (2020)
6. He, X., Liao, L., Zhang, H., Nie, L., Hu, X., Chua, T.: Neural collaborative filtering. In: Proceedings of the 26th International Conference on World Wide Web, pp. 173–182 (2017)
7. Hu, G., Zhang, Y., Yang, Q.: CoNet: collaborative cross networks for cross-domain recommendation. In: Proceedings of the 27th ACM International Conference on Information and Knowledge Management, pp. 667–676 (2018)
8. Kang, S., Hwang, J., Lee, D., Yu, H.: Semi-supervised learning for cross-domain recommendation to cold-start users. In: Proceedings of the 28th ACM International Conference on Information and Knowledge Management, pp. 1563–1572 (2019)
9. Koren, Y., Bell, R., Volinsky, C.: Matrix factorization techniques for recommender systems. Computer **42**, 30–37 (2009)
10. Kula, M.: Metadata embeddings for user and item cold-start recommendations. In: Proceedings of the 2nd Workshop on New Trends on Content-Based Recommender Systems co-located with 9th ACM Conference on Recommender Systems, pp. 14–21 (2015)
11. Li, P., Tuzhilin, A.: DDTCDR: deep dual transfer cross domain recommendation. In: Proceedings of the 13th International Conference on Web Search and Data Mining, pp. 331–339 (2020)
12. Liu, M., Li, J., Li, G., Pan, P.: Cross domain recommendation via bi-directional transfer graph collaborative filtering networks. In: Proceedings of the 29th ACM International Conference on Information & Knowledge Management, pp. 885–894 (2020)
13. Man, T., Shen, H., Jin, X., Cheng, X.: Cross-domain recommendation: an embedding and mapping approach. In: Proceedings of the 26th International Joint Conference on Artificial Intelligence, pp. 2464–2470 (2017)
14. Mnih, A., Salakhutdinov, R.R.: Probabilistic matrix factorization. In: Proceedings of the 21th International Conference on Neural Information Processing Systems, pp. 1257–1264 (2007)
15. Pan, W., Liu, N.N., Xiang, E.W., Yang, Q.: Transfer learning to predict missing ratings via heterogeneous user feedbacks. In: Proceedings of the 22nd International Joint Conference on Artificial Intelligence, pp. 2318–2323 (2011)
16. Pan, W., Xiang, E., Liu, N., Yang, Q.: Transfer learning in collaborative filtering for sparsity reduction. In: Proceedings of the 24th AAAI Conference on Artificial Intelligence. vol. 24, pp. 230–235 (2010)
17. Recht, B., Re, C., Wright, S., Niu, F.: Hogwild!: a lock-free approach to parallelizing stochastic gradient descent. In: Proceedings of the 24th International Conference on Neural Information Processing Systems, vol. 24, pp. 693–701 (2011)
18. Rendle, S., Freudenthaler, C., Gantner, Z., Schmidt-Thieme, L.: BPR: Bayesian Personalized Ranking from Implicit Feedback. In: Proceedings of the 25th Conference on Uncertainty in Artificial Intelligence, pp. 452–461 (2009)
19. Wang, J., Lv, J.: Tag-informed collaborative topic modeling for cross domain recommendations. Knowl. Based Syst. **203**, 106119 (2020)

20. Wang, X., He, X., Wang, M., Feng, F., Chua, T.: Neural graph collaborative filtering. In: Proceedings of the 42nd International ACM SIGIR Conference on Research and Development in Information Retrieval, pp. 165–174 (2019)
21. Zhang, Q., Hao, P., Lu, J., Zhang, G.: Cross-domain recommendation with semantic correlation in tagging systems. In: 2019 International Joint Conference on Neural Networks, pp. 1–8 (2019)
22. Zhao, C., Li, C., Xiao, R., Deng, H., Sun, A.: CATN: cross-domain recommendation for cold-start users via aspect transfer network. In: Proceedings of the 43rd International ACM SIGIR Conference on Research and Development in Information Retrieval, pp. 229–238 (2020)
23. Zhu, F., Wang, Y., Chen, C., Zhou, J., Li, L., Liu, G.: Cross-domain Recommendation: Challenges, Progress, and Prospects. arXiv preprint arXiv:2103.01696 (2021)

ColBERT-FairPRF: Towards Fair Pseudo-Relevance Feedback in Dense Retrieval

Thomas Jaenich$^{(\boxtimes)}$, Graham McDonald, and Iadh Ounis

University of Glasgow, Glasgow, UK
t.jaenich.1@research.gla.ac.uk,
{graham.mcdonald,iadh.ounis}@glasgow.ac.uk

Abstract. Pseudo-relevance feedback mechanisms have been shown to be useful in improving the effectiveness of search systems for retrieving the most relevant items in response to a user's query. However, there has been little work investigating the relationship between pseudo-relevance feedback and fairness in ranking. Indeed, using the feedback from an initial retrieval to revise a query can in principle also allow to optimise objectives beyond relevance, such as the fairness of the search results. In this work, we show how a feedback mechanism based on the successful ColBERT-PRF model can be used for retrieving fairer search results. Therefore, we propose a novel fair feedback mechanism for multiple representation dense retrieval (ColBERT-FairPRF), which enhances the distribution of exposure over groups of documents in the search results by fairly extracting the feedback embeddings that are added to the user's query representation. To fairly extract representative embeddings, we apply a clustering approach since traditional methods based on counting are not applicable in the dense retrieval space. Our results on the 2021 TREC Fair Ranking Track test collection demonstrate the effectiveness of our method compared to ColBERT-PRF, with statistical significant improvements of up to ~19% in Attention Weighted Ranked Fairness. To the best of our knowledge, ColBERT-FairPRF is the first query expansion method for fairness in multiple representation dense retrieval.

1 Introduction

Extending a user's initial query with additional informative terms has been shown to be a useful mechanism for improving the effectiveness of search systems. Many approaches from the literature, for example [8,10,17,18], use pseudo-relevance feedback (PRF) for identifying such informative terms and expanding the user's query. Essentially, PRF approaches expand the initial query by appending it with the terms that are expected to be the most informative from the top-ranked documents (often referred to as the pseudo-relevant set) in an initial ranked retrieval. Recently, novel PRF approaches for dense retrieval, such as ColBERT-PRF [15], have been shown to be successful for improving the relevance of search results. In particular, ColBERT-PRF is the state-of-the-art PRF

© The Author(s), under exclusive license to Springer Nature Switzerland AG 2023
J. Kamps et al. (Eds.): ECIR 2023, LNCS 13981, pp. 457–465, 2023.
https://doi.org/10.1007/978-3-031-28238-6_36

model in multiple representation dense retrieval, where contextual embeddings for each token of a document and a query are leveraged. However, there has been little work investigating the relationship between PRF in dense retrieval and fairness in ranking. Indeed, using representative and discriminative embeddings from an initial retrieval to revise a query can in principle also allow to optimise objectives beyond relevance, such as the fairness of the search results. In this work, we investigate the effects of PRF in dense retrieval, in terms of the exposure that particular groups of documents, for example documents that discuss different geographic locations, receive in the search results. We propose a novel fair PRF mechanism for multiple representation dense retrieval, named ColBERT-FairPRF, that is effective for enhancing the fairness of the distribution of exposure over groups of documents in a ranking. In particular, we show that selecting feedback terms from each of the groups individually leads to the groups receiving a fairer exposure in the search results, without significantly affecting relevance. Our experiments on the 2021 TREC Fair Ranking Track test collection demonstrate the effectiveness of our proposed approach compared to ColBERT-PRF in providing a fair exposure to particular groups of documents. Our results show a significant improvement of ~19% in Attention Weighted Ranked Fairness [13] compared to ColBERT-PRF, without any significant decreases in utility as measured by nDCG. To the best of our knowledge, ColBERT-FairPRF is the first PRF method to investigate fairness in multiple representation dense retrieval. This work presents a first investigation into fair PRF for dense retrieval. Therefore we leave the comparison of ColBERT-FairPRF to other non-PRF based fairness approaches to future work.

2 Related Work

Pseudo-Relevance feedback (PRF) has been shown to be effective in increasing the retrieval performance in terms of relevance in many previous works, for example [8,10,17,18]. Traditional PRF models, such as Rocchio's algorithm [12], the RM3 relevance language model [1], or the Bo1 model from the Divergence from Randomness framework [2] used statistical information about the distributions of terms in the pseudo-relevant set and the larger collection of documents (that the pseudo-relevant set is retrieved from) for selecting informative terms. However, with the emergence of large pre-trained language models such as BERT [4], contextualised embeddings, known as *dense* representations, have become widely used for representing queries and documents. As such, the emergence of BERT-like retrieval approaches have sparked the development of new PRF methods for dense retrieval. In particular, ColBERT-PRF [15] which is based on ColBERT [9] has been shown to be effective for improving the retrieval performance of its original model. Colbert-PRF expands the user's initial query with additional contextual embeddings from the pseudo-relevant set. In this work, we investigate, how multiple representation dense retrieval based on ColBERT-PRF can be leveraged to create search results that are both relevant to the query and also provide a fair exposure to multiple *groups* of documents. A group is a set of documents that share a particular fairness characteristic of interest, for example documents that are about a particular geographic location.

There is little previous work that investigated the relationship between fairness and PRF. Shariq et al. [14] introduced a cluster-based PRF mechanism for decreasing bias in the search results set by increasing the findability of documents. Relatedly, Wilkie et al. [16] investigated how revising a user's query can impact the retrievability of documents. More recently, Bigdeli et al. [3] introduced a bias-aware PRF framework for selecting neutral (i.e., non-biased) documents to form the pseudo-relevant set of documents that the informative terms are selected from. Differently from the works of [3,14,16] that mainly focused on the investigation and mitigation of bias with PRF, in this work our focus is on providing groups of documents a fair exposure to the user by using PRF. Specifically, we want to leverage PRF to produce a fair exposure distribution over particular groups of documents. The importance of this task has been highlighted by the TREC Fair Ranking Track [5]. Therefore, we use the 2021 TREC Fair Ranking Track test collection to investigate if PRF can be leveraged to increase the fairness of multiple representation dense retrieval.

3 ColBERT-FairPRF

We propose to extend the PRF mechanism of ColBERT-PRF [15] to provide a fair exposure in the search results for multiple groups of documents. To score documents w.r.t. a query, q, and generate a ranked list of results, ColBERT-PRF (and ColBERT [9]) deploy an approximate nearest neighbour search to find the nearest document token embeddings for each of the query token embeddings, q_e. The token similarities are then summed to calculate the similarity of a document to a query. Our approach ColBERT-PRF [15] builds on ColBERT by selecting useful embeddings from the pseudo-relevant set and appending them to the initial query representation.

To provide a fair exposure to the documents from a set of fairness groups, G, our ColBERT-FairPRF approach selects feedback embeddings from the highest ranked documents of *each* of the individual groups in the pseudo-relevant set. Selecting embeddings from the groups individually ensures that all of the groups that we want to be fair to are represented in the expanded query. In particular, we want to generate a re-ranking, $rank_R$, of the initial ColBERT ranking, $rank_C$, where each of the groups $g \in G$ receives a fair exposure in $rank_R$. To do this, we first identify specific feedback documents for each of the groups $g \in G$. For each group, g, we identify the k most relevant documents from the group g in $rank_C$ to form the pseudo-relevant set, PRS_g, for the group g. Next, we deploy the ColBERT-PRF approach to identify the most representative embeddings, $\{v_1, \ldots, v_k\}$, for each of the sets PRS_g, for $g \in G$. In particular, following [15], for each PRS_g we deploy a simple KMeans clustering approach to obtain k representative centroid embeddings from the feedback documents for group g.

Having identified the k most representative centroid embeddings for PRS_g, we need to (1) identify the term embeddings in PRS_g that are expected to be the most useful for representing the group g, (2) append these useful (and representative) embeddings to the initial query, q, to form the expanded query q_e, and (3) re-score the documents in the initial ranking, $rank_C$, to form the re-ranked (and final) ranking $rank_R$. To do this, we leverage ColBERT-PRF [15] to identify, for each of the k centroids, the closest term embedding, t_i, to the centroid embedding v_i. For further details about the centroid creation, we refer the reader to the original ColBERT-PRF [15] work. The identified k term embeddings for each of the groups $g \in G$ are then appended to the query representation as the expansion embeddings, E_e, to form the expanded query representation, q_e.

Finally, each of the documents, d_i, in $rank_C$ are re-scored with respect to the expanded query, q_e, to form the final re-ranked search results, as follows:

$$s(q_e, d_i) = \sum_{i=1}^{|q_e|} \max_{j=1,\dots,|d|} \phi_{q_i}^T \phi_{d_j} + \beta \sum_{e_i \in E_e} \max_{j=1,\dots,|d|} e_i \phi_{d_j} \qquad (1)$$

where $|q_e|$ is the number of embeddings in the initial query, $|d|$ is the number of embeddings in d_i, ϕ_{q_i} is an embedding for a token in $|q_e|$, ϕ_{d_j} is an embedding for a token in d_i, E_e are the expansion embeddings for q_e and $\beta > 0$ controls the contribution of the expansion embeddings to q_e. The score $s(q, d)$ for a document d given a query q is obtained by summing the maximum similarities between the query token embeddings and the document token embeddings ($\max_{j=1,\dots,|d|} \phi_{q_i}^T \phi_{d_j}$) with the maximum similarities between the expansion embeddings and the document token embeddings ($\max_{j=1,\dots,|d|} e_i \phi_{d_j}$).

4   Experimental Setup

In this work, we aim to answer the following two research questions:

- **RQ1**: Can our ColBERT-FairPRF improve the fairness of multiple representation dense retrieval?
- **RQ2**: What is the effect of improving fairness using ColBERT-FairPRF on the relevance of the search results?

Test Collection: We use the test collection of the TREC 2021 Fair Ranking Track [5]. The test collection consists of approximately 6.5 million articles from the English language Wikipedia. Each of the articles has associated group labels for two fairness characteristics, namely geographic location and gender. In our experiments, we use the geographic location labels as our fairness groups since all of the articles have geographic labels, whereas the coverage of gender labels is relatively small since they are only available for biographical articles. There are 48 evaluation queries representing Wikipedia topics, such as *mathematics* and *finance*. Relevance assessments were obtained through TREC pooling.

Indexing and Retrieval: We parse the document collection to remove the Wiki-Markup using PyAutoCorpus[1] and index the collection using the ColBERTIndexer[2] from PyTerrier [11].

Baselines: We compare our proposed ColBERT-FairPRF approach against the ColBERT [9] and ColBERT-PRF [15] approaches. We deploy the default parameters for ColBERT-PRF [15], i.e. we select 3 feedback documents and identify 24 representative centroids, before appending 10 embeddings to the query. We set ColBERT-PRF's β value to 1, thereby giving the expansion embeddings the maximum possible influence. For our proposed ColBERT-FairPRF approach, we use the same parameter values as ColBERT-PRF, however they are applied per group. We leave investigating the influence of the β parameter to future work.

Metrics: Following the TREC Fair Ranking Track [5], as our fairness metric, we report Attention Weighted Ranked Fairness (AWRF) [13], calculated as $AWRF(F) = \Delta(\epsilon(F), \hat{p})$, where Δ is the difference between the actual group exposure, ϵ, and the target exposure, $\hat{p}$. We use the official target exposure distributions, $\hat{p}$, from the test collection. For relevance, we report $nDCG$ [7]. Both measures are computed over the top k documents. We evaluate at $k=[10,20,50]$.

5   Results

Table 1 presents the results of our ColBERT-FairPRF approach compared to ColBERT and ColBERT-PRF, in terms of the mean Attention Weighted Ranked Fairness (AWRF) and nDCG over all of the queries. For both AWRF and nDCG, a higher score indicates a better performance.

Table 1. Mean nDCG and AWRF calculated over the 48 queries. The * indicates that the proposed approach significantly outperform both baselines (t-test, p<0.001). There are no significant differences in terms of nDCG.

Approach	nDCG@10	AWRF@10	nDCG@20	AWRF@20	nDCG@50	AWRF@50
ColBERT	**0.401**	0.586	**0.350**	0.627	**0.282**	0.675
ColBERT-PRF	0.360	0.568	0.330	0.614	0.262	0.670
ColBERT-FairPRF	0.362	**0.703***	0.314	**0.747***	0.266	**0.766***

To answer **RQ1**, we analyse the Attention Weighted Ranked Fairness. From Table 1, we observe that the differences between the approaches are clearly visible. ColBERT-PRF and ColBERT produce similar results for mean AWRF at all considered values of k, e.g. at $k=50$ both approaches achieve approximately 0.67. It is apparent that at all considered values of k, ColBERT-FairPRF outperforms both of the baselines in terms of fairness with a maximum increase

---

[1] https://github.com/seanmacavaney/pyautocorpus.
[2] https://github.com/terrierteam/pyterrier_colbert.

of up to ~19% at k=10. This observation is supported by the results of significance testing (t-test, p<0.001) showing significant improvements over both of the baselines in terms of AWRF. To investigate **RQ2**, the utility of the rankings need to be considered. By analysing the performance of the approaches in terms of nDCG, we observe that only marginal differences are visible between the approaches. In particular, ColBERT achieves the highest nDCG scores followed by Colbert-FairPRF for k=10 and k=50. The marginal difference in utility is confirmed by the statistical significance test. Indeed, there are no significant differences between ColBERT-FairPRF and the baselines in terms of nDCG.

To further understand the behaviour of the approaches, we examine the per query performances on both nDCG and AWRF. Figure 1 shows the AWRF scores of the approaches for every query. The queries are ordered by decreasing AWRF of the ColBERT-FairPRF. Accordingly, the query on which ColBERT-FairPRF performs the best is at the left side of the plot. From Fig. 1(a) we can clearly see that ColBERT-FairPRF has the highest scores on the majority of the queries compared to the baselines. In particular, on 43 out of 48 queries our approach achieves the highest AWRF score. On three queries Colbert-PRF is the highest scoring approach. ColBERT has the fairest ranking for two queries.

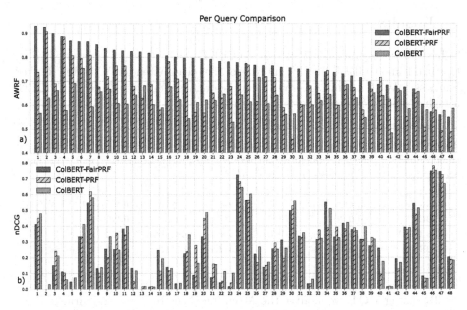

Fig. 1. Per query effectiveness in terms of (a) AWRF and (b) nDCG. In both of the plots, the queries are ordered by decreasing AWRF score for ColBERT-FairPRF.

Figure 1(b) shows the nDCG scores per query. To allow an easy comparison with Fig. 1(a) the queries in both plots are aligned in the same order. From comparing the plots (a) and (b) in Fig. 1, we note that there is not a clear relationship between the amount of gain in fairness, plot (a), and the increase/decrease or overall relevance for a query, plot (b). For example, for queries 12 and 16 ColBERT-FairPRF manages to increase both fairness and relevance compared to both of the baselines. This shows that it is possible to increase fairness while also improving the relevance of the results. As future work, we will investigate using query performance prediction [6] approaches to further tailor ColBERT-FairPRF to automatically weigh the PRF components on a per query basis to further optimise fairness and relevance. From observing the results for the individual queries, we note that, ColBERT-FairPRF manages to increase both fairness and relevance when the initial query terms are relatively coherent, for example for the queries "acting, actor, actress", "doctor, physician, surgeon" and "education, literacy", compared to less coherent queries such as "internet, company, online, business". We will also investigate this interesting research direction in future work.

Table 2. Example expansions for ColBERT PRF and ColBERT-FairPRF.

Original Query	'disco','discotheque','nightclub','deejay','dj','remix','dance music
ColBERT-PRF	'gibbons','proposition','stereo','mix', #mat','disco','1975','8','new','which'

Fairness Group	Colbert-FairPRF
Unknown	'thieves','collin','remixes','mix','disco','dj','got','de','united','which'
Africa	'bays','gee','vision','mt','#oa','financial','dance','secondary','active','fr'
Antarctica	'#ater','#dur','antarctica','antarctica','disco','ci','m','named','#r','south'
Asia	'honda','apartment','resignation','creating','dj','hop','#shi','#shi','#2','age'
Europe	'djs','afro','disco','#ty','together','late','#es','used','music','#k'
Latin America	'flores','reggae','#itt','rican','#mus','duo','dj','#ja','#de','music'
North America	'gibbons','bronx','leonard','#rra','disco','dj','hip','36','video','york'
Oceania	'botany','edwin','dj','hip','#ren','dance','zealand','23','station','-'

To further clarify how Colbert-FairPRF expands a query, we present a detailed investigation of the query "Disco Music". Table 2 gives an overview of how the query is enriched by the different approaches. ColBERT-PRF creates 10 new query tokens that are used to expand the original query. Colbert-FairPRF applies the same approach for every geographic group and creates 10 query tokens per group. The lower portion of Table 2 shows the identified expansion tokens per group. It is noticeable that for some groups, the expansion embeddings are related to the geographic location. For example, the group Antarctica contains the token "antarctica", while the group Northern America contains "bronx" and "york" that point to distinct geographic locations. By adding these tokens to the query embeddings, our proposed approach boosts the initially underrepresented groups and brings their exposure closer to the desired target exposure. In future work, adjusting the weights of the individual tokens can be used to further target any underexposed groups. Moreover, different β values can be explored to achieve further improvements in fairness, as well as utility.

6   Conclusions

In this work we have explored the potential of pseudo-relevance feedback (PRF) in multiple representation dense retrieval to provide a fair exposure to different groups of documents in a ranking. We proposed ColBERT-FairPRF, a new PRF approach that expands a user's query with representative embeddings for each of the fairness groups. Our experiments on the TREC 2021 Fair Ranking Test collection show that our approach is able to significantly improve the fairness of exposure (+ ~19%) in a ranking without decreasing the ranking's utility. We believe that this initial exploration of the relationship between dense retrieval PRF and fairness will open up interesting future research directions.

References

1. Abdul-Jaleel, N., et al.: UMASS at TREC 2004: Novelty and hard. Computer Science Department Faculty Publication Series p. 189 (2004)
2. Amati, G., Van Rijsbergen, C.J.: Probabilistic models of information retrieval based on measuring the divergence from randomness. ACM Trans. Inf. Syst. (TOIS) 20(4), 357–389 (2002)
3. Bigdeli, A., Arabzadeh, N., Seyedsalehi, S., Zihayat, M., Bagheri, E.: On the orthogonality of bias and utility in ad hoc retrieval. In: Proceedings of SIGIR (2021)
4. Devlin, J., Chang, M.W., Lee, K., Toutanova, K.: BERT: pre-training of deep bidirectional transformers for language understanding. arXiv preprint arXiv:1810.04805 (2018)
5. Ekstrand, M., McDonald, G., Raj, A., Johnson, I.: Overview of the TREC 2021 fair ranking track. In: Proceedings of TREC (2022)
6. He, B., Ounis, I.: Query performance prediction. Inf. Syst. 31(7), 585–594 (2006)
7. Järvelin, K., Kekäläinen, J.: Cumulated gain-based evaluation of IR techniques. ACM Trans. Inf. Syst, 20(4), 422–446 (2002)
8. Keikha, A., Ensan, F., Bagheri, E.: Query expansion using pseudo relevance feedback on Wikipedia. J. Intell. Inf. Syst. 50(3), 455–478 (2018)
9. Khattab, O., Zaharia, M.: Colbert: efficient and effective passage search via contextualized late interaction over BERT. In: Proceedings of SIGIR (2020)
10. Lv, Y., Zhai, C.: Positional relevance model for pseudo-relevance feedback. In: Proceedings of SIGIR (2010)
11. Macdonald, C., Tonellotto, N.: Declarative experimentation in information retrieval using PyTerrier. In: Proceedings of ICTIR (2020)
12. Rocchio, J.: Relevance feedback in information retrieval. In: The Smart Retrieval System-Experiments in Automatic Document Processing, pp. 313–323 (1971)
13. Sapiezynski, P., Zeng, W., Robertson, R., Mislove, A., Wilson, C.: Quantifying the impact of user attention on fair group representation in ranked lists. In: Compilation Process of WWW (2019)
14. Shariq, B., Andreas, B.: Improving retrievability of patents with cluster-based pseudo-relevance feedback document selection. In: Proceedings of CIKM (2009)
15. Wang, X., Macdonald, C., Tonellotto, N., Ounis, I.: Pseudo-relevance feedback for multiple representation dense retrieval. In: Proceedings of ICTIR (2021)
16. Wilkie, C., Azzopardi, L.: Best and fairest: an empirical analysis of retrieval system bias. In: Proceedings of ECIR (2014)

17. Xu, Y., Jones, G.J., Wang, B.: Query dependent pseudo-relevance feedback based on Wikipedia. In: Proceedings of SIGIR (2009)
18. Yan, R., Hauptmann, A., Jin, R.: Multimedia search with pseudo-relevance feedback. In: Proceedings of ICIVR (2003)

C²LIR: Continual Cross-Lingual Transfer for Low-Resource Information Retrieval

Jaeseong Lee[iD], Dohyeon Lee[iD], Jongho Kim[iD], and Seung-won Hwang[✉][iD]

Seoul National University, Seoul, South Korea
{tbvj5914,waylight3,jongh97,seungwonh}@snu.ac.kr

Abstract. This paper proposes a method to train information retrieval (IR) model for a low-resource language with a small corpus and no parallel sentences. Although neural IR models based on pretrained language models (PLMs) have shown high performance in high-resource languages (HRLs), building PLM for LRLs is challenging. We propose C²LIR, a method to build a high-performing neural IR model for LRL, with dictionary-based pretraining objectives for cross-lingual transfer from HRL. Experiments on the monolingual and cross-lingual IR in diverse low-resource scenarios show the effectiveness and data efficiency of C²LIR.

Keywords: Low-resource language · Neural IR · Cross-lingual transfer

1 Introduction

Although the pretrained language model (PLM) shows promising results in information retrieval [8], building IR models for low-resource languages (LRL) is challenging since they are data-hungry. A large-scale corpus and relevant query-document pairs are required to train such IR models. One can consider translating LRL queries and documents into a high-resource language (HRL), i.e., English, but it also requires sufficient parallel sentences.

An alternative direction requires no parallel sentences: mDPR [3,19] leverages a multilingual Pretrained Language Model (mPLM) containing the target LRL. However, due to 'the curse of multilinguality' [17] mPLM is limited to about 100 major languages, leaving most of 6500+ languages unseen. We solve this problem by injecting unseen LRL into a PLM trained with HRL. Figure 1 compares our method with the possible injection approaches. First, **Retraining** (Fig. 1a) trains again from scratch [14] by adding an LRL corpus to the original corpus, i.e., HRL+LRL. As a more efficient alternative, **Continual Pretraining** (Fig. 1b) [2,16], avoids expensive retraining by keeping the existing PLM and extending the pretraining process with LRL. However, the corpus size of LRL is inevitably smaller, thus their performance is reportedly degraded due to such imbalances [17].

J. Lee and D. Lee—Equal Contribution.

© The Author(s), under exclusive license to Springer Nature Switzerland AG 2023
J. Kamps et al. (Eds.): ECIR 2023, LNCS 13981, pp. 466–474, 2023.
https://doi.org/10.1007/978-3-031-28238-6_37

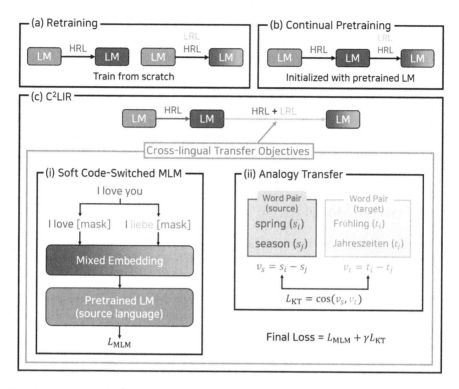

Fig. 1. Overview of C²LIR and the baselines. HRL: Pretrain PLM with high resource language. LRL: Pretrain PLM with low resource language.

To overcome such resource imbalance, we propose **C²LIR** (Fig. 1c). We design cross-lingual pretraining objectives with a word-to-word dictionary. A naïve baseline is to create cross-lingual resources through code-switching [12]. Instead of requiring an expensive parallel resource such as 'I love you' and 'Ich liebe dich', it replaces a subset of words to the target language. Though, this new code-switched sentence is a lower quality cross-lingual resource compared to its translated German pair because: (i) it loses the language-specific knowledge from the embedding of the original word 'love'. (ii) in case of word ambiguity caused by polysemy or homonym, simple word-based replacement would change the sentence semantics. For example, the word 'spring' may indicate both a season and a curved metal, such that replacing the translation of one in place for another would become faulty.

To overcome the downsides of code-switching, we propose novel pretraining objectives (Fig. 1c), namely (i) **soft code-switched MLM** and (ii) **analogy transfer** : (i) we mix each source word and target word according to the ratio in the embedding level to preserve the semantics from the original word. (ii) with HRL analogy pair and a dictionary, we employ analogy alignment as another pretraining objective. Doing so, the relationship between LRL's analogy pair,

(Frühling, Jahreszeiten), can be learned from HRL's pretrained relationships, (spring, season).

We pretrain the model with the proposed method and verify it over a diverse range of scenarios, concerning available cross-lingual resources during fine-tuning. Two virtual languages are used to fit our scenario to ensure both the low-resource setting for training and reliability for evaluation. Experimental results on XOR-Retrieve [3] and Mr. TyDi [19] show that C²LIR improves the performance of the neural IR model despite ten times smaller training resources used than mDPR. In addition, C²LIR is still effective even if the resource is extremely scarce.

2   Proposed Model

As motivated in Fig. 1(c), our model adds two cross-lingual pretraining objectives to *Continual Pretraining* [16]: (i) soft code-switched masked language modeling (MLM) objective and (ii) analogy transfer objective.

Soft Code-Switched MLM Objective. To prevent losing language-specific information from code-switching, we adopt a soft interpolation of the embeddings of the words.

First, given a sentence S_t in the target language t, we substitute some random tokens with [MASK] token. Let S_t^m be the masked sentence. Next, we replace the words in S_t^m with the counterparts in the source language s using the provided dictionary, to get a code-switched sentence S_s^m. Then, we tokenize the translated words again. We denote the embedding vectors of a sentence from each the source language and the target language as h_s and h_t. Finally, instead of using h_t as input for masked language modeling directly, the mixture of the two embeddings is used as follows:

$$h_{cs} = \lambda_{cs} \cdot h_t + (1 - \lambda_{cs}) \cdot h_s \quad (0 \le \lambda_{cs} \le 1) \tag{1}$$

where λ_{cs} is sampled from a beta distribution $\beta(\alpha, \alpha)$. We normalize the length of subword embeddings, by applying *1-to-1* length normalization [11]. s Note that in such a way, the masking rate is preserved. We can process similarly when a given sentence is in the source language s, to obtain h_{cs}.

We formulate the soft code-switched MLM objective as follows:

$$L_{\text{MLM}}(\theta_{\text{LM}}) = \sum_{i \in \mathbf{m}} -\log p_{\theta_{\text{LM}}}(w_i | h_{cs}) \tag{2}$$

Analogy Transfer Objective. To overcome the ambiguity in code-switching, we add analogy transfer as a pretraining objective. It aligns the analogy pair from LRL, using the HRL analogy pair and dictionary.

Given analogy word pairs in two languages $(w_{s,i}, w_{s,j})$ and $(w_{t,i}, w_{t,j})$, we minimize the cosine similarity between the vector offset of word embedding pairs,

$e_{s,i} - e_{s,j}$ and $e_{t,i} - e_{t,j}$. The average of subword embeddings is used as the word embedding. We formulate the loss function as follows:

$$L_{AT} = -\frac{1}{N} \sum \cos(e_{s,i} - e_{s,j}, e_{t,i} - e_{t,j}) \qquad (3)$$

where N is the number of word pairs it utilizes per batch.

We propose the final loss as the combination of the two:

$$L = L_{MLM} + \gamma \cdot L_{AT} \qquad (4)$$

3   Experiments

Simulating Unseen LRLs. Real unseen languages lack a high-quality evaluation dataset to the best of our knowledge. Therefore we simulate unseen LRLs following Liu et al. [12], but assuming scarcer resources. We randomly sample 80K Wikipedia sentences and use MUSE dictionary [10]. To construct word analogy pairs, we utilize English analogy word pairs [15] and substitute the words with MUSE dictionary. As a result, *s-Korean* and *s-Bengali* are simulated from Korean and Bengali. The two selected are the lowest-resourced languages in the evaluation dataset and dictionary we use when sorted by the corpus size.

Tasks. First, we consider *cross-lingual IR* with XOR-retrieve [3], whose queries are written in an LRL but retrieval corpus is written in English. We build models with (*lrl-en*) and without (*lrl-en-0shot*) the train dataset. Second, we tackle *monolingual IR* with Mr. TyDi [19], whose queries and retrieval corpus are both written in an LRL. Unlike *lrl-en*, we believe train data is less likely to exist in LRL, thus we build a model without train dataset (*lrl-lrl-0shot*).[1]

Methods. We compare our approach with the following baselines to inject LRL into English PLM:[2]

- Dictionary-base translation (*DT*): Translation using cross-lingual dictionary following Conneau et al. [6].
- *Retraining (mDPR)*: We pretrain bilingual BERT [7] from scratch to build custom mDPR [3,19].
- *Continual Pretraining (CP)*: We pretrain English PLM further with bilingual corpus. The detailed construction is similar to Wang et al. [16].
- *CP-cs*: We conduct *CP* with code-switched corpus.
- C^{2}LIR: We apply our proposed soft code-switched MLM objective and analogy transfer objective, upon *CP*.

---

[1] XOR-Retrieve train set contains just 2.5k LRL queries, where the average query length is less than 10 words. Mr. Tydi contains LRL documents aligned with LRL queries, which are far unlikely to exist. Thus we discard the train dataset of Mr. Tydi.

[2] Although we can also apply C^{2}LIR on another PLM, such as mBERT, we experiment with English PLM. Comparison can be found in Table 4.

Table 1. Comparison between C²LIR and baselines. For *lrl-en* and *lrl-en-0shot*, we report R2 (R@2kt[%]) and R5 (R@5kt[%]) following [3]. For *lrl-lrl-0shot*, we report MRR (MRR@100[%]) and R (R@100[%]) following [19].

Unseen LRL		s-Korean						s-Bengali					
		lrl-en		lrl-lrl-0shot		lrl-en-0shot		lrl-en		lrl-lrl-0shot		lrl-en-0shot	
Method	Iter	R2	R5	MRR	R	R2	R5	R2	R5	MRR	R	R2	R5
C²LIR	0.1M	**32.3**	**40.7**	**31.1**	**57.4**	**13.7**	**21.8**	**44.7**	**55.3**	46.8	**87.8**	30.9	**41.8**
CP-cs	0.1M	27.4	40.0	29.6	57.1	12.6	16.8	39.1	49.7	46.3	86.0	**31.3**	**41.8**
CP	0.1M	22.1	26.3	27.2	55.5	13.7	16.5	33.6	42.1	38.5	79.3	17.8	28.9
mDPR	1M	21.8	28.8	27.9	56.9	13.0	21.4	35.5	43.1	42.6	79.7	21.4	36.2
DT	-	16.5	24.7	22.4	52.1	–	–	28.3	38.6	33.1	72.5	–	–

Implementation and Hyperparameters. We use BERT-base model [7] as English PLM. We build 32k wordpieces [18] for target unseen LRL.

In the pretraining stage, we use both LRL and English corpus.[3] For the source language corpus, we use Wikipedia articles, and BookCorpus [20]. We set $\alpha = 0.75, \gamma = 10$ in Eq. 4. Hyperparameters are similar to Devlin et al. [7]. We pretrain for 1M steps for *Retraining*, and 0.1M steps for C²LIR or *CP*. For *CP-cs*, we use word replacement probability of 30% following Liu et al. [12].

For *lrl-en*, we first fine-tune on Natural Questions (NQ) [9] following Asai et al. [3]. Then we fine-tune with XOR-Retrieve for 245 epochs, to make similar iterations to Asai et al. [3]. For *lrl-en-0shot*, we use NQ dataset only with code-switched queries, where word replacement probability is 90%. We report R@2kt and R@5kt on the dev set of XOR-Retrieve since the test set is not publicly available.

For *lrl-lrl-0shot*, as we do not use LRL monolingual IR training data (such as Mr.TyDi. training data), we utilize code-switching and transfer from cross-lingual IR for better adaptation. We fine-tune on code-switched queries and documents from NQ, then on XOR-Retrieve with code-switched documents. Finally, we follow the sparse-dense hybrid method with the default evaluation settings from Mr.TyDi.[4] and report MRR@100 and R@100 on the test set of Mr.TyDi.

3.1 Result and Analysis

Effectiveness of C²LIR C²LIR outperforms baselines (Table 1). In detail, (1) C²LIR outperforms *CP-cs* and *DT*, which supports that the proposed pretraining objectives overcome the downside of code-switching. (2) During the pretraining stage, C²LIR efficiently reduces the computation budget to adapt to the new target language by 10x. While *CP* and *DT* fail to achieve comparable performance with *Retraining (mDPR)*, C²LIR outperforms *Retraining* baseline.

Data Efficiency in Extreme Setting. To show C²LIR remains effective even when available dictionary size is very scarce (Table 2), we shift to a more aggressive low-resource setting. We select the top n% words in the MUSE dictionary

---

[3] We allow 10 times more English sentences than LRL, based on preliminary experiments to select the upsample ratio of the LRL corpus.

[4] https://github.com/castorini/mr.tydi/tree/4281b6515a.

words by the occurrences in the English corpus. In lrl-lrl-0shot and lrl-en-0shot, we choose 10% of words, that is approximately 2000-3000 words for each language. In lrl-en scenario, aware of other baselines not using any dictionary, we severely limit the size of the dictionary. Only 1% of the words are selected to generate about 200-300 word pairs.

Table 2 shows that C^2LIR outperforms the other baselines, even with far scarcer resources. Especially, we emphasize 1) In lrl-lrl-0shot, C^2LIR outperforms BM25 baseline, while *CP* doesn't. 2) In lrl-en, C^2LIR outperforms *mDPR*, even when 1% of the dictionary is used.

Table 2. Comparison between C^2LIR and baselines on s-Korean with limited dictionary. We use 10% of dictionary if not stated. (*: Does not use dictionary, i.e., same as Table 1. †: uses 1% of dictionary)

Method	lrl-en		lrl-lrl-0shot		lrl-en-0shot	
	R2	R5	MRR	R	R2	R5
C^2LIR	24.9†	31.9†	28.4	55.2	6.7	13.7
CP	22.1*	26.3*	26.0	54.4	8.4	12.3
mDPR	21.8*	28.8*	27.6	54.8	9.5	11.9
BM25	–	–	25.9*	54.8*	–	–

Table 3. Ablating one objective drops performance from C^2LIR.

SCS	AT	< C^2LIR
✓	✗	9/12
✗	✓	11/12

Table 4. Average score of C^2LIR on s-Korean, varying initial PLM.

	En	M
lrl-en	36.5	33.0
lrl-lrl-0shot	44.3	42.5
lrl-en-0shot	17.7	17.7

Ablation Study. We conduct an ablation study on Soft Code-Switched MLM (SCS) and Analogy Transfer (AT). Among 12 (LRL, metric) pairs from Table 1, ablating SCS or AT drops the performance of C^2LIR in 9, 11 pairs respectively (Table 3). This explains that both preserving language-specific knowledge and dissolving ambiguities are necessary objectives to understand LRL.

EnBERT vs mBERT. We justify why we used English PLM (EnBERT) as initial PLM rather than multilingual PLM (mBERT), in the main experiments. Following mBERT [14], we build a multilingual PLM, called *mBERT7*, with 7 languages to maximize the performance of English [5]. Table 4 discloses that applying C^2LIR starting from EnBERT (En) is better or comparable than starting from mBERT7 (M).

4 Related Work

4.1 Injecting LRLs

Retraining (Fig. 1a). A compute-intensive way to support more languages is to retrain a PLM from scratch, by extending the training corpus to include those written in the target language [14]. In the field of Information retrieval, mDPR [3, 19] leverages PLM incorporating multiple languages into one shared architecture. We show that mDPR has limited performance in low-resource languages and requires cross-lingual pretraining as an alternative (Sect. 3).

Continual Pretraining (Fig. 1b). As an efficient alternative for Retraining, continuing the pretraining procedure by extending the original vocabulary of a PLM [2,4,16] reduces the training cost. Transliteration [13] or pre-designed embeddings [1], if available for a new language, can be leveraged for higher efficiency. Our proposed method can be orthogonally applied to these augmentations.

4.2 Cross-lingual Pretraining Objectives

Without parallel sentences, translating with only a dictionary [6] is possible, but it showed limited performance. Code-switching is an alternative pretraining objective [12] in the inadequate-resource scenario. However, as explained in Sect. 1, compared to parallel sentences, code-switched ones have the following limitations: (i) losing the language-specific signals of the original word, and (ii) not properly handling ambiguity caused by homonyms or polymorphs.

Our proposed objectives focus on addressing the two limitations. Though the existing soft code-switching [11] scheme also targeted the limitations, it assumes the existence of PLM, so the embedding of the target language must exist. Ours differs from theirs as we do not require pretrained target embedding.

5   Conclusion

This paper proposes new cross-lingual pretraining objectives to efficiently support unseen low-resource language for information retrieval. Experimental results show the effectiveness and data efficiency of C²LIR in diverse scenarios.

Acknowledgement. This work was supported by Institute of Information & communications Technology Planning & Evaluation (IITP) grant funded by the Korea government(MSIT) [NO.2021-0-01343, Artificial Intelligence Graduate School Program (Seoul National University)]. This research was supported by the MSIT(Ministry of Science and ICT), Korea, under the ITRC(Information Technology Research Center) support program(IITP-2023-2020-0-01789) supervised by the IITP(Institute for Information & Communications Technology Planning & Evaluation). We would like to thank Google's TPU Research Cloud (TRC) program for providing Cloud TPUs.

References

1. Ansell, A., et al.: MAD-G: Multilingual adapter generation for efficient cross-lingual transfer. In: Findings of the Association for Computational Linguistics: EMNLP 2021, pp. 4762–4781. Association for Computational Linguistics, Punta Cana, Dominican Republic (Nov 2021). https://doi.org/10.18653/v1/2021.findings-emnlp.410
2. Artetxe, M., Ruder, S., Yogatama, D.: On the cross-lingual transferability of monolingual representations. In: Proceedings of the 58th Annual Meeting of the Association for Computational Linguistics, pp. 4623–4637. Association for Computational Linguistics, Online (Jul 2020). https://doi.org/10.18653/v1/2020.acl-main.421

3. Asai, A., Kasai, J., Clark, J., Lee, K., Choi, E., Hajishirzi, H.: XOR QA: cross-lingual open-retrieval question answering. In: Proceedings of the 2021 Conference of the North American Chapter of the Association for Computational Linguistics: Human Language Technologies, pp. 547–564. Association for Computational Linguistics, Online (Jun 2021). https://doi.org/10.18653/v1/2021.naacl-main.46

4. Chau, E.C., Lin, L.H., Smith, N.A.: parsing with multilingual bert, a small corpus, and a small treebank. In: Findings of the Association for Computational Linguistics: EMNLP 2020, pp. 1324–1334. Association for Computational Linguistics, Online (Nov 2020). https://doi.org/10.18653/v1/2020.findings-emnlp.118

5. Conneau, A., et al.: Unsupervised cross-lingual representation learning at scale. In: Proceedings of the 58th Annual Meeting of the Association for Computational Linguistics. pp. 8440–8451. Association for Computational Linguistics, Online (Jul 2020). https://doi.org/10.18653/v1/2020.acl-main.747

6. Conneau, A., Lample, G., Ranzato, M., Denoyer, L., Jégou, H.: Word translation without parallel data. arXiv preprint arXiv:1710.04087 (2017)

7. Devlin, J., Chang, M.W., Lee, K., Toutanova, K.: BERT: pre-training of deep bidirectional transformers for language understanding. In: Proceedings of the 2019 Conference of the North American Chapter of the Association for Computational Linguistics: Human Language Technologies, Volume 1 (Long and Short Papers), pp. 4171–4186. Association for Computational Linguistics, Minneapolis, Minnesota (Jun 2019). https://doi.org/10.18653/v1/N19-1423

8. Karpukhin, V., et al.: Dense passage retrieval for open-domain question answering. In: Proceedings of the 2020 Conference on Empirical Methods in Natural Language Processing (EMNLP), pp. 6769–6781. Association for Computational Linguistics, Online (Nov 2020). https://doi.org/10.18653/v1/2020.emnlp-main.550

9. Kwiatkowski, T., et al.: Natural Questions: a benchmark for question answering research. Trans. Assoc. Comput. Linguist. **7**, 452–466 (2019)

10. Lample, G., Conneau, A., Ranzato, M., Denoyer, L., Jégou, H.: Word translation without parallel data. In: International Conference on Learning Representations (Feb 2018)

11. Lee, D., Lee, J., Lee, G., Chun, B.g., Hwang, S.w.: SCOPA: Soft code-switching and pairwise alignment for zero-shot cross-lingual transfer. In: Proceedings of the 30th ACM International Conference on Information & Knowledge Management, CIKM 2021, pp. 3176–3180. Association for Computing Machinery, New York (Oct 2021). https://doi.org/10.1145/3459637.3482176

12. Liu, Z., Winata, G.I., Fung, P.: Continual mixed-language pre-training for extremely low-resource neural machine translation. In: Findings of the Association for Computational Linguistics: ACL-IJCNLP 2021, pp. 2706–2718. Association for Computational Linguistics, Online (Aug 2021). https://doi.org/10.18653/v1/2021.findings-acl.239

13. Muller, B., Anastasopoulos, A., Sagot, B., Seddah, D.: When being unseen from mbert is just the beginning: handling new languages with multilingual language models. In: Proceedings of the 2021 Conference of the North American Chapter of the Association for Computational Linguistics: Human Language Technologies, pp. 448–462. Association for Computational Linguistics, Online (Jun 2021). https://doi.org/10.18653/v1/2021.naacl-main.38

14. Google Research: BERT (2019). https://github.com/google-research/bert/blob/eedf5716ce1268e56f0a50264a88cafad334ac61/multilingual.md

15. Ushio, A., Espinosa-Anke, L., Schockaert, S., Camacho-Collados, J.: BERT is to NLP what AlexNet is to CV: Can Pre-Trained Language Models Identify Analogies? In: Proceedings of the ACL-IJCNLP 2021 Main Conference. Association for Computational Linguistics (2021)

16. Wang, Z., K, K., Mayhew, S., Roth, D.: Extending multilingual bert to low-resource languages. In: Findings of the Association for Computational Linguistics: EMNLP 2020, pp. 2649–2656. Association for Computational Linguistics, Online (Nov 2020). https://doi.org/10.18653/v1/2020.findings-emnlp.240

17. Wu, S., Dredze, M.: Are all languages created equal in multilingual BERT? In: Proceedings of the 5th Workshop on Representation Learning for NLP, pp. 120–130. Association for Computational Linguistics, Online (Jul 2020). https://doi.org/10.18653/v1/2020.repl4nlp-1.16

18. Wu, Y., et al.: Google's Neural Machine Translation System: Bridging the Gap between Human and Machine Translation. arXiv:1609.08144 [cs] (Oct 2016)

19. Zhang, X., Ma, X., Shi, P., Lin, J.: Mr. TyDi: A Multi-lingual Benchmark for Dense Retrieval. arXiv:2108.08787 [cs] (Aug 2021)

20. Zhu, Y., et al.: Aligning books and movies: towards story-like visual explanations by watching movies and reading books. In: 2015 IEEE International Conference on Computer Vision (ICCV), pp. 19–27. IEEE, Santiago, Chile (Dec 2015). https://doi.org/10.1109/ICCV.2015.11

Joint Extraction and Classification of Danish Competences for Job Matching

Qiuchi Li and Christina Lioma[⊠]

University of Copenhagen, Universitetsparken 1, 2100 Copenhagen, Denmark
{qiuchi.li,c.lioma}@di.ku.dk

Abstract. The matching of competences, such as skills, occupations or knowledges, is a key desiderata for candidates to be fit for jobs. Automatic extraction of competences from CVs and Jobs can greatly promote recruiters' productivity in locating relevant candidates for job vacancies. This work presents the first model that jointly extracts and classifies competence from Danish job postings. Different from existing works on skill extraction and skill classification, our model is trained on a large volume of annotated Danish corpora and is capable of extracting a wide range of danish competences, including skills, occupations and knowledges of different categories. More importantly, as a single BERT-like architecture for joint extraction and classification, our model is lightweight and efficient at inference. On a real-scenario job matching dataset, our model beats the state-of-the-art models in the overall performance of Danish competence extraction and classification, and saves over 50% time at inference.

Keywords: Competence extraction and classification · Job matching · Danish BERT

1 Introduction

Job matching, also known as person-job fit or job-resume matching, is a crucial and challenging scenario in job recruitment where matchers need to search suitable candidates for job vacancies from a huge pool of candidate profiles. The booming increase in job vacancies on recruitment platforms creates a high demand for prompt and accurate identification of matched candidates, placing a great burden to recruiters [9]. The absence of such matching systems will cause financial losses to both job seekers and companies [2].

Competences, specifically skills, knowledges or occupations, serve as one of the most important criteria for judging the relevant candidates [1]. Accurate and prompt extraction of competences from Jobs and CVs can promote accurate matching of relevant candidates and liberate recruiters' from their burden. While existing works [4,7,10,14,17] have mainly attempted to match job and candidates by their representation as a whole, the extraction of competences

© The Author(s), under exclusive license to Springer Nature Switzerland AG 2023
J. Kamps et al. (Eds.): ECIR 2023, LNCS 13981, pp. 475–483, 2023.
https://doi.org/10.1007/978-3-031-28238-6_38

can support a clear presentation of reasons of matching along with the matching result. However, the job matching context poses greater challenges to competence extraction algorithms in the following aspects:

- **Extraction accuracy.** A recruiter is often not knowledgeable to the industry related to the matching task, and relies on the extracted competences for finding the relevant candidates. Therefore, the extracted competence should be of high quality to support real matching scenario.
- **Fine-grained categories.** A competence can be expressed in different ways. Apart from exact term matching, fine-grained categorization of extracted competences should be devised to account for this issue and promote finding more relevant candidates.
- **Efficiency.** For productivity, a competence extraction model should be able to generate prompt response to incoming jobs and CVs.

Machine learning algorithms have been developed for automatic identification [6,11], extraction [3,5,12,15] and classification [16] of competences from Job postings or CVs. These works mainly target English or Chinese job postings. For Danish Jobs, however, the research on competence extraction is limited by the lack of available annotated data. Zhang et al. [16] investigated Danish competence classification with distant supervision. On their collected tiny-scale dataset of 60 Danish job postings, few-shot learning and cross-language transfer learning led to decent performance. However, the extraction of Danish competences is still an unsolved task.

We frame the task of Danish competence extraction as a token classification task, and propose a novel model for jointly extracting and classifying Danish competences for job matching. The model is based on pre-trained text encoder for Danish job postings [15] and maps the encoded sentence representations to produce named entity recognition (NER) labels for competence extraction and multi-class labels for competence classification. The model is trained on around 200,000 sentences from Danish Jobs and CVs with annotations of European Skills, Competences, Qualifications and Occupations (ESCO) [13]. Different from [16], we include a wider range of competences in the ESCO taxonomy, broadly covering the main categories of skills, occupations and knowledges. The model is jointly trained from annotation labels of both tasks, and extracts and classifies competences in separate steps at prediction. Our model achieves improved accuracy over the best existing practices on fine-grained competence extraction and classification, and takes only half of the prediction time.

2 Task Definition and Data Description

The task is to extract and classify Danish Competences from job postings. Specifically, the input is a Danish sentence or a sequence of tokens $\mathcal{X} = \{x_1, x_2, ..., x_N\}$, and the output is a list of (text span, class) tuples $\mathcal{Y} = \{(s_1, c_1), (s_2, c_2), ..., (s_K, c_K)\}$. Each span $s_k = \{x_i\}_{i=k_{start}}^{k_{end}}$ is a continuous subsequence of tokens of $\mathcal{X}$, and $c_k \in \mathcal{C}$ is the class label of s_k that belongs to a pre-set collection of labels. Under this notation, SKILLSPAN [15] targets at establishing the mapping from $\mathcal{X}$ to $\mathcal{Y}^S = \{s_1, s_2, ..., s_K\}$, while KOMPE-TENCER [16] manages to predict c_k for each input s_k. We seek to directly learn the mapping $\mathcal{X} \rightarrow \mathcal{Y}$ from annotated data with a single model.

We proposed to jointly extract and classify Danish competences. The extraction of competences entities is formulated as a named entity recognition (NER) task, where a 3-class label is predicted for each token: [**O, I, B**]. **B** marks the beginning of an entity, **I** refers to the inner part of an entity, and **O** stands for a non-entity token. A "**BII...I**" pattern indicates a multi-token entity. The classification of entities is formulated as a multi-class classification task.

An important ingredient to our model is the resource for Danish competences. For this purpose, we rely on the European Skills, Competences, Qualifications and Occupations (ESCO) taxonomy, which contains a total number of 16898 skills, occupations, knowledges in 28 different languages. Each ESCO entity has a textual description and associated to 4-leveled annotations. This work aims at extracting text spans that are considered as ESCO entities, and further classify them into top-level categories in the ESCO taxonomy, include 10 occupation categories ($C_0 - C_9$), 8 skill categories ($S_1 - S_8$), 2 language skill categories (L_0, L_1), 11 knowledge categories ($K_{00} - K_{10}$), as well as 6 transversal skills and competences ($T_1 - T_6$). We also include three labels (C_{-1}, K_{-1}, S_{-1}) for non-ESCO occupations, knowledges and skills.

We apply our model on a collection of annotated sentences from the Jobindex[1] database. Jobindex is a job portal located in Denmark. It originally targeted at the Danish market and has expanded to have sites in 3 other countries. The sentences come from an abundance of Danish jobs and candidate profiles - see Fig. 1 for an illustration of the main text fields in each of them. For jobs, the sentences come from its textual descriptions. From candidate profiles, we extract sentences from educational and work experience. We apply exact phrase matching to detect ESCO entities in each sentence, split the sentence into a sequence of tokens, and insert NER labels and class labels based on the extracted ESCO entities.

3   Our Model

We build a single model to tackle ESCO entity extraction and classification. As shown in Fig. 2, NER labels and ESCO class labels are produced based on a multi-layer Transformer text encoder. The model jointly learns from annotated labels of both tasks in the training step, but produces the labels in a sequential manner at prediction.

---

[1] https://www.jobindex.dk/.

Fig. 1. Ingredients of a job posting and a candidate profile in the Jobindex database. The texts are translated to English for a better understanding.

3.1   Text Encoder

We take existing Danish BERT models for the text encoder, and formulate the extraction and classification of ESCO labels as a fine-tuning task. Danish BERT (**DaBERT**)[2] is a publicly available BERT model trained by Certainly[3] on 9.5GB Danish texts. Zhang et al. [16] obtained a Danish BERT for job recommendation context, namely **DaJobBERT**, by further training **DaBERT** on 24.5M Danish job posting sentences for one epoch. **DaJobBERT** is reported to have superior few-shot performance over **DaBERT** [16]. Both encoders are included in this work.

3.2   ESCO Detection and Classification

Token-wise NER labels are produced for detecting ESCO entities. We construct simple feed-forward neural network f_{NER}, containing a single hidden layer with Tanh as the activation function, for mapping each encoded token vector to a 3-class NER label.

For ESCO classification, all tokens in the extracted ESCO entities are aggregated to a fixed-dimensional vector. Specfically, we take the average of the encoded token vectors for each entity span, and pass it to a feed-forward network f_{CLS} to produce its ESCO class label.

3.3   Joint Training

A training sample is annotated with NER labels for all tokens and ESCO class labels for the ESCO entities. ESCO detection loss $\mathcal{L}_{NER}$ is the average cross-entropy loss over all tokens against the golden NER labels in a sentence.

---

[2] https://huggingface.co/Maltehb/danish-bert-botxo.
[3] https://certainly.io/.

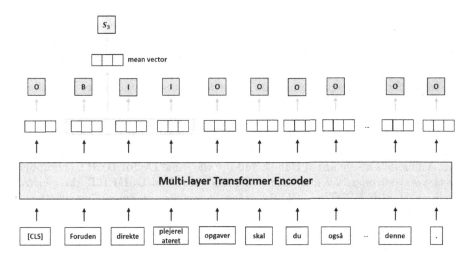

Fig. 2. The architecture of the joint ESCO extraction and classification model.

For ESCO classification loss, we pass the average token vectors of each **golden entity** to f_{CLS}. and compute the average cross-entropy loss between the output logits and true ESCO labels over all entities as the ESCO classification loss $\mathcal{L}_{CLS}$. The overall training loss is a combination of both losses controlled by a linear weight λ:

$$\mathcal{L} = \mathcal{L}_{CLS} + \lambda \mathcal{L}_{NER}. \tag{1}$$

Both the text encoder and feed-forward networks (f_{CLS}, f_{NER}) are learned by minimizing the loss $\mathcal{L}$ with a standard back-propagation algorithm.

3.4 Two-Step Prediction

The model extracts and classifies ESCO entities in separate steps for an input sentence at prediction. First, the sentence is passed to the text encoder and NER network f_{NER} to compute an NER label for each token. The entities are extracted accordingly: we detect the **B** labels in the tokens, and at take the longest **BII...I** sequence as the entities for each **B** label. Then, the encoded tokens for each extracted entity are passed f_{CLS} to produce ESCO class labels.

4 Experiment

Data. We evaluate different models for ESCO extraction and classification on job and candidate texts in the Jobindex database. A total number of 217661 sentences are obtained. We split the data into training, validation and test sets at a ratio of 8:1:1.

Models. In addition to our model, a list of BERT-based models are included. Due to the absence of existing models for the same purpose, we aggregate SKILLSPAN [15] and KOMPETENCER [16] into a two-step pipeline as the state-of-the-art model. As another two-step approach, we train a separate model based on the competence extraction and classification architecture part of our model. The idea is to check if jointly rendering two tasks leads to improved model capacity. Furthermore, we include an intuitive single-model strategy that views the joint extraction and classification of ESCO competences as an end-to-end NER model, where class-specific NER labels are produced, such as **B-S$_1$**, **I-L$_0$**, etc. All models above are in Danish and use the same **DaJobBERT** checkpoint as the text encoder. We also train our model based on **DaBERT**, the general-purpose Danish BERT model. By doing so we aim at examining whether a domain-adapted Danish BERT can further enhance our model capacity.

All models have a 12-layer, 12-attention-head structure with a model size of 768 and an intermediate size of 3072. We use AdamW [8] as the optimizer. We train on the training dataset for one epoch and do early stopping in terms of the validation loss. We set $\lambda = 0.1$ to place more attention of our model on competence classification. All models are implemented in the Hugging Face toolkit[4] under PyTorch 1.9.0 and trained on a GPU server with 4 T A100 40GB cards.

Metrics. The same effectiveness metrics is applied to all models. For competence extraction, we compute the F1 score for judging whether a token belongs to an ESCO entity. For competence classification, the weighted macro-F1 score is computed following the established practice [16]. It is a weighted average of F1 scores for all classes considering the support for each class. To check the model efficiency in a real job matching scenario, we also test the average processing time on each job, over the same set of 100 random job postings.

5 Results and Discussions

Overall Performance. As shown in Table 1, all models have similar capacity in extracting ESCO entities. In comparison, SKILLSPAN achieved an F1 score between 0.55 and 0.65 on a relatively small dataset of around 15000 English job sentences [15]. The relatively high ESCO extraction performance proves that large-scale annotated data contributes to strong ESCO extraction capabilities. For classification of ESCO labels, however, our model beats the existing works by a remarkable margin. In addition, the SOTA solution takes twice as much storage, and takes slightly more than twice as much as our model due to the intermediate data processing steps between the two models. All the above advantages makes our model an obviously better candidate under the job matching scenario.

Effect of Joint Learning. The huge gap in the competence classification performance between single-model and two-model approaches reveals a huge positive influence of ESCO extraction to ESCO classification, which could be well

---
[4] https://huggingface.co/.

Table 1. Size, inference time and effectiveness metrics for models in the experiment. For effectiveness metrics, we compute the average and standard deviation of each value over 5 runs of different random seeds.

Model name	Model size	Inference	ESCO extraction (F1)	ESCO classification (Weighted Macro F1)
SKILLSPAN + KOMPETENCER	862.12M	1.72 s	0.864 ± 0.002	0.436 ± 0.003
Detection + Classification	858.39M	2.14 s	0.865 ± 0.003	0.337 ± 0.002
All-class NER	429.35M	0.67 s	0.856 ± 0.001	0.588 ± 0.000
Our model	431.54M	0.73 s	0.860 ± 0.001	0.623 ± 0.001
Ur model - DaBERT	431.54M	0.70 s	0.841 ± 0.000	0.627 ± 0.003

captured by the joint learning architecture. Compared to a single NER prediction task for all classes, our structure is a better proposal. Since the all-class NER view leads to a doubling of class numbers, the class imbalance and data sparsity issue are likely to bring negative impact to competence classification.

Effect of domain-adapted Danish BERT. Building our model based on **DaBERT** encoder yields close performance to on the **DaJobBERT** encoder. This is to the contrary of the observations in [16], where **DaJobBERT** significantly outperformed **DaBERT** in the few-shot setting. We have demonstrated that, n sufficient in-domain data, different pre-trained Danish BERT encoders have minimal influence to competence extraction and classification.

6 Conclusion

We present a novel model for jointly extracting and classification Danish competences for Job Matching. On a large collection of annotated samples, this model excels at extracting competences of fine-grained categories, in over 50% less time compared to the SOTA approach. The strong effectiveness and efficiency makes it better at tackling the requirements of job matching.

This work is limited to Danish language and the model is not evaluated on a publicly available dataset. We will examine if similar findings hold on publicly available English job postings. We also plan to integrate this model to an automatic job recommendation framework to directly study its impact on a real job matching scenario.

Acknowledgement. This research was supported by the Innovation Fund Denmark, grant no. 0175-000005B. We are grateful for Jobindex's support on providing the data and setting up the experiment.

References

1. Bogers, T., Kaya, M.: an exploration of the information seeking behavior of recruiters. In: Proceedings of the RecSys in HR 2021 Workshop, pp. 11–18 (2021)
2. Cardoso, A., Mourão, F., Rocha, L.: The matching scarcity problem: When recommenders do not connect the edges in recruitment services. Expert Syst. Appli. **175**, 114764 (2021). https://doi.org/10.1016/j.eswa.2021.114764, https://www.sciencedirect.com/science/article/pii/S0957417421002050
3. Chernova, M.: Occupational skills extraction with FinBERT. Master's thesis, Metropolia University of Applied Sciences (2020)
4. Dave, V.S., Zhang, B., Al Hasan, M., AlJadda, K., Korayem, M.: A Combined Representation Learning Approach for Better Job and Skill Recommendation. In: Proceedings of the 27th ACM International Conference on Information and Knowledge Management, CIKM 2018, pp. 1997–2005. Association for Computing Machinery, New York (Oct 2018). https://doi.org/10.1145/3269206.3272023
5. Gugnani, A., Misra, H.: Implicit skills extraction using document embedding and its use in job recommendation. In: Proceedings of the AAAI Conference on Artificial Intelligence, vol. 34(08), pp. 13286–13293 (Apr 2020). https://doi.org/10.1609/aaai.v34i08.7038, https://ojs.aaai.org/index.php/AAAI/article/view/7038
6. Jia, S., Liu, X., Zhao, P., Liu, C., Sun, L., Peng, T.: Representation of job-skill in artificial intelligence with knowledge graph analysis. In: 2018 IEEE Symposium on Product Compliance Engineering - Asia (ISPCE-CN), pp. 1–6 (2018). https://doi.org/10.1109/ISPCE-CN.2018.8805749
7. Li, Q., Lioma, C.: Template-based recruitment email generation for job recommendation (2022). https://doi.org/10.48550/arXiv.2212.02885 arXiv: 2212.02885
8. Loshchilov, I., Hutter, F.: Fixing weight decay regularization in adam. CoRR abs/arXiv: 1711.05101 (2017)
9. Montuschi, P., Gatteschi, V., Lamberti, F., Sanna, A., Demartini, C.: Job recruitment and job seeking processes: how technology can help. IT Professional **16**(5), 41–49 (2014). https://doi.org/10.1109/MITP.2013.62
10. Qin, C., et al.: An enhanced neural network approach to person-job fit in talent recruitment. ACM Trans. Inf. Syst. **38**(2), 15:1–15:33 (2020). https://doi.org/10.1145/3376927
11. Sayfullina, L., Malmi, E., Kannala, J.: Learning representations for soft skill matching, arXiv: 1807.07741 (2018)
12. Tamburri, D.A., Heuvel, W.J.V.D., Garriga, M.: Dataops for societal intelligence: a data pipeline for labor market skills extraction and matching. In: 2020 IEEE 21st International Conference on Information Reuse and Integration for Data Science (IRI), pp. 391–394 (2020). https://doi.org/10.1109/IRI49571.2020.00063
13. le Vrang, M., Papantoniou, A., Pauwels, E., Fannes, P., Vandensteen, D., De Smedt, J.: ESCO: Boosting job matching in Europe with semantic interoperability. Computer **47**(10), 57–64 (2014). https://doi.org/10.1109/MC.2014.283
14. Yan, R., Le, R., Song, Y., Zhang, T., Zhang, X., Zhao, D.: Interview Choice Reveals Your Preference on the Market: To Improve Job-Resume Matching through Profiling Memories. In: Proceedings of the 25th ACM SIGKDD International Conference on Knowledge Discovery & Data Mining, KDD 2019 pp. 914–922. Association for Computing Machinery, New York (Jul 2019). https://doi.org/10.1145/3292500.3330963

15. Zhang, M., Jensen, K., Sonniks, S., Plank, B.: SkillSpan: Hard and soft skill extraction from English job postings. In: Proceedings of the 2022 Conference of the North American Chapter of the Association for Computational Linguistics: Human Language Technologies, pp. 4962–4984. Association for Computational Linguistics, Seattle, United States (Jul 2022). https://doi.org/10.18653/v1/2022.naacl-main.366, https://aclanthology.org/2022.naacl-main.366
16. Zhang, M., Jensen, K.N., Plank, B.: Kompetencer: Fine-grained skill classification in danish job postings via distant supervision and transfer learning. In: Proceedings of the Language Resources and Evaluation Conference. pp. 436–447. European Language Resources Association, Marseille, France (June 2022). https://aclanthology.org/2022.lrec-1.46
17. Zhu, C., et al.: Person-Job Fit: Adapting the Right Talent for the Right Job with Joint Representation Learning. arXiv:1810.04040 [cs] (Oct 2018)

A Study on FGSM Adversarial Training for Neural Retrieval

Simon Lupart[✉] and Stéphane Clinchant

Naver Labs Europe, Meylan, France
{simon.lupart,stephane.clinchant}@naverlabs.com

Abstract. Neural retrieval models have acquired significant effectiveness gains over the last few years compared to term-based methods. Nevertheless, those models may be brittle when faced to typos, distribution shifts or vulnerable to malicious attacks. For instance, several recent papers demonstrated that such variations severely impacted models performances, and then tried to train more resilient models. Usual approaches include synonyms replacements or typos injections – as data-augmentation – and the use of more robust tokenizers (characterBERT, BPE-dropout). To further complement the literature, we investigate in this paper adversarial training as another possible solution to this robustness issue. Our comparison includes the two main families of BERT-based neural retrievers, i.e. dense and sparse, with and without distillation techniques. We then demonstrate that one of the most simple adversarial training techniques – the Fast Gradient Sign Method (FGSM) – can improve first stage rankers robustness and effectiveness. In particular, FGSM increases models performances on both in-domain and out-of-domain distributions, and also on queries with typos, for multiple neural retrievers.

Keywords: Neural IR · Robustness · Adversarial training

1 Introduction

Stochastic Gradient Descent (SGD) is the main optimization method in Machine Learning, enabling to effectively optimize neural networks with millions of parameters. Despite the great performances from SGD, neural networks models still suffer from robustness issues when face to distributions shifts or noise. The seminal work of Goodfellow et al. [6] showed, for instance, how to manipulate model-predictions – in an adversarial way – by adding gradient-targeted perturbations in images at the pixel level. Their approach, the Fast Gradient Sign Method (FGSM), was the first and simpler algorithm to perform such attack. While this opened the way to possible stronger attacks, it was shown in the meantime, that the same techniques could also be used to train more robust and resilient models. Beyond original attacks, Adversarial Training (AT) could be used to increase model robustness, as a regularization or data-augmentation [11].

While in the field of Information Retrieval (IR), several works demonstrated that Pre-trained Language Models (PLM) based architectures had the same

J. Kamps et al. (Eds.): ECIR 2023, LNCS 13981, pp. 484–492, 2023.
https://doi.org/10.1007/978-3-031-28238-6_39

robustness issues in zero-shot and noisy environments [15,20,25], it seems to our knowledge that one of the simple adversarial training technique – FGSM-AT – has not been evaluated for first stage rankers. As an initial study, we consider in this work FGSM-AT, both to increase model robustness both for in and out-of-domain. Then, we apply FGSM-AT on domain adaptation scenarios, to further analyse AT in environments with fewer annotated samples. Overall, this paper investigates the following Research Questions (RQ):

- RQ1: How performances change on in-domain and out-of-domain distributions with FGSM Adversarial Training?
- RQ2: Does FGSM Adversarial Training increase performances in environments with noise in queries such as typos?
- RQ3: Is FGSM Adversarial Training beneficial for domain adaptation?

2   Related Works

There is an abundant literature on adversarial methods, which can be grouped in mainly two families: the white-box and the black-box methods. In the white-box settings, one assumes full access to the model and can therefore compute models gradients (e.g., FGSM, PGD [1,6,12]), in difference to the black-box settings, where gradients are hidden from the attacker. In particular for the black-box case, attacks thus rely on various heuristic techniques, by iterating on the models inputs/outputs. While white box settings apply well in Computer Vision, examples of black box attacks are more common in NLP due to the discrete nature of words. For instance, BERT-Attack [9], iteratively replaces words by their synonyms – using a MLM BERT head – to find possible replacement-words that could trigger the model to make wrong predictions. To further specify the literature on adversarial methods, some works purely focus on malignant objectives [3], while others try to overcome the weaknesses of current architectures (Adversarial Training). As an example of the former, Carlini et al. [3] show that by poisoning a minimal fraction of the training set, we could control the prediction of particular test samples.

With the emergence of PLM-based models in IR (dense bi-encoder, SPLADE, ColBERT [4,8,18]) replacing old term-based approaches (BM25 [17]), some literature also appeared on adversarial methods in IR. In the current literature, the first works focus on malignant attacks, also known as Search Engine Optimization (SEO) [21–23]. Applied to IR, the goal becomes to either promote or demote the rank of a particular document or set of documents. As a leverage, existing works usually add several tokens in a document, that are optimized to modify its rank for a given query, or a set of queries. Distillation being also very commonly used in IR [7], grey-box approaches also appeared. In their work, [10] present the idea as to first distil a model – on which we would not have access to the gradient – into a copy, and then attack through the gradient of the copy.

Although the literature on SEO is already rich, it appears that adversarial training in IR is very limited, to our knowledge. Zhuang et al. [24] used data-augmentation on typos to make models more robust to typos. Later, the

same authors proposed a dedicated architecture for typos [25]: their model used
CharacterBERT, and a smoothing technique they called Self-Teaching, which
forces the model to predict the same score for a given (query, document) pair
with/without the typos. In the meantime, Sidiropoulos et al. [19] also experimented with data-augmentation and contrastive losses between queries with and
without typos, and had similar results. In the following of the paper, we aim at
applying the same methods with perturbations directly injected in the embedding space – in difference to previous works that worked at the token level – and
also with adversarial perturbations.

3   Adversarial Training

This section introduces adversarial training for first stage rankers in IR, using
the most simple approach, i.e. FGSM-AT. Standard training in IR usually uses a
contrastive $\texttt{InfoNCE}$ loss [13] on triplets $T_i = (q_i, d_i^+, d_i^-)$, which aims at increasing the similarity between the query and the positive document, while reducing
it for the negative documents. It can be seen as minimizing the loss:

$$\mathcal{L}_{\texttt{InfoNCE}}(T_i) = -\frac{e^{s(q_i, d_i^+)}}{e^{s(q_i, d_i^+)} + \sum_j e^{s(q_i, d_{i,j}^-)}}$$

Now in an adversarial training scenario, each triplet is perturbed by an $\epsilon_i =
(\epsilon_{i_q}, \epsilon_{i_{d+}}, \epsilon_{i_{d-}})$, containing independent perturbations for the query, and each
of the documents (applied on the inputs embeddings). Then, to ensure that
the model would predict the same scores in a local vicinity around any training triplet, FGSM-AT minimizes the joint objective containing the original and
adversarial losses as follows[1]:

$$\mathcal{L}_{\text{total}}(T_i) = \mathcal{L}_{\texttt{InfoNCE}}(T_i) + \mathcal{L}_{\text{adv}}(T_i + \epsilon_i)$$

$$\epsilon_i = \text{argmax}_{||r||_2 \leq ||r_{\max}||_2} \mathcal{L}_{\text{adv}}(T_i + r)$$

where $\mathcal{L}_{\text{adv}}$ is the adversarial loss, either the original $\mathcal{L}_{\texttt{InfoNCE}}$ ranking loss – which
is the case we consider for the following of the paper – or a measure of divergence
on scores directly (e.g., Kullback Leibler Divergence between the distributions
of scores). Note that adversarial training can be defined with a norm (here $||.||_2$)
and an upper-bound on the norm (here $r_{\max}$). With FGSM-AT, the min-max
optimization process is simplified by approximating ϵ_i in one step, computing the
gradient with respect to the input at T_i, and taking the direction that maximizes
it. The norm of the perturbation is also constant ($||\epsilon_i||_2 = ||r_{\max}||_2$):

$$\epsilon_i = -r_{\max}\frac{g_i}{||g_i||_2} \qquad g_i = \nabla_{T_i}\mathcal{L}_{\text{adv}}(T_i).$$

Intuitively, this helps the model to smooth the representation space, and act
as a regularization. From another perspective, this can also be seen as a data
augmentation, as we simply create one new sample for each original sample.

[1] Note that FGSM-AT can be applied on any loss, and thus generalizes to the margin-
MSE loss for the case of distillation [7].

4   Experiments

We compare two neural retrieval architectures: *(i)* a **dense bi-encoder**, which uses dot products to compute similarities between the `mean` tokens representations of queries and documents [8], *(ii)* and **SPLADE** – as a sparse bi-encoder – which represents them as high-dimensional bag-of-words vectors [5]. Both models are trained on MS MARCO. The dense bi-encoder is trained for 5 epochs, over the full set of 500k queries, in batches of size 16 with 32 negatives per query (with the hard-negatives released by [5] from SPLADE-Cocondenser). For SPLADE, we use standard training for 150k steps, with batches of size 128, and only one BM25 negative per query. For distillation, the dense bi-encoder uses a released msmarco-hard-negatives dataset[2] hosted on the Transformers library [16] where negatives were scored by a larger reranker, while SPLADE uses its own negatives [5] also scored by a reranker. We kept the same batch sizes and numbers of negatives during distillation than previously. Both models are trained with In-Batch-Negatives and a learning rate of 2e−5 with linear scheduler. To add FGSM-AT, we start from the previous best checkpoint, and resume training for resp. 2 epochs or 60k iterations with the targeted perturbations. This follows the settings from [14], with the motivation that FGSM-AT or noise injection can help to recover from a sharp minimum. Also, including FGSM-AT for only the last steps reduces training cost in comparison to FGSM-AT from scratch (each step of FGSM-AT being twice longer than a regular step). The value of the perturbation norm $r_{\max}$ is fixed to 0.01 (best value from 0.1, 0.01, 0.001 in our initial study). As additional baselines for FGSM-AT, we compare it with a ϵ–**random** baseline that adds a perturbation in a random direction of the embedding space (instead of the one given by the gradient), and a token level baseline **token-random**, that replace 15% of the original tokens with another random token. The models are then benchmarked on the MS MARCO collection [2], with both the original MS MARCO dev queries and TREC DL 2019/2020 judgements. To evaluate on out-of-distribution condition, we use the 13 available datasets from the BEIR benchmark [20]. Metrics are the default ones, MRR@10 and Recall@1k for MS MARCO, and nDCG@10 for TREC and BEIR datasets.

4.1   RQ1: How Performances Change on In-Domain and Out-of-Do-Main Distributions with FGSM Adversarial Training?

Table 1 reports the general comparison of FGSM-AT in-domain (MS MARCO and TREC), together with the mean nDCG@10 score out-of-domain on BEIR. The first six rows report the performances on both the dense bi-encoder and SPLADE without distillation, while for lower rows, the comparison is made on models trained with distillation. Without distillation, the random baselines first reveals that FGSM-AT is more effective than random noise injection (both at the token level or in the embeddings). We also notice the high improvements from

---

[2] https://huggingface.co/datasets/sentence-transformers/msmarco-hard-negatives.

Table 1. In-domain performances on MS MARCO dev et TREC DL tracks, and out-of-domain average performance on the 13 BEIR datasets. We report scores for both standard negative training (-N) and distillation training (-D). Results with † indicates p-values < 0.05 on paired t-test.

Dataset ($\rightarrow$)	MS MARCO dev		TREC DL 2019		TREC DL 2020		BEIR(13)
Models ($\downarrow$)	MRR@10	R@1k	nDCG@10	R@1k	nDCG@10	R@1k	nDCG@10
bi-encoder -N	33.24	95.75	65.93	**76.05**	66.08	81.11	39.54
+token-random	33.21	95.42	65.61	75.12	66.26	79.76	39.09
+ϵ-random	32.98	95.52	64.99	75.51	65.34	79.66	38.68
+FGSM	**35.49**†	**96.38**†	**69.24**†	75.57	**68.89**†	**81.32**	**41.63**
SPLADE -N	34.59	96.50	68.56	79.65	67.55	83.60	43.97
+FGSM	**36.02**†	**96.97**†	**70.13**	**82.44**†	**69.18**	**86.61**†	**45.51**
bi-encoder -D	37.13	**97.43**	71.08	**81.18**	69.68	**83.95**	**45.13**
+FGSM	**37.49**†	97.20	**71.42**	79.80	**70.32**	82.97	44.90
SPLADE -D	37.05	**97.89**	72.99	85.62	70.05	88.77	**49.55**
+FGSM	**37.51**†	97.78	**73.49**	**86.40**	**70.89**	**88.84**	49.41

FGSM for the dense model (+2.25 on MS MARCO), but also for the sparse model (+1.43) in-domain. For models with distillation, we observe that there is an improvement on MS MARCO dev MRR@10, in particular for SPLADE-D, but this improvement is more contested for the dense bi-encoder. For fair comparison, we mention here that we kept similar FLOPS (with and without FGSM) in the case of SPLADE: with FLOPS of 1.3 for the negative training models (-N), and FLOPS of 1.0 for the distilled models (-D)[3]. Now looking at performances out-of-domain, we have a high increase on models without distillation with FGSM on the 13 BEIR-datasets. This gain seems to saturate for the distilled models. Overall, FGSM has a very similar behaviour on the dense and sparse architectures, as a proof of its consistency.

From the observations made on the distilled models, we hypothesis that FGSM-AT and distillation have both a similar label smoothing effect: through the distilled scores and the MSE-loss for distillation, and through the adversarial perturbations for FGSM. This would explain the mixed gains in this case, and why performance increases do not add up. However, note that FGSM-AT smooths representations without requiring external knowledge from a reranker, in difference to distillation.

4.2   RQ2: Does FGSM Adversarial Training Increase Performances in Environments with Noise in Queries Such as Typos?

For the second research question, we examine the effect of adversarial training on queries with typos. To do so, we evaluate our models on the queries varia-

---

[3] Having FLOPS values around 1.0 is a common practice with SPLADE to have a good efficiency-effectiveness trade-off.

Table 2. Robustness to variation from the query-variation generator dataset - on TREC DL 2019 queries (nDCG@10). D-N and D-D are resp. the non-distil/distil version of dense bi-encoder. The same notation is used for SPLADE with S-N and S-D.

#	Q-Variation	D-N	+FGSM	S-N	+FGSM	D-D	+FGSM	S-D	+FGSM
a	Original	65.93	**69.24**†	68.586	**70.13**	71.08	**71.42**	72.99	**73.49**
b	RandomChar	38.82	**41.93**	44.14	**46.76**	45.52	**47.33**	49.66	48.20
c	NeighbChar	36.35	**41.18**†	45.63	44.31	48.86	**49.11**	51.28	**53.04**
d	QWERTYChar	34.43	**40.87**†	43.73	43.07	46.29	**48.01**	49.23	**49.71**
e	RMStopWords	63.2	**66.90**†	69.07	**70.18**	70.46	**70.90**	71.53	**71.63**
f	T5DescToTitle	59.34	**63.10**†	61.69	61.02	64.28	**65.92**	64.68	**64.77**
g	RandomOrder	65.81	**67.62**	67.58	**69.46**	70.76	**71.36**	71.58	**72.01**
h	BackTransla	58.06	**61.29**	56.77	**59.76**	61.36	**63.78**	60.95	**64.26**
i	T5QQP	63.84	**64.62**	64.91	64.36	**69.50**	68.36	66.31	**67.73**
j	WordEmbSyn	60.30	**63.34**	57.59	**61.92**	67.67	**69.82**	69.06	**69.42**
k	WordNetSyn	45.31	**60.58**†	62.43	**64.25**	61.62	**63.41**†	61.67	60.32
l	**Average**	53.73	**58.25**	58.38	**59.57**	61.60	**62.68**	62.65	**63.14**

tions dataset [15], based on TREC DL 19. Table 2 contains variations in queries that do not change the semantic of the original query, but apply noise on it, with **typos** (rows b/c/d), **paraphrasing** (h/i/j/k) and changes in the **word ordering** (g) or the **naturality** (e/f). First, independently from FGSM-AT, we can observe the important drops in all categories, especially for the typos. On typos, SPLADE seems to be naturally more robust than the dense (+4.65 in average without distillation), even-though drops are really important for both models. The better performances of SPLADE may be due to the natural robustness brought by the MLM head.

Now on FGSM-AT, our observation is that, while FGSM-AT is a general method (not a priori focus on one type of noise), it helps in almost all cases. In particular we see gains on paraphrasing for all models, and even on queries with typos for the dense. Due to the small number of queries, lots of p-values are over 0.05, however, by computing the mean per category, we observe – while not reported in the table – that D-N, S-N and D-D have statistically significant increases for paraphrasing (p-values < 0.05). This suggest that representations of models trained with FGSM-AT are more robust, and queries with the same intent will be closer to each other.

4.3 RQ3: Is FGSM Adversarial Training Beneficial for Domain Ada-Ptation?

As a final research question, we consider the case of scarce training data, through the example of domain adaptation, to investigate if FGSM-AT could mitigate overfitting of pre-trained IR models. For this experiment, we start from the

Table 3. Domain adaptation comparison on BEIR Datasets for the dense bi-encoder.

Dataset	Fever		FiQA		NFCorpus	
	nDCG@10	R@100	nDCG@10	R@100	nDCG@10	R@100
Zero-shot	76.98	93.54	29.38	58.08	29.17	25.35
Finetuning	84.46	**95.78**	**32.66**	**61.75**	38.05	46.43
+FGSM	**87.10**	95.45	29.75	61.69	**39.42**	**48.00**

previous distilled dense bi-encoder (D-D) trained on MS MARCO, and finetune it with negative training triplets from résp. datasets from the BEIR benchmark (as for the experiments in Sect. 4.1, we sampled 32 negatives per query from SPLADE-Cocondenser, and also used lower learning rates for adaptation). Only few of BEIR datasets have actual train/dev/test sets which is why we perform our experiment on Fever, FiQA and NFCorpus (containing resp. 110k, 5.5k and 2.6k training queries). Training is done in 100 epochs for FiQA and NFCorpus, and 10 epochs for Fever, with the best checkpoint being selected using the dev set. Training sets being relatively small, we need to train models with a high number of epochs, which is our motivation for using FGSM-AT on this particular settings to smooth representations. Another motivation is that training a reranker for distillation is challenging with only few training samples, and also distillation would require to retrain a reranker for each of the new domain, which is expensive.

Table 3 reports the finetuning results. First, we notice that the distilled bi-encoder – initially trained on MS MARCO with distillation – is able to learn from the new BEIR annotations, in particular on Fever and NFCorpus (+7.48 and +8.88 nDCG@10 resp.), and overall that FGSM-AT prevents the models from overfitting. Results of FGSM-AT are different on FiQA, but this dataset is also the one on which models have the most struggle to learn from the new annotations (gains from only +2.28), so the different behaviour may be due to poor training data, more than FGSM-AT in itself.

5   Conclusion

In this study, we experimented with FGSM to train first stage rankers. Our experiments revealed that a simple regularization on the embedding space could increase the in-domain performances on MS MARCO, especially for models trained without distillation, on which it additionally strengthen the generalization capacities. Besides, FGSM-AT enables a better adaptation to new domains, even on top of distilled models. In future work, we plan to investigate adversarial training directly on rerankers to see if improvements on rerankers could transfer during distillation. Finally, we hope our study would encourage the community to reconsider this baseline method when dealing with robustness issues.

References

1. Athalye, A., Carlini, N., Wagner, D.: Obfuscated gradients give a false sense of security: circumventing defenses to adversarial examples (2018). https://doi.org/10.48550/ARXIV.1802.00420. https://arxiv.org/abs/1802.00420
2. Bajaj, P., et al.: MS MARCO: a human generated machine reading comprehension dataset (2018)
3. Carlini, N., Terzis, A.: Poisoning and backdooring contrastive learning (2021). https://doi.org/10.48550/ARXIV.2106.09667. https://arxiv.org/abs/2106.09667
4. Formal, T., Lassance, C., Piwowarski, B., Clinchant, S.: SPLADE v2: sparse lexical and expansion model for information retrieval (2021). https://doi.org/10.48550/ARXIV.2109.10086. https://arxiv.org/abs/2109.10086
5. Formal, T., Lassance, C., Piwowarski, B., Clinchant, S.: From distillation to hard negative sampling: making sparse neural IR models more effective (2022). https://doi.org/10.48550/ARXIV.2205.04733. https://arxiv.org/abs/2205.04733
6. Goodfellow, I.J., Shlens, J., Szegedy, C.: Explaining and harnessing adversarial examples (2015)
7. Hofstätter, S., Althammer, S., Schröder, M., Sertkan, M., Hanbury, A.: Improving efficient neural ranking models with cross-architecture knowledge distillation (2020). https://doi.org/10.48550/ARXIV.2010.02666. https://arxiv.org/abs/2010.02666
8. Karpukhin, V., et al.: Dense passage retrieval for open-domain question answering (2020)
9. Li, L., Ma, R., Guo, Q., Xue, X., Qiu, X.: BERT-ATTACK: adversarial attack against BERT using BERT (2020). https://doi.org/10.48550/ARXIV.2004.09984. https://arxiv.org/abs/2004.09984
10. Liu, J., et al.: Order-Disorder: imitation adversarial attacks for black-box neural ranking models (2022). https://doi.org/10.48550/ARXIV.2209.06506. https://arxiv.org/abs/2209.06506
11. Ma, X., Nogueira dos Santos, C., Arnold, A.O.: Contrastive fine-tuning improves robustness for neural rankers (2021). https://doi.org/10.48550/ARXIV.2105.12932. https://arxiv.org/abs/2105.12932
12. Madry, A., Makelov, A., Schmidt, L., Tsipras, D., Vladu, A.: Towards deep learning models resistant to adversarial attacks (2017). https://doi.org/10.48550/ARXIV.1706.06083. https://arxiv.org/abs/1706.06083
13. van den Oord, A., Li, Y., Vinyals, O.: Representation learning with contrastive predictive coding (2018). https://doi.org/10.48550/ARXIV.1807.03748. https://arxiv.org/abs/1807.03748
14. Orvieto, A., Kersting, H., Proske, F., Bach, F., Lucchi, A.: Anticorrelated noise injection for improved generalization (2022). https://doi.org/10.48550/ARXIV.2202.02831. https://arxiv.org/abs/2202.02831
15. Penha, G., Câmara, A., Hauff, C.: Evaluating the robustness of retrieval pipelines with query variation generators (2021). https://doi.org/10.48550/ARXIV.2111.13057. https://arxiv.org/abs/2111.13057
16. Reimers, N., Gurevych, I.: Sentence-BERT: sentence embeddings using Siamese BERT-networks (2019). https://doi.org/10.48550/ARXIV.1908.10084. https://arxiv.org/abs/1908.10084
17. Robertson, S.: The probabilistic relevance framework: BM25 and beyond. Found. Trends® Inf. Retrieval **3**(4), 333–389 (2009). http://scholar.google.de/scholar.bib?q=info:U4l9kCVIssAJ:scholar.google.com/&output=citation&hl=de&as_sdt=2000&as_vis=1&ct=citation&cd=1

18. Santhanam, K., Khattab, O., Saad-Falcon, J., Potts, C., Zaharia, M.: ColBERTv2: effective and efficient retrieval via lightweight late interaction (2021). https://doi.org/10.48550/ARXIV.2112.01488. https://arxiv.org/abs/2112.01488

19. Sidiropoulos, G., Kanoulas, E.: Analysing the robustness of dual encoders for dense retrieval against misspellings. In: Proceedings of the 45th International ACM SIGIR Conference on Research and Development in Information Retrieval. ACM, July 2022. https://doi.org/10.1145/3477495.3531818

20. Thakur, N., Reimers, N., Rücklé, A., Srivastava, A., Gurevych, I.: BEIR: a heterogenous benchmark for zero-shot evaluation of information retrieval models (2021). https://doi.org/10.48550/ARXIV.2104.08663. https://arxiv.org/abs/2104.08663

21. Wang, Y., Lyu, L., Anand, A.: BERT rankers are brittle: a study using adversarial document perturbations (2022). https://doi.org/10.48550/ARXIV.2206.11724. https://arxiv.org/abs/2206.11724

22. Wu, C., et al.: Certified robustness to word substitution ranking attack for neural ranking models (2022)

23. Wu, C., Zhang, R., Guo, J., de Rijke, M., Fan, Y., Cheng, X.: PRADA: practical black-box adversarial attacks against neural ranking models (2022). https://doi.org/10.48550/ARXIV.2204.01321. https://arxiv.org/abs/2204.01321

24. Zhuang, S., Zuccon, G.: Dealing with typos for BERT-based passage retrieval and ranking (2021). https://doi.org/10.48550/ARXIV.2108.12139. https://arxiv.org/abs/2108.12139

25. Zhuang, S., Zuccon, G.: CharacterBERT and self-teaching for improving the robustness of dense retrievers on queries with typos. In: Proceedings of the 45th International ACM SIGIR Conference on Research and Development in Information Retrieval. ACM, July 2022. https://doi.org/10.1145/3477495.3531951

Dialogue-to-Video Retrieval

Chenyang Lyu$^{(\boxtimes)}$, Manh-Duy Nguyen, Van-Tu Ninh, Liting Zhou,
Cathal Gurrin, and Jennifer Foster

School of Computing, Dublin City University, Dublin, Ireland
{chenyang.lyu2,manh.nguyen5,van.ninh2}@mail.dcu.ie,
{liting.zhou,cathal.gurrin,jennifer.foster}@dcu.ie

Abstract. Recent years have witnessed an increasing amount of dialogue/conversation on the web especially on social media. That inspires the development of dialogue-based retrieval, in which retrieving videos based on dialogue is of increasing interest for recommendation systems. Different from other video retrieval tasks, dialogue-to-video retrieval uses structured queries in the form of user-generated dialogue as the search descriptor. We present a novel dialogue-to-video retrieval system, incorporating structured conversational information. Experiments conducted on the AVSD dataset show that our proposed approach using plain-text queries improves over the previous counterpart model by 15.8% on R@1. Furthermore, our approach using dialogue as a query, improves retrieval performance by 4.2%, 6.2%, 8.6% on R@1, R@5 and R@10 and outperforms the state-of-the-art model by 0.7%, 3.6% and 6.0% on R@1, R@5 and R@10 respectively.

Keywords: Dialog-based retrieval · Dialogue search query · Conversational information

1 Introduction

The aim of a video retrieval system is to find the best matching videos according to the queries provided by the users [5, 8, 20, 25, 26]. Video retrieval has significant practical value as the vast volume of videos on the web has triggered the need for efficient and effective video search systems. In this paper, we focus on improving the performance of video retrieval systems by combining both textual descriptions of the target video with interactive dialogues between users discussing the content of the target video.

Previous work on video retrieval applied a CNN-based architecture [12, 16, 18] combined with an RNN network [3] to handle visual features and their time-series information [2, 30, 32]. Meanwhile, another RNN model was employed to embed a textual description into the same vector space as the video, so that their similarity could be computed in order to perform the retrieval [2, 26, 32]. Due to the huge impact of the transformer architecture [29] in both text and

---

C. Lyu, M.-D. Nguyen and V.-T. Ninh—Contributed equally.

© The Author(s), under exclusive license to Springer Nature Switzerland AG 2023
J. Kamps et al. (Eds.): ECIR 2023, LNCS 13981, pp. 493–501, 2023.
https://doi.org/10.1007/978-3-031-28238-6_40

image modalities, this network has also been widely applied in the video retrieval research field, obtaining improvements over previous approaches [4,9,13,17,22].

Current video retrieval research, however, mainly focuses on plain text queries such as video captions or descriptions. The need to search videos using queries with complex structures becomes more important when the initial simple text query is ambiguous or not sufficiently well described to find the correct relevant video. Nevertheless, there are only a few studies that focus on this problem [23,24]. Madusa et al. [23] used a dialogue, a sequence of questions and answers about a video, as a query to perform the retrieval because this sequential structure contains rich and detailed information. Specifically, starting with a simple initial description, a video retrieval model would return a list of matching videos from which a question and its answer were generated to create an extended dialogue. This iterative process continued until the correct video was found. Unlike the model of Maeoki et al. [24] which applied a CNN-based encoder and an LSTM [14] to embed data from each modality and to generate questions and answers, Madusa et al's system, VIRED [23], applied Video2Sum [28] to convert a video into a textual summary which can be used with the initial query to get the generated dialogue with the help of a BART model [19].

In this paper, we focus on a less-studied aspect of video retrieval: dialogue-to-video retrieval where the search query is a user-generated dialogue that contains structured information from each turn of the dialogue. The need for dialogue-to-video retrieval derives from the increasing amount of online conversations on social media, which inspires the development of effective dialogue-to-video retrieval systems for many purposes, especially recommendation systems [1,11,33]. Different from general text-to-video retrieval, dialogue-to-video uses user-generated dialogues as the search query to retrieve videos. The dialogue contains user discussion about a certain video, which provides dramatically different information than a plain-text query. This is because during the interaction between users in the dialogue, a discussion similar to the following could happen "A: *The main character of that movie was involved in a horrible car accident when he was 13.* B: *No, I think you mean another character.*". Such discussion contains subtle information about the video of interest and thus cannot be treated as a plain-text query.

Therefore, to incorporate the conversational information from dialogues, we propose a novel dialogue-to-video retrieval approach. In our proposed model, we sequentially encode each turn of the dialogue to obtain a dialogue-aware query representation with the purpose of retaining the dialogue information. Then we calculate the similarity between this dialogue-aware query representation and individual frames in the video in order to obtain a weighted video representation. Finally, we use the video representation to compute an overall similarity score with the dialogue-aware query. To validate the effectiveness of our approach, we conduct dialogue-to-video experiments on a benchmark dataset AVSD [1]. Experimental results show that our approach achieves significant improvements over previous state-of-the-art models including FIT and VIRED [4,23,24].

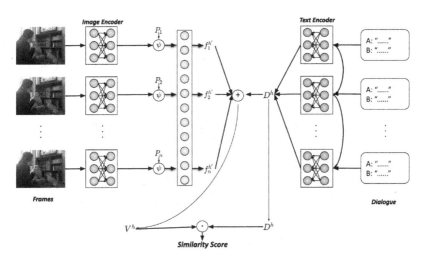

Fig. 1. The architecture of our proposed approach.

2   Methodology

In this section, we describe how our dialogue-to-video retrieval system works. Our retrieval system consists of two major components: 1) a ***temporal-aware video encoder*** responsible for encoding the image frames in video with temporal information. 2) a ***dialogue-query encoder*** responsible for encoding the dialogue query with conversational information. As shown in Fig. 1, our model receives video-query pairs and produces similarity scores. Each video consists of n frames: $V = \{f_1, f_2, ..., f_n\}$ and each dialogue query is composed of m turns of conversation: $D = \{d_1, d_2, ..., d_m\}$.

In the *video encoder*, we encode each frame f_i to its visual representation f_i^h. Then we incorporate temporal information to the corresponding frame representation and feed them into a stacked MULTI-HEAD-ATTENTION module, yielding temporal frame representation $f_i^{h'}$. In the *dialogue-query encoder*, we sequentially encode D by letting $d_i^h = \text{TEXT-ENCODER}(d_{i-1}^h, d_i)$ in order to produce a dialogue-history-aware dialogue representation. We then obtain the final dialogue-query representation by fusing all d_i^h: $D^h = g(d_1^h, ..., d_m^h)$ where g represents our fusion function. After obtaining D^h, we use it to calculate similarities with each frame $f_i^{h'}$, which are then used to obtain a video representation V^h based on the weighted summation of all $f_i^{h'}$. Finally, we obtain the dialogue-to-video similarity score using the dot-product between D^h and V^h.

2.1   Temporal-Aware Video Encoder

Our *temporal-aware video encoder*, which is built on Vision Transformer [7] firstly encodes each frame f_i to its visual representation:

$$f_i^h = \text{IMAGE-ENCODER}(f_i) \tag{1}$$

Then we inject the positional information of the corresponding frame in the video to the frame representation and feed it to the MULTI-HEAD-ATTENTION module:

$$f_i^{h'} = \text{MULTI-HEAD-ATTENTION}([f_1^p, \ldots\ldots, f_n^p]) \tag{2}$$

where f_i^p is the frame representation with positional information $f_i^p = \psi(f_i^h, p_i)$ and p_i is the corresponding positional embedding. Practically, we add *absolute* positional embedding vectors to frame representation as in BERT [6]: $f_i^p = f_i^h + p_i$. Finally, we obtain the temporal-aware video representation $V^{h'} = \{f_1^{h'}, \ldots\ldots, f_n^{h'}\}$.

2.2   Dialogue-Query Encoder

The dialogue-query encoder is responsible for encoding the dialogue-query $D = \{d_1, d_2, \ldots\ldots, d_m\}$:

$$d_i^h = \text{TEXT-ENCODER}(d_{i-1}^h, d_i) \tag{3}$$

where TEXT-ENCODER is a Transformer-based encoder model [6,27,29] in our experiments. Then we fuse all d_i^h to obtain a dialogue-level representation D^h for the dialogue-query:

$$D^h = g(d_1^h, \ldots\ldots, d_m^h) \tag{4}$$

2.3   Interaction Between Video and Dialogue-Query

To calculate the similarity score between each V and D, we firstly compute the similarity scores between dialogue-query D^h and each frame $f_i^{h'}$. Then we obtain a weighted summation of all frames $f_i^{h'}$ as the video representation V^h:

$$V^h = \sum_{i=1}^{n} c_i f_i^h \tag{5}$$

$$c_i = \frac{e^{\phi(D^h, f_i^h)}}{\sum_{j=1}^{n} e^{\phi(D^h, f_j^h)}} \tag{6}$$

The final similarity score is obtained by dot-product between D^h and V^h:
$s = D^h(V^h)^T$

2.4   Training Objective

We perform in-batch contrastive learning [10,15]. For a batch of N video-dialogue pairs $\{(V_1, D_1), \ldots, (V_N, D_N)\}$, the dialogue-to-video and video-to-dialogue match loss are:

$$L_{d2v} = -\frac{1}{N} \sum_{i=1}^{N} \frac{e^{D_i^h (V_i^h)^T}}{\sum\limits_{j=1}^{N} e^{D_i^h (V_j^h)^T}} \tag{7}$$

$$L_{v2d} = -\frac{1}{N} \sum_{i=1}^{N} \frac{e^{D_i^h (V_i^h)^T}}{\sum\limits_{j=1}^{N} e^{D_j^h (V_i^h)^T}} \tag{8}$$

The overall loss to be minimized during the training process is $L = (L_{d2v} + L_{v2d})/2$.

3   Experiments

3.1   Dataset

We conduct our experiments on the popular video-dialogue dataset: AVSD [1].[1] In AVSD, each video is associated with a 10-round dialogue discussing the content of the corresponding video. We follow the standard dataset split of AVSD [1,24], 7,985 videos for training, 863 videos for validation and 1,000 videos for testing.

3.2   Training Setup

Our implementation is based on CLIP [27] from Huggingface [31]. CLIP is used to initialize our IMAGE-ENCODER and TEXT-ENCODER. For performance and efficiency consideration, we employ ViT-B/16 [27] as our image encoder.[2] We train our system with a learning rate of 1×10^{-5} for 10 epochs, with a batch size of 16. We use a maximum gradient norm of 1. The optimizer we used is AdamW [21], for which the ϵ is set to 1×10^{-8}. We perform early stopping when the performance on validation set degrades. We employ R@K, Median Rank and Mean Rank as evaluation metrics [1].

3.3   Results

We present our experimental results on the test set of AVSD [1] in Table 1, where we also show the results of recent baseline models including: 1) LSTM [24], an LSTM-based interactive video retrieval model; 2) FIT [4], a Transformer-based

---

[1] https://video-dialog.com.
[2] https://openai.com/blog/clip/.

Table 1. Experimental results on AVSD dataset

	Use dialogue	R@1	R@5	R@10	MedRank	MeanRank
LSTM [24]	✓	4.2	13.5	22.1	N/A	119
FIT [4]	✗	5.6	18.4	27.5	25	95.4
FIT + Dialogue [4]	✓	10.8	28.9	40	18	58.7
VIRED [23]	✓	24.9	49.0	60.8	6.0	30.3
D2V + Script	✗	21.4	45.9	57.5	9.0	39.8
D2V + Summary	✗	23.4	48.5	59.1	6.0	33.5
D2V + Dialogue	✓	**25.6**	**52.1**	**65.1**	**5.0**	**28.9**

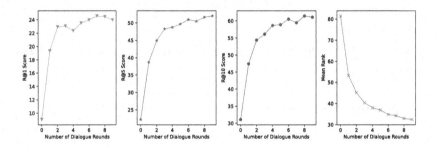

Fig. 2. Effect of dialogue rounds

text-to-video retrieval model using the video summary as the search query; 3) FIT [4] + Dialogue, the FIT model with dialogue in AVSD [1] as the search query[3]; 4) VIRED [23], a video retrieval system based on FIT and CLIP [27] using the dialogue summary as the initial query and model-generated dialogue as an additional query. In Table 1, our model is named D2V (**D**ialogue-to-**V**ideo). We also include the results of our system using the video caption (script in AVSD dataset) – D2V+SCRIPT – and the dialogue summary (summary in AVSD dataset) as the search query – D2V+SUMMARY.

The results in Table 1 show that our proposed approach, D2V, achieves superior performance compared to previous models. First, D2V+SCRIPT with plain-text video caption input outperforms its counterpart FIT by a large margin (15.8 R@1 improvement) and even obtains significant improvements (by 10.6 R@1) over FIT using dialogue as input. That shows the effectiveness of our proposed model architecture. Second, D2V+DIALOGUE significantly outperforms D2V+SCRIPT and D2V+SUMMARY by 3.2 R@1 and 2.2 R@1 respectively, which demonstrates the benefit of incorporating dialogue as a search query. The results in Table 1 show that the dialogue does indeed contain important information about the video content and demonstrates the plausibility of using dialogue as a search query.

---

[3] We concatenate all the rounds of dialogue as plain text to serve as the search query.

Effect of Dialogue Rounds. We investigate the effect of dialogue rounds on the retrieval performance. The results on the validation set of AVSD are shown in Fig. 2, where we use a varying number of dialogue rounds (from 1 to 10) when retrieving videos. We observe a consistent improvement with an increasing number of dialogue rounds. The results show that with more rounds of dialogue, we can obtain better retrieval performance. The improvement brought by increasing the dialogue rounds is more significant especially in the early stage (when using 1 round of dialogue versus 3 rounds).

4 Conclusion

In this paper, we proposed a novel dialogue-to-video retrieval model which incorporates conversational information from dialogue-based queries. Experimental results on the AVSD benchmark dataset show that our approach with a plaintext query outperforms previous state-of-the-art models. Moreover, our model using dialogue as a search query yields further improvements in retrieval performance, demonstrating the importance of utilising dialogue information.

Acknowledgements. This work was funded by Science Foundation Ireland through the SFI Centre for Research Training in Machine Learning (18/CRT/6183). We thank the reviewers for their helpful comments.

References

1. Alamri, H., et al.: Audio-visual scene-aware dialog. In: Proceedings of the IEEE Conference on Computer Vision and Pattern Recognition (2019)
2. Anne Hendricks, L., Wang, O., Shechtman, E., Sivic, J., Darrell, T., Russell, B.: Localizing moments in video with natural language. In: ICCV (2017)
3. Bahdanau, D., Cho, K.H., Bengio, Y.: Neural machine translation by jointly learning to align and translate. In: 3rd International Conference on Learning Representations, ICLR 2015 (2015)
4. Bain, M., Nagrani, A., Varol, G., Zisserman, A.: Frozen in time: a joint video and image encoder for end-to-end retrieval. In: IEEE International Conference on Computer Vision (2021)
5. Cheng, X., Lin, H., Wu, X., Yang, F., Shen, D.: Improving video-text retrieval by multi-stream corpus alignment and dual softmax loss. arXiv:2109.04290 (2021)
6. Devlin, J., Chang, M.W., Lee, K., Toutanova, K.: BERT: pre-training of deep bidirectional transformers for language understanding. In: Proceedings of the 2019 Conference of the North American Chapter of the Association for Computational Linguistics: Human Language Technologies, Volume 1 (Long and Short Papers), Minneapolis, Minnesota, pp. 4171–4186. Association for Computational Linguistics, June 2019. https://doi.org/10.18653/v1/N19-1423. https://www.aclweb.org/anthology/N19-1423
7. Dosovitskiy, A., et al.: An image is worth 16 × 16 words: transformers for image recognition at scale. In: International Conference on Learning Representations (2020)

8. Dzabraev, M., Kalashnikov, M., Komkov, S., Petiushko, A.: MDMMT: multido-main multimodal transformer for video retrieval. In: CVPR (2021)
9. Gabeur, V., Sun, C., Alahari, K., Schmid, C.: Multi-modal transformer for video retrieval. In: Vedaldi, A., Bischof, H., Brox, T., Frahm, J.-M. (eds.) ECCV 2020. LNCS, vol. 12349, pp. 214–229. Springer, Cham (2020). https://doi.org/10.1007/978-3-030-58548-8_13
10. Gao, T., Yao, X., Chen, D.: SimCSE: simple contrastive learning of sentence embeddings. In: Empirical Methods in Natural Language Processing (EMNLP) (2021)
11. He, F., et al.: Improving video retrieval by adaptive margin. In: Proceedings of the 44th International ACM SIGIR Conference on Research and Development in Information Retrieval, pp. 1359–1368 (2021)
12. He, K., Zhang, X., Ren, S., Sun, J.: Deep residual learning for image recognition. In: Proceedings of the IEEE Conference on Computer Vision and Pattern Recognition, pp. 770–778 (2016)
13. Hezel, N., Schall, K., Jung, K., Barthel, K.U.: Efficient search and browsing of large-scale video collections with vibro. In: Þór Jónsson, B., et al. (eds.) MMM 2022. LNCS, vol. 13142, pp. 487–492. Springer, Cham (2022). https://doi.org/10.1007/978-3-030-98355-0_43
14. Hochreiter, S., Schmidhuber, J.: Long short-term memory. Neural Comput. **9**(8), 1735–1780 (1997)
15. Karpukhin, V., et al.: Dense passage retrieval for open-domain question answering. In: Empirical Methods in Natural Language Processing (EMNLP) (2020)
16. Krizhevsky, A., Sutskever, I., Hinton, G.E.: ImageNet classification with deep convolutional neural networks. Commun. ACM **60**(6), 84–90 (2017)
17. Le, T.-K., Ninh, V.-T., Tran, M.-K., Healy, G., Gurrin, C., Tran, M.-T.: AVSeeker: an active video retrieval engine at VBS2022. In: Þór Jónsson, B., et al. (eds.) MMM 2022. LNCS, vol. 13142, pp. 537–542. Springer, Cham (2022). https://doi.org/10.1007/978-3-030-98355-0_51
18. LeCun, Y., Bottou, L., Bengio, Y., Haffner, P.: Gradient-based learning applied to document recognition. Proc. IEEE **86**(11), 2278–2324 (1998)
19. Lewis, M., et al.: BART: denoising sequence-to-sequence pre-training for natural language generation, translation, and comprehension. In: Proceedings of the 58th Annual Meeting of the Association for Computational Linguistics, pp. 7871–7880. Association for Computational Linguistics, July 2020. https://doi.org/10.18653/v1/2020.acl-main.703. https://www.aclweb.org/anthology/2020.acl-main.703
20. Liu, Y., Albanie, S., Nagrani, A., Zisserman, A.: Use what you have: video retrieval using representations from collaborative experts. arXiv:1907.13487 (2019)
21. Loshchilov, I., Hutter, F.: Decoupled weight decay regularization. In: International Conference on Learning Representations (2019). https://openreview.net/forum?id=Bkg6RiCqY7
22. Luo, H., et al.: CLIP4Clip: an empirical study of CLIP for end to end video clip retrieval and captioning. Neurocomputing **508**, 293–304 (2022)
23. Madasu, A., Oliva, J., Bertasius, G.: Learning to retrieve videos by asking questions. arXiv preprint arXiv:2205.05739 (2022)
24. Maeoki, S., Uehara, K., Harada, T.: Interactive video retrieval with dialog. In: Proceedings of the IEEE/CVF Conference on Computer Vision and Pattern Recognition Workshops, pp. 952–953 (2020)
25. Miech, A., Laptev, I., Sivic, J.: Learning a text-video embedding from incomplete and heterogeneous data. arXiv:1804.02516 (2018)
26. Mithun, N.C., Li, J., Metze, F., Roy-Chowdhury, A.K.: Learning joint embedding with multimodal cues for cross-modal video-text retrieval. In: ICMR (2018)

27. Radford, A., et al.: Learning transferable visual models from natural language supervision. In: International Conference on Machine Learning, pp. 8748–8763. PMLR (2021)

28. Song, Y., Chen, S., Jin, Q.: Towards diverse paragraph captioning for untrimmed videos. In: Proceedings of the IEEE/CVF Conference on Computer Vision and Pattern Recognition, pp. 11245–11254 (2021)

29. Vaswani, A., et al.: Attention is all you need. In: Proceedings of the 31st International Conference on Neural Information Processing Systems, NIPS 2017, USA, pp. 6000–6010. Curran Associates Inc. (2017). http://dl.acm.org/citation.cfm?id=3295222.3295349

30. Venugopalan, S., Rohrbach, M., Donahue, J., Mooney, R., Darrell, T., Saenko, K.: Sequence to sequence-video to text. In: Proceedings of the IEEE International Conference on Computer Vision, pp. 4534–4542 (2015)

31. Wolf, T., et al.: Transformers: state-of-the-art natural language processing. In: Proceedings of the 2020 Conference on Empirical Methods in Natural Language Processing: System Demonstrations, pp. 38–45. Association for Computational Linguistics, October 2020. https://doi.org/10.18653/v1/2020.emnlp-demos.6. https://aclanthology.org/2020.emnlp-demos.6

32. Yang, X., Zhang, T., Xu, C.: Text2Video: an end-to-end learning framework for expressing text with videos. IEEE Trans. Multimedia **20**(9), 2360–2370 (2018)

33. Zheng, Y., Chen, G., Liu, X., Sun, J.: MMChat: multi-modal chat dataset on social media. In: Proceedings of the 13th Language Resources and Evaluation Conference. European Language Resources Association (2022)

Time-Dependent Next-Basket Recommendations

Sergey Naumov[1,3], Marina Ananyeva[1,2], Oleg Lashinin[1(✉)], Sergey Kolesnikov[1], and Dmitry I. Ignatov[2]

[1] Tinkoff, 2-Ya Khutorskaya Ulitsa, 38A, bld. 26, Moscow 127287, Russian Federation
fotol@bk.ru
[2] National Research University Higher School of Economics, 20 Myasnitskaya St, Moscow 101000, Russian Federation
[3] Moscow Institute of Physics and Technology, Institutskiy Pereulok, 9, Dolgoprudny, Moscow Oblast 141701, Russian Federation

Abstract. There are various real-world applications for next-basket recommender systems. One of them is guiding a website user who wants to buy anything toward a collection of items. Recent works demonstrate that methods based on the frequency of prior purchases outperform other deep learning algorithms in terms of performance. These techniques, however, do not consider timestamps and time intervals between interactions. Additionally, they often miss the time period that passes between the last known basket and the prediction time. In this study, we explore whether such knowledge could improve current state-of-the-art next-basket recommender systems. Our results on three real-world datasets show how such enhancement may increase prediction quality. These findings might pave the way for important research directions in the field of next-basket recommendations.

Keywords: Recommender systems · Next-basket recommendations · Time-dependent recommendations

1 Introduction

Next-basket recommender systems (NBR) have been actively studied in the research community [19,24]. The developed methods may employ a variety of data sources, including past user purchases [2,8,11,12], current session click history [1,21], and other user and item attributes [3,10,17,26]. However, state-of-the-art approaches [8,12] usually do not take into account timestamps of interactions. Even though they weigh the baskets according to their order of appearance, they are still not (1) time-aware approaches nor (2) able to generate time-dependent recommendations.

The contribution of D.I. Ignatov to the paper was done within the framework of the HSE University Basic Research Program.

Time-aware recommender models can extract additional information from historical interaction timestamps [27]. If the model does not consider them, it treats the baskets as equidistant. In practise, time gaps are very important when modelling user behaviour. Small gaps between baskets could result in greater dependence on recent interactions in subsequent baskets. According to [18], large time gaps could be a sign of weaker connections between user behaviour in the past and present.

Time-dependent recommendations can change depending on when the predictions were made [25]. In non-time-dependent approaches [2,8,12], users' representations are calculated at the time of the last known basket. However, the user's interests could change if some time elapses between the last known interactions and the prediction time [18]. Fortunately, we have the ability to update representations to reflect the current moment in time. We can use the times when test baskets were purchased in offline experiments. Alternatively, we can use the time period when a user sees recommendations in an online scenario. The key concept is that recommendations change over time, even when a user does not further interact with any items. As a result, these models are known as time-dependent ones [25,27].

Recent works have emphasised the superiority of straightforward frequency-based approaches in the next-basket recommendations [12,16,24]. Unfortunately, the majority of cutting-edge algorithms lack time features. One of them is TIFU-KNN [12], which uses purchase frequency to make recommendations based on the purchases of the target user's neighbours. In this paper, we add time information to TIFU-KNN. Specifically, the main contribution can be listed as follows:

- We modify TIFU-KNN, a state-of-the-art approach for next-basket recommendations, to make it (1) time-aware and (2) time-dependent
- We conduct comprehensive experiments to demonstrate how such a straightforward change can enhance the quality of recommendations on three real-world datasets.

2   Related Work

Different approaches have been applied to solve next-basket recommendations. Previously published works employ Markov Chains [30], Recurrent Neural Networks [2,11,22,30], Attention mechanisms [23,31], Graph Neural Networks [31], and frequency-based approaches [8,12]. Frequency-based methods perform better [19,24] than other methods, despite deep neural networks' great success in other research areas. It emphasises the significance of enhancing frequency-based models.

The addition of time features to recommender models is another area of study. Time is used as additional information by time-aware models to model user interests [4,25,27]. For instance, the well-known SASRec [15] has been improved by the TiSASRec [18], which has a time interval-aware Self-Attention mechanism. Recent independent studies [6,13,20] have revealed that TiSASRec typically outperforms SASRec in terms of quality. Time-dependent models' predictions can differ depending on the current time context (hour, day of the week, or month) [7,9,29] or the time before the most recent interaction [5,14,32]. The authors of [28] introduced DRM that can dynamically change next-basket recommendations based on the user's current time context. Although it makes sense to use time as a feature for both training and predictions, there are not many time-dependent next-basket approaches.

3   Original and Modified Versions of TIFU-KNN

Original TIFU-KNN is a KNN-based non-DL method described in [12]. Among non-deep-learning models, it currently displays the best results in the next-basket recommendation task [19,24]. The baskets are separated into nearly equal-sized groups. It allows to introduce an additional global time-decayed factor. Within the group, each basket has a unique ordinal number. Utilisation of two different weights simultaneously is the key concept; baskets are weighted within groups, and groups are weighted among themselves. The weight of each basket in the group varies depending on when it was purchased $r_b(i) = r_b^i$ (r_b power i), and $i = 0, 1, \ldots, B(g) - 1$ is the index number from the most recent basket in the group to the earliest basket, $B(g)$ is the number of baskets in the group g. Similarly, earlier groups of baskets have smaller weight $r_g(j) = r_g^j$ (r_g power j), $j = 0, 1, \ldots, G - 1$ from the most recent group to the earliest group.

We consider a I-sized multi-hot vector v_b that represents a basket b, where I is the number of items. If a basket b contains an item i then the corresponding component equals 1, and otherwise equals 0. If the group vector v_g is thus obtained as a weighted average vector of the baskets v_b, and the user vector v_u is taken into consideration as a weighted average vector of the groups v_g:

$$v_g = \sum_{i=0}^{B(g)-1} \frac{r_b(i) \cdot v_{b_i}}{B(g)}, \quad v_u = \sum_{j=0}^{G-1} \frac{r_g(j) \cdot v_{g_j}}{G}, \tag{1}$$

where r_b is the time-decayed ratio within a group, r_g is the time-decayed ratio across groups, $B(g)$ is the number of baskets in group g, and G is the number of groups, v_{b_i} is the vector of the i-th basket, v_{g_j} is the vector of the j-th group, v_u is the user's final vector representation.

The average of the vectors v_u of the k closest users is also calculated for each user's nearest neighbours vector $v_{nn}(v_u)$ (using Euclidean distance).

$$KNN(v_u) = \{v_{u_0}, v_{u_1}, v_{u_2}, \ldots, v_{u_{K-1}}\}, \quad v_{nn}(v_u) = \sum_{i=0}^{K} \frac{KNN(v_u)[i]}{K}. \tag{2}$$

The prediction vector $P(u)$ for each user is the weighted sum of the user's own vector v_u and the nn-vector $v_{nn}(v_u)$:

$$P(u) = \alpha \cdot v_u + (1 - \alpha) \cdot v_{nn}(v_u), \tag{3}$$

where α is the balance coefficient between two parts. $P(u)$ is used to calculate the final recommendations.

Time-Aware TIFU-KNN (TIFU-KNN-TA) is easily attainable with a few adjustments. Each user's entire purchase history is divided into equal time segments of gs days. $ts_{last}(u)$ corresponds to the timestamp of the last train basket of user u. Then the first group's baskets are distributed between $ts_{last}(u) - gs$ and $ts_{last}(u)$. Interactions between $ts_{last}(u) - 2 \cdot gs$ and $ts_{last}(u) - gs$ form the second group. Group timestamp restrictions for user u are as follows:

$$group_m(u) : \ (ts_{last}(u) - (m+1) \cdot gs, \ ts_{last}(u) - m \cdot gs), \quad m = 0, 1, 2, \ldots \quad (4)$$

As a result, each user's group size in days is fixed, but the number of baskets in each group and the total number of groups can vary. For the group $r_g(j)$, the attenuation is still the same as it is in the default TIFU-KNN (Eq. 1). On the other hand, the group's r_b basket coefficient has changed. Instead of the basket number, the exponent is now the number of days until the group's end (or natural logarithm of the number of days, depending on the hyperparameter use_log). Let us denote the right limit of the group g from Eq. 4 as $rl(g)$, and the timestamp of the basket b as $ts(b)$:

$$\Delta ts(b, g) = rl(g) - ts(b), \quad (5)$$

$$r_b(\Delta ts) = r_b^{\Delta ts(b,g)} \quad \text{or} \quad r_b(\Delta ts) = r_b^{\ln (\Delta ts(b,g))}, \quad (6)$$

$$v_g = \sum_{i=0}^{B(g)} \frac{r_b(\Delta ts) \cdot v_{b_i}}{B(g)}. \quad (7)$$

Time-dependent TIFU-KNN (TIFU-KNN-TD) has two differences from Time-aware TIFU-KNN. During the prediction stage, a timestamp of the next basket $ts_{test}(u)$ is served to the model for each user. In offline experiments, this could be the moment when a user buys test or validation baskets. Additionally, we can use time of predictions if the experiments are online. In order to create groups of baskets for the purpose of calculating the user vector v_u^{new}, $ts_{test}(u)$ is used instead of the maximum timestamp $ts_{last}(u)$ from the train. User u now has the following group timestamp limitations:

$$group_m(u) : \ (ts_{test}(u) - (m+1) \cdot gs, \ ts_{test}(u) - m \cdot gs), \quad (8)$$

Thus, on the validation and test stages, the model has new vectors v_u^{new} for each user. However, the nearest neighbour representations are determined for $ts_{last}(u)$. This is done to prevent the need to continually recalculate the vectors of all nearby users. As a result, we calculate vectors v_u^{new} for the target user u using the current moment of time. However, the vectors for nearest neighbours are only based on training stage.

$$KNN(v_u^{new}) = \{v_{u_0}, v_{u_1}, v_{u_2}, \ldots, v_{u_{K-1}}\},$$

$$v_{nn}(v_u^{new}) = \sum_{i=0}^{K} \frac{KNN(v_u^{new})[i]}{K}, \quad (9)$$

$$P(u) = \alpha \cdot v_u^{new} + (1 - \alpha) \cdot v_{nn}(v_u^{new}). \quad (10)$$

Table 1. Dataset statistics after preprocessing.

Dataset	#users	#items	#baskets	#baskets per user	#items per basket	#items per user
Dunnhumby	2471	8644	251361	101.72	7.71	381.09
Tafeng	14006	13674	94274	6.73	6.34	37.61
Instacart	19999	26677	629067	31.45	9.94	100.22

4   Experiments

We have provided experiments to answer the following research questions:

- **RQ1:** Can we increase the quality of recommendations by taking time intervals into account instead of basket numbers?
- **RQ2:** Does the consideration of the time of prediction improve the quality of recommendations?

4.1   Datasets

We make use of the three open source datasets for the Next Basket Recommendation problem to ensure the reproducibility of our research:

- **Dunnhumby**[1] includes transactions of 2,500 households at a retailer over a two-year period. A basket is a collection of all the items that were purchased in a single transaction.
- **TaFeng**[2] includes four months of Chinese grocery store transactions. Each basket contains the user's daily purchases.
- **Instacart**[3] it contains a sample of over 3 million grocery orders from over 200,000 users with an average of 4 to 100 orders from each user. Every order is considered to be one basket.

From each dataset, we remove users with fewer than three baskets and items bought by fewer than five users. We sample 20,000 Instacart users and 10,000 Dunnhumby items before filtering. Table 1 displays the statistics of the datasets after prepossessing. Every dataset was divided into a training, validation, and test set for our experiments. For each user, the training part consists of all baskets except the final one. The remaining baskets are split in half, with 50% going to the test part and 50% to the validation part.

---

[1] https://www.kaggle.com/datasets/frtgnn/dunnhumby-the-complete-journey.
[2] https://www.kaggle.com/datasets/chiranjivdas09/ta-feng-grocery-dataset.
[3] https://www.kaggle.com/competitions/instacart-market-basket-analysis/data.

Table 2. Results of our models compared against the baselines. The best and second best performing models are indicated by boldface and underline, respectively. ▲% shows our models' improvements over the best baseline.

Dataset	Metric	Baselines				Ours	
		G-Pop	GP-Pop	UP-CF@r	TIFU-KNN	TIFU-KNN-TA (▲%)	TIFU-KNN-TD (▲%)
DHB	Recall@5	0.1379	0.2326	0.2397	0.2491	**0.2572** (3.3)	<u>0.2570</u> (3.2)
	nDCG@5	0.1229	0.2222	0.2294	0.2355	**0.2433** (3.3)	<u>0.2422</u> (2.8)
	Recall@10	0.1359	0.2473	0.2611	0.2709	**0.2760** (1.9)	<u>0.2743</u> (1.3)
	nDCG@10	0.1158	0.2188	0.2298	0.2384	**0.2439** (2.3)	<u>0.2425</u> (1.7)
TaFeng	Recall@5	0.0815	0.1026	0.1244	0.1403	<u>0.1415</u> (0.9)	**0.1448** (3.2)
	nDCG@5	0.0895	0.0979	0.1121	<u>0.1347</u>	0.1341 (-0.4)	**0.1393** (3.4)
	Recall@10	0.0841	0.1260	0.1537	0.1632	<u>0.1642</u> (0.6)	**0.1673** (2.5)
	nDCG@10	0.0877	0.1047	0.1227	<u>0.1406</u>	0.1401 (-0.4)	**0.1453** (3.3)
Instacart	Recall@5	0.1092	0.4070	0.4371	0.4524	<u>0.4541</u> (0.4)	**0.4559** (0.8)
	nDCG@5	0.1183	0.4238	0.4527	0.4668	<u>0.4691</u> (0.5)	**0.4725** (1.2)
	Recall@10	0.0969	0.4000	0.4276	<u>0.4476</u>	0.4469 (-0.2)	**0.4496** (0.4)
	nDCG@10	0.1051	0.4039	0.4320	0.4484	<u>0.4493</u> (0.2)	**0.4526** (0.9)

4.2   Baseline Methods

In order to ensure that our research is thorough, we also include the following baselines:

- **G-Pop:** this baseline just recommends the most frequent items in the dataset.
- **GP-Pop:** for each user, the most frequently purchased items are recommended first, followed by the most frequent items in the entire dataset.
- **UP-CF@r:** a hybrid of the recency-aware user-wise popularity and user-wise collaborative filtering presented in [8].

4.3   Metrics

We calculate **Recall** and **nDCG**, which have been used in previous NBR studies, to assess the effectiveness of our methods. Based on the average basket size in the datasets Table 1, we picked values of 5 and 10 for the *topk* parameter.

4.4   Experiment Settings

We search for the optimal parameters using Optuna[4] with 300 trials for each model, optimising Recall@10. The random seeds are all fixed. We make the experiment code available online, including hyperparameter search spaces[5].

---

[4] https://optuna.org.

[5] https://github.com/sergunya17/time_dependent_nbr.

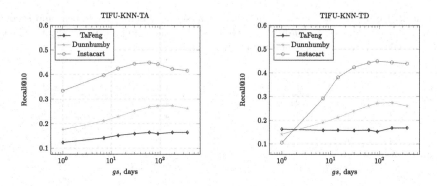

Fig. 1. Recall@10 w.r.t. different *gs* values across all included datasets.

4.5   Results

Table 2 answer both **RQ1** and **RQ2**. TIFU-KNN-TA outperforms all included baselines on Dunnhumby and Instacart and shows similar performance on other datasets (**RQ1**). This demonstrates the value of replacing ordinal number weighting of baskets with weighting based on the amount of time between interactions.

Moreover, TIFU-KNN-TD outperforms all included algorithms both on TaFeng and Instacart. Additionally, it has higher quality on original TIFU-KNN in all experiments (**RQ2**). Finally, TIFU-KNN is better than UP-CF@r on all metrics and datasets which is in line with [2]. As we can see, our modifications improved quality of recommendations by introducing time features both for training and prediction stages.

It is important to note the dependence on hyperparameter values. The two novel hyperparameters for the suggested methods are *use_log* and *gs*. Our experimental findings across all included datasets indicate that the quality is either unchanged or slightly improved when the logarithm is used. Additionally, the quality of recommendations can be completely affected by varying the value of parameter *gs*. While fixing the remaining values in the optimal configuration for each model and dataset, we varied the value of hyperparameter *gs*. Figure 1 shows the results.

5   Conclusion

In this study, we demonstrated the importance of providing time-dependent and time-aware next-basket recommendations. We made some minor adjustments to the state-of-the-art TIFU-KNN next-basket recommender system to show the impact of time context. On three real-world datasets, the quality of next-basket predictions was improved by merely substituting basket number for interaction weighting using timestamp-based descent. We believe that these findings will spur additional study into the creation of time-sensitive next-basket recommendation techniques for both training and prediction phases.

References

1. Ananyeva, M., Lashinin, O., Ivanova, V., Kolesnikov, S., Ignatov, D.I.: Towards interaction-based user embeddings in sequential recommender models. In: Vinagre, J., Al-Ghossein, M., Jorge, A.M., Bifet, A., Peska, L. (eds.) Proceedings of the 5th Workshop on Online Recommender Systems and User Modeling, ORSUM@RecSys 2022. CEUR Workshop Proceedings, vol. 3303. CEUR-WS.org (2022). http://ceur-ws.org/Vol-3303/paper10.pdf
2. Ariannezhad, M., Jullien, S., Li, M., Fang, M., Schelter, S., de Rijke, M.: ReCANet: a repeat consumption-aware neural network for next basket recommendation in grocery shopping. In: Proceedings of the 45th International ACM SIGIR Conference, SIGIR 2022, pp. 1240–1250. ACM (2022)
3. Bai, T., Nie, J.Y., Zhao, W.X., Zhu, Y., Du, P., Wen, J.R.: An attribute-aware neural attentive model for next basket recommendation. In: The 41st International ACM SIGIR Conference on Research & Development in Information Retrieval, pp. 1201–1204 (2018)
4. Campos, P.G., Díez, F., Cantador, I.: Time-aware recommender systems: a comprehensive survey and analysis of existing evaluation protocols. User Model. User-Adap. Interact. **24**(1), 67–119 (2014). https://doi.org/10.1007/s11257-012-9136-x
5. Cao, J., et al.: Deep structural point process for learning temporal interaction networks. In: Oliver, N., Pérez-Cruz, F., Kramer, S., Read, J., Lozano, J.A. (eds.) ECML PKDD 2021. LNCS (LNAI), vol. 12975, pp. 305–320. Springer, Cham (2021). https://doi.org/10.1007/978-3-030-86486-6_19
6. Chen, Z., Zhang, W., Yan, J., Wang, G., Wang, J.: Learning dual dynamic representations on time-sliced user-item interaction graphs for sequential recommendation. In: Proceedings of the 30th ACM International Conference on Information & Knowledge Management, pp. 231–240 (2021)
7. Cho, J., Hyun, D., Kang, S., Yu, H.: Learning heterogeneous temporal patterns of user preference for timely recommendation. In: Proceedings of the Web Conference 2021, pp. 1274–1283 (2021)
8. Faggioli, G., Polato, M., Aiolli, F.: Recency aware collaborative filtering for next basket recommendation. In: Proceedings of the 28th ACM Conference on User Modeling, Adaptation and Personalization, pp. 80–87 (2020)
9. Fan, Z., Liu, Z., Zhang, J., Xiong, Y., Zheng, L., Yu, P.S.: Continuous-time sequential recommendation with temporal graph collaborative transformer. In: Proceedings of the 30th ACM International Conference on Information & Knowledge Management, pp. 433–442 (2021)
10. Fouad, M.A., Hussein, W., Rady, S., Philip, S.Y., Gharib, T.F.: An efficient approach for rational next-basket recommendation. IEEE Access **10**, 75657–75671 (2022)
11. Hu, H., He, X.: Sets2Sets: learning from sequential sets with neural networks. In: Proceedings of the 25th ACM SIGKDD International Conference on Knowledge Discovery & Data Mining, pp. 1491–1499 (2019)
12. Hu, H., He, X., Gao, J., Zhang, Z.L.: Modeling personalized item frequency information for next-basket recommendation. In: Proceedings of the 43rd International ACM SIGIR Conference on Research and Development in Information Retrieval, pp. 1071–1080 (2020)
13. Huang, Z., Ma, J., Dong, Y., Foutz, N.Z., Li, J.: Empowering next POI recommendation with multi-relational modeling. arXiv preprint arXiv:2204.12288 (2022)

14. Ji, W., Wang, K., Wang, X., Chen, T., Cristea, A.: Sequential recommender via time-aware attentive memory network. In: Proceedings of the 29th ACM International Conference on Information & Knowledge Management, pp. 565–574 (2020)
15. Kang, W.C., McAuley, J.: Self-attentive sequential recommendation. In: 2018 IEEE International Conference on Data Mining (ICDM), pp. 197–206. IEEE (2018)
16. Kolesnikov, S., Lashinin, O., Pechatov, M., Kosov, A.: TTRS: Tinkoff transactions recommender system benchmark. arXiv preprint arXiv:2110.05589 (2021)
17. Lashinin, O., Ananyeva, M.: Next-basket recommendation constrained by total cost. In: Vinagre, J., et al. (eds.) Proceedings of the 5th Workshop on Online Recommender Systems and User Modeling, ORSUM@RecSys 2022. CEUR Workshop Proceedings, vol. 3303. CEUR-WS.org (2022). http://ceur-ws.org/Vol-3303/brainstorming1.pdf
18. Li, J., Wang, Y., McAuley, J.: Time interval aware self-attention for sequential recommendation. In: Proceedings of the 13th International Conference on Web Search and Data Mining, pp. 322–330 (2020)
19. Li, M., Jullien, S., Ariannezhad, M., de Rijke, M.: A next basket recommendation reality check. arXiv preprint arXiv:2109.14233 (2021)
20. Liu, C., et al.: C^2-Rec: an effective consistency constraint for sequential recommendation. arXiv preprint arXiv:2112.06668 (2021)
21. Moreira, G.S.P., Rabhi, S., Ak, R., Kabir, M.Y., Oldridge, E.: Transformers with multi-modal features and post-fusion context for e-commerce session-based recommendation. arXiv preprint arXiv:2107.05124 (2021)
22. Qin, Y., Wang, P., Li, C.: The world is binary: contrastive learning for denoising next basket recommendation. In: Proceedings of the 44th International ACM SIGIR Conference on Research and Development in Information Retrieval, pp. 859–868 (2021)
23. Ren, P., Chen, Z., Li, J., Ren, Z., Ma, J., De Rijke, M.: RepeatNet: a repeat aware neural recommendation machine for session-based recommendation. In: Proceedings of the AAAI Conference on Artificial Intelligence, vol. 33, pp. 4806–4813 (2019)
24. Shao, Z., Wang, S., Zhang, Q., Lu, W., Li, Z., Peng, X.: A systematical evaluation for next-basket recommendation algorithms. arXiv preprint arXiv:2209.02892 (2022)
25. Shi, Y., Larson, M., Hanjalic, A.: Collaborative filtering beyond the user-item matrix: a survey of the state of the art and future challenges. ACM Comput. Surv. (CSUR) 47(1), 1–45 (2014)
26. Van Maasakkers, L., Fok, D., Donkers, B.: Next-basket prediction in a high-dimensional setting using gated recurrent units. Expert Syst. Appl. 212, 118795 (2023)
27. Vinagre, J., Jorge, A.M., Gama, J.: An overview on the exploitation of time in collaborative filtering. Wiley Interdisc. Rev. Data Min. Knowl. Discov. 5(5), 195–215 (2015)
28. Wang, P., Zhang, Y., Niu, S., Guo, J.: Modeling temporal dynamics of users' purchase behaviors for next basket prediction. J. Comput. Sci. Technol. 34(6), 1230–1240 (2019). https://doi.org/10.1007/s11390-019-1972-2
29. Wu, Y., Li, K., Zhao, G., Xueming, Q.: Personalized long-and short-term preference learning for next poi recommendation. IEEE Trans. Knowl. Data Eng. 34(4), 1944–1957 (2020)
30. Yu, F., Liu, Q., Wu, S., Wang, L., Tan, T.: A dynamic recurrent model for next basket recommendation. In: Proceedings of the 39th International ACM SIGIR Conference, SIGIR 2016, pp. 729–732. Association for Computing Machinery, New York (2016)

31. Yu, L., Sun, L., Du, B., Liu, C., Xiong, H., Lv, W.: Predicting temporal sets with deep neural networks. In: Proceedings of the 26th ACM SIGKDD International Conference on Knowledge Discovery & Data Mining, pp. 1083–1091 (2020)
32. Zhu, Y., et al.: What to do next: modeling user behaviors by time-LSTM. In: IJCAI, vol. 17, pp. 3602–3608 (2017)

Investigating the Impact of Query Representation on Medical Information Retrieval

Georgios Peikos[1]([envelope]) [ID], Daria Alexander[2,3] [ID], Gabriella Pasi[1] [ID],
and Arjen P. de Vries[2] [ID]

[1] University of Milano-Bicocca, Milan, Italy
{georgios.peikos,gabriella.pasi}@unimib.it
[2] Radboud University, Nijmegen, The Netherlands
daria.alexander@ru.nl, arjen@cs.ru.nl
[3] Spinque, Utrecht, The Netherlands

Abstract. This study investigates the effect that various patient-related information extracted from unstructured clinical notes has on two different tasks, i.e., patient allocation in clinical trials and medical literature retrieval. Specifically, we combine standard and transformer-based methods to extract entities (e.g., drugs, medical problems), disambiguate their meaning (e.g., family history, negations), or expand them with related medical concepts to synthesize diverse query representations. The empirical evaluation showed that certain query representations positively affect retrieval effectiveness for patient allocation in clinical trials, but no statistically significant improvements have been identified in medical literature retrieval. Across the queries, it has been found that removing negated entities using a domain-specific pre-trained transformer model has been more effective than a standard rule-based approach. In addition, our experiments have shown that removing information related to family history can further improve patient allocation in clinical trials.

Keywords: Medical information retrieval · Information extraction · Query reformulation

1 Introduction

The widespread adoption of electronic health records (EHRs) in hospitals and ambulatory care settings has created a significant amount of information that can also be exploited for research purposes [16,18]. As EHRs are created to support multiple functionalities, they are a source of diverse patient-related information, such as demographics, diagnoses, family history, current and past medical problems, medications, vital signs, and laboratory data [16]. However, as their primary purpose is to support clinical care rather than research, EHRs can be unstructured (e.g., narrative clinical notes), and contain several textual peculiarities (e.g., medical jargon) and negations. As a result, when clinical notes are

J. Kamps et al. (Eds.): ECIR 2023, LNCS 13981, pp. 512–521, 2023.
https://doi.org/10.1007/978-3-031-28238-6_42

being used as inputs (i.e., queries) in medical Information Retrieval (IR) tasks, their format and content impose several challenges that need to be addressed.

A significant challenge is associated with the presence of negated content, and information on family history [11], the impact of which on clinical trial retrieval has been investigated in previous studies [8,19]. In particular, Koopman and Zuccon showed that in the patient retrieval task, negated content should be negatively weighted, rather than simply being removed. Another challenge is associated with the issue of vocabulary mismatch that may affect retrieval effectiveness. In this direction, Agosti et al. investigated whether query expansion and reduction techniques can improve effectiveness in medical literature and clinical trials retrieval [4].

This work enriches previous works [4,19] by conducting further experiments that combine novel transformer-based methods with standard rule-based ones, for entity extraction and semantic meaning disambiguation. In addition, it aims at gathering new practical insights on the impact of various query representations across the studied medical retrieval tasks, by exploiting additional patient related information, such as a patient's historical information. We specifically aim to answer the following research questions:

1. Does the presence of various medical entities of a clinical note have an impact on the overall retrieval effectiveness?
2. How does the presence of negated content affect retrieval performance?
3. How does the presence of sentences with non-identified medical entities impact retrieval performance?
4. How does the presence of family history and/or patient's historical information affect retrieval performance?
5. What is the impact of medical entity expansion, using a knowledge base, on retrieval performance?

To answer these questions, we employ and combine several state-of-the-art methods for entity extraction and semantic meaning disambiguation, synthesize distinct queries, and compare their effectiveness in terms of achieved retrieval performance across four benchmark collections.

2   Related Work

Query representation plays a crucial role in both of the studied retrieval tasks as it is highly connected with the overall retrieval performance [4,14,19,30,35]. Regarding the clinical trial retrieval task introduced in TREC 2021 [35], most of the proposed systems [3] extract medical conditions, procedures or drugs, and expand them using the Unified Medical Language System (UMLS) [10]. In addition, the retrieval approach proposed in [29] leverages a zero-shot neural query synthesis method that generates multiple queries from an EHR, which are independently used for retrieval. When combined with a neural point-wise re-ranker, this approach achieves a P@10 score of 0.59. Similar query representations have been used for medical literature retrieval in the 2014 and 2015 TREC Clinical

Decision Support (CDS) tracks [30,31,34]. As reported in [34] and [9,31], the top performing retrieval approaches in the corresponding 2014 and 2015 collections achieve P@10 scores equal to 0.39 and 0.52.

In the literature, other information extraction approaches have been introduced whose effect on retrieval effectiveness in the studied tasks has not been investigated yet. Specifically, to extract drugs and dosages, initial works leveraged rule-based methods [6,21,36], while others have introduced neural architectures [25,38,39]. Recently, pre-trained language models, such as BioBERT have been proven effective for medical entity extraction [23]. In addition, identifying negated content in clinical notes is an essential task that has been widely tackled using the ConText algorithm [17]. However, recently a pre-trained transformer-based model has been fine-tuned in domain-specific data to identify negations [5]. Furthermore, the family history identification task has also been investigated in the literature [13,17,33], while its effect on the retrieval performance, under a certain retrieval task, has been outlined in [19]. Lastly, identifying patient related historical information has also been explored in the literature [7,17].

3   Methodology and Experimental Setup

An overview of the proposed methodology for information extraction, entity meaning disambiguation, and entity expansion is presented in Fig. 1. In detail, given a patient's clinical note (i.e., verbose query representation), we synthesized several query representations by combining the displayed methods (Fig. 1). Then, each of the synthesized queries has been used for retrieving both relevant clinical trials and relevant medical literature.

To extract medical problems, treatments, and tests we employed a pre-trained transformer-based NER model [1] trained on the n2c2 dataset introduced in [37]. We synthesized queries based on all of the possible entity combinations and used them for retrieval. In the literature, several publicly available libraries and models exist for drug, dosage and disease extraction. From these, following previous studies [22,39], and based on the impact that the synthesized query has on retrieval performance, we employed and evaluated SciSpacy [28], Stanza [39] and BioBERT [2,24], out of which Stanza proved to perform the most robust in both tasks. Also, clinical notes often contain sentences that do not mention medical entities. These sentences may mention patient related habits, such as smoking, physical activity, among others. In our experiments we investigated the effect that these sentences have on the retrieval performance, as it is possible that these sentences only contain noisy information.

Having extracted the mentioned medical entities, we set out to identify those that are negated, related to family history or refer to historical patient information. To identify negations, we compared the broadly used ConText algorithm [12,17] that relies on regular expressions with the pre-trained transformer-based model introduced in [5], which is fine-tuned on negation assertion in clinical notes. For the identification of family history and historical information, we employed the ConText algorithm implemented in medSpacy [15] that allows for

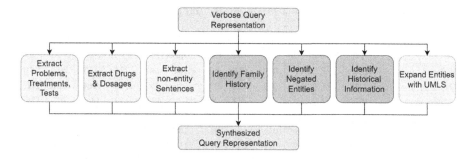

Fig. 1. Overview of our methodology, where information extraction methods are outlined in blue, entity meaning disambiguation methods in green, and entity expansion in purple. (Color figure online)

multi-token regular expressions to be used for case-specific semantic meaning disambiguation. As a result, it is feasible to disambiguate the semantic meaning (negation, family history, historical information) for all of the extracted entities from the previously mentioned methods. Finally, for all the extracted entities, we employed the UMLS [10] to retain the original entities and expand them with aliases, UMLS concepts, and definitions, inspired by [4].

To conduct the retrieval experiments with the created synthesized queries, we employed PyTerrier [27] with its default indexing parameters (i.e., porter-stemming and stopword removal)[1] and the BM25 model [32]. The empirical evaluation of the clinical trial retrieval task was performed using the test collection provided by the TREC 2021 Clinical Trials track [35] and another collection introduced by Koopman and Zuccon [20]. Regarding the medical literature retrieval, we used the TREC 2014 and TREC 2015 Clinical Decision Support collections [31,34]. For indexing the TREC collections, we used *ir-datasets* [26] and indexed all of the available document fields. Finally, the retrieval performance was evaluated based on precision-oriented measures following the official TREC guidelines for each retrieval task. The source code that implements our methodology is publicly available online[2].

4   Experimental Results and Analysis

Table 1 presents the obtained retrieval effectiveness scores across the four employed collections. Due to space limitations, in this work we report only the best performing synthesized query representations. The statistical significance is tested against the effectiveness achieved by the verbose query representation according to a paired t-test with Bonferroni multiple testing correction, at significance levels $0.05(°)$.

---

[1] All indexing parameter combinations were evaluated, however these parameters lead to greater retrieval performance.

[2] https://github.com/inf_extraction_med_ir.

Table 1. Retrieval effectiveness achieved by the top-performing synthesized queries.

Query Repr.	TREC Clinical			Clinical [20]			TREC CDS 2014			TREC CDS 2015		
	Bpref	P@5	P@25	Bpref	P@5	P@25	Bpref	P@5	P@25	Bpref	P@5	P@25
Verbose query	.184	.291	.211	.065	.050	.023	.153	.307	.224	.153	.363	**.279**
$Q1_{prob\_treat\_test}$	.211°	.323	.218	.077	.032	.021	.146	.280	.216	**.160**	**.363**	.276
$Q2_{drug\_dis}$	.196	.192	.167	.073	.046	.016	.139	.233	.179	.096	.207	.136
$Q3_{un\_comb\_Q1\_Q2}$	**.214°**	.299	**.227**	.084	**.054**	.025	.156	.287	**.241**	.153	.304	.264
$Q4_{non\_neg\_Q1\_trans}$	**.214°**	.323	.218	.082	.029	.023	.148	.293	.211	.156	.341	.261
$Q5_{non\_neg\_Q1\_con}$	.205	.291	.201	.074	.036	.020	.150	.287	.200	.151	.333	.241
$Q6_{comb\_Q4\_no\_ent}$	.206	.304	.220	.083	.036	.021	.149	.320	.212	.155	.341	.262
$Q7_{comb\_Q3\_no\_ent}$	.212°	.304	.225	**.090**	.050	**.026**	**.160**	**.333**	.239	.150	.304	.261
$Q8_{Q4\_rem\_fam\_hist}$	.207	.312	.206	.087	.014	.017	.148	.287	.189	.149	.311	.219
$Q9_{Q4\_rem\_fam}$	.212°	**.331**	.216	.084	.025	.023	.150	.307	.209	.158	.348	.240
$Q10_{Q4\_rem\_hist}$	.205	.304	.202	.083	.021	.018	.148	.267	.187	.149	.319	.231
$Q11_{Q9\_exp\_def}$	.183	.213	.143	.089	**.054**	.017	.121	.180	.135	.125	.163	.159
$Q12_{Q9\_exp\_alia}$	.182	.208	.143	.089	.050	.016	.124	.200	.135	.126	.200	.154
Human adhoc	-	-	-	.094	.071	.034	.282	.367	.307	.251	.422	.307

RQ1: Does the presence of various medical entities of a clinical note have an impact on the overall retrieval effectiveness? Experimental findings suggest that $Q1_{prob\_treat\_test}$, which contains the concatenated text of a patient's extracted problems, treatments, and tests (extracted using the transformer-based model [1]), improves retrieval performance (for both tasks) compared to $Q2_{drug\_dis}$, which contains the concatenated text of a patient's identified disease and drug using Stanza. A performance increase was achieved when these representations were combined by taking their union and keeping the unique terms ($Q3_{un\_comb\_Q1\_Q2}$).

RQ2: How does the presence of negated content affect retrieval performance? Representations $Q4_{non\_neg\_Q1\_trans}$ and $Q5_{non\_neg\_Q1\_con}$ contain the non-negated entities of $Q1_{prob\_treat\_test}$. The changes in the bpref measure suggest that queries resulting from the pre-trained transformer model [5] ($Q4_{non\_neg\_Q1\_trans}$) are more effective than those based on the ConText algorithm ($Q5_{non\_neg\_Q1\_con}$). Generally, removing negated entities improves retrieval effectiveness (bpref increases) in clinical trial retrieval, but not in medical literature retrieval.

RQ3: How does the presence of sentences with non-identified medical entities impact retrieval performance? Regarding the presence of sentences with non-identified medical entities, $Q7_{comb\_Q3\_no\_ent}$ appears to be the representation that leads to improvements over the baseline in three collections. This representation contains the non-negated entities of $Q3_{un\_comb\_Q1\_Q2}$ combined with those sentences that do not contain any identified medical entity. Similarly, $Q6_{comb\_Q4\_no\_ent}$ combines $Q4_{non\_neg\_Q1\_trans}$ with those non-entity sentences. The findings suggest that these sentences contain essential information rather than noise.

RQ4: How does the presence of family history and/or patient's historical information affect retrieval performance? Here, we observed the effects of removing from $Q4_{non\_neg\_Q1\_trans}$ both entities related to the family history and patient's historical information ($Q8_{Q4\_rem\_fam\_hist}$), removing only family history ($Q9_{Q4\_rem\_fam}$), and removing only historical information ($Q10_{Q4\_rem\_hist}$). The obtained representations seem to positively affect clinical trial retrieval, for instance, removing information related to family history leads to greater retrieval precision. However, similarly to [19], we also noticed that removing historical information identified by ConText may lead to errors, for example, when a clinical note contains only past medical information.

RQ5: What is the impact of medical entity expansion, using a knowledge base, on retrieval performance? We found out that query expansion with aliases, medical concepts and concept definitions did not improve retrieval performance. However, from all the evaluated representations, $Q11_{Q9\_exp\_def}$, which expands $Q9_{Q4\_rem\_fam}$ with aliases, medical concepts and definitions, and $Q12_{Q9\_exp\_alia}$, which expands it with aliases and medical concepts, yield greater retrieval performance.

A general observation that can be drawn from Table 1 is that none of the synthesized queries outperformed the human-generated ad-hoc queries (when available). Four medical assessors have created these ad-hoc queries, as described in [20], while for some topics, multiple short queries have been created. In this work, we concatenate these short queries and keep the unique terms to create a single query representation. All in all, the reported results suggest that the evaluated query representations improve the performance in clinical trial retrieval but not in medical literature retrieval.

4.1   A Qualitative Example

Figure 2 provides a qualitative example of an EHR (i.e., verbose query representation) from the TREC 2021 Clinical Trials collection, to illustrate the information extracted from it. Specifically, the terms highlighted in blue are those that compose the $Q1_{prob\_treat\_test}$ representation. The sentences that do not contain medical entities are mentioned in yellow, while in red we mention those medical entities that have not been identified by the employed method, despite the fact

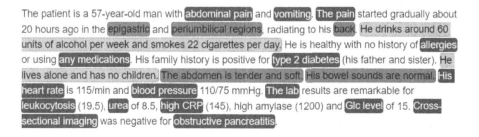

Fig. 2. Verbose query representation of TREC Clinical, topic 21.

that these terms describe the patient's condition. The $Q4_{non\_neg\_Q1\_trans}$ representation, contains all terms in blue, excluding the terms *allergies, any medications, cross-sectional imaging,* and *obstructive pancreatitis* that have been identified as negations. By combining the sentences in yellow with the query representation $Q4_{non\_neg\_Q1\_trans}$, one can obtain the $Q6_{comb\_Q4\_no\_ent}$ representation for this topic. By removing the identified medical entity *type 2 diabetes*, one can obtain the $Q8_{Q4\_rem\_fam\_hist}$ and $Q9_{Q4\_rem\_fam}$ representations, as for this topic no term has been identified as patient's historical information. Lastly, by expanding the $Q9_{Q4\_rem\_fam}$ representation using UMLS aliases or definitions one can obtain the last two query representation for this topic, as mentioned in Table 1.

5   Conclusions and Future Research

Based on the preliminary results, we draw the following conclusions. Firstly, exploiting a proper query representation that contains extracted medical entities improves retrieval performance, especially in the task of patient allocation in clinical trials. These improvements are greater when combining different approaches. Secondly, the results suggest that transformer-based models for negation identification, fine-tuned in domain-specific data, outperformed the standard rule-based approach. Thirdly, we showed that in this retrieval task, important patient information, such as a patient's habits, is not identifiable by current transformer-based models as these sentences do not contain medical entities. However, these sentences contain essential information. Removing family information improves early precision in clinical trial retrieval but not in medical literature retrieval. However, removing historical information is not as effective due to the employed identification method. Lastly, our results suggest that entity expansion with UMLS does not improve effectiveness in the considered retrieval tasks. Based on our findings, our future work will be focused on exploiting more accurate methods for extracting historical information and information from those sentences that do not contain medical entities. Also, as the obtained results, in terms of precision, did not show any statistically significant improvements, future work will focus on handling the importance of the extracted content on a per-query basis.

Acknowledgements. This work was supported by the EU Horizon 2020 ITN/ETN on Domain Specific Systems for Information Extraction and Retrieval (H2020-EU.1.3.1., ID: 860721).

References

1. Bert-base-uncased clinical NER. https://huggingface.co/samrawal/bert-base-uncased_clinical-ner. Accessed 12 Oct 2022
2. BioBert. https://github.com/alvaroalon2/bio-nlp/tree/master/models. Accessed 17 Oct 2022

3. The Thirtieth Text REtrieval Conference (TREC 2021) Proceedings. https://trec. nist.gov/pubs/trec30/trec2021.html. Accessed 03 Oct 2022

4. Agosti, M., Nunzio, G.M.D., Marchesin, S.: An analysis of query reformulation techniques for precision medicine. In: Piwowarski, B., Chevalier, M., Gaussier, É., Maarek, Y., Nie, J., Scholer, F. (eds.) Proceedings of the 42nd International ACM SIGIR Conference on Research and Development in Information Retrieval, SIGIR 2019, Paris, France, 21–25 July 019, pp. 973–976. ACM (2019). https://doi.org/ 10.1145/3331184.3331289

5. van Aken, B., Trajanovska, I., Siu, A., Mayrdorfer, M., Budde, K., Loeser, A.: Assertion detection in clinical notes: medical language models to the rescue? In: Proceedings of the Second Workshop on Natural Language Processing for Medical Conversations. Association for Computational Linguistics (2021). https:// aclanthology.org/2021.nlpmc-1.5

6. Akkasi, A., Varoğlu, E., Dimililer, N.: Chemtok: a new rule based tokenizer for chemical named entity recognition. BioMed Res. Int. (2016). https://doi.org/10. 1155/2016/4248026

7. Alfattni, G., Peek, N., Nenadic, G.: Extraction of temporal relations from clinical free text: a systematic review of current approaches. J. Biomed. Inf. **108**, 103488 (2020). https://doi.org/10.1016/j.jbi.2020.103488

8. Averbuch, M., Karson, T.H., Ben-Ami, B., Maimon, O., Rokach, L.: Context-sensitive medical information retrieval. In: Fieschi, M., Coiera, E.W., Li, Y.J. (eds.) MEDINFO 2004 - Proceedings of the 11th World Congress on Medical Informatics, San Francisco, California, USA, 7–11 September 2004. Studies in Health Technology and Informatics, vol. 107, pp. 282–286. IOS Press (2004). https://doi.org/10. 3233/978-1-60750-949-3-282

9. Balaneshinkordan, S., Kotov, A., Xisto, R.: WSU-IR at TREC 2015 clinical decision support track: joint weighting of explicit and latent medical query concepts from diverse sources. In: Voorhees, E.M., Ellis, A. (eds.) Proceedings of the Twenty-Fourth Text REtrieval Conference, TREC 2015, Gaithersburg, Maryland, USA, 17–20 November 2015. NIST Special Publication, vol. 500–319. National Institute of Standards and Technology (NIST) (2015), http://trec.nist.gov/pubs/trec24/ papers/wsu_ir-CL.pdf

10. Bodenreider, O.: The unified medical language system (umls): integrating biomedical terminology. Nucleic acids Res. **32**(suppl_1), D267–D270 (2004)

11. Chapman, W.W., Bridewell, W., Hanbury, P., Cooper, G.F., Buchanan, B.G.: Evaluation of negation phrases in narrative clinical reports. In: AMIA 2001, American Medical Informatics Association Annual Symposium, Washington, DC, USA, 3–7 November 2001. AMIA (2001). https://knowledge.amia.org/amia-55142-a2001a-1. 597057/t-001-1.599654/f-001-1.599655/a-021-1.600074/a-022-1.600071

12. Chapman, W.W., Bridewell, W., Hanbury, P., Cooper, G.F., Buchanan, B.G.: A simple algorithm for identifying negated findings and diseases in discharge summaries. J. Biomed. Inf. **34**(5), 301–310 (2001)

13. Dai, X., Rybinski, M., Karimi, S.: Searchehr: a family history search system for clinical decision support. In: Demartini, G., Zuccon, G., Culpepper, J.S., Huang, Z., Tong, H. (eds.) CIKM 2021: The 30th ACM International Conference on Information and Knowledge Management, Virtual Event, Queensland, Australia, 1–5 November 2021, pp. 4701–4705. ACM (2021). https://doi.org/10.1145/3459637. 3481986

14. Dhayne, H., Kilany, R., Haque, R., Taher, Y.: Emr2vec: bridging the gap between patient data and clinical trial. Comput. Ind. Eng. **156**, 107236 (2021). https://doi. org/10.1016/j.cie.2021.107236

15. Eyre, H., et al.: Launching into clinical space with medspacy: a new clinical text processing toolkit in python. In: AMIA Annual Symposium Proceedings, vol. 2021, p. 438. American Medical Informatics Association (2021)

16. Gliklich, R.E., Leavy, M.B., Dreyer, N.A.: Tools and technologies for registry inter-operability, registries for evaluating patient outcomes: a user's guide, addendum 2 (2019)

17. Harkema, H., Dowling, J.N., Thornblade, T., Chapman, W.W.: Context: an algorithm for determining negation, experiencer, and temporal status from clinical reports. J. Biomed. Inf. **42**(5), 839–851 (2009)

18. Hersh, W.R.: Adding value to the electronic health record through secondary use of data for quality assurance, research, and surveillance. Clin. Pharmacol. Ther. **81**, 126–128 (2007)

19. Koopman, B., Zuccon, G.: Understanding negation and family history to improve clinical information retrieval. In: Geva, S., Trotman, A., Bruza, P., Clarke, C.L.A., Järvelin, K. (eds.) The 37th International ACM SIGIR Conference on Research and Development in Information Retrieval, SIGIR 2014, Gold Coast, QLD, Australia - 06–11 July 2014, pp. 971–974. ACM (2014). https://doi.org/10.1145/2600428.2609487

20. Koopman, B., Zuccon, G.: A test collection for matching patients to clinical trials. In: Perego, R., Sebastiani, F., Aslam, J.A., Ruthven, I., Zobel, J. (eds.) Proceedings of the 39th International ACM SIGIR conference on Research and Development in Information Retrieval, SIGIR 2016, Pisa, Italy, 17–21 July 2016, pp. 669–672. ACM (2016). https://doi.org/10.1145/2911451.2914672

21. Krallinger, M., Leitner, F., Rabal, O., Vazquez, M., Oyarzabal, J., Valencia, A.: Chemdner: The drugs and chemical names extraction challenge. J. Cheminf. **7**, 1–11 (2015)

22. Leaman, R., Islamaj, R., Lu, Z.: The overview of the NLM-Chem BioCreative VII track: full-text chemical identification and indexing in PubMed articles. In: BioCreative VII Challenge Evaluation Workshop, pp. 108–113 (2021)

23. Lee, J., Yoon, W., Kim, S., Kim, D., Kim, S., Ho So, C., Kang, J.: Biobert: a pre-trained biomedical language representation model for biomedical text mining. Bioinformatics **36**(4), 1234–1240 (2020). https://doi.org/10.1093/bioinformatics/btz682

24. Lee, J., et al.: Biobert: a pre-trained biomedical language representation model for biomedical text mining. Bioinform. **36**(4), 1234–1240 (2020). https://doi.org/10.1093/bioinformatics/btz682

25. Luo, L., et al.: An attention-based bilstm-crf approach to document-level chemical named entity recognition. Bioinformatics (Oxford, England) **34** (2017). https://doi.org/10.1093/bioinformatics/btx761

26. MacAvaney, S., Yates, A., Feldman, S., Downey, D., Cohan, A., Goharian, N.: Simplified data wrangling with ir_datasets. In: SIGIR (2021)

27. Macdonald, C., Tonellotto, N.: Declarative experimentation ininformation retrieval using pyterrier. In: Proceedings of ICTIR 2020 (2020)

28. Neumann, M., King, D., Beltagy, I., Ammar, W.: ScispaCy: fast and robust models for biomedical natural language processing. In: Proceedings of the 18th BioNLP Workshop and Shared Task, pp. 319–327. Association for Computational Linguistics, Florence, Italy, August 2019. https://doi.org/10.18653/v1/W19-5034, https://www.aclweb.org/anthology/W19-5034

29. Pradeep, R., Li, Y., Wang, Y., Lin, J.: Neural query synthesis and domain-specific ranking templates for multi-stage clinical trial matching. In: Proceedings of the 45th International ACM SIGIR Conference on Research and Development in Information Retrieval, pp. 2325–2330. SIGIR 2022, Association for Computing Machinery, New York, NY, USA (2022). https://doi.org/10.1145/3477495.3531853

30. Roberts, K., Simpson, M.S., Demner-Fushman, D., Voorhees, E.M., Hersh, W.R.: State-of-the-art in biomedical literature retrieval for clinical cases: a survey of the TREC 2014 CDS track. Inf. Retr. J. **19**(1-2), 113–148 (2016). https://doi.org/10.1007/s10791-015-9259-x

31. Roberts, K., Simpson, M.S., Voorhees, E.M., Hersh, W.R.: Overview of the TREC 2015 clinical decision support track. In: Voorhees, E.M., Ellis, A. (eds.) Proceedings of The Twenty-Fourth Text REtrieval Conference, TREC 2015, Gaithersburg, Maryland, USA, 17–20 November 2015. NIST Special Publication, vol. 500–319. National Institute of Standards and Technology (NIST) (2015). http://trec.nist.gov/pubs/trec24/papers/Overview-CL.pdf

32. Robertson, S.E., Walker, S., Jones, S., Hancock-Beaulieu, M., Gatford, M.: Okapi at TREC-3. In: Harman, D.K. (ed.) Proceedings of The Third Text REtrieval Conference, TREC 1994, Gaithersburg, Maryland, USA, November 2–4, 1994. NIST Special Publication, vol. 500–225, pp. 109–126. National Institute of Standards and Technology (NIST) (1994). http://trec.nist.gov/pubs/trec3/papers/city.ps.gz

33. Rybinski, M., Dai, X., Singh, S., Karimi, S., Nguyen, A., et al.: Extracting family history information from electronic health records: natural language processing analysis. JMIR Med. Inf. **9**(4), e24020 (2021)

34. Simpson, M.S., Voorhees, E.M., Hersh, W.R.: Overview of the TREC 2014 clinical decision support track. In: Voorhees, E.M., Ellis, A. (eds.) Proceedings of The Twenty-Third Text REtrieval Conference, TREC 2014, Gaithersburg, Maryland, USA, 19–21 November 2014. NIST Special Publication, vol. 500–308. National Institute of Standards and Technology (NIST) (2014). https://trec.nist.gov/pubs/trec23/papers/overview-clinical.pdf

35. Soboroff, I.: Overview of trec 2021. In: 30th Text REtrieval Conference. Gaithersburg, Maryland (2021)

36. Tikk, D., Solt, I.: Improving textual medication extraction using combined conditional random fields and rule-based systems, journal of the american medical informatics association. J. Am. Med. Inf. Assoc. **17**, 540–544 (2010). https://doi.org/10.1136/jamia.2010.004119

37. Uzuner, Ö., South, B.R., Shen, S., DuVall, S.L.: 2010 i2b2/va challenge on concepts, assertions, and relations in clinical text. J. Am. Med. Inf. Assoc. **18**(5), 552–556 (2011). https://doi.org/10.1136/amiajnl-2011-000203

38. Xu, B., Xiufeng, S., Zhao, Z., Zheng, W.: Leveraging biomedical resources in bilstm for drug drug interaction extraction. IEEE Access 1 (2018). https://doi.org/10.1109/ACCESS.2018.2845840

39. Zhang, Y., Zhang, Y., Qi, P., Manning, C.D., Langlotz, C.P.: Biomedical and clinical English model packages for the Stanza Python NLP library. J. Am. Med. Inf. Assoc. **28**(9), 1892–1899 (2021)

Where a Little Change Makes a Big Difference: A Preliminary Exploration of Children's Queries

Maria Soledad Pera[1]([✉])(iD), Emiliana Murgia[2](iD), Monica Landoni[3](iD), Theo Huibers[4](iD), and Mohammad Aliannejadi[5](iD)

[1] Web Information Systems, TU Delft, Delft, The Netherlands
m.s.pera@tudelft.nl
[2] Università degli Studi di Milano, Bicocca, Italy
emiliana.murgia@unimib.it
[3] Università della Svizzera Italiana, Lugano, Switzerland
monica.landoni@usi.ch
[4] University of Twente, Enschede, The Netherlands
t.w.c.huibers@utwente.nl
[5] University of Amsterdam, Amsterdam, The Netherlands
m.aliannejadi@uva.nl

Abstract. This paper contributes to the discussion initiated in a recent SIGIR paper describing a gap in the information retrieval (IR) literature on query understanding–where they come from and whether they serve their purpose. Particularly the connection between query variability and search engines regarding consistent and equitable access to all users. We focus on a user group typically underserved: *children*. Using preliminary experiments (based on logs collected in the *classroom* context) and arguments grounded in children IR literature, we emphasize the importance of dedicating research efforts to interpreting queries formulated by children and the information needs they elicit. We also outline open problems and possible research directions to advance knowledge in this area, not just for children but also for other often-overlooked user groups and contexts.

Keywords: Queries · Children · Search · Query processing

1 Introduction

In their recent SIGIR perspective paper, Alaofi et al. [2] spotlight a crucial gap in the Information Retrieval (IR) literature regarding *understanding where queries come from*; that is, why they are worded a certain way or whether *query*

---

[1] "Multiple queries can represent a single information need" [8]. In this context, query variability refers to the various keyword or phrase combinations searchers can employ to articulate their requirements when faced with the same information need [2,8].

All authors contributed equally to the discussions presented in this work.

variability[1] can affect the search process, as not all queries lead us to useful information. Expanding on this discussion, we bring attention to a user group often underserved in the IR realm: **children**. Native users of search engines (SE) [17] who are in the process of acquiring vocabulary and domain knowledge, children struggle with translating their information needs into queries that prompt SE to retrieve and rank resources that are actually about what they were looking for [21,45,58]. They have in-development (cognitive) skills and an affinity to search tools that differ from adults and thus deserve actions tailored to them.

Literature on children and their interactions with IR tools is relatively limited [33]; more so from the IR perspective—the human-computer interaction community has long recognized children as important actors and has allocated efforts to outlining user experiences with and for children. Of note, strategies that simplify query formulation using images [50] or spelling suggestions specifically responding to children's query misspellings [20]. Several probabilistic, lexical, and neural-based models offer children query suggestions [5,7,48,56,57,63]. How children (re)formulate queries, along with SE performance in response to children's queries, have also been explored from diverse perspectives, including relevance, suitability, and emotion [6,12,13,34,41,59,60,60]. For the most part, the children IR community has focused on understanding search behavior and system performance. However, it has seldom considered factors that may influence how children choose the keywords to initiate the search process and the cascading effect on the results they see.

In this work, we discuss insights emerging from an initial exploration of query variations formulated by children in the **classroom** context. Given the preliminary stage of our exploration and to control *scope*, we limit our analysis to *children ages 9 to 11 in the classroom context*. Whenever possible, and to offer context to our findings, we discuss alignment with observations from experiments and literature concerning commonly-studied searchers (adults) reported in [2]. Further, we outline future directions for this research area inherent to the user group and context under scrutiny. Focusing on children enables the inspection of a range of issues, as they are not biased by previous experiences or keep them undisclosed because of social pressure-shame; on the contrary, they are more open given their limited digital and literacy skills, as reported by involved observers, teachers, and parents [22,52]. We argue that studying children could be a means to better understand other user groups experiencing similar issues relating to (lack of) access to information. With that, we invite the IR community to leverage discussions in this manuscript that add to those in [2], and together use them as a blueprint to study this area further.

2   Preliminary Exploration

To decipher children's queries, where they come from, and what social, linguistic, and cognitive factors, among others, influence their formulation, we probe query variability and their effect on search results.

Contrary to [2], we cannot turn to known **test collections** like TREC-8 Query Track [14], UQV-100 [9] or ClueWeb12C [15], as they do not explicitly capture the interactions with SE of non-traditional user groups, such as children. To bypass this limitation, we reached out to Landoni et al., who shared the logs produced for the studies presented in [39,41]. Data collection took place using the same protocol in three different classrooms in two Italian-speaking countries. We were permitted to use anonymized data (stored in a secure location) for research purposes. As part of regular instruction, children engaged in online inquiry tasks related to subjects common to the primary school curriculum, e.g. science and history. Search prompts for these tasks invited children to discover resources explaining for example current environmental concerns or how to recognize different types of volcanoes. Further, some questions were fact-based (e.g., "Where was ancient Rome founded?") and others open-ended (e.g., "How were the pyramids built?").

This resulted in the logs we use as a test collection, called CQL, comprised of topics (search prompts), queries, and the corresponding SERP (up to the 10th result) generated using Bing (https://www.bing.com) (advertisements which are often present in SERP were excluded; to prevent user profiling, each query induced a new browsing instance). Each SERP result is labeled as (non-) relevant by expert educators. Overall, CQL includes 345 queries across 64 topics and 1,538 unique labeled URLs. In the context of this work, as in [2], we define **variations** as the set of queries formulated to address the same user information need (i.e., queries generated in response to the same search prompt—topic.).

We associate each SERP result with a **reading level**. For this, we used Python's Textstat (https://pypi.org/project/textstat/), a library for readability prediction of texts in Italian based on *Flesch Reading Ease*. As reported in [4,44], there is no consensus on the "best" or "more suitable" readability formula to use when determining the text complexity of texts; more so for texts written in languages beyond English. Consequently, we use Flesch Reading Ease for readability estimation, given its popularity and broad adoption in the literature [26,28]. We also append the **emotion** inferred for each SERP result using Python's FEEL-IT (https://pypi.org/project/feel-it/), which is based on the Italian BERT model UmBERTo, fine-tuned to predict four emotions: *anger, fear, joy,* and *sadness* [11]. We adopted FEEL-IT, a state-of-the-art strategy specifically designed and empirically proven to be effective when applied to Italian text [10,11]. Due to the preliminary nature of this work, we bypassed manual assessment for reading level and emotion labeling in favor of automated strategies. Moreover, the reading level and emotion of each SERP result were inferred from its title and snippet text sample. Although using snippets as a proxy for the content of the corresponding full page has been shown to be a viable alternative [3,53], we expect to examine the whole text of SERP results in future iterations of this study.

Variability. We first look at whether children, like adults [2], adopt a range of alternatives to express the same underlying information need. Analysis of CQL reveals query variations for 38 (of the 64) topics, with an average of 7 query variations per topic. We depict in Fig. 1 (top) variation counts grouped by topic, which range from 2 to 18, with a median of 5.5. To ease visualization, we excluded from the figure topics for which no variations were found. It is worth noting that while these results verify variability exists, as stated in [2], it is still unknown what causes these variations among children or how to design SE that can "alleviate or potentially exploit" [2] this variability to better serve them.

Commercial SE. To explore the effectiveness and consistency of search results, we probe CQL utilizing multiple lenses. We use MRR and nDCG@5 to investigate disparity in relevant resources instigated by query variations; with Rank Bias Overlap (RBO) [62] we measure the consistency of retrieved results across variations. As in [2], to compute RBO, we compare the SERP generated for any pair of query variations for a given topic, which we then average. Visible from Fig. 1 is that, except for a handful of topics, query variability causes fluctuations in performance (MRR and nDCG@5, resp.). It is also apparent (from RBO) that (even minor) changes to how queries are expressed or the terminology used can yield dissimilar result sets.

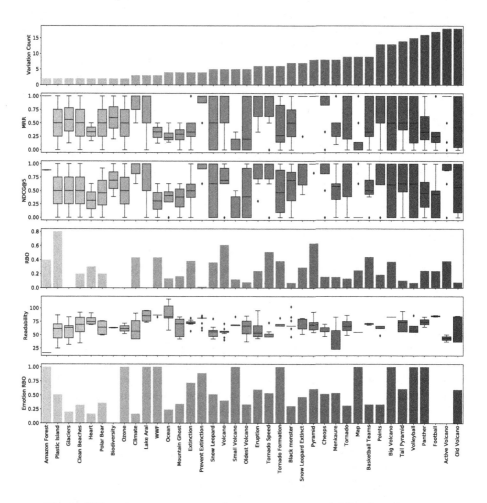

Fig. 1. Different measures are used to assess query variability based on CQL.

We see discrepancies in the position of the first relevant result retrieved, the number of relevant results positioned earlier in the ranking, and the result sets retrieved. These initial outcomes exemplify the impact of query variability on SE (also demonstrated in [2]), evidencing inequity on how SE serve searchers simply by them using different terms to address essentially the same topic. For example, 'plastic island' and 'plastic island place' (translations from the original Italian) are similar, yet they yield different results, with the top-1 being relevant for the former and not for the latter.

Readability. With children at the center, we must consider the potential connection between query variability and text complexity. For this, we gauge the readability level of the first result retrieved for each query variation. As discussed in [6,29], children tend to linearly explore SERP, starting from the top. This is why it is crucial to determine if even slight variations negatively impact children, i.e., lead to results that may address the intended information need but are above a level children can read and understand, rendering them useless. In the case of Flesch Reading Ease, the higher the score, the easier the corresponding text is to read. A score of 80 aligns with the expected reading level for 10- to 11-year-olds. Query variability visibly alters the results young searchers are exposed to (Fig. 1-Readability). For most topics, the readability score is far below the one they can comprehend. This emphasizes the need for IR research on boosting children's abilities to formulate queries (via scaffolding or novel query formulation strategies) that can ease the reach of suitable results, as well as incorporating readability as another measure of relevance for children. The latter can leverage foundational knowledge resulting from existing research in the medical domain that argues in favor of considering readability (among other factors) as a criterion for optimization beyond topicality [19].

Emotions. Recent studies examine the emotion profile of commercial SE, i.e., the emotions inferred from results retrieved for queries formulated by diverse user groups [36,42,46]. This motivated examining changes in emotions, if any, that are the direct result of query variability. We again rely on RBO as a proxy to capture the (dis)similarity of emotions observed among top results (the first and perhaps only children will engage with) generated in response to query variations. The higher the RBO, the more homogeneous the emotions inferred. As illustrated in Fig. 1 (bottom), query variations spur results conferring the same emotion for very few topics. As mentioned in [42], the affective lens is one of importance to expand upon, particularly in the context of younger children without the presence of the expert-in-the-loop (parents or teachers, depending on the context of the search), as well as other non-traditional user groups, such as those afflicted with mental health disorders [46].

3   Directions Inherent from Children and the Classroom

Here, we examine the research directions that resulted from reported initial findings (Sect. 2), and how these in turn call for further study areas to be investigated to obtain a holistic understanding of where children's queries come from.

Alaofi et al. [2] anticipate that understanding where queries come from requires revisiting *information-seeking models*. This is also true for children's queries, as literature about the information-seeking behavior of this user group is limited and seldom grounded on theoretical models [16]. Further, to our knowledge, existing theoretical models do not explicitly account for users for whom the concepts of uncertainty and aboutness might be challenging to grasp [37,45].

There are no *test collections* to guide children IR advancements. Developing collections representative of children with different skills and abilities, even within the same age range, capturing a variety of topics would be indispensable. In turn, this would enable the collaboration among researchers and practitioners on the creation of benchmarks to compare the effect of query variations across user groups. At the same time, the need to protect children's *privacy* [1,54,61] somehow interferes with gathering, building, and curating collections to enable researchers to study how starting from the same prompts describing a search task, different queries are formed, with only very few of them retrieving resources that are safe, readable, relevant, and trustworthy for children to use.

Lessons learned from children IR literature indicate that, as mentioned in [2], context and cognition factors, mediated by age, could shape children's query variability. We believe that the *roles* children play when searching (which are not mutually exclusive) can also impact their keyword selection to engage with the search process [23,24,38]. This is prompted by the mapping between cognitive bias in IR already identified in the literature for traditional user groups [27] and search roles observed among children [24,42]. We look at *the classroom* as providing context and social support to children, as well as the necessary scaffolding while developing media literacy. An example is *query elicitation from teachers* and how assisted searchers (children who depend on guidance to have a successful search experience) [38], outperform online searchers working in isolation. Similarly, Rutter et al. [51] look at how communication between teachers and children helps them to better express their information needs and retrieve useful information. we need to better understand how the choice of keywords used by teachers in the formulation of a search task together with those shared in a class discussion can result in several query variations and identify those bringing to *safe, useful, understandable, and trustworthy* results.

Further, the *interplay between distraction and reformulation of a particular query*, is summed up in the role of the distracted searcher [22,38], easily attracted by other activities and quickly abandoning the search task. *Task complexity and formulation* are crucial in children's search experience [51]. The quest for the right complexity to equally avoid boredom and frustration relies on teachers' expertise and their ability to match children's ever-changing interests and skills.

The spread of online *mis/disinformation* is something to be attentive to when it comes to children [32], who are known to be easily influenced and less critical than adults [30,49]. The rule-bound searcher [22,39] believes *fixedness* is a way to keep safe by starting from the same trusted source, often Wikipedia, and repeating the same query with no reformulations. Instead, as children's safety is paramount, they need to be actively trained in formulating queries that can deter the retrieval of misinformation, recognizing trustworthy sources, and developing the ability to judge the quality of the results they are offered [49]. Query autocompletion to support formulation in this context might not always be effective [31]. Further, there is a lack of research on what constitutes dis/misinformation for children beyond fake news [43,55]. These are some of the reasons why it is of utmost importance to allocate research so that SE can "cope with query variations that have been 'nudge' towards misinformation" [2].

Input *modality* is another influential factor [2], more so for young searchers. Researchers already note changes in how children express their queries depending upon the interface they interact with (a text search box, a voice-driven search like Alexa, or personalized conversational agents) [7,35,39,52]. We wonder whether those distinguishing traits would remain if we were to study children addressing the same search prompts on different devices or if children's perceptions of technology would bias their formulation [18].

4   Concluding Remarks

Equity in IR technology is a complex problem. The IR community has risen to this challenge with works on fairness, bias, and accessibility [25,47,64], but there is still much to be done. With their call for SE to *"provide more consistent, accurate and relevant search results regardless of the searcher's framing of the query"*, Alaofi et al. [2] expand the discourse in this area by highlighting query variability and its potential impact on equitable information access. Inspired by their work, and aiming to bring attention to young searchers, we inspected queries produced by children ages 9 to 11 in the classroom, i.e., in a specific state of cognitive and linguistic development, and captured factors that can contribute to the discussion.

With this work, focused on facets specific to young searchers, we hope to add to the comprehensive picture started in [2] on how searchers select the keywords they use to initiate a search, and how in turn query variability could hinder access to information. Paraphrasing Bilal "valuable findings from work related to children IR could serve as another layer towards advancing knowledge in mainstream IR" [40]. Consequently, we posit that this work can encourage similar and perhaps more rigorous investigations once benchmarks will be available to run comparisons across user groups. These investigations will enable us to learn more about children (beyond the ages we study) as well as other user groups and contexts underserved in IR literature for which it is critical that the queries they employ mitigate mis/disinformation, including searchers affected with mental health disorders, those with low literacy, and language learners (e.g. refugees seeking online resources). This will require multidisciplinary teams with expertise beyond IR to ascertain the various factors that make query variability so crucial.

References

1. Agesilaou, A., Kyza, E.A.: Whose data are they? elementary school students' conceptualization of data ownership and privacy of personal digital data. Int. J. Child-Comput. Interact. **33**, 100462 (2022)
2. Alaofi, M., et al.: Where do queries come from? In: Amigó, E., Castells, P., Gonzalo, J., Carterette, B., Culpepper, J.S., Kazai, G. (eds.) SIGIR 2022: The 45th International ACM SIGIR Conference on Research and Development in Information Retrieval, Madrid, Spain, 11–15 July 2022, pp. 2850–2862. ACM (2022). https://doi.org/10.1145/3477495.3531711
3. Allen, G., et al.: BiGBERT: classifying educational web resources for kindergarten-12th grades. In: Hiemstra, Djoerd, Moens, Marie-Francine., Mothe, Josiane, Perego, Raffaele, Potthast, Martin, Sebastiani, Fabrizio (eds.) ECIR 2021. LNCS, vol. 12657, pp. 176–184. Springer, Cham (2021). https://doi.org/10.1007/978-3-030-72240-1_13
4. Allen, G., Milton, A., Wright, K.L., Fails, J.A., Kennington, C., Pera, M.S.: Supercalifragilisticexpialidocious: Why using the "right" readability formula in children's web search matters. In: European Conference on Information Retrieval. pp. 3–18. Springer (2022)
5. Anuyah, O., Fails, J.A., Pera, M.S.: Investigating query formulation assistance for children. In: Giannakos, M.N., Jaccheri, L., Divitini, M. (eds.) Proceedings of the 17th ACM Conference on Interaction Design and Children, IDC 2018, Trondheim, Norway, June 19–22, 2018. pp. 581–586. ACM (2018). https://doi.org/10.1145/3202185.3210779, https://doi.org/10.1145/3202185.3210779
6. Anuyah, O., Milton, A., Green, M., Pera, M.S.: An empirical analysis of search engines' response to web search queries associated with the classroom setting. Aslib J. Inf. Manag. **72**(1), 88–111 (2020). https://doi.org/10.1108/AJIM-06-2019-0143, https://doi.org/10.1108/AJIM-06-2019-0143
7. Azpiazu, I.M., Dragovic, N., Anuyah, O., Pera, M.S.: Looking for the movie seven or sven from the movie frozen?: A multi-perspective strategy for recommending queries for children. In: Shah, C., Belkin, N.J., Byström, K., Huang, J., Scholer, F. (eds.) Proceedings of the 2018 Conference on Human Information Interaction and Retrieval, CHIIR 2018, New Brunswick, NJ, USA, March 11–15, 2018. pp. 92–101. ACM (2018). https://doi.org/10.1145/3176349.3176379, https://doi.org/10.1145/3176349.3176379
8. Bailey, P., Moffat, A., Scholer, F., Thomas, P.: User variability and ir system evaluation. In: Proceedings of The 38th International ACM SIGIR conference on research and development in Information Retrieval. pp. 625–634 (2015)
9. Bailey, P., Moffat, A., Scholer, F., Thomas, P.: UQV100: A test collection with query variability. In: Perego, R., Sebastiani, F., Aslam, J.A., Ruthven, I., Zobel, J. (eds.) Proceedings of the 39th International ACM SIGIR conference on Research and Development in Information Retrieval, SIGIR 2016, Pisa, Italy, July 17–21, 2016. pp. 725–728. ACM (2016), https://doi.org/10.1145/2911451.2914671
10. Bellodi, E., Bertagnon, A., Gavanelli, M.: Comparing emotion and sentiment analysis tools on italian anti-vaccination for covid-19 posts. In: Proceedings of the Sixth Workshop on Natural Language for Artificial Intelligence (NL4AI 2022) co-located with 21th International Conference of the Italian Association for Artificial Intelligence (AI* IA 2022) (2022)

11. Bianchi, F., Nozza, D., Hovy, D.: FEEL-IT: Emotion and Sentiment Classification for the Italian Language. In: Proceedings of the 11th Workshop on Computational Approaches to Subjectivity, Sentiment and Social Media Analysis. Association for Computational Linguistics (2021)

12. Bilal, D.: Ranking, relevance judgment, and precision of information retrieval on children's queries: Evaluation of google, yahoo!, bing, yahoo! kids, and ask kids. J. Assoc. Inf. Sci. Technol. **63**(9), 1879–1896 (2012). https://doi.org/10.1002/asi. 22675, https://doi.org/10.1002/asi.22675

13. Bilal, D., Gwizdka, J.: Children's query types and reformulations in google search. Inf. Process. Manag. **54**(6), 1022–1041 (2018). https://doi.org/10.1016/j.ipm.2018. 06.008, https://doi.org/10.1016/j.ipm.2018.06.008

14. Buckley, C., Walz, J.A.: The TREC-8 Query track. In Proceedings of the 8th Text REtrieval Conference (1999), http://trec.nist.gov/pubs/trec8/papers/ qtrack.pdf

15. Dai, Z., Callan, J.: Context-aware document term weighting for ad-hoc search. In: Huang, Y., King, I., Liu, T., van Steen, M. (eds.) WWW '20: The Web Conference 2020, Taipei, Taiwan, April 20–24, 2020. pp. 1897–1907. ACM / IW3C2 (2020), https://doi.org/10.1145/3366423.3380258

16. Dania, B.: Theoretical applications in children and youth information behavior research: 1999–2019. Proceedings of the Association for Information Science and Technology **59**(1), 11–22 (2022)

17. Danovitch, J.H.: Growing up with google: How children's understanding and use of internet-based devices relates to cognitive development. Human Behavior and Emerging Technologies **1**(2), 81–90 (2019)

18. Desai, S., Twidale, M.: Is alexa like a computer? a search engine? a friend? a silly child? yes. In: 4th Conference on Conversational User Interfaces. pp. 1–4 (2022)

19. van Doorn, J., Odijk, D., Roijers, D.M., de Rijke, M.: Balancing relevance criteria through multi-objective optimization. In: Proceedings of the 39th International ACM SIGIR conference on Research and Development in Information Retrieval. pp. 769–772 (2016)

20. Downs, B., Pera, M.S., Wright, K.L., Kennington, C., Fails, J.A.: Kidspell: Making a difference in spellchecking for children. Int. J. Child Comput. Interact. **32**, 100373 (2022). https://doi.org/10.1016/j.ijcci.2021.100373, https://doi.org/10. 1016/j.ijcci.2021.100373

21. Dragovic, N., Azpiazu, I.M., Pera, M.S.: "is sven seven?": A search intent module for children. In: Perego, R., Sebastiani, F., Aslam, J.A., Ruthven, I., Zobel, J. (eds.) Proceedings of the 39th International ACM SIGIR conference on Research and Development in Information Retrieval, SIGIR 2016, Pisa, Italy, July 17–21, 2016. pp. 885–888. ACM (2016), https://doi.org/10.1145/2911451.2914738

22. Druin, A., Foss, E., Hatley, L., Golub, E., Guha, M.L., Fails, J., Hutchinson, H.: How children search the internet with keyword interfaces. In: Proceedings of the 8th International conference on interaction design and children. pp. 89–96 (2009)

23. Foss, E., Druin, A.: Children's internet search: Using roles to understand children's search behavior. Synthesis Lectures on information concepts, retrieval, and services **6**(2), 1–106 (2014)

24. Foss, E., Druin, A., Brewer, R., Lo, P., Sanchez, L., Golub, E., Hutchinson, H.: Children's search roles at home: Implications for designers, researchers, educators, and parents. Journal of the American Society for Information Science and Technology **63**(3), 558–573 (2012)

25. Gao, R., Shah, C.: Addressing bias and fairness in search systems. In: Proceedings of the 44th international ACM SIGIR conference on research and development in information retrieval. pp. 2643–2646 (2021)

26. Ginesti, G., Sannino, G., Drago, C.: Board connections and management commentary readability: the role of information sharing in italy. Corporate Governance: The international journal of business in society (2017)
27. Gomroki, G., Behzadi, H., Fattahi, R., Salehi Fadardi, J.: Identifying effective cognitive biases in information retrieval. Journal of Information Science p. 01655515211001777 (2021)
28. Grego, G., Spina, S., Danilo, R., et al.: Predicting readability of texts for italian l2 students: A preliminary study. In: ALTE (2017). Learning and assessment: making the connections-Proceedings of the ALTE 6th International Conference, 3–5 May 2017. pp. 272–278. ALTE (2017)
29. Gwizdka, J., Bilal, D.: Analysis of children's queries and click behavior on ranked results and their thought processes in google search. In: Proceedings of the 2017 conference on conference human information interaction and retrieval. pp. 377–380 (2017)
30. Hämäläinen, E.K., Kiili, C., Marttunen, M., Räikkönen, E., González-Ibáñez, R., Leppänen, P.H.: Promoting sixth graders' credibility evaluation of web pages: an intervention study. Computers in Human Behavior **110**, 106372 (2020)
31. Hiemstra, D.: Reducing misinformation in query auto-completions. In: Wagner, A. (ed.) OSSYM 2020: Second International Symposium on Open Search Technology, 12–14 October, 2020, Web Meeting hosted by CERN, Geneva, Switzerland, pp. 1–4. SI, Zenodo (2020)
32. Howard, P.N., Neudert, L.M., Prakash, N., Vosloo, S.: Digital misinformation/disinformation and children. UNICEF. Retrieved on February 20, 2021 (2021)
33. Huibers, T., Landoni, M., Murgia, E., Pera, M.S.: IR for children 2000–2020: Where are we now? In: Diaz, F., Shah, C., Suel, T., Castells, P., Jones, R., Sakai, T. (eds.) SIGIR '21: The 44th International ACM SIGIR Conference on Research and Development in Information Retrieval, Virtual Event, Canada, July 11–15, 2021. pp. 2689–2692. ACM (2021). https://doi.org/10.1145/3404835.3462822, https://doi.org/10.1145/3404835.3462822
34. Jochmann-Mannak, H., Huibers, T., Sanders, T.: Children's information retrieval: beyond examining search strategies and interfaces. In: 2nd BCS IRSG Symposium: Future Directions in Information Access 2008 2. pp. 64–72 (2008)
35. Kammerer, Y., Bohnacker, M.: Children's web search with google: the effectiveness of natural language queries. In: proceedings of the 11th International Conference on Interaction Design and Children. pp. 184–187 (2012)
36. Kazai, G., Thomas, P., Craswell, N.: The emotion profile of web search. In: Proceedings of the 42nd international ACM SIGIR conference on research and development in information retrieval. pp. 1097–1100 (2019)
37. Kuhlthau, C.C.: A principle of uncertainty for information seeking. Journal of documentation (1993)
38. Landoni, M., Huibers, T., Aliannejadi, M., Murgia, E., Pera, M.S.: Getting to know you: Search logs and expert grading to define children's search roles in the classroom. In: DESIRES. pp. 44–52 (2021)
39. Landoni, M., Matteri, D., Murgia, E., Huibers, T., Pera, M.S.: Sonny, cerca! evaluating the impact of using a vocal assistant to search at school. In: Crestani, F., Braschler, M., Savoy, J., Rauber, A., Müller, H., Losada, D.E., Bürki, G.H., Cappellato, L., Ferro, N. (eds.) Experimental IR Meets Multilinguality, Multimodality, and Interaction - 10th International Conference of the CLEF Association, CLEF 2019, Lugano, Switzerland, September 9–12, 2019, Proceedings. Lecture Notes in Computer Science, vol. 11696, pp. 101–113. Springer (2019), https://doi.org/10.1007/978-3-030-28577-7_6

40. Landoni, M., Murgia, E., Huibers, T., Pera, M.S.: Report on the 1st ir for children 2000–2020: where are we now?(ir4c) workshop at sigir 2021: the need to spotlight research on children information retrieval. In: ACM SIGIR Forum. vol. 55, pp. 1–7. ACM New York, NY, USA (2022)

41. Landoni, M., Pera, M.S., Murgia, E., Huibers, T.: Inside out: Exploring the emotional side of search engines in the classroom. In: Kuflik, T., Torre, I., Burke, R., Gena, C. (eds.) Proceedings of the 28th ACM Conference on User Modeling, Adaptation and Personalization, UMAP 2020, Genoa, Italy, July 12–18, 2020. pp. 136–144. ACM (2020), https://doi.org/10.1145/3340631.3394847

42. Landoni, M., Pera, M.S., Murgia, E., Huibers, T.: Inside out: Exploring the emotional side of search engines in the classroom. In: Proceedings of the 28th ACM conference on user modeling, adaptation and personalization. pp. 136–144 (2020)

43. Loos, E., Ivan, L.: Special issue "fighting fake news: A generational approach" (2022)

44. Madrazo Azpiazu, I., Pera, M.S.: An analysis of transfer learning methods for multilingual readability assessment. In: Adjunct Publication of the 28th ACM Conference on User Modeling, Adaptation and Personalization. pp. 95–100 (2020)

45. Maron, M.E.: On indexing, retrieval and the meaning of about. Journal of the american society for information science **28**(1), 38–43 (1977)

46. Milton, A., Pera, M.S.: What snippets feel: Depression, search, and snippets. In: 1st Joint Conference of the Information Retrieval Communities in Europe (CIRCLE). CEUR Workshop Proceedings, 2621 (2020)

47. Olteanu, A., Garcia-Gathright, J., de Rijke, M., Ekstrand, M.D., Roegiest, A., Lipani, A., Beutel, A., Olteanu, A., Lucic, A., Stoica, A.A., et al.: Facts-ir: fairness, accountability, confidentiality, transparency, and safety in information retrieval. In: ACM SIGIR Forum. vol. 53, pp. 20–43. ACM New York, NY, USA (2021)

48. Pera, M.S., Ng, Y.: Using online data sources to make query suggestions for children. Web Intell. **15**(4), 303–323 (2017). https://doi.org/10.3233/WEB-170367, https://doi.org/10.3233/WEB-170367

49. Pilgrim, J., Vasinda, S.: Fake news and the "wild wide web": A study of elementary students' reliability reasoning. Societies **11**(4), 121 (2021)

50. Polajnar, T., Glassey, R., Azzopardi, L.: Juse: a picture dictionary query system for children. In: Ma, W., Nie, J., Baeza-Yates, R., Chua, T., Croft, W.B. (eds.) Proceeding of the 34th International ACM SIGIR Conference on Research and Development in Information Retrieval, SIGIR 2011, Beijing, China, July 25–29, 2011. pp. 1281–1282. ACM (2011). https://doi.org/10.1145/2009916.2010160, https://doi.org/10.1145/2009916.2010160

51. Rutter, S., Clough, P.D., Toms, E.G.: Using classroom talk to understand children's search processes for tasks with different goals. Information Research: An International Electronic Journal **24**(1), n1 (2019)

52. Rutter, S., Ford, N., Clough, P.: How do children reformulate their search queries? Information Research: An International Electronic Journal **20**(1), n1 (2015)

53. Shen, D., Chen, Z., Yang, Q., Zeng, H.J., Zhang, B., Lu, Y., Ma, W.Y.: Webpage classification through summarization. In: Proceedings of the 27th annual international ACM SIGIR conference on Research and development in information retrieval. pp. 242–249 (2004)

54. Sun, K., Sugatan, C., Afnan, T., Simon, H., Gelman, S.A., Radesky, J., Schaub, F.: "they see you're a girl if you pick a pink robot with a skirt": A qualitative study of how children conceptualize data processing and digital privacy risks. In: Proceedings of the 2021 CHI Conference on Human Factors in Computing Systems. pp. 1–34 (2021)

55. Sundin, O., Francke, H.: In search of credibility: pupils' information practices in learning environments. Information Research: An International Electronic Journal **14**(4), n4 (2009)

56. Torres, S.D., Hiemstra, D., Weber, I., Serdyukov, P.: Query recommendation for children. In: Chen, X., Lebanon, G., Wang, H., Zaki, M.J. (eds.) 21st ACM International Conference on Information and Knowledge Management, CIKM'12, Maui, HI, USA, October 29 - November 02, 2012. pp. 2010–2014. ACM (2012). https://doi.org/10.1145/2396761.2398562, https://doi.org/10.1145/2396761.2398562

57. Torres, S.D., Hiemstra, D., Weber, I., Serdyukov, P.: Query recommendation in the information domain of children. J. Assoc. Inf. Sci. Technol. **65**(7), 1368–1384 (2014). https://doi.org/10.1002/asi.23055, https://doi.org/10.1002/asi.23055

58. Torres, S.D., Weber, I.: What and how children search on the web. In: Macdonald, C., Ounis, I., Ruthven, I. (eds.) Proceedings of the 20th ACM Conference on Information and Knowledge Management, CIKM 2011, Glasgow, United Kingdom, October 24–28, 2011. pp. 393–402. ACM (2011), https://doi.org/10.1145/2063576.2063638

59. Vanderschantz, N., Hinze, A.: A study of children's search query formulation habits. In: Hall, L.E., Flint, T., O'Hara, S., Turner, P. (eds.) HCI 2017 - Digital make-believe. Proceedings of the 31st International BCS Human Computer Interaction Conference, BCS HCI 2017, University of Sunderland, St Peter's campus, Sunderland, UK, 3–6 July 2017. Workshops in Computing, BCS (2017). https://doi.org/10.14236/ewic/HCI2017.7, https://doi.org/10.14236/ewic/HCI2017.7

60. Vanderschantz, N., Hinze, A.: Children's query formulation and search result exploration. International Journal on Digital Libraries **22**(4), 385–410 (2021). https://doi.org/10.1007/s00799-021-00316-9

61. Vasiliki, C., Stephane, C., Rosanna, D.G., Riina, V., Marina, E.P., Ignacio, S.M.J., Emilia, G.G., et al.: Artificial intelligence and the rights of the child: Towards an integrated agenda for research and policy. Tech. rep, Joint Research Centre (Seville site) (2022)

62. Webber, W., Moffat, A., Zobel, J.: A similarity measure for indefinite rankings. ACM Transactions on Information Systems (TOIS) **28**(4), 1–38 (2010)

63. Wood, A., Ng, Y.: Orthogonal query recommendations for children. In: Anderst-Kotsis, G. (ed.) Proceedings of the 18th International Conference on Information Integration and Web-based Applications and Services, iiWAS 2016, Singapore, November 28–30, 2016. pp. 298–302. ACM (2016). https://doi.org/10.1145/3011141.3011220, https://doi.org/10.1145/3011141.3011220

64. Yu, R.: Improving Knowledge Accessibility on the Web-from Knowledge Base Augmentation to Search as Learning. Ph.D. thesis (2020)

Visconde: Multi-document QA with GPT-3 and Neural Reranking

Jayr Pereira[1,2](✉)[iD], Robson Fidalgo[2][iD], Roberto Lotufo[1][iD],
and Rodrigo Nogueira[1][iD]

[1] NeuralMind, Campinas, Brazil
{jayr.pereira,roberto,rodrigo.nogueira}@neuralmind.ai
[2] Centro de Informática, Universidade Federal de Pernambuco, Recife, Brazil
{jap2,rdnf}@cin.ufpe.br

Abstract. This paper proposes a question-answering system that can answer questions whose supporting evidence is spread over multiple (potentially long) documents. The system, called Visconde, uses a three-step pipeline to perform the task: decompose, retrieve, and aggregate. The first step decomposes the question into simpler questions using a few-shot large language model (LLM). Then, a state-of-the-art search engine is used to retrieve candidate passages from a large collection for each decomposed question. In the final step, we use the LLM in a few-shot setting to aggregate the contents of the passages into the final answer. The system is evaluated on three datasets: IIRC, Qasper, and StrategyQA. Results suggest that current retrievers are the main bottleneck and that readers are already performing at the human level as long as relevant passages are provided. The system is also shown to be more effective when the model is induced to give explanations before answering a question. Code is available at https://github.com/neuralmind-ai/visconde.

1 Introduction

In recent years, question-answering (QA) tasks that use relatively short contexts (e.g., a paragraph) have seen remarkable progress in multiple domains [14,15]. However, in many cases, the necessary information to answer a question is spread over multiple documents or long ones [7,9,23,36]. To solve this task, QA models are based on a pipeline comprised of a retriever and a reader component [8, 13,28,30]. Most of these approaches rely on fine-tuning large language models on supervised datasets, which may be available for a variaty of domains. Other approaches use Transformers for long sequences like LongT5 [10] to process the context document and the question at once [29,33], which might not scale to longer sequences (e.g., documents with hundreds of pages).

The few-shot capability of LLMs may reduce the costs for solving QA tasks, as it allows one to implement QA systems for different domains without needing a specific annotated dataset. In addition, recent studies showed that adding a chain-of-thought (CoT) reasoning step before answering significantly improves

J. Kamps et al. (Eds.): ECIR 2023, LNCS 13981, pp. 534–543, 2023.
https://doi.org/10.1007/978-3-031-28238-6_44

LLMs' zero or few-shot effectiveness on diverse QA benchmarks [16]. In this work, we propose Visconde,[1] a QA system that combines a state-of-the-art retriever and a few-shot (CoT) approach to induce an LLM to generate the answer as a generative reader. The retriever is a multi-stage pipeline that uses BM25 [27] to select candidate documents followed by a monoT5 reranker [22]. The reader uses GPT-3 [2] in a few-shot setting that reason over the retrieved records to produce an answer. We induce CoT by asking the model to explain how the evidence documents can answer the question. Our system rivals state-of-the-art supervised models in three datasets: IIRC, Qasper, and StrategyQA.

Our main contribution is to show that current multi-document QA systems are close to human-level performance *as long as ground truth contexts are provided as input to the reader.* When a SOTA retriever selects the context, we observe a significant drop in effectiveness. Thus, we argue future work on multi-document QA should focus on improving retrievers.

2   Related Work

Most approaches for multi-document QA are typically based on a retriever followed by a reader component [37]. The retriever aims to select relevant documents for given a question, while the reader seeks to infer the final answer from them. Recent studies used dense retrievers [7, 30] or commercial search engines [17] for this task. For the reader component, some studies used sequence-to-sequence models to generate natural language answers [12, 18, 34], numerical reasoning models adapted to reason also over text [7, 8, 30], or LLMs to aggregate information from multiple documents [17, 21].

Recent work enriches this pipeline by adding components to perform query decomposition [1, 4, 6, 11, 24] or evidence retrieved by a web search engine [17, 21, 24]. Our work is similar to these, but we focus on evaluating the limitations of this method and found that the retrieval component needs more work.

3   Our Method: Visconde

Visconde is a multi-document QA system that has three main steps: *Question decomposition, Document Retrieval,* and *Aggregation.* As illustrated in Fig. 1, the system first decomposes the user question when necessary and searches for relevant documents to answer the subquestions. The retrieved documents are the basis for generating an explanation and a final answer using an LLM.

Question Decomposition: We use GPT-3 (text-davinci-002) with five in-context examples for question decomposition. In Fig. 1 we show an example of a question that needs to be decomposed (**Q**) extracted from the IIRC dataset [7] and the subquestions generated by the model (**Q1** and **Q2**). The five examples

---

[1] The name is a homage to *Visconde de Sabugosa* a fictional character invented by Monteiro Lobato that is a corn cob doll whose wisdom comes from reading books.

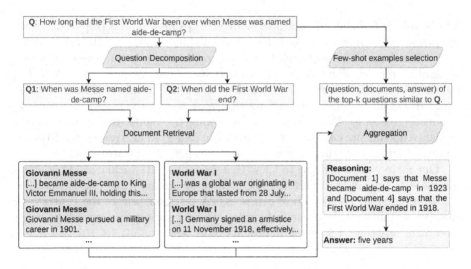

Fig. 1. Visconde QA flow.

used as few-shot in the prompt were randomly selected from the training set of StrategyQA dataset [9], which has questions and decomposed subquestions.

Document Retrieval: For document retrieval, we used a strategy divided into three main steps: 1) document indexing – we create an inverted index using Pyserini [19]; 2) candidates retrieval – we use the Pyserini implementation of the BM25 algorithm to retrieve candidate documents. 3) document reranking – we rerank the top-1000 documents retrieved using a sequence-to-sequence model designed for reranking, the monoT5 model [22], an adaptation of the T5 model [25].[2] monoT5 receives as input a sequence with the document and the query, and a softmax function is applied only on the logits calculated by T5 to the tokens `true` and `false`. The log probability of the token `true` is used as the document relevance score given the question. The output is a list of documents ranked by the relevance scores.

Aggregation: We use GPT-3 (text-davinci-002) as a few-shot learner for the aggregation step. Different studies have shown that the effectiveness of LLM can be improved by inducing it to first generate reasoning steps before answering a question [3,16,31,32]. We use CoT to induce the LLM to reason over multiple documents and answer a question, as shown in the example in Fig. 2. In our prompt, each example has a list of context documents (e.g., [Document 1]), a question, an evidence paragraph, and the answer. The context documents of the target example are the top-k documents from the retriever step. When the question is decomposed, we use the top-k documents from each subquestion. For the target example, an evidence paragraph is not provided, and the LLM must generate it, as well as a final answer.

---

[2] We used the 3 billion parameters version, whose checkpoint is available at https://huggingface.co/castorini/monot5-3b-msmarco-10k.

– Few-shot examples omitted due to space –

Example 5:

[Document 1]: Title: Giovanni Messe. Content: [...] became aide-de-camp to King Victor Emmanuel III, holding this post from 1923 to 1927.

– Documents omitted due to space –

[Document 5]: Title: World War I. Content: [...] Germany signed an armistice on 11 November 1918, effectively ending the war.

Question: How long had the First World War been over when Messe was named aide-de-camp?

Evidence: **According to [Document 5], the First World War ended on November 11th, 1918. According to [Document 1], Messe was named aide-de-camp on 1923. Therefore, the First World War had been over for 5 years when Messe was named aide-de-camp.**

Answer: **5 years**

Fig. 2. Reasoning prompt example. The **bold** text is the model's output.

We tested two approaches for prompt construction: 1) using static prompts with a pre-defined list of examples; and 2) using dynamic prompts by selecting in-context examples that are similar to the test example. For dynamic prompts, we encode the questions from the training dataset using SentenceTransformers [26] and apply a KNN algorithm to find the k most similar to the test question, as Liu et al. [20] did for other tasks.[3]

4 Experiments

4.1 IIRC

The Incomplete Information Reading Comprehension (IIRC) dataset [7] consists of information-seeking questions that require retrieving the necessary information missing from the original context. Each original context is a paragraph from the English Wikipedia, which comes with a set of links to other Wikipedia pages.

Pre-processing: we used the dynamic prompt described in Sect. 3. For this, we automatically generated reasoning paragraphs for 10% of the IIRC training set (1340 questions) using GPT-3. In addition, we processed the context articles provided by the dataset to create a searchable index.

Procedure: following the framework depicted in Fig. 1, we first decomposed the questions from the IIRC test set. We performed document retrieval on a database of Wikipedia documents provided in the dataset. In the aggregation step, we applied four methods: 1) using GPT-3 without CoT and providing the

---

[3] We used this model as our sentence encoder: *sentence-transformers/msmarco-bert-base-dot-v5*.

links and the ground truth contexts, i.e., skipping the document retrieval step; 2) using the reasoning step with the links and ground truth contexts; 3) using reasoning step over the intersection of retrieved documents and the documents cited by the main context; and 4) reasoning over the documents retrieved from the entire Wikipedia subset provided by the dataset.

4.2 Qasper

Qasper [5] is an information-seeking QA dataset over academic research papers. The task consists of retrieving the most relevant evidence paragraph for each question and answering the question.

Procedure: we did not apply query decomposition because the questions in this dataset are closed-ended, i.e., they do not require decomposition as they are grounded in a single paper of interest [5]. For example, the question "How is the text segmented?" only makes sense concerning its grounded paper. Besides, we skipped the BM25 step for document retrieval as the monoT5 reranker can score each paragraph in the paper in a reasonable time. The document retrieval step consists of reranking the paper's paragraphs based on the question and choosing the top five as context documents. We did not notice any advantage in using a dynamic prompt in this dataset.

4.3 StrategyQA

StrategyQA [9] is a dataset focused on open-domain questions that require reasoning steps. This dataset has three tasks: 1) question decomposition, measured using a metric called SARI, generally used to evaluate automatic text simplification systems [35]; 2) evidence paragraph retrieval, measured as the recall of the top ten retrieved results; and 3) question answering, measured in terms of accuracy.

Pre-processing: we did not generate reasoning paragraphs for the training examples since the context comprises long paragraphs that exceed the model input size limit (4000 tokens). We processed the context articles provided by the dataset to create a searchable index, by splitting the articles into windows of three sentences each.

Procedure: we applied question decomposition and performed retrieval using the approach described in Sect. 3. We used the top five retrieved documents for each decomposed question as the context in the reading step.

4.4 Results

In Table 1, we present the results of our experiments. First, we show the results obtained in the IIRC dataset. Our approach outperforms the baselines (i.e., Ferguson et al. [7]'s) in different settings: 1) Using the gold context searched by humans (Gold Ctx); 2) Searching for context in the links the dataset provides (Linked pages); and 3) Searching for contexts in the entire dataset. We report

Table 1. Visconde and similar methods results on IIRC, Qasper and StrategyQA.

IIRC		
Model	F1	EM
Human	88.4	85.7
Finetuned		
Ferguson et al. [7]	31.1	27.7
Ferguson et al. [7] Linked pages	32.5	29.0
Ferguson et al. [7] Gold Ctx	70.3	65.6
PReasM (pretrain + finetuning) [30] Gold Ctx	-	73.3
PReasM (pretrain + finetuning) [30]	-	47.2
Sup_{A+QA} (supervised) [8]	51.6	-
Few-shot		
Visconde (4-shot dynamic prompt) Gold Ctx and CoT	84.2	74.7
Visconde (4-shot dynamic prompt) Gold Ctx	80.3	70.0
Visconde (4-shot static prompt) Gold Ctx	74.3	62.7
Visconde (4-shot dynamic prompt) Linked pages	48.2	40.7
Visconde (4-shot dynamic prompt) CoT	47.9	40.0

Qasper						
Model	Extractive	Abstractive	Boolean	Unanswerable	Evidence F1	Answer F1
Human	58.9	39.7	79.0	69.4	71.6	60.9
LED-base	30.0	15.0	68.9	45.0	29.9	33.6
SOTA sup. [33]	-	-	-	-	-	53.1
Visconde	52.3	21.7	86.2	48.3	38.5	49.1

StrategyQA			
Model	Acc	Recall@10	SARI
Human	87.00	0.586	-
Baseline	63.60	0.195	-
Leaderboard's SOTA	69.80	0.537	0.555
GOPHER-280B OB_{Google}^{PoE} [17]	66.20	-	-
Visconde (1-shot, static prompt, gold evidences)	73.80	-	-
Visconde (1-shot, static prompt)	69.43	0.331	0.570

Visconde's results with and without CoT and using a static prompt instead of a dynamic one. The dynamic prompt leads to better performance in this dataset. When using the gold contexts, Visconde approaches human performance in terms of F1. However, when the system has to search for context, performance decreases. Also, the system performs better when using CoT. With CoT,

Visconde tends to perform better in questions requiring basic arithmetic operations, which is consistent with the literature [16,32]. By inspecting the model output in relation to the expected answer, we noticed that in some cases the system answers questions marked as unanswerable by the human annotators. This may occur 1) because the retriever found a relevant document containing the answer or 2) because GPT-3 answers the question even when the necessary information is not in the context. Different ways to write numerals may also affect the results. For example, the model might answer "five years", while "5 years" is expected.

For Qasper, we present the results in terms of F1 for the answer and the evidence. The LED-base model is the baseline [5]. The SOTA model is the model proposed by Xiong et al. [33]. Visconde outperformed the baseline but did not surpass the SOTA model. Xiong et al. [33]'s model is a long Transformer finetuned on the task. Regarding evidence F1, our system outperforms the baseline, but there is still a gap between our performance and human performance. Visconde had a high performance on the boolean questions but a low score on the abstractive ones. The table shows that even the human result is lower for the abstractive question than other types.

For StrategyQA, we present the results in terms of answer accuracy, evidence recall@10, and SARI for question decomposition. Automated methods are still far from human performance. Our approach outperforms the baselines presented in the paper [9] in terms of answer accuracy and evidence recall@10. We also outperform the leaderboard's SOTA model[4] in the quality of the questions decomposition measured with SARI. However, we did not surpass SOTA's recall@10 in retrieving the appropriate evidence paragraphs and coming close in the answer accuracy. We also outperform Lazaridou et al. 's approach [17], which also uses a few-shot LLM.

5   Conclusion

This paper describes a system for multi-document question answering that uses a passage reranker to retrieve documents and large language models to reason over them and compose an answer. Our system rivals state-of-the-art supervised models in three datasets: IIRC, Qasper, and StrategyQA. Our results suggest that using GPT-3 as a reader is close to human-level performance as long as relevant passages are provided, while current retrievers are the main bottleneck. We also show that inducing the model to give explanations before answering a question improves effectiveness.

Acknowledgments. This research was partially supported by Fundação de Amparo à Pesquisa do Estado de São Paulo (FAPESP) (project id 2022/01640-2) and by Coordenação de Aperfeiçoamento de Pessoal de Nível Superior (CAPES) (Grant code: 88887.481522/2020-00). We also thank Centro Nacional de Processamento de Alto Desempenho (CENAPAD-SP) and Google Cloud for computing credits.

---

[4] https://leaderboard.allenai.org/strategyqa/submissions/public. Accessed on July 20, 2022.

References

1. Boerschinger, B., et al.: Boosting search engines with interactive agents. Transactions on Machine Learning Research (2022). https://openreview.net/pdf?id=0ZbPmmB61g
2. Brown, T., et al.: Language models are few-shot learners. In: Larochelle, H., Ranzato, M., Hadsell, R., Balcan, M., Lin, H. (eds.) Advances in Neural Information Processing Systems, vol. 33, pp. 1877–1901. Curran Associates, Inc. (2020). https://proceedings.neurips.cc/paper/2020/file/1457c0d6bfcb4967418bfb8ac142f64a-Paper.pdf
3. Creswell, A., Shanahan, M.: Faithful reasoning using large language models. arXiv preprint arXiv:2208.14271 (2022)
4. Das, R., Dhuliawala, S., Zaheer, M., McCallum, A.: Multi-step retriever-reader interaction for scalable open-domain question answering (2019). https://doi.org/10.48550/ARXIV.1905.05733, https://arxiv.org/abs/1905.05733
5. Dasigi, P., Lo, K., Beltagy, I., Cohan, A., Smith, N.A., Gardner, M.: A dataset of information-seeking questions and answers anchored in research papers. In: Proceedings of the 2021 Conference of the North American Chapter of the Association for Computational Linguistics: Human Language Technologies, pp. 4599–4610. Association for Computational Linguistics, June 2021. https://doi.org/10.18653/v1/2021.naacl-main.365, https://aclanthology.org/2021.naacl-main.365
6. Feldman, Y., El-Yaniv, R.: Multi-hop paragraph retrieval for open-domain question answering. In: Proceedings of the 57th Annual Meeting of the Association for Computational Linguistics. pp. 2296–2309. Association for Computational Linguistics, Florence, Italy, July 2019. https://doi.org/10.18653/v1/P19-1222, https://aclanthology.org/P19-1222
7. Ferguson, J., Gardner, M., Hajishirzi, H., Khot, T., Dasigi, P.: IIRC: a dataset of incomplete information reading comprehension questions. in: proceedings of the 2020 conference on empirical methods in Natural Language Processing (EMNLP), pp. 1137–1147. Association for Computational Linguistics, November 2020. https://doi.org/10.18653/v1/2020.emnlp-main.86, https://aclanthology.org/2020.emnlp-main.86
8. Ferguson, J., Hajishirzi, H., Dasigi, P., Khot, T.: Retrieval data augmentation informed by downstream question answering performance. In: Proceedings of the Fifth Fact Extraction and VERification Workshop (FEVER), pp. 1–5. Association for Computational Linguistics, Dublin, Ireland, May 2022. https://doi.org/10.18653/v1/2022.fever-1.1, https://aclanthology.org/2022.fever-1.1
9. Geva, M., Khashabi, D., Segal, E., Khot, T., Roth, D., Berant, J.: Did Aristotle use a laptop? A question answering benchmark with implicit reasoning strategies. Trans. Assoc. Comput. Linguist. 9, 346–361 (2021). https://doi.org/10.1162/tacl_a_00370, https://doi.org/10.1162/tacl_a_00370
10. Guo, M., et al.: Longt5: efficient text-to-text transformer for long sequences (2021). https://doi.org/10.48550/ARXIV.2112.07916, https://arxiv.org/abs/2112.07916
11. Huebscher, M.C., Buck, C., Ciaramita, M., Rothe, S.: Zero-shot retrieval with search agents and hybrid environments (2022). https://doi.org/10.48550/ARXIV.2209.15469, https://arxiv.org/abs/2209.15469
12. Izacard, G., Grave, E.: Leveraging passage retrieval with generative models for open domain question answering. In: Proceedings of the 16th Conference of the European Chapter of the Association for Computational Linguistics: Main Volume, pp. 874–880. Association for Computational Linguistics, April 2021. https://doi.org/10.18653/v1/2021.eacl-main.74, https://aclanthology.org/2021.eacl-main.74

13. Karpukhin, V., et al.: Dense passage retrieval for open-domain question answering. In: Proceedings of the 2020 Conference on Empirical Methods in Natural Language Processing (EMNLP), pp. 6769–6781. Association for Computational Linguistics, November 2020. https://doi.org/10.18653/v1/2020.emnlp-main.550, https://aclanthology.org/2020.emnlp-main.550

14. Khashabi, D., et al.: Unifiedqa: crossing format boundaries with a single QA system (2020)

15. Khashabi, D., Kordi, Y., Hajishirzi, H.: Unifiedqa-v2: stronger generalization via broader cross-format training. arXiv preprint arXiv:2202.12359 (2022)

16. Kojima, T., Gu, S.S., Reid, M., Matsuo, Y., Iwasawa, Y.: Large language models are zero-shot reasoners (2022). https://doi.org/10.48550/ARXIV.2205.11916, https://arxiv.org/abs/2205.11916

17. Lazaridou, A., Gribovskaya, E., Stokowiec, W., Grigorev, N.: Internet-augmented language models through few-shot prompting for open-domain question answering. arXiv preprint arXiv:2203.05115 (2022)

18. Lewis, P., et al.: Retrieval-augmented generation for knowledge-intensive NLP tasks. In: Larochelle, H., Ranzato, M., Hadsell, R., Balcan, M., Lin, H. (eds.) Advances in Neural Information Processing Systems, vol. 33, pp. 9459–9474. Curran Associates, Inc. (2020). https://proceedings.neurips.cc/paper/2020/file/6b493230205f780e1bc26945df7481e5-Paper.pdf

19. Lin, J., Ma, X., Lin, S.C., Yang, J.H., Pradeep, R., Nogueira, R.: Pyserini: a python toolkit for reproducible information retrieval research with sparse and dense representations. In: Proceedings of the 44th International ACM SIGIR Conference on Research and Development in Information Retrieval, pp. 2356–2362. SIGIR 2021, Association for Computing Machinery, New York, NY, USA (2021). https://doi.org/10.1145/3404835.3463238

20. Liu, J., Shen, D., Zhang, Y., Dolan, B., Carin, L., Chen, W.: What makes good in-context examples for gpt-3? (2021). https://doi.org/10.48550/ARXIV.2101.06804, https://arxiv.org/abs/2101.06804

21. Nakano, R., et al.: Webgpt: browser-assisted question-answering with human feedback (2021). https://doi.org/10.48550/ARXIV.2112.09332, https://arxiv.org/abs/2112.09332

22. Nogueira, R., Jiang, Z., Pradeep, R., Lin, J.: Document ranking with a pre-trained sequence-to-sequence model. In: Findings of the Association for Computational Linguistics: EMNLP 2020, pp. 708–718. Association for Computational Linguistics, November 2020. https://doi.org/10.18653/v1/2020.findings-emnlp.63, https://aclanthology.org/2020.findings-emnlp.63

23. Perez, E., Lewis, P., Yih, W.T., Cho, K., Kiela, D.: Unsupervised question decomposition for question answering. In: Proceedings of the 2020 Conference on Empirical Methods in Natural Language Processing (EMNLP), pp. 8864–8880 (2020)

24. Press, O., Zhang, M., Min, S., Schmidt, L., Smith, N.A., Lewis, M.: Measuring and narrowing the compositionality gap in language models. arXiv preprint arXiv:2210.03350 (2022)

25. Raffel, C., et al.: Exploring the limits of transfer learning with a unified text-to-text transformer. J. Mach. Learn. Res. **21**(140), 1–67 (2020)

26. Reimers, N., Gurevych, I.: Sentence-bert: sentence embeddings using siamese bert-networks (2019). https://doi.org/10.48550/ARXIV.1908.10084, https://arxiv.org/abs/1908.10084

27. Robertson, S.E., Walker, S., Jones, S., Hancock-Beaulieu, M., Gatford, M.: Okapi at trec-3. In: TREC (1994)

28. Sachan, D.S., Lewis, M., Yogatama, D., Zettlemoyer, L., Pineau, J., Zaheer, M.: Questions are all you need to train a dense passage retriever. arXiv preprint arXiv:2206.10658 (2022)
29. Tay, Y., et al.: Unifying language learning paradigms (2022). https://doi.org/10.48550/ARXIV.2205.05131, https://arxiv.org/abs/2205.05131
30. Trivedi, H., Balasubramanian, N., Khot, T., Sabharwal, A.: Teaching broad reasoning skills via decomposition-guided contexts (2022). https://doi.org/10.48550/ARXIV.2205.12496, https://arxiv.org/abs/2205.12496
31. Wang, X., et al.: Self-consistency improves chain of thought reasoning in language models (2022). https://doi.org/10.48550/ARXIV.2203.11171, https://arxiv.org/abs/2203.11171
32. Wei, J., et al.: Chain of thought prompting elicits reasoning in large language models (2022). https://doi.org/10.48550/ARXIV.2201.11903, https://arxiv.org/abs/2201.11903
33. Xiong, W., Gupta, A., Toshniwal, S., Mehdad, Y., Yih, W.T.: Adapting pretrained text-to-text models for long text sequences (2022). https://doi.org/10.48550/ARXIV.2209.10052, https://arxiv.org/abs/2209.10052
34. Xiong, W., et al.: Answering complex open-domain questions with multi-hop dense retrieval (2020). https://doi.org/10.48550/ARXIV.2009.12756, https://arxiv.org/abs/2009.12756
35. Xu, W., Napoles, C., Pavlick, E., Chen, Q., Callison-Burch, C.: Optimizing statistical machine translation for text simplification. Trans. Assoc. Comput. Linguist. **4**, 401–415 (2016). https://doi.org/10.1162/tacl_a_00107, https://doi.org/10.1162/tacl_a_00107
36. Yang, Z., et al.: Hotpotqa: a dataset for diverse, explainable multi-hop question answering. In: Proceedings of the 2018 Conference on Empirical Methods in Natural Language Processing, pp. 2369–2380 (2018)
37. Zhu, F., Lei, W., Wang, C., Zheng, J., Poria, S., Chua, T.S.: Retrieving and reading: a comprehensive survey on open-domain question answering (2021). https://doi.org/10.48550/ARXIV.2101.00774, https://arxiv.org/abs/2101.00774

Towards Detecting Interesting Ideas
Expressed in Text

Bela Pfahl[ID] and Adam Jatowt[✉][ID]

University of Innsbruck, Innsbruck, Austria
bela.pfahl@student.uibk.ac.at, adam.jatowt@uibk.ac.at

Abstract. In recent years, product and project ideas are often sourced from public competitions, where anyone can enter their own solutions to an open-ended question. While copious ideas can be gathered in this way, it becomes difficult to find the most promising results among all entries. This paper explores the potential of automating the detection of interesting ideas and studies the effect of various features of ideas on the prediction task. A BERT-based model is built to rank ideas by their predicted interestingness, using text embeddings from idea descriptions and the concreteness, novelty as well as the uniqueness of ideas. The model is trained on a dataset of OpenIDEO idea competitions. The results show that language models can be used to speed up finding promising ideas, but care must be taken in choosing a suitable dataset.

Keywords: Detecting interesting ideas · Ideas ranking · Text novelty · Text concreteness

1 Introduction

Ideas are an essential part of technological and societal progress and innovation. Good constructive or conceptual ideas usually come at the beginning of every successful project or product. Open communities, where the public is invited to contribute their creative solutions to various challenges, have become popular in recent years for finding such ideas [16,19]. Many profit and nonprofit organizations use open online platforms to publicly collect users' answers to open-ended, exploratory questions or propositions for concrete products and product improvements. Collecting many ideas in this way is easy, but finding interesting entries in the resulting collection is a time-consuming and non-trivial task. What makes up a good – and therefore interesting – idea depends on a number of factors, such as novelty, feasibility, relevance and specificity [4]. If one could automatically rank a set of new, unseen ideas based on their predicted interestingness, it would save time and increase efficiency in analyzing them. This paper aims to explore to what extent automatic ranking is possible and find out which features of an idea are important for its perceived quality.

For our analysis, we construct a supervised regression model based on a neural network architecture using DistilBERT's contextual text embeddings and

J. Kamps et al. (Eds.): ECIR 2023, LNCS 13981, pp. 544–552, 2023.
https://doi.org/10.1007/978-3-031-28238-6_45

additional textual features. These include the concreteness of idea descriptions, their uniqueness in the competition and their novelty compared to existing ideas on the internet. The constructed model is trained and evaluated on a dataset of 21 OpenIDEO[1] [6] online idea competitions, containing nearly 5,000 ideas that have been expert rated and interacted with by a community of idea writers. We later discuss the unique properties and biases of the considered dataset and their effects on the models' predictions.

2   Related Work

The problem of measuring the interestingness of text has been addressed before in the field of recommender systems [1], knowledge discovery [8] and in analyzing interesting user content on the web [10] or in archival document collections [7]. There exist few works that we are aware of that deal with the interestingness or quality of ideas. Dasgupta and Dey [17] propose a model to rank textual ideas based on their innovativeness and Baba et al. [20] show an approach to assist idea selection using crowd raters.

Most relevant to our approach is the work of Ahmed et al. [2], who tackle the task of classifying high quality ideas using a supervised approach on a similar dataset of OpenIDEO idea competitions. They aim to improve the ranking of ideas and explore which aspects of an idea have the most impact on its chances of winning competitions. A set of features containing the amount of received likes, text descriptors such as long words or vocabulary size, the readability, coherence and the uniqueness of an idea is extracted from each idea's textual description, on which a random forest classifier is then trained. They find that the number of comments, sentences and long words had the most impact on the classification results. Further, they conclude that most features, 298 of 319, had no impact on the result at all. Among these were all coherence and semantic measurements. Most other predictive features, such as the number of sentences, long words, syllables or complex words, had a high correlation with the text length of the idea descriptions. The authors have in a later paper extended their ranking approach to also consider the diversity of found ideas [3].

In contrast to the work by Ahmed et al., which uses an ordinary classification approach, our model utilizes contextual text embeddings created by DistilBERT [14] and computes the uniqueness and novelty of ideas in relation to other ideas and prior ideas from the internet using sBERT [12]. Further, we focus on predicting the interestingness of ideas based on textual features alone, using community feedback as a label, rather than as a feature.

3   Methodology

Model. To explore if the perceived interestingness of an idea can – to some extent – be evaluated automatically, while also studying the importance of different features, we construct a supervised regression model that ranks ideas by

---

[1] https://www.openideo.com/.

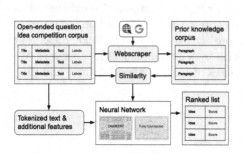

Fig. 1. Overview of the model's architecture

Normal vs Smoothed distribution Histogram for column mixed_score

Fig. 2. Distribution of the mixed score label with a Gaussian filter.

Table 1. Overview of used features

Feature	Description
Embeddings	Text embeddings of ideas' title, subtitle or description, obtained using DistilBERT
Concreteness	Measured using the mean and median of word concreteness scores obtained from the MRC Psycholinguistic Database
Uniqueness	Semantic similarity score relative to other ideas in the same competition, measured using sBERT
Novelty	Semantic similarity score relative to prior knowledge scraped from the internet using the Google search engine, also measured using sBERT

their predicted interestingness. An overview of the model's architecture can be seen in Fig. 1. For training the model, a dataset containing ideas together with a ground truth quality value is used. Embeddings of the ideas descriptions are obtained using DistilBERT [14], which are then combined with additional features extracted from the descriptions and used as input for a fully connected linear network. The additional features include concreteness values, which are calculated for each idea using a bootstrapped version of the MRC Psycholinguistic Database [11]. Concreteness scores are retrieved for all words in the idea that are found in the database, of which the mean and median is then computed. Further, uniqueness and novelty scores are obtained by calculating the semantic similarity of ideas in the dataset to each other, as well as to the prior knowledge respectively, using sBERT [12]. To build a collection of prior ideas, we scraped all top-10 returned search results on the Google search engine for the open-ended question of each competition available in our dataset that was used as a query. Then, we extracted textual content from those collected search results as a proxy for prior ideas related to the input question. The sBERT embeddings of all the target ideas in the dataset and their corresponding extracted prior ideas were then compared using cosine similarity to create two distance matrices for uniqueness and novelty. A uniqueness and novelty score is computed for each idea using the Local Outlier Factor (LOF) [5]. Both the concreteness score of an idea and its estimated uniqueness and novelty, calculated according to the above methods, serve as features for the classification model. An overview of the used features is given in Table 1. The model is implemented using PyTorch[2], building on the DistilBERT implementation provided by the HuggingFace library[3] and is trained on a training dataset with the Adam Optimizer [9].

Metrics. The models performance is evaluated with the normalized discounted cumulative gain (nDCG) and a thresholded cumulative accuracy profile (CAP) [18], similar as used in [21], on a list of outputs ranked by the predicted scores. In CAP, the cumulative score of all entries in the ranked list up to each index k is plotted as a curve, showing how many of the 'good' ideas can be found by reviewing the output list up to index k (The cumulative accuracy (CA)). The accuracy ratio (AR) describes how much of the potential performance is achieved. It is computed by dividing the area between the best theoretical curve and the achieved curve by the area between the best the theoretical curve and the random performance curve. The CAP is thresholded, i.e. removing all ideas below a score of 0.5, to make the difference between curves more meaningful.

4   Experiments

Dataset. We perform our experiments on a dataset of 21 OpenIDEO competitions[4] containing a total of $4,857$ written ideas answering different open-ended questions (e.g., "How might we get products to people without generating plastic waste?"). The dataset's ideas vary strongly in length and writing quality, ranging from single sentences to whole project descriptions. Some ideas are selected by a jury of experts to win each competition and receive monetary prices, chosen from candidate 'refinement' ideas, where the authors get additional feedback. During the competitions, ideas are edited frequently and other users can comment and like them. On average, a competition contains about 3.7% top ideas. To combat the strong class imbalance, we construct a mixed score label by linearly combining winning ideas, refinement ideas, comments and likes, normalized by the maximum values in each competition. This is possible because all three are well correlated (Point-biserial correlation between top ideas, likes is 0.522; top ideas, comments is 0.543; likes, comments is 0.817). To further alleviate the imbalanced distribution, we apply Label Distribution Smoothing (LDS) [15], where we calculate a smoothed label distribution using a Gaussian kernel and adjust the training loss by the density of the labels, emphasizing accuracy in rare labels. This should improve the model's ability to find high-scoring ideas. A distribution of the mixed score label and applied LDS can be seen in Fig. 2. The dataset is split randomly into 80% training set and 20% evaluation set ideas.

---

[2] https://pytorch.org/.

[3] https://huggingface.co/docs/transformers/model_doc/distilbert#transformers. DistilBertModel/.

[4] https://www.openideo.com/challenges.

Table 2. Performance comparison. Top: Bias baselines. Mid: Full models. Bottom: Ablation models.

Model/Baseline	NDCG@300	nDCG@all	AR	CA@0.1	CA@0.5	CA@0.8
Random order	0.42	0.77	0.0	0.1	0.5	0.8
Description length in words	0.65	0.85	0.62	0.45	**0.86**	**0.98**
Subtitle length in words	0.49	0.80	0.13	0.17	0.55	0.90
Title/Subtitle has 'Update'	0.60	0.85	0.27	0.45	0.57	0.85
description_all_features	**0.67**	**0.86**	**0.65**	**0.51**	**0.86**	0.97
subtitle_all_features	0.61	0.83	0.50	0.42	0.79	0.94
subtitle_no_updates_all_features	0.57	0.82	0.25	0.24	0.64	0.84
subtitle_no_features	0.60	0.83	0.48	0.45	0.75	0.95
subtitle_concreteness	0.65	0.85	0.53	0.46	0.78	0.93
subtitle_outlier	0.62	0.84	0.54	0.47	0.84	0.95
subtitle_novelty	0.60	0.83	0.54	0.41	0.83	0.97

Biases and Model Versions. While conducting the experiments, we found out that the OpenIDEO competitions have a strong bias between the text length of ideas descriptions or the word "update" occurring in the title or subtitle and their mixed score label. This is likely because authors that receive feedback from the community or jury wrote already promising ideas that they then edit even further (such ideas are then annotated with the word "update"). As this bias prevents learning in the model, we introduce three separate model versions: description_all_features uses the title and description, subtitle_all_features uses the title and subtitle and subtitle_no_updates_all_features uses the title and subtitle, with all 'update' related strings filtered out. The length of title and subtitle are not correlated to the label. To study the effect of different features on the model, an ablation study is performed on the subtitle model: The model subtitle_no_features uses the title, subtitle and no additional features, the models subtitle_concreteness, subtitle_outlier and subtitle_novelty the named feature, respectively. All models were trained for 10 epochs with stochastic gradient descent.

Results. The results for all three models can be seen in Table 2 and Fig. 3. In addition to the model results, Table 2 also lists the scores achieved when sorting all ideas in random order, by the description length in words, the subtitle length in words, and, with all ideas containing the string 'Update' in title or subtitle ordered before all other ideas. The model description_all_features, which relies on the descriptions using all additional features, performs the best. However, a similar performance can be achieved with the text length alone. The models subtitle_all_features and subtitle_no_updates_all_features, which both only use title and subtitle, perform worse, but the results are achieved without the text length bias present. The ablation study shows that the additional features do not have a statistically relevant effect on the performance of the model, as can be seen by comparing the different model versions in the last section of Table 2.

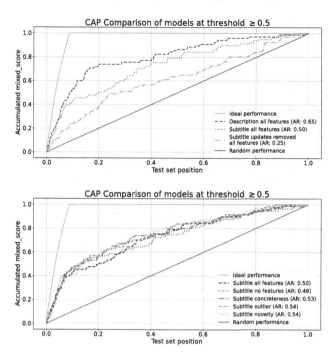

Fig. 3. CAP for all models. Top: Models using all features. Bottom: Ablation study on subtitle model. (X-axis: Relative position in the testset. Y-axis: Accumulated mixed score.)

The experiments show that it is possible to train a supervised model that, to some extent, can predict expert and community feedback based on textual idea descriptions. However, the performance is highly dependent on the quality and structure of the dataset used. The model version `subtitle_no_updates_all_features` performed considerably worse than the versions using the biased text length and update titles and subtitles. The big performance difference shows that each bias in the dataset can provide an easy-to-utilize, implicit feature on which the model bases its predictions instead of learning more meaningful patterns. The ablation studies' results support this observation, as the other additional features had little to no effect on the model's performance. This suggests, that the concreteness, novelty and outlierness had no measurable impact on the jury's decisions and community engagement, which was better predicted by textual semantics encoded into DistilBERT embeddings. Because the used OpenIDEO dataset contains ideas in many different stages, where some ideas were selected as promising early on and had enough opportunity to be subsequently improved, it is unclear if the features are not measurable on the given dataset or generally unsuited for predicting interesting ideas.

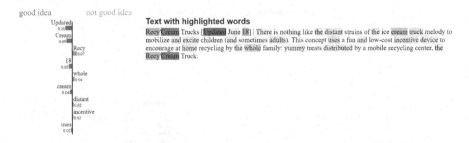

Fig. 4. LIME result for True Positive example answering the question "How might we establish better recycling habits at home?". Predicted score: 0.64, Ground truth score: 0.98.

Finally, an example decision of the model on a true positive sample can be seen in the LIME [13] analysis in Fig. 4, where the update phrase in the title is the strongest indicator for the model. Vivid words such as "cream", "truck" or "yummy" had a positive impact, while abstract terms like "incentive device" had a detrimental effect.

5   Conclusions

This paper explored if and how language models can be utilized to find interesting ideas in large collections and studied the effect of different features of an idea on its perceived interestingness. Our model uses DistilBERT embeddings of textual idea descriptions, concreteness, outlier and novelty scores. The results show that the model is able to rank ideas substantially better than using a random order, therefore indicating that automated methods can indeed help in finding interesting ideas. The concreteness, outlier and novelty features had little to no impact on the model's decisions.

To improve the method in future work, finding a better and bigger dataset that captures ideas in an earlier, more homogeneous state appears to be very important. With such a dataset, it would be possible to investigate further why the extracted additional features of concreteness, novelty and outlierness had little to no effect. Furthermore, additional metrics for calculating the novelty, uniqueness and concreteness of ideas, as well as additional features should be investigated.

Acknowledgment. The dataset consisting of 21 OpenIDEO competitions used in the paper has been provided by the research group for Innovation & Entrepreneurship (https://www.uibk.ac.at/smt/innovation-entrepreneurship/) at the Department of Strategic Management, Marketing and Tourism of the University of Innsbruck. Parts of the experiments have been conducted while Bela Pfahl was employed as a student research assistant in the group.

References

1. Adamopoulos, P., Tuzhilin, A.: On unexpectedness in recommender systems: Or how to better expect the unexpected. ACM Trans. Intell. Syst. Technol. (TIST) **5**(4), 1–32 (2014)
2. Ahmed, F., Fuge, M.: Capturing winning ideas in online design communities. In: Proceedings of the 2017 ACM Conference on Computer Supported Cooperative Work and Social Computing, ACM, New York, NY, USA (2017). https://doi.org/10.1145/2998181.2998249
3. Ahmed, F., Fuge, M.: Ranking ideas for diversity and quality. J. Mech. Des. **140**(1) (2018). https://doi.org/10.1115/1.4038070
4. Blohm, I., Riedl, C., Füller, J., Leimeister, J.M.: Rate or trade? identifying winning ideas in open idea sourcing. Inf. Syst. Res. **27**(1), 27–48 (2016). https://doi.org/10.1287/isre.2015.0605
5. Breunig, M.M., Kriegel, H.P., Ng, R.T., Sander, J.: Lof. In: Dunham, M. (ed.) Proceedings of the 2000 ACM SIGMOD International Conference on Management of Data, pp. 93–104. ACM Conferences, ACM, New York, NY (2000). https://doi.org/10.1145/342009.335388
6. Fuge, M., Tee, K., Agogino, A., Maton, N.: Analysis of collaborative design networks: a case study of openideo. J. Comput. Inf. Sci. Eng. **14**(2) (2014). https://doi.org/10.1115/1.4026510
7. Jatowt, A., Hung, I.-C., Färber, M., Campos, R., Yoshikawa, M.: Exploding TV sets and disappointing laptops: suggesting interesting content in news archives based on surprise estimation. In: Hiemstra, D., Moens, M.-F., Mothe, J., Perego, R., Potthast, M., Sebastiani, F. (eds.) ECIR 2021. LNCS, vol. 12656, pp. 254–269. Springer, Cham (2021). https://doi.org/10.1007/978-3-030-72113-8_17
8. Kuznetsov, S.O., Makhalova, T.: On interestingness measures of formal concepts. Inf. Sci. **442–443**, 202–219 (2018). https://doi.org/10.1016/j.ins.2018.02.032
9. Loshchilov, I., Hutter, F.: Fixing weight decay regularization in adam. CoRR abs/1711.05101 (2017)
10. Naveed, N., Gottron, T., Kunegis, J., Alhadi, A.C.: Bad news travel fast: a content-based analysis of interestingness on twitter. In: Proceedings of the 3rd International Web Science Conference, pp. 1–7 (2011)
11. Paetzold, G., Specia, L.: Inferring psycholinguistic properties of words. In: Proceedings of the 2016 Conference of the North American Chapter of the Association for Computational Linguistics: Human Language Technologies. Association for Computational Linguistics, Stroudsburg, PA, USA (2016). https://doi.org/10.18653/v1/n16-1050
12. Reimers, N., Gurevych, I.: Sentence-BERT: sentence embeddings using siamese BERT-networks. arXiv (2019). https://doi.org/10.48550/arXiv.1908.10084
13. Ribeiro, M.T., Singh, S., Guestrin, C.: Why should i trust you?": Explaining the predictions of any classifier. In: Proceedings of the 22nd ACM SIGKDD International Conference on Knowledge Discovery and Data Mining, San Francisco, CA, USA, 13–17 August 2016, pp. 1135–1144 (2016)
14. Sanh, V., Debut, L., Chaumond, J., Wolf, T.: Distilbert, a distilled version of bert: smaller, faster, cheaper and lighter. CoRR abs/1910.01108 (2019)
15. Sun, Y., Wong, A.K.C., Kamel, M.S.: Classification of imbalanced data: a review. Int. J. Pattern Recogn. Artif. Intell. **23**(04), 687–719 (2009)
16. Terwiesch, C., Xu, Y.: Innovation contests, open innovation, and multiagent problem solving. Manage. Sci. **54**(9), 1529–1543 (2008). https://doi.org/10.1287/mnsc.1080.0884

17. Dasgupta, T., Dey, L.: Automatic scoring for innovativeness of textual ideas (2016)
18. van der Burgt, M.: Calibrating low-default portfolios, using the cumulative accuracy profile. J. Risk Model Validation **1**(4), 17–33 (2008)
19. Wahl, J., Füller, J., Hutter, K.: What's the problem? how crowdsourcing and text-mining may contribute to the understanding of unprecedented problems such as covid-19. R&D Manage. **52**(2), 427–446 (2022). https://doi.org/10.1111/radm.12526
20. Baba, Y., Li, J., Kashima, H.: Crowdea: multi-view idea prioritization with crowds. In: Proceedings of the AAAI Conference on Human Computation and Crowdsourcing, vol. 8, pp. 23–32 (2020). https://ojs.aaai.org/index.php/hcomp/article/view/7460
21. Zhang, Y., Siriaraya, P., Kawai, Y., Jatowt, A.: Predicting time and location of future crimes with recommendation methods. Knowledge-Based Systems **210**, 106503 (2020). https://doi.org/10.1016/j.knosys.2020.106503

Towards Linguistically Informed Multi-objective Transformer Pre-training for Natural Language Inference

Maren Pielka[1(✉)], Svetlana Schmidt[1,2], Lisa Pucknat[1,3], and Rafet Sifa[1]

[1] Fraunhofer IAIS, Schloss Birlinghoven, 53757 Sankt Augustin, Germany
Maren.Pielka@iais.fraunhofer.de
[2] Ruhr-Universität Bochum, Universitätsstraße 150, 44801 Bochum, Germany
[3] Rheinische Friedrich-Wilhelms Universität Bonn,
Regina-Pacis-Weg 3, 53113 Bonn, Germany

Abstract. We introduce a linguistically enhanced combination of pre-training methods for transformers. The pre-training objectives include POS-tagging, synset prediction based on semantic knowledge graphs, and parent prediction based on dependency parse trees. Our approach achieves competitive results on the Natural Language Inference task, compared to the state of the art. Specifically for smaller models, the method results in a significant performance boost, emphasizing the fact that intelligent pre-training can make up for fewer parameters and help building more efficient models. Combining POS-tagging and synset prediction yields the overall best results.

1 Introduction

Understanding entailment and contradictions is a particularly hard task for any machine learning (ML) model. The system has to deeply comprehend the semantics of natural language, and have access to some amount of background knowledge that is often helpful in understanding many real-world statements. Current solutions still show deficits with respect to both criteria. At the same time, state-of-the-art language models such as GPT-3 [3] and XLM-RoBERTa [4,10] rely heavily on massive amounts of data for pre-training, and are quite resource-extensive due to their large number of parameters.

To address those shortcomings, we present a linguistically enhanced approach for multi-objective pre-training of transformer models. We inject extra knowledge into the model by pre-training for three additional language modelling tasks, one of which is a novel approach. Specifically, we utilize external information about part of speech tags, syntactic parsing, and semantic relations between words. Our main contribution can be summarized as follows:

We aim to become independent of huge data resources for pre-training, and having to train models with a large number of parameters, by injecting as much external knowledge to the model as possible. This goal is being quantified by

J. Kamps et al. (Eds.): ECIR 2023, LNCS 13981, pp. 553–561, 2023.
https://doi.org/10.1007/978-3-031-28238-6_46

evaluating our model on the Stanford Natural Language Inference (SNLI) [2] data set. We compare different implementations of the transformer architecture (BERT [7] and XLM-RoBERTa [4,10]), with the aim to show that the smaller model, BERT, is able to perform competitively when being enhanced with additional knowledge during pre-training.

Our approach does not require any additional data or annotations for pre-training. It is pre-trained for the additional tasks on the same data set that it is later being fine-tuned on. The labels for the word-level pre-training are generated in a semi-supervised fashion, by utilizing existing models and knowledge bases for those tasks. We therefore argue, that we can achieve close to state-of-the-art performance with a comparatively small (BERT-base) model and minimal additional effort in terms of data and computation time.

2 Related Work

In Natural Language Inference (NLI), first introduced by [5], one has to determine whether a given *hypothesis* can be inferred from a given *premise*, or whether it contradicts the premise. Further, a new research field emerged from the NLI task named Contradiction Detection (CD). Multiple languages, besides the commonly used English language, such as Persian [18], Spanish [20], and German [14,15,21,22] were studied. We follow up on the latter research, in which a portion of the SNLI dataset was machine-translated into German. They found that RNNs handle machine-translated data quite well, with difficulties in complicated sentence structures, translation artifacts, and understanding of world knowledge. A fine-tuned XLM-RoBERTa model seemed to be most promising with regard to the difficulties mentioned above. Still, qualitative exploration [13] has shown that among other things, the model struggles with prepositional references, incomplete sentences as well as antonyms and homonyms, which gave rise to enhance the model with lower level linguistic tasks.

BERT [7] and XLM-RoBERTa [4,10] are among the state of the art transformer-based encoder models for text classification tasks. They use the pre-training objectives of Masked Language Modeling and Next Sentence Prediction (only BERT) to obtain a large amount of language understanding in an unsupervised way. Dependency Injected Bidirectional Encoder Representations from Transformers (DIBERT) [24] utilizes a third pre-training objective called parent prediction injecting syntactic structures of dependency trees.

The approach of integrating the external semantic knowledge into a transformer model was presented by [1]. In their work, WordNet embeddings are combined with the BERT architecture in two ways: during *external combination* the outputs of WordNet and BERT are combined for the additional classification and in *internal inclusion* the WordNet representations are integrated into the internal BERT architecture. The resulting models were evaluated on four GLUE [25] datasets for Sentiment Analysis, Linguistic Acceptability, Sentence Similarity and Natural Language Inference tasks [1].

A similar approach has been introduced by [27]. They pre-train a BERT model on five different, linguistics aware tasks such as POS-tagging, semantic role labeling and syntactic parsing, achieving competitive results on GLUE benchmark tasks. The main difference between this work and ours is that we focus on minimizing the amount of pre-training data and model parameters, utilizing the same data sets for both custom pre-training and fine-tuning. In addition, we introduce the novel synset prediction objective. Unlike the approach of [1], we utilize only one synset extracted for each word in the data.

3 Data

The Stanford Natural Language Inference (SNLI) data set was introduced by Bowman et al. [2]. It is the largest collection of human-generated premise and hypothesis pairs for the NLI task with over 570,000 examples. The data was collected in a crowd-source campaign. Workers were instructed to devise hypotheses inspired by premise image captions. These sentences should entail, contradict or not relate to the original caption. In a final effort, sentence pairs were labeled by different annotators with one of three labels - *entailment*, *contradiction* or *neutral* (if hypothesis does not relate to premise). The gold label for each pair was chosen based on a majority vote of annotators.

4 Methodology: Pretraining Methods

Injecting syntactic and semantic information into the architecture is achieved by training with different pre-training objectives. All of our pre-training objectives are word-based, meaning that we utilize the output vector mapping to the corresponding input-token for these tasks instead of the special [CLS] token, which is commonly used for sentence level classification tasks. All of our labels are generated in a semi-supervised manner. We take advantage of already present and well working architectures to predict labels for POS-tagging and dependency parsing and create labels for different synsets with the nltk wordnet[1] interface supporting the WordNet[2] [8] lexical database.

4.1 POS-tagging

The main objective of part-of-speech (POS)-tagging is to predict the syntactic function of a word in a sentence. Words can have different meanings in different contexts. Therefore, POS-tagging is used, among other things, to identify the context in which a word occurs. The used tagset includes common parts of speech such as adjective, noun and verb, but also finer graduations such as numerical and symbol words. We extract labels from spaCys implementation for POS-tagging [9]. Among the common POS-tags are: NOUN (noun), VERB (verb), ADJ (adjective), ADV (adverb), DET (determiner), PRON (pronoun). The full list can be found at the spaCy repository[3].

---

[1] https://www.nltk.org/howto/wordnet.html.

[2] https://wordnet.princeton.edu/.

[3] https://github.com/explosion/spaCy/blob/master/spacy/glossary.py.

The following example shows semi-supervised generated tags for a tokenized sentence from the SNLI dataset. An underscore corresponds to the beginning of a word. As POS-tags are associated with complete words, but some words are being split into multiple tokens during tokenization, each input-token is assigned the POS-tag for the complete word. So, tokens for words that are split up by the tokenizer all map to the same POS-tag.

```
_A _person _on  _a _horse _jump  s  _over _a _broken _down _air  plan   e    .
DET NOUN  ADP DET NOUN VERB VERB ADP DET  VERB   ADP  NOUN NOUN NOUN PUNCT
```

4.2 Parent Prediction

For parent prediction (PP) [24] the parent of each word is predicted. The parent is deduced from a corresponding dependency tree of the sentence, which was created using the NLP library Stanza [17]. The dependency tree provides information about the syntactic dependency relation between words. Each word is assigned to exactly one other word, so each word has precisely one parent. The central clause, i.e. the root clause without parent, is a (finite) verb.

4.3 Synset Prediction

In order to enhance the model with semantic knowledge, we take advantage of the WordNet [8,12] knowledge graph. WordNet is the lexical database for the English language. The nouns, verbs, adjectives and adverbs in WordNet are organized in groups, based on their semantic similarity. Those groups are called synsets (synonyms sets). One synset represents one distinct concept, thus one synset can contain several lexical units, where each of the lexical units represents one meaning of a word. Since words have several meanings, they can be associated with several synsets. For instance, the synset for the word *lady* in the sentence *"The lady is weeding her garden."* contains three possible meanings, as it can be seen below.

Synset('lady.n.01'), Synset('dame.n.02'), Synset('lady.n.03')

The main objective of this pre-training task is the prediction of labels representing semantic knowledge. We extract the synsets from WordNet for nouns, verbs and adjectives. The WordNet[4] nltk corpus reader is used for the extraction of synsets. The first synset in a set of synsets represents the most common meaning of a word. Thus, we utilize the first synset for semantic representation of a word. For example, for the word *lady* the synset *Synset('lady.n.01')* is chosen. We argue that since most words have a unique meaning, this approach is a reasonable heuristic, even though it will introduce a small amount of noise by assigning the wrong synset to uncommon words. To our best knowledge,

---

[4] https://www.nltk.org/_modules/nltk/corpus/reader/wordnet.html.

it is the first attempt to utilise the synsets for pre-training the model with semantic knowledge.

The following example shows the tokenized sentence from above and the corresponding labels. Similar to the example in 4.1 the label for a complete word is assigned to each of the subword tokens, just as in case with _we ed ing.

_The _lady _is _we   ed   ing _her _garden

no_syn lady_n_01 be_v_01 weed_v_01 weed_v_01 weed_v_01 no_syn garden_n_01

5   Experiments and Results

In the next section, we describe the experimental setup and further evaluate our proposed pre-training objectives quantitatively and qualitatively. We do not use any additional data, other than the SNLI training set, and prolong the overall training only by a few epochs.

The main model[5] is based on a BERT architecture with approximately 110M parameters, 12-layers, 12 attention heads and a hidden state of size 768. A simple feed-forward layer is used for classification and shared across each output vector or, in case of finetuning, for the special [CLS] token. The BERT model has been pre-trained for the Masked Language Modeling and Next Sentence Prediction tasks on a large corpus of English data from books [28] and Wikipedia. Further, we compare to a large XLM-RoBERTa[6] with 355M parameters. Binary Cross Entropy Loss in combination with AdamW optimizer [11] is used for all experiments. For pre-training a learning rate of 6e-5 is used. For fine-tuning we utilize a learning rate of 5e–6.

Evaluating our main model, the overall best results are achieved when we pre-train for POS-tagging and synset prediction, yielding a significant performance boost over the baseline model (see Table 1). This proofs that linguistically informed pre-training does in fact help the model to capture additional knowledge that is helpful for the classification task. Apparently, not all combinations of pre-training methods work equally well. For example, combining all three approaches yields slightly worse results than combining only POS-tagging and parent prediction, or POS-tagging and synset prediction. A possible explanation for this behavior is that the model "forgets" previously learned knowledge, if it is trained for multiple tasks in a row. It is yet to be explored, whether it would help the model if the objectives were applied subsequently to specific layers of the transformer.

---

[5] https://huggingface.co/bert-base-cased.

[6] https://huggingface.co/xlm-roberta-large.

Table 1. Performance comparison for different pre-training configurations on the SNLI test set, in percent. The abbreviations stand for: POS=POS-tagging, PP=Parent Prediction, Syn=Synset prediction.

Pretraining configuration	Acc.	F1-Score (Cont.)	F1-Score (Ent.)	F1-Score (Neut.)
No additional pretraining	88.6	91.6	89.7	84.5
POS	90.0	92.4	90.9	86.7
PP	89.5	92.1	90.4	85.9
POS+PP	90.2	92.8	91.1	86.5
Syn	89.9	92.3	90.8	86.6
POS+Syn	**90.4**	**93.2**	**91.7**	**86.7**
PP+Syn	89.9	92.6	90.6	86.3
POS+PP+Syn	89.9	92.5	90.7	86.4

Table 2. Performance comparison for different model architectures on the SNLI test set, in percent. We compare our approaches with (POS+Syn) and without pre-training to the current best result on the data set by [26].

Configuration	Base model	Num. param.	Acc.	F1 (Cont.)	F1 (Ent.)	F1 (Neut.)
Current SOTA (EFL)	Roberta-large	355 M	93.1	n.a	n.a	n.a
No add. pretraining	Xlm-roberta-large	345 M	91.5	94.5	92.1	87.7
POS+Syn	SXlm-roberta-large	345 M	91.5	94.5	92.0	88.1
No add. pretraining	Bert-base-cased	110 M	88.6	91.6	89.7	84.5
POS+Syn	Bert-base-cased	110 M	90.4	93.2	91.1	86.7

Comparing the different model architectures (Table 2), it is apparent that adding further pre-training tasks helps the smaller models achieve competitive results compared to xlm-roberta-large, while it does not yield a huge performance boost for the large model itself. In order to prove that the improvement is significant and due to pre-training tasks, we compare the mean performance for five training and evaluation runs of the model architectures with additional pre-training and without it in Table 3. While the difference between xlm-roberta-large performances with and without additional pre-training is almost not noticeable, the mean of evaluation results of the smaller model with additional pre-training shows improvement. This suggests that enhancing the smaller models with additional knowledge could make them competitive, and thereby not having to rely on extensive computational resources. At the same time, both models achieve results that are comparable to the current state of the art [23, 26].

Table 3. Mean with standard deviation of different model architectures performance on the SNLI test set, in percent. Each of the models was evaluated five times and the mean was calculated over all five evaluation results per setting.

Configuration	Base model	Mean Acc.	Mean F1 (Cont.)	Mean F1 (Ent.)	Mean F1 (Neut.)
No add. pretraining	Xlm-roberta-large	91.8(±0.10)	94.5(±0.08)	88.4(±0.10)	87.6(±0.15)
POS+Syn	Xlm-roberta-large	91.8(±0.06)	94.5(±0.09)	88.6(±0.12)	87.8(±0.09)
No add. pretraining	Bert-base-cased	89.5(±0.10)	91.8(±0.08)	86.0(±0.10)	84.3(±0.15)
POS+Syn	Bert-base-cased	89.6(±0.06)	92.0(±0.09)	86.1(±0.12)	84.5(±0.09)

6   Conclusion and Outlook

We presented a combination of linguistically enhanced pre-training methods for transformers. The experimental results illustrate that the performance of the transformer models on the NLI task can be improved by enhancing the models with syntactic and semantic knowledge. The novel method of synset prediction shows that enriching transformer models with semantic knowledge positively affects the ability of the models to learn semantic correlations in data. Moreover, it is not required to utilize a large transformer model for handling the task of detecting contradictions, entailments or neutral expressions. Another important advantage of our approach is that the improvement can be achieved with no additional training data. Part of this research has already successfully been applied in an industry context, for finding contradictions in financial reports [6], showing that an informed approach also facilitates domain adaptation.

A significant limitation of our work is the rule-based annotation procedure for the synset prediction task, utilizing always the first (most probable) synset extracted from WordNet as a label for a given word. This is clearly not ideal, as it introduces some noise, and less common synsets are not represented in the labels. Nevertheless, the results show that we can already achieve a performance improvement by using this simplified approach. It would be an interesting direction of research, to treat this problem as a machine learning task on its own and train a dedicated model, which would most likely enhance the downstream performance even further. This, of course, would require a certain amount of manual annotations. In this regard, it could also be meaningful to reduce the number of predicted synsets by grouping them together into clusters or hypernym groups, which would make the learning problem easier and less sparse.

Future work includes extending the approach to other, less research-covered languages such as German, Italian or Arabic. We also aim to further reduce the model size by integrating more external knowledge. One idea would be data augmentation methods with the goal to align the languages in feature space, similar to the approach presented by [16]. Another direction of research is training on prototypical examples, as suggested by [19]. Those could be created using linguistic rules, thus reducing the amount of hand-annotated training data and teaching the model the essential rules of contradiction and entailment.

Finally, we also plan to apply the pre-training approaches to other tasks, such as toxicity detection or relation extraction from financial documents.

Acknowledgments. We thank the anonymous reviewers for their valuable feedback. This research has been funded by the Federal Ministry of Education and Research of Germany and the state of North-Rhine Westphalia as part of the Lamarr-Institute for Machine Learning and Artificial Intelligence, LAMARR22B.

References

1. Barbouch, M., Verberne, S., Verhoef, T.: WN-BERT: Integrating wordnet and BERT for lexical semantics in natural language understanding. Comput. Linguist. Netherlands J. **11**, 105–124 (2021)
2. Bowman, S., Angeli, G., Potts, C., Manning, C.: A large annotated corpus for learning natural language inference. In: Proceeding of EMNLP (2015)
3. Brown, T.B., et al.: Language models are few-shot learners. CoRR abs/2005.14165 (2020). https://arxiv.org/abs/2005.14165
4. Conneau, A., et al.: Unsupervised cross-lingual representation learning at scale. arXiv preprint arXiv:1911.02116 (2019)
5. Dagan, I., Glickman, O., Magnini, B.: The PASCAL recognising textual entailment challenge. In: Quiñonero-Candela, J., Dagan, I., Magnini, B., d'Alché-Buc, F. (eds.) MLCW 2005. LNCS (LNAI), vol. 3944, pp. 177–190. Springer, Heidelberg (2006). https://doi.org/10.1007/11736790_9
6. Deußer, T., et al.: Contradiction detection in financial reports. In: Proceeding of NLDL 2023 (2023)
7. Devlin, J., Chang, M.W., Lee, K., Toutanova, K.: Bert: Pre-training of deep bidirectional transformers for language understanding. arXiv preprint arXiv:1810.04805 (2018)
8. Fellbaum, C.: WordNet: An Electronic Lexical Database. Bradford Books, Denver (1998)
9. Honnibal, M., Montani, I., Van Landeghem, S., Boyd, A.: spacy: Industrial-strength natural language processing in python (2020)
10. Liu, Y., et al.: Roberta: a robustly optimized BERT pretraining approach. CoRR abs/1907.11692 (2019). http://arxiv.org/abs/1907.11692
11. Loshchilov, I., Hutter, F.: Decoupled weight decay regularization. arXiv preprint arXiv:1711.05101 (2017)
12. Miller, G.A.: Wordnet: a lexical database for English. Commun. ACM **38**(11), 39–41 (1995)
13. Pielka, M., Rode, F., Pucknat, L., Deußer, T., Sifa, R.: A linguistic investigation of machine learning based contradiction detection models: an empirical analysis and future perspectives. In: Proceedings of ICMLA 2022 (2022)
14. Pielka, M., et al.: Tackling contradiction detection in German using machine translation and end-to-end recurrent neural networks. In: Proceedings of ICPR 2020 (2021)
15. Pucknat, L., Pielka, M., Sifa, R.: Detecting contradictions in German text: a comparative study. In: 2021 IEEE Symposium Series on Computational Intelligence (SSCI), pp. 01–07. IEEE (2021)
16. Pucknat, L., Pielka, M., Sifa, R.: Towards informed pre-training for critical error detection in English-German. In: Proceedings of LWDA 2022 (2022)

17. Qi, P., Zhang, Y., Zhang, Y., Bolton, J., Manning, C.D.: Stanza: a python natural language processing toolkit for many human languages. arXiv preprint arXiv:2003.07082 (2020)
18. Rahimi, Z., ShamsFard, M.: Contradiction detection in persian text. arXiv preprint arXiv:2107.01987 (2021)
19. von Rueden, L., Houben, S., Cvejoski, K., Bauckhage, C., Piatkowski, N.: Informed pre-training on prior knowledge. arXiv preprint arXiv:2205.11433 (2022)
20. Sepúlveda-Torres, R., Bonet-Jover, A., Saquete, E.: Here are the rules: ignore all rules': automatic contradiction detection in Spanish. Appl. Sci. **11**(7), 3060 (2021)
21. Sifa, R., et al.: Towards automated auditing with machine learning. In: Proceedings of the ACM Symposium on Document Engineering 2019, pp. 1–4 (2019)
22. Sifa, R., Pielka, M., Ramamurthy, R., Ladi, A., Hillebrand, L., Bauckhage, C.: Towards contradiction detection in German: a translation-driven approach. In: Proceedings of IEEE SSCI 2019 (2019)
23. Sun, Z., et al.: Self-explaining structures improve NLP models. arXiv preprint arXiv:2012.01786v2 (2020)
24. Wahab, A., Sifa, R.: Dibert: dependency injected bidirectional encoder representations from transformers. In: Proceedings of IEEE SSCI 2021 (2021)
25. Wang, A., Singh, A., Michael, J., Hill, F., Levy, O., Bowman, S.R.: Glue: a multi-task benchmark and analysis platform for natural language understanding. arXiv preprint arXiv:1804.07461 (2018)
26. Wang, S., Fang, H., Khabsa, M., Mao, H., Ma, H.: Entailment as few-shot learner (2021)
27. Zhou, J., Zhang, Z., Zhao, H.: LIMIT-BERT : linguistic informed multi-task BERT. CoRR abs/1910.14296 (2019)
28. Zhu, Y., Kiros, R., Zemel, R., Salakhutdinov, R., Urtasun, R., Torralba, A., Fidler, S.: Aligning books and movies: towards story-like visual explanations by watching movies and reading books (2015). https://doi.org/10.48550/ARXIV.1506.06724, https://arxiv.org/abs/1506.06724

Dirichlet-Survival Process: Scalable Inference of Topic-Dependent Diffusion Networks

Gaël Poux-Médard$^{(\boxtimes)}$ (ID), Julien Velcin (ID), and Sabine Loudcher (ID)

Université de Lyon, Lyon 2, ERIC UR 3083,
5 Avenue Pierre Mendès France, 69676 Bron Cedex, France
{gael.poux-medard,julien.velcin,sabine.loudcher}@univ-lyon2.fr

Abstract. Information spread on networks can be efficiently modeled by considering three features: documents' content, time of publication relative to other publications, and position of the spreader in the network. Most previous works model up to two of those jointly, or rely on heavily parametric approaches. Building on recent Dirichlet-Point processes literature, we introduce the Houston (Hidden Online User-Topic Network) model, that jointly considers all those features in a non-parametric unsupervised framework. It infers dynamic topic-dependent underlying diffusion networks in a continuous-time setting along with said topics. It is unsupervised; it considers an unlabeled stream of triplets shaped as *(time of publication, information's content, spreading entity)* as input data. Online inference is conducted using a sequential Monte-Carlo algorithm that scales linearly with the size of the dataset. Our approach yields consequent improvements over existing baselines on both cluster recovery and subnetworks inference tasks.

Keywords: Spreading process · Network inference · Clustering · Bayesian nonparametrics

1 Introduction

1.1 Overview of the Contribution

Over the last decades, information spread patterns have become more and more complicated. The volume of data that flows on social networks keeps increasing every day that passes, and results in complex diffusion processes that can be described by many factors. However, recent advances suggest that documents complex diffusion processes can be efficiently modeled considering only three variables: their publication date (when), the publisher (who) and their semantic content (what). The idea of considering these three factors is not novel. However, most of the models that tackle diffusion problems tend to consider up to two of these, but seldom the three parameters.

J. Kamps et al. (Eds.): ECIR 2023, LNCS 13981, pp. 562–570, 2023.
https://doi.org/10.1007/978-3-031-28238-6_47

We introduce the Houston model, that tackles the problem by jointly infer-ring clusters of textual documents spreading online *and* the subnetworks they spread on. In this context, a cluster is a set of documents that share similar semantic content *and* similar diffusion patterns ; the associated subnetwork is a set of nodes whose edges represent the probabilities for any element of a clus-ter to spread between two nodes. Our method builds on recent Dirichlet-Point processes advances [9, 18, 23, 24]. To the best of our knowledge, it is the first model that considers semantic content, publication dynamics and the network of spreading documents in an online, non-parametric and unsupervised way.

Firstly, we briefly review existing works on topic-aware diffusion networks inference Sect. 1.2. We then introduce Dirichlet-Point processes in general in Sect. 2.1, detail which Dirichlet process and which Point process are considered Sect. 2.2, to finally build the final Dirichlet-Survival Process in Sect. 2.3. The Sequential Monte-Carlo optimization algorithm is finally discussed, and applied to synthetic datasets in Sect. 3.

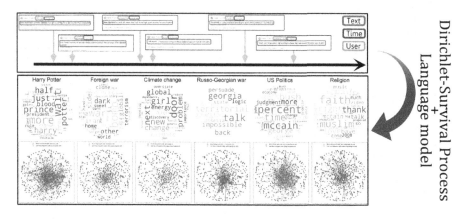

Fig. 1. From a stream of textual documents, we model the underlying topic-dependent diffusion subnetworks. Inference is unsupervised, non-parametric and conducted online, meaning data is processed sequentially. Results in the bottom row come from the application of our method to the Memetracker dataset [17]. Nodes colors represent traditional medias (red) and blog (blue). (Color figure online)

1.2   Related Works

It has been underlined on several occasions that efficiently modeling informa-tion diffusion involves accounting for the network's structure [16, 22], publication times [8, 12] and documents' content [10, 15]. Some approaches consider sequen-tially all three factors. Typically, they first infer topics based on documents content, and only then they use this information to infer the latent diffusion subnetworks [7, 10, 15, 26, 28, 29]. These approaches run with a lower computa-tional complexity at the cost of a lesser accuracy, as shown in the experimental

section. The work most closest to ours [4] is, to our knowledge, the only one
that jointly models documents' content, dynamics and structure. It develops an
unsupervised topic-dependent network inference method. The approach breaks
down the topic-aware diffusion into two factors: each node is assumed to have a
given sensitivity to a topic, and a certain authority on them. Given this assumption, the authors develop a parametric prior on the probability for a diffusion
cascade to belong to a given topic. The textual content (or side information) is
then accounted for using a homogeneous Poisson textual model [19], combined
with the above prior. The model is optimized using an EM algorithm. However,
the optimization algorithm is not designed for online optimization –data cannot
be added sequentially–, and topics optimization is parametric –the number of
topics must be provided.

2 Model

2.1 Background

To answer these limitations, we build a Dirichlet-Survival process. Dirichlet-
Point processes –of which the Dirichlet-Survival process is a special case– are
created by merging Dirichlet processes with Point processes. The method has
been explored by combining Hawkes processes to several variants of Dirichlet processes (hierarchical [18], mixed membership [27], powered [23], multivariate [24]).
Besides, in [14], the authors show that a large part of the literature on underlying
diffusion network inference [8,12–14,21,28] can be expressed as special cases of a
counting point process. The method allows to infer dynamic underlying diffusion
networks using convex optimization tools.

However, no work considered the combination of Dirichlet-Point processes to
other processes than the Hawkes process. Our approach using Survival analysis
explores this new connection: we design an optimization algorithm (Sequential
Monte Carlo) for online non-parametric topics-aware diffusion subnetworks inference (the number of topics/subnetworks does not have to be chosen in advance).
This results in the Dirichlet-Survival prior.

2.2 Dirichlet Process and Survival Analysis

Dirichlet Process. The Dirichlet process is used as a non-parametric prior
distribution over clusters in many clustering algorithms. It can be written as
follows:

$$P(s_i = k | \{s_m\}_{m=1,...,n-1}, \alpha_0) = \begin{cases} \frac{N_k}{\alpha_0 + \sum_k^K N_k} & \text{if k} = 1, ..., \text{K} \\ \frac{\alpha_0}{\alpha_0 + \sum_k^K N_k} & \text{if k} = \text{K+1} \end{cases} \quad (1)$$

where s_i is a variable that represents the cluster of the ith observation, $N_k = |\{s_i | s_i = k\}_{i=1,...,n-1}|$ the population of cluster k, K the total number of non-
empty clusters and α_0 a concentration hyper-parameter. The choice of $K + 1$
means a new cluster is opened and K in increased by 1. Note that references
[23,24] use the powered version of this process [25].

Network Inference Model. The edges of topic-dependent networks are inferred using the NetRate model [12], which is part of a broad literature on underlying spreading networks inference [10,12–14,28]. In particular in [14], the authors demonstrate that all these models can be expressed as special cases of a counting point process. These processes take a collection of independent times-tamped diffusion cascades $\vec{c} = \{(u_i^c, t_i^c)\}_i$ as input, where u_i^c is the node on which the ith event occurred and t_i^c the time at which it happened in cascade c. The process is entirely characterized by a hazard function $H(t_i^c | t_j^c, \alpha_{u_j^c, u_i^c})$, which is the instantaneous infection rate of u_i^c at time t_i^c by u_j^c previously infected at time t_j^c, given it infection did not happen before t_i. In this paper, we express the hazard function as a constant $H(t|t_i, \alpha) = \alpha$, implying by definition that the probability of an event *non* happening before a time t given t_i decays exponentially as $e^{-\alpha(t-t_i)}$. The associated convex likelihood of α can be found in [12] (Eq. 7).

2.3   Dirichlet-Survival Process

In [9] the authors define the Dirichlet-Hawkes process by replacing the integer counts in Eq. 1 by the intensity of a Hawkes process. It can be interpreted as replacing integers counts in Dirichlet Processes by non-integer time-dependent counts, encoded by the intensity of the point process. Here, we consider the hazard rate of the NetRate model instead to account for networks structure. Each node is associated to its own temporal point process, and counts are replaced by the number of times any neighbour has been infected, weighted according to time and to edges strength. Using the methodology introduced in [9] and substituting the Hawkes process by the hazard rate of a survival model [14], we make a yet unexplored bridge between Dirichlet processes and Survival analysis. We remind that [14] reformulates the work of [8,12,13,28] in terms of Survival analysis and associated counting processes; we settled on using NetRate here, but any of these models would fit as well in our approach. The point process nature of survival analysis discussed in [14] makes this extension sound with respect to previous works on Dirichlet-Point processes [9,18,23,27].

Let $\mathbf{A}^{(k)}$ be the adjacency matrix of the subnetwork associated to cluster k, whose entries are $\alpha_{i,j}^{(k)}$. We define $(u_j^c, t_j^c)^{(k)}$ as an event of cascade c observed on node u_j at t_j attributed to subnetwork $A^{(k)}$. We write the history of events in cascade c attributed to the subnetwork k as $\mathcal{H}_{i,c}^{(k)} = \{(u_j^c, t_j^c)^{(k)}\}_{j:t_j < t_i}$. We note $\mathcal{H}_{i,c} = \{\mathcal{H}_{i,c}^{(k)}\}_k$ and $\mathbf{A} = \{\mathbf{A}^{(k)}\}_k$. We consider a new event from cascade c observed on node u_i^c at time t_i^c. At this point, the new event is not yet associated to any subnetwork. We write the Dirichlet-Survival prior probability for the new event to belong to subnetwork k:

$$P(s_i = k | \mathcal{H}_{\mathbf{i,c}}, \mathbf{A}, \lambda_0) = \begin{cases} \dfrac{\lambda_0^{(k)} + \sum_{\mathcal{H}_{i,c}^{(k)}} H(t_i^c | t_j^c, \alpha_{u_j, u_i}^{(k)})}{\lambda_0^{(K+1)} + \sum_k^K \lambda_0^{(k)} + \sum_{\mathcal{H}_{i,c}^{(k)}} H(t_i^c | t_j^c, \alpha_{u_j, u_i}^{(k)})} & \text{if } k = 1, ..., K \\[3em] \dfrac{\lambda_0^{(K+1)}}{\lambda_0^{(K+1)} + \sum_k^K \lambda_0^{(k)} + \sum_{\mathcal{H}_{i,c}^{(k)}} H(t_i^c | t_j^c, \alpha_{u_j, u_i}^{(k)})} & \text{if } k = K+1 \end{cases}$$

$$(2)$$

We introduced a new parameter $\lambda_0 = \{\lambda_0^{(k)}\}_{k=1,...,K+1}$, which translates the probability for a new observation not to have been triggered by any neighbour. It represents the probability that an event of cluster k is exogenous [15,20].

The Dirichlet-Survival prior is coupled to a sequential language model. For simplicity, we consider the bag-of-words Dirichlet-Multinomial model, as in [9, 18,23]; note that more refined sequential language models are also fit to our approach (Dynamic Topic Model [6], online LDA [3], online PLSA [5], etc.).

The input data is a stream of events. Each event takes the form of a triplet (u_i^c, t_i^c, v_i^c), where c is the cascade an event has been observed in, u_i^c is the node corresponding to the event, t_i^c is its publication time, and v_i^c represents its textual content (e.g. words in a tweet or in a news article). By combining the Dirichlet-Survival prior to the textual likelihood, we get the posterior distribution of the ith observation belonging to cluster (or subnetwork) k as:

$$P(s_i | v_i^c, \mathbf{N}, \mathcal{H}_{\mathbf{i,c}}, \mathbf{A}, \theta_0, \lambda_0) \propto \underbrace{P(v_i^c | s_i, \mathbf{N}, \theta_0)}_{\text{Dirichlet-Multinomial}} \times \underbrace{P(s_i | \mathcal{H}_{\mathbf{i,c}}, \mathbf{A}, \lambda_0)}_{\text{Dirichlet-Survival prior (Eq. 2)}} \qquad (3)$$

where $\vec{N}$ contains the words counts within each cluster, v_i^c contains the words count in document i, and θ_0 the concentration parameter of the model.

Finally, inference is conducted using a Sequential Monte Carlo algorithm similar to [9,18,23]. We perform several parallel runs on the same data stream. Within each run, each new observation in the stream is assigned to a cluster according to Eq. 3. The adjacency matrix $\mathbf{A}$ is then updated by optimizing the convex likelihood associated to the NetRate point process (Eq. 7 in [12]). Finally, we compute the likelihood of the language model for each run; runs that have a likelihood lesser than a threshold are discarded and replaced by more likely ones. The process is repeated until the end of the data stream. According to this algorithm, Eq. 1, and introducing a cutoff on the exponential hazard function (observations older than a time t_{old} are ignored), the optimization runs in $\mathcal{O}(N_{obs} N_{runs}(N_{nodes} + K))$ where N_{part} is the number of particles, N_{nodes} is the maximum network size and K the number of clusters (typically $N_{runs} \ll K \ll N_{nodes}$). Inference hence scales linearly with the size of the dataset.

We point out that the Dirichlet-Survival process is not about refining complex diffusion models such as [4,7,26]. Instead, it introduces a different angle for tackling content-aware diffusion problems. This new angle allows for unsupervised, non-parametric and online inference.

3   Experiments

3.1   Data and Experimental Setup

All data, codes and results are available in open access[1]. We consider 3 different network types of 500 nodes each: power-law (**PL**) [2], random Erdös-Renye (**ER**) [11] and a real network of hyperlinks between political blogs (**Blogs**) [1]. From each network, we randomly sample 5 subnetworks of 250 nodes and assign random weights α between 0 and 1 to their edges. Each of the generated subnetworks is used to propagate one given cluster of information. We then simulate infection cascades on each subnetwork according to the exponential NetRate model. Finally, we associate 5 words drawn from a vocabulary of size 100 to each so-generated event according to its associated subnetwork (or cluster). We generate a total of 55,000 events $\{(u_i^c, t_i^c, v_i^c)\}_{i,c}$ for each network.

Our hyperparameters are $\theta_0 = 0.1$ and $\lambda_0^{(k)} = 0.001 \; \forall k$. The SMC algorithm considers 4 parallel runs. We consider a constant hazard rate $H(t_i|t_j, \alpha_{j,i}) = \alpha_{j,i}$, so the probability of a new event *not* happening decays exponentially with time.

Table 1. Results on clusters (NMI, ARI) and edges (AUC-ROC, F1, MAE) retrieval.

		Houston	NRxDM	DHP	NetRate
PL	NMI	**0.809**	0.669	0.449	–
	ARI	**0.688**	0.330	0.063	–
	AUC-ROC	**0.807**	0.719	–	0.731
	F1	**0.199**	0.106	–	0.005
	MAE	**0.267**	0.338	–	0.460
ER	NMI	**0.787**	0.711	0.638	–
	ARI	**0.631**	0.488	0.411	–
	AUC-ROC	**0.849**	0.800	–	0.659
	F1	**0.263**	0.176	–	0.005
	MAE	**0.229**	0.278	–	0.481
Blogs	NMI	**0.750**	0.668	0.372	–
	ARI	**0.609**	0.365	0.023	–
	AUC-ROC	0.701	0.613	–	**0.710**
	F1	**0.168**	0.087	–	0.005
	MAE	**0.374**	0.444	–	0.499

3.2   Results

We compare to 3 similar baselines used as ablation tests: **Dirichlet-Hawkes process (DHP)** [9] clusters textual data by using temporal dynamics, and does not consider structure; **NetRate** [12] infers a dynamic network based on

---

[1] https://github.com/GaelPouxMedard/HOUsToN.

observed cascades without considering their content; **NetRate x Dirichlet-Multinomial (NRxDM)** first uses textual information to infer clusters, and only then infers the underlying subnetwork for each cluster, in the same fashion as [10,15]. For the record, Houston runs in $\mathcal{O}(N_{obs}N_{runs}(N_{nodes} + K))$, whereas NRxDM runs in $\mathcal{O}(N_{obs}N_{runs}K + N_{nodes}N_{obs}K)$, DHP runs in $\mathcal{O}(N_{obs}N_{runs}K)$ and NetRate runs in $\mathcal{O}(N_{nodes}N_{obs})$. When applicable, we evaluate on a classification task (scores NMI and ARI with respect to the clusters used for data generation) and a network inference task (AUC-ROC, F1 and MAE on the true edges, same metrics as in [12]).

We see in Table 1 that Houston consistently outperforms methods that do not consider jointly text, time and structure of the network. To summarize, NRxDM only considers textual information to build clusters, making the network inference miss a great deal of temporal and structural information. DHP considers textual information and temporal dynamics, but misses the structural information. NetRate does not consider textual data and infers the network based on temporal dynamics only. Houston bridges the gap between these models, by making a joint use of textual, temporal and structural information.

As an illustration of what Dirichlet-Processes can yield on real-world data, we draft its application to the Memetracker dataset [17] in Fig. 1 (bottom). We retrieve the diffusion network associated to meme clusters and observe diverse spreading dynamics. Topics spread in distinct parts of the global network, and mostly do so through a reduced set of densely connected nodes, as shown in [13].

4   Conclusion

In this paper, we propose the Dirichlet-Survival process as an alternative way to jointly model textual, temporal and structural information in spreading processes. Ablation tests demonstrate the relevance of the proposed approach. As a prior, the Dirichlet-Survival process can add a dynamic network dimension to any sequential Bayesian model; it could be coupled to models that account for any type of clustering (e.g. images, time series, labels), or simply more refined language models. Its introduction opens new perspectives on traditional machine learning problems, including topic-dependent spreading processes on networks.

References

1. Adamic, L.A., Glance, N.: The political blogosphere and the 2004 U.S. election: Divided they blog. In: Proceedings of the 3rd International Workshop on Link Discovery, pp. 36–43. LinkKDD 2005, Association for Computing Machinery, New York, NY, USA (2005)
2. Albert, R., Barabási, A.L.: Statistical mechanics of complex networks. Rev. Mod. Phys. **74**, 47–97 (2002)

3. AlSumait, L., Barbará, D., Domeniconi, C.: On-line lda: adaptive topic models for mining text streams with applications to topic detection and tracking, pp. 3–12 (2008). https://doi.org/10.1109/ICDM.2008.140

4. Barbieri, N., Manco, G., Ritacco, E.: Survival factorization on diffusion networks. In: Machine Learning and Knowledge Discovery in Databases, pp. 684–700 (2017). https://doi.org/10.1007/978-3-319-71249-9_41

5. Bassiou, N.K., Kotropoulos, C.L.: Online plsa: Batch updating techniques including out-of-vocabulary words. IEEE Trans. Neural Netw. Learn. Syst. 25(11), 1953–1966 (2014). https://doi.org/10.1109/TNNLS.2014.2299806

6. Blei, D.M., Lafferty, J.D.: Dynamic topic models. In: Proceedings of the 23rd International Conference on Machine Learning. p. 113–120. ICML 2006, Association for Computing Machinery, New York, NY, USA (2006). https://doi.org/10.1145/1143844.1143859

7. Choudhari, J., Dasgupta, A., Bhattacharya, I., Bedathur, S.: Discovering topical interactions in text-based cascades using hidden markov hawkes processes, pp. 923–928 (2018). https://doi.org/10.1109/ICDM.2018.00112

8. Du, N., Song, L., Smola, A., Yuan, M.: Learning networks of heterogeneous influence. In: NIPS, vol. 4, pp. 2780–2788, January 2012

9. Du, N., Farajtabar, M., Ahmed, A., Smola, A., Song, L.: Dirichlet-hawkes processes with applications to clustering continuous-time document streams. In: 21th ACM SIGKDD International Conference on Knowledge Discovery and Data Mining (2015). https://doi.org/10.1145/2783258.2783411

10. Du, N., Song, L., Woo, H., Zha, H.: Uncover topic-sensitive information diffusion networks. In: Proceedings of the Sixteenth International Conference on Artificial Intelligence and Statistics, AISTATS. JMLR Workshop and Conference Proceedings, vol. 31, pp. 229–237. JMLR.org (2013)

11. Erdős, P., Rényi, A.: On the evolution of random graphs. In: Publication of The Mathematical Institute of The Hungarian Academy of Sciences, pp. 17–61 (1960)

12. Gomez-Rodriguez, M., Balduzzi, D., Schölkopf, B.: Uncovering the temporal dynamics of diffusion networks. In: ICML, pp. 561–568 (2011)

13. Gomez-Rodriguez, M., Leskovec, J., Schoelkopf, B.: Structure and dynamics of information pathways in online media. In: WSDM (2013)

14. Gomez-Rodriguez, M., Leskovec, J., Schölkopf, B.: Modeling information propagation with survival theory. In: ICML, vol. 28, p. III-666–III-674 (2013)

15. He, X., Rekatsinas, T., Foulds, J.R., Getoor, L., Liu, Y.: Hawkestopic: a joint model for network inference and topic modeling from text-based cascades. In: ICML (2015)

16. Larremore, D., Carpenter, M., Ott, E., Restrepo, J.: Statistical properties of avalanches in networks. Phys. Rev. E 85, 066131 (2012). https://doi.org/10.1103/PhysRevE.85.066131

17. Leskovec, J., Backstrom, L., Kleinberg, J.: Meme-tracking and the dynamics of the news cycle. In: Proceedings of the 15th ACM SIGKDD International Conference on Knowledge Discovery and Data Mining, pp. 497–506. KDD 2009, Association for Computing Machinery, New York, NY, USA (2009). https://doi.org/10.1145/1557019.1557077

18. Mavroforakis, C., Valera, I., Gomez-Rodriguez, M.: Modeling the dynamics of learning activity on the web. In: Proceedings of the 26th International Conference on World Wide Web, pp. 1421–1430. WWW 2017 (2017)

19. Mei, Q., Fang, H., Zhai, C.: A study of poisson query generation model for information retrieval, pp. 319–326 (2007). https://doi.org/10.1145/1277741.1277797

20. Myers, S.A., Zhu, C., Leskovec, J.: Information diffusion and external influence in networks. In: Proceedings of the 18th ACM SIGKDD International Conference on Knowledge Discovery and Data Mining, pp. 33–41. KDD 2012, Association for Computing Machinery, New York, NY, USA (2012). https://doi.org/10.1145/2339530.2339540

21. Nickel, M., Le, M.: Modeling sparse information diffusion at scale via lazy multivariate hawkes processes. In: Proceedings of the Web Conference 2021, pp 706–717. WWW 2021, Association for Computing Machinery, New York, NY, USA (2021). https://doi.org/10.1145/3442381.3450094

22. Poux-Médard, G., Pastor-Satorras, R., Castellano, C.: Influential spreaders for recurrent epidemics on networks. Phys. Rev. Res. **2**, 023332 (2020). https://doi.org/10.1103/PhysRevResearch.2.023332

23. Poux-Médard, G., Velcin, J., Loudcher, S.: Powered hawkes-dirichlet process: challenging textual clustering using a flexible temporal prior. In: 2021 IEEE International Conference on Data Mining (ICDM), pp. 509–518 (2021)

24. Poux-Médard, G., Velcin, J., Loudcher, S.: Multivariate powered dirichlet-hawkes process. In: ECIR (2023)

25. Poux-Médard, G., Velcin, J., Loudcher, S.: Powered dirichlet process for controlling the importance of "rich-get-richer" prior assumptions in bayesian clustering. ArXiv (2021)

26. Suny, P., Li, J., Mao, Y., Zhang, R., Wang, L.: Inferring multiplex diffusion network via multivariate marked hawkes process. ArXiv abs/1809.07688 (2018)

27. Tan, X., Rao, V.A., Neville, J.: The Indian buffet hawkes process to model evolving latent influences. In: UAI (2018)

28. Wang, L., Ermon, S., Hopcroft, J.E.: Feature-enhanced probabilistic models for diffusion network inference. In: Flach, P.A., De Bie, T., Cristianini, N. (eds.) ECML PKDD 2012. LNCS (LNAI), vol. 7524, pp. 499–514. Springer, Heidelberg (2012). https://doi.org/10.1007/978-3-642-33486-3_32

29. Yang, S.H., Zha, H.: Mixture of mutually exciting processes for viral diffusion. In: Dasgupta, S., McAllester, D. (eds.) Proceedings of the 30th International Conference on Machine Learning. Proceedings of Machine Learning Research, vol. 28, pp. 1–9 (2013)

Consumer Health Question Answering Using Off-the-Shelf Components

Alexander Pugachev[1]([✉]), Ekaterina Artemova[2], Alexander Bondarenko[3] [iD],
and Pavel Braslavski[1,4] [iD]

[1] HSE University, Moscow, Russia
apugachev@hse.ru
[2] Center for Information and Language Processing (CIS), MaiNLP Lab., LMU
Munich, Munich, Germany
[3] Friedrich-Schiller-Universität Jena, Jena, Germany
[4] Ural Federal University, Yekaterinburg, Russia

Abstract. In this paper, we address the task of open-domain health
question answering (QA). The quality of existing QA systems heavily
depends on the annotated data that is often difficult to obtain, especially in the medical domain. To tackle this issue, we opt for PubMed
and Wikipedia as trustworthy document collections to retrieve evidence.
The questions and retrieved passages are passed to off-the-shelf question
answering models, whose predictions are then aggregated into a final
score. Thus, our proposed approach is highly data-efficient. Evaluation
on 113 health-related yes/no question and answer pairs demonstrates
good performance achieving AUC of 0.82.

Keywords: Health question answering · Medical information retrieval

1 Introduction

People actively seek answers to health-related questions online [13,15]. However,
about half of top-ranked search engines' results may provide incorrect answers
to such questions [6,30,31]. Consequently, there have been many research efforts
to improve health-related search by ranking documents higher that contain relevant and correct information using, for example, a trustworthiness predictor
or explicit expert relevance feedback [16,34]. Also, TREC Health Misinformation Track [9,10] addressed the task of ranking documents returned to health-
related queries according to three dimensions: usefulness, credibility, and correctness. Submitted solutions utilized a wide range of IR and NLP methods such
as: (1) the fusion of domain-specific representation models with neural quality
estimators [27], (2) ensembles of BERT-based classifier built w.r.t. each target
dimension [33], (3) continuous active learning to collect the datasets aimed at
training T5-based classifier [1], and (4) axiomatic re-ranking [5], etc.

In this work, we take another perspective and move from the ranking task
to open-domain question answering (OpenQA). Medical and health QA is an

J. Kamps et al. (Eds.): ECIR 2023, LNCS 13981, pp. 571–579, 2023.
https://doi.org/10.1007/978-3-031-28238-6_48

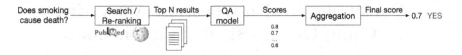

Fig. 1. Our proposed three-step open-domain question answering pipeline for health-related yes/no questions.

area of active research; a recent survey [19] provides a comprehensive overview of the field, including methods and datasets. For a historical perspective, we also refer the interested reader to a survey of pre-neural network methods in biomedical QA [3].

OpenQA aims to find an answer in a large document collection [26,35]. Traditionally, OpenQA pipeline has two components: (1) a *retriever* that returns relevant documents (or their parts – e.g. paragraphs) for the question and (2) a QA model (also referred to as *reader*) that infers the answer from the question-document pair obtained in the previous stage. In our study, we follow this architecture, but in contrast to medical QA systems consisting of dedicated components (see, for instance, [11]), we build our system from ready-to-use third-party blocks. At the retrieval stage, we do not assess information sources' credibility, but restrict the evidence search to PubMed and Wikipedia, both found to be reliable information sources in health and medical domains [21,23]. We use existing search APIs to retrieve documents, thus sparing indexing and ranker training. At the *reading* stage we use freely available QA models trained on existing data – either from the general or medical domain – thus making our approach very data-efficient.

The complete pipeline of our approach is presented in Fig. 1. Given a health-related question like "Does smoking cause death?", we first retrieve relevant documents from either PubMed or Wikipedia. Next, we use question answering models that output prediction probabilities of an answer for every pair of the question and retrieved document in the top-ranked results. In the final step, we aggregate the scores obtained for each question-document pair into a single final answer score.

To test our approach, we use a collection of 113 yes/no health questions like "Does celandine help with cancer?" with ground-truth expert answer annotations. Since, on the one hand, we address a binary prediction task, and on the other hand, we want our approach to inform the asker about to what degree the answer is conclusive, we use AUC as an evaluation measure. Our experiments show that using a mash-up of existing tools and solutions and sparing tailored training data achieves satisfactory results. The most effective combination of Google search over Wikipedia and RoBERTa model fine-tuned on general-domain BoolQ dataset achieves an AUC score of 0.82. Our proposed approach can serve as a strong baseline for health-related yes/no question answering. Our code and data are publicly available on GitHub.[1]

---

[1] https://github.com/apugachev/consumer-health-qa

Table 1. The upper part of the table describes 113 test questions (keywords are in **bold**). The bottom part provides examples and statistics of BoolQ, PubMedQA, and BioASQ datasets the readers were trained on. Note that PubMedQA contains 55 questions with the answer *maybe*, while the rest of the data – only *yes/no* answers.

Source	#Questions (y/n)	Example
TREC [2,9]	84 (42/42)	Can **dupixent** treat **eczema**?
Yandex [6]	15 (7/8)	Does **celandine** help with **cancer**?
HBT [4]	14 (12/2)	Does **smoking** cause **death**?
BoolQ [8]	12,697 (7,907/4,790)	Is there a treatment for the bubonic plague?
PubMedQA [18]	500 (276/55/169)	Do mitochondria play a role in remodeling lace?
BioASQ [29]	742 (611/131)	Does metformin interfere thyroxine absorption?

2   Data

Document Collections. The idea of our approach is that we do not need to search for medical information in the wild but instead focus on trustworthy collections: PubMed and Wikipedia. PubMed[2] is a large collection of biomedical literature, comprised of 34 million items at the time of writing. Wikipedia[3] is a large online encyclopedia driven by massive community efforts. At the time of writing, English Wikipedia contains more than 6.5 million articles and is considered a valuable resource of health-related information [28]. An advantage of Wikipedia in the context of our study is that articles about diseases often contain a summary of related treatments, side effects, as well as misbeliefs.

Questions and Answers. The primary source of the test questions used for the evaluation of our proposed approach is the TREC 2019 Decision Track [2] and TREC 2021 Health Misinformation Track [9] data. All the questions have a similar structure: they ask whether a treatment/medicine is helpful for a disease/condition. The test suites contain a question, its corresponding keyword query, a narrative, an answer (*helpful/unhelpful*), and a link to a respective medical publication as evidence for the answer. In our experiments, we make use only of a query, a question, and an answer. To ensure a higher diversity of the test data, we added 15 questions from the *Yandex* log translated from Russian into English and provided with a grounded answer [6]. Finally, we added 14 questions from a study dealing with health beliefs in Twitter (hereafter *HBT*) [4]. These questions are generated from the verified statements from the paper and depart from the rest of the test questions following the *Does X help Y?* pattern and its variations. Since TREC 2019 data is the only one that contains questions with *inconclusive* labels, we removed such questions from the test set. In total, the final test set contains 113 questions, 94 of which are provided with PubMed document IDs as answer evidence (see details and examples in Table 1).

---

[2] https://pubmed.ncbi.nlm.nih.gov/.
[3] https://en.wikipedia.org/.

3   Approach

Evidence Search. We experiment with three different PubMed retrievers:
(1) native PubMed search [24], (2) Google search over PubMed (implemented
as a custom search engine restricted to the PubMed domain), and (3) Google's
BioMed Explorer.[4] Native PubMed search [14] implements two-stage ranking:
first, documents are retrieved based on BM25 scores and then re-ranked using a
Lambda-MART-based model [7]. BioMed Explorer combines term- and BERT-
based retrieval and is trained on a mix of human-annotated and automatically
generated data from biomedical and general domains.[5] In each case, we perform separate searches for keyword and question query variants. In the case of
PubMed keyword search, we create a conjunctive (AND) query and restrict the
search to TITLE and ABSTRACT fields, thus aiming for high-precision results.
In the rest of the configurations, we run a default search with the query as a
string. Then, we fetch titles and abstracts of up to top 10 results for subsequent
processing.

We search Wikipedia with keyword and question queries using (1) Wikipedia
API[6] with default parameters and (2) Google custom search engine restricted
to the English Wikipedia domain. We fetch up to 10 articles and split them
into paragraphs (each paragraph is combined with the original Wikipedia page
title), lemmatize, and rank based on query term occurrences. Top 10 ranked
paragraphs are then passed to the readers.

Question Answering. We employ three third-party question answering models: a RoBERTa-large model fine-tuned on BoolQ[7] and two BioLinkBERT-large
models – fine-tuned on PubMedQA and BioASQ.[8] RoBERTa-large [22] is a
Transformer-based model with 355M parameters that follows BERT's [12] learning regime with some optimizations. BoolQ [8] is a QA dataset consisting of
general-domain yes/no questions from Google search log, Wikipedia context
paragraphs, and ground-truth answers. BoolQ is categorized into topics, the
topic closest to the medical domain is "Nature/Science", which comprises about
20% of the dataset. Fine-tuned RoBERTa achieves an accuracy of 0.86 on the
BoolQ test set, which is a good trade-off between the model's performance and
size. LinkBERT [32] is also a BERT-like model with a document relation prediction as an auxiliary learning objective. BioLinkBERT-large with 340M parameters is pre-trained on PubMed corpus with citation links that demonstrated
state-of-the-art on PubMedQA and BioASQ subsets of the BLURB benchmark
for biomedical NLP [17] at the time of publication – 0.73 and 0.95 accuracy
points, respectively. PubMedQA [18] is dataset with PubMed abstracts containing 1K expert-annotated yes/maybe/no questions along with a larger portion of
unlabeled and automatically generated items. BioLinkBERT model that we use

---

[4] https://g.co/research/biomedexplorer/.

[5] https://ai.googleblog.com/2020/05/an-nlu-powered-tool-to-explore-covid-19.html.

[6] https://wikipedia.readthedocs.io/en/latest/code.html.

[7] https://huggingface.co/apugachev/roberta-large-boolq-finetuned.

[8] https://github.com/michiyasunaga/LinkBERT.

in our work utilizes only expert-annotated data for training. BioASQ[9] is a yearly challenge on biomedical question answering. At the time of writing the BioASQ training data comprises 4,719 questions of four types: factoid, yes/no, summary, and list questions. BioLinkBERT model employed in our study leverages only a subset of yes/no questions from the BioASQ 2019 edition [25].

The statistics of the data that was used for fine-tuning the readers are summarized in the bottom part of Table 1. The BoolQ dataset is significantly larger than the two medical QA datasets. Moreover, BoolQ questions from real users are "simpler", than the more specialized questions from PubMedQA and BioASQ. We pass questions and up to 10 retrieved PubMed abstracts or Wikipedia paragraphs to the readers that then return a continuous value from 0 to 1 (0 corresponds to a "no" answer and 1 – to "yes"). If no evidence was retrieved, we assign an inconclusive 0.5 score to the question answer.

Score Aggregation. Finally, we use three score aggregation methods: (1) the final score is derived solely from the top 1 evidence document, (2) plain average over the top 10 results (or less, if fewer documents are returned), and (3) weighted average (weights linearly decrease with the increased rank and sum up to one).

4   Results and Discussion

Table 2 reports the results of different configurations of our approach to answering health-related yes/no questions. Although we did not perform a thorough component-based evaluation, we can make some observations about the quality of components in our pipeline based on indirect indicators. For instance, Google retrieved the highest number of evidence documents from PubMed among top 10 results (see 'Hits' column in Table 2). However, we cannot unequivocally interpret these numbers — they can signal a higher search quality or also a search bias in the test collection: TREC annotators might have used Google or another major search engine to find evidence (most of our test questions come from the TREC tracks). We also applied the three readers to the PubMed abstracts available for 94 out of 113 questions (these abstracts come from the original data [2,6,9] and were manually selected by human annotators; one abstract per question). Readers fine-tuned on BoolQ, PubMedQA, and BioASQ achieved 0.88, 0.65, and 0.80 AUC points, respectively. These scores can be regarded as an upper limit estimate for these QA models applied to PubMed abstracts, i.e., the decline in the final scores can be attributed to retrievers' deficiencies. However, one should compare these values with caution, since evidence PubMed abstracts are available not for all questions, and the human bias in selecting these evidence documents may also play a role.

Using PubMed or Wikipedia only leads to a reduced recall and sometimes to no results at all (see '#0' column in Table 2). For example, three out of five search configurations in our experiments failed to find any documents for the

---

[9] http://www.bioasq.org/.

Table 2. AUC scores for different configurations. Hits: number of evidence PubMed documents in top 10 results; #0: number of queries with no evidence results. Final score aggregations variants: based on top 1 document, plain (avg), and weighted average (wavg) over top-ranked documents. Off-the-shelf readers: RoBERTa-large fine-tuned on BoolQ, BioLinkBERT models fine-tuned on PubMedQA and BioASQ. The overall best result is in **bold**, best results for each retriever are underlined.

	Retriever	Query	Hits	#0	RoBERTa-large (BoolQ)			BioLinkBERT (PubMedQA)			BioLinkBERT (BioASQ)		
					top 1	avg	wavg	top 1	avg	wavg	top 1	avg	wavg
PubMed	PubMed	keywords	10	31	0.65	_0.66_	_0.66_	0.61	0.62	0.61	0.56	0.58	0.58
		question	7	42	0.61	0.62	0.62	0.57	0.58	0.58	0.58	0.48	0.48
	Google	keywords	56	1	_0.79_	0.73	0.73	0.59	0.66	0.68	0.71	0.65	0.69
		question	39	2	0.71	0.76	0.77	0.64	0.64	0.67	0.65	0.57	0.62
	BioMed Explorer	keywords	41	0	0.72	0.75	0.74	0.63	0.64	0.66	0.58	0.74	0.71
		question	39	0	0.74	_0.77_	_0.77_	0.60	0.69	0.69	0.72	0.71	0.73
Wikipedia	Wikipedia	keywords	–	26	_0.81_	0.77	0.78	0.68	0.65	0.68	0.58	0.57	0.59
		question	–	56	0.65	0.68	0.68	0.57	0.56	0.58	0.48	0.48	0.47
	Google	keywords	–	19	0.80	0.79	**0.82**	0.63	0.72	0.68	0.55	0.57	0.57
		question	–	16	0.75	0.75	0.75	0.57	0.67	0.64	0.52	0.53	0.56

TREC 2021 question "Can I get rid of a pimple overnight by applying toothpaste?" and its corresponding keyword query "toothpaste pimple overnight", while Web results for these queries are abundant. Overall, native searchers of PubMed and Wikipedia suffer the most from the no returned results. This is due to restrictions imposed on a PubMed query search and its poor ability to handle question-like queries. Post-processing Wikipedia search results and re-ranking aiming for a higher precision also lead to a lower recall.

The evaluation results (see Table 2) show that RoBERTa fine-tuned on BoolQ significantly outperforms BioLinkBERT models in all configurations. We can conclude that the volume of data for fine-tuning the reader is more important than the in-domain pre-training of the language model. The best results are achieved on Wikipedia documents that often contain relevant information formulated in plain language. The impact of using up to top 10 retrieved results compared to just top 1 document is somewhat mixed: in some cases accounting for documents beyond top 1 improves results, but in other cases, the effect is the opposite. Overall, BioLinkBERT fine-tuned on PubMedQA outperforms its counterpart fine-tuned on BioASQ. General-domain Google search scores higher than other retrievers and Wikipedia is a more useful document collection for consumer health QA in our settings. Using keyword query searches often result in higher evaluation scores of our QA pipeline, although in the case of PubMed with BioMed Explorer (the latter is marketed as a question answering system) question queries outperform keyword variants in the majority of cases.

The highest evaluation results in our experiments are obtained using keyword queries (provided by human annotators), which can be seen as a limitation of our approach since we do not use automatic conversion of questions to queries. However, most natural language questions in the test collection can be transformed into keyword queries automatically by filtering out verbs, determiners, prepositions, and sometimes adverbs as the 'pimple–toothpaste' example suggests (see examples in Table 1).

5 Conclusion

Our solution exploits evidence search in PubMed and Wikipedia for open-domain health question answering. In our approach, we use different search tools to retrieve evidence documents and ready-to-use question answering models. Coupled with simple score aggregation heuristics, this combination delivers satisfactory results – best configuration achieves AUC of 0.82 on 113 test yes/no questions. The proposed approach does not use annotated data directly and does not require training on the target data or task. Thus, it can be considered a strong baseline. However, the main limitation of our work is a small test set, such that the evaluation results and the conclusions should be taken with a grain of salt.

There is ample room for future improvements within our proposed pipeline. In the future, we plan to elaborate on search results post-processing. In particular, we plan to investigate if evidence *sentences* in contrast to paragraphs can help to achieve better results. We also plan to explore if medical thesauri can help to increase recall of the Wikipedia search. Increasing the number and types of test questions, probably gleaning them from various existing question answering datasets, is another interesting avenue for future work.

Acknowledgments. A. Pugachev's research was supported in part through computational resources of HPC facilities at HSE University [20]. A. Bondarenko's work was supported by the Deutsche Forschungsgemeinschaft (DFG) in the project "ACQuA 2.0: Answering Comparative Questions with Arguments" (project 376430233) as part of the priority program "RATIO: Robust Argumentation Machines" (SPP 1999). P. Braslavski's work was supported in part by the Ministry of Science and Higher Education of the Russian Federation (project 075-02-2022-877).

References

1. Abualsaud, M., et al.: UWaterlooMDS at the TREC 2021 health misinformation track. In: Proceedings of the Thirtieth REtrieval Conference Proceedings (TREC 2021). National Institute of Standards and Technology (NIST), Special Publication (2021)
2. Abualsaud, M., Lioma, C., Maistro, M., Smucker, M.D., Zuccon, G.: Overview of the TREC 2019 decision track. In: Proceedings of the Twenty-Eigth Text REtrieval Conference (TREC 2019) (2019)

3. Athenikos, S.J., Han, H.: Biomedical question answering: a survey. Comput. Methods Program. Biomed. **99**(1), 1–24 (2010)
4. Bhattacharya, S., Tran, H., Srinivasan, P.: Discovering health beliefs in Twitter. In: Proceedings of the Information Retrieval and Knowledge Discovery in Biomedical Text, Papers from the 2012 AAAI Fall Symposium. AAAI Technical Report, vol. FS-12-05. AAAI (2012)
5. Bondarenko, A., et al.: Webis at TREC 2021: deep learning, health misinformation, and podcasts tracks. In: The Thirtieth REtrieval Conference Proceedings (TREC 2021). National Institute of Standards and Technology (NIST), Special Publication (2021)
6. Bondarenko, A., Shirshakova, E., Driker, M., Hagen, M., Braslavski, P.: Misbeliefs and biases in health-related searches. In: Proceedings of the 30th ACM International Conference on Information and Knowledge Management (CIKM 2021), pp. 2894–2899. ACM (2021)
7. Burges, C.J.: From RankNet to LambdaRank to LambdaMART: an overview. Technical Report, Microsoft Research Technical Report MSR-TR-2010-82 (2010)
8. Clark, C., Lee, K., Chang, M.W., Kwiatkowski, T., Collins, M., Toutanova, K.: BoolQ: exploring the surprising difficulty of natural yes/no questions. In: Proceedings of the 2019 Conference of the North American Chapter of the Association for Computational Linguistics: Human Language Technologies, Volume 1 (Long and Short Papers), pp. 2924–2936 (2019)
9. Clarke, C.L.A., Maistro, M., Smucker, M.D.: Overview of the TREC 2021 health misinformation track. In: Proceedings of the Thirtieth Text REtrieval Conference, TREC 2021. NIST Special Publication, (NIST) (2021)
10. Clarke, C.L.A., Rizvi, S., Smucker, M.D., Maistro, M., Zuccon, G.: Overview of the TREC 2020 health misinformation track. In: Proceedings of the Twenty-Ninth Text REtrieval Conference, TREC 2020. NIST Special Publication, (NIST) (2020)
11. Demner-Fushman, D., Mrabet, Y., Ben Abacha, A.: Consumer health information and question answering: helping consumers find answers to their health-related information needs. J. Am. Med. Inf. Assoc. **27**(2), 194–201 (2019)
12. Devlin, J., Chang, M.W., Lee, K., Toutanova, K.: BERT: pre-training of deep bidirectional transformers for language understanding. In: Proceedings of the 2019 Conference of the North American Chapter of the Association for Computational Linguistics: Human Language Technologies, Volume 1 (Long and Short Papers), pp. 4171–4186. Association for Computational Linguistics (2019)
13. Finney Rutten, L.J., Blake, K.D., Greenberg-Worisek, A.J., Allen, S.V., Moser, R.P., Hesse, B.W.: Online health information seeking among US adults: measuring progress toward a healthy people 2020 objective. Public Health Rep. **134**(6), 617–625 (2019)
14. Fiorini, N., et al.: Best match: new relevance search for PubMed. PLoS Biol. **16**(8), e2005343 (2018)
15. Fox, S., Duggan, M.: Health online 2013. Health **2013**, 1–55 (2013)
16. Fröbe, M., Günther, S., Bondarenko, A., Huck, J., Hagen, M.: Using keyqueries to reduce misinformation in health-related search results. In: Proceedings of the 2nd Workshop Reducing Online Misinformation through Credible Information Retrieval 2022 co-located with The 44th European Conference on Information Retrieval ECIR 2022. CEUR Workshop Proceedings, vol. 3138, pp. 1–10. CEUR-WS.org (2022)
17. Gu, Y., et al.: Domain-specific language model pretraining for biomedical natural language processing. ACM Trans. Comput. Healthcare (HEALTH) **3**(1), 1–23 (2021)

18. Jin, Q., Dhingra, B., Liu, Z., Cohen, W., Lu, X.: PubMedQA: a dataset for biomedical research question answering. In: Proceedings of the 2019 Conference on Empirical Methods in Natural Language Processing and the 9th International Joint Conference on Natural Language Processing (EMNLP-IJCNLP), pp. 2567–2577 (2019)

19. Jin, Q., et al.: Biomedical question answering: a survey of approaches and challenges. ACM Comput. Surv. (CSUR) 55(2), 1–36 (2022)

20. Kostenetskiy, P., Chulkevich, R., Kozyrev, V.: HPC resources of the higher school of economics. J. Phys. Conf. Ser. 1740(1), 012050 (2021)

21. Laurent, M.R., Vickers, T.J.: Seeking health information online: does Wikipedia matter? J. Am. Med. Inf. Assoc. 16(4), 471–479 (2009)

22. Liu, Y., et al.: RoBERTa: a robustly optimized BERT pretraining approach. arXiv preprint arXiv:1907.11692 (2019)

23. Morshed, T., Hayden, S.: Google versus PubMed: comparison of google and PubMed's Search tools for answering clinical questions in the emergency department. Ann. Emerg. Med. 75(3), 408–415 (2020)

24. National Center for Biotechnology Information (US), Bethesda (MD): Entrez Programming Utilities Help (2010)

25. Nentidis, A., Bougiatiotis, K., Krithara, A., Paliouras, G.: Results of the seventh edition of the BioASQ challenge. in: Joint European Conference on Machine Learning and Knowledge Discovery in Databases, pp. 553–568 (2020)

26. Prager, J.: Open-domain question answering. Found. Trends Inf. Retrieval 1(2), 91–231 (2007)

27. Schlicht, I.B., de Paula, A.F.M., Rosso, P.: UPV at TREC health misinformation track 2021 ranking with SBERT and quality estimators. CoRR abs/2112.06080 (2021)

28. Smith, D.A.: Situating Wikipedia as a health information resource in various contexts: a scoping review. PloS One 15(2), e0228786 (2020)

29. Tsatsaronis, G., et al.: An overview of the BioASQ large-scale biomedical semantic indexing and question answering competition. BMC Bioinf. 16(1), 1–28 (2015)

30. White, R.: Beliefs and biases in web search. In: Jones, G.J.F., Sheridan, P., Kelly, D., de Rijke, M., Sakai, T. (eds.) Proceedings of the 36th International Conference on Research and Development in Information Retrieval (SIGIR 2013), pp. 3–12. ACM (2013)

31. White, R.W., Hassan, A.: Content bias in online health search. ACM Trans. Web (TWEB) 8(4), 1–33 (2014)

32. Yasunaga, M., Leskovec, J., Liang, P.: LinkBERT: pretraining language models with document links. In: Proceedings of the 60th Annual Meeting of the Association for Computational Linguistics (Volume 1: Long Papers), pp. 8003–8016 (2022)

33. Zhang, B., Naderi, N., Jaume-Santero, F., Teodoro, D.: DS4DH at TREC health misinformation 2021: multi-dimensional ranking models with transfer learning and rank fusion. arXiv preprint arXiv:2202.06771 (2022)

34. Zhang, D., Tahami, A.V., Abualsaud, M., Smucker, M.D.: Learning trustworthy web sources to derive correct answers and reduce health misinformation in search. In: Proceedings of the 45th International Conference on Research and Development in Information Retrieval (SIGIR 2022), pp. 2099–2104. ACM (2022)

35. Zhu, F., Lei, W., Wang, C., Zheng, J., Poria, S., Chua, T.S.: Retrieving and reading: a comprehensive survey on open-domain question answering. arXiv preprint arXiv:2101.00774 (2021)

MOO-CMDS+NER: Named Entity Recognition-Based Extractive Comment-Oriented Multi-document Summarization

Vishal Singh Roha[1]([✉]) [iD], Naveen Saini[2] [iD], Sriparna Saha[1] [iD],
and Jose G. Moreno[3,4] [iD]

[1] Indian Institute of Technology Patna, Bihar, India
vishal.roha95@gmail.com, sriparna@iitp.ac.in
[2] Indian Institute of Information Technology, Lucknow, Uttar Pradesh, India
nsaini1988@gmail.com
[3] University of La Rochelle, L3i, 17000 La Rochelle, France
jose.moreno@irit.fr
[4] University of Toulouse, IRIT, Toulouse, France

Abstract. In this work, we propose an unsupervised extractive summarization framework for generating good quality summaries which are supplemented by the comments posted by the end-users. Using the evolutionary multi-objective optimization concept, different objective functions for assessing the quality of a summary, like diversity and the relevance of sentences in relation to comments, are optimized simultaneously. In the literature, named entity recognition (NER) has been shown to be useful in the summarization process. The current work is the first of its kind where we have introduced a new objective function that utilizes the concept of NER in news documents and user comments to score the news sentences. To test how well the new objective function works, different combinations of the NER-based objective function with already existing objective functions were tested on the English and French datasets using ROUGE 1, 2, and SU4 F1-scores. We have also investigated the abstractive and compressive summarization approaches for our comparative analysis. The code of the proposed work is available at the github repository https://github.com/vishalsinghroha/Unsupervised-Comment-based-Multi-document-Extractive-Summarization.

Keywords: Unsupervised learning · Multi-document summarization · Named entity recognition · User-comments · Evolutionary algorithm · Multi-objective optimization · Information retrieval

1 Introduction

Nowadays, the internet contains an exponential amount of text content that must be summarized precisely in a limited number of words to keep up with the latest information [1,2,12,16,23]. Numerous studies for document(s) summarization utilizing comments have been done in the literature. A deep learning framework called reader-aware summary generator (RASG) [8] employs a

J. Kamps et al. (Eds.): ECIR 2023, LNCS 13981, pp. 580–588, 2023.
https://doi.org/10.1007/978-3-031-28238-6_49

sequence-to-sequence architecture that includes a copy and attention mechanism. It uses a semantic alignment scoring between each word in a news item and the related user comments to produce an abstractive summary of the news document in order to capture the reader-focused component of the user comments. On the other hand, reader-aware multi-document summarization (RA-MDS) [15] uses a variational auto-encoder (VAEs) based multi-document summarization (MDS) framework for the production of compressive summaries. It determines the weights of news documents based on user comments utilizing unigrams, bigrams, and entities. However, the fundamental problem with deep learning models is that they need a lot of training data. Hu et al. [10] described an extractive comment-based summarization system that is based on graphs. Three relations-topic, quotation, and mention-connect user comments to three relation graphs. Following that, each user comment's relevance is calculated using two unsupervised techniques. Other classical graph-based methods, such as LexRank [7] and TexRank [18], rate the news sentences in an unsupervised setting using the page rank algorithm. However, both methods assign lengthier sentences higher scores, which leads to redundant information in the resultant summary.

In [22], an approach, namely, *MOO-CMDS*, was proposed that aims to summarize multi-documents and utilizes user comments posted by the end-user. In *MOO-CMDS*, there were two phases for the summary generation task: (a) identification of the useful/relevant comments with respect to the news documents; (b) generation of summary utilizing multi-objective optimization (MOO) based evolutionary framework [26]. In the second phase, four different objective functions evaluating the quality of the summary are simultaneously optimized to improve the quality of the summary. The potential of named entity recognition (NER) [14] has already been demonstrated in the literature in various information extraction tasks such as sentiment analysis and neural machine translation [6,9,14]. Therefore, in the current article, we have extended the work of [22] by introducing a NER-based objective function. In other words, we aim to develop an unsupervised multi-document extractive summarization framework using NER in conjunction with already existing objective functions discussed in [22]. We have named our approach as *MOO-CMDS+NER*, where MOO and CMDS stand for multi-objective optimization and comment-oriented multi-document summarization, respectively.

The proposed objective function (Named entity score: η_5) utilizes the named entity information to identify the essential aspects present in the news sentences and user comments, such as people's names, geographical names, monetary values, brands, and more, to assign scores to the news sentences. Further, two different versions of η_5 are explored in our work. We have computed η_5 in two different ways, namely, version 1 and version 2, to explore its ability to score the news sentences. In version 1, the frequency of Named Entity Terms (NETs) present in the news sentences is calculated by their total occurrences in the user comments to score the news sentences. In version 2, the news sentences are scored on the basis of whether the NETs of a news sentence are present in the user comments or not.

Thus, key contributions of our current work are two-fold: (a) investigating the named entity recognition-based objective function in *MOO-CMDS*; (b) comparative analysis of abstractive [4] and compressive summarization [20] approaches with our extractive summarization framework. Also, the quality of the summary generated by using the proposed objective function is independent of the language and topic used. To validate this point, two distinct datasets belonging to different languages namely English and French are used. Both datasets contain multiple news documents along with their corresponding comments. To evaluate the performance of the new objective with the already used objective functions in *MOO-CMDS* and the other extractive, abstractive, and compressive summarization frameworks, ROUGE-1, 2, and SU4 F-score [11,17] are used. Based on the observation, we found that the proposed objective function has shown an average improvement of (a) 2.95%, (b) 11.15%%, and (c) 32.19% over the extractive, abstractive, and compressive baselines, respectively. A detailed discussion of results is done in Sect. 3.

2   Proposed MOO-CMDS+NER Approach

The current work proposes an unsupervised extractive multi-document summarization framework that utilizes the comments available with them, to automatically construct the summary. In order to optimize the summary quality, our framework uses a multi-objective evolutionary framework [26] similar to *MOO-SMDS*. From now onwards, we will call our proposed framework *MOO-CMDS+NER*. The four different objective functions used in *MOO-CMDS*, which were simultaneously optimized to improve the summary quality are (i) diversity (η_1) to avoid redundancy in the summary; (ii) user-attention score (η_2) which takes into account the useful comments in calculating the news sentence relevance; (iii) density-based score (η_3) which considers the syntactic and semantic weights of words along with the identified useful comments to calculate the news' sentence score; (iv) user-attention with syntactic score (η_4) is similar to η_3, but it assigns more importance to syntactic weight. For their mathematical definition, readers can refer to MOO-CDMS [22]. Along with these objective functions, in *MOO-CMDS+NER*, we propose two different versions of a new objective function (Named entity score (NES): η_5) that examines NER capabilities in our extractive comment-oriented summary generation task. After a single run of *MOO-CMDS+NER*, our multi-objective optimization-based approach generates a variety of alternative summaries. Finally, a single best summary can be selected by the user depending on his/her choice.

The flow of our *MOO-CMDS+NER* is the same as that of *MOO-CMDS* having different modules (a) extraction of useful comments; (b) calculation of objective functions; and (c) summary generation using identified useful comments. Therefore, we are not discussing the architecture in detail due to a length restriction. The subsequent section will highlight the details of the NER-based objective function.

2.1 Named Entity Recognition-Based Objective Function (η_5)

Named Entity Recognition (NER) [14] is a promising task in the field of information extraction and aims to quickly identify essential aspects of a text, such as geographical names, people's names, monetary values, brands, and more. The news sentence scores are calculated in this case using two separate versions that leverage a phrase's named entity terms (NETs). In the first version, the score/weight of a news sentence is computed by finding the frequency of the common NETs present in the news sentence and the corresponding user comments. Equation 1 is used to compute the weight of qth word of jth sentence in ith document (i.e., $entity\_score_{v_{ij}^q}$).

$$entity\_score_{v_{ij}^q} = wf_{i,j}^c + 1 \tag{1}$$

where v_{ij}^q denotes the qth word of jth sentence in ith document, $wf_{l,k}^c$ represents the frequency (total number of occurrence) of cth NET in the jth news sentence of ith document appeared across all the corresponding user (useful) comments. There might be a case in which a news sentence contains NETs not present in user comments; then, in order to assign weights to those news sentences, we have added 1 to $wf_{l,k}^c$ (to each named entity term of the news sentence). Suppose two news sentences do not have NETs present in user comments, then the sentence with more NETs will be assigned a higher score. Now, the final score of each news sentence is computed by summing the weights of all NETs of a news sentence obtained from Eq. 1 divided by the number of unique entities present across all the news document sentences. This step is represented by using Eq. 2.

$$new_{d_{i,j}} = \sum_{q=1}^{|D_{ij}|} entity\_score_{v_{ij}^q} / |DU_i| \tag{2}$$

where $new_{d_{i,j}}$ denotes the named entity weight of the jth news sentence of the ith document, $|D_{ij}|$ is the number of words in jth sentence of ith document, $d_{i,j}$ is the jth sentence of ith document, and $|DU_i|$ represents the total number of unique entities in the ith document. Then η_5 can be computed using Eq. 3.

$$\eta_5 = \sum_{j=1}^{|D|} new_{d_{i,j}} / |D| \tag{3}$$

where, $|\mathcal{D}|$ is the number of sentences in the output summary. In the 2nd version, rather than computing the frequencies of NETs present in the news sentences of a document, we will instead find whether the NET of a news sentence is present in the corresponding user comments or not. If a NET of a news sentence is present in a user comment, it will be assigned a score of 3; otherwise, 1. This step is mathematically represented by Eq. 4.

$$entity\_score_{v_{ij}^q} = \begin{cases} 3, & \text{if } v_{ij}^q \in DU_i \cap CU_i \\ 1, & \text{otherwise} \end{cases} \tag{4}$$

where $entity\_score_{v_{ij}^q}$ and v_{ij}^q hold the same meaning as in Eq. 1 and $DU_i \cap CU_i$ represents the intersection set of NETs of the ith document and the associated user comments. There might be cases when the NETs of a news sentence are not present in the corresponding user comments. So, to assign scores to those news sentences, each NET of that news sentence is provided a score of 1. Now, to assign a score to NETs present in both the news sentence and user comments, a bunch of different values in the range of 2 to 5 are utilized, and after analysis, the value of 3 is found to be the best. Therefore, a score of 3 is assigned to provide a higher weightage to the NETs, which are not present in user comments. Similarly, the score of each news sentence is calculated by summing the weights of words obtained from Eq. 5 divided by the number of unique entities obtained from the intersection of the corresponding document and user comments. Equation 5 represents the mathematical calculations involved in this step.

$$new_{v_{i,j}} = \sum_{q=1}^{|D_{ij}|} entity\_score_{v_{ij}^q}/|DU_i \cap CU_i| \tag{5}$$

where $new_{v_{i,j}}$ has the same meaning as in Eq. 2. Finally, as for the 1st version, η_5 is computed using Eq. 3.

3   Experimental Setup and Comparative Results

3.1   Datasets, Evaluation Metrics, and Parameter Setting

To check the effectiveness of our proposed *MOO-CMDS+NER*, two distinct datasets belonging to English and French are utilized. The first is the RA-MDS dataset, which is freely available in English and consists of news documents and their corresponding comments on 45 different news topics. The second dataset is in French language and available at GitHub repository[1]. It contains 40 different topics/themes, each with user comments. For performance evaluation, we have employed the ROUGE-N F-score [11,17], where N takes the values of 1, 2, or SU4. All used parameter values are left unchanged from our baseline paper [22].

3.2   Comparison Methods

For comparative analysis, we have compared our proposed method with extractive *MOO-CMDS* [22], in conjunction with (a) six extractive summarization methods (two graph-based: TextRank [18] and LexRank [7], one topic-based: Centroid [5], and three transformer-based models: GPT-2 [21], XLNet [24], and BERT [19]); (b) three transformer-based abstractive summarization methods: BART [13], PEGASUS [25], and Longformer [3]; (c) two compressive-based summarization methods, CLTS [20] and RA-MDS [20].

3.3   Comparative Results

Table 2 includes the results obtained by the extensive study done on both the versions of objective 5, i.e., η_5, along with other objective functions, on

---

[1] https://github.com/vishalsinghroha/FrenchDatasetwithcomments.

Table 1. Comparison of ROUGE scores between different models for English and French datasets. Here, '-' means results are not provided in the reference paper. MOO-CDMS contains result of '$\eta_1 + \eta_2 + \eta_3 + \eta_4$'. Improvements in terms of points attained by our proposed MOO-CMDS+NER using a combination of '$\eta_1 + \eta_2 + \eta_5(version2)$' with respect to different baselines are presented in ().

Type of Summarization	System	English Dataset			French Dataset		
		ROUGE 1	ROUGE 2	ROUGE SU4	ROUGE 1	ROUGE 2	ROUGE SU4
Extractive	TextRank	0.3308 (+0.1457)	0.1075 (+0.1541)	0.1367 (+0.1299)	0.0838 (+0.2146)	0.0241 (+0.1413)	0.0289 (+0.1299)
	LexRank	0.4040 (+0.0725)	0.1678 (+0.0938)	0.1866 (+0.0800)	0.0798 (+0.2186)	0.0174 (+0.1480)	0.0253 (+0.1335)
	Centroid	0.3987 (+0.0778)	0.1713 (+0.0903)	0.1890 (+0.0776)	0.0053 (+0.2931)	0.0005 (+0.1649)	0.0009 (+0.1579)
	BERT	0.3721 (+0.1044)	0.1378 (+0.1238)	0.1626 (+0.1040)	0.0869 (+0.2115)	0.0180 (+0.1474)	0.0278 (+0.1310)
	GPT-2	0.3971 (+0.0794)	0.1441 (+0.1175)	0.1707 (+0.0959)	0.0920 (+0.2064)	0.0222 (+0.1432)	0.0316 (+0.1272)
	XLNet	0.3930 (+0.0866)	0.1529 (+0.1087)	0.1752 (+0.0914)	0.0855 (+0.2129)	0.0222 (+0.1432)	0.0302 (+0.1286)
Abstractive	BART	0.4452 (+0.0313)	0.2284 (+0.0332)	0.2383 (+0.0283)	0.0703 (+0.2281)	0.0253 (+0.1401)	0.0272 (+0.1316)
	Pegasus	0.3661 (+0.1104)	0.1599 (+0.1017)	0.1757 (+0.0909)	0.0815 (+0.2169)	0.0244 (+0.1410)	0.0304 (+0.1284)
	Longformer	0.3428 (+0.1337)	0.1268 (+0.1348)	0.1419 (+0.1247)	0.0855 (+0.2129)	0.0222 (+0.1432)	0.0302 (+0.1286)
Compressive	CLTS	0.3735 (+0.1030)	0.1352 (+0.1264)	0.1558 (+0.1108)	0.0879 (+0.2105)	0.0879 (+0.0775)	0.0234 (+0.1354)
	RA-MDS	0.4430 (+0.0335)	0.1710 (+0.0906)	0.1960 (+0.0706)	-	-	-
Extractive	MOO-CMDS	0.4680(+0.0085)	0.2513(+0.0103)	0.2590(+0.0076)	0.2898(+0.0086)	0.1519(+0.0135)	0.1484(+0.0104)
	MOO-CMDS+NER	0.4765	0.2616	0.2666	0.2984	0.1654	0.1588

both the English and French dataset. While Table 1 reports the comparative results between our method. From Table 1, it can be observed that our *MOO-CMDS+NER* approach optimizing '$\eta_1 + \eta_2 + \eta_5(verison2)$' outperforms all the other approaches by a great margin for both the English and the French dataset. Further, all the combinations of both versions of η_5 are performing significantly better than the other methods. For the English dataset, version 1 of η_5 outperforms the extractive (MOO-CDMS using combination of '$\eta_1 + \eta_2 + \eta_3 + \eta_4$') by 0.74%, 4.54%, and 3.36%, abstractive (BART) by 5.9%, 15.0%, and 12.3%, and compressive (RA-MDS) by 6.4%, 53.6%, and 36.6%, in terms of ROUGE 1, 2 and SU4 F1-scores, respectively, when it is optimized along with η_1, η_2, and η_4. For the same dataset (English), when version 2 of η_5 is optimized with all other remaining objectives, it outperforms the MOO-CDMS by 1.82%, 4.09%, and 2.93%, BART by 7.03%, 14.54%, and 11.88%, and RA-MDS by 7.56%, 52.98%, and 36.02%, in terms of ROUGE 1,2 and SU4 F1-scores, respectively. Additionally, for the English dataset '$\eta_1 + \eta_5$' outperforms (a) '$\eta_1 + \eta_2$' by 3.09%, 2.17%, and 2.43% for version 1 and 2.65%, 4.73%, and 4.60% for version 2, and (b) '$\eta_1 + \eta_4$' by 3.69%, 8.32%, and 6.94% for version 1 and 3.25%, 11.03%, and 9.21% for version 2, respectively.

For the French dataset, versions 1 and 2 of η_5 have outperformed the extractive, abstractive, and compressive baselines by (a) 0.0086, 0.0135, and 0.0104, (b) 0.2129, 0.1432, and 0.1286, and (c) 0.2105, 0.0775, and 0.1354 points. This difference is due to the fact that the French dataset's gold summary is only a few phrases long and quite concise. However, some models provide greater importance to longer phrases, which lowers the ROUGE F1 score. The performances of '$\eta_1 + \eta_5$' for both versions of η_5 are also comparable to the combination of '$\eta_1 + \eta_2$', '$\eta_1 + \eta_3$', and '$\eta_1 + \eta_4$' which are (a) 0.2657, 0.1295, and 0.1211, (b) 0.2444, 0.1096. and 0.1063, and (c) 0.2715, 0.1330, and 0.1313 in terms of ROUGE F1-score for

Table 2. ROUGE scores obtained after conducting ablation study on different objective functions. Note: 'R' indicates ROUGE and the best ROUGE scores.

combination	English Dataset						French Dataset					
	η_5(Version 1)			η_5(Version 2)			η_5(Version 1)			η_5(Version 2)		
	R 1	R 2	R SU4	R 1	R 2	R SU4	R 1	R 2	R SU4	R 1	R 2	R SU4
$\eta_1 + \eta_5$	0.4597	0.2356	0.2449	0.4577	0.2415	0.2501	0.2671	0.1311	0.1352	0.2542	0.1106	0.1163
$\eta_1 + \eta_2 + \eta_5$	0.4689	0.2549	0.2625	**0.4765**	**0.2616**	**0.2666**	**0.3037**	**0.1628**	**0.1566**	**0.2984**	**0.1654**	**0.1588**
$\eta_1 + \eta_3 + \eta_5$	0.4642	0.2394	0.2505	0.4634	0.2396	0.2516	0.2783	0.1365	0.1343	0.2976	0.1521	0.1503
$\eta_1 + \eta_4 + \eta_5$	0.4555	0.2385	0.2487	0.4601	0.2405	0.2498	0.2726	0.1263	0.1225	0.267	0.1257	0.1259
$\eta_1 + \eta_2 + \eta_3 + \eta_5$	0.4656	0.2443	0.2524	0.4701	0.2512	0.2562	0.2791	0.1466	0.137	0.2725	0.1425	0.1339
$\eta_1 + \eta_2 + \eta_4 + \eta_5$	**0.4715**	**0.2627**	**0.2677**	0.4685	0.2569	0.2642	0.2836	0.1497	0.1476	0.2869	0.1492	0.1448
$\eta_1 + \eta_3 + \eta_4 + \eta_5$	0.4711	0.2452	0.2539	0.4687	0.2595	0.2674	0.2657	0.1225	0.1200	0.2780	0.1332	0.1292
$\eta_1 + \eta_2 + \eta_3 + \eta_4 + \eta_5$	0.4655	0.2619	0.2650	0.4732	0.2610	0.2661	0.2861	0.1516	0.1410	0.2973	0.1517	0.1426

the French dataset. Table 2 further demonstrates that the combination of objectives η_1, η_2, and η_5 consistently yields the highest ROUGE-N scores across both datasets and different versions of η_5. As a result, by comparing the performance of the two versions of η_5, we can conclude that both versions are equally effective and that either one of them may be utilized for the summary generation task. Thus, we conclude that by efficiently identifying the common syntactic patterns between the news documents and their corresponding user comments, NER aids in enhancing the quality of summary generation.

4   Conclusions and Future Works

In the current work, we have introduced a NER-based objective function that aids in summarizing multi-documents in an unsupervised way using their associated comments. Further, to improve the performance of our approach, we investigate two different versions of the NER-based objective function. The results obtained on English and French datasets clearly demonstrate the effectiveness of both versions of the NER-based objective function on the summary generation task. Therefore, either of the versions can be used with the combination of other objective functions for the summary generation task. In the future, we'd like to extend our proposed framework into a cross-lingual environment.

Acknowledgements. Dr. Sriparna Saha gratefully acknowledges the Young Faculty Research Fellowship (YFRF) Award, supported by Visvesvaraya Ph.D. Scheme for Electronics and IT, Ministry of Electronics and Information Technology (MeitY), Government of India, being implemented by Digital India Corporation (formerly Media Lab Asia) for carrying out this research. Dr. Naveen Saini acknowledge the postdoctoral program of the CIMI LabEx and the support received from Indian Institute of Information Technology Lucknow, India. Dr. Jose G Moreno acknowledges TERMITRAD (2020-2019-8510010) and ANR-MEERQAT (ANR-19-CE23-0028) projects.

References

1. Alami, N., Meknassi, M., En-nahnahi, N.: Enhancing unsupervised neural networks based text summarization with word embedding and ensemble learning. Expert Syst. Appl. **123**, 195–211 (2019)
2. Anand, D., Wagh, R.: Effective deep learning approaches for summarization of legal texts. J. King Saud Univ.-Comput. Inf. Sci. **34**(5), 2141–2150 (2019)
3. Beltagy, I., Peters, M.E., Cohan, A.: Longformer: the long-document transformer. arXiv preprint arXiv:2004.05150 (2020)
4. Bing, L., Li, P., Liao, Y., Lam, W., Guo, W., Passonneau, R.J.: Abstractive multi-document summarization via phrase selection and merging. arXiv preprint arXiv:1506.01597 (2015)
5. Blei, D.M., Ng, A.Y., Jordan, M.I.: Latent dirichlet allocation. J. Mach. Learn. Res **3**, 993–1022 (2003)
6. Boroş, E., et al.: Alleviating digitization errors in named entity recognition for historical documents. In: Proceedings of the 24th Conference on Computational Natural Language Learning, pp. 431–441 (2020)
7. Erkan, G., Radev, D.R.: Lexrank: graph-based lexical centrality as salience in text summarization. J. Artif. Intell. Res. **22**, 457–479 (2004)
8. Gao, S., Chen, X., Li, P., Ren, Z., Bing, L., Zhao, D., Yan, R.: Abstractive text summarization by incorporating reader comments. In: Proceedings of the AAAI Conference on Artificial Intelligence, vol. 33, pp. 6399–6406 (2019)
9. Goyal, A., Gupta, V., Kumar, M.: A deep learning-based bilingual Hindi and Punjabi named entity recognition system using enhanced word embeddings. Knowl.-Based Syst. **234**, 107601 (2021)
10. Hu, M., Sun, A., Lim, E.P.: Comments-oriented document summarization: understanding documents with readers' feedback. In: Proceedings of the 31st Annual International ACM SIGIR Conference on Research and Development in Information Retrieval, pp. 291–298 (2008)
11. Jain, R., Mavi, V., Jangra, A., Saha, S.: Widar-weighted input document augmented rouge. arXiv preprint arXiv:2201.09282 (2022)
12. Jangra, A., Saha, S., Jatowt, A., Hasanuzzaman, M.: Multi-modal summary generation using multi-objective optimization. In: Proceedings of the 43rd International ACM SIGIR Conference on Research and Development in Information Retrieval, pp. 1745–1748 (2020)
13. Lewis, et al.: Bart: denoising sequence-to-sequence pre-training for natural language generation, translation, and comprehension. arXiv preprint arXiv:1910.13461 (2019)
14. Li, J., Sun, A., Han, J., Li, C.: A survey on deep learning for named entity recognition. IEEE Trans. Knowl. Data Eng. **34**(1), 50–70 (2020)
15. Li, P., Bing, L., Lam, W.: Reader-aware multi-document summarization: an enhanced model and the first dataset. arXiv preprint arXiv:1708.01065 (2017)
16. Li, P., Wang, Z., Lam, W., Ren, Z., Bing, L.: Salience estimation via variational auto-encoders for multi-document summarization. In: Thirty-First AAAI Conference on Artificial Intelligence (2017)
17. Lin, C.Y.: Rouge: A package for automatic evaluation of summaries. In: Text Summarization Branches Out, pp. 74–81 (2004)
18. Mihalcea, R., Tarau, P.: Textrank: Bringing order into text. In: Proceedings of the 2004 Conference on Empirical Methods in Natural Language Processing, pp. 404–411 (2004)

19. Miller, D.: Leveraging bert for extractive text summarization on lectures. arXiv preprint arXiv:1906.04165 (2019)
20. Pontes, E.L., Huet, S., Torres-Moreno, J.M., Linhares, A.C.: Compressive approaches for cross-language multi-document summarization. Data Knowl. Eng. **125**, 101763 (2020)
21. Radford, A., Wu, J., Child, R., Luan, D., Amodei, D., Sutskever, I., et al.: Language models are unsupervised multitask learners. OpenAI Blog **1**(8), 9 (2019)
22. Roha, V.S., Saini, N., Saha, S., Moreno, J.G.: Unsupervised framework for comment-based multi-document extractive summarization. In: Proceedings of the Genetic and Evolutionary Computation Conference, pp. 574–582 (2022)
23. Saini, N., Saha, S., Jangra, A., Bhattacharyya, P.: Extractive single document summarization using multi-objective optimization: exploring self-organized differential evolution, grey wolf optimizer and water cycle algorithm. Knowl.-Based Syst. **164**, 45–67 (2019)
24. Yang, Z., Dai, Z., Yang, Y., Carbonell, J., Salakhutdinov, R.R., Le, Q.V.: Xlnet: generalized autoregressive pretraining for language understanding. In: Advances in Neural Information Processing Systems, vol. 32 (2019)
25. Zhang, J., Zhao, Y., Saleh, M., Liu, P.: Pegasus: pre-training with extracted gap-sentences for abstractive summarization. In: International Conference on Machine Learning, pp. 11328–11339. PMLR (2020)
26. Zhou, A., Qu, B.Y., Li, H., Zhao, S.Z., Suganthan, P.N., Zhang, Q.: Multiobjective evolutionary algorithms: a survey of the state of the art. Swarm Evol. Comput. **1**(1), 32–49 (2011)

Don't Raise Your Voice, Improve Your Argument: Learning to Retrieve Convincing Arguments

Sara Salamat[1(✉)], Negar Arabzadeh[2], Amin Bigdeli[1], Shirin Seyedsalehi[1], Morteza Zihayat[1], and Ebrahim Bagheri[1]

[1] Toronto Metropolitan University, Toronto, ON, Canada
{sara.salamat,abigdeli,shirin.seyedsalehi,
mzihayat,bagheri}@torontomu.ca
[2] University of Waterloo, Waterloo, ON, Canada
narabzad@uwaterloo.ca

Abstract. The Information Retrieval community has made strides in developing neural rankers, which have show strong retrieval effectiveness on large-scale gold standard datasets. The focus of existing neural rankers has primarily been on measuring the relevance of a document or passage to the user query. However, other considerations such as the convincingness of the content are not taken into account when retrieving content. We present a large gold standard dataset, referred to as CoRe, which focuses on enabling researchers to explore the integration of the concepts of convincingness and relevance to allow for the retrieval of relevant yet persuasive content. Through extensive experiments on this dataset, we report that there is a close association between convincingness and relevance that can have practical value in how convincing content are presented and retrieved in practice.

1 Introduction

There has been an increasing attention on mining and identifying argumentative structures from monologues (micro-level) and dialogues (macro-level) in the context of discussion forums and social networks [2,3,12,18–20], which are often referred to as *argument mining*. The works in the argument mining literature explore various tasks including argument detection [3], argument component classification [12], as well as inter and intra argument relation identification [2], to name a few. The major objective of these tasks is to identify arguments, understand their structure and model their relations with each other within a formal argumentation framework [16]. A specific strand of research in this area has focused on identifying, modeling, and predicting *persuasiveness* of arguments. These works are interested in determining what types of arguments and what forms of argumentative structures are capable of convincing the target audience [4,6,8,15,17,26]. There have been a variety of methods that focus on argument persuasion (convincingness) including those that leverage surface textual, social interaction, and argument-related features for ranking arguments [22], as well as others that adopt an end-to-end approach for modeling convincingness [6].

© The Author(s), under exclusive license to Springer Nature Switzerland AG 2023
J. Kamps et al. (Eds.): ECIR 2023, LNCS 13981, pp. 589–598, 2023.
https://doi.org/10.1007/978-3-031-28238-6_50

Table 1. Broad areas of related work.

Reference	Task		
	Argument mining	Convincingness	Relevance
[2,6,12,16,19,20]	✓	✗	✗
[4,6,8,15,17,22,26]	✗	✓	✗
[5,21]	✗	✗	✓
Our work	✗	✓	✓

Other researchers have ventured into modeling argument quality [5,21]. While researchers have explored various aspects of argumentative structures, to the best of our knowledge, the *notion of convincingness* of content has not been explored within the context of Information Retrieval (IR). We believe that it is important to understand the process behind the effective retrieval of convincing content because as discussed by Vecchi et al. [20], a careful treatment of such content could be used for social good in areas such as retrieving factual and convincing information for purposes including countering misinformation.

The work in the literature, shown in Table 1, can broadly be classified as those that (1) perform argument mining, (2) measure content convincingness, and (3) determine content relevance. We note that there are no earlier works that have considered the retrieval of convincing information. In other words, retrieval tasks are often focused on optimizing relevance without necessarily taking convincingness of content into account. As such, our work in this paper is among the first to explore how IR ranking models can capture and incorporate the notion of convincingness and integrate it into the retrieval process. Our objective is to rank documents to be *both relevant and convincing*. We systematically curate and publicly release a gold standard of queries and relevant documents, each of which comes with an explicit degree of convincingness. We benefited from the Change My View (CMV) subreddit (r/changemyview) in order to capture content convincingness. The CMV subreddit allows users to exchange information with each other on specific topics with the hope of changing each others' opinion, and to explicitly specify how much and to what extent their opinions have changed. We consider content that have changed the opinion of a larger number of users to be more convincing.

Based on the curated dataset, we explore whether it would be possible to learn the notion of convincingness through training different neural ranking models. The idea is that given recent state-of-the-art neural rankers are becoming increasingly better at learning the concept of relevance when shown pairs of queries and their relevant documents, we hypothesize that it might be possible to learn the concept of convincingness by using a similar strategy. Through extensive experiments, we make an important observation that the concepts of relevance and convincingness are (at least on the CMV subreddit) highly correlated phenomena. We find that highly relevant documents to a query are those that are considered to be the most convincing for the users. Our findings align very closely with those of researchers in *cognitive psychology* [13] who have shown

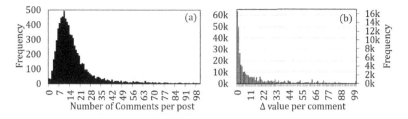

Fig. 1. (a) Distribution of the number of comments per post in CoRe; (b) Distribution of Δ values for the comments. Number of comments with Δ=0 is scaled with the left y-axis and the number of comments with non-zero Δ are scaled with the right y-axis.

that people tend to be convinced more easily when presented with highly relevant information. We show that retrieving documents that are highly relevant would lead to the retrieval of highly convincing ones. This observation suggests that relevance could be a significant contributing factor to convincingness; and therefore, users who would like to persuade others would need to focus their arguments on highly relevant content.

The **contributions of our work** can be summarized as follows: (1) We collect and publicly release a dataset, referred to as *CoRe (Convincing Retrieval)*, which includes 7,937 topics along with subsequent arguments on each topic that have explicit labels for their convincingness at 5 levels; (2) We adopt state-of-the-art neural rankers to learn concepts of relevance and convincingness using our CoRe dataset in order to rank content based on both criteria; (3) We systematically show that the concepts of relevance and convincingness are highly correlated where a retrieval process that maximizes the likelihood of relevance will also be effective for retrieving convincing content.

Reproducibility : The CoRe dataset is publicly available: https://github.com/sara-salamat/CoRe.

2   The Convincing Retrieval (CoRe) Dataset

Most gold standard datasets for the ad hoc retrieval task capture the concept of relevance between a query and its related documents. The objective of our work in this paper is to additionally introduce the concept of convincingness in order to facilitate the process of retrieving relevant and convincing content. To curate such a dataset, we leverage the popular subreddit known as the Change My View subreddit. This subreddit is a community, with over 1.5 million members, on which users post their opinions on a particular topic and challenge others to convince them to change their viewpoints. The community works based on a scoring system, called deltas (Δ), which provides the means to assign credits to convincing arguments. Users are expected to reply to the comment that has changed at least one aspect of their opinion by rewarding it a delta (Δ) and explain how they were convinced to change their opinion [1]. The more convincing a comment is, the more deltas it will receive.

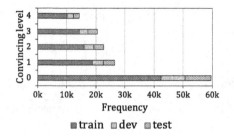

Fig. 2. Convincingness frequency in CoRe.

Table 2. CoRe dataset statistics.

# comments	153,755
# posts	7,937
# users	46,419
Avg length of posts' content	330.93 words
Avg length of posts' title	14.42 words
Avg length of comments	120.27 words

Table 3. CoRe dataset train/dev/test set statistics.

	Train	Dev	Test
Number of posts	5,555	1,189	1,193
Average number of comments	18.26	17.36	18.06
Median number of comments	13	12	13
Average number of Δ per comment	33.58	35.37	33.42
Median number of Δ per comment	1	2	2
Average number Δ per post	613.6	614.19	603.91
Median number Δ per post	515	509	516

In order to gather our CoRe gold standard, we collected all posts and comments published on CMV for a period of 15 months starting from January 2021. To avoid *recency bias*, we did not include any posts that were still active as the deltas on their comments may not have yet reached a steady state. Furthermore, in order to avoid *topical bias*, we did not prioritize the collection of any topics and all content were collected as available on CMV. Table 2 shows the statistics of the content included in our CoRe dataset. For each post, we obtained all of its first level responses and considered them to be the relevant documents for that post. We consider this to be a reasonable assumption since according to CMV rules, any irrelevant responses to the post will be removed by the CMV administrators. As shown in Fig. 1(a), the majority of the posts received between 7 to 14 comments, i.e., the majority of topics in our gold standard have between 7–14 relevant documents. Furthermore, for each of the comments, we collected their delta values whose distribution is depicted in Fig. 1(b). As seen in the Figure, from 153k comments in CoRe, 63k (41%) of these comments did not receive any deltas indicating that no user on CMV considered them to be convincing.

We map delta values into five different levels where comments with no Δ are placed in level zero and are considered not to be convincing at all. The other four levels consist of comments with increasing convincingness with 1–5, 6–20, 21–100 and 100 and more deltas, respectively. We have created splits of the CoRe dataset so it can be used for training neural models by randomly assigning 70% of the posts to the train set, 15% to the development set and 15% to the test set. Table 3 shows the statistics of the data in each split, which have a similar distribution in terms of number of posts, comments and average number of Δ per comment and per post. Additionally in Fig. 2, we depict the frequency of comments placed in the different levels as well as how comments with varying levels of convincingness are placed in different splits. We have ensured that convincingness levels retain a similar ratio in each split. CoRe is structured in TREC format where each CMV post is a query, its first-level comments are its relevant documents, and the convincingness level of each comment is related to its deltas.

3   Evaluation Tasks

We introduce two independent retrieval tasks for the CoRe dataset, namely (1) **relevance ranking**: to retrieve and rank all relevant comments to a post, and (2) **convincingness ranking**: to rank-order the comments of a post based on their degree of convincingness. **The Relevance Ranking Task.** The goal of this task is to perform ad hoc retrieval on the CoRe dataset. Given a query q, the goal of an ad hoc retriever is to use method M to retrieve a ranked list of documents D_q from a collection of items (i.e., C) such that $M(q, C) = D_q$. Given q, D_q is compared to a judged set of items R_q to evaluate the performance of M. In the context of CoRe, each post is considered to be a query, which needs to be satisfied through a retrieval method M based on the set of all comments in the corpus. Each post p is accompanied with a set of comments $C_p = \{C_{p^1}, C_{p^2}, ...C_{p^n}\}$. Given p, all the comments in C_p are considered as relevant, i.e., C_i is only relevant to p if $C_i \in C_p$, and comments not in C_p are considered to be irrelevant to p. The goal of the relevance ranking task is to identify a ranked list of comments D_p for a given post p from a collection of comments using retrieval method M, i.e., $D_p = M(p, C)$.

In order to operationalize M, we employ widely-used bi-encoder-based dense neural retrievers, which have shown promising performance on other tasks [7, 10,11,23–25]. Neural rankers need to be trained on a gold dataset. For this purpose, we adopt two strategies: **(1)** In the first strategy, we train the ranker on a completely different relevance judgment dataset, which is non-overlapping with CoRe. The reason for this is that we would like to ensure that the ranker only learns the concept of relevance and does not have a chance to observe the concept of convincingness (as present in CoRe). To this end, we adopt the MS MARCO dataset, which consists of over 500k queries and their relevant judgment documents. **(2)** In the second strategy, we train the ranker on the training split of the CoRe dataset; however, when using this split, we only consider comments that are related to each post as being relevant and ignore the convincingness levels of the comments when training the ranker. The reason for this is that the goal of relevance ranking is to rank comments based on their relevance to the post.

When training the rankers, for each post p, pairs of (p, C_i) are positive samples if $C_i \in C_p$, otherwise, (p, C_i) is a negative sample. The ranker is trained to predict the label for each (p, C_i). We set the maximum sequence length to 300, the number of training epochs to 30 and learning rate to 2e-5. We use Faiss [9] for efficient indexing. Table 5 illustrates the performance of the rankers based on which we make several observations: **(a)** Consistent with earlier findings on base language models for neural rankers, the best performance on relevance ranking is seen when BERT is used [14]. **(b)** The first strategy that uses MS MARCO to train the ranker is more effective for relevance ranking, which shows that relevance learnt on a different corpus is transferable to CoRe; and, **(c)** In two of the language models with the largest number of parameters, i.e., BERT and RoBERTa, the model trained on CMV shows weaker performance compared to one trained on MS MARCO. On the other hand, on the smaller language model, i.e., DistilBERT, the model trained on CMV shows better performance. This can be due to the need for a large number of training samples to tune language models with a large number of parameters, i.e., BERT and RoBERTa.

The Convincingness Ranking Task. The objective of the second task is to learn the concept of convincingness and rank comments according to their degree of convincingness. Formally stated, given a pair of post p, and comment C_i where $C_i \in C_p$, i.e., C_i is a comment related to post p, our goal is to learn the level of convincingness of a pair (p, C_i) while minimizing the difference between the predicted convincingness level through function $S(p, C_i)$ with its actual level of

Table 4. Results on the convincing ranking task.

	Training		Evaluation metric	
Model	Dataset	Task	Recall@10	ndcg@10
DistilBERT	MS MARCO	relevance	0.739	0.688
	CoRe	relevance	0.732	0.674
	CoRe	convincing	0.738	0.689
BERT	MS MARCO	relevance	**0.741**	0.697
	CoRe	relevance	**0.741**	0.695
	CoRe	convincing	**0.741**	**0.699**
RoBERTa	MS MARCO	relevance	0.738	0.684
	CoRe	relevance	**0.741**	0.696
	CoRe	convincing	0.729	0.655

Table 5. Results on relevance retrieval task.

	Training dataset	Recall		MAP		nDCG	
		@10	@100	@10	@100	@10	@100
DistilBERT	MS MARCO	0.212	0.394	0.164	0.204	0.384	0.381
	CoRe	0.234	0.454	0.185	0.234	0.414	0.424
BERT	MS MARCO	**0.260**	**0.466**	**0.213**	**0.266**	**0.462**	**0.454**
	CoRe	0.236	0.462	0.183	0.233	0.409	0.424
RoBERTa	MS MARCO	0.227	0.414	0.183	0.228	0.415	0.404
	CoRe	0.192	0.378	0.144	0.179	0.351	0.357

convincingness. To learn the representation for function S, we adopt two strategies: (1) In the first strategy, we benefit from the convincingness levels in our CoRe dataset to learn which comments are convincing in the context of the post. We train a bi-encoder based dense-retriever architecture discussed in the first task to train a model based on comment convincingness levels available in CoRe. In contrast to the first task where there were only two relevance levels, here we are dealing with five levels of convincingness. (2) In the second strategy, we use the same neural rankers that were trained for the relevance ranking task to estimate the convincingness of a comment. We use rankers that have learnt the concept of relevance to rank comments based on their convincingness to investigate whether there are any meaningful relationships between the concepts of relevance and convincingness.

Based on results of the convincingness ranking task shown in Table 4, our most notable observation is that regardless of whether the training task was on relevance or convincingness ranking, the results of the convincingness ranking task is similar (0.741) regardless of whether the neural rankers were trained on the MS MARCO or the CoRe datasets. This is an important finding as it shows the neural rankers trained on MS MARCO for relevance ranking are competitive with those rankers trained on CoRe for convincingness. This might indicate that relevance and convincingness are correlated.

4   In-depth Analysis

In order to take an in-depth look into a possible correlation between relevance and convincingness, we first compare the rankings produced by models that were trained on MS MARCO with their counterparts trained on the convincingness levels in CoRe. Then, we compare the rankings produced by both approaches through a *stratified* strategy.

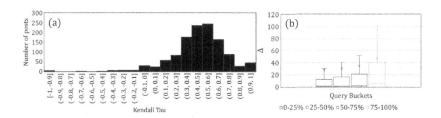

Fig. 3. (a) Distribution of Δ values; (b) distribution of Kendall Tau values.

Association Between Relevance and Convincingness. To assess the degree of association between the two concepts, we compare the retrieved list of comments for a given post when retrieved using the two different strategies, once using rankers trained on relevance and once through rankers trained on convincingness. We employ the Kendall Tau rank correlation to evaluate the correlation between the predicted scores for each of the comments in the retrieved lists for every post. Figure 3(a) presents the histogram of the Kendall Tau correlation values. The Figure shows how correlated the ranked list of comments from the BERT model trained on MS MARCO is to the BERT model trained on CoRe. For each post, the closer the value of Kendall Tau is to one, the higher the correlation between the two retrieved lists would be. From the Figure, we observe that the majority of comments experience a strong correlation (over 0.3), which indicates that the performance of the ranker trained on relevance is quite correlated with a ranker trained on CoRe on an individual post level. This shows that, at least on the CoRe dataset, the concepts of convincingness and relevance are correlated with each other.

Stratified Comparison of Relevance and Convincingness. We study the relationship between the performance of queries' relevance-based retrieval and their comments' convincingness. To do so, we categorize queries into 4 equally-size buckets based on the percentile of their performance (recall@100) where the worst performing queries are in the 0–25% bucket and the 75–100% bucket includes 25% of the highest performing queries. We plot the distribution of deltas associated with the comments on each post under each bucket, which are shown in Fig. 3(b). As shown, the best-performing query bucket, i.e., the yellow bucket, consists of comments with higher degrees of convincingness compared to lower performing query buckets. This finding shows that the easier the query is, the more convincing its comments are and vice versa, i.e., the comments for the hardest queries gain the lowest number of deltas compared to better performing buckets of queries. We find that in CoRe, relevance and convincingness are correlated, which means a ranker that has been effectively trained for relevance retrieval could be an effective out-of-the-box ranker for convincingness retrieval.

5   Concluding Remarks

In this paper, we have introduced the task of convincing IR and offered a systematically collected dataset, called CoRe. The dataset allows the community to explore the retrieval of persuasive content. Based on extensive experiments, we find that the concepts of relevance and convincingness may be correlated, which suggests that, at least in the context of the CMV subreddit, convincing content are those that are relevant to the topic of the query. This reinforces findings in cognitive psychology that indicate people are more likely to be convinced when they are presented with highly relevant content.

References

1. Change my view (cmv) (2018). https://www.reddit.com/r/changemyview/wiki/deltasystem/
2. Chakrabarty, T., Hidey, C., Muresan, S., McKeown, K., Hwang, A.: AMPER-SAND: argument mining for PERSuAsive oNline discussions. In: Proceedings of the 2019 Conference on Empirical Methods in Natural Language Processing and the 9th International Joint Conference on Natural Language Processing (EMNLP-IJCNLP), pp. 2933–2943. Association for Computational Linguistics, Hong Kong, China, November 2019. https://doi.org/10.18653/v1/D19-1291, https://aclanthology.org/D19-1291
3. Cheng, L., Wu, T., Bing, L., Si, L.: Argument pair extraction via attention-guided multi-layer multi-cross encoding. In: Proceedings of the 59th Annual Meeting of the Association for Computational Linguistics and the 11th International Joint Conference on Natural Language Processing (Volume 1: Long Papers), pp. 6341–6353 (2021)
4. Dayter, D., Messerli, T.C.: Persuasive language and features of formality on the r/changemyview subreddit. Internet Pragmatics 5(1), 165–195 (2022)
5. Dumani, L., Schenkel, R.: Quality-aware ranking of arguments. In: Proceedings of the 29th ACM International Conference on Information & Knowledge Management, pp. 335–344 (2020)
6. Dutta, S., Das, D., Chakrabarty, T.: Changing views: Persuasion modeling and argument extraction from online discussions. Inf. Process. Manage. 57(2), 102085 (2020)
7. Gao, J., Xiong, C., Bennett, P., Craswell, N.: Neural approaches to conversational information retrieval. arXiv preprint arXiv:2201.05176 (2022)
8. Habernal, I., Gurevych, I.: Which argument is more convincing? analyzing and predicting convincingness of web arguments using bidirectional lstm. In: Proceedings of the 54th Annual Meeting of the Association for Computational Linguistics (Volume 1: Long Papers), pp. 1589–1599 (2016)
9. Johnson, J., Douze, M., Jégou, H.: Billion-scale similarity search with GPUs. IEEE Trans. Big Data 7(3), 535–547 (2019)
10. Karpukhin, V., et al.: Dense passage retrieval for open-domain question answering. arXiv preprint arXiv:2004.04906 (2020)
11. Lin, J., Nogueira, R., Yates, A.: Pretrained transformers for text ranking: bert and beyond. Synth. Lect. Hum. Lang. Technol. 14(4), 1–325 (2021)
12. Lugini, L., Litman, D.: Contextual argument component classification for class discussions. arXiv e-prints pp. arXiv-2102 (2021)
13. Maio, G.R., Hahn, U., Frost, J.M., Kuppens, T., Rehman, N., Kamble, S.: Social values as arguments: similar is convincing. Front. Psychol. 5, 829 (2014)
14. Reimers, N., Gurevych, I.: Sentence-bert: sentence embeddings using siamese bert-networks. arXiv preprint arXiv:1908.10084 (2019)
15. Simpson, E., Gurevych, I.: Finding convincing arguments using scalable bayesian preference learning. Trans. Assoc. Comput. Linguist. 6, 357–371 (2018)
16. Stab, C., Gurevych, I.: Parsing argumentation structures in persuasive essays. Comput. Linguist. 43(3), 619–659 (2017)
17. Tan, C., Niculae, V., Danescu-Niculescu-Mizil, C., Lee, L.: Winning arguments: interaction dynamics and persuasion strategies in good-faith online discussions. In: Proceedings of the 25th International Conference on World Wide Web, pp. 613–624 (2016)

18. Trabelsi, A., Zaiane, O.R.: Finding arguing expressions of divergent viewpoints in online debates. In: Proceedings of the 5th Workshop on Language Analysis for Social Media (LASM), pp. 35–43 (2014)
19. Tran, N., Litman, D.: Multi-task learning in argument mining for persuasive online discussions. In: Proceedings of the 8th Workshop on Argument Mining, pp. 148–153 (2021)
20. Vecchi, E.M., Falk, N., Jundi, I., Lapesa, G.: Towards argument mining for social good: a survey. In: Proceedings of the 59th Annual Meeting of the Association for Computational Linguistics and the 11th International Joint Conference on Natural Language Processing (Volume 1: Long Papers), pp. 1338–1352 (2021)
21. Wachsmuth, H., Stein, B., Ajjour, Y.: "Pagerank" for argument relevance. In: Proceedings of the 15th Conference of the European Chapter of the Association for Computational Linguistics: Volume 1, Long Papers, pp. 1117–1127 (2017)
22. Wei, Z., Liu, Y., Li, Y.: Is this post persuasive? ranking argumentative comments in online forum. In: Proceedings of the 54th Annual Meeting of the Association for Computational Linguistics (Volume 2: Short Papers), pp. 195–200 (2016)
23. Xiong, L., et al.: Approximate nearest neighbor negative contrastive learning for dense text retrieval. CoRR abs/2007.00808 (2020). https://arxiv.org/abs/2007.00808
24. Yang, W., Zhang, H., Lin, J.: Simple applications of BERT for ad hoc document retrieval. CoRR abs/1903.10972 (2019). http://arxiv.org/abs/1903.10972
25. Yu, S., Liu, Z., Xiong, C., Feng, T., Liu, Z.: Few-shot conversational dense retrieval. In: Proceedings of the 44th International ACM SIGIR Conference on Research and Development in Information Retrieval, pp. 829–838 (2021)
26. Zeng, J., Li, J., He, Y., Gao, C., Lyu, M., King, I.: What Changed Your Mind: The Roles of Dynamic Topics and Discourse in Argumentation Process, p. 1502–1513. Association for Computing Machinery, New York, NY, USA (2020), https://doi.org/10.1145/3366423.3380223

Learning Query-Space Document Representations for High-Recall Retrieval

Sara Salamat[1(⊠)], Negar Arabzadeh[2], Fattane Zarrinkalam[3], Morteza Zihayat[1], and Ebrahim Bagheri[1]

[1] Toronto Metropolitan University, Toronto, ON, Canada
{sara.salamat,mzihayat,bagheri}@torontomu.ca
[2] University of Waterloo, Waterloo, ON, Canada
narabzad@uwaterloo.ca
[3] University of Guelph, Guelph, ON, Canada
fzarrink@uoguelph.ca

Abstract. Recent studies have shown that significant performance improvements reported by neural rankers do not necessarily extend to a diverse range of queries. There is a large set of queries that cannot be effectively addressed by neural rankers primarily because relevant documents to these queries are not identified by first-stage retrievers. In this paper, we propose a novel document representation approach that represents documents within the query space, and hence increases the likelihood of recalling a higher number of relevant documents. Based on experiments on the MS MARCO dataset as well as the hardest subset of its queries, we find that the proposed approach shows synergistic behavior to existing neural rankers and is able to increase recall both on MS MARCO dev set queries as well as the hardest queries of MS MARCO.

1 Introduction

There have been recent works in the literature that have shown the approach adopted by neural ranking models that captures relevance through learning document and query representations that maximizes the similarity of relevant queries and documents and minimizes the relevance of dissimilar ones does not necessarily scale to a full range of different query types [8,11,28,29]. For instance, Arabzadeh et al. found that regardless of the underlying neural ranking architecture, neural rankers are not able to satisfy a large group of queries (an average precision of zero) within the MS MARCO collection. These queries were referred to as MS MARCO *Chameleons*. The long-tailed performance of state-of-the-art neural ranking models on gold standard collections, such as MS MARCO, indicates that it is important to identify ways through which all queries, especially those that are hard for neural ranking models, can be handled effectively.

One of the main observations about hard queries is that they not only struggle with poor precision but also struggle with low recall. In essence, the poor recall on such queries can also explain the low precision since due to the heavy computation cost of neural models for full-collection retrieval, most existing neural ranking models are specifically devoted to re-ranking a set of candidates retrieved by a first-stage retriever [14]. In any full ranking stack, whether it is industrial [3,13,26] or research-oriented

J. Kamps et al. (Eds.): ECIR 2023, LNCS 13981, pp. 599–607, 2023.
https://doi.org/10.1007/978-3-031-28238-6_51

[7,24], the goal of the first stage of the retrieval a.k.a *the recall stage*, is to collect all potential relevant documents w.r.t the query using computationally cheap and efficient methods. Further, in the ranking stack, the retrieved pool of candidates will get re-ranked with more expensive, complex, and accurate rerankers [10,12,16,20]. Hence, the main objective of the first stage of a full-ranking stack is to efficiently provide a high-recall pool of document candidates. As such, neural ranking models will only be able to show improved precision if relevant documents are already retrieved and included in the list of documents retrieved by the first-stage retriever. However, in practice, first-stage retrievers struggle with finding a sufficient number of relevant documents for harder queries (low recall), which translates into poor precision by neural rankers.

Existing research has hinted at the fact that the low recall can be due to the difficulty associated with learning appropriate representations for hard queries, their relevant documents, or both [22,30]. In other words, inappropriate query or document representations can significantly impact recall. For example, Bagheri et al. [2] have shown that the choice of document representation can have a notable impact on recall. As such, there have been approaches that explore ways through which more effective query and document representations can be learned. Nogueira et al. [17] have been among the first to explore how document representations could be slightly modified to improve retrieval effectiveness. They found that appending documents with artificially-generated queries from that document using a transformer architecture can lead to noticeable performance improvement. Similarly, Dai and Callan advocated for the idea of learning document term weights that could then lead to a more effective weighted document representation and hence more effective retrieval [5].

Inspired by such studies that have shown the impact of document representation on recall, in this paper, we aim specifically for high-recall retrieval, especially for harder queries. We hypothesize that harder queries with poor recall are those whose relevant documents' representations are not similar to the query itself and, as such, the first-stage retriever is not able to retrieve the relevant documents in the first stage. In such cases, the relevant documents lack any notable resemblance to their relevant query; therefore, we propose to fully replace the original document with a more concise representation of that document. This representation is derived by learning a transformer architecture that learns to generate a query from a document when trained on a collection of gold query-document pairs. In our approach, the original document is replaced by a query-inspired representation of that document, which has the following benefits: (b1) given the new document representation is in the form of a query, learning embeddings that would match the query and the new document representation could be easier; and, (b2) the new document representation is a reformulation of the document but in query space; hence increases the chances of being effectively matched with the relevant query.

In order to evaluate our work, we conduct our experiments on the MS MARCO passage collection [15] and show that our proposed concise document representation so called as $q2q$ (Query to Query-space representation) is able to retrieve a non-overlapping set of relevant documents compared to the original first-stage retrievers. We show that by systematically integrating the results of our work with that of the first-stage retrievers, we are able to improve recall significantly on the queries from the MS

MARCO development set. We also show that such an improvement is not only observed over all of the queries but the improvements are much more substantial on the harder MS MARCO queries known as MS MARCO Chameleons.

Reproducibility: We publicly release the code, data, and trained models on https://github.com/sara-salamat/queryspace-representation.

2 Proposed Approach

The objective of our work is to facilitate high recall in first-stage retrievers by an alternative document representation that is closer to query representations. We propose to replace each document with its corresponding query representation where the query representation is generated by a transformer trained specifically on query-document pairs. On this basis, our approach consists of two steps: 1) learning alternative document representations; and 2) training a neural ranking model to learn the association between the query representation and the reformulated document representations.

Step 1. Learning Alternative Document Representations: In order to learn alternative document representations that are closer to the query space, we are inspired by methods such as Doc2Query, which modify document representations by appending additional query terms to the document. Unlike these methods and instead of expanding the document, we are interested in fully replacing the document representation with one in the query space. We believe such an approach will ensure that the document space is sufficiently close to the query space to lead to improved recall. To this end, we adopt a transformer architecture to generate queries from an input document. More formally, we let $\mathcal{G}$ be a query translation function, which is trained to generate queries from an input document. With $\mathcal{G}$, we will be able to generate query representations for each document in the document corpus $\mathcal{D}$ such that each generated query would be able to efficiently retrieve the document it was generated from. Therefore, given a document $d \in \mathcal{D}$, and the translation function $\mathcal{G}$, we generate $\hat{q}_d$ as $\hat{q}_d = \mathcal{G}(d)$.

It has been shown that $\mathcal{G}$ can be efficiently learned [17] by fine-tuning a transformer [23] based on a relevant judgment dataset. Simply put, based on the association between existing queries and their associated relevant documents, the transformer will learn to generate queries for a given document. Such a fine-tuned transformer will act as $\mathcal{G}$, and since the translation function is not deterministic, we can generate multiple queries for each document by translating the document several times. Hence, we can generate a query set $\hat{Q}_d$ per document $d \in \mathcal{D}$. Ideally, $\hat{Q}_d$ can be interpreted as a set of all queries that can be answered by document d. Moreover, for each document d, we define the query-to-query representation of document d as $q2q(d)$ through the concatenation of its corresponding generated query set, as follows:

$$q2q(d) = concat(\hat{q}_i)|\hat{q}_i \in \hat{Q}_d$$

In our work, we propose to use $q2q(d)$ as the alternative representation for document d.

Step 2. Training Neural Ranker based on Alternative Document Representation: Similar to the training strategy adopted for neural ranking models (dense retrievers), given a query q and its set of relevant documents R_q^+, we fine-tune a large pre-trained language model to maximize the similarity between representations of a query and the documents. In essence, a neural ranking model learns a mapping function ϕ, which maximizes the similarity of the representation of the query ($\phi(q)$) and its relevant documents ($\phi(R_q^+)$) and minimizes the similarity of the representation of the query and its irrelevant documents $\phi(R_q^-)$. Such a mapping function is often obtained by fine-tuning contextualized language models such as BERT [6]. In the context of our work, we fine-tune neural ranking architectures to maximize the similarity between the representation for q and the reformulated representation of a document based on its set of generated queries $q2q(R_q^+)$. This neural ranking model learns the new representations of the query and documents by maximizing the similarity of $\phi(q)$ and $\phi(q2q(R_q^+))$ minimizing the similarity with Negative sampled documents $\phi(q2q(R_q^-))$.

3   Experiments

Dataset. We evaluate our proposed approach on the MS MARCO passage collection [15]. We trained both our generation function and our neural ranking models on the MS MARCO training set and evaluated them on the 6,980 small MS MARCO dev set queries, which are intended for evaluation purposes.

Query Sets. We perform the evaluation on the small MS MARCO dev set queries as well as the set of its poorly performing queries, a.k.a. MS MARCO Chameleons [1]. The MS MARCO Chameleons consists of three sets: 1) Veiled Chameleons ("Hard" set); 2) Pygmy Chameleons ("Harder" set); and, 3) Lesser Chameleon ("Hardest" set). We refer the interested reader to [1] for more details on the Chameleons sets.

Query Translation Function. To generate alternative document representations, we fine-tuned a T5 transformer based on the query-document pairs of the MS MARCO train set. We ran experiments by representing documents through a set of k corresponding queries where $k \in \{5, 10, 20, 30\}$. We explore the impact of k in our experiments.

Dense Retriever. For the neural ranking model, we adopt the widely-used Sentence-BERT (SBERT) [18], which has shown to have strong retrieval performance and low computational overhead. To have a fair comparison, and due to computational limitations, we performed fine-tuning for 5 epochs on the MS MARCO train set with lr = $2e - 5$. Our model uses Multiple Negatives Ranking Loss (MNRL) [9]. We used five negative samples for each query and used DistilBERT [19] as our base model for training.

Baselines. We compare our work with two state of the art document representation techniques, namely DeepCT [5] and Doc2Query [17]. DeepCT learns a weighted representation of the document based on a neural attention mechanism, while Doc2Query expands the initial document representation with additional query-related terms.

4   Results and Findings

Impact of the Number of Generated Queries. In Fig. 1(a), we report the performance of our proposed approach in terms of recall@k where $k \in \{10, 20, 100, 200, 500, 1000\}$ on MS MARCO dev set when representing the document with N-generated queries where $N \in \{5, 10, 20, 30\}$. As shown in this Figure, we observe that the number of queries used to form the alternative document representation has a notable impact on performance. This is especially noticeable as we increase the number of queries from 5

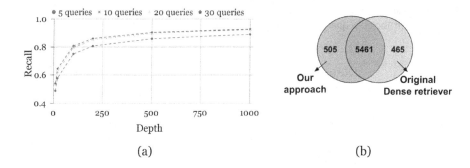

Fig. 1. (a) Performance of our proposed representation in terms of recall at different cutoffs. (b) The number of retrieved relevant documents on top-100 retrieved documents.

to 20. However, adding more queries after 20 does not lead to any statistically significant improvements in performance (paired t-test with 95% confidence interval). This observation is aligned with other document expansion work [17] where the authors also reported that after appending 20 queries to the document, there were no significant improvements observed in retrieval performance. Thus, for the rest of the experiments reported in this paper, we report the results with documents represented by 20 queries.
Performance Comparison. To evaluate the performance of our proposed approach, we compare its performance to that of the base dense retriever in Table 1. As shown, the performance of the two models is quite competitive and similar to each other in terms of recall at different depths. We note that the original dense retriever is performing slightly better at different depths. However, upon further in-depth inspection of performance, we find that while the models have comparable quantitative performance, they do not necessarily have overlapping retrieval performance in practice. In other words, the similar measured performance is not due to a similar retrieval at the query level since the two models are showing retrieval effectiveness on non-overlapping sets of relevant documents. Figure 1(b) exhibits this performance where the number of unique relevant documents as well as the number of overlapping relevant documents retrieved by two rankers are shown. As seen, there are 505 unique relevant documents that are retrieved by our method that are not identified by the base retriever and similarly 465 unique relevant documents that were not identified by our method while there are 5,461 shared relevant documents between the two methods. This is a clear indication of synergistic behavior between the two models. As we will show later in our experiments, our method has been able to identify relevant documents for harder queries that are not retrieved by the base method. As such, as proposed in literature [4,21,27], we adopt the pairwise reciprocal rank fusion between the original runs and our approach to interpolate the two runs and benefit from the complementary behavior of the two models.

Table 1 shows the results of the integration of our method with the base retriever using the pairwise reciprocal rank fusion. From Table 1, we observe that (1) selecting the pool of candidates from the combined pool of retrieved documents from the base retriever as well as our proposed approach leads to a constant increase in recall. The observed differences are statistically significant on all query sets at all depths (paired t-test with 95% confidence interval). (2) As noted earlier in the paper, first stage retriever

Table 1. Recall values of our proposed approach at different cut-offs on MS MARCO dev set as well as the Chameleons query subsets (hard, harder, and hardest).

	Retrieval method	Recall cut-off					
		10	20	100	200	500	1000
MS MARCO Dev Set	SBERT	0.5457	0.6423	0.8053	0.8549	0.9003	0.9259
	Doc2query	0.4502	0.543	0.7193	0.786	0.8536	0.8919
	DeepCT	0.4761	0.5725	0.7537	0.8097	0.872	0.9035
	$q2q$	0.5291	0.6244	0.7996	0.8509	0.898	0.9197
	Interpolated $q2q$	0.5731	0.6691	0.847	0.8926	0.9334	0.9500
	%Improvement	**5.02%**	**4.17%**	**5.18%**	**4.41%**	**3.68%**	**2.60%**
Veiled (hard)	SBERT	0.2065	0.3344	0.6153	0.7085	0.7996	0.8491
	Doc2query	0.1203	0.2071	0.4642	0.5777	0.7075	0.7785
	deepct	0.0872	0.2046	0.5137	0.618	0.7381	0.8012
	$q2q$	0.2038	0.3277	0.6200	0.7123	0.7995	0.8436
	Interpolated $q2q$	0.2354	0.3638	0.6797	0.7714	0.849	0.8887
	%Improvement	**14.00%**	**8.79%**	**10.47%**	**8.88%**	**6.18%**	**4.66%**
Pygmy (harder)	SBERT	0.1441	0.2616	0.5602	0.6674	0.7731	0.8289
	Doc2query	0.0695	0.136	0.3944	0.5173	0.6606	0.7432
	deepct	0.0369	0.1303	0.445	0.5596	0.6948	0.7674
	$q2q$	0.1499	0.2587	0.5706	0.6756	0.7761	0.8273
	Interpolated $q2q$	0.1677	0.2866	0.626	0.7341	0.8258	0.8725
	%Improvement	**16.38%**	**9.56%**	**11.75%**	**9.99%**	**6.82%**	**5.26%**
Lesser (hardest)	SBERT	0.0871	0.1806	0.4818	0.6051	0.7214	0.7871
	Doc2query	0.0269	0.0627	0.2889	0.4123	0.5778	0.6818
	deepct	0.0012	0.0605	0.3437	0.4692	0.6307	0.7159
	$q2q$	0.095	0.1799	0.4907	0.6133	0.7302	0.7949
	Interpolated $q2q$	0.1035	0.2021	0.5396	0.6718	0.7812	0.8399
	%Improvement	**18.83%**	**11.90%**	**12.00%**	**11.02%**	**8.29%**	**6.71%**

methods and in general neural rankers struggle to satisfy hard queries especially those represented in the MS MARCO Chameleons dataset. We report performance of our work on the three variations of the Chameleons dataset. It is important to note that while our approach leads to a noticeable improvement of ~5% on recall@100 on the whole MS MARCO dev set, this improvement is at least ~10% on the Chameleons dataset (2x higher than the overall dataset). This is a significant observation, since as reported in [17], most queries in Chameleons showed an average precision of zero indicating that neural rankers are not able to retrieve any relevant documents for these queries. Therefore, a significant boost in the number of relevant documents returned by the first stage retriever has the potential to impact their overall retrieval effectiveness in the next stage. The statistically significant improvement over recall, especially on the Chameleons dataset, is an indication that our proposed representation is quite effective for hard queries. (3) Finally, when comparing our work with two state-of-the-art document representation baseline methods, namely, DeepCT and Doc2Query, we find that our proposed $q2q$ method shows a better performance compared to both of the methods, with and without interpolation, on all four query sets at various cut-off points.

We note that in order to study the generalizability of our approach, we replicated the experiments with dense retrievers using other base language models, e.g., miniLM-L6-v2 [25]. While noting that the results were consistent with the above-mentioned findings, due to limited space, we have included these results in our Github repository.

5   Concluding Remarks

Our work in this paper builds on observations from the literature that have shown neural rankers are not as equally effective across a range of queries, i.e., while they significantly improve the performance of a subset of queries, they fail to satisfy others. We tend to improve the performance of the hardest queries for state-of-the-art neural rankers by attempting to provide high-recall at the first-stage retrieval. We observe that neural rankers struggle to learn suitable representations to connect hard queries to their relevant documents. As such, we propose to learn query-like representations for documents and show that training a dense retriever on the generated alternative document representations would be more effective for connecting queries to documents that would otherwise not be matched. The experiments confirm that our proposed representation $q2q$ is able to retrieve non-overlapping relevant documents compared to the original dense retrievers. Thus, integrating the original dense retriever runs with documents retrieved based on our proposed representation can increase the recall of the first stage retriever by 5% overall on MS MARCO dev set queries and over 10% on the hardest MS MARCO queries.

References

1. Arabzadeh, N., Mitra, B., Bagheri, E.: Ms marco chameleons: challenging the ms marco leaderboard with extremely obstinate queries. In: Proceedings of the 30th ACM International Conference on Information and Knowledge Management, pp. 4426–4435 (2021)
2. Bagheri, E., Ensan, F., Al-Obeidat, F.: Impact of document representation on neural ad hoc retrieval. In: Proceedings of the 27th ACM International Conference on Information and Knowledge Management, pp. 1635–1638. CIKM 2018, Association for Computing Machinery, New York, NY, USA (2018). https://doi.org/10.1145/3269206.3269314
3. Chen, Q., Zhao, H., Li, W., Huang, P., Ou, W.: Behavior sequence transformer for e-commerce recommendation in alibaba (2019)
4. Cormack, G.V., Clarke, C.L.A., Buettcher, S.: Reciprocal rank fusion outperforms condorcet and individual rank learning methods. In: Proceedings of the 32nd International ACM SIGIR Conference on Research and Development in Information Retrieval, pp. 758–759. SIGIR 2009, Association for Computing Machinery, New York, NY, USA (2009). https://doi.org/10.1145/1571941.1572114
5. Dai, Z., Callan, J.: Context-aware sentence/passage term importance estimation for first stage retrieval. arXiv preprint arXiv:1910.10687 (2019)
6. Devlin, J., Chang, M.W., Lee, K., Toutanova, K.: Bert: pre-training of deep bidirectional transformers for language understanding. arXiv preprint arXiv:1810.04805 (2018)

7. Gallagher, L., Chen, R.C., Blanco, R., Culpepper, J.S.: Joint optimization of cascade ranking models. In: Proceedings of the Twelfth ACM International Conference on Web Search and Data Mining, pp. 15–23. WSDM 2019, Association for Computing Machinery, New York, NY, USA (2019). https://doi.org/10.1145/3289600.3290986

8. Gao, L., Dai, Z., Chen, T., Fan, Z., Van Durme, B., Callan, J.: Complementing lexical retrieval with semantic residual embedding. arXiv preprint arXiv:2004.13969 (2020)

9. Henderson, M.L., et al.: Efficient natural language response suggestion for smart reply. CoRR abs/1705.00652 (2017). http://arxiv.org/abs/1705.00652

10. Jones, K.S., Walker, S., Robertson, S.E.: A probabilistic model of information retrieval: development and comparative experiments: part 2. Inform. Process. Manage. **36**(6), 809–840 (2000)

11. Karpukhin, V., et al.: Dense passage retrieval for open-domain question answering. arXiv preprint arXiv:2004.04906 (2020)

12. Lafferty, J., Zhai, C.: Document language models, query models, and risk minimization for information retrieval. In: Proceedings of the 24th Annual International ACM SIGIR Conference on Research and Development in Information Retrieval, pp. 111–119 (2001)

13. Liu, S., Xiao, F., Ou, W., Si, L.: Cascade ranking for operational e-commerce search. In: Proceedings of the 23rd ACM SIGKDD International Conference on Knowledge Discovery and Data Mining (2017). https://doi.org/10.1145/3097983.3098011

14. MacAvaney, S., Nardini, F.M., Perego, R., Tonellotto, N., Goharian, N., Frieder, O.: Efficient document re-ranking for transformers by precomputing term representations. CoRR abs/2004.14255 (2020). https://arxiv.org/abs/2004.14255

15. Nguyen, T., et al.: Ms marco: a human generated machine reading comprehension dataset. In: CoCo@ NIPs (2016)

16. Nogueira, R., Yang, W., Cho, K., Lin, J.: Multi-stage document ranking with bert. arXiv preprint arXiv:1910.14424 (2019)

17. Nogueira, R., Yang, W., Lin, J., Cho, K.: Document expansion by query prediction. arXiv preprint arXiv:1904.08375 (2019)

18. Reimers, N., Gurevych, I.: Sentence-bert: sentence embeddings using siamese bert-networks. arXiv preprint arXiv:1908.10084 (2019)

19. Sanh, V., Debut, L., Chaumond, J., Wolf, T.: Distilbert, a distilled version of BERT: smaller, faster, cheaper and lighter. CoRR abs/1910.01108 (2019). http://arxiv.org/abs/1910.01108

20. Schütze, H., Manning, C.D., Raghavan, P.: Introduction to Information Retrieval, vol. 39. Cambridge University Press, Cambridge (2008)

21. Shehata, D., Arabzadeh, N., Clarke, C.L.: Early stage sparse retrieval with entity linking. In: Proceedings of the 31st ACM International Conference on Information and Knowledge Management, pp. 4464–4469 (2022)

22. Singhal, A., et al.: Modern information retrieval: a brief overview. IEEE Data Eng. Bull. **24**(4), 35–43 (2001)

23. Vaswani, A., et al.: Attention is all you need. CoRR abs/1706.03762 (2017). http://arxiv.org/abs/1706.03762

24. Wang, L., Lin, J.J., Metzler, D.: A cascade ranking model for efficient ranked retrieval. In: Proceedings of the 34th International ACM SIGIR Conference on Research and Development in Information Retrieval (2011)

25. Wang, W., Wei, F., Dong, L., Bao, H., Yang, N., Zhou, M.: Minilm: deep self-attention distillation for task-agnostic compression of pre-trained transformers. Adv. Neural Inform. Process. Syst. **33**, 5776–5788 (2020)

26. Wang, Z., Zhao, L., Jiang, B., Zhou, G., Zhu, X., Gai, K.: Cold: towards the next generation of pre-ranking system (2020)

27. Willett, P.: Combination of similarity rankings using data fusion. J. Chem. Inform. Model. **53**(1), 1–10 (2013)

28. Xiong, L., et al.: Approximate nearest neighbor negative contrastive learning for dense text retrieval. arXiv preprint arXiv:2007.00808 (2020)
29. Zhan, J., Mao, J., Liu, Y., Zhang, M., Ma, S.: Repbert: contextualized text embeddings for first-stage retrieval. arXiv preprint arXiv:2006.15498 (2020)
30. Zhang, H., Abualsaud, M., Ghelani, N., Smucker, M.D., Cormack, G.V., Grossman, M.R.: Effective user interaction for high-recall retrieval: Less is more. In: Proceedings of the 27th ACM International Conference on Information and Knowledge Management, pp. 187–196 (2018)

Investigating Conversational Search Behavior for Domain Exploration

Phillip Schneider[1]([✉]) [iD], Anum Afzal[1] [iD], Juraj Vladika[1] [iD], Daniel Braun[2] [iD],
and Florian Matthes[1] [iD]

[1] Technical University of Munich, Munich, Germany
{phillip.schneider,anum.afzal,juraj.vladika,matthes}@tum.de
[2] University of Twente, Enschede, Netherlands
d.braun@utwente.nl

Abstract. Conversational search has evolved as a new information retrieval paradigm, marking a shift from traditional search systems towards interactive dialogues with intelligent search agents. This change especially affects exploratory information-seeking contexts, where conversational search systems can guide the discovery of unfamiliar domains. In these scenarios, users find it often difficult to express their information goals due to insufficient background knowledge. Conversational interfaces can provide assistance by eliciting information needs and narrowing down the search space. However, due to the complexity of information-seeking behavior, the design of conversational interfaces for retrieving information remains a great challenge. Although prior work has employed user studies to empirically ground the system design, most existing studies are limited to well-defined search tasks or known domains, thus being less exploratory in nature. Therefore, we conducted a laboratory study to investigate open-ended search behavior for navigation through unknown information landscapes. The study comprised of 26 participants who were restricted in their search to a text chat interface. Based on the collected dialogue transcripts, we applied statistical analyses and process mining techniques to uncover general information-seeking patterns across five different domains. We not only identify core dialogue acts and their interrelations that enable users to discover domain knowledge, but also derive design suggestions for conversational search systems.

Keywords: Conversational interfaces · Exploratory search · Dialogue study

1 Introduction

Driven by major advances in natural language processing, the ubiquitous availability of conversational agents ushered in a new era for human-computer interfaces. In consequence, interactions between humans and machines shift towards the medium of language [8,9]. Even though modern conversational agents have a broad skill set in following task-oriented commands or engaging in short chit-chat conversations, their information-seeking capabilities are predominantly confined

© The Author(s), under exclusive license to Springer Nature Switzerland AG 2023
J. Kamps et al. (Eds.): ECIR 2023, LNCS 13981, pp. 608–616, 2023.
https://doi.org/10.1007/978-3-031-28238-6_52

to answering factoid questions. A limitation of the question-answering paradigm is its inherent dependence on the users' prior knowledge, which is the prerequisite for being able to ask meaningful questions in the first place [3,7]. Especially in exploratory search scenarios, where users with unclear information goals are confronted with unfamiliar domains, it is necessary to support search behaviors that go beyond simple query-response interactions [20]. Hence, there is a growing research interest in multi-turn conversational search systems. While some scholars approach this topic by developing theories and conceptual frameworks [1,11,14], others perform laboratory studies in combination with dialogue analysis to ground models of search behavior in empirical observations [16,18,19].

However, most existing laboratory studies focus only on experimental setups with search scenarios that are not exploratory in nature but constrained by predefined information needs and search tasks. Therefore, we designed a study for answering the research question: *What is the characteristic dialogue structure of information-seeking conversations for domain exploration?* As far as we know, our experiment is the first to collect transcripts of completely open-ended exploratory search dialogues in five domains. Our contributions are twofold: (i) We publish an annotated corpus of exploratory search dialogues with five domain datasets.[1] (ii) We identify core dialogue acts and domain-independent dialogue flow patterns which can inform the design of conversational systems.

2   Related Work

In the literature on conversational systems, dialogue analysis is common research practice. It facilitates the conceptual understanding of human behavior by means of examining communication patterns, information flows, or vocabulary choice [4,21]. Concerning information search, dialogue analysis has been applied to characterize information-seeking conversations and develop theoretical models [17]. While some researchers gather dialogue data from natural settings, such as reference interviews or online support platforms [6,10,12], others conduct controlled laboratory studies to set up an artificial search scenario, as is the case in our experiment. Vtyurina et al. [19] performed an experiment to explore users' preferences in solving search tasks with three kinds of assistants: a chatbot, a human, and a perceived automatic system simulated by a human. A study carried out by Trippas et al. [15] investigated how users communicate in an audio-only search setting. Both experiments had clearly defined search tasks and information needs assigned to the participants, which were not exploratory but goal-oriented. Another related work is from Vakulenko et al. [18], where students engaged in conversations for exploratory browsing to find a specific dataset in an open data portal. In contrast, we do not restrict the search to a predefined task, but instead, we only instructed participants to explore an unknown dataset. A further distinction is that we record multiple dialogue sessions across five domains and compare general interaction patterns of exploratory search.

---

[1] Repository: https://github.com/sebischair/conversational-domain-exploration-data.

3 Method

A total of 26 participants took part in the study, which was conducted in English. The participants were university students recruited from a teaching course. All students had previous experience with chat interfaces, but no prior knowledge of the datasets they were instructed to explore. We scraped five publicly available datasets from the internet. All of them have a relational structure, in which a set of items is characterized by a set of attributes. The tabular datasets were selected by two aspects. For one thing, they had to be licensed under Creative Commons BY-SA 3.0 or BY-SA 4.0, and for another, they had to contain enough interesting data items for an engaging conversation. Ultimately, we acquired one dataset for each of the following domains: geography, history, media, nutrition, and sports. Table 3 in the Appendix lists each of the five datasets along with a short description of their content.

The experimental setup consisted of 26 chat sessions between two participants, where one participant acted as an information seeker and the other as an information provider. Based on personal preference, every student was given one dataset in the form of a spreadsheet as an information source for the provider role. We grouped the students into pairs with two distinct datasets, ensuring mutual interest in each other's domain. Each pair was assigned to a text-based chatroom. Seekers were only instructed to explore and inquire information about the unknown dataset of their partner, but no concrete search task was specified. After one session of 15 min, students in the seeker role completed a feedback survey. It contained two free-form questions about unmet information needs as well as suggestions to improve the search experience. After completing the feedback survey, the participants switched roles and started with their second chat session regarding the other domain.

After running the experiment, the dialogue scripts were annotated. This task of dialogue act annotation identifies the function or goal of a given utterance [13]. Two researchers independently labeled each message with a speech act and corresponding dialogue act. To assess the reliability of the inter-annotator agreement, we calculated Cohen's Kappa coefficient [5]. The annotations of speech and dialogue acts had coefficients of 0.93 and 0.86, respectively. As suggested by Cohen, Kappa values from 0.81 to 1.00 indicate almost perfect agreement. To come up with a suitable group of dialogue acts, we started with an initial set derived from the widely known taxonomy of Bach and Harnish [2]. Through regular discussions, we clarified ambiguous labels and resolved disagreements between the annotators. Thereby, the set of used dialogue acts evolved through adding or removing certain acts to better fit the dialogue corpus.

For examining the annotated corpus, we calculated various descriptive statistics. Furthermore, we employed process mining techniques since they have been successfully applied to discover sequential patterns from event logs in various data formats, including conversational transcripts. We chose a Python-based state-of-the-art process mining library called PM4Py.[2]

---

[2] PM4Py process mining package: https://pm4py.fit.fraunhofer.de.

4   Results and Discussion

Statistical Analysis. We performed a statistical analysis to describe the occurrence of linguistic constituents like speech or dialogue acts. The annotated dialogue corpus contains 669 individual messages from 26 sessions. Table 1 lists the most important summary statistics. Looking at the different domains, we see that the minimum and maximum message count varies greatly, ranging from a chat with 8 messages to a more extensive chat with 62 messages. Participants exchanged on average 25.7 text messages with each other, which is significantly higher than in the more goal-directed conversational search experiment from Vakulenko et al. [18]. Considering the verbosity of the messages, the mean length of the utterances is 46.4 characters. The standard deviation of text length was unusually large for history, due to a single very long message. Besides this outlier, no irregularities in the dataset were found. It can be noted that history and sports are the domains with not only the lowest message count and the smallest average of messages per session, but they also have the shortest messages on average when excluding the outlier in the history domain.

Table 1. Summary statistics of dialogue corpus.

Aspect	Geography	History	Media	Nutrition	Sports	Overall
Number of messages	169	98	156	153	93	669
Number of sessions	6	5	5	5	5	26
Min-max messages per session	11–41	8–29	20–62	21–45	16–22	8–62
Average messages per session	28.2	19.6	31.2	30.6	18.6	25.7
Average characters per message	47.3	48.5	45.0	49.0	40.9	46.4

More insights about the linguistic elements of the dialogue transcripts can be gained by comparing the distribution of speech acts. As proposed by Bach and Harnish [2], there are four groups of illocutionary speech acts: Constatives express an intention to convey information. Directives express the intention to get the addressee to do something. Commissives are acts of obligating oneself to do something.

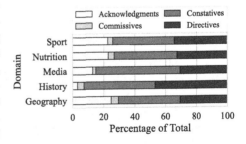

Fig. 1. Distribution of speech acts.

Acknowledgments express mutual understanding or attitudes that are expected on particular occasions. Figure 1 illustrates the distribution of observed speech acts across the five domains. In total, constatives (44.2%) and directives (34.1%) are most predominant, accounting for over three-quarters of all speech acts. Acknowledgments and commissives make up 18.1% and 3.6%, respectively. Considering all domains, constatives have a higher occurrence than directives, followed by acknowledgments and commissives. The history sessions deviate from this rule since the ratios of constatives and directives are almost equal and there are only very few acknowledgments.

Table 2. Distribution of dialogue acts sorted by relative frequency.

Speech act	Dialogue act	Percentage	Definition
Directives	Request	32.0%	Express a general information need
Constatives	Describe	19.3%	Provide a description of an information item
Acknowledgments	Acknowledge	8.1%	Express that an utterance was understood
Constatives	List	7.0%	List multiple information items from the data
Constatives	Rank	6.7%	Rank information items by a given metric
Acknowledgments	Greet	5.8%	Open the conversation with an initial greeting
Constatives	Count	4.6%	Count the number of a specified set of items
Acknowledgments	Thank	4.2%	Express gratitude with regard to a response
Constatives	Verify	3.7%	Verify if the dataset contains a specific item
Commissives	Offer	3.0%	Offer options for information exploration
Constatives	Accept	2.4%	Agree with a suggested exploration direction
Directives	Clarify	2.1%	Ask a clarifying question for a given utterance
Commissives	Promise	0.6%	Promise to perform a requested action
Constatives	Reject	0.4%	Disagree with a suggested exploration direction

Dialogue acts are specialized speech acts that depend on the conversation setting. Thus, they allow for more granular linguistic analysis. We present an overview of all identified dialogue acts in Table 2. Expressing an information request is the most frequent act, which accounts for every third utterance, whereas describing a data item ranks second with 19.3%. Together, these two dialogue acts represent already half of all interactions. Other information-providing dialogue acts, such as listing, ranking, verifying, or counting, claim a significant share as well. It is also noteworthy that after a suggested exploration direction, seekers acted receptive, accepting over five times more often than rejecting. With regard to the speech act of acknowledgments, Table 2 shows that acknowledging and thanking are important functions for communicating feedback. As can be seen from the relatively few offer (3.0%) and promise (0.6%) dialogue acts, the participants used commissives rarely. The same holds true for asking clarification questions. Overall, the vast majority of utterances serve the functions of requesting information, providing information, and giving feedback.

Process Mining Analysis. Aside from identifying dialogue acts, we applied an inductive miner algorithm to discover the underlying core process for exploratory search. Figure 2 depicts the extracted domain-independent dialogue sequence flow which manifested itself in all five domains. The nodes correspond to the ten most frequent dialogue acts. The arrows show the direction of the conversation flow. Lastly, there are diamond-shaped gateway symbols, an exclusive gateway (X) breaks the flow into mutually exclusive flows, and an inclusive gateway (+) represents concurrent flows. Concurrency means that the order of dialogue acts varied between the analyzed conversation transcripts. From Figure 2, we can discern that the dialogue logs consist of three concurrent main loops. One is for requesting information, a second is for describing information items, and a third incorporates a sequence of the remaining dialogue acts. The two former loops are indispensable since they are used at least once in every conversation, whereas the third loop allows for skipping specific dialogue acts.

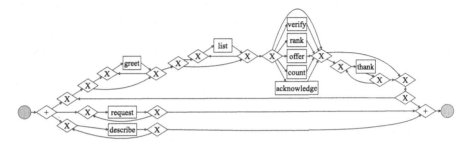

Fig. 2. Process model of extracted dialogue sequence flow.

Considering the dialogue transcripts across all domains, we observed that most chat sessions started with a greeting and an ensuing question from the information seeker. This request usually came in the form of asking what the dataset is about in general, followed by asking about its different dimensions. The information provider tried to fulfill these requests by listing column names or exemplary data records. On that basis, the seeker proceeded with asking more detailed questions, often about aggregated, or sorted data items, which relate to count and rank dialogue acts. In the related study from Vakulenko et al. [18], similar dialogue acts for browsing through relational data structures were identified. These acts help information seekers to better understand the scope of the dataset by getting to know the existing attributes along with their facets. When the provider had trouble finding an answer, the provider sometimes offered alternative options. The seeker, in turn, gave feedback by acknowledging utterances or expressing gratitude for relevant responses. The participants commonly ended their chats by thanking each other.

Discussion of Design Suggestions. Our analysis demonstrated the key role of certain dialogue acts in domain exploration scenarios. Effective search agents must be able to emulate these communication functions and handle the dialogue flow outlined hereinafter. Concerning the interplay between constatives and directives, we found that the observed dialogues are seeker-driven with every third dialogue act being a request. Information providers addressed these requests by first listing or describing metadata, along with counting, ranking, and verifying information items. This iterative interaction process supported seekers in building a mental model of the dataset's dimensionality, which was essential for asking more complex follow-up questions. This is in line with Belkin's Anomalous State of Knowledge model (ASK) [3] because it is only possible for seekers to formulate an information need if they have first perceived a knowledge gap. Providers were guided in their informative responses through the seekers' acknowledgments.

A similar picture emerged from the feedback surveys of the participants. Overall, they had a positive impression of their chat sessions, although when asked about suggestions for improvement, they pointed towards both the chat interface and the dialogue interaction. For instance, they wanted visual data summaries in the chat window, clickable buttons with predefined replies, or timestamped messages. Regarding the interaction, they demanded a short introduction of the dataset, i.e., metadata, right at the start of the session. This prior knowledge was an essential requirement for engaging in a deeper exploration of the domain. Also, they criticized one-word answers, wishing for more descriptive responses and question suggestions from the provider. Some of these user preferences were also observed by Vtyurina et al. [19]. In consequence, it seems evident that exploratory conversational search depends on a proactive search agent, which can quickly familiarize the user with the explored domain.

5   Conclusion and Future Work

In this paper, we presented a laboratory study designed to investigate conversational search behavior in the context of discovering unfamiliar domains. Our empirical analysis not only revealed core dialogue acts that information seekers use, but also a domain-independent dialogue act flow sequence. In addition, we believe that our derived design suggestions are vital for a user-centered design of conversational agents. Two major limitations of our study arise due to having a biased participant sample of only university students and focusing exclusively on tabular datasets. In future work, we plan to use our insights to build and evaluate exploratory search agents for more diverse user groups with the capability to retrieve information from different data sources and structures.

Acknowledgments. This work has been supported by the German Federal Ministry of Education and Research (BMBF) Software Campus grant 01IS17049.

A   Appendix

Table 3. Overview of the five domain datasets used in the experiment.

Domain	# Rows	# Columns	Short description
Geography	98	5	Geographic information about nature parks
History	11341	17	Biographic data about historical figures
Media	500	5	Data about time travel literature, films, and TV series
Nutrition	285	9	Nutritional values of common food products
Sports	77	6	Data about international football records

References

1. Azzopardi, L., Dubiel, M., Halvey, M., Dalton, J.: Conceptualizing agent-human interactions during the conversational search process. In: The Second International Workshop on Conversational Approaches to Information Retrieval (2018)
2. Bach, K., Harnish, R.: Linguistic Communication and Speech Acts. MIT Press (1979). https://doi.org/10.2307/2184680
3. Belkin, N.J.: Anomalous states of knowledge as a basis for information retrieval. Can. J. Inform. Sci. **5**, 133–143 (1980)
4. Bunt, H.: Dynamic interpretation and dialogue theory. Struct. Multimodal Dial. **2**, 139–166 (2000). https://doi.org/10.1075/z.99.10bun
5. Cohen, J.: A coefficient of agreement for nominal scales. Educ. Psychol. Measure. **20**(1), 37–46 (1960). https://doi.org/10.1177/001316446002000104
6. Daniels, P.J., Brooks, H.M., Belkin, N.J.: Using problem structures for driving human-computer dialogues. In: Recherche d'Informations Assistée par Ordinateur, pp. 645–660. Centre de hautes études internationales d'informatique documentaire (1985). https://doi.org/10.5555/3157680.3157726
7. Furnas, G.W., Landauer, T.K., Gomez, L.M., Dumais, S.T.: The vocabulary problem in human-system communication. Commun. ACM **30**(11), 964–971 (1987). https://doi.org/10.1145/32206.32212
8. Klopfenstein, L.C., Delpriori, S., Malatini, S., Bogliolo, A.: The rise of bots: a survey of conversational interfaces, patterns, and paradigms. In: Proceedings of the 2017 Conference on Designing Interactive Systems, pp. 555–565 (2017). https://doi.org/10.1145/3064663.3064672
9. The Conversational Interface. Springer, Cham (2016). https://doi.org/10.1007/978-3-319-32967-3
10. Qu, C., Yang, L., Croft, W.B., Trippas, J.R., Zhang, Y., Qiu, M.: Analyzing and characterizing user intent in information-seeking conversations. In: The 41st International ACM Sigir Conference On Research and Development In Information Retrieval, pp. 989–992 (2018). https://doi.org/10.48550/arXiv.1804.08759
11. Radlinski, F., Craswell, N.: A theoretical framework for conversational search. In: Proceedings of the 2017 Conference on Conference Human Information Interaction and Retrieval, pp. 117–126 (2017). https://doi.org/10.1145/3020165.3020183
12. Saracevic, T., Spink, A., Wu, M.M.: Users and intermediaries in information retrieval: what are they talking about? In: Jameson, A., Paris, C., Tasso, C. (eds.) User Modeling. CISM, vol. 383, pp. 43–54. Springer (1997). https://doi.org/10.1007/978-3-7091-2670-7_6
13. Sinclair, J.M., Sinclair, J.M., Coulthard, M., et al.: Towards an analysis of discourse: The English Used By Teachers and Pupils. Oxford University Press, USA (1975). https://doi.org/10.2307/3585455
14. Thomas, P., Czerwinksi, M., McDuff, D., Craswell, N.: Theories of conversation for conversational IR. ACM Trans. Inform. Syst. **39**(4), 1–23 (2021). https://doi.org/10.1145/3439869
15. Trippas, J.R., Spina, D., Cavedon, L., Joho, H., Sanderson, M.: Informing the design of spoken conversational search: perspective paper. In: Proceedings of the 2018 Conference On Human Information Interaction & Retrieval, pp. 32–41 (2018). https://doi.org/10.1145/3176349.3176387
16. Trippas, J.R., Spina, D., Thomas, P., Sanderson, M., Joho, H., Cavedon, L.: Towards a model for spoken conversational search. Inform. Process. Manage. **57**(2), 102162 (2020). https://doi.org/10.1016/j.ipm.2019.102162

17. Vakulenko, S., Kanoulas, E., De Rijke, M.: A large-scale analysis of mixed initiative in information-seeking dialogues for conversational search. ACM Trans. Inform. Syst. **39**(4), 1–32 (2021). https://doi.org/10.48550/ARXIV.2104.07096
18. Vakulenko, S., Savenkov, V., de Rijke, M.: Conversational browsing. arXiv preprint arXiv:2012.03704 (2020). https://doi.org/10.48550/arXiv.2012.03704
19. Vtyurina, A., Savenkov, D., Agichtein, E., Clarke, C.L.: Exploring conversational search with humans, assistants, and wizards. In: Proceedings of the 2017 Chi Conference Extended Abstracts on Human Factors in Computing Systems, pp. 2187–2193 (2017). https://doi.org/10.1145/3027063.3053175
20. White, R.W., Roth, R.A.: Exploratory search: beyond the query-response paradigm. Synth. Lect. Inform. Concepts Retrieval Serv. **1**(1), 1–98 (2009). https://doi.org/10.2200/S00174ED1V01Y200901ICR003
21. Yankelovich, N.: Using natural dialogs as the basis for speech interface design. In: Human Factors and Voice Interactive Systems, pp. 255–290. Springer, Boston (2008). https://doi.org/10.1007/978-0-387-68439-0_9

Evaluating Humorous Response Generation to Playful Shopping Requests

Natalie Shapira[1(✉)], Oren Kalinsky[2], Alex Libov[2], Chen Shani[3],
and Sofia Tolmach[2]

[1] Bar-Ilan University, Ramat Gan, Israel
nd1234@gmail.com
[2] Amazon Science, Tel Aviv, Israel
{orenk,alibov,sofiato}@amazon.com
[3] The Hebrew University of Jerusalem, Jerusalem, Israel
Chen.shani@mail.huji.ac.il

Abstract. AI assistants are gradually becoming embedded in our lives, utilized for everyday tasks like shopping or music. In addition to the everyday utilization of AI assistants, many users engage them with playful shopping requests, gauging their ability to understand – or simply seeking amusement. However, these requests are often not being responded to in the same playful manner, causing dissatisfaction and even trust issues.

In this work, we focus on equipping AI assistants with the ability to respond in a playful manner to irrational shopping requests. We first evaluate several neural generation models, which lead to unsuitable results – showing that this task is non-trivial. We devise a simple, yet effective, solution, that utilizes a knowledge graph to generate template-based responses grounded with commonsense. While the commonsense-aware solution is slightly less diverse than the generative models, it provides better responses to playful requests. This emphasizes the gap in commonsense exhibited by neural language models.

User: Buy me the moon!
Assistant:(Default) I'm sorry, you can't buy that.
        (Desired) We only sell the moon when it is blue.

Fig. 1. Task illustrative example. The user playfully asks the AI assistant for a non-shoppable item. The goal is to provide a better response than the default.

1 Introduction

AI assistants such as Amazon's Alexa, Apple's Siri, Google Assistant, and Microsoft's Cortana are becoming increasingly popular. Users commonly expect the assistants to support a wide range of human capabilities, sometimes beyond their original intended tasks – such as carrying a conversation or responding to humor.

N. Shapira and C. Shani—Work was done during an internship at Amazon.
Except for the first author, the rest of the authors follow the ABC of surnames.

J. Kamps et al. (Eds.): ECIR 2023, LNCS 13981, pp. 617–626, 2023.
https://doi.org/10.1007/978-3-031-28238-6_53

There has been a recent trend in the field of artificial intelligence toward addressing challenges related to social skills and commonsense [6, 8, 19, 21, 26, 33], including the challenging area of generative commonsense [10].

We focus here on the shopping scenario, where users playfully ask to purchase non-shoppable items (e.g., **"Buy me the moon"**). Recently, [22] analyzed Alexa traffic, showing this is a prevalent use-case. Moreover, they sketched a detection approach, leaving the problem of appropriate and scalable response generation as an open question. Responding to playful requests falls under the wider field of computational humor, considered an AI-complete problem [24] (Fig. 1).

Contemporary AI assistants rely on hand-curated responses (**"We only sell the moon when it is blue"**), which are hard to scale. Moreover, they provide a stopgap solution only for a handful of requests, missing many playful opportunities. For any detected playful request with no prepared response, a catch-all answer such as **"I'm sorry, you can't buy that"** can be applied. However, we strive to improve over this laconic, non-playful response.

To provide a scalable and suitable response mechanism, we first define the task of generating responses to playful shopping requests (see Sect. 3). We explore different methods for *automatically generating* satisfactory responses (Sect. 4). Motivated by neural-models' poor performance, we devise a simple knowledge graph and template based solution (Sect. 5). We hypothesize that incorporating commonsense will positively surprise users, as they do not expect it from a computer. Indeed, the commonsense-aware approach provides more suitable responses to playful requests. This is yet another example of the gap in commonsense that neural models exhibit, emphasizing its importance [19–21].

Our main contributions are: we 1) define a novel humor generation task of responding to playful shopping requests; 2) evaluate modern generative LMs on this task; and 3) devise a simple, yet effective, approach that leverages commonsense through templates and achieves state-of-the-art results.

We release our data to facilitate research in the challenging field of humor generation, and in particular on this novel task[1].

2   Related Work

In line with a recent review [1], we distinguish between two humor generation approaches, templates and neural networks.

Template-based systems often rely on external knowledge such as corpora or knowledge graphs [3, 15, 25, 29]. Their Achilles heel is low diversity, as they produce similar content with almost the same wording or repetitive humor mechanism.

Only a few humor-generation systems utilized neural-based solutions [12, 18, 32]. None of them tackled a question-answering scenario or involved a zero-shot setting with prompt engineering, as we present here. We note that although neural-based methods have achieved state-of-the-art results in many NLP tasks [5], generating *humorous* text is still in its infancy.

---

[1] https://registry.opendata.aws/shopping-humor-generation/.

As for evaluation, a standardized methodology is still missing. The majority of studies used human judges to rate the generated texts on a 1–5 Likert scale. [28] proposed to measure the frequency of humorous outputs in the system's output (using a threshold on the numeric scores). [2] suggested a humor variation of the Turing test, in which an automatically generated pun is considered humorous if annotators are unable to tell it apart from human-generated puns.

We note that the evaluation effort focused on the quality of each generated output on its own, neglecting the systems' output diversity (ability to generate a variety of humorous texts). One notable example is the automated diversity, measured using the ratio of distinct uni- and bi-grams in the output [32]. More recently, [1] suggested measuring it in terms of syntactical and lexical features, as well as the joking mechanism.

Due to the highly subjective and individual nature of humor, previous research shows significant disagreement between different annotators [4,15,24, 31] - a finding that we reproduce here.

3 Problem Definition

We define a *non-shoppable item* as any object, entity, or concept that is impossible to purchase online via an AI assistant (e.g., love, brain, galaxy). We assume a detector exists and focus on *response generation*.

Given a *non-shoppable item*, we wish to automatically generate an *appropriate* response, such that: 1) the response expresses the assistant's understanding that the item is impossible to purchase, 2) it is sensible within the conversation's playful context, and 3) it has correct grammar. A *playful* response will also be humorous. The generic "I'm sorry, you can't buy that" is the *minimal* appropriate response, as it is very broad, laconic, and not playful (but satisfies all the requirements of an appropriate response).

While this definition refers to a single response, we extend the problem to also address diversity, which is the ability of the system to produce a variety of outputs for a given input.

4 Methods

In line with the division of [1], we explored two approaches towards automatically generating appropriate responses to non-shoppable requests: 1) off the shelf Generative Language Models (GLM), and 2) combining hand-crafted templates with commonsense knowledge graphs.

4.1 Generative Language Models (GLMs)

Large Language Models (LMs), neural-based, are widely used for text generation, where the input prompts affect their output drastically [11]. We employ T5-3B [17] and GPT-2 [16] in a zero-shot learning setup.

As input to these models, we curated general freestyle shopping-oriented response prompts (e.g., "It is impossible to buy *<non-shoppable-item>* because"). Additionally, we constructed commonsense-aware prompts based on the relations in ConceptNet, a commonsense knowledge graph [23]. This was done in an attempt to inject some commonsense-based wit into the neural LMs. For example, we constructed the prompt "You want to buy <non-shoppable-item>? Let me check if I have it next to".[2]

Fig. 2. ConceptNet template example.

4.2   Knowledge-Graph Templates

We chose 30 relations from ConceptNet and manually created two templates – one for each edge direction. For example, consider the AtLocation relation: When the *non-shoppable-item* is in the start we produced *"Just a second, I'm going to <connected-item> to get it for you"*, while when the *non-shoppable-item* is at the end, the template is *"Sorry, I'm all out. Maybe just <connected-item>...?"* (See Fig. 2). We also employed additional filtering steps removing concepts that are too similar to the non-shoppable item, or that are problematic (using age of acquisition score, readability score, and profanities filters).[3]

5   Evaluation

We first set out to create a dataset of non-shoppable items.[4] We used Amazon Mechanical Turk (MTurk) to generate a list of 100 non-shoppable items[5]. Specifically, we requested workers to be playful and write items that are impossible to buy via an AI assistant. All items were manually approved by our team. We denote this set as 100-NSI.

---

[2] The full list is included in the code repository. T5 had 95 prompts, and GPT-2 had 89 (the prompts that were suffix-based are irrelevant to GPT-2 that attends to the prefix. Top-K = 50, Top-P = 0.95, Beam width = 10, Max length GPT-2 = 50 T5-3B = 20.

[3] The full list of relations, templates, and filtering logic is included in the code repository.

[4] The dataset of non-shoppable items and responses are included in the code repository.

[5] Workers were paid 5 cents per generated non-shoppable item.

5.1 Response Sets

We used 100-NSI to automatically construct responses using the methods presented in Sect. 4.

1. **ConceptNet Templates:** For each 100-NSI item, we automatically generated all possible responses (4,198 in total). We then randomly chose 1,000 responses for further evaluation.
2. **T5:** For each 100-NSI item, we automatically generated up to 10 responses per prompt out of the 95 prompts (43,056 in total). We then randomly chose 10 random responses for each item, ending with 1,000 responses.
3. **GPT2:** For each 100-NSI item, we automatically generated up to 10 responses per prompt out of the 89 prompts (78,913 in total). We then randomly chose 10 random responses for each item, ending with 1,000 responses.
4. **Hand-Crafted:** Hand-curated responses created by a team of experts from a major commercial voice AI assistant. A total of 33 responses for 24 items.
5. **Random:** We constructed a baseline of arbitrary 2–20 words sentences that contain the non-shoppable items. Sentences extracted from C4 corpus [17].

Table 1. Evaluation of the different response generation methods. The "Better than Default" column denotes the percentage of responses that were classified as better than the minimal appropriate response ("I'm sorry, you can't buy that"). Diversity classified according to [1]. Automated diversity is measured using the ratio of distinct unigrams and bigrams divided by the total number of words in the whole set, according to [32]. $N = 1000$ for all methods except for humans ($N = 33$). Results show that while the gap to human performance is still quite large, the commonsense-based templates outperform neural GLMs. However, it lacks diversity.

Method	Better than default	Diversity	Auto. diversity (Unigram)	Auto. diversity (Bigram)	Dispute
Hand-crafted	78.79%	3	57.83%	90.12%	60.61%
ConceptNet templates	**38.60%**	2	**16.47%**	37.34%	45.40%
T5-3B	15.30%	**2.5**	9.76%	27.59%	27.40%
GPT-2	9.20%	**2.5**	**16.47%**	**49.54%**	28.10%
Random	2.70%	3	32.17%	79.57%	14.40%

5.2 Annotation Task

We used MTurk to evaluate the quality of generated responses. The task placed the worker as a user initiating a non-shoppable request taken from 100-NSI (e.g., "I want to buy sleep") and the AI assistant's generated response (e.g., "Oh, you want to sleep? maybe you need to go to bed", generated by ConceptNet templates).

We then asked workers to evaluate the response (three workers per response). Due to the subjectivity of the task, we added an anchor – the rating was compared to the generic "I'm sorry, you can't buy that" response, which is the *minimal* appropriate response according to our definition (see Sect. 3).[6] Thus, we asked workers whether the responses they received is *a better experience than* "I'm sorry, you can't buy that" on the following Likert scale: really bad (−2), bad (−1), same experience (0), good (1), very good (2). The three ratings for each response were then averaged.

We note that this annotation is far from trivial due to: 1) its subjective nature, and 2) the difficulty in assessing generative output [27]. To account for these caveats we deployed several methods to raise the quality of annotation such as qualification tests and planted test questions with bad responses (See Sect. 6).

5.3   Results and Discussion

Results can be seen in Table 1. While all automatic generation methods are still far from the human baseline, they exceed the random baseline, proving they can generate appropriate responses. The template-based approach outperforms generative LM methods: 38.6% of ConceptNet template responses rated better than "I'm sorry, you can't buy that", utterly beating the generative models that reached only 15.3% and 9.2% (T5 and GPT-2, respectively). Interestingly, 14.4% of ConceptNet template responses were better scored than average human responses (compared to only 3.6% and 2.1% for T5 and GPT-2 respectively).

Diversity. To avoid distortion originated by a system having only a handful of responses, we also measure diversity. We use the measures proposed by [32] to compare the different methods' diversity (see Table 1). Results show that the GLM approach is more diverse than the templates one. Moreover, the gap to human-curated responses is still quite large.

A Note About Task's Difficulty. We used a 5-point Likert scale. A response is classified as in dispute if at least one of the three annotators rated it as positive ($\{1, 2\}$) and one as negative ($\{-1, -2\}$). In line with previous works, we see high disagreement (24%). Interestingly, we found a highly significant positive correlation between the response's mean score (on the Likert scale) and disagreement ($R = 0.48$, p-value < 0.001). Meaning, bad responses are easy to agree upon, whereas good responses were more likely to be in dispute. While such an analysis was not done before, we hypothesize this is a general finding, since responses can often be objectively bad, while a good experience is often subjective and personal-taste dependent.

---

[6] Preliminary experiments showed that annotators tended to rank responses with a discourse issue as worse than the baseline response (−1/−2).

Table 2 presents the top and bottom-ranked samples from the user study. On one hand, there are cases where both ConceptNet templates and T5-3B generate great responses. Yet, some responses appear to be nonsense due to issues in the knowledge graph or poor performance of the models which were not tuned on this task. Finally, the table presents two cases of a strong dispute. It seems that in these cases, the AI assistant appears cynical or rude, and we expect this type of humor to be highly subjective.

To conclude, evaluating responses to playful shopping requests is not a trivial task. However, when treated carefully, it is possible to achieve meaningful and insightful results. Our experiment shows that unlike some other natural language generation tasks, in which Generative LMs can outperform any other approach, here they still fall short. While template-based approaches show improved results, they are still far from human performance. Moreover, they are slightly less diverse compared to LMs. Further analysis shows that it is easier for annotators to agree in the bad-response region compared to the good-response region. The proposed method can be applied to a specific problem and may be relevant to other similar problems, but it does not provide a universal solution for humorous problems.

6   Ethical Considerations

We identify two ethical concerns – crowd worker wages and hurtful response generation.

Crowd Worker Wages. Payment for crowd workers was above the minimum wage in the US ($0.03 for a median work time of 7 seconds per hit; $15.43 per hour). Qualification-test included a warning about the rejection policy and candidates confirmed that they understood and agreed. The planted test questions were simple and clearly unambiguous, as five judges had 100% agreement and no concern someone else would disagree. For example, the sentence *"and leagues create custom wish Gamers also select participate. Own to may their cups and teams they the"* is clearly broken and expected to be rated as a "very bad" experience (we allowed for "bad" as well). In addition, we used a large safety margin (at least four errors) before auto-rejecting. Some edge cases were manually reassessed by looking at the errors, error rate (total hits) and a sample of other hits. Overall, we rejected 5/75 workers (18.2% of the hits).

Hurtful Response Generation. T5 and GPT-2 can be potentially dangerous in manifesting undesirable biases and representational harm [7,9,14,30]. ConceptNet, which is mostly generated by humans, may also reflect social biases [13]. User safety should be kept in mind before applying those methods in production, providing safety measures against responses proposing harmful actions. For example, the "sleep" request response: "Oh, you want to sleep? Maybe you need alcohol", or another example based on ConceptNet is the "happiness" request response: "Oh, you want happiness? Maybe you need to look better.".

Table 2. Examples of responses generated by the different methods, presenting the top-ranked and bottom ranked in the human evaluation, as well as a few with a strong dispute.

Experience	User	AI assistant's response	Method
Good	Order space	I have to warn you, it has zero pressure	ConceptNet
	Do you sell a street	Would you like it to be one-way or two-way?	
	Buy me the moon	Sorry, I'm all out. Maybe just the stars...?	T5-3B
	Do you sell laugh?	We have no jokes left in our stock	
Bad	Buy a smile	Are you motivated by heather told a joje?	ConceptNet
	Buy me a governor	Hmm... no, they wont help you, they could not a governor	T5-3B
	Shop for budget	Do you have a great camera?	GPT-2
	Do you sell a farm?	I'll grab some iced tea for you on Friday	
Dispute	Can I buy laziness?	Check with your local community service center	GPT-2
	Can I buy a thought?	So I guess you're not really into art	

7   Conclusion

In this work, we focus on equipping AI assistants with the ability to *respond* in a playful manner to irrational shopping questions. We compare between neural Generative LMs and a simple, yet effective, template-based solution, grounded with commonsense. Results show that generating appropriate responses is a non-trivial task – both in terms of the generation itself and in objectively measuring the output. Interestingly, the template-based approach outperforms state-of-the-art Generative LMs, providing further evidence on the importance of commonsense in constrained text generation. However, we note that the gap to human-expert responses is still quite large. Inspired with previous works showing that integrating real-world knowledge graphs can improve over the vanilla LM approach, an interesting line of research would be to constrain and guide LM generation by commonsense knowledge graphs.

References

1. Amin, M., Burghardt, M.: A survey on approaches to computational humor generation. In: Proceedings of the The 4th Joint SIGHUM Workshop on Computational Linguistics for Cultural Heritage, Social Sciences, Humanities and Literature, pp. 29–41 (2020)
2. Binsted, K., Ritchie, G.: Computational rules for generating punning riddles (1997)

3. Dybala, P., Ptaszynski, M., Higuchi, S., Rzepka, R., Araki, K.: Humor prevails!-implementing a joke generator into a conversational system. In: Wobcke, W., Zhang, M. (eds.) Australasian Joint Conference on Artificial Intelligence. LNCS, vol. 5360, pp. 214–225. Springer, Cham (2008). https://doi.org/10.1007/978-3-540-89378-3_21

4. Goldberg, K., Roeder, T., Gupta, D., Perkins, C.: Eigentaste: a constant time collaborative filtering algorithm. Inform. Retriev. **4**(2), 133–151 (2001)

5. Goldberg, Y.: Neural network methods for natural language processing. Synth. Lect. Hum. Lang. Technol. **10**(1), 1–309 (2017)

6. Hessel, J., et al.: Do androids laugh at electric sheep? humor "understanding" benchmarks from the new yorker caption contest. arXiv preprint arXiv:2209.06293 (2022)

7. Kirk, H.R., et al.: Bias out-of-the-box: an empirical analysis of intersectional occupational biases in popular generative language models. Adv. Neural Inform. Process. Syst. **34** (2021)

8. Le, M., Boureau, Y.L., Nickel, M.: Revisiting the evaluation of theory of mind through question answering. In: Proceedings of the 2019 Conference on Empirical Methods in Natural Language Processing and the 9th International Joint Conference on Natural Language Processing (EMNLP-IJCNLP), pp. 5872–5877 (2019)

9. Liang, P.P., Wu, C., Morency, L.P., Salakhutdinov, R.: Towards understanding and mitigating social biases in language models. In: International Conference on Machine Learning, pp. 6565–6576. PMLR (2021)

10. Lin, B.Y., et al.: Commongen: A constrained text generation challenge for generative commonsense reasoning. arXiv preprint arXiv:1911.03705 (2019)

11. Liu, P., Yuan, W., Fu, J., Jiang, Z., Hayashi, H., Neubig, G.: Pre-train, prompt, and predict: a systematic survey of prompting methods in natural language processing. arXiv preprint arXiv:2107.13586 (2021)

12. Luo, F., Li, S., Yang, P., Chang, B., Sui, Z., Sun, X., et al.: Pun-gan: generative adversarial network for pun generation. arXiv preprint arXiv:1910.10950 (2019)

13. Mehrabi, N., Zhou, P., Morstatter, F., Pujara, J., Ren, X., Galstyan, A.: Lawyers are dishonest? quantifying representational harms in commonsense knowledge resources. arXiv preprint arXiv:2103.11320 (2021)

14. Nadeem, M., Bethke, A., Reddy, S.: Stereoset: Measuring stereotypical bias in pretrained language models. arXiv preprint arXiv:2004.09456 (2020)

15. Petrovic, S., Matthews, D.: Unsupervised joke generation from big data. In: ACL (2), pp. 228–232 (2013)

16. Radford, A., Wu, J., Child, R., Luan, D., Amodei, D., Sutskever, I., et al.: Language models are unsupervised multitask learners. OpenAI blog **1**(8), 9 (2019)

17. Raffel, C., et al.: Exploring the limits of transfer learning with a unified text-to-text transformer. arXiv preprint arXiv:1910.10683 (2019)

18. Ren, H., Yang, Q.: Neural joke generation. Final Project Reports of Course CS224n (2017)

19. Sakaguchi, K., Bras, R.L., Bhagavatula, C., Choi, Y.: Winogrande: an adversarial winograd schema challenge at scale. Commun. ACM **64**(9), 99–106 (2021)

20. Sap, M., LeBras, R., Fried, D., Choi, Y.: Neural theory-of-mind? on the limits of social intelligence in large LMS. arXiv preprint arXiv:2210.13312 (2022)

21. Sap, M., Rashkin, H., Chen, D., LeBras, R., Choi, Y.: Socialiqa: commonsense reasoning about social interactions. arXiv preprint arXiv:1904.09728 (2019)

22. Shani, C., Libov, A., Tolmach, S., Lewin-Eytan, L., Maarek, Y., Shahaf, D.: "alexa, what do you do for fun?" characterizing playful requests with virtual assistants. arXiv preprint arXiv:2105.05571 (2021)

23. Speer, R., Chin, J., Havasi, C.: Conceptnet 5.5: an open multilingual graph of general knowledge. In: Thirty-first AAAI Conference on Artificial Intelligence (2017)
24. Stock, O., Strapparava, C.: Hahacronym: humorous agents for humorous acronyms (2003)
25. Stock, O., Strapparava, C.: Hahacronym: a computational humor system. In: Proceedings of the ACL Interactive Poster and Demonstration Sessions, pp. 113–116 (2005)
26. Talmor, A., et al.: Commonsenseqa 2.0: Exposing the limits of AI through gamification. arXiv preprint arXiv:2201.05320 (2022)
27. Tevet, G., Habib, G., Shwartz, V., Berant, J.: Evaluating text gans as language models. arXiv preprint arXiv:1810.12686 (2018)
28. Valitutti, A.: How many jokes are really funny? In: Human-Machine Interaction in Translation: Proceedings of the 8th International NLPCS Workshop, vol. 41, p. 189. Samfundslitteratur (2011)
29. Valitutti, A., Doucet, A., Toivanen, J.M., Toivonen, H.: Computational generation and dissection of lexical replacement humor. Natl. Lang. Eng. **22**(5), 727–749 (2016)
30. Weidinger, L., et al.: Ethical and social risks of harm from language models. arXiv preprint arXiv:2112.04359 (2021)
31. Winters, T., Nys, V., Schreye, D.D.: Automatic joke generation: learning humor from examples. In: Streitz, N., Konomi, S. (eds.) International Conference on Distributed, Ambient, and Pervasive Interactions. LNCS, vol. 10922, pp. 360–377. Springer, Cham (2018). https://doi.org/10.1007/978-3-319-91131-1_28
32. Yu, Z., Tan, J., Wan, X.: A neural approach to pun generation. In: Proceedings of the 56th Annual Meeting of the Association for Computational Linguistics (Volume 1: Long Papers), pp. 1650–1660 (2018)
33. Zellers, R., Holtzman, A., Bisk, Y., Farhadi, A., Choi, Y.: Hellaswag: can a machine really finish your sentence? arXiv preprint arXiv:1905.07830 (2019)

Joint Span Segmentation and Rhetorical Role Labeling with Data Augmentation for Legal Documents

T. Y. S. S Santosh[(✉)], Philipp Bock, and Matthias Grabmair

School of Computation, Information, and Technology,
Technical University of Munich, Munich, Germany
{santosh.tokala,philipp.bock,matthias.grabmair}@tum.de

Abstract. Segmentation and Rhetorical Role Labeling of legal judgements play a crucial role in retrieval and adjacent tasks, including case summarization, semantic search, argument mining etc. Previous approaches have formulated this task either as independent classification or sequence labeling of sentences. In this work, we reformulate the task at span level as identifying spans of multiple consecutive sentences that share the same rhetorical role label to be assigned via classification. We employ semi-Markov Conditional Random Fields (CRF) to jointly learn span segmentation and span label assignment. We further explore three data augmentation strategies to mitigate the data scarcity in the specialized domain of law where individual documents tend to be very long and annotation cost is high. Our experiments demonstrate improvement of span-level prediction metrics with a semi-Markov CRF model over a CRF baseline. This benefit is contingent on the presence of multi sentence spans in the document.

Keywords: Rhetorical Role Labeling · Semi-Markov CRF · Data augmentation

1 Introduction

Rhetorical Role Labeling (RRL) of legal documents involves segmenting a document into semantically coherent chunks and assigning a label to the chunk that reflects its function in the legal discourse (e.g., preamble, fact, evidence, reasoning). RRL for long legal case documents is a precursor task to several downstream tasks, such as case summarization [5,9,12,22], fact-based semantic case search [21], argument mining [25] and judgement prediction [12].

Prior works in RRL on legal judgements have regarded the task either as straightforward classification of sentences without modeling any contextual dependency between them [1,25] or as sequence labeling [3,8,12,27]. Initial works [5,9,22] performed RRL using hand-crafted features as part of a summarization pipeline. Savelka et al. [24] employed a CRF on hand-crafted features to segment US court decisions into functional and issue specific parts. Similarly, Walker et al. [25] used engineered features for RRL on US Board of Veterans' Appeals (BVA) decisions. With the rise of deep learning, Yamada et al. [27],

J. Kamps et al. (Eds.): ECIR 2023, LNCS 13981, pp. 627–636, 2023.
https://doi.org/10.1007/978-3-031-28238-6_54

Ghosh et al. [8], Paheli et al. [3] and Ahmad et al. [1] employed deep learning based BiLSTM-CRF models for RRL on Japanese civil rights judgements, Indian Supreme Court opinions, UK supreme court judgements and the US BVA corpus respectively. More recently, Kalamkar et al. [12] benchmarked RRL on Indian legal documents using a Hierarchical Sequential Labeling Network model (HSLN). The corpus they used claims to be the largest available corpus of legal documents annotated with rhetorical sentence roles.

In this work we approach RRL on legal documents with the observation that the texts of judgement are not only very long, but also often contain large sections of the same sentence type (e.g. explanations of case facts). We hence build models that segment the document into thematically coherent sets of contiguous sequence of sentences (which we refer to as *spans*) and assign them labels. We also hypothesize that modeling documents at this span level can also help to capture certain types of contexts effectively that may be spread across long sequences of sentences that can be collapsed into a much smaller number of thematically coherent spans. For example, when case documents are to be retrieved according to certain types of information, then aggregating that content from a small number of topical blocks across a long document is intuitive. At the same type, we explore how this assumption of topical continuity in the law can help RRL models learn better from small amounts of training data.

To tackle this problem as sequential span classification, we apply semi-Markov Conditional Random Field (CRF) [23], which have been proposed to jointly handle span segmentation and labeling. Semi-Markov CRFs have been used in various tasks such as Chinese word segmentation [16,17], named entity recognition [2,31,32], character-level parts of speech labelling [13], phone recognition [19], chord recognition [20], biomedical abstract segmentation [28] and piano transcription [29]. Most previous works dealt with shorter input sequences and thus contained smaller span lengths, which allows for a convenient upper bound on the maximum length of a span. In this work, we assess the performance of semi-Markov CRFs on legal judgements, which are usually very long and also possess a potentially large range of labels, making this setup even more challenging.

Obtaining sufficiently large amounts of annotated data for deep learning models in specialized domains like the law is very expensive as it requires expert annotators. To mitigate this data scarcity, we explore three strategies of data augmentation (DA) such as random deletion of words, back translation and swapping of sentences within a span. DA techniques which are common in computer vision field, has witnessed growing interest in NLP tasks due to the twin challenge of large annotated data for neural networks and expensive data annotation in low-resource domains [6]. In sum, this paper contributes the casting RRL of legal judgments as a sequential span classification task and associated experiments with semi-Markov CRFs on existing public datasets. We also explore three data augmentation strategies to assess their impact on the task. Our experiments demonstrate that our semi-Markov CRF model performs better compared to a CRF baseline on documents characterized by multi-sentence spans.[1]

---

[1] Our code is available at https://github.com/TUMLegalTech/Span-RRL-ECIR23.

2 Method

Our hierarchical semi-Markov CRF model takes the judgement document $x = \{x_1, x_2, \ldots, x_m\}$ as input, where $x_i = \{x_{i1}, x_{i2}, \ldots, x_{in}\}$ and outputs the rhetorical role label sequence $l = \{l_1, l_2, \ldots, l_m\}$ with $l_i \in L$. x_i and x_{jp} denote i^{th} sentence and p^{th} token of jth sentence, respectively. m and n denote the number of sentences and tokens in the ith sentence respectively. l_i is the rhetorical role corresponding to sentence x_i and L denotes set of pre-defined rhetorical role labels.

2.1 Hierarchical Semi-Markov CRF Model

Our model contains a semi-Markov CRF component [23] built on top of a Hierarchical Sequential Labeling Network model [11] with the following layers:

Encoding Layers: Following [12], we encode each sentence with BERT-BASE [14] to obtain token level representations $z_i = \{z_{i1}, z_{i2}, \ldots, z_{in}\}$. These are passed through a Bi-LSTM layer [10] followed by an attention pooling layer [30] to obtain sentence representations $s = \{s_1, s_2, \ldots, s_m\}$.

$$u_{it} = \tanh(W_w z_{it} + b_w) \quad \& \quad \alpha_{it} = \frac{\exp(u_{it} u_w)}{\sum_s \exp(u_{is} u_w)} \quad \& \quad s_i = \sum_{t=1}^{n} \alpha_{it} u_{it} \quad (1)$$

where W_w, b_w, u_w are trainable parameters.

Context Enrichment Layer: The sentence representations s are passed through a Bi-LSTM to obtain contextualized sentence representations $c = \{c_1, c_2, \ldots, c_m\}$, which encode contextual information from surrounding sentences.

Classification Layer: A semi-Markov CRF takes the sequence of sentence representations c and segments it into labeled spans $k = \{k_1, \ldots, k_{|s|}\}$ with $k_j = (a_j, b_j, y_j)$ where a_j and b_j are the starting and ending position of the sentences in the jth span, and y_j is the corresponding rhetorical role label of the jth span. $|s|$ denotes the total number of spans where $\sum_{l=1}^{|s|}(b_j - a_j + 1) = m$. We model the conditional probability through a semi-Markov CRF which jointly tackles the span segmentation and label assignment for a span as follows:

$$p(y|c) = \frac{1}{Z(c)} \exp(\sum_{j=1}^{|s|} F(k_j, c) + A(y_{j-1}, y_j)) \quad (2)$$

$$\text{where} \quad Z(c) = \sum_{k' \in K} \exp(\sum_j F(k'_j, c) + A(y_{j-1}, y_j)) \quad (3)$$

where $F(k_j, c)$ is the score assigned for span k_j (i.e., for interval $[a_j, b_j]$ belonging to label y_j based on span input c) and $A(y_{j-1}, y_j)$ is the transition score of the labels of two adjacent spans. $Z(c)$ denotes the normalization factor computed as the sum over the set of all possible spans K against c. The score $F(k_j, c)$ is computed using a learnable weight and bias matrix.

$$F(k_j, c) = W^T . f(k_j, c) + b \qquad (4)$$

where W and b denote trainable parameters and $f(k_j, c)$ represents span representation of jth span derived from c.

To obtain the span representations $f(k_j, c)$, we pass the sentence-level representations c for the sentences in the given span k_j through a BiLSTM layer initially to capture the context of the span. Then we obtain the span representation $f(k_j, c)$ as the concatenation of the first two and final two sentences vectors, and the mean of the sentences in the span. In case of shorter spans, we repeat the same sentence to match the dimension.

We maximize the above defined conditional log-likelihood to estimate the parameters and train the model end-to-end. We perform inference using the Viterbi decoding algorithm [7] to obtain the best possible span sequence along with its label assignment. These computations are done in logarithmic space to avoid numerical instability. In traditional semi-Markov CRF which are applied to relatively shorter sequences in the previous works, the assumption is that that there exists no transition between the same rhetorical labels. However, due to the long input data and a larger range of potential label spans, we relax this assumption as we can deal with a certain maximum span length due to computational constraints as it involves quadratic complexity.

2.2 Data Augmentation

The main goal of Data Augmentation in low resource settings is to increase the diversity of training data which in turn helps the model to generalize better on test data. In this regard, we implement the following three Data Augmentation techniques as preliminary analysis and leave the exploration of more advanced techniques as a future work.

Word deletion [26] is a noise based method that deletes words within a sentence at random. The augmented data differs from the original without affecting the rhetorical role of the sentence as the rhetorical role of the sentence can be derived from the other words present in the sentence. This helps the model to derive better contextual understanding of the sentence rather than relying on word-level surface features.

In **back-translation** [18], we translate the original text at sentence level into other languages and then back to the original language to obtain augmented data. Unlike word level methods, this method does not not directly deal with individual words but rewrites the whole sentence. This makes the model robust to any writing style based spuriously correlated features and learn the semantic information conveyed by the text.

Sentence swapping [4] is based on the notion that a minor change in order of sentences is still readable for humans. We restrict swapping of sentences to those within a single span, which preserves the overall discourse flow of the document.

While some discontinuities will be introduced, the text remains content complete and rhetorical roles do not change. This helps the model to learn the discourse flow of the document and makes the model overcome the limitation of having transition between same spans as described in the previous sub-section.

3 Experiments and Discussion

Datasets: We experiment on two datasets - (i) BUILDNyAI dataset [12] consisting of judgement documents from the Indian supreme court, high court and district courts. It consists of publicly available train and validation splits with 184 and 30 documents, respectively, annotated with 12 different rhetorical role labels along with 'None'. As test dataset is not publicly available, we split and use training dataset for both training and validation and test it on the validation partition; (ii) the BVA PTSD dataset [25] consists of 25 decisions[2] by the U.S. Board of Veterans' Appeals (BVA) from appealed disability claims by veterans for service-related post-traumatic stress disorder (PTSD). We use 19 documents for training and validation, and 6 as test. They are annotated with 5 rhetorical roles along with 'None'.

Baselines: We compare our method, *HSLN-spanCRF+DA (data augmentation)* against the following variants : *HSLN-CRF* (normal CRF, no DA), *HSLN-spanCRF* (spanCRF, no DA) and *HSLN-CRF+DA* (normal CRF with DA).

Metrics: We use both span-macro-F1 and span-micro-F1, which is computed based on match of span-by-span labels[3] (i.e., it encompasses both segmentation into exact spans as well their labeling). We also report span-segmentation-F1 which only evaluates on segmentation of spans ignoring the label. We further evaluate at the sentence level using micro-F1 and macro-F1 following previous works [12].

Implementation Details: We use the hyperparameters of [12] for the HSLN model. For the semi-Markov CRF, we obtain the the maximum segment length using validation set and set it to 30 and 4 for BUILDNyAI and BVA datasets respectively. We used a batch size of 1 and trained our model end-to-end using Adam [15] optimizer with a learning rate of 1e-5. For data augmentation, we employed a maximum word deletion rate of 20%. For back-translation, we used English, German and Spanish as the sequence of languages. We augmented the dataset once using each DA technique and thus models with DA component were trained with four times the size of training dataset.

---

[2] The dataset actually contains 75 decisions, out of which only 25 documents have annotation label for every sentence.

[3] We post-process and merge the same consecutive labels to obtain the span labels.

Table 1. Model performance on BUILDNyAI and BVA datasets

| | BUILDNyAI | | | | | BVA PTSD | | | | |
| | Span | | | Sentence | | Span | | | Sentence | |
Model	s-mic.	s-mac.	s-seg	mic.	mac.	s-mic.	s-mac.	s-seg	mic.	mac.
CRF	0.31	0.28	0.33	0.80	0.60	0.67	0.58	0.71	0.81	0.74
spanCRF	0.38	0.35	0.39	0.76	0.56	0.67	0.56	0.69	0.78	0.72
CRF + DA	0.32	0.32	0.34	**0.82**	**0.63**	0.72	0.64	**0.75**	**0.85**	**0.81**
spanCRF + DA	**0.40**	**0.36**	**0.41**	0.81	0.58	**0.73**	**0.65**	**0.75**	0.83	0.80

Performance Evaluation: Table 1 reports the performance of our model and its variants on the two datasets. On BUILDNyAI, we observe that spanCRF performs better compared to a normal CRF in span-level metrics (statistically significant ($p \leq 0.05$) using McNemar Test), with a drop at the sentence-level. With the addition of data augmentation (DA), both CRF and spanCRF performance improves. However, the increase is larger for spanCRF's sentence level metrics (statistically significant ($p \leq 0.05$) using McNemar Test). This can be attributed to spanCRF having to compute the optimal segmentation path over all the possible paths, which requires enough data to learn and generalize better. On the other hand, on the BVA PTSD dataset, spanCRF did not show a significant impromavement compared to normal CRF. This is because 73.8% of the spans in BVA dataset (BUILDNyAI: 31%) have length 1 and the mean span length is 1.85 (BUILDNyAI: 6.81) which does not allow spanCRF to show its potential. However, the trend towards a beneficial effect of data augmentation persists.

Effect of Maximum Span Length: We create variants of spanCRF by varying the maximum span length. First section in Table 2 shows that increasing the span length improved the performance on span-level metrics with a marginal drop at the sentence-level. We choose 30 as the maximum span length due to the computational resource constraints and our very long judgment documents.

Effect of Span Representation: We experiment with various span representations such as *grConv* [13] (Gated Recursive Convolutional Neural Networks), *simple* [28] involving concatenation of first and last sentence representation in span. We also create a variant of our proposed span representation by removing the BiLSTM (*ours w/o BiLSTM*). From second section in Table 2, we observe a performance drop without the BiLSTM layer (both at span- and sentence-level) indicating the importance of capturing context specifically at the span level to obtain good representations. We notice less improvement with *grConv*, which can also be attributed to its high number of parameters for our low data condition. Though *simple* achieves an improvement in span-level metrics, it shows a huge drop in sentence-level performance.

Table 2. First and second section indicates the effect of max span length (w/o DA) and different span feature representations (w/o DA) on BUILDNyAI

	Span			Sentence	
Model	s-mic.	s-mac.	s-seg	mic.	mac.
CRF (len = 1)	0.31	0.28	0.33	**0.80**	**0.60**
spanCRF (len = 5)	0.33	0.30	0.34	0.68	0.45
spanCRF (len = 10)	0.34	0.32	0.36	0.71	0.48
spanCRF (len = 20)	0.36	0.33	0.37	0.73	0.52
spanCRF (len = 30)	**0.38**	**0.35**	**0.39**	0.76	0.56
CRF (no span)	0.31	0.28	0.33	**0.80**	**0.60**
Span CRF (ours)	**0.38**	**0.35**	**0.39**	0.76	0.56
Span CRF (ours w/o BiLSTM)	0.36	0.32	0.37	0.75	0.55
Span CRF (grConv)	0.32	0.30	0.34	0.74	0.51
Span CRF (simple)	0.34	0.33	0.36	0.72	0.52

Ablation on Data Augmentation Strategies : We observe the effect of each data augmentation strategy in isolation. From Table 3, we observe that, in the case of CRF, each of the augmentation strategies boosted performance at the sentence-level by a considerable margin. With all three augmentation strategies combined, CRF witnessed a considerable jump, indicating the complementarity between the strategies. Similarly, we observe an improvement with each data augmentation strategy in case of spanCRF, and the greatest increase when using all three strategies combined.

Table 3. Different data augmentations on CRF and spanCRF on BUILDNyAI

	Span						Sentence			
	s-mic.		s-mac.		s-seg		mic.		mac.	
Model	CRF	sp.CRF	CRF	sp.CRF	CRF	sp.CRF	CRF	sp.CRF	CRF	sp.CRF
No Augmentation	0.31	0.38	0.28	0.35	0.33	0.39	0.80	0.76	0.60	0.56
+ Swapping	**0.32**	0.39	0.30	**0.36**	**0.34**	0.40	**0.82**	0.80	0.62	**0.58**
+ Deletion	**0.32**	0.39	0.30	**0.36**	**0.34**	0.40	0.81	0.78	0.61	**0.58**
+ Back translation	**0.32**	**0.40**	0.31	**0.36**	**0.34**	0.40	0.81	0.77	0.62	0.57
+ All three DA	**0.32**	**0.40**	**0.32**	**0.36**	**0.34**	**0.41**	**0.82**	**0.81**	**0.63**	**0.58**

4 Conclusion

Our experiments demonstrate that while semi-Markov CRFs help to boost the predictions at the span level, data augmentation strategies can mitigate data scarcity and improve the performance both at sentence- and span-levels, albeit conditioned on the documents exhibiting patterns of longer passages of the same rhetorical type. While this is typical for legal judgments, it is not universal. In the future, we hence would like to combine the complimentary sentence- and span-level methods. We would also like to explore different data augmentation strategies to alleviate the bottle neck of limited annotated data and expensive data annotation, especially in these specialized domains.

References

1. Ahmad, S.R., Harris, D., Sahibzada, I.: Understanding legal documents: classification of rhetorical role of sentences using deep learning and natural language processing. In: 2020 IEEE 14th International Conference on Semantic Computing (ICSC), pp. 464–467. IEEE (2020)
2. Arora, R., Tsai, C.T., Tsereteli, K., Kambadur, P., Yang, Y.: A semi-markov structured support vector machine model for high-precision named entity recognition. In: Proceedings of the 57th Annual Meeting of the Association for Computational Linguistics, pp. 5862–5866 (2019)
3. Bhattacharya, P., Paul, S., Ghosh, K., Ghosh, S., Wyner, A.: Deeprhole: deep learning for rhetorical role labeling of sentences in legal case documents. Artific. Intell. Law 1–38 (2021)
4. Dai, X., Adel, H.: An analysis of simple data augmentation for named entity recognition. In: Proceedings of the 28th International Conference on Computational Linguistics, pp. 3861–3867 (2020)
5. Farzindar, A., Lapalme, G.: Letsum, an automatic legal text summarizing. In: Legal Knowledge and Information Systems: JURIX 2004, the Seventeenth Annual Conference, vol. 120, p. 11. IOS Press (2004)
6. Feng, S.Y., et al.: A survey of data augmentation approaches for NLP. In: Findings of the Association for Computational Linguistics: ACL-IJCNLP 2021, pp. 968–988 (2021)
7. Forney, G.D.: The viterbi algorithm. Proc. IEEE **61**(3), 268–278 (1973)
8. Ghosh, S., Wyner, A.: Identification of rhetorical roles of sentences in indian legal judgments. In: Legal Knowledge and Information Systems: JURIX 2019: The Thirty-second Annual Conference. vol. 322, p. 3. IOS Press (2019)
9. Hachey, B., Grover, C.: Extractive summarisation of legal texts. Artific. Intell. Law **14**(4), 305–345 (2006)
10. Hochreiter, S., Schmidhuber, J.: Long short-term memory. Neural Comput. **9**(8), 1735–1780 (1997)
11. Jin, D., Szolovits, P.: Hierarchical neural networks for sequential sentence classification in medical scientific abstracts. In: Proceedings of the 2018 Conference on Empirical Methods in Natural Language Processing, pp. 3100–3109 (2018)
12. Kalamkar, P., et al.: Corpus for automatic structuring of legal documents. In: LREC (2022)
13. Kemos, A., Adel, H.: Neural semi-markov conditional random fields for robust character-based part-of-speech tagging. In: Proceedings of NAACL-HLT, pp. 2736–2743 (2019)
14. Kenton, J.D.M.W.C., Toutanova, L.K.: Bert: Pre-training of deep bidirectional transformers for language understanding. In: Proceedings of NAACL-HLT, pp. 4171–4186 (2019)

15. Kingma, D.P., Ba, J.: Adam: a method for stochastic optimization. In: Bengio, Y., LeCun, Y. (eds.) 3rd International Conference on Learning Representations, Conference Track Proceedings, ICLR 2015, 7–9 May 2015, San Diego, CA, USA (2015)
16. Kong, L., Dyer, C., Smith, N.A.: Segmental recurrent neural networks. In: Bengio, Y., LeCun, Y. (eds.) 4th International Conference on Learning Representations, Conference Track Proceedings, 2–4 May 2016. ICLR 2016. San Juan, Puerto Rico (2016)
17. Liu, Y., Che, W., Guo, J., Qin, B., Liu, T.: Exploring segment representations for neural segmentation models. In: Proceedings of the Twenty-Fifth International Joint Conference on Artificial Intelligence, pp. 2880–2886 (2016)
18. Lowell, D., Howard, B., Lipton, Z.C., Wallace, B.C.: Unsupervised data augmentation with naive augmentation and without unlabeled data. In: Proceedings of the 2021 Conference on Empirical Methods in Natural Language Processing, pp. 4992–5001 (2021)
19. Lu, L., Kong, L., Dyer, C., Smith, N.A., Renals, S.: Segmental recurrent neural networks for end-to-end speech recognition. Interspeech **2016**, 385–389 (2016)
20. Masada, K., Bunescu, R.C.: Chord recognition in symbolic music using semi-markov conditional random fields (2017)
21. Nejadgholi, I., Bougueng, R., Witherspoon, S.: A semi-supervised training method for semantic search of legal facts in canadian immigration cases. In: JURIX, pp. 125–134 (2017)
22. Saravanan, M., Ravindran, B., Raman, S.: Automatic identification of rhetorical roles using conditional random fields for legal document summarization. In: Proceedings of the Third International Joint Conference on Natural Language Processing: Volume-I (2008)
23. Sarawagi, S., Cohen, W.W.: Semi-markov conditional random fields for information extraction. Adv. Neural Inform. Process. Syst. **17** (2004)
24. Savelka, J., Ashley, K.D.: Segmenting us court decisions into functional and issue specific parts. In: JURIX, pp. 111–120 (2018)
25. Walker, V.R., Pillaipakkamnatt, K., Davidson, A.M., Linares, M., Pesce, D.J.: Automatic classification of rhetorical roles for sentences: comparing rule-based scripts with machine learning. In: ASAIL@ ICAIL (2019)
26. Wei, J., Zou, K.: Eda: easy data augmentation techniques for boosting performance on text classification tasks. In: Proceedings of the 2019 Conference on Empirical Methods in Natural Language Processing and the 9th International Joint Conference on Natural Language Processing (EMNLP-IJCNLP), pp. 6382–6388 (2019)
27. Yamada, H., Teufel, S., Tokunaga, T.: Neural network based rhetorical status classification for japanese judgment documents. In: Legal Knowledge and Information Systems, pp. 133–142. IOS Press (2019)
28. Yamada, K., Hirao, T., Sasano, R., Takeda, K., Nagata, M.: Sequential span classification with neural semi-markov crfs for biomedical abstracts. In: Findings of the Association for Computational Linguistics: EMNLP 2020, pp. 871–877 (2020)
29. Yan, Y., Cwitkowitz, F., Duan, Z.: Skipping the frame-level: Event-based piano transcription with neural semi-CRFS. Adv. Neural Inform. Process. Syst. **34**, 20583–20595 (2021)
30. Yang, Z., Yang, D., Dyer, C., He, X., Smola, A., Hovy, E.: Hierarchical attention networks for document classification. In: Proceedings of the 2016 Conference of the North American Chapter of the Association for Computational Linguistics: Human Language Technologies, pp. 1480–1489 (2016)

31. Ye, Z., Ling, Z.H.: Hybrid semi-markov CRF for neural sequence labeling. In: Proceedings of the 56th Annual Meeting of the Association for Computational Linguistics (Volume 2: Short Papers), pp. 235–240 (2018)
32. Zhuo, J., Cao, Y., Zhu, J., Zhang, B., Nie, Z.: Segment-level sequence modeling using gated recursive semi-markov conditional random fields. In: Proceedings of the 54th Annual Meeting of the Association for Computational Linguistics (Volume 1: Long Papers), pp. 1413–1423 (2016)

Trigger or not Trigger: Dynamic Thresholding for Few Shot Event Detection

Aboubacar Tuo[(✉)] [iD], Romaric Besançon[iD], Olivier Ferret[iD],
and Julien Tourille[iD]

Université Paris-Saclay, CEA, List, 91120 Palaiseau, France
{aboubacar.tuo,romaric.besancon,olivier.ferret,julien.tourille}@cea.fr

Abstract. Recent studies in few-shot event trigger detection from text address the task as a word sequence annotation task using prototypical networks. In this context, the classification of a word is based on the similarity of its representation to the prototypes built for each event type and for the "non-event" class (also named null class). However, the "non-event" prototype aggregates by definition a set of semantically heterogeneous words, which hurts the discrimination between trigger and non-trigger words. We address this issue by handling the detection of non-trigger words as an out-of-domain (OOD) detection problem and propose a method for dynamically setting a similarity threshold to perform this detection. Our approach increases f-score by about 10 points on average compared to the state-of-the-art methods on three datasets.

Keywords: Few-shot learning · Event Detection · Prototypical networks

1 Introduction

Event Detection (ED) is an important task in Information Extraction (IE) that aims at extracting instances of given types of events from text [14,15]. This extraction consists in identifying event *triggers*, which are words or phrases that explicitly indicate the presence of an event in a sentence. For example, in the sentence *"John D. Idol will [**take over**] as Chief Executive."*, a *"Start-Position"* event is triggered by the phrase *"take over"*. Supervised machine learning approaches for ED have been extensively studied in past years, including feature-based methods [11], convolutional neural networks [14], recurrent neural networks [13], and graph-based models [12,16,28]. However, all these approaches rely on large-scale annotated datasets for training, which are generally difficult to obtain. Few-Shot Event Detection (FSED) has therefore received a great interest in recent years with the emergence of few-shot learning methods, notably

---

This publication was made possible by the use of the FactoryIA supercomputer, financially supported by the Île-de-France Regional Council.

J. Kamps et al. (Eds.): ECIR 2023, LNCS 13981, pp. 637–645, 2023.
https://doi.org/10.1007/978-3-031-28238-6_55

via metric learning [6,20,21,25], and the development of pre-trained language models capable of transferring their language knowledge to novel tasks.

FSED has been implemented in several forms: *event identification*, which determines if a word in a sentence is a trigger according to a type of event [1,2], *event classification*, whose objective is to choose the type of event associated with an already identified trigger in a sentence [4,8–10,19], and *event detection*, which achieves these two steps jointly [3,23].

These research efforts cast FSED as a sequence labeling task, turned into a word classification problem addressed by using prototypical networks [20], which are particularly suited to few-shot learning. In this context, a prototype is built for each type of event and the "non-event" class (also named null class) [3,23,29]. However, the intrinsic heterogeneity of the "non-event" prototype makes the discrimination between trigger and non-trigger words based on the similarity with prototypes difficult. To solve this problem, we formulate FSED as an out-of-domain detection problem [18]. We consider the words of the null class as out-of-domain examples and learn a dynamic similarity threshold so that these examples are not associated with any event class.

In summary, our contribution is threefold: (1) we propose a new way to handle the null class in FSED; (2) we define a new model for FSED using prototypical networks and a contrastive loss; (3) we compute a dynamic decision threshold using the Empirical Cumulative Distribution Function (ECDF). We experiment on several datasets and set a new state-of-the-art performance for the task.

2   Our Approach

2.1   Problem Definition

We formulate FSED as a N-ways k-shots episodic learning [25] with prototypical networks. We cast the trigger detection task as a sequence labeling problem at word level and use the Inside-Outside-Beginning (IOB) format following [3] and [23]. At each episode, a *support set* is drawn from the labeled data. It contains N event types and k annotated sentences per type (k being small, e.g. from 1 to 10). A second set of sentences, called *query set*, is used to make predictions based on the annotated sentences of the support set. Each sentence can contain one or more event types characterized by a trigger. The identification of both the event type and the position of the trigger is performed by assigning a label to each word. Triggers are annotated with the corresponding event tag and words that are not part of a trigger are labeled with the "O" tag. This amounts to a word-level multi-class classification.

We build a prototype for each class from the examples in the support set by taking the average of the embeddings of the k triggers of this class. Then, we classify each word of the query set according to its similarity with these prototypes. During training, these similarities are used in an objective function to update the model's weights. However, this formulation implies having a prototype for the class "O", which is built in practice by gathering words that are not semantically homogeneous. Inspired by few-shot out-of-domain detection

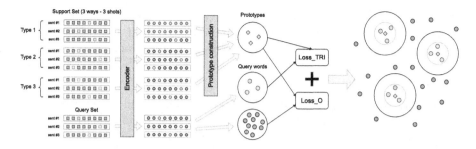

Fig. 1. Overview of the model

research efforts [17,22], we avoid building the "O" prototype and propose an approach based on a dynamic threshold adapted to each sentence by using the ECDF of the words of this sentence.

2.2   Model

Figure 1 presents an overview of our model. We detail its components hereafter.

Encoder. This component takes a sentence as input and produces a contextual representation for each word. Given a sentence $x = w_1, \ldots, w_L$, of length L, the encoder provides $\bar{e} = e_1, \ldots, e_L$, where e_i is the embedding of the word w_i. We use the BERT-Weighted encoder proposed by [23], which provides the best results for FSED.

Prototype Module. This component builds a prototype for each class of events by averaging the representations of its support trigger words and classifies the query words according to their similarities to these prototypes. In contrast to [23] and [3], we do not build a prototype for the null class but rely on a similarity threshold to decide whether a word is part of the null class or not.

Training. The objective function commonly used in prototypical networks is the cross entropy loss. Here, we propose a contrastive loss, which is more suitable for metric learning. Unlike cross entropy, whose objective is to learn to predict a label or values from an input, contrastive losses predict the relative similarity between inputs. This objective is more appropriate in our case as we are interested in making the triggers of each class of events closer to their prototypes than to the "O" examples. For a given class y, the objective function has two terms:

- **TRI-Loss:** a term to match triggers with their prototypes:

$$\mathcal{L}_{TRI}(\bar{e}, y) = \sum_{j \neq k} \max(0, \mathcal{M}_0 - s(e_{tr}, c^j) + s(e_{tr}, c^y)) \tag{1}$$

- **O-Loss:** a term to separate the "null words" e_i from the prototype c^y.

$$\mathcal{L}_O(\bar{e}, y) = \max(0, \max_{i \neq tr}(s(e_i, c^y) - \mathcal{M}_1)) \tag{2}$$

where $s(.)$ is a cosine similarity function, $\mathbf{c}^y$ the embedding of the prototype for the class y, and $\mathbf{e_{tr}}$ and $\mathbf{e_{i \neq tr}}$ respectively the trigger word and the null words of the query example. $\mathcal{M}_0$ and $\mathcal{M}_1$ are hyperparameters modeling the margins. The model is then trained end-to-end by minimizing $\mathcal{L} = \mathcal{L}_{TRI} + \mathcal{L}_O$ through a back-propagation step at each episode.

Classification and Null Class Processing. The standard approach with prototypical networks is to classify each word according to its similarity with the class prototypes. In our out-of-trigger model, since we do not have a prototype for the null class, we must rely on a threshold above which a word can be assigned to a specific class and under which the word is discarded as a non-event. Typically, in works such as [22] and [17], a global threshold is defined using the distribution of similarity values on a validation set. However, in our case, we observed empirically that the distributions of similarity values between a trigger and the prototypes vary too much from one sentence to another (see Fig. 2a). Hence, we cannot set a global threshold.

To tackle this problem, we propose to search for the probability corresponding to the optimal threshold by using the ECDF on the maximum similarity values. This allows us to obtain a dynamic threshold that is specific to the considered event mention. More precisely, given that the similarities of the triggers are higher than those of the null words, we assume that, for a given sentence, the similarities of the triggers will only be present above a certain quantile (quite high) in the distribution of the similarities. We also assume that this quantile is fairly stable even if it does not correspond to the same similarity value from one sentence to another. In practice, for a given query sentence, we select the most similar sentence in the support set and we search for the threshold giving the best f-score for the classification of all the words of this sentence. We vary the threshold between the minimum and maximum similarities and adopt the one that maximizes the f-score on the selected sentence. Then, we determine the probability corresponding to this threshold using its ECDF. Finally, we determine the optimal threshold for the query example from its ECDF and the previously determined probability. However, since the probabilities directly computed from the ECDF depend on the number of words in the sentences, we linearly interpolate the ECDF over a larger number of points before estimating the probabilities, which allows us to artificially make all sentences the same length (we use 512 points). In the example of Fig. 2b, "sentence 1" (support sentence) has its optimal threshold at 0.71 corresponding to a probability of 0.97. We then report this probability on the ECDF of "sentence 2" (query sentence) to obtain its optimal threshold, which is equal to 0.92.

3   Experiments

3.1   Evaluation Framework and Hyperparameters

We experiment on the ACE 2005 [26], MAVEN [27], and FewEvent [4] datasets. We use the splits of [2] for ACE 2005 and MAVEN and of [3] for FewEvent.

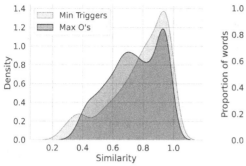 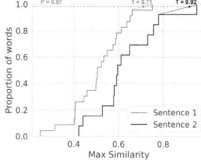

(a) Min triggers vs max "O"s similarities   (b) ECDF for two different sentences

Fig. 2. a shows that trigger and "O" words are not separable with a global threshold because the distribution of the minimal values of triggers' similarities (Min triggers) and the distribution of the maximal values of "O"s' similarities (Max "O"s) overlap too much. b shows, from the ECDF of two example sentences, that the optimal similarity varies from one sentence to another.

In all cases, the test and training sets contain distinct classes so that during evaluation, the model has to deal with new classes it has never seen before.

We adopt the N ways, k shots episodic evaluation, which consists in building episodes with N classes and k annotated examples per class. In the standard episodic evaluation setting [25], the test sets are sampled such that all the classes are equally distributed, which clearly does not correspond to the event mention distribution in real data. As a result, the reported performance scores do not reflect the effectiveness of these models when they are applied to a new domain. We adopt the more realistic configuration of [29], which builds the support set with $N * K$ examples and uses all other examples in the test set as queries.

For the experiments, we used the pre-trained BERT-base model as backbone and adopted the "Weighted" strategy of [23] to obtain contextual word embeddings. We adopt a maximum sequence length of 128 tokens, a batch size of 1, a learning rate of 1e−5 and 30,000 N ways, k shots episodes to train the model. The hyperparameters $\mathcal{M}_0 = 1$ and $\mathcal{M}_1 = 0.4$ were obtained on the validation set among the values {0.2, 0.4, 0.6, 0.8, 1}.

3.2   Results and Analysis

We compare our approach, **OUTFIT** (i.e OUT oF trIgger deTection), with four other models from the literature in a 5-ways 5-shots configuration. **PA-CRF** [3] and [23] are state-of-the-art models that compute a prototype for the null class. **PA-CRF** captures transition probabilities between IOB labels in a few-shot framework based on Conditional Random Fields [7] whereas [23] proposes a better exploitation of BERT layers for the encoder. **HCL-TAT** [30] is also a prototype-free model for the null class based on a decision threshold equal

642   A. Tuo et al.

Table 1. FSED mean and standard deviation of f-scores over five runs. † stands for the results from the original paper. * denotes statistically significant improvements over the best of our three reference models using the Almost Stochastic Order test [5] as implemented by [24].

	Model	ACE 2005	MAVEN	FewEvent
5-ways, 5-shots	PROTO	49.2 ± 1.2	51.6 ± 1.4	53.6 ± 0.7
	PA-CRF [3]	64.0 ± 0.6	65.2 ± 0.3	65.3 ± 2.0
	[23]'s model	66.4 ± 1.8	67.1 ± 1.5	67.4 ± 1.1
	HCL-TAT† [30]	–	–	66.9 ± 0.7
	OUTFIT (ours)	$\mathbf{74.0^*} \pm \mathbf{1.1}$	$\underline{76.9} \pm 1.1$	$\mathbf{79.6^*} \pm \mathbf{4.2}$
	– PoS tags	$\underline{72.2} \pm 2.2$	$\mathbf{77.5^*} \pm \mathbf{0.8}$	$\underline{77.9} \pm 3.9$
	– contrastive	66.5 ± 5.7	63.1 ± 12.6	75.9 ± 5.4
	– weighted	59.2 ± 3.6	50.0 ± 2.3	70.9 ± 2.7
	Oracle threshold	82.5 ± 1.9	87.2 ± 1.1	84.1 ± 0.5
1w,5s	FS-Causal† [2]	76.9 ± 1.4	55.0 ± 0.4	–
	OUTFIT	80.9 ± 2.9	81.1 ± 1.1	79.1 ± 2.1

to the average similarities during an episode. We compare these methods to a baseline prototypical model that builds a prototype for the null class, uses the cross-entropy loss and a BERT encoder (**PROTO**). **FS-Causal** [2] leverages a causal intervention on the triggers by taking into account their context to solve the trigger curse problem. Since their reported results are only evaluated class by class, it corresponds to a 1 way k shots configuration. We also add results corresponding to the optimal threshold found directly on the query set instances (**Oracle threshold**), which gives an indication of the best result that can be achieved with our approach.

In preliminary experiments, we noted that the precision ($\approx 65\%$) was relatively low compared to the recall ($\approx 80\%$), indicating that the model was identifying too many words as triggers. To increase the precision, we filtered the predictions by their parts-of-speech, keeping only the tags most commonly associated with event triggers in the training set (verb, adverb, and noun).

The results of our evaluation are reported in Table 1. Our method (OUTFIT) sets a new state-of-the-art performance with an average increase of 10 points in f1-score for all three datasets in the 5 ways, 5 shots setting. The ablation study suggests that the weighted encoder and the contrastive loss, combined with our new formulation, play important roles in the global performance of the model. More specifically, we can note that the contrastive loss strongly helps decrease the variance of the results. We also think that this loss, combined with our threshold finding strategy, contributes to the strong difference of performance with HCL-TAT while our objectives are initially close. As our preliminary experiments suggested, filtering candidate triggers according to their PoS tags allows

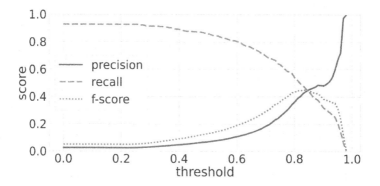

Fig. 3. Scores using a global threshold, on the FewEvent dataset.

increasing the performance by a few points for two datasets. In the 1 way, 5 shots setting, our model also improves performance compared to FS-Causal for the two datasets with results for FS-Causal. This result first shows that the improvement brought by our proposal is not limited to one setting. Furthermore, it is also interesting because considering new types of events one by one is the more general strategy for the adaptation to a new domain in which the number of types of events is not known in advance.

Finally, Fig. 3 shows the scores for a global threshold ranging from 0 to 1 on the FewEvent dataset, in the 5 ways, 5 shots setting. We notice that the best f-score that can be obtained with a global threshold is around 0.45. This clearly justifies the interest in finding a dynamic threshold rather than a global threshold as used in [22] and [17]. However, the significant gap between the oracle and our model suggests that our approach could be further improved.

4   Conclusion and Perspectives

In this paper, we address FSED as an Out-of-Domain Detection task using prototypical networks. This method avoids building a prototype for the null class, which is inherently heterogeneous, and provides a dynamic threshold to decide whether a word is a trigger or not. Experimental results suggest that this new formulation provides an important performance boost compared to other state-of-the-art methods. To the best of our knowledge, this is the first research effort that cast FSED as a sequence labeling task while treating the null class as an out-of-domain detection problem.

We believe that our method could be applied to other sequence labeling tasks and will investigate more particularly its application to Event Argument Extraction and Named Entity Recognition as future work.

References

1. Bronstein, O., Dagan, I., Li, Q., Ji, H., Frank, A.: Seed-based event trigger labeling: how far can event descriptions get us? In: ACL-IJCNLP, pp. 372–376 (2015). https://doi.org/10.3115/v1/P15-2061
2. Chen, J., Lin, H., Han, X., Sun, L.: Honey or poison? Solving the trigger curse in few-shot event detection via causal intervention. In: Proceedings of EMNLP, Punta Cana, Dominican Republic, pp. 8078–8088. Association for Computational Linguistics, November 2021. https://doi.org/10.18653/v1/2021.emnlp-main.637
3. Cong, X., Cui, S., Yu, B., Liu, T., Yubin, W., Wang, B.: Few-shot event detection with prototypical amortized conditional random field. In: Findings of ACL-IJCNLP, pp. 28–40, August 2021. https://doi.org/10.18653/v1/2021.findings-acl.3
4. Deng, S., Zhang, N., Kang, J., Zhang, Y., Zhang, W., Chen, H.: Meta-learning with dynamic-memory-based prototypical network for few-shot event detection. In: WSDM, Houston, TX, USA, pp. 151–159, January 2020. https://doi.org/10.1145/3336191.3371796
5. Dror, R., Shlomov, S., Reichart, R.: Deep dominance - how to properly compare deep neural models. In: ACL, Florence, Italy, pp. 2773–2785, July 2019. https://doi.org/10.18653/v1/P19-1266
6. Geng, R., Li, B., Li, Y., Zhu, X., Jian, P., Sun, J.: Induction networks for few-shot text classification. In: Proceedings of the 2019 Conference on Empirical Methods in Natural Language Processing and the 9th International Joint Conference on Natural Language Processing (EMNLP-IJCNLP), Hong Kong, China, pp. 3904–3913. Association for Computational Linguistics, November 2019. https://doi.org/10.18653/v1/D19-1403
7. Lafferty, J.D., McCallum, A., Pereira, F.C.N.: Conditional random fields: probabilistic models for segmenting and labeling sequence data. In: ICML, San Francisco, CA, USA, pp. 282–289 (2001)
8. Lai, V., Dernoncourt, F., Nguyen, T.H.: Learning prototype representations across few-shot tasks for event detection. In: EMNLP, pp. 5270–5277 (2021)
9. Lai, V.D., Nguyen, T.: Extending event detection to new types with learning from keywords. In: W-NUT 2019, Hong Kong, China, pp. 243–248, November 2019. https://doi.org/10.18653/v1/D19-5532
10. Lai, V.D., Nguyen, T.H., Dernoncourt, F.: Extensively matching for few-shot learning event detection. In: Workshop NUSE, pp. 38–45 (2020). https://doi.org/10.18653/v1/2020.nuse-1.5
11. Li, Q., Ji, H., Huang, L.: Joint event extraction via structured prediction with global features. In: ACL, Sofia, Bulgaria, pp. 73–82, August 2013
12. Liu, X., Luo, Z., Huang, H.: Jointly multiple events extraction via attention-based graph information aggregation. In: EMNLP, pp. 1247–1256 (2018). https://doi.org/10.18653/v1/D18-1156
13. Nguyen, T.H., Cho, K., Grishman, R.: Joint event extraction via recurrent neural networks. In: NAACL-HLT, San Diego, California, pp. 300–309 (2016). https://doi.org/10.18653/v1/N16-1034
14. Nguyen, T.H., Grishman, R.: Event detection and domain adaptation with convolutional neural networks. In: ACL-IJCNLP. Beijing, China, pp. 365–371 (2015). https://doi.org/10.3115/v1/P15-2060
15. Nguyen, T.H., Grishman, R.: Event detection and domain adaptation with convolutional neural networks. In: Proceedings of the 53rd Annual Meeting of the Association for Computational Linguistics and the 7th International Joint Conference on Natural Language Processing, Beijing, China, pp. 365–371. Association for Computational Linguistics, July 2015. https://doi.org/10.3115/v1/P15-2060

16. Nguyen, T.H., Grishman, R.: Graph convolutional networks with argument-aware pooling for event detection. In: Thirty-Second AAAI Conference on Artificial Intelligence (2018)
17. Nimah, I., Fang, M., Menkovski, V., Pechenizkiy, M.: ProtoInfoMax: prototypical networks with mutual information maximization for out-of-domain detection. In: Findings of the Association for Computational Linguistics: EMNLP, pp. 1606–1617. Association for Computational Linguistics, November 2021. https://doi.org/10.18653/v1/2021.findings-emnlp.138
18. Schölkopf, B., Platt, J.C., Shawe-Taylor, J., Smola, A.J., Williamson, R.C.: Estimating the support of a high-dimensional distribution. Neural Comput. **13**(7), 1443–1471 (2001). https://doi.org/10.1162/089976601750264965
19. Shen, S., Wu, T., Qi, G., Li, Y.F., Haffari, G., Bi, S.: Adaptive knowledge-enhanced Bayesian meta-learning for few-shot event detection. In: Findings of ACL-IJCNLP, pp. 2417–2429, August 2021. https://doi.org/10.18653/v1/2021.findings-acl.214
20. Snell, J., Swersky, K., Zemel, R.: Prototypical networks for few-shot learning. In: Advances in Neural Information Processing Systems, vol. 30 (2017)
21. Sung, F., Yang, Y., Zhang, L., Xiang, T., Torr, P.H.S., Hospedales, T.M.: Learning to compare: relation network for few-shot learning. In: 2018 IEEE/CVF Conference on Computer Vision and Pattern Recognition, pp. 1199–1208 (2018)
22. Tan, M., et al.: Out-of-domain detection for low-resource text classification tasks. In: EMNLP-IJCNLP, pp. 3566–3572. Association for Computational Linguistics, November 2019. https://doi.org/10.18653/v1/D19-1364
23. Tuo, A., Besançon, R., Ferret, O., Tourille, J.: Better exploiting BERT for few-shot event detection. In: Rosso, P., Basile, V., Métais, E., Meziane, F. (eds.) NLDB 2022. LNCS, vol. 13286, pp. 291–298. Springer, Cham (2022). https://doi.org/10.1007/978-3-031-08473-7_26
24. Ulmer, D.: Deep-significance: easy and better significance testing for deep neural networks, March 2021. https://doi.org/10.5281/zenodo.4638709. https://github.com/Kaleidophon/deep-significance
25. Vinyals, O., Blundell, C., Lillicrap, T., Kavukcuoglu, K., Wierstra, D.: Matching networks for one shot learning. In: Advances in Neural Information Processing Systems, vol. 29 (2016)
26. Walker, C., Strassel, S., Julie Medero, K.M.: ACE 2005 multilingual training corpus (2006). https://doi.org/10.35111/mwxc-vh88
27. Wang, X., et al.: MAVEN: a massive general domain event detection dataset. In: Proceedings of the 2020 Conference on Empirical Methods in Natural Language Processing (EMNLP), pp. 1652–1671. Association for Computational Linguistics, November 2020. https://doi.org/10.18653/v1/2020.emnlp-main.129
28. Yan, H., Jin, X., Meng, X., Guo, J., Cheng, X.: Event detection with multi-order graph convolution and aggregated attention. In: EMNLP-IJCNLP, pp. 5766–5770 (2019)
29. Yang, Y., Katiyar, A.: Simple and effective few-shot named entity recognition with structured nearest neighbor learning. In: EMNLP, pp. 6365–6375. Association for Computational Linguistics, November 2020. https://doi.org/10.18653/v1/2020.emnlp-main.516
30. Zhang, R., Wei, W., Mao, X.L., Fang, R., Chen, D.: HCL-TAT: a hybrid contrastive learning method for few-shot event detection with task-adaptive threshold. In: Findings of the Association for Computational Linguistics: EMNLP, pp. 1808–1819. Association for Computational Linguistics (2022)

The Impact of a Popularity Punishing Hyperparameter on ItemKNN Recommendation Performance

Robin Verachtert[1,2]($\boxtimes$) (iD), Jeroen Craps[1] (iD), Lien Michiels[1,2] (iD),
and Bart Goethals[1,2,3] (iD)

[1] Froomle NV, Antwerp, Belgium
robin.verachtert@froomle.com
[2] University of Antwerp, Antwerp, Belgium
{lien.michiels,bart.goethals}@uantwerpen.be
[3] Monash University, Melbourne, Australia

Abstract. Collaborative filtering techniques have a tendency to amplify popularity biases present in the training data if no countermeasures are taken. The ItemKNN algorithm with conditional probability-inspired similarity function has a hyperparameter α that allows one to counteract this popularity bias. In this work, we perform a deep dive into the effects of this hyperparameter in both online and offline experiments, with regard to both accuracy metrics and equality of exposure. Our experiments show that the hyperparameter can indeed counteract popularity bias in a dataset. We also find that there exists a trade-off between countering popularity bias and the quality of the recommendations: Reducing popularity bias too much results in a decrease in click-through rate, but some counteracting of popularity bias is required for optimal online performance.

Keywords: Recommendation systems · AB test · Nearest neighbour

1 Introduction

Collaborative filtering algorithms are widely used for recommendation systems. To make predictions of what users may like, they rely on past preferences for items expressed by users. These preferences can, for example, be expressed by interacting with an item. Collaborative filtering methods can suffer from a 'rich get richer' effect when they fail to address the popularity bias in the data. For example, when some items are visited more often by users, the recommendation algorithm is also more likely to recommend them. This bias towards already popular items is generally considered undesirable, and many solutions have been proposed to address this bias [e.g. 1,17,24]. Even some of the earlier works on collaborative filtering were mindful of this inherent popularity bias. When Deshpande and Karypis [7] proposed the ItemKNN algorithm, they added a hyperparameter α to their conditional probability-inspired similarity function with

J. Kamps et al. (Eds.): ECIR 2023, LNCS 13981, pp. 646–654, 2023.
https://doi.org/10.1007/978-3-031-28238-6_56

the explicit purpose of discounting popular items that may otherwise dominate recommendations. Recent works have shown that despite advances in the field, ItemKNN and other nearest neighbour-based methods are still competitive, provided they are well-tuned [9,10,16,22]. Because of their inherent scalability, they remain popular methods in production environments.

In this work, we investigate how different values of the hyperparameter α impact performance and equality of exposure, as a measure of popularity bias, in both offline and online experiments with ItemKNN on three news datasets.

We answer the following three research questions:

- **RQ1:** How does the hyperparameter α impact the equality of exposure?
- **RQ2:** How does the hyperparameter α impact accuracy and CTR results?
- **RQ3:** Do the offline and online results agree?

Our work is done in the context of the popular item-to-item recommendation paradigm, recommending similar items in the context of another item, which we will refer to as *context item*. We focus our work on the news domain, as they have a specific interest in combatting popularity bias for ethical reasons, and, of course, because our partners agreed to perform the online tests discussed in this work. All data processed in these experiments was collected in accordance with GDPR: Users consented to receive personalised recommendations, as well as to have their data analysed and to participate in AB testing.

We find that the hyperparameter α can be used to increase the equality of exposure. Secondly, we find that it is necessary to seek a trade-off between equality of exposure and recommendation quality. We leave a thorough investigation into this trade-off for future work. Finally, we note that our offline and online results do not align due to the inherent popularity bias persisted in the offline evaluation [4].

2   Related Work

Popularity bias has been extensively studied in the context of recommender systems [e.g. 1,17,24]. Although the effect of popularity bias on ItemKNN has been studied [2], to the best of our knowledge, the impact of the hyperparameter α on popularity bias has not. In the original work by Deshpande and Karypis [7], the impact of α is evaluated solely in terms of MRR and HitRate, both accuracy measures. Recent work by Pellegrini et al. [19] suggests that not recommending popular items makes recommendations more personalised and can positively impact the recommender system's performance.

ItemKNN remains a popular and competitive baseline, despite recent advances in recommendation algorithms [9,10,16,22].

Due to their scalability, neighbourhood-based methods such as ItemKNN remain a popular choice in production settings [3,8,15,20]. Therefore, a thorough investigation of how the popularity bias can be countered is of great practical relevance.

Offline and online results often do not correlate [4,11,21], although some works have achieved success [12,18]. Popularity bias is an important factor in this failure to correlate and thus we investigate its impact in this work [4].

3   Experimental Setup

In this work, we focus on the item-to-item recommendation problem. The recommendation system needs to recommend users new items while they are currently visiting an item page on the website. The item the user is visiting is the only information the system uses to generate recommendations.

The dataset $\mathcal{D}$ consists of triplets (u, i, t) where $u \in U$ is the user, $i \in I$ is the item, and $t \in \mathbb{N}$ is the timestamp of when user u interacted with item i. Then the recommendation for user u is a function: $\Phi(\mathcal{D}_u^l)$, where $\mathcal{D}_u$ is the list of items that the user has seen and $\mathcal{D}_u^l$ is the last item that the user has seen.

Algorithm. We use the ItemKNN algorithm, with the similarity between items computed using the conditional probability-inspired similarity function, defined as

$$sim(i, j) = \frac{|\{u|i, j \in \mathcal{D}_u\}|}{|\{u|i \in \mathcal{D}_u\}| \cdot |\{u|j \in \mathcal{D}_u\}|^\alpha}$$

Here, i is a context item, j is a target item and α is a hyperparameter that punishes popular items in the similarity computation [7].

Specific values for α can be linked to other similarity measures. When $\alpha = 1$ it provides the same recommendations as the lift similarity measure. In the specific case of item-to-item recommendations, $\alpha = 0.5$ leads to the same recommendations as cosine similarity.

Metrics. To evaluate the exposure of articles, we measure both the item-space coverage and the Gini coefficient as suggested in previous works on evaluation [6,13]. Coverage computes the percentage of the available catalogue recommended at least once during an experiment, while the Gini coefficient gives more insight into the recommendation distribution by measuring the inequalities in the number of recommendations each item in the catalogue receives. To evaluate the accuracy of the recommendations, we measured normalised discounted cumulative gain (NDCG) [14], recall [13] and mean reciprocal rank [13]. For brevity, we report only the NDCG results in this paper. Both other accuracy metrics support the same findings. In online trials, we evaluate the quality of the recommendations by click-through rate (CTR).

Datasets. For our experiments, we use three different newspaper websites as our testing platforms, referred to as NP1, NP2 and NP3. The statistics of online traffic and offline exports on these websites can be found in Table 1. Offline datasets are constructed by selecting events from an eight-day window on the website.

Table 1. Statistics of websites used in the online tests.

Website	Users (per day)	Articles read (per day)	Clicks (per day)	$|U|$	$|I|$	$|\mathcal{D}|$	Gini coeff.
NP1	300K	1M	25K	410 843	2 382	4 049 944	0.79
NP2	200K	800K	14K	234 839	2 404	2 852 956	0.77
NP3	1M	4M	160K	1 215 900	5 531	13 842 991	0.88

Offline Experiments. In our offline experiments, we closely mimic the online setup. The first day of our eight-day dataset is used to make sure that we always have a full day of training data when training a model. The second day is used for optimising other hyperparameters than α. The last six days are used for evaluation. Models are trained, following the online setting, on a single day of training data. During optimisation and evaluation, we expect the model to predict a user's last event between 10 AM and 2 PM on each day, using their second to last event in that window as the context item. The measurements from each of the six evaluation days are averaged and reported in this paper.

As our online tests show three items to the user, we also evaluate the offline metrics on the top three recommendations. We ran our experiments for $\alpha \in \{0, 0.1, 0.2, 0.3, 0.4, 0.5, 0.6, 0.7, 0.8, 0.9, 1\}$. For our online tests, we selected $\alpha \in \{0, 0.2, 0.5, 0.7, 1\}$, as they resulted in different exposure distributions. For brevity, we only report results for these values of α.

Online Experiments. Recommendations were displayed in a horizontal list of three items, just after the end of an article. The models for both the control and treatment groups are re-trained every 15 minutes, using a day of training data. This training window was optimised following the procedure defined by Verachtert et al. [22]. In order to evaluate the impact of α in a real and dynamic environment, we have performed a sequence of trials. In each of these trials, a control group of 75% of the users received recommendations using $\alpha = 0.5$. The treatment group (25% of users) received recommendations using a different $\alpha \in \{0, 0.2, 0.7, 1\}$ for each trial period. As it is not possible to compare the CTR between treatment groups, we instead use the lift in CTR for each treatment group compared to the control group during each trial.

4 Experiments

RQ1: How Does the Hyperparameter α Impact the Equality of Exposure? In Table 2 we show that increasing α leads to higher coverage and to more equal exposure between items. Increasing α from 0.7 to 1.0 does lead to only minor improvements in the Gini coefficient and to a reduction of coverage in two datasets.

In Fig. 1 we look beyond the metrics and inspect how the α hyperparameter impacts how often items are recommended on the NP3 website. Items are sorted by popularity, from most popular to least popular along the x-axis. When α is 0, almost all recommendations are from the most popular items. As the value of the hyperparameter increases, more and more different items are recommended, until the distribution shifts when α is 1, and mostly unpopular items are recommended. This insight explains the slight decrease in coverage for some of the datasets, and why the Gini coefficient did not decrease further when increasing α to the max. These distribution plots, also show that none of the α settings provides true equality of exposure, as the middle section of items is always under-recommended, compared to popular or unpopular items depending on the value of α.

Table 2. Coverage and Gini coefficient results for each of the hyperparameter configurations in the online experiments.

α	Coverage@3 (%)					Gini coeff.				
	0.0	0.2	0.5	0.7	1.0	0.0	0.2	0.5	0.7	1.0
NP1	71	87	94	**97**	95	0.91	0.89	0.83	0.79	**0.76**
NP2	57	78	93	**94**	**94**	0.92	0.90	0.83	0.78	**0.76**
NP3	78	94	97	**100**	99	0.91	0.90	0.80	**0.70**	**0.70**

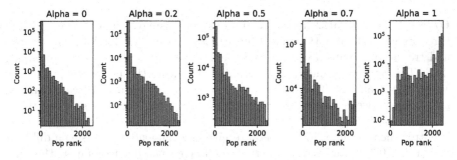

Fig. 1. Number of times items are recommended on the NP3 website experiment, ranked by popularity. The lowest rank is the most visited item.

RQ2: How Does the Hyperparameter α Impact Accurracy and CTR Results?
In Table 3, we show the NDCG@3 for each of the settings of α in our offline tests and the lift in CTR during the online tests.

In the offline experiments increasing the α hyperparameter beyond 0.2 leads to a decrease in performance. As less popular items are recommended, accuracy suffers. Online we find a similar result, higher values of α do not correlate with a higher CTR. However, maximal online performance is reached with the control setting of $\alpha = 0.5$.

So, while a higher α results in a higher coverage and a lower Gini coefficient, both the click-through rate and the NDCG show a decrease in performance when we increase α too much. In our news use-cases, exposure equality and countering popularity bias need to be balanced with recommendation performance. Popular items are relevant to many users, and so if we want to showcase more, less popular, items, we might need to accept a performance decline.

Table 3. NDCG@3 (offline) results and CTR (online) results. CTR results are relative performance compared to the control setting ($\alpha = 0.5$).

α	NDCG@3 (%)					CTR lift (%)				
	0.0	0.2	0.5	0.7	1.0	0.0	0.2	0.5	0.7	1.0
NP1	8.52	**9.15**	7.68	5.27	0.70	-6.40	-4.05	**0**	-4.82	-21.26
NP2	5.54	**6.43**	6.41	4.47	1.07	-3.28	-1.28	**0**	-6.12	-26.45
NP3	6.44	**7.02**	6.48	4.16	0.40	-6.87	-3.91	**0**	-6.93	-31.60

RQ3: Do the Offline and Online Results Agree? In the offline results, the optimal setting for all datasets is $\alpha = 0.2$. However, in our online results, $\alpha = 0.2$ is not optimal, instead $\alpha = 0.5$ is the optimal setting.

Our datasets, like many news datasets, show an unbalanced reading behaviour, indicated by the high Gini coefficient in Table 1. Users read the most popular items much more often than the other items. This popularity bias leads to higher performance in offline results for algorithms with more popularity bias (lower α). However, in the production setting, recommending mostly popular items leads to recommending popular items not related to the context item. Users looking for related articles do not click on these popularity-based recommendations. These results follow the common finding, due to popularity bias offline and online results do not align nicely. However, we can see the value of the offline experimentation in the performance of the $\alpha = 1$ setting. The bad offline performance is reflected in the online results.

5 Conclusion

We find that while the hyperparameter α is able to counteract popularity bias, it is only a proxy for true exposure equality. Therefore, further research is required on how to combat the popularity bias of the ItemKNN algorithm. Secondly, we note that our offline and online results do not align, due to the inherent popularity bias in typical offline evaluation [4,5,23]. Our findings suggest that it is worthwhile to opt for suboptimal offline test results in terms of accuracy, but with a lower Gini index. However, a trade-off should be sought between fair exposure and user experience. We leave a thorough investigation of this trade-off and a framework for determining the setting most likely to perform best in online tests for future work. Finally, we note that our results are limited to the news domain. We see no reason to believe that our findings will not generalize to other domains, as they were not dependent on specific characteristics of the news context. However, it is our aim to replicate these findings in other domains, provided we find partners to perform these trials with.

References

1. Abdollahpouri, H., Burke, R., Mobasher, B.: Controlling popularity bias in learning-to-rank recommendation. In: Proceedings of the Eleventh ACM Conference on Recommender Systems, RecSys 2017, pp. 42–46. Association for Computing Machinery, New York (2017). https://doi.org/10.1145/3109859.3109912. ISBN 9781450346528
2. Abdollahpouri, H., Mansoury, M., Burke, R., Mobasher, B.: The unfairness of popularity bias in recommendation. In: CEUR Workshop Proceedings, vol. 2440 (2019). https://ceur-ws.org/Vol-2440/paper4.pdf
3. Bambini, R., Cremonesi, P., Turrin, R.: A recommender system for an IPTV service provider: a real large-scale production environment. In: Ricci, F., Rokach, L., Shapira, B., Kantor, P.B. (eds.) Recommender Systems Handbook, pp. 299–331. Springer, Boston (2011). https://doi.org/10.1007/978-0-387-85820-3_9
4. Beel, J., Genzmehr, M., Langer, S., Nürnberger, A., Gipp, B.: A comparative analysis of offline and online evaluations and discussion of research paper recommender system evaluation. In: Proceedings of the International Workshop on Reproducibility and Replication in Recommender Systems Evaluation, RepSys 2013, pp. 7–14. Association for Computing Machinery, New York (2013). https://doi.org/10.1145/2532508.2532511. ISBN 9781450324656
5. Beel, J., Langer, S.: A comparison of offline evaluations, online evaluations and user studies in the context of research-paper recommender systems. In: Kapidakis, S., Mazurek, C., Werla, M. (eds.) Research and Advanced Technology for Digital Libraries, pp. 153–168. Springer Cham (2015). https://doi.org/10.1007/978-3-319-24592-8_12. ISBN 978-3-319-24592-8
6. Castells, P., Hurley, N., Vargas, S.: Novelty and diversity in recommender systems. In: Ricci, F., Rokach, L., Shapira, B. (eds.) Recommender Systems Handbook, pp. 603–646. Springer, New York (2022). https://doi.org/10.1007/978-1-0716-2197-4_16. ISBN 978-1-0716-2197-4
7. Deshpande, M., Karypis, G.: Item-based top-n recommendation algorithms. ACM Trans. Inf. Syst. **22**(1), 143–177 (2004). https://doi.org/10.1145/963770.963776. ISSN 1046–8188
8. Eksombatchai, C., et al.: Pixie: a system for recommending 3+ billion items to 200+ million users in real-time. In: Proceedings of the 2018 World Wide Web Conference, WWW 2018, pp. 1775–1784. International World Wide Web Conferences Steering Committee, Republic and Canton of Geneva, CHE (2018). https://doi.org/10.1145/3178876.3186183. ISBN 9781450356398
9. Ferrari Dacrema, M., Boglio, S., Cremonesi, P., Jannach, D.: A troubling analysis of reproducibility and progress in recommender systems research. ACM Trans. Inf. Syst. **39**(2) (2021). https://doi.org/10.1145/3434185. ISSN 1046–8188
10. Ferrari Dacrema, M., Cremonesi, P., Jannach, D.: Are we really making much progress? A worrying analysis of recent neural recommendation approaches. In: Proceedings of the 13th ACM Conference on Recommender System, RecSys 2019, pp. 101–109. Association for Computing Machinery, New York (2019). https://doi.org/10.1145/3298689.3347058. ISBN 9781450362436

11. Garcin, F., Faltings, B., Donatsch, O., Alazzawi, A., Bruttin, C., Huber, A.: Offline and online evaluation of news recommender systems at Swissinfo.ch. In: Proceedings of the 8th ACM Conference on Recommender Systems, RecSys 2014, pp. 169–176. Association for Computing Machinery, New York (2014). https://doi.org/10.1145/2645710.2645745. ISBN 9781450326681

12. Gruson, A., et al.: Offline evaluation to make decisions about playlist recommendation algorithms. In: Proceedings of the Twelfth ACM International Conference on Web Search and Data Mining, WSDM 2019, pp. 420–428. Association for Computing Machinery, New York (2019). https://doi.org/10.1145/3289600.3291027. ISBN 9781450359405

13. Gunawardana, A., Shani, G., Yogev, S.: Evaluating recommender systems. In: Ricci, F., Rokach, L., Shapira, B. (eds.) Recommender Systems Handbook, pp. 547–601. Springer, New York (2022). https://doi.org/10.1007/978-1-0716-2197-4_15 ISBN 978-1-0716-2197-4

14. Järvelin, K., Kekäläinen, J.: cumulated gain-based evaluation of IR techniques. ACM Trans. Inf. Syst. **20**(4), 422–446 (2002). https://doi.org/10.1145/582415.582418. ISSN 1046–8188

15. Kersbergen, B., Sprangers, O., Schelter, S.: Serenade - low-latency session-based recommendation in e-commerce at scale. In: Proceedings of the 2022 International Conference on Management of Data, SIGMOD 2022, pp. 150–159. Association for Computing Machinery, New York (2022). https://doi.org/10.1145/3514221.3517901. ISBN 9781450392495

16. Ludewig, M., Mauro, N., Latifi, S., Jannach, D.: Performance comparison of neural and non-neural approaches to session-based recommendation. In: Proceedings of the 13th ACM Conference on Recommender Systems, RecSys 2019, pp. 462–466. Association for Computing Machinery, New York (2019). https://doi.org/10.1145/3298689.3347041. ISBN 9781450362436

17. Mansoury, M., Abdollahpouri, H., Pechenizkiy, M., Mobasher, B., Burke, R.: Feedback loop and bias amplification in recommender systems. In: Proceedings of the 29th ACM International Conference on Information and Knowledge Management, CIKM 2020, pp. 2145–2148. Association for Computing Machinery, New York (2020). https://doi.org/10.1145/3340531.3412152. ISBN 9781450368599

18. Mei, M.J., Zuber, C., Khazaeni, Y.: A lightweight transformer for next-item product recommendation. In: Proceedings of the 16th ACM Conference on Recommender Systems, RecSys 2022, pp. 546–549. Association for Computing Machinery, New York (2022). https://doi.org/10.1145/3523227.3547491. . ISBN 9781450392785

19. Pellegrini, R., Zhao, W., Murray, I.: Don't recommend the obvious: estimate probability ratios. In: Proceedings of the 16th ACM Conference on Recommender Systems, RecSys 2022, pp. 188–197. Association for Computing Machinery, New York (2022). https://doi.org/10.1145/3523227.3546753. ISBN 9781450392785

20. Rehorek, T., Biza, O., Bartyzal, R., Kordik, P., Povalyev, I., Podstavek, O.: Comparing offline and online evaluation results of recommender systems. In: REVEAL 2018: Proceedings of the Workshop on Offline Evaluation for Recommender Systems (2018). https://users.fit.cvut.cz/rehorto2/files/comparing-offline-online.pdf

21. Rossetti, M., Stella, F., Zanker, M.: Contrasting offline and online results when evaluating recommendation algorithms. In: Proceedings of the 10th ACM Conference on Recommender Systems, RecSys 2016, pp. 31–34. Association for Computing Machinery, New York (2016). https://doi.org/10.1145/2959100.2959176. ISBN 9781450340359

22. Verachtert, R., Michiels, L., Goethals, B.: Are we forgetting something? Correctly evaluate a recommender system with an optimal training window. In: Proceedings of the Perspectives on the Evaluation of Recommender Systems Workshop 2022, Seattle, WA, USA. CEUR-WS.org, September 2022
23. Zangerle, E., Bauer, C.: Evaluating recommender systems: survey and framework. ACM Comput. Surv. **55**(8) (2022). https://doi.org/10.1145/3556536. ISSN 0360–0300
24. Zhu, Z., He, Y., Zhao, X., Caverlee, J.: Popularity bias in dynamic recommendation. In: Proceedings of the 27th ACM SIGKDD Conference on Knowledge Discovery and Data Mining, KDD 2021, pp. 2439–2449. Association for Computing Machinery, New York (2021). https://doi.org/10.1145/3447548.3467376. ISBN 9781450383325

Neural Ad-Hoc Retrieval Meets Open Information Extraction

Duc-Thuan Vo[1], Fattane Zarrinkalam[2]([✉]), Ba Pham[3], Negar Arabzadeh[4], Sara Salamat[1], and Ebrahim Bagheri[1]

[1] Toronto Metropolitan University, Toronto, Canada
{thuanvd,sara.salamat,bagheri}@ryerson.ca
[2] University of Guelph, Guelph, Canada
fzarrink@uoguelph.ca
[3] University of Toronto, Toronto, Canada
ba.pham@theta.utoronto.ca
[4] University of Waterloo, Waterloo, Canada
narabzad@uwaterloo.ca

Abstract. This paper presents the idea of systematically integrating relation triples derived from Open Information Extraction (OpenIE) with neural rankers in order to improve the performance of the ad-hoc retrieval task. This is motivated by two reasons: (1) to capture longer-range semantic associations between keywords in documents, which would not otherwise be immediately identifiable by neural rankers; and (2) identify closely mentioned yet semantically unrelated content in the document that could lead to a document being incorrectly considered to be relevant for the query. Through our extensive experiments on three widely used TREC collections, we show that our idea consistently leads to noticeable performance improvements for neural rankers on a range of metrics.

1 Introduction

Ad-hoc Information Retrieval (IR) is focused on identifying and ranking relevant documents given a user information need expressed in the form of a query [6,10]. While effective in practice, traditional keyword-based IR systems [11,23] can face challenges such as *vocabulary mismatch* or *terminological ambiguity* when the user chooses to use dissimilar query terms from those used in the document collection while formulating their meaning/intent. To address these challenges, there have been recent works in *semantics-enabled* IR that integrate information from knowledge graphs to model documents and queries through a set of concepts (aka knowledge graph entities) [1,5,16]. Researchers have shown that while knowledge graph-based methods are not necessarily always stronger than their keyword-based counterparts, they do offer significant performance improvements on subsets of queries that are difficult for keyword-based methods and hence exhibit synergistic impact, which is valuable in practice [6,17].

More recently, neural ranking models have been proposed for ad-hoc retrieval that automatically learn distributed representations of queries and documents

J. Kamps et al. (Eds.): ECIR 2023, LNCS 13981, pp. 655–663, 2023.
https://doi.org/10.1007/978-3-031-28238-6_57

Q: Paul Dirac Occupation
D_1: Paul was a scientist, who works …Cambridge.
D_2: Dirac recalls the time when John began his career as scientist…Cambridge.

Fig. 1. A sample query and two related document snippets.

that can effectively capture relevance relations (*representation-based*) [18], or to model the query-document relevance directly from their word-level interactions (*interaction-based*) [4,22]. Such neural approaches have shown to be effective in the context of ad-hoc retrieval when large scale training data is available [4]. However, although they are able to capture semantic associations between terms, they are not designed to capture longer-range semantic association between terms and entities within queries and documents [13]. For example, in Fig. 1, both documents, i.e., D_1 and D_2, employ semantically similar terminology which are highly relevant to the terms in the query. However, D_2 is clearly not related to the query primarily because it is not related to the subject of the query, although it does contain the term *'Dirac'* that relates to terms such as *'career'* and *'scientist'* that are semantically related to the term *'occupation'* in the query. Neural ranking models would typically rank D_2 higher than D_1, despite it not being related to the query, because the query term *'Dirac'*, which is mentioned in D_2, is more specific compared to the term *'Paul'*.

Motivated by such examples, our work is focused on strengthening existing neural ranking models by capturing longer-range semantic associations between terms within a document and minimizing the impact of potentially irrelevant yet syntactically related terms. We propose to achieve this objective by exploiting Open Information Extraction (OpenIE) [7,12,14] to capture relation information from documents and integrating them with neural ranking models. OpenIE enables us to extract relation triples from textual corpora. For example in Fig. 1, $< 'Paul', \ 'was', \ 'a \ scientist'>$ and $< 'John', \ 'began \ his \ career \ as', \ 'scientist'>$ are two triples extracted by OpenIE from D_1 and D_2, respectively. We observe that while D_2 contains *'Dirac'*, OpenIE identifies a lack of a relationship between this term and *'scientist'* and therefore does not include it in a relation triple. Therefore, integrating OpenIE with a neural ranking model would allow the model to realize that there is a direct relation between *'Paul'* and *'scientist'* in D_1, which does not exist between *'Dirac'* and *'scientist'* in D_2. Therefore, allowing the neural ranking model to learn that D_1 is relevant to the query while D_2 is not. The most significant contributions of our work are as follows:

1. We propose to integrate OpenIE with both representation-based and interaction-based neural ranking models;
2. We show that longer-range semantic associations and contextual information can be extracted by OpenIE and incorporated in neural ranking models;
3. Through systematically evaluating on TREC collections, we show our approach improves the performance of state-of-the-art neural ranking models.

2 Methodology

We define two main research questions that will be systematically studied in this paper. In these research questions, we will investigate whether integrating OpenIE with neural ranking models can show improved retrieval performance against baselines with which OpenIE was integrated. **RQ1** will explore the performance of *representation-based* models and **RQ2** will evaluate *interaction-based* models on the queries of the benchmark datasets. Therefore, we first use OpenIE to automatically extract triples representing basic propositions or assertions from text. Specifically, given a document d composed of a sequence of m sentences, i.e., $d = <S_1, ..., S_m>$, we apply OpenIE over the sentences, to extract its facts in the form of a sequence of triples $\theta_S = <T_1, T_2, ..., T_p>$. Each triple is in the form of ternary relation $T_i = <a_{i1}, p_i, a_{i2}>$, where p_i denotes the predicate that shows a semantic relation between the first argument a_{i1}, and the second one a_{i2}. Finally, each document d is represented as a sequence of triples θ_d by concatenating the triples extracted from each sentence, i.e., $\theta_d = <T_1, T_2, ..., T_K>$, where $K \leq m \times p$. Then, we integrate OpenIE outputs into neural ranking models, i.e., representation-based and interaction-based models, as follows:

2.1 Integrating OpenIE with Representation-Based Ranking Models

Given a query q and a document d, for representation-based ranking models, in the embedding layer, we use two independent identical neural network models Γ_q and Γ_d to map q and d into feature vectors v_q and v_d, respectively. We apply a Siamese architecture in which Γ_q and Γ_d are identical [3]. To integrate the extracted triples from OpenIE into the embedding layer of the models, given a document d represented as a sequence of triples, i.e., $\theta_d = <T_1, T_2, ..., T_K>$, we consider each triple $T_i = <a_{i1}, p_i, a_{i2}>$ as a sequence of terms where each term is an argument or a predicate. Then, we embed each term instead of each word as the initial embeddings for Γ_d. For example, the triple $<$'a bombing', 'happened in', 'Iraq'$>$ is considered as three terms. One possible advantage of considering the extracted triple's terms by OpenIE instead of unigrams in the embedding layer is that it can both capture the local information and leverage rich contextual information from the whole document. Given the extracted feature vectors of query and document, v_q and v_d, in the matching layer, the relevance score of the query-document pair (q, d) is calculated by the matching function learnt between the query and document spaces.

2.2 Integrating OpenIE with Interaction-Based Ranking Models

In order to incorporate the extracted triples from OpenIE with an interaction-based model, given a document d represented as a set of K triples, i.e., $\theta_d = T_1, T_2, ..., T_K$, we first convert each triple $T_i = <a_{i1}, p_i, a_{i2}>$ to its equivalent sequence of words, i.e., $<w_1, w_2, ..., w_Y>$ by concatenating the words in each of the arguments and the predicate. The flattened form of each triple generates a sentence. For example, the triple $<$'Paul', 'was', 'a scientist'$>$ is converted to the sentence 'Paul was a scientist'. Then, we develop a reworked version of each document d, denoted by d^r, by representing it as a set of sentences derived from the flattened relation tuples. Given a query q and a tuple-based document d^r, we first employ an embedding layer to map each word $w \in q$ or d^r into an L-dimension embedding v_w. Then, in the translation layer we build a translation matrix TM between q and d^r based on the word embedding pairwise similarity between words mentioned in the query and the document [2]. Finally, we calculate the final ranking score from the matrix TM by first applying a feature extractor $\phi()$ on TM, and then combining the features by a ranking layer to produce the final ranking score: $f(q,d) = tanh(a^T \phi(TM) + b)$ where a and b are the ranking parameters that need to be learnt and $tanh()$ is the activation function to control the range of ranking scores.

3 Empirical Evaluation

3.1 Datasets and Experimental Setups

Dataset. To answer our research questions, we conducted experiments on three widely used TREC collections, namely the ClueWeb09-B, ClueWeb12-B and Gov2. We conducted our experiments on both *Title* and *description* fields as our queries and the results were consistent, but for the sake of space, we only report the results on Titles. Further, our results are reported based on a 5-fold cross-validation for each collection.

OpenIE. Without loss of generality, we employed LS3RyIE [19] as a clause-based framework that focuses on the use of syntactic and dependency parsing.

Embedding Layer. For pre-trained embeddings, we used the publicly available Wikipedia-based GloVe embeddings, with an embedding dimension of 100 [15].

Neural Ranking Baselines. We selected two state-of-the-art baselines from each category of neural ranking models as follows: (a) *Representation-based models:* (1) MV-LSTM [20] adopts Bi-LSTM to produce positional sequence representations with two hidden vectors to indicate the meaning of the whole sentence from two directions for capturing the local information as well as the global information in the sentences. (2) Match-LSTM [21] employs neural attention models to derive attention weighted vector representations of the premise and perform word-by-word matching of the hypothesis with the premise. (b) *Interaction-based models:* (1) Conv-KNRM [4] extends KNRM by applying CNNs to represent n-grams of different lengths in a unified embedding space. Its remaining architecture is identical to KNRM. Both Conv-KNRM and KNRM use kernel pooling on interaction features to compute similarity scores.(2) KNRM [22] employs kernel-pooling to produce soft-match signals between words and then learns word embeddings and the ranking layer in an end-to-end fashion. For the implementation of the neural ranking models, we used the MatchZoo framework [9].

Table 1. Performance of neural ranking models on ClueWeb 09-B.

		MRR	MAP	nDCG@10	nDCG@20	P@10	P@20
Representation	MV-LSTM	51.85	40.05	47.15	46.36	48.02	45.93
	MV-LTSM + IE	54.97 (6.02%)	41.09 (2.56%)	48.93 (3.78%)	49.94 (7.72%)	49.13 (3.16%)	49.54 (6.96%)
	Match-LSTM	35.39	23.84	22.41	23.19	24.87	24.23
	Match-LSTM + IE	39.23 (10.85%)	25.25 (5.91%)	26.28 (17.26%)	26.73 (15.26%)	26.88 (8.08%)	26.07 (7.59%)
Interaction	Conv-KNRM	34.21	29.75	28.70	27.93	31.65	30.56
	Conv-KNRM + IE	39.95 (16.77%)	34.56 (16.17%)	36.75 (28.05%)	31.45 (12.63%)	41.24 (30.29%)	38.74 (26.28%)
	KNRM	36.42	29.59	28.53	29.69	30.93	30.18
	KNRM + IE	42.14 (15.71%)	34.04 (15.03%)	35.78 (25.41%)	31.02 (4.49%)	40.56 (31.16%)	37.56 (24.48%)

3.2 Findings

To answer RQs 1 and 2, we evaluate the effect of integrating OpenIE into two neural ranking baseline models from representation-based models (RQ1) and two baselines from interaction-based models (RQ2). The results of each neural ranking baseline and its integrated version with OpenIE on ClueWeb 09-B, ClueWeb 12-B and Gov2 are reported in Tables 1, 2 and 3, respectively.

By comparing each neural ranking baseline and its integrated version with OpenIE (e.g., Conv-KNRM vs. Conv-KNRM+IE), we observe that integrating OpenIE into all the baselines leads to improvements in the performance of the baselines on all the three TREC collections and in terms of all the evaluation metrics. For example, on ClueWeb09-B in terms of MAP, the Conv-KNRM+IE method improves the Conv-KNRM method by a margin of 16.17%. We can conclude that OpenIE effectively contributes to improving both representation-based and interaction-based neural ranking models.

We also observe that, in most cases, improvements to interaction-based models (i.e., Conv-KNRM and KNRM) is greater compared to the improvements made to representation-based models (i.e., MV-LSTM and Match-LSTM). For example, on ClueWeb09-B in terms of MAP, the interaction-based neural ranking models, i.e., Conv-KNRM+IE and KNRM+IE, improve Conv-KNRM and KNRM by a margin of 16.17% and 15.03%, respectively. However, MV-LSTM+IE and Match-LSTM+IE, which are representation-based models, outperform their corresponding baselines by a margin of 2.56% and 5.91%, respectively, which is less compared to the improvements observed on the representation-based models. This indicates that integrating OpenIE with neural ranking models is more effective for interaction-based models compared to representation-based models. However although, MV-LSTM+IE, which is a representation-based method, is the best performing model on the TREC collections with MAP of $\geq 39.81\%$, the interaction-based Conv-KNRM+IE and KNRM+IE models are in the second and third rank with the score of $\geq 34.56\%$ and $\geq 33.47\%$, respectively. Our findings here are consistent with those reported in [8], which show interaction-based neural models often exhibit a superior performance compared to representation-based models in the context of ad-hoc retrieval as interaction-based models can capture more precise matching signals between words in the document and query spaces as they are trained to align the two spaces. Overall, we recommend integrating OpenIE methods into both interaction-based and representation-based models for ad-hoc retrieval as improvements on both types are noticeable.

Table 2. Performance of neural ranking models on ClueWeb 12-B.

		MRR	MAP	nDCG@10	nDCG@10	P@10	P@20
Representation	MV-LSTM	48.93	42.18	43.77	45.81	43.25	43.00
	MV-LSTM + IE	54.51 (11.4%)	44.24 (4.65%)	47.05 (7.49%)	47.47 (3.62%)	46.23 (6.89%)	44.74 (4.04%)
	Match-LSTM	38.48	33.54	29.56	31.78	35.20	34.87
	Match-LTSM + IE	39.43 (2.46%)	34.58 (3.10%)	31.54 (6.69%)	32.56 (2.45%)	36.58 (3.92%)	36.10 (3.52%)
Interaction	Conv-KNRM	41.64	38.08	35.94	37.36	39.59	38.49
	Conv-KNRM + IE	53.61 (28.74%)	43.41 (13.99%)	39.04 (28.32%)	43.41 (16.19%)	45.09 (13.89%)	44.80 (15.07%)
	KNRM	48.86	43.12	39.01	40.73	38.59	38.50
	KNRM + IE	49.89 (2.11%)	44.23 (2.62%)	43.01 (10.25%)	43.25 (6.18%)	42.10 (9.09%)	41.87 (8.75%)

Table 3. Performance of neural ranking models on Gov2.

		MRR	MAP	nDCG@10	nDCG@20	P@10	P@20
Representation	MV-LSTM	37.89	41.25	42.45	40.24	38.36	37.89
	MV-LSTM + IE	38.79 (7.35%)	43.13 (2.37%)	43.82 (4.56%)	41.91 (2.46%)	39.67 (2.46%)	38.79 (2.46%)
	Match-LSTM	48.36	33.60	34.63	36.36	33.95	32.41
	Match-LTSM + IE	50.17 (3.74%)	34.40 (4.29%)	35.95 (3.81%)	37.80 (3.96%)	35.84 (5.56%)	34.19 (5.49%)
Interaction	Conv-KNRM	50.54	35.68	38.47	38.56	38.56	35.85
	Conv-KNRM + IE	55.23 (9.27%)	38.05 (4.96%)	39.46 (6.64%)	40.23 (2.57%)	40.23 (4.33%)	38.36 (7.00%)
	KNRM	51.25	35.11	40.21	39.78	39.56	35.87
	KNRM + IE	54.87 (7.06%)	36.56 (4.13%)	41.85 (4.07%)	41.25 (3.69%)	40.84 (3.24%)	36.64 (2.14%)

4   Concluding Remarks

This paper explores the positive impact of OpenIE on neural ad-hoc retrieval. We propose that longer-range semantic associations within a document as well as contextual information that can be derived from relation triples extracted by OpenIE techniques can improve the performance of neural ranking models. This paper shows how OpenIE relation triples can be incorporated into interaction-based and representation-based neural ranking models. Based on our experiments on the three TREC collections, we have shown that the consideration of relation triples leads to consistent performance improvement over the baseline neural ranking models on a range of metrics including MRR, nDCG and MAP. As a part of our future work, we will be examining the potentially synergistic impact between neural ranking methods when trained directly on document text and when trained on relation triples as proposed in this paper. We also intend to study how OpenIE techniques can contribute to transformer-based neural ranking models.

References

1. Aliannejadi, M., Zamani, H., Crestani, F., Croft, W.B.: Asking clarifying questions in open-domain information-seeking conversations. In: Piwowarski, B., Chevalier, M., Gaussier, É., Maarek, Y., Nie, J., Scholer, F. (eds.) Proceedings of the 42nd International ACM SIGIR Conference on Research and Development in Information Retrieval, SIGIR 2019, Paris, France, 21–25 July 2019, pp. 475–484. ACM (2019). https://doi.org/10.1145/3331184.3331265
2. Berger, A.L., Lafferty, J.D.: Information retrieval as statistical translation. In: Gey, F.C., Hearst, M.A., Tong, R.M. (eds.) SIGIR 1999: Proceedings of the 22nd Annual International ACM SIGIR Conference on Research and Development in Information Retrieval, Berkeley, CA, USA, 15–19 August 1999, pp. 222–229. ACM (1999). https://doi.org/10.1145/312624.312681
3. Bromley, J., Guyon, I., LeCun, Y., Säckinger, E., Shah, R.: Signature verification using a Siamese time delay neural network. In: Cowan, J.D., Tesauro, G., Alspector, J. (eds.) Advances in Neural Information Processing Systems 6 [7th NIPS Conference], Denver, Colorado, USA, pp. 737–744. Morgan Kaufmann (1993). http://papers.nips.cc/paper/769-signature-verification-using-a-siamese-time-delay-neural-network
4. Dai, Z., Xiong, C., Callan, J., Liu, Z.: Convolutional neural networks for soft-matching n-grams in ad-hoc search. In: Chang, Y., Zhai, C., Liu, Y., Maarek, Y. (eds.) Proceedings of the Eleventh ACM International Conference on Web Search and Data Mining, WSDM 2018, Marina Del Rey, CA, USA, 5–9 February 2018, pp. 126–134. ACM (2018). https://doi.org/10.1145/3159652.3159659
5. Dietz, L., Kotov, A., Meij, E.: Utilizing knowledge graphs for text-centric information retrieval. In: Collins-Thompson, K., Mei, Q., Davison, B.D., Liu, Y., Yilmaz, E. (eds.) The 41st International ACM SIGIR Conference on Research and Development in Information Retrieval, SIGIR 2018, Ann Arbor, MI, USA, 08–12 July 2018, pp. 1387–1390. ACM (2018). https://doi.org/10.1145/3209978.3210187
6. Ensan, F., Bagheri, E.: Document retrieval model through semantic linking. In: de Rijke, M., Shokouhi, M., Tomkins, A., Zhang, M. (eds.) Proceedings of the Tenth ACM International Conference on Web Search and Data Mining, WSDM 2017, Cambridge, UK, 6–10 February 2017, pp. 181–190. ACM (2017). https://doi.org/10.1145/3018661.3018692

7. Gashteovski, K., Yu, M., Kotnis, B., Lawrence, C., Niepert, M., Glavas, G.: BenchIE: a framework for multi-faceted fact-based open information extraction evaluation. In: Muresan, S., Nakov, P., Villavicencio, A. (eds.) Proceedings of the 60th Annual Meeting of the Association for Computational Linguistics (Volume 1: Long Papers), ACL 2022, Dublin, Ireland, 22–27 May 2022, pp. 4472–4490. Association for Computational Linguistics (2022). https://doi.org/10.18653/v1/2022.acl-long.307

8. Guo, J., Fan, Y., Ai, Q., Croft, W.B.: A deep relevance matching model for ad-hoc retrieval. In: Mukhopadhyay, S., et al. (eds.) Proceedings of the 25th ACM International Conference on Information and Knowledge Management, CIKM 2016, Indianapolis, IN, USA, 24–28 October 2016, pp. 55–64. ACM (2016). https://doi.org/10.1145/2983323.2983769

9. Guo, J., Fan, Y., Ji, X., Cheng, X.: Matchzoo: A learning, practicing, and developing system for neural text matching. In: Piwowarski, B., Chevalier, M., Gaussier, É., Maarek, Y., Nie, J., Scholer, F. (eds.) Proceedings of the 42nd International ACM SIGIR Conference on Research and Development in Information Retrieval, SIGIR 2019, Paris, France, 21–25 July 2019, pp. 1297–1300. ACM (2019). https://doi.org/10.1145/3331184.3331403

10. Haddad, D., Ghosh, J.: Learning more from less: Towards strengthening weak supervision for ad-hoc retrieval. In: Piwowarski, B., Chevalier, M., Gaussier, É., Maarek, Y., Nie, J., Scholer, F. (eds.) Proceedings of the 42nd International ACM SIGIR Conference on Research and Development in Information Retrieval, SIGIR 2019, Paris, France, 21–25 July 2019, pp. 857–860. ACM (2019). https://doi.org/10.1145/3331184.3331272

11. Karimzadehgan, M., Zhai, C.: Estimation of statistical translation models based on mutual information for ad hoc information retrieval. In: Crestani, F., Marchand-Maillet, S., Chen, H., Efthimiadis, E.N., Savoy, J. (eds.) Proceeding of the 33rd International ACM SIGIR Conference on Research and Development in Information Retrieval, SIGIR 2010, Geneva, Switzerland, 19–23 July 2010, pp. 323–330. ACM (2010). https://doi.org/10.1145/1835449.1835505

12. Kolluru, K., Mohammed, M., Mittal, S., Chakrabarti, S., Mausam: Alignment-augmented consistent translation for multilingual open information extraction. In: Muresan, S., Nakov, P., Villavicencio, A. (eds.) Proceedings of the 60th Annual Meeting of the Association for Computational Linguistics (Volume 1: Long Papers), ACL 2022, Dublin, Ireland, 22–27 May 2022, pp. 2502–2517. Association for Computational Linguistics (2022). https://doi.org/10.18653/v1/2022.acl-long.179

13. MacAvaney, S., Feldman, S., Goharian, N., Downey, D., Cohan, A.: ABNIRML: analyzing the behavior of neural IR models. Trans. Assoc. Comput. Linguistics **10**, 224–239 (2022). https://doi.org/10.1162/tacl_a_00457

14. Mausam, Schmitz, M., Soderland, S., Bart, R., Etzioni, O.: Open language learning for information extraction. In: Tsujii, J., Henderson, J., Pasca, M. (eds.) Proceedings of the 2012 Joint Conference on Empirical Methods in Natural Language Processing and Computational Natural Language Learning, EMNLP-CoNLL 2012, Jeju Island, Korea, 12–14 July 2012, pp. 523–534. ACL (2012), https://aclanthology.org/D12-1048/

15. Pennington, J., Socher, R., Manning, C.D.: Glove: Global vectors for word representation. In: Moschitti, A., Pang, B., Daelemans, W. (eds.) Proceedings of the 2014 Conference on Empirical Methods in Natural Language Processing, EMNLP 2014, a meeting of SIGDAT, a Special Interest Group of the ACL, Doha, Qatar, 25–29 October 2014, pp. 1532–1543. ACL (2014). https://doi.org/10.3115/v1/d14-1162

16. Reinanda, R., Meij, E., de Rijke, M.: Knowledge graphs: an information retrieval perspective. Found. Trends Inf. Retr. **14**(4), 289–444 (2020). https://doi.org/10.1561/1500000063
17. Shehata, D., Arabzadeh, N., Clarke, C.L.A.: Early stage sparse retrieval with entity linking. In: Hasan, M.A., Xiong, L. (eds.) Proceedings of the 31st ACM International Conference on Information and Knowledge Management, Atlanta, GA, USA, 17–21 October 2022, pp. 4464–4469. ACM (2022). https://doi.org/10.1145/3511808.3557588
18. Shen, Y., He, X., Gao, J., Deng, L., Mesnil, G.: Learning semantic representations using convolutional neural networks for web search. In: Chung, C., Broder, A.Z., Shim, K., Suel, T. (eds.) 23rd International World Wide Web Conference, WWW 2014, Companion Volume, Seoul, Republic of Korea, 7–11 April 2014, pp. 373–374. ACM (2014). https://doi.org/10.1145/2567948.2577348
19. Vo, D., Bagheri, E.: Self-training on refined clause patterns for relation extraction. Inf. Process. Manag. **54**(4), 686–706 (2018). https://doi.org/10.1016/j.ipm.2017.02.009
20. Wan, S., Lan, Y., Guo, J., Xu, J., Pang, L., Cheng, X.: A deep architecture for semantic matching with multiple positional sentence representations. In: Schuurmans, D., Wellman, M.P. (eds.) Proceedings of the Thirtieth AAAI Conference on Artificial Intelligence, Phoenix, Arizona, USA, 12–17 February 2016, pp. 2835–2841. AAAI Press (2016). http://www.aaai.org/ocs/index.php/AAAI/AAAI16/paper/view/11897
21. Wang, S., Jiang, J.: Learning natural language inference with LSTM. In: Knight, K., Nenkova, A., Rambow, O. (eds.) NAACL HLT 2016, The 2016 Conference of the North American Chapter of the Association for Computational Linguistics: Human Language Technologies, San Diego California, USA, 12–17 June 2016, pp. 1442–1451. The Association for Computational Linguistics (2016). https://doi.org/10.18653/v1/n16-1170
22. Xiong, C., Dai, Z., Callan, J., Liu, Z., Power, R.: End-to-end neural ad-hoc ranking with kernel pooling. In: Kando, N., Sakai, T., Joho, H., Li, H., de Vries, A.P., White, R.W. (eds.) Proceedings of the 40th International ACM SIGIR Conference on Research and Development in Information Retrieval, Shinjuku, Tokyo, Japan, 7–11 August 2017, pp. 55–64. ACM (2017). https://doi.org/10.1145/3077136.3080809
23. Yang, P., Lin, J.: Reproducing and generalizing semantic term matching in axiomatic information retrieval. In: Azzopardi, L., Stein, B., Fuhr, N., Mayr, P., Hauff, C., Hiemstra, D. (eds.) ECIR 2019. LNCS, vol. 11437, pp. 369–381. Springer, Cham (2019). https://doi.org/10.1007/978-3-030-15712-8_24

Augmenting Graph Convolutional Networks with Textual Data for Recommendations

Sergey Volokhin[1]([✉]), Marcus D. Collins[2], Oleg Rokhlenko[2], and Eugene Agichtein[1,2]

[1] Emory University, Atlanta, GA, USA
svolokh@emory.edu
[2] Amazon, Seattle, WA, USA
{collmr,olegro,eugeneag}@amazon.com

Abstract. Graph Convolutional Networks have recently shown state-of-the-art performance for collaborative filtering-based recommender systems. However, many systems use a pure user-item bipartite interaction graph, ignoring available additional information about the items and users. This paper proposes an effective and general method, *TextGCN*, that utilizes rich textual information about the graph nodes, specifically user reviews and item descriptions, using pre-trained text embeddings. We integrate those reviews and descriptions into item recommendations to augment graph embeddings obtained using LightGCN, a SOTA graph network. Our model achieves a 7–23% statistically significant improvement over this SOTA baseline when evaluated on several diverse large-scale review datasets. Furthermore, our method captures semantic signals from the text, which are not available when using graph connections alone.

Keywords: Graph Convolutional Networks · Product recommendations · Textual augmentation

1 Introduction

Graph neural network (GNN) approaches to recommendation models have grown in popularity in recent years [20], which is natural since so much of the information in these systems is easily mapped to a graph structure. While there is still some controversy over whether graph-embedding methods outperform more conventional recommendation systems [4], the appeal of GNN systems is strong. It has long been clear that side information and additional knowledge, typically social connections between users, or structured knowledge about items, enhance any recommendation system [19]. However, the use of *unstructured* information about items or users has lagged, despite the availability of vast quantities of unstructured text in the form of user reviews and item descriptions. We are aware of only a couple of examples where such unstructured information has been used in GNN recommender systems [16].

© The Author(s), under exclusive license to Springer Nature Switzerland AG 2023
J. Kamps et al. (Eds.): ECIR 2023, LNCS 13981, pp. 664–675, 2023.
https://doi.org/10.1007/978-3-031-28238-6_58

Our intuition is that unstructured review text and item descriptions capture a great deal of semantic and behavioral information unavailable from the purely topological structure of a user-item interaction graph. We posit that this unstructured text may also contain information that can't be found in conventional knowledge graphs either. For instance, particular users may express what they like about items differently. We not only want to find similar users in terms of *what* items they like or what actors or characters, or attributes they seem to gravitate towards. We want to find similar users in terms of *how* they describe those items and attributes.

At the same time, many GNN recommender systems are increasingly complex, while in at least some cases, it has been shown that the sophisticated mixing and attention mechanisms used might even hinder recommendation accuracy [9]. Therefore, we seek to take the simplest approach that we can find to incorporate unstructured review and item description data into a GNN framework. We will show that a simple means of incorporating unstructured text into a GNN recommender improves the performance of a popular baseline system, LightGCN [9], by a similar amount as much more sophisticated approaches. In summary, our contributions are:

1. We explore ways to augment interaction-based Graph Recommender Systems with textual information for improved node representation and introduce a simple and general approach for integrating both graphical and textual representations of users and items.
2. We experimentally demonstrate the effectiveness of our combined model for recommendation performance, with 6–21% improvements across the evaluation metrics.

We also release the code[1] we have written to the scientific community for transparency and reproducibility.

2 Related Work

Recommending items to users is a naturally graph-oriented problem, and Graph Convolution Networks have recently achieved significant gains in recommender systems [20]. A few of these systems attempt to incorporate additional information about the users or items in the graph, with some success. We discuss them in this section to put our contributions in context.

Many systems that use additional information about entities or users do so by augmenting the user-item graph with additional nodes. For instance, TGCN [1] includes a third class of nodes called *tags* which encode additional structured metadata. KGAT [17] includes in the graph categorical entities connected to items (for example, actor or director entities are connected to movie items). Mei *et al.* [12] choose instead to connect additional entity nodes directly to users, constructing an interaction graph with user-entity and user-item edges. KCAN [3] also attempts to encode knowledge graphs alongside the user-item

---

[1] https://github.com/sergey-volokhin/TextGCN.

interaction graph but uses much more complex methods to incorporate that information into both user and item representations; it achieves mostly minor improvements over KGAT.

MKGAT (for Multi-modal KGAT) [16], is the most similar approach to ours. MKGAT first encodes the user-item interaction graph to produce user and item embeddings. It then concatenates those with embeddings of text and images. These first two steps are similar to our approach. However, MKGAT then adds an additional graph attention network. We use much simpler linear layers or gradient-boosted decision trees to combine the text and interaction graph vectors into a regression and achieve higher relative improvements over our baselines. MKGAT does not use review or product description text but uses the text associated with entities to build their representation. Moreover, none of the additional information is used directly to augment user representations.

Unlike all methods described above, our approach uses additional information to augment both user and item representations.

While it is well known that side information improves graph recommender performance [16,17,20], our model is distinguished in two ways. First, we use raw text from reviews and descriptions, which requires no processing. Second, we have taken a much simpler approach to incorporate that text, which proves equally effective in terms of the relative improvement of each recommendation metric. For instance, MKGAT [16] compares its base model with and without text inputs and finds that using unstructured text improves recall by 3.1% and nDCG by 3.5%. As shown below, the simple approach of TextGCN improves recall @20 by 8–18% and nDCG@20 by 10–31% over LightGCN.

3   Methodology

In this section, we describe our TextGCN framework, which uses the original LightGCN framework as a starting point. First, we discuss two baseline improvements, not involving text, which we have applied to both LightGCN and TextGCN. We sought to improve on the original baseline to help confirm that the improvements we observe when including text indeed are due to the additional information contained in the text over what is available in the graph. Next, we describe how the textual information is used to improve upon the baselines in our proposed model.

3.1   Non-text Related Improvements

The LightGCN paper [9] primarily focused on simplifying the general GNN architecture and left several optimizations unexplored.

Activation Function. We replaced the softplus activation function used in LightGCN with SELU [11] when calculating the Bayesian Personalized Ranking (BPR) loss [15]. Using SELU yielded better results for both our baseline models and those incorporating text features. Therefore every result in this work was computed using SELU as an activation function.

Negative Sampling. The original LightGCN work randomly sampled one negative example for each positive example per user. We conducted several experiments with more complex sampling functions and report below results with the best of these, Dynamic Negative Sampling (DNS) [21], which improved all metrics. DNS first ranks all the items and then selects negatives with the lowest score. Details about these experiments are shown in Sect. 5.1.

3.2 TextGCN

Table 1. Notation used in this paper

Symbol	Definition
d_i	Vector or textual description of item i
$r_{u,i}$	Vector of review written by user u about item i
$\mathcal{N}_x$	Set of neighbors of node x
$\mathbf{t}_x$	Textual representation of node x
$\mathbf{e}_x$	LightGCN vector representing node x

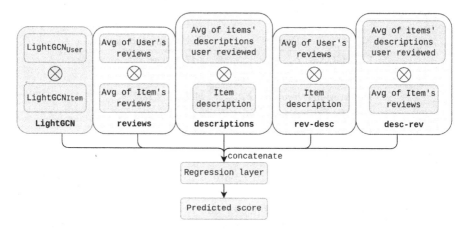

Fig. 1. Architecture of the proposed model. We use 5 features, 4 of which involve textual information (reviews and item descriptions). The LightGCN model is pretrained on user-item interaction graph and frozen, so the first feature does not change when training the regression layer. $\otimes$ represents dot product

Figure 1 shows the architecture of the model. First, we train and freeze the LightGCN model. Then we combine textual representations of items and users to create 4 additional features. The features are mathematically defined in Table 2 using notations from Table 1. Finally, we train a regression layer which predicts the scores for user-item pairs.

We experimented with using both reviews and items' descriptions (see Sect. 4 for a description of the data sources) in both the user and item node representations.

We can represent *users* by using the average of the reviews they have written (Eq. 1) or by using the average of the descriptions of the items they have reviewed (Eq. 2):

$$\mathbf{t}_u = \frac{\sum_{j \in \mathcal{N}_u} r_{u,j}}{|\mathcal{N}_u|} \equiv avg(r_u) \tag{1}$$

$$\mathbf{t}_u = \frac{\sum_{j \in \mathcal{N}_u} d_j}{|\mathcal{N}_u|} \equiv avg_u(d) \tag{2}$$

We can represent *items* by using either average of the reviews written about them (Eq. 3) or using their descriptions (Eq. 4):

$$\mathbf{t}_i = \frac{\sum_{v \in \mathcal{N}_i} r_{v,i}}{|\mathcal{N}_i|} \equiv avg(r_i) \tag{3}$$

$$\mathbf{t}_i = d_i \tag{4}$$

Table 2. All features used in the TextGCN model, and the average weights of the corresponding neurons in the regression layers across all datasets.

Name	Feature	Weight in pred. layer
LightGCN	$\mathbf{e}_u \cdot \mathbf{e}_i$	1.47 ± 0.08
Reviews	$avg(r_u) \cdot avg(r_i)$	8.66 ± 4.45
Descriptions	$avg_u(d) \cdot d_i$	24.44 ± 2.27
Rev-desc	$avg(r_u) \cdot d_i$	-12.80 ± 2.79
Desc-rev	$avg_u(d) \cdot avg(r_i)$	-7.23 ± 4.66

We use those representations to create features for the model, the list of features that the model uses can be found in Table 2. Each feature is constructed by applying dot product on different user and item representations, and is then fed into a regression layer which estimates the score for that user-item pair. For instance, the feature *rev-desc* is the dot product of the average vector of the user's embedded reviews and the vector of the item's embedded description: $avg(r_u) \cdot d_i$.

3.3   Semantic Similarity

In our experiments mixing unstructured text with user-item interaction graphs, we need to compute the similarity between the textual representations of items and users. We used the Sentence-Transformers [14] framework of SOTA sentence and text embeddings to achieve that. Specifically, we use the "all-MiniLM-L6-v2"[2] model trained for semantic search and clustering tasks. Although larger models could further improve the quality, optimizing the specific language model is out of the scope of our work.

4   Data

Previous research has made use of the well-known Amazon Reviews data from 2014 [8] in the Books domain (herein "Books'14"), and we used that data, among others, to validate our code. However, the Books'14 data does not include textual item metadata like descriptions. Therefore, we use a newer version of that data for our experiments, Amazon Reviews 2018 [13], also in the Books domain (herein, "Books'18"), which includes textual item descriptions. Table 3 lists the statistics of the data.

Books'18 has a different distribution than Books'14, so for consistency, we sub-sampled Books'18 to have a similar ratio of users to items as in Books'14 (herein "Sampled'18") and used that as the final dataset in Books domain. Despite those three datasets being very similar in nature and structure, the results we have obtained for Sampled'18 were four times better than results we obtained for Books'14 on 3/4 metrics.

We calculated several sparsity-related metrics on each dataset included in Table 3 to investigate this discrepancy. These metrics demonstrate that Books'18 is much sparser than Books'14. While sub-sampling Books'18 does bring the number of users, items, and total samples closer to Books'14, centrality measures remain smaller than those for Books'14. That seems counter-intuitive: higher edge sparsity should result in lower recommendation performance. Nonetheless, this observation shows that it is important not to compare results across different generations of review data or even across different subsets of a single generation of data.

Sparsity metrics and other statistics for data from the other domains we use in this work ("Toys and Games", "Movies", and "Electronics") are also shown in Table 3.

Table 3. Data statistics (centralities are multiplied by 10^4)

Data	#users	#items	#samples	Sparsity	Degree centr.		Eigenvector centr.	
					Mean	Median	Mean	Median
Books'14	53k	92k	2.9M	99.938%	2.29	1.39	11.48	4.13
Books'18	174k	96k	4M	97.605%	0.86	0.45	5.12	0.89
Sampled'18	92k	58k	2.7M	99.949%	1.79	1.06	7.98	1.43
Movies	268k	78k	3.1M	99.961%	1.09	0.50	9.80	4.55
Toys	64k	32k	0.75M	99.963%	1.32	0.83	12.36	5.40
Electronics	139k	40k	2.1M	99.984%	0.43	0.23	4.45	1.62

---

[2] https://huggingface.co/sentence-transformers/all-MiniLM-L6-v2.

5   Experiments

We first describe the baselines we use for comparison and then the results of our experiments adding unstructured text representations to those baselines. The results of all models described in this section are shown in Table 4.

5.1   Baselines

Collaborative Filtering Baseline. The first baseline does not use graphs and works as a sanity check to ensure that a basic CF model does not outperform our much more complex approach. We use the "implicit" [6] Python library to build several CF systems from the user-item interaction matrix. The results of the best CF system–BayesianPersonalizedRanking in all cases–are shown in Table 4, marked as "CF (BPR)".

Graph-Based Baselines. We experiment with LightGCN [9] and several derivatives of it as baselines. LightGCN uses 3 graph propagation layers, with a simple mean aggregation over neighbor nodes, normalized symmetrically by the degree of each node. The final node representation is a simple average over the three layers' outputs (the formulas are available in the original paper [9]).

Single Layer. In the original LightGCN paper, the best model for the Books'14 data uses only the outputs from the final (i.e., third) layer. However, in our experiments, it did not perform as well as the version which takes the average of all layers. The authors called this the "Single" variation. We put the results for it in Table 4.

Alternate Aggregators. Before selecting LightGCN as our base model, we evaluated several other Graph Convolutional Networks, all of which are available in the Python `torch_geometric` [5] package (GCN [10], GAT (v1 [18] and v2), GraphSAGE [7]), however, all of them performed worse than LightGCN. Results are shown in Table 4.

Dynamic Negative Sampling (DNS). The authors of the LightGCN paper have noted [9] that more advanced negative sampling techniques could improve Light-GCN, and we decided to try one such sampling method to evaluate whether the improvements obtained using additional unstructured text would still appear when using improved sampling methods. Following [21], we rank 1000 random items for each user, pick the 40 lowest ranked items, then pick 5 random positives, and train on the Cartesian product of those 2 sets (200 samples per user).
Per Sect. 3.1, we use 1-to-1 sampling, and the activation function in the BPR loss is SELU. Results for all those models can be found in Table 4.

Table 4. All the Baseline and Experimental models that were run on Sampled'18 dataset. Names of TextGCN models reflect which layer is used on top of LightGCN for the final score prediction. 'Linear' is the TextGCN Baseline model with a linear top layer. Metrics averaged over 5 runs

Model	Recall	Precision	Hit rate	nDCG
CF (BPR)	0.1422	0.0391	0.4662	0.1029
Graph baselines				
SAGE	0.0963	0.0263	0.3479	0.0674
GAT	0.1366	0.0359	0.4452	0.0984
GATv2	0.1384	0.0364	0.4503	0.0993
GCN	0.1419	0.0374	0.4584	0.1029
"Single"	0.1162	0.0318	0.4081	0.0833
LightGCN	0.1690	0.0455	0.5210	0.1244
LightGCN w DNS	0.1813	0.0490	0.5467	0.1353
TextGCN				
XGBoost	0.1539	0.0372	0.4736	0.1075
GBDT	0.1749	0.0453	0.5308	0.1304
Linear	0.1833	0.0460	0.5308	0.1350
Linear w DNS	**0.1923**	**0.0485**	**0.5481**	**0.1428**

5.2   TextGCN

We leverage the semantic information in encoded item descriptions and reviews by combining it with the final node vectors from the LightGCN network. The formulas for all features are described in Table 2.

We have experimented with combining different representations. For example, we can represent users by concatenating their LightGCN node vector with the user-averaged item description vector and represent items similarly computing $(\mathbf{e}_u \| avg_u(d)) \cdot (\mathbf{e}_i \| d_i)$ as a feature. However, we found these were highly correlated with other existing features and so degraded the performance. We have omitted these from the table.

Finally, we freeze the LightGCN embeddings when training this final regressor. We also experimented with back-propagating the error signal back through the LightGCN model (unfreezing the model), but we achieved better results with frozen graph and text embeddings. Therefore we show only experiments with frozen embeddings below.

All models are run for 1000 epochs or until they converge and do not show any improvement for 75 consecutive epochs. Evaluation is performed every 25 epochs. All the experiments are run on 5 random vertically-sampled folds of the data: each training fold has all the users and is of size 80% of the whole dataset, and results are averaged.

Prediction Layer

Linear Layer. In this version, we use a simple dense linear layer on top of the features shown in Table 2. Despite the simplicity, it already significantly outperforms the baseline. Furthermore, we can add complexity by introducing hidden layers and increasing their sizes. For example, we added a 16-node hidden layer, which improved the results by an additional ≈1%.

Gradient Boosted Decision Trees. We experimented with gradient-boosted decision trees (GBDT) and XGBoost [2] regressors, however, they also performed worse than LightGCN (results in Table 4).

6   Results and Discussion

Table 4 shows variants of TextGCN that we described in Sect. 5.2. Surprisingly, a simple linear model did best here, and XGBoost performed worse than LightGCN. It proved too easy to overfit the ranking model to the data, and it failed to generalize well. So we use Linear version of TextGCN to run all further experiments.

Our main results for all datasets are shown in Table 5. When we compare each model with text to the corresponding baseline without text, the model incorporating text features is superior to the baseline on every metric on all the datasets. TextGCN improves recall@20 by 13.19%, precision@20 by 12.12%, hit rate by 11.19%, and nDCG by 20.25% on average over LightGCN. Adding Dynamic Negative Sampling boosts the performance of both the LightGCN and TextGCN, however, the improvement gets smaller: recall@20 by 8.03%, precision@20 by 7.02%, hit rate by 6.27%, and nDCG by 10.39%. This supports the conclusion that the actual text of user reviews or descriptions contains useful information beyond what is available in the user-item graph itself (Table 6).

Table 2 shows which features the models use, as well as the average weights in the prediction layer for each feature for all TextGCN models we have trained. Those weights can act as proxies for feature importances and we can draw conclusions from them. We notice that the highest weights are for the "description" feature, which calculates the similarity between the user's average item description vector and the candidate item vector. We speculate that this is because users are attracted to similar aspects across products. Such aspects (say "lightweight") might be implicit in a user-item interaction graph if enough users interacted with the *same* set of products that shared that feature. However, using the text descriptions to represent both users and items appears to be more effective and direct.

Two other textual features, using different textual representations from each user and item (*rev-desc* and *desc-rev*) have negative importance, which suggests that there is no direct useful link between the description given to the product by the seller and the reviews written by the users.

Table 5. Main results for all 4 domains. All metrics @20. All results are statistically significant ($p \ll 0.001$)

Model	Sampled'18 (Books)				Toys			
	Recall	Precis	Hit	nDCG	Recall	Precis	Hit	nDCG
LightGCN	0.1700	0.0429	0.5044	0.1222	0.0988	0.0107	0.1787	0.0571
TextGCN	0.1833	0.0460	0.5308	0.1350	0.1160	0.0125	0.2064	0.0693
LightGCN w DNS	0.1822	0.0463	0.5290	0.1330	0.1114	0.0122	0.2035	0.0662
TextGCN w DNS	0.1923	0.0485	0.5481	0.1428	0.1268	0.0136	0.2258	0.0762
	Movies				Electronic			
	Recall	Precis	Hit	nDCG	Recall	Precis	Hit	nDCG
LightGCN	0.1575	0.0195	0.3074	0.0939	0.0543	0.0058	0.1087	0.0299
TextGCN	0.1723	0.0212	0.3321	0.1107	0.0640	0.0067	0.1261	0.0392
LightGCN w DNS	0.1789	0.0227	0.3475	0.1123	0.0703	0.0078	0.1432	0.0474
TextGCN w DNS	0.1895	0.0239	0.3644	0.1241	0.0750	0.0082	0.1513	0.0514

Table 6. Average relative improvement of TextGCN over the corresponding baseline across all datasets

Metric	w/o DNS	w DNS
Recall	+13.19%	+8.03%
Precision	+12.12%	+7.02%
Hit	+11.19%	+6.27%
nDCG	+20.25%	+10.39%

7 Conclusions

In this work, we have established that knowledge from unstructured text can be exploited to improve recommendations using Graph Neural Networks. This text captures information not present in the user-item interaction graph. By examining the computed feature importances of our ranking models, we identified that using the item description text to augment both user and item representations had the strongest positive influence on recommendation metrics. Furthermore, we have shown that we can efficiently augment lightweight graph embeddings with this text and substantially improve recommendation performance without complex models to combine the graph and textual representations.

References

1. Chen, B., et al.: TGCN: tag graph convolutional network for tag-aware recommendation. In: Proceedings of the 29th ACM International Conference on Information & Knowledge Management, CIKM 2020, pp. 155–164. Association for Computing Machinery, New York (2020). https://doi.org/10.1145/3340531.3411927

2. Chen, T., Guestrin, C.: XGBoost: a scalable tree boosting system. In: Proceedings of the 22nd ACM SIGKDD International Conference on Knowledge Discovery and Data Mining, KDD 2016, pp. 785–794. ACM, New York (2016). https://doi.org/10.1145/2939672.2939785

3. Demartini, G., et al.: Conditional graph attention networks for distilling and refining knowledge graphs in recommendation. In: Proceedings of the 30th ACM International Conference on Information & Knowledge Management, pp. 1834–1843 (2021). https://doi.org/10.1145/3459637.3482331

4. Deng, Y.: Recommender systems based on graph embedding techniques: a review. IEEE Access **10**, 51587–51633 (2022). https://doi.org/10.1109/access.2022.3174197

5. Fey, M., Lenssen, J.E.: Fast graph representation learning with PyTorch geometric. In: ICLR Workshop on Representation Learning on Graphs and Manifolds (2019)

6. Frederickson, B.: Fast python collaborative filtering for implicit datasets (2017). https://github.com/benfred/implicit

7. Hamilton, W., Ying, Z., Leskovec, J.: Inductive representation learning on large graphs. In: Guyon, I., et al. (eds.) Advances in Neural Information Processing Systems, vol. 30. Curran Associates, Inc. (2017). https://proceedings.neurips.cc/paper/2017/file/5dd9db5e033da9c6fb5ba83c7a7ebea9-Paper.pdf

8. He, R., McAuley, J.: Ups and downs: modeling the visual evolution of fashion trends with one-class collaborative filtering. In: proceedings of the 25th International Conference on World Wide Web, pp. 507–517 (2016). https://doi.org/10.1145/2872427.2883037

9. He, X., Deng, K., Wang, X., Li, Y., Zhang, Y., Wang, M.: LightGCN: simplifying and powering graph convolution network for recommendation. In: Proceedings of the 43rd International ACM SIGIR Conference on Research and Development in Information Retrieval, pp. 639–648 (2020). https://doi.org/10.1145/3397271.3401063

10. Kipf, T.N., Welling, M.: Semi-supervised classification with graph convolutional networks. arXiv preprint arXiv:1609.02907 (2016). https://doi.org/10.48550/arXiv.1609.02907

11. Klambauer, G., Unterthiner, T., Mayr, A., Hochreiter, S.: Self-normalizing neural networks. In: Guyon, I., et al. (eds.) Advances in Neural Information Processing Systems, vol. 30. Curran Associates, Inc. (2017). https://proceedings.neurips.cc/paper/2017/file/5d44ee6f2c3f71b73125876103c8f6c4-Paper.pdf

12. Mei, D., Huang, N., Li, X.: Light graph convolutional collaborative filtering with multi-aspect information. IEEE Access **9**, 34433–34441 (2021). https://doi.org/10.1109/access.2021.3061915

13. Ni, J., Li, J., McAuley, J.: Justifying recommendations using distantly-labeled reviews and fine-grained aspects. In: Proceedings of the 2019 Conference on Empirical Methods in Natural Language Processing and the 9th International Joint Conference on Natural Language Processing (EMNLP-IJCNLP), pp. 188–197 (2019). https://doi.org/10.18653/v1/D19-1018

14. Reimers, N., Gurevych, I.: Sentence-BERT: sentence embeddings using siamese BERT-networks. In: Proceedings of the 2019 Conference on Empirical Methods in Natural Language Processing. Association for Computational Linguistics (2019). https://arxiv.org/abs/1908.10084

15. Rendle, S., Freudenthaler, C., Gantner, Z., Schmidt-Thieme, L.: BPR: Bayesian personalized ranking from implicit feedback. In: Proceedings of the Twenty-Fifth Conference on Uncertainty in Artificial Intelligence, UAI 2009, pp. 452–461. AUAI Press, Arlington (2009). https://doi.org/10.48550/arXiv.1205.2618

16. Sun, R., et al.: Multi-modal knowledge graphs for recommender systems. In: Proceedings of the 29th ACM International Conference on Information & Knowledge Management, pp. 1405–1414 (2020). https://doi.org/10.1145/3340531.3411947
17. Teredesai, A., et al.: KGAT: knowledge graph attention network for recommendation. In: Proceedings of the 25th ACM SIGKDD International Conference on Knowledge Discovery & Data Mining, pp. 950–958 (2019). https://doi.org/10.1145/3292500.3330989
18. Veličković, P., Cucurull, G., Casanova, A., Romero, A., Lio, P., Bengio, Y.: Graph attention networks. arXiv preprint arXiv:1710.10903 (2017). https://doi.org/10.48550/arXiv.1710.10903
19. Wang, S., et al.: Graph learning based recommender systems: a review. In: Zhou, Z.H. (ed.) Proceedings of the Thirtieth International Joint Conference on Artificial Intelligence, IJCAI 2021, pp. 4644–4652. IJCAI International Joint Conference on Artificial Intelligence, International Joint Conferences on Artificial Intelligence (2021). https://doi.org/10.24963/ijcai.2021/630. 30th International Joint Conference on Artificial Intelligence, IJCAI 2021; Conference date: 19-08-2021 Through 27-08-2021
20. Wu, S., Sun, F., Zhang, W., Xie, X., Cui, B.: Graph neural networks in recommender systems: a survey. ACM Comput. Surv. (2022). https://doi.org/10.1145/3535101
21. Zhang, W., Chen, T., Wang, J., Yu, Y.: Optimizing top-N collaborative filtering via dynamic negative item sampling. In: Proceedings of the 36th International ACM SIGIR Conference on Research and Development in Information Retrieval, SIGIR 2013, pp. 785–788. Association for Computing Machinery, New York (2013). https://doi.org/10.1145/2484028.2484126

Utilising Twitter Metadata for Hate Classification

Oliver Warke[1]([✉])(iD), Joemon M. Jose[1](iD), and Jan Breitsohl[2](iD)

[1] School of Computing Science, University of Glasgow, Glasgow, UK
o.warke.1@research.gla.ac.uk
[2] Adam Smith Business School, University of Glasgow, Glasgow, UK

Abstract. Social media has become an essential daily feature of people's lives. Social media platforms provide individuals wishing to cause harm with an open, anonymous, and far-reaching channel. As a result, society is experiencing a crisis concerning hate and abuse on social media. This paper aims to provide a better method of identifying these instances of hate via a custom BERT classifier which leverages readily available metadata from Twitter alongside traditional text data. With Accuracy, F1, Recall and Precision scores of 0.85, 0.75, 0.76, and 0.74, the new model presents a competitive performance compared to similar state-of-the-art models. The increased performance of models within this domain can only benefit society as they provide more effective means to combat hate on social media.

Keywords: Hate · Social media · Deep learning · Metadata

1 Introduction

When considering society's interaction on the web, it is impossible to ignore the widespread use of social media [2]. These platforms provide individuals wishing to cause harm with an open, anonymous, and far-reaching channel for the rapid dissemination of user generated content. When a user is provided with anonymity it can create a sense of invulnerability; facilitating bullying and hate speech [4]. This is the case for both positive and negative content, with research showing how a user's intent impacts how far-reaching the content is [19]. Bullying and hate have proven negative impacts on personal [15] and community levels [26], with research showing "social media usage is associated with lower task performance, increased technostress, and lower happiness" [7]. As a result, society is experiencing a crisis with regards to hate and abuse on social media [33]. It is therefore imperative to gain control and reduce this crisis, one avenue which can be utilised is the automated detection of hate on social media.

Whilst there has been lots of work on detecting hate on social media, little has fully utilised the available user-generated content, instead focusing on text data. Alongside text data, there are numerous numerical and categorical features relating to the tweet object and author. Previous work has researched this metadata in the context of hate but falls short of investigating state-of-the-art models. This paper presents a comprehensive evaluation of the metadata features and

J. Kamps et al. (Eds.): ECIR 2023, LNCS 13981, pp. 676–684, 2023.
https://doi.org/10.1007/978-3-031-28238-6_59

various methods for combining them with text to use within classification models. Any increase in classification performance within this domain and a greater knowledge of the available uses of social media data can only benefit society.

2 Literature Review

Social media promotes interaction, connectivity, learning, and many other positive features. In contrast, social media has also been linked to depression, anxiety, stress, anger, and many forms of hate [5]. The presence of hate on social media has been gaining traction in traditional media channels, with Time Magazine devoting a front page to the culture of social media hate [25]. Also seen by the recent racial abuse of football players after the Euros 2022 competition [14]. Many social media platforms fail to employ effective countermeasures against hate. Twitter is one of the largest platforms and most criticised with regards to user generated content, with researchers and wider society recognising it as a venue for hate, toxicity, and bullying [12,18]. It is therefore paramount to gain a better understanding of hate speech on social media, which has been recognised by other researchers. The Oxford Internet Institute [16] state; "To develop effective responses to hate speech, including through education, it is essential to better monitor and analyse the phenomenon by drawing on clear and reliable data... this also means better understanding the occurrence, virulence and reach of online hate speech." Despite the extensive work surrounding hate on social media, researchers have used various terminology to describe the phenomenon alongside 'hate' such as 'bullying' [6], 'abuse' [23], and 'toxicity' [34]. We would also like to note that 'hate' and 'hate speech' are often used interchangeably within research. We define 'online hate' or 'hate' as an extreme interaction intended to cause harm to an individual or group, while 'hate speech' is defined as an extreme interaction intended to cause harm to an individual or group because of a specific characteristic. In this paper the dataset uses a 'hate' label to refer to what we consider to be hate speech.

Hate detection is a necessary first step in the fight against hate before developing effective responses. Text-based classification models are increasingly common due to the short text messages predominantly seen on social media and advancement of models which excel with short text; such as BERT [9,13,21]. Given hate's sensitive nature, automatic hate identification must be as precise as possible. Misidentifying benign content as hate and failing to identify hate both result in undesirable outcomes [17]. Research frequently leverages techniques to improve hate classification. For example, Ayo et al. [3] research combining text with multiple features such as word and sentence level embeddings. Vosoughi et al. [32] curate a dataset for sentiment classification featuring metadata, then combining these features using a Bayesian approach. The best model outperforming other models by more than 3%. Miró-Llinares et al. [20] use metadata to identify hate but only use a random forest classifier. They conclude that tweet metadata is more beneficial to classification performance than account metadata but raise concerns about the trade off between utilising metadata and computational power required. Despite these works, there is a lack of comprehensive evaluation of twitter metadata and its use within hate classification. Given the previous discussion of BERT's success, there is a need for comparison between text, metadata, and combined classification approaches.

3   Methodology

3.1   Dataset and Feature Selection

We used Twitter due to the ease of data access [1], the extensive range of datasets related to the hate domain, and the high presence of hate on the platform. Numerous datasets contain hate, abuse, bullying, and other forms of extreme negative behaviours [31]. Despite this there are few containing twitter metadata or the resources to obtain it. However, one dataset that included Twitter IDs was the Founta et al. dataset [11]. It was uniquely suited to our work as it enabled the collection of twitter metadata associated with each tweet rather than limiting research to the dataset's existing information. This however, also poses an ethical concern regarding user consent. Within our work, usernames were never accessed and user IDs were not used or stored beyond the gathering of account metadata such as followers and account creation time. We would also like to note that all tweets involved were publicly available on twitter and gathered through the platform's academic research API. Additionally, no users were individually targeted for analysis.

Founta et al. carried out a rigorous annotation process, ensuring robust data quality. The labels included in the final dataset were 'normal', 'spam', 'abusive' and 'hateful'. They defined Abusive Language as "Any strongly impolite, rude or hurtful language using profanity, that can show a debasement of someone or something, or show intense emotion". Hate speech as "Language used to express hatred towards a targeted individual or group, or is intended to be derogatory, to humiliate, or to insult the members of the group, based on attributes such as race, religion, ethnic origin, sexual orientation, disability, or gender". This paper disregarded the spam tweets within the dataset as we were uninterested in that behaviour. Therefore, the final dataset for this research contained 9,039 'normal', 4,094 'abusive', and 1,778 'hateful' labelled tweets.

Twitter provides numerous features for each tweet; however, not all features are useful in aiding the classification model. Therefore, we had to deduce which features benefit the classification model and which have a negative or negligible impact. Some features were discarded straight away, such as 'geo' for the tweet object, as very few tweets had a value. The final lists of tweet and account metadata features evaluated were; 'retweets', 'replies', 'likes', and 'quotes' and 'followers', 'following', 'tweet count', 'listed count', and 'account creation time'. Singh et al. [30] advocate the use of normalisation for classifier features, finding that normalisation improves classification performance in most cases, although some normalisation methods were not effective. They also note that the optimal normalisation method was subjective to the classification task. We implemented Quantile Transformation, using scikit-learn's [24] Quantile Transformer class to transform the highly skewed distribution to a normal distribution. Other research papers support the use of normalisation [27,29].

3.2 Classifier

To combine BERT's text classification with metadata we explored various methods. The first method was appending the metadata to the tweet text with added context. "You are a bloody idiot." becomes "You are a bloody idiot. This tweet has 3 replies and 12 likes. The user has 147 followers". The concept being that given BERT's context-aware capabilities, it would be able to make use of the features with added context. The second uses the final BERT classification as an input feature in a random forest model. Random Forest was tested due to it's prior use in the hate research domain when using metadata. The third is using a linear layer to combine the BERT output logits with the numerical features. The combined features are first passed through a linear layer and then a sigmoid layer which applies a sigmoid function to the output reducing it to either 1 or 0. Out of these combination methods the best performing was BERT with the linear layer.

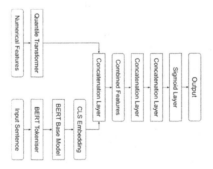

Fig. 1. MetaBert model diagram

To establish a baseline for model performance comparison we explored several state-of-the-art models used within the hate domain, including BERT. The first is HateBERT [8], a model derived from BERT that focuses on 'offensive', 'abusive' and 'hateful' language. HateBERT features intensive pre-training on this language before being deployed for fine-tuning on domain-specific tasks, the authors specifically highlight its portability between datasets. The second model was DistilBERT, a lightweight variation of BERT which has been proven to have competitive performance, with Sanh et al. [28] finding that DistilBERT retains the majority of its performance across difference tasks whilst being 60% faster. We elected to benchmark DistilBERT for its successful use in hate detection by Mutanga et al. [22]. The inclusion of previously utilised models within a domain-specific context provides more significance to metaBERT's performance within the hate domain.

4   Results and Discussion

Table 1. Table showing the feature permutation scores

Retweets	Replies	Likes	Quotes	Followers	Following	Tweet count	Listed count	Account creation time
0.07	0.02	0.04	0.01	0.17	0.17	0.23	0.15	0.15

Table 2. Table showing the metadata feature performance

Metric	Tweet RF	Account RF	Combined RF
Accuracy	0.65	0.73	0.72
Precision	0.29	0.64	0.65
Recall	0.69	0.67	0.66
F1-score	0.42	0.65	0.65

The feature permutation analysis and the random forest classification results show the metadata features have varying impacts on classification models Table 1. The tweet metadata had less impact than account metadata features, whilst the classification metrics for the account metadata RF model far outperform the Tweet metadata RF model Table 2. This is contradictory to the work by Miró-Llinares et al. [20], who found tweet metadata was superior for identifying hate. The poor performance of the tweet metadata model could be attributed to the large volume of zero values present in tweet metadata (retweet (68%), replies (85%), likes (74%), and quotes (96%)). In their work Miró-Llinares et al. eliminated tweets with a percentage of null values larger than 25–30%. This is not a realistic representation of data on Twitter. Not all hateful tweets will have rich metadata associated with them, but that does not exclude them from being hateful behaviour. Any classification system has to be evaluated with realistic null values.

All models were trained and tested over the publicly available Founta et al. hate dataset [11]. The train-test split was 80% and 20% respectively. This resulted in 11,928 training and 2,983 testing data points. All models were trained with batch sizes of 32, four epochs, and one of the four learning rates recommended by the BERT authors (3e-4, 1e-4, 5e-5, 3e-5). The best-performing model was metaBERT, with the highest accuracy (0.85) and the highest F1-score (0.75) Table 3. However, metaBert does not significantly outperform any of the other state-of-the-art models. We conducted a statistical analysis of the results of metaBERT and BERT using McNemar's test [10]. McNemar's test captures the errors made by two models, testing for a significant difference between the two models. The resultant p-value was 0.924, failing to reject the null hypothesis; that the two models have a similar proportion of error rates on the same test set. This indicates that whilst the metadata features can be useful within the classification;

they do not produce a significant increase in performance. Additionally, all the pure metadata models were outperformed by all text and text + metadata models. We can therefore state that tweet text is more useful within classification than tweet and account metadata. When including extra features such as metadata, there must be a consideration of the associated extra computational costs. In this case, results show that despite the metadata improving performance, there is not a large enough improvement to justify the metadata included within the model.

Table 3. Table showing the overall performance of the models

Metric	Meta BERT	BERT	HateBERT	DistilBERT
Accuracy	0.85	0.84	0.84	0.83
Precision	0.76	0.75	0.74	0.72
Recall	0.74	0.73	0.74	0.68
F1-score	0.75	0.74	0.74	0.69

5   Conclusion

This paper presents a novel investigation into the use of Twitter metadata for hate classification. Given the crisis of hate on social media, any contribution to this domain is important due to the need to understand and target ways to reduce hate. Metadata is readily available for any classification task which uses Twitter data; we recommend that researchers at least explore using metadata in their work. This paper finds that not all metadata features are equally impactful within the models, account metadata had a greater impact on classification performance than tweet metadata. This is not necessarily true for all classification tasks, and each feature's impact may be subject to the level of user interaction within that feature. The competitive performance of the metaBERT model against other state-of-the-art models proves that Twitter metadata is a valuable resource which should not be ignored in favour of pure tweet text data within classification tasks. However, given that the performance improvements gained were not significant, researchers should also evaluate the computational cost of using metadata features. Whilst they may produce higher performance in classification metrics, any benefits gained may be outweighed by the extra computational cost.

5.1   Limitations and Future Work

Few Twitter hate datasets contain metadata or the ability to access it. As such, we were unable to replicate the results for additional datasets, which would be hugely beneficial for metadata research. Additionally, the data was four years old and many data points had been lost to time. Another limitation of the work is the use of only three classes to represent the spectrum of hateful behaviour. Hate is very nuanced, with many different forms. An extensive multiclass model with classes across the full spectrum of hateful behaviour allows for more significant analysis and insight into the phenomenon.

Future work should investigate metadata features, evaluating their performance and computational costs, across different domains. With image classification becoming more prominent within the field, a complete multimodal classification model could be employed utilising text, metadata, and images. Research could also examine the different forms of hate, introducing a multiclass dataset covering these various forms from a multimodal perspective.

References

1. Twitter api for academic research | products | twitter developer platform. https://developer.twitter.com/en/products/twitter-api/academic-research
2. Auxier, B., Anderson, M.: Social media use in 2021. Pew Research Center **1**, 1–4 (2021)
3. Ayo, F.E., Folorunso, O., Ibharalu, F.T., Osinuga, I.A.: Machine learning techniques for hate speech classification of twitter data: state-of-the-art, future challenges and research directions. Comput. Sci. Rev. **38**, 100311 (2020)
4. Barlett, C.P., DeWitt, C.C., Maronna, B., Johnson, K.: Social media use as a tool to facilitate or reduce cyberbullying perpetration: a review focusing on anonymous and nonanonymous social media platforms. Violence Gender **5**(3), 147–152 (2018)
5. Best, P., Manktelow, R., Taylor, B.: Online communication, social media and adolescent wellbeing: a systematic narrative review. Child Youth Serv. Rev. **41**, 27–36 (2014)
6. Bretschneider, U., Peters, R.: Detecting cyberbullying in online communities (2016)
7. Brooks, S.: Does personal social media usage affect efficiency and well-being? Comput. Hum. Behav. **46**, 26–37 (2015)
8. Caselli, T., Basile, V., Mitrović, J., Granitzer, M.: HateBERT: Retraining BERT for abusive language detection in English. arXiv preprint arXiv:2010.12472 (2020)
9. Dai, X., Karimi, S., Hachey, B., Paris, C.: Cost-effective selection of pretraining data: a case study of pretraining BERT on social media. arXiv preprint arXiv:2010.01150 (2020)
10. Dietterich, T.G.: Approximate statistical tests for comparing supervised classification learning algorithms. Neural Comput. **10**(7), 1895–1923 (1998)
11. Founta, A.M., et al.: Large scale crowdsourcing and characterization of twitter abusive behavior. In: Twelfth International AAAI Conference on Web and Social Media (2018)
12. Frenkel, S., Conger, K.: Hate speech's rise on twitter is unprecedented, researchers find, December 2022. https://www.nytimes.com/2022/12/02/technology/twitter-hate-speech.html
13. Heidari, M., Jones, J.H.: Using BERT to extract topic-independent sentiment features for social media bot detection. In: 2020 11th IEEE Annual Ubiquitous Computing, Electronics & Mobile Communication Conference (UEMCON), pp. 0542–0547. IEEE (2020)
14. Holden, M., Phillips, M.: England's black players face racial abuse after euro 2020 defeat, July 2021. https://www.reuters.com/world/uk/uk-pm-johnson-condemns-racist-abuse-england-soccer-team-2021-07-12/

15. Horner, S., Asher, Y., Fireman, G.D.: The impact and response to electronic bullying and traditional bullying among adolescents. Comput. Hum. Behav. **49**, 288–295 (2015)
16. Institute, O.I.: UNESCO, on Genocide Prevention, U.N.O., the Responsibility to Protect: Addressing hate speech on social media: contemporary challenges (2021). https://unesdoc.unesco.org/ark:/48223/pf0000379177
17. Isaksen, V., Gambäck, B.: Using transfer-based language models to detect hateful and offensive language online. In: Proceedings of the Fourth Workshop on Online Abuse and Harms, pp. 16–27 (2020)
18. Masud, S., et al.: Hate is the new infodemic: a topic-aware modeling of hate speech diffusion on twitter. In: 2021 IEEE 37th International Conference on Data Engineering (ICDE), pp. 504–515. IEEE (2021)
19. Mathew, B., Dutt, R., Goyal, P., Mukherjee, A.: Spread of hate speech in online social media. In: Proceedings of the 10th ACM Conference on Web Science, pp. 173–182 (2019)
20. Miró-Llinares, F., Moneva, A., Esteve, M.: Hate is in the air! but where? Introducing an algorithm to detect hate speech in digital microenvironments. Crime Sci. **7**(1), 1–12 (2018)
21. Mozafari, M., Farahbakhsh, R., Crespi, N.: A BERT-based transfer learning approach for hate speech detection in online social media. In: Cherifi, H., Gaito, S., Mendes, J.F., Moro, E., Rocha, L.M. (eds.) COMPLEX NETWORKS 2019. SCI, vol. 881, pp. 928–940. Springer, Cham (2020). https://doi.org/10.1007/978-3-030-36687-2_77
22. Mutanga, R.T., Naicker, N., Olugbara, O.O.: Hate speech detection in twitter using transformer methods. Int. J. Adv. Comput. Sci. Appl. **11**(9) (2020)
23. Park, J.H., Fung, P.: One-step and two-step classification for abusive language detection on twitter. arXiv preprint arXiv:1706.01206 (2017)
24. Pedregosa, F., et al.: Scikit-learn: machine learning in python. J. Mach. Learn. Res. (2011)
25. Person, J.S.: How trolls are ruining the internet, August 2016. https://time.com/4457110/internet-trolls/
26. Saha, K., Chandrasekharan, E., De Choudhury, M.: Prevalence and psychological effects of hateful speech in online college communities. In: Proceedings of the 10th ACM Conference on Web Science, pp. 255–264 (2019)
27. Sanchez, L., He, J., Manotumruksa, J., Albakour, D., Martinez, M., Lipani, A.: Easing legal news monitoring with learning to rank and BERT. In: Jose, J.M., et al. (eds.) ECIR 2020. LNCS, vol. 12036, pp. 336–343. Springer, Cham (2020). https://doi.org/10.1007/978-3-030-45442-5_42
28. Sanh, V., Debut, L., Chaumond, J., Wolf, T.: DistilBERT, a distilled version of BERT: smaller, faster, cheaper and lighter. arXiv preprint arXiv:1910.01108 (2019)
29. Sefara, T.J.: The effects of normalisation methods on speech emotion recognition. In: 2019 International Multidisciplinary Information Technology and Engineering Conference (IMITEC), pp. 1–8. IEEE (2019)
30. Singh, D., Singh, B.: Investigating the impact of data normalization on classification performance. Appl. Softw. Comput. **97**, 105524 (2020)
31. Vidgen, B., Derczynski, L.: Directions in abusive language training data, a systematic review: garbage in, garbage out. PLoS ONE **15**(12), e0243300 (2020)

32. Vosoughi, S., Zhou, H., Roy, D.: Enhanced twitter sentiment classification using contextual information. arXiv preprint arXiv:1605.05195 (2016)
33. Walther, J.B.: Social media and online hate. Curr. Opin. Psychol. (2022)
34. Wijesiriwardene, T., et al.: ALONE: a dataset for toxic behavior among adolescents on twitter. In: Aref, S., et al. (eds.) SocInfo 2020. LNCS, vol. 12467, pp. 427–439. Springer, Cham (2020). https://doi.org/10.1007/978-3-030-60975-7_31

Evolution of Filter Bubbles and Polarization in News Recommendation

Han Zhang[1], Ziwei Zhu[2]([⊠]), and James Caverlee[1]([⊠])

[1] Texas A&M University, 400 Bizzell St., College Station, USA
zh89118877@tamu.edu, caverlee@cse.tamu.edu
[2] George Mason University, 4400 University Drive, Fairfax, USA
zzhu20@gmu.edu

Abstract. Recent work in news recommendation has demonstrated that recommenders can over-expose users to articles that support their pre-existing opinions. However, most existing work focuses on a static setting or over a short-time window, leaving open questions about the long-term and dynamic impacts of news recommendations. In this paper, we explore these dynamic impacts through a systematic study of three research questions: 1) How do the news reading behaviors of users change after repeated long-term interactions with recommenders? 2) How do the inherent preferences of users change over time in such a dynamic recommender system? 3) Can the existing SOTA static method alleviate the problem in the dynamic environment? Concretely, we conduct a comprehensive data-driven study through simulation experiments of political polarization in news recommendations based on 40,000 annotated news articles. We find that users are rapidly exposed to more extreme content as the recommender evolves. We also find that a calibration-based intervention can slow down this polarization, but leaves open significant opportunities for future improvements

Keywords: Filter bubble · Recommender system · Dynamic

1 Introduction

It has been demonstrated by recent work [12,14] that *personalized news recommender systems* can over-expose users to news articles supporting their pre-existing opinions. With increasing reliance on personalized recommendations to consume news from digital news apps [2,6], such a filter bubble phenomenon paves the way for continued (and potentially increased) intellectual segregation and political polarization.

While these important studies have demonstrated the problem of filter bubbles and political polarization, most existing work [1,15,16,18] focuses on the problem under a static or short-term setting, leaving open questions about the *long-term and dynamic impacts* of news recommendations. For example, how fast do these filter bubbles form? Does polarization oscillate? Or is it fixed? Can interventions alleviate this polarization? Hence, in this work, we conduct

J. Kamps et al. (Eds.): ECIR 2023, LNCS 13981, pp. 685–693, 2023.
https://doi.org/10.1007/978-3-031-28238-6_60

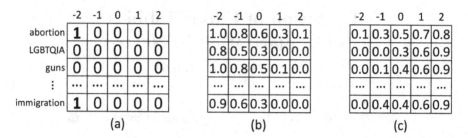

Fig. 1. (a) shows an article matrix. (b) shows a preference matrix for a 'solid liberal' user. (c) shows a preference matrix for a 'core conservative' user.

a systematic study to investigate the long-term and dynamic impacts of news recommender systems organized around three key research questions: 1) How do the news reading behaviors of users change after repeated long-term interactions with recommenders? 2) How do the inherent preferences of users change over time in such a dynamic recommender system? 3) Can a SOTA intervention method alleviate the problem in the dynamic environment?

Concretely, we conduct an extensive data-driven study through simulations of news recommendations based on 40,000 annotated news articles to study the impacts of news recommenders. To uncover how the recommender influences the news reading behaviors of users and intensifies polarization over time, we consider that the political preferences of users can be influenced by recommended and read news. Unsurprisingly, we find that users are rapidly exposed to more extreme content as the recommender evolves and the inherent political preferences of users become increasingly radical. Moreover, we also observe that users read more and more extreme content even if they are immune to recommendation influence and keep their inherent political opinions invariant. Last, we further conduct experiments with a calibration-based method [19], which is a SOTA static method for addressing filter bubbles. We find that such a calibration-based intervention can slow down this polarization but still leaves open significant opportunities for future improvements.

2   Dynamic Experiment Setup

In this section, we first introduce our framework for studying dynamic news recommendation, including the dataset, the experimental process, and the metrics.

2.1   Dataset

We use a variation of the dataset from [12], which consists of a collection of 40,000 news articles and a set of 500 users. The 40,000 articles are with annotations of their topics and political stances. Specifically, there are 14 topics: *abortion, environment, guns, health care, immigration, LGBTQIA, taxes, technology, trade, Trump impeachment, US military, welfare, US 2020 election, and*

Algorithm 1: Dynamic News Recommendation

---

1 **Bootstrap:** Randomly expose 10 articles from each topic (140 in total) to each user, and collect initial clicks $\mathcal{D}$, and train the first model ψ by $\mathcal{D}$;

2 **for** $t = 1 : 40,000$ **do**

3   Randomly choose a user u_t as the current visiting user;

4   Recommend 5 articles to the current user u_t by ψ;

5   Collect new clicks and add them to $\mathcal{D}$;

6   Update preference matrix of user u_t;

7   **if** $t\%200 == 0$ **then**

8   Retrain ψ by $\mathcal{D}$;

---

racism. Each article can cover one or more topics. For political stance, each article is labeled as one of $\{-2, -1, 0, 1, 2\}$, which spans the ideological spectrum from extreme liberal (-2) to extreme conservative ($+2$). There are 8,000 articles for each political stance. We can use a binary utility matrix $\mathbf{A}_i \in \{0, 1\}^{14 \times 5}$ to represent the topic and stance for an article i. Figure 1(a) shows an example of an article related to abortion and immigration with a political stance of -2.

The user set is simulated based on the Pew survey of U.S. political typologies [7], which summarizes 9 political typologies in the U.S. and their opinions toward different topics. We consider the five most representative typologies: *solid liberal* (extreme liberal), *opportunity democrats* (lean toward liberal), *bystanders* (mild group), *market skeptic republicans* (lean toward conservative), and *core conservatives* (extreme conservative). For each typology, we generate 100 users, where each user can be represented by a preference matrix $\mathbf{U}_u \in \mathbb{R}^{14 \times 5}$ to represent the user's political stances toward different topics. The larger $\mathbf{U}_u(t, s)$ is, the more likely user u holds an opinion of stance s toward the topic t. Figure 1(b) shows an example preference matrix of a 'solid liberal' user and Fig. 1(c) shows an example preference matrix for a 'core conservative' user.

With the utility matrices for news articles and preference matrices of users, we can determine the preference of a user for an article by vectorizing their corresponding matrices and then taking the dot product to calculate the preference score between them. We can further use this preference score to determine user-read-article behaviors. The higher the preference score is, the more likely a user is to click and read the article. More details about how news articles are annotated, how user preference matrices are generated from the Pew survey, and the user click model can be found in [12].

2.2 Dynamic Recommendation Process

Next, we conduct a dynamic recommendation experiment to study how users are impacted by a personalized news recommender. The detailed experimental process is presented in Algorithm 1. We first conduct a bootstrap step to collect initial click data from all users by randomly showing 140 articles (10 articles from each topic) and then training the first recommendation model with the

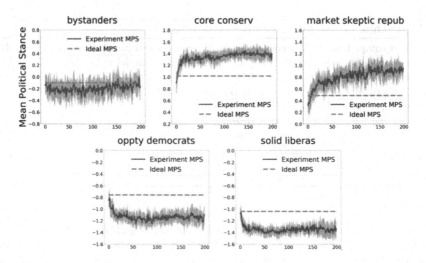

Fig. 2. MPS changes over time for five user groups (c = 0).

initial click data. Then, we run the dynamic experiment for 40,000 iterations. At each iteration, a random user will come and ask for recommendations of 5 articles. The user will iterate all the 5 articles and determine whether click and read them. The interaction data will be stored for further model training. We retrain the model after every 200 iterations, resulting in 200 experiment epochs. In this work, we use the fundamental Matrix Factorization (MF) [12] model as the core approach to deliver recommendations.

Moreover, in the real world, users' preferences can be influenced by recommendations exposed to them. If an article was recommended and read by a user, the corresponding opinions of the user will be reinforced, and the user is more likely to click articles with the same political stances and topics in the future. So, we model these dynamics by changing preference matrices of users corresponding to what articles are exposed and read by users. We first define an influence parameter c to determine to what degree users can be influenced by recommendations. Then, every time a user u is exposed to an article i, if u clicks and reads i, we update the preference matrix $\mathbf{U}_u$ of u by $\mathbf{U}_u \leftarrow \mathbf{U}_u + c \cdot \mathbf{A}_i$. A larger c means that people are more susceptible to the recommendation influence.

2.3   Evaluation Metrics

To show how recommendations influence user behavior, we calculate the Mean Political Stance (MPS) for iteration t: $MPS_t = \sum_{p=1}^{5} y_{u_t,p} \cdot stance(p) / \sum_{p=1}^{5} y_{u_t,p}$, where we iterate the 5 recommended articles (from top position $p = 1$ to the end $p = 5$), and if u_t clicks and reads article at position p, $y_{u_t,p} = 1$, otherwise $y_{u_t,p} = 0$. We calculate the average political stance of articles read by the user at interaction t, and $stance(p)$ returns the political stance of an article at position p. We report the average MPS for each user group in each experiment epoch and show how it evolves over 200 epochs.

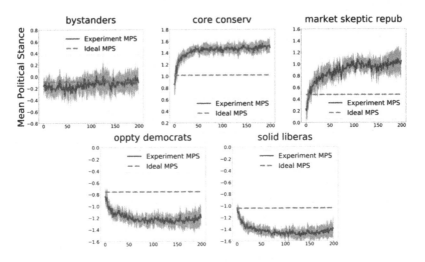

Fig. 3. MPS changes over time for five user groups (c = 0.03).

We also calculate the User Mean Political Stance (UMPS) for each user: $UMPS = \sum_{s=-2}^{2} \sum_{t=1}^{14} s \cdot \mathbf{U}_u(t,s)$ to directly show the evolution of inherent user preference. The UMPS reflects the current user preference. After each epoch, we calculate the average UMPS of each user group and show how their inherent preference change over 200 epochs.

3   Experimental Results

We empirically study three key research questions: (RQ1) In such a dynamic recommendation process, will users be exposed to and read more and more similar articles with more extreme political stances? (RQ2) Will users be influenced by these recommendations and become more and more radical over time? and (RQ3) Can an existing intervention method alleviate the filter bubble problem? All experiments are repeated 10 times, and we report the averaged results.

RQ1: Evolution of User News Reading Behaviors. First, we study how do the news reading behaviors of users change after repeated long-term interactions with the recommender. We report the averaged MPS of each user group to depict the pattern of news reading behaviors, and show the changing of behavior patterns with influence parameter $c = 0$ in Fig. 2 and with $c = 0.03$ in Fig. 3. The x-axis in these figures represents the experiment epochs, each of which contains 200 interactions. In the figures, besides the MPS in each epoch during the experiment, we also plot the MPS during the bootstrap step for each user group, which indicates the true initial political stance of each user group and can be regarded as the ideal MPS we want to achieve for dynamic recommendation.

From the result, we can see that even though user political preference remains static (the influence parameter is set to be 0), the absolute value of MPS of

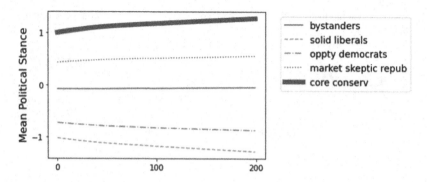

Fig. 4. The UMPS of different user groups change over time (c = 0.03).

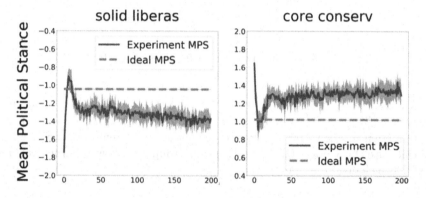

Fig. 5. Changing of MPS with the Calibrated Recommendation method (c = 0.03)

different groups except the 'bystanders' group becomes larger. In other words, even if users are immune to the influence of recommendations and keep their political preferences invariant, they will still read more and more extreme news. After we add the influence parameter into the experiment and compare the results in Fig. 2 and Fig. 3, we can observe even more severe trend of reading extreme news: except for the 'bystanders' group, the other four groups become more and more deviated from the ideal MPS, demonstrating the rapid trend of radicalization and polarization of users.

RQ2: Evolution of User Preference. Next, we unveil how the inherent political preferences of users evolve over time. Here, we measure the averaged UMPS for each user group to indicate the current inherent preference of the user group and show the changing of UMPS over time to depict how the user preference is influenced by recommendations. In Fig. 4, we show the empirical result with c = 0.03, which clearly illustrates that except for 'bystanders', all other user groups become more and more radical. That is to say, if the exposed recommendations can change users' inherent opinions, users will move toward more extreme stances after long-term interactions with the new recommender.

RQ3: Effectiveness of Intervention. Last, we study how well a SOTA static method for mitigating filter bubble performs for the dynamic recommendation. Here, we conduct experiments to evaluate the performance of the Calibrated Recommendation method [19], which is one of the SOTA static methods for addressing filter bubbles. The Calibrated method re-ranks the recommendation list from the recommender so that the re-ranked list contains a distribution that follows an ideal distribution (the distribution learned from the bootstrap step in our case). We show the results for the "solid liberals" and "core conservatives" groups with $c = 0.03$ in Fig. 5. From these results, we can observe that the Calibrated method can only slow down the polarization process, but it cannot prevent the trend of radicalization and polarization. Hence, we conclude that the calibration method produces very limited effects in such a dynamic scenario motivating efforts for more effective methods.

4 Related Work

Filter bubble is a long-standing problem for recommender systems, widely studied in many large-scale platforms like Twitter, Facebook, and YouTube [3,4,9,13,17]. One of the major reasons raising filter bubbles is the nature of recommendation algorithms to deliver content that users are more likely to click on to maximize utility [5,8,11]. Such a problem of filter bubbles can lead to damaged user experience and intensify intellectual segregation and polarization in society [10]. Specifically, a recent work [12] analyzes and compares how different algorithms form filter bubbles and expose more extreme content to users in a news recommender system. However, most prior work is focused on short-term and static scenarios, which motivates us to explore the long-term and dynamic nature of filter bubbles in this work.

5 Conclusion and Future Work

In this paper, we conduct a comprehensive data-driven study through simulation experiments of political polarization in news recommendations based on 40,000 annotated news articles. Specifically, we answer three research questions: 1) How do the news reading behaviors of users change after repeated long-term interactions with recommenders? 2) How do the inherent preferences of users change over time in such a dynamic recommender system? 3) How effective can the existing SOTA intervention method alleviate the problem in the dynamic environment? We find that users are rapidly exposed to more extreme content and become more radical as the system evolves. We also find that a calibration-based intervention slows down this polarization, but leaves open significant opportunities for future improvements

Acknowledgements. This work is supported in part by NSF grants IIS-1939716 and IIS-1909252.

References

1. Abdollahpouri, H., Mansoury, M., Burke, R., Mobasher, B.: The unfairness of popularity bias in recommendation. arXiv preprint arXiv:1907.13286 (2019)
2. Agarwal, D., Chen, B.C., Elango, P.: Explore/exploit schemes for web content optimization. In: 2009 Ninth IEEE International Conference on Data Mining, pp. 1–10. IEEE (2009)
3. Bakshy, E., Messing, S., Adamic, L.A.: Exposure to ideologically diverse news and opinion on Facebook. Science **348**(6239), 1130–1132 (2015)
4. Barberá, P., Jost, J.T., Nagler, J., Tucker, J.A., Bonneau, R.: Tweeting from left to right: Is online political communication more than an echo chamber? Psychol. Sci. **26**(10), 1531–1542 (2015)
5. Chu, W., Park, S.T.: Personalized recommendation on dynamic content using predictive bilinear models. In: Proceedings of the 18th International Conference on World Wide Web, pp. 691–700 (2009)
6. Das, A.S., Datar, M., Garg, A., Rajaram, S.: Google news personalization: scalable online collaborative filtering. In: Proceedings of the 16th International Conference on World Wide Web, pp. 271–280 (2007)
7. Doherty, C., Kiley, J., Johnson, B.: Political typology reveals deep fissures on the right and left: conservative republican groups divided on immigration, 'openness'. Pew Research Center (2017)
8. Dumais, S., Joachims, T., Bharat, K., Weigend, A.: Sigir 2003 workshop report: implicit measures of user interests and preferences. In: ACM SIGIR Forum, vol. 37, pp. 50–54. ACM New York (2003)
9. Eady, G., Nagler, J., Guess, A., Zilinsky, J., Tucker, J.A.: How many people live in political bubbles on social media? evidence from linked survey and twitter data. SAGE Open **9**(1), 2158244019832705 (2019)
10. Epstein, R., Robertson, R.E.: The search engine manipulation effect (seme) and its possible impact on the outcomes of elections. Proc. Natl. Acad. Sci. **112**(33), E4512–E4521 (2015)
11. Johnson, C.C.: Logistic matrix factorization for implicit feedback data. Adv. Neural. Inf. Process. Syst. **27**(78), 1–9 (2014)
12. Liu, P., Shivaram, K., Culotta, A., Shapiro, M.A., Bilgic, M.: The interaction between political typology and filter bubbles in news recommendation algorithms. In: Proceedings of the Web Conference 2021, pp. 3791–3801 (2021)
13. Min, S.J., Wohn, D.Y.: All the news that you don't like: cross-cutting exposure and political participation in the age of social media. Comput. Hum. Behav. **83**, 24–31 (2018)
14. Pariser, E.: The filter bubble: what the Internet is hiding from you. Penguin UK (2011)
15. Park, Y.J., Tuzhilin, A.: The long tail of recommender systems and how to leverage it. In: Proceedings of the 2008 ACM Conference on Recommender Systems, pp. 11–18 (2008)
16. Rodriguez, C.G., Moskowitz, J.P., Salem, R.M., Ditto, P.H.: Partisan selective exposure: the role of party, ideology and ideological extremity over time. Transl. Issues Psychol. Sci. **3**(3), 254 (2017)
17. Shapiro, M.A., Park, H.W.: More than entertainment: Youtube and public responses to the science of global warming and climate change. Soc. Sci. Inf. **54**(1), 115–145 (2015)

18. Steck, H.: Item popularity and recommendation accuracy. In: Proceedings of the Fifth ACM Conference on Recommender Systems, pp. 125–132 (2011)
19. Steck, H.: Calibrated recommendations. In: Proceedings of the 12th ACM Conference on Recommender Systems, pp. 154–162 (2018)

Capturing Cross-Platform Interaction for Identifying Coordinated Accounts of Misinformation Campaigns

Yizhou Zhang[1], Karishma Sharma[2], and Yan Liu[1(✉)]

[1] University of Southern California, Los Angeles, CA 90007, USA
{zhangyiz,yanliu.cs}@usc.edu
[2] Amazon, Sunnyvale, CA 94089, USA
karish.sharma24@gmail.com

Abstract. Recent years have witnessed the increasing abuse of coordinated accounts on multiple social media platforms. Such accounts are usually operated by misinformation campaigns to manipulate the public opinions on different platforms jointly. However, existing methods mainly focus on detecting such accounts by capturing the coordinated activities within a single platform. As a result, their performances are limited as they can not make use of the information from other platforms. In this work, we propose that capturing cross-platform coordinated activities can bring a significant boost to identifying the accounts operated by misinfromation campaigns. To leverage such information in a practical way, we design a novel **Conditional Gaussian-distribution Basis** to extract cross-platform correlation from **Coordinated Activity Set**, which can be easily acquired. Experimental results indicate that our methodology outperform baselines and its own variants that can not leverage cross-platform information.

1 Introduction

Recent researches reveal the existence of active coordinated accounts, which are usually operated by a disinformation campaign such as Internet Research Agency (IRA) [10,11,14], on multiple platforms. Due to the different statistical properties of the accounts on different platforms, the difficulty of detection on different platforms also varies. We analyze those accounts interacting (post, comment or share) with the sampled information[1] posted or interested by IRA. In this case, on Reddit among 5k accounts interacted with the information targeted by IRA, only 96 of them are coordinated accounts. In contrary, on Twitter this ratio is 312 out of 2025 accounts, which is much more balanced than on Reddit and leads to much easier training of machine learning based coordination detector.

An intuitive solution to address the above issues is to exploit cross-platform information about coordination. Cross-platform coordination has been reported by recent researches [6,8,17]. For example [11] reveals that IRA first posts candidate contents on Reddit to evaluate their influence and then selectively spreads

---

[1] The information here mean the posts on Reddit and tweets on Twitter.

© The Author(s), under exclusive license to Springer Nature Switzerland AG 2023
J. Kamps et al. (Eds.): ECIR 2023, LNCS 13981, pp. 694–702, 2023.
https://doi.org/10.1007/978-3-031-28238-6_61

those contents with high influence on Twitter. As a result, we can use the information on those **"easier"** and well-studied platforms (like Twitter, we denote such platforms as **aid platform**) to boost the detection on **"hard"** platforms (like Reddit, we denote such platforms as **target platform**).

However, incorporating cross-platform information is highly challenging, because the existing methods for cross-platform social media analysis often assume that (1) there is an underlying mapping between accounts from different platforms and (2) we can acquire an approximately accurate mapping (known as *social network alignment*) [4,9,20]. Such assumptions are not realistic for coordinated account detection. First, existing tools for social network alignment are still far from sufficiently accurate. Even for the state-of-the-art model, when provided with 70% supervision, the precision@5 is still lower than 60% [4]. Second, but more importantly, most of the coordinated accounts are social bots or controlled by human operators of misinformation campaigns or organizations [10]. In such a case, the underlying mapping may follow a different distribution from normal users (the human operator case) or not exist (the social bot case).

In this paper, to boost the coordination detection on target platforms with cross-platform information in a more practical way, we propose to make use of **Coordinated Activity Set**, which consists of the activities conducted by the coordinated accounts of the same misinformation campaign on the aid platform. Compared to social network alignment, coordinated activity set gets rid of the assumption of underlying cross-platform account mapping. To obtain an accurate coordinated activity set, all we need is a sufficiently precise single-platform coordination detector on the aid platforms. This can be easily satisfied because the precision of unsupervised state-of-the-art coordination detector on well-studied and easier platforms like Twitter can easily surpass 90% [15]. And in a slightly looser semi-supervised learning setting where only 5% labelled data is provided, even some simple baselines like Label Propagation Algorithm can achieve a precision of 88% [21].

To capture the correlation information in coordinated activity set, we design an activity trace based deep neural detector incorporated with **Conditional Gaussian-distribution Basis (CGB)**. In this model, the activity trace of an account on the target platform is first encoded as a single-platform representation by a neural encoder. Then the encoding is forwarded into CGB, where each basis is a conditional Gaussian distribution of a coordinated activity on the aid platform given the representation. With the learnt distributions, we can calculate the probability density of each coordinated event on every basis and aggregate them to get a cross-platform feature, which will be fused with the single-platform representation for detection. A theoretical guarantee on sufficient expressive power of CGB ensures it to capture any complicated cross-platform interaction when enough parameters are provided, without assuming the specific strategies of the coordinated accounts on either platform. In general, our contributions include:

- We propose a new direction for detecting coordinated accounts on social media: we can boost the coordination detector by modeling cross-platform interaction in **coordinated activity set**, which can be easily acquired.
- We design **conditional Gaussian-distribution basis**, which has a theoretic guarantee on sufficient expressive power to capture complicated cross-platform coordinated activities distributed in continuous time.
- Experiments show that our model outperform existing baselines which only consider single-platform information.

2   Related Work

The abuse of coordinated accounts to manipulate public opinions has raised people's concern on the credibility of information on social media [14–16]. Such coordinated accounts, usually operated by misinformation campaigns, spread or influence the spread of information to control the visibility of specific narratives. The earliest reported case of coordination is the intervene of Internet Research Agency on the U.S. 2016 President Election [10]. To address this challenge, researchers try to design detection algorithms including two main directions:

Individual Features. Some researches reveal that coordinated accounts of the same misinformation campaigns may appear some shared characteristics such as the linguistic features [1], metadata [7,19] and the pattern of activity traces [10]. By applying unsupervised or supervised learning, we can train a model to extract such individual features and identify the coordinated accounts.

Collective Behaviours. In addition to the above individual features, some researches also propose to detect malicious accounts by capturing the collective behaviours of them. There are two popular ways. The first one is to represent the interaction of accounts by a graph based on prior knowledge and hand-crafted metrics such as time synchronization and co-appearance [2,5]. Then a graph clustering or adjacency matrix decomposition followed by a detector will be conducted to identify the coordinated accounts. However, such methods usually rely on the quality of the prior knowledge. To address this challenge, recent researches propose the second way, which is to apply deep learning to learn account representations by maximizing the data likelihood [15,21].

3   Preliminary

In this section, we introduce the definitions for the task of detecting coordinated accounts in social networks by capturing cross-platform coordination.

Definition 1: Activity Profile. An activity profile of an account is the sequence of all events of this account ordered in time, which can be formulated as $C_s = [t_1, t_2, ..., t_n]$. Each timestamp t_i corresponds to an activity by the account. The activities represent account actions on the network such as posting original content, re-sharing, replying, or reacting to other posts.

Definition 2: Coordinated Activity Set. A coordinated activity set of a misinformation campaign is the set of all activities conducted by the accounts

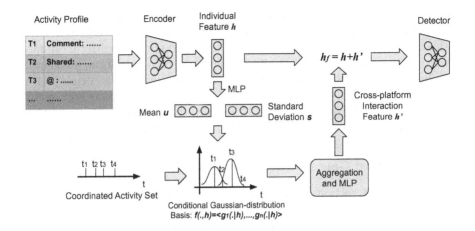

Fig. 1. The overview of our proposed method.

belonging to this campaign, which can be formulated as $S = \{t_1, t_2, ..., t_n\}$. Each timestamp t_i corresponds to an activity by an account.

Task Definition: Cross-Platform Interaction based Coordination Detection. This task aims at training a machine learning model that can exploit both single-platform and cross-platform information to identify the undiscovered coordinated accounts on social media. We denote the platform where our model will be applied to conduct detection as **target platform**. Input data includes:

- The activity profile of the account to be classified (coordinated account or normal user) on the target platform
- A known coordinated activity set of the same misinformation campaign on another platform. We denoted this platform as **aid platform**. The activity set on the aid platform will be accessible during training and testing.

4 Proposed Method

Figure 1 shows the main pipeline of the proposed method. In this framework, an time-series encoder (could be an RNN or a Transformer [12, 13, 15, 18, 22]) first encodes the activity profile to a representation vector h. Then another neural module takes h and the coordinated activity set on aid platform as input and output a vector h' which encodes the cross-platform-interaction information. After that, we fuse h and h' to acquire the final representation h_f and forward it into the multi-layer perceptron for prediction:

$$h_f = h + h', \quad \hat{p}(y|h_f) = \sigma(MLP_p(h_f)) \tag{1}$$

where y is the label (coordinated account or normal user) the σ is the sigmoid function. The whole model can be end-to-end trained with a **binary entropy**

loss function in a supervised manner. The key step in the above pipeline is to encode the cross-platform-interaction information as h' via the neural module. In this work, given an coordinated activity on the aid platform with timestamp t and the representation h of an activity profile on the target platform, we quantify their interaction as $p(t|h)$, which is the conditional probability density of an event on time t given h. To capture the above interaction, the neural module must have sufficient expressive power to encode $p(t|h)$. To this end, we designed an interaction extractor based on conditional Gaussian-distribution basis.

4.1   Conditional Gaussian-Distribution Basis

To extract the interaction between an event at time t on the aid platform and the representation h of an account on the target platform, we encode them as a vector with a set of Conditional Gaussian-distribution Basis (CGB) $f(t, h)$:

$$f(t, h) = < g_1(t|h), g_2(t|h), \ldots, g_n(t|h) > \tag{2}$$

where each $g_j(t|h)$ is a 1-dimensional Gaussian distribution whose parameters (mean μ_j and standard deviation s_j) are dynamically computed by a MLP:

$$g_j(t|h) = \frac{1}{s_j(h)\sqrt{2\pi}} \exp\left(-\frac{(t - \mu_j(h))^2}{2(s_j(h))^2}\right) \tag{3}$$

$$< \mu_1(h), \ldots >, < \log s_1(h), \ldots >= MLP_g(h) \tag{4}$$

Then we aggregate all $f(t_i, h)$ for all events i in the coordinated activity set and then fed forward the aggregation into another MLP to acquire the final h' to be fused with h. In this paper, for the aggregation, we apply top k pooling followed by a summation on each dimension separately.

4.2   Expressive Power of Conditional Gaussian-Distribution Basis

An intuitive concern to the conditional Gaussian-distribution basis is that the a set of Gaussian distribution might be too simple to capture complicated interaction between t and h. To address the above concern, in this section, we will provide a theoretic guarantee of CGB's expressive power:

Theorem 1. *(Dasgupta, 2008, Theorem 33.2 [3]) Let $q(x)$ be a continuous probability density where $x \in \mathbb{R}$. For any continuous probability density function $p(x)$ with $x \in \mathbb{R}$ and any $\epsilon > 0$, there exists a number of components $K \in \mathbb{N}$, mixture weight vector $w \in \mathbb{R}^K$ satisfying $\sum_{i=1}^{K} w_i = 1$, mean vector $\mu \in \mathbb{R}^K$ and scale vector $s \in \mathbb{R}^K$ such that for the mixture distribution $\hat{p}(x) = \sum_{i=1}^{K} w_i q(\frac{x - \mu_i}{s_i})$, we have $|p(x) - \hat{p}(x)| < \epsilon$*

If we apply the standard Gaussian distribution as $q(x)$, then every Gaussian-distribution basis with parameter μ_i and s_i corresponds to a $q(\frac{x - \mu_i}{s_i})$. Thus, we can easily get the following corollary:

Table 1. Results on detection of coordinated disinformation campaigns of Russian (IRA) interference in US Elections.

Method	AP	AUC	F1	Prec	Rec	Macro
AMDN	15.0	78.8	15.9	9.2	57.9	54.3
AMDN-HAGE	16.5	80.5	16.7	9.8	56.5	55.1
RNN	84.9	94.8	87.2	94.4	81.0	93.5
Transformer	87.6	99.7	93.0	91.0	95.2	96.5
Ours (RNN)	86.6	92.3	90.0	**96.5**	85.7	93.8
Ours (Transformer)	**98.8**	**99.9**	**95.5**	91.3	**100.0**	**97.7**

Corollary 1. *For any $p(t|h)$ and any error bound $\epsilon > 0$, when the number of Gaussian-distribution basis n is large enough, there exist a mixture weight vector $w(h)$ and a set of Gaussian-distribution basis $< g_1(t|h), ..., g_n(t|h) >$ such that:*

$$|p(t|h) - \sum_{j=1}^{K} w_j(h)g_j(t|h)| < \epsilon \tag{5}$$

The above corollary reveal the universal approximation ability of Gaussian-distribution basis when correct $w(h)$ is provided. Meanwhile, note that the $w(h)$ is decided only by h. Therefore, h has already catch the necessary information for $w(h)$ and all the information about the interaction are contained by the $g_j(t|h)$.

5 Experiments

In this section, we evaluate our model on detecting coordinated accounts operated by Internet Research Agency (IRA), a well-known misinformation campaign. An investigation by the U.S. Congress verified that IRA tried to manipulate the U.S. 2016 President Election. Coordinated accounts related to IRA have been found on multiple popular social media platforms, such as Twitter and Reddit. In this paper, we apply Reddit as the target platform and Twitter as the aid platform because the automatic coordination detection on Twitter have been explored by many previous researches [10,14–16,21].

Dataset: The dataset we applied in this work contain two components: 5k activity profiles of accounts on Reddit (among them 96 are coordinated accounts [11]), and Coordinated activity set of IRA accounts on Twitter [10]. The whole coordinated activity set is accessible during both training and testing stage. As for the activity profiles, we split them to 60%/20%/20% for training/validation/testing sets. Due to the unbalance of different categories, we apply a weighted binary cross entropy loss to allocate more weights to the coordinated accounts.

Baselines: We mainly compare our method with two kinds of baselines. **AMDN** and **AMDN-HAGE** [15] learn representations in an unsupervised manner and separately learn a supervised detector. **RNN** (LSTM) and **Transformer** learn a sequence classification model in an end-to-end manner.

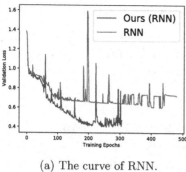

(a) The curve of RNN.

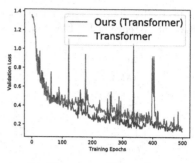

(b) The curve of Transformer.

Fig. 2. The validation-loss curve during training. The y-axis is the validation loss and the x-axis is the training epoch. The blue line in Fig. 2a ends early due to the early stopping. (Color figure online)

We report the Average Precision (AP), Area Under ROC Curve (AUC), F1 score, Precision, Recall and Macro F1 score [15]. For our model, we evaluate two versions applying **RNN** and **Transformer** as the encoders respectively. For hyper-parameters of our model and the end-to-end baseline, we keep the hyper-parameters of the encoder and detector the same and only fine-tune the hyper-parameters on the stacked module (CGB).

The performance of the two models based on unsupervised representation learning is significantly lower. We suggest that this is because on Reddit the ratio of coordinated account is too low. As a result, the coordinated behaviours are covered by the interaction between normal users. Without supervised signal, the model can hardly learn meaningful features that help detecting coordinated accounts. Also, compared to the end-to-end baselines, our methods achieves significantly better performance. We also present the validation loss curve in Fig. 2. As we can see, the validation loss of our model drops faster and lower, especially for RNN case, indicating that CGB helps the model generalize better.

6   Conclusion

In this work, we proposed that capturing the cross-platform interaction from coordinated activity set is beneficial for coordinated account detection. To enable the detector to capture such information, we design a conditional Gaussian-distribution basis, which has a theoretic guarantee to express any complicated interaction given sufficient parameter size. Experiment results show that cross-platform interaction captured by our novel design can bring a significant boost.

Acknowledgement. The work in this paper is supported by NSF Research Grant IIS-2226087. The views and conclusions in this paper are of the authors and should not be interpreted as representing the social policies of the funding agency, or U.S. Government. Yizhou Zhang is also partially supported by the Annenberg Fellowship of the University of Southern California. We are sincerely thankful to our anonymous reviewers for their feedback, comments and suggestions.

References

1. Addawood, A., Badawy, A., Lerman, K., Ferrara, E.: Linguistic cues to deception: Identifying political trolls on social media. In: Proceedings of the International AAAI Conference on Web and Social Media, vol. 13, pp. 15–25 (2019)
2. Cao, Q., Yang, X., Yu, J., Palow, C.: Uncovering large groups of active malicious accounts in online social networks. In: Proceedings of the 2014 ACM SIGSAC Conference on Computer and Communications Security, pp. 477–488 (2014)
3. DasGupta, A.: Asymptotic theory of statistics and probability, vol. 180. Springer, New York (2008) https://doi.org/10.1007/978-0-387-75971-5
4. Gao, S., Zhang, Z., Su, S., Philip, S.Y.: Reborn: transfer learning based social network alignment. Inf. Sci. **589**, 265–282 (2022)
5. Gupta, S., Kumaraguru, P., Chakraborty, T.: MalReG: detecting and analyzing malicious retweeter groups. In: Proceedings of the ACM India Joint International Conference on Data Science and Management of Data, pp. 61–69 (2019)
6. Horawalavithana, S., Ng, K.W., Iamnitchi, A.: Twitter Is the megaphone of cross-platform messaging on the white helmets. In: Thomson, R., Bisgin, H., Dancy, C., Hyder, A., Hussain, M. (eds.) SBP-BRiMS 2020. LNCS, vol. 12268, pp. 235–244. Springer, Cham (2020). https://doi.org/10.1007/978-3-030-61255-9_23
7. Im, J., et al.: Still out there: modeling and identifying Russian troll accounts on twitter. In: 12th ACM Conference on Web Science, pp. 1–10, (2020)
8. Horawalavithana, S., Iamnitchi, A.: NG Kin Wai: Twitter, facebook and youtube against the white helmets, Multi-platform information operations (2021)
9. Li, C., et al.: Adversarial learning for weakly-supervised social network alignment. In: Proceedings of the AAAI Conference on Artificial Intelligence, vol. 33, pp. 996–1003 (2019)
10. Luceri, L., Giordano, S., Ferrara, E.: Detecting troll behavior via inverse reinforcement learning: a case study of Russian trolls in the 2016 us election. In: Proceedings of the International AAAI Conference on Web and Social Media, vol. 14, pp. 417–427 (2020)
11. Lukito, J.: Coordinating a multi-platform disinformation campaign: internet research agency activity on three us social media platforms, 2015 to 2017. Polit. Commun. **37**(2), 238–255 (2020)
12. Mei, H., Eisner, J.M.: The neural hawkes process: a neurally self-modulating multivariate point process. In: Advances in Neural Information Processing Systems, vol. 30 (2017)
13. Rumelhart, D.E., Hinton, G.E., Williams, R.J.: Learning representations by back-propagating errors. Nature **323**(6088), 533–536 (1986)
14. Sharma, K., Seo, S., Meng, C., Rambhatla, S., Liu, Y.: Covid-19 on social media: analyzing misinformation in twitter conversations. arXiv preprint arXiv:2003.12309 (2020)
15. Sharma, K., Zhang, Y., Ferrara, E., Liu, Y.: Identifying coordinated accounts on social media through hidden influence and group behaviours. arXiv preprint arXiv:2008.11308 (2020)
16. Sharma, K., Zhang, Y., Liu, Y.: Covid-19 vaccines: characterizing misinformation campaigns and vaccine hesitancy on twitter. arXiv preprint arXiv:2106.08423 (2021)
17. Starbird, K., Arif, A., Wilson, T.: Disinformation as collaborative work: surfacing the participatory nature of strategic information operations. In: Proceedings of the ACM on Human-Computer Interaction, vol. 3(CSCW), pp. 1–26 (2019)

18. Vaswani, A., et al.: Attention is all you need. In: Advances in Neural Information Processing Systems, vol. 30 (2017)
19. Zannettou, S., Caulfield, T., De Cristofaro, E., Sirivianos, M., Stringhini, G., Blackburn, J.: Disinformation warfare: understanding state-sponsored trolls on twitter and their influence on the web. In: Companion Proceedings of the 2019 World Wide Web Conference, pp. 218–226 (2019)
20. Zhang, J., et al.: Mego2vec: embedding matched ego networks for user alignment across social networks. In: Proceedings of the 27th ACM International Conference on Information and Knowledge Management, pp. 327–336 (2018)
21. Zhang, Y., Sharma, K., Liu, Y.: VIGDET: knowledge informed neural temporal point process for coordination detection on social media. In: Advances in Neural Information Processing Systems, vol. 34 (2021)
22. Zuo, S., Jiang, H., Li, Z., Zhao, T., Zha, H.: Transformer hawkes process. In: International Conference on Machine Learning, pp. 11692–11702. PMLR (2020)

Author Index

Printed in the United States
by Baker & Taylor Publisher Services